Scattering Techniques Applied to Supramolecular and Nonequilibrium Systems

NATO ADVANCED STUDY INSTITUTES SERIES

A series of edited volumes comprising multifaceted studies of contemporary scientific issues by some of the best scientific minds in the world, assembled in cooperation with NATO Scientific Affairs Division.

Series B. Physics

Recent Volumes in this Series

Volume 68 – Superconductor Materials Science: Metallurgy, Fabrication, and Applications
edited by Simon Foner and Brian B. Schwartz

Volume 69 – Photovoltaic and Photoelectrochemical Solar Energy Conversion
edited by F. Cardon, W. P. Gomes, and W. Dekeyser

Volume 70 – Current Topics in Elementary Particle Physics
edited by K. H. Mütter and K. Schilling

Volume 71 – Atomic and Molecular Collision Theory
edited by Franco A. Gianturco

Volume 72 – Phase Transitions: *Cargèse 1980*
edited by Maurice Lévy, Jean-Claude Le Guillou, and Jean Zinn-Justin

Volume 73 – Scattering Techniques Applied to Supramolecular and Nonequilibrium Systems
edited by Sow-Hsin Chen, Benjamin Chu, and Ralph Nossal

Volume 74 – Rigorous Atomic and Molecular Physics
edited by G. Velo and A. S. Wightman

Volume 75 – Nonlinear Phenomena in Physics and Biology
edited by Richard H. Enns

Volume 76 – Metal Hydrides
edited by Gust Bambakidis

Volume 77 – Nonlinear Phenomena at Phase Transitions and Instabilities
edited by T. Riste

This series is published by an international board of publishers in conjunction with NATO Scientific Affairs Division

A Life Sciences B Physics	Plenum Publishing Corporation London and New York
C Mathematical and Physical Sciences	D. Reidel Publishing Company Dordrecht, Boston, and London
D Behavioral and Social Sciences E Applied Sciences	Sijthoff & Noordhoff International Publishers Alphen aan den Rijn, The Netherlands, and Germantown, U.S.A.

Scattering Techniques Applied to Supramolecular and Nonequilibrium Systems

Edited by

Sow-Hsin Chen

Massachusetts Institute of Technology
Cambridge, Massachusetts

Benjamin Chu

State University of New York
Stony Brook, New York

and

Ralph Nossal

National Institutes of Health
Bethesda, Maryland

PLENUM PRESS • NEW YORK AND LONDON

Published in cooperation with NATO Scientific Affairs Division

Library of Congress Cataloging in Publication Data

NATO Advanced Study Institute on Scattering Techniques Applied to Supramolecular and Nonequilibrium Systems (1980 : Wellesley College)
Scattering techniques applied to supramolecular and nonequilibrium systems.

(NATO advanced study institutes series. Series B, Physics ; v. 73)
Proceedings of the institute held Aug. 3-12, 1980, at Wellesley College, Wellesley, Mass.
"Published in cooperation with NATO Scientific Affairs Division."
Bibliography: p.
Includes index.
1. Scattering (Physics)—Congresses. 2. Biophysics—Technique—Congresses. I. Chen, Sow-Hsin, 1935- . II. Chu, Benjamin. III. Nossal, Ralph. IV. North Atlantic Treaty Organization. Division of Scientific Affairs. V. Title. VI. Series.
QH505.N34 1980 574.19'285 81-13767
AACR2

ISBN-13: 978-1-4684-4063-8 e-ISBN-13: 978-1-4684-4061-4
DOI: 10.1007/978-1-4684-4061-4

Proceedings of a NATO Advanced Study Institute on Scattering Techniques Applied to Supramolecular and Nonequilibrium Systems, held August 3-12, 1980, at Wellesley College, Wellesley, Massachusets

A Division of Plenum Publishing Corporation
233 Spring Street, New York, N.Y. 10013

PREFACE

This Advanced Study Institute was held at Wellesley College, Wellesley, MA., from 3 to 12 August 1980. It followed by four years the second "Capri School on Photon Correlation Spectroscopy". During the intervening period there had been many new applications of dynamic light scattering techniques to the study of systems whose properties depend either on collective molecular interactions or on the formation or activity of supramolecular structures. Consequently, emphasis at this conference was on light scattering studies of subjects such as dynamical correlations in dense polymer solutions, phase transitions in gels, spinodal decomposition of binary fluids, Bénard instabilities in nonequilibrium fluids, the formation of micelles and phospholipid vesicles, and movements of the molecular assemblies of muscle tissue. The instructional programme also included tutorial lectures on two complementary spectroscopic techniques which have benefited from dramatic advances in instrumentation, these being small angle X-ray (SAXS) and small angle neutron (SANS) scattering. Strong cold neutron and synchrotron X-ray sources have become available, and data now can be acquired rapidly with newly developed position-sensitive detectors. Several reviews of recent applications of SAXS and SANS were also provided.

The organizers of the ASI hoped to provide a forum for theoreticians and experimentalists to assess advances in fields which, although related, were sufficiently different that a great deal of unfamiliar information could be communicated. The ordering of the papers in this volume closely approximates that of the talks presented at the Advanced Study Institute. The reader will discover that diverse phenomena of physical, chemical and biological interest were discussed.

Part I of this book contains reviews of the basic physical principles underlying X-ray, neutron, and dynamic light scattering. Also included are discussions of facilities and instrumentation currently available for performing SANS and SAXS experiments, and summaries of procedures for analyzing quasielastic light scattering data. Part II pertains to polymer and colloidal systems, containing papers on polymer solutions, gels, micelles, and phospholipid vesicles. Part III concerns non-equilibrium systems and includes papers on critical behavior, nucleation and spinodal decomposition, and convective hydrodynamic instabilities. In addition, reviews of some highly complex mathematical modelling of non-equilibrium stochastic behavior are presented. Part IV contains papers describing several SANS and dynamic light scattering studies of the structure and motions of biological systems, among which are investigations of membrane and virus structure, mobility of micro-organisms, and structure and dynamics of muscle cells. Finally, Part V contains short presentations of twenty-seven contributed posters on such topics as new experimental techniques, hydrogen-bonded liquids, polymer solutions, colloidal suspensions, micelles and microemulsions, biomolecules and biological organelles and cells.

We are particularly fortunate to have had a group of expert lecturers who not only delivered excellent talks in the classroom but also gave generously of their time in the after-hour discussions. They are warmly thanked for their hard work in providing such interesting and stimulating lecture notes.

The United States Organizing Committee consisted of: S.H. Chen, MIT, Cambridge, MA (Director); B. Chu, SUNY, Stony Brook, L.I., N.Y.; and R. Nossal, NIH, Bethesda, MD. The European Advisory Committee was: J.P. Boon, Université Libre, Bruxelles, Belgium; and P. Bergé, CENS, Gif-sur-Yvette, France. Principal support for the Institute was provided by the NATO Scientific Affairs Division. The numerous warm and stimulating conversations between participants who oftentimes had to invent special variations of established languages in order to communicate attested to the wisdom of expending funds in this manner. A supplemental combined grant was received from the Division of Computer Research and Technology, the Division of Research Services, the National Institute of Dental Research, and the Fogarty International Center at the National Institutes of Health. These funds furnished support for participants from non-NATO countries. Generous contributions also were obtained from the following instrument manufacturers: Langley-Ford Instruments; Malvern Scientific Corporation; Nicomp Instruments; Science Research Systems of Troy; and Technology for Energy Corporation. The latter funds were used to provide a concert and other amenities which enabled the participants to live and work harmoniously for almost two weeks.

Located near Boston on a beautiful 500 acre site, Wellesley College afforded the academic and recreational attributes necessary for an intense, extended, working relationship between the participants. We thank the College administration for their many efforts to make our visit fruitful and enjoyable. In addition to lecture halls and conference rooms, facilities were provided for an instrument show and an ongoing display of posters relating to contributed papers. Also, after attending formal evening lectures, we were able to retire to the "Great Hall" of our dormitory where discussions oftentimes continued into the early hours of the morning.

Special mention must be made of the assistance given by Paul Wang and Brigitte Herpigny, who performed many of the numerous daily tasks relating to the organization and smooth operation of the conference. Their patience was severely tested, but they nonetheless performed with admirable good humor. Lastly, we acknowledge the cheerful assistance of Anne and Cathy Chen, who typed this book and whose help made possible its timely completion.

May 1981

S.H. Chen
B. Chu
R. Nossal

Cambridge, Massachusetts
Stony Brook, New York
Bethesda, Maryland

CONTENTS

PART I - ADVANCES IN INSTRUMENTATION AND TECHNIQUES

PART V - CONTRIBUTED PAPERS

A. Techniques

B. Hydrogen-Bonded Liquids

C. Polymer Solutions

D. Colloids, Microemulsions, and Micelles

G. Cells

PART I

Advances In Instrumentation And Techniques

INTERACTION OF THERMAL NEUTRONS AND PHOTONS WITH MATTER

Sow-Hsin Chen

Department of Nuclear Engineering
Massachusetts Institute of Technology
Cambridge, Massachusetts 02139

CONTENTS

1. RADIATION SCATTERING AS A TOOL FOR CONDENSED MATTER SPECTROSCOPY

Spectroscopy is concerned with measurement of the structure and dynamics of the ground state or low lying excited states of a condensed matter. A radiation is useful as a tool for spectroscopy if it couples weakly (in a sense to be discussed later) to the many body system. Because in this case the double differential cross-section (per unit solid angle, per unit energy transfer) for the radiation scattering can be written schematically as (see Figure 1):

$$\frac{d^2\sigma}{d\Omega d\omega} \sim \left(\frac{d\sigma}{d\Omega}\right)_o \sum_{i,f} P_i \left| \langle f \,|\sum_{\ell=1}^{n} e^{i(\underline{k}_1-\underline{k}_2)\cdot\underline{r}_\ell}|\, i\rangle \right|^2 \delta(\hbar\omega - E_i + E_f) \tag{1}$$

where the first factor $(d\sigma/d\Omega)_o$ refers to differential scattering cross-section from the basic unit of scattering in the system and

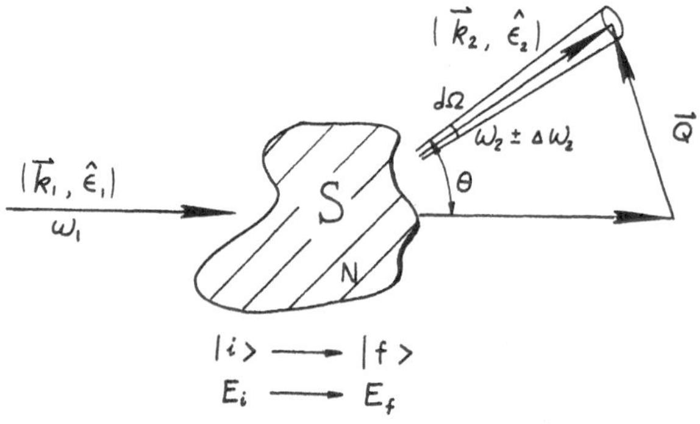

Figure 1. Scattering Geometry.

the second factor, usually called the dynamic structure factor, represents the time dependent structure of the system as seen by the radiation. This clear separation of the basic scattering problem, as embodies in the first factor, from the dynamic structure of the system itself, is only possible when the radiation couples weakly to the system and therefore the use of Born-approximation in deriving Equation (1) is valid. Both thermal neutrons and photons with energy up to x-ray region satisfy this criterion and thus are useful as probes for condensed matter structures. The dynamic structure factor contains two parameters related to the energy and momentum of the probe, namely the momentum transfer

$$\hbar\underline{Q} = \hbar\,(\underline{k}_1 - \underline{k}_2) \tag{2}$$

and the energy transfer

$$\hbar\omega = \hbar\,(\omega_1 - \omega_2) \tag{3}$$

to the system. In general the double differential cross-section depends on all four parameters: k_1, ω_1, k_2, ω_2, namely, both incident and scattered wavevectors and frequencies. But this complexity is reduced greatly when the Born approximation is applicable. In this case the dynamic structure factors depend, aside from state of the system, only on two external parameters $\underline{Q}$ and ω. This has an immediate significant implication in experimental consideration. We shall explain later, in Section 6, that qualitatively speaking when the radiation imparts momentum $\hbar \underline{Q}$ and energy $\hbar\omega$ to the system it effectively probes the structure and dynamics of the system with a spatial resolution of $R = Q^{-1}$ and time resolution of $\tau = \omega^{-1}$. For the purpose of this Study Institute, we are mostly interested in situations where $nk_1 \simeq nk_2 = k$, and n is the index of refraction in the medium. In this case Q can be expressed in terms of wave number k and scattering angle θ as:

$$Q = 2k \sin \frac{\theta}{2} \tag{4}$$

and we shall call it the Bragg wavenumber.

In order to probe the system spatial structure at different levels, one would like to change Q accordingly. For instance, to detect a periodic structure of $d \simeq 10\text{Å}$, one needs to have $Q_o \sim 2\pi/d = 0.628\text{Å}^{-1}$. One can use cold neutrons of wave length $\lambda = 4\text{Å}$ ($k = 2\pi/\lambda = 1.57\text{Å}^{-1}$) and work at a scattering angle around $\theta \simeq 23°$ so that $2k \sin\theta/2 = 2 \times 1.57 \sin(23°/2) \simeq 0.628\text{Å}^{-1}$; alternatively one can use x-rays of wavelength $\lambda = 0.62\text{Å}$ and work at an angle around $\theta \sim 3.6°$; or one can even use γ-rays of $\lambda = 0.03\text{Å}$ (412 keV γ-rays from Au^{198}, and work at an angle around $\theta \sim 0.17°$.

Similar consideration applies to the frequency or time resolution. Since the relevant variable is the difference of incident and scattered energies one can tune the dynamic range of the probe by varying accuracies of the energy difference measurement. A good example is a recent neutron scattering study of polymer chain dynamics by Higgins, Nicholson and Hayter,[1] using the so-called spin-echo technique. Using 8Å neutrons ($\hbar\omega_1 = 1279\ \mu$eV) they were able to measure energy difference of 0.01 μeV in the scattered neutrons and thus were able to probe the chain dynamics at a time scale of 0.1 μsec over the distances scale of up to 30Å ($Q \sim 0.03\text{Å}^{-1}$). Figure 2 summarizes the present status of neutron and photon spectroscopy relevant to the discussions in this Institute in terms of its (Q, ω) and (R, τ) resolutions. The important point to notice is the regions of overlap of the three methods because of the recent advances in techniques and instrumentations.

The primary purpose of this lecture is to derive the important relations given in Equation (1) and to identify in each case, thermal neutron, x-ray and light scattering, the corresponding basic scattering unit which contributes to the factor $(d\sigma/d\Omega)_o$. We shall comment on meanings of the dynamic structure factor using some examples in the final section.

2. THERMAL NEUTRON SCATTERING

Consider a combined system of a material medium, describable by Hamiltonian H_s, and a neutron with kinetic energy $P^2/2M$ and

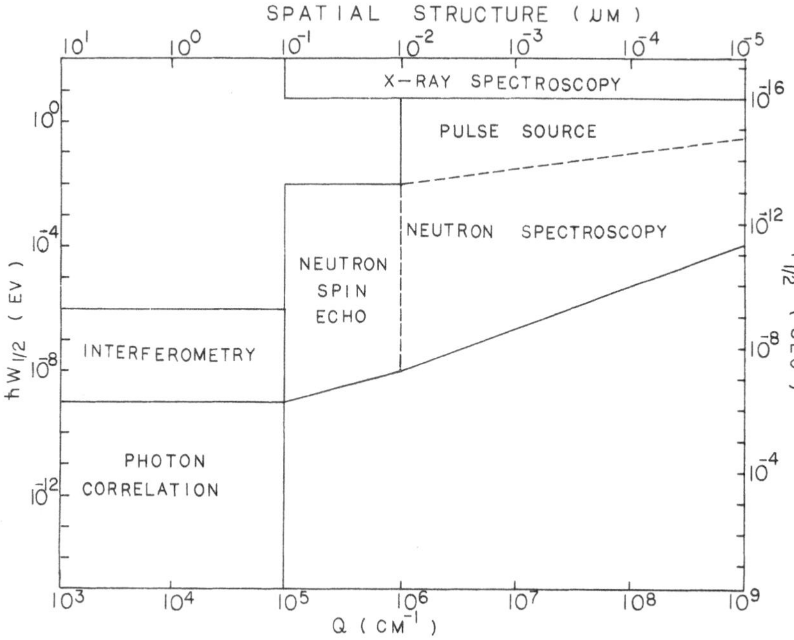

Figure 2. (Q,ω) and (R, τ) resolutions for each scattering techniques. Notice the contact between neutron and photon spectroscopy made possible by advant of the neutron spin-echo spectroscopy.

their mutual interaction V. The time independent Schrodinger equation for the system is

$$(\frac{P^2}{2M} + H_s + V)\,\Phi(\underline{r}, \{R\}) = E\,\Phi(\underline{r}, \{R\}). \tag{5}$$

We denote by $\underline{r}$ position of the neutron and $\{\underline{R}\}$ the collection of coordinates of all particles in the system. The material system has a set of stationary states $\{|n\rangle\}$ defined by

$$H_s|n\rangle = E_n|n\rangle \tag{6}$$

$$\langle n|n'\rangle = \delta_{nn'} \tag{7}$$

Using this set of stationary states we expand the total wave function as

$$\Phi(\underline{r}, \{\underline{R}\}) = \sum_n \psi_n(\underline{r})\,|n\rangle \tag{8}$$

Substituting Equation (8) into Equation (5) and taking a scalar product with $\langle n|$ on both sides, we obtain a wave equation for the neutron in the presence of the material system:

$$(\nabla^2 + K_n^2)\,\psi_n(\underline{r}) = \frac{2M}{\hbar^2}\sum_{n'} \langle n|V|n'\rangle\,\psi_{n'}(\underline{r}) \tag{9}$$

where $\hbar^2K_n^2/2M = E-E_n$ is the kinetic energy of the neutron in the system.

The neutron wave Equation (9) can be used in two ways. The first application is to consider propagation of neutrons in the material medium. For this application we take the system to be in the ground state, i.e., $|n\rangle = |n'\rangle = |0\rangle$, and introduce an optical potential $U(\underline{r}) = \langle o|V|o\rangle$. Then writing $\psi_o \equiv \psi$ and $K_o \equiv K$, Equation (9) reduces to a wave equation

$$(\nabla^2 + K^2)\,\psi(\underline{r}) = \frac{2M}{\hbar^2}\,U(\underline{r})\,\psi(\underline{r}). \tag{10}$$

One normally takes the optical potential to be a pseudo potential[2]

$$U(\underline{r}) = \frac{2\pi\hbar^2}{M}\sum_{\ell=1}^{n} b_\ell\,\delta(\underline{r}-\underline{R}_\ell). \tag{11}$$

A simple example is to take a homonuclear system with idential bound atom scattering length $b_i = b$,[3] and work out the index of refraction of neutrons in the medium. Write

$$U(\underline{r}) = \frac{2\pi\hbar^2 b}{M}\sum_{\ell=1}^{n}\delta(\underline{r}-\underline{R}_\ell)$$

$$= \frac{2\pi\hbar^2 b}{M}\,n(\underline{r}) \tag{12}$$

where $n(\underline{r})$ is the local number density of nuclei. Putting Equation (12) in Equation (10) we have

$$(\nabla^2 + K^2)\,\psi(\underline{r}) = 4\pi b n(\underline{r})\,\psi(\underline{r}). \tag{13}$$

Take a plane wave $\psi(\underline{r}) = \exp[i\underline{K}'\cdot\underline{r}]$ propagating in an optically homogeneous medium where we can set $n(\underline{r}) = n$ (this is valid when no Bragg condition is satisfied). We have from Equation (13) a relation

$$- K'^2 + K^2 = 4\pi bn \tag{14}$$

From Equation (14) the index of refraction follows as

$$n \equiv \frac{K'}{K} = \sqrt{1 - \frac{4\pi bn}{K^2}} = 1 - \frac{bn\lambda^2}{2\pi} . \tag{15}$$

Take a typical case of nickel for which $n = 9.13 \times 10^{22}/cm^3$ and $b = 1.03 \times 10^{-12}$cm. We have for 4Å neutrons $bn\lambda^2/2\pi \sim 2.4 \times 10^{-5}$. Index of refraction of a material with $b > 0$ is therfore slightly optically rare with respect to neutron wave. Notice the λ^2 dependençe in Equation (15). This means for ultra cold neutrons of λ = 800Å (speed ∿5M/sec), n = 0.04 and the total reflection from the surface of Ni film of up to $\theta_c = 87.7°$ is possible.

If one takes into account the periodic variation of $n(\underline{r})$ in a crystalline solid then one can proceed to work out the dynamic theory of neutron diffraction in a perfect crystal.[4]

The second application is to solve Equation (9) for scattered wave function. Scattering of neutrons in general induces excitation or de-excitation of the medium, so the matrix element $\langle n|V|n'\rangle$ of the interaction potential has to be evaluated between the initial and final states of the scattering system. For calculation of scattering cross-section, we are interested in the asymptotic wave function at distances large compared to the system size.

First, solve for the Green's function

$$(\nabla^2 + K_n^2)\, G_n(|\underline{r} - \underline{r}'|) = -\delta(\underline{r} - \underline{r}') \tag{16}$$

This gives

$$G_n(|r - r'|) = \frac{e^{iK_n|\underline{r}-\underline{r}'|}}{4\pi|\underline{r}-\underline{r}'|} \xrightarrow[r \gg r']{} \frac{e^{iK_n r}}{4\pi r}\, e^{iK_n\hat{\underline{r}}\cdot\underline{r}'} \tag{17}$$

Constructing an inhomogeneous solution from Equation (17) and adding to it the homogeneous solution which represents the incident plane wave, we get

$$\psi_n(\underline{r}) = \delta_{nn_o} e^{i\underline{K}\cdot\underline{r}} - \frac{e^{iK_n r}}{r}\left(\frac{M}{2\pi h^2}\right)\sum_{n'}\int_V d^3r' e^{-i\underline{K}_n\cdot\underline{r}'} V_{nn'}(r')\psi_{n'}(\underline{r}') \tag{18}$$

where $\underline{K} \equiv \underline{k}_1$ denotes the incident wavevector, $K_n \equiv k_2 = K_n \hat{r}$ denotes the scattered wavevector and the integration is over the volume of the system.

To go further from here one makes the Fermi-approximation which essentially consists of two parts: make the Born approximation by taking

$$\psi_{n'}(\underline{r}') \simeq \delta_{n'n_o} e^{i\underline{K}\cdot\underline{r}'} \tag{19}$$

and simultaneously using the pseudo-potential

$$V(\underline{r}) = \frac{2\pi\hbar^2}{M} \sum_{\ell=1}^{N} b_\ell \delta(\underline{r} - \underline{R}_\ell) \tag{20}$$

to get

$$\psi_n(\underline{r}) = \delta_{nn_o} e^{i\underline{K}\cdot\underline{r}} - \frac{e^{iK_n r}}{r} \left(\frac{M}{2\pi\hbar^2}\right) \int_V d^3r' e^{i(\underline{K}-\underline{K}_n)\cdot\underline{r}'} V_{nn_o}(\underline{r}'). \tag{21}$$

There are a number of discussions in the literature with regard to the Fermi-approximation which we shall not go into here.[5] The accuracy of this approximation is estimated to be of the order of 0.1% in the case of a hydrogen atom bound in a molecule. Taking a typical neutron-nuclear potential of depth $-V_o \sim 36\text{MeV}$ and width $r_o \sim 2 \times 10^{-13}\text{cm}$, validity of the Fermi approximation rest on the following considerations:

A. $Kr_o \ll 1$ (in fact $\sim 10^{-4}$) -- a condition for validity of low energy scattering from a bound nucleus where a scattering length b is sufficient to characterize the cross-section.

B. For a square well potential $b \propto V_o r_o^3$ and one can redefine fictitious potential parameters $\bar{V}_o, \bar{r}_o$, such that not only the scattering length is preserved, i.e., $\bar{V}_o \bar{r}_o^3 = V_o r_o^3$, but at the same time the condition for validity of the Born approximation is also satisfied, i.e., $m\bar{V}_o \bar{r}_o^3 / \hbar^2 \ll 1$.

C. The fictitious range of nuclear force $\bar{r}_o$ chosen above can still be much smaller than amplitude of the zero point vibration A of the bound nucleus in the molecule.

Physically speaking, even though neutron-nuclear interaction is so strong that the Born approximation is not applicable, a fortunate situation arises because the interaction is such a short range that one can smear out the interaction range considerably so as to decrease its strength in such a way that the Born approximation can be used. In practice A $\sim 10^{-9}$cm and $\bar{r}_o$ can be taken to be $100r_o = 2 \times 10^{-11}$ cm. We now define a scattering amplitude $f_{nn_o}(\theta)$ by writing Equation (21) in an asymtotic form

$$\psi_n(\underline{r}) = \delta_{nn_o} e^{i\underline{K}\cdot\underline{r}} + f_{nn_o}(\theta)\frac{e^{iK_n r}}{r} \tag{22}$$

and identify

$$f_{nn_o}(\theta) = -\sum_{\ell=1}^{N} b_\ell \langle n|e^{i(\underline{k}_1-\underline{k}_2)\cdot\underline{R}_\ell}|n_o\rangle . \tag{23}$$

First term in Equation (22) represents the incident wave ψ_{inc} while the second term the scattered wave ψ_{sc} in region $\theta \neq 0$. Differential cross-section for elastic scattering can then be obtained by calculating

$$d\sigma = \text{(Number of neutrons elastically scattered into } d\Omega \text{ in the direction } \hat{k}_2\text{/sec)} \div \text{(incident neutron flux)}$$

$$= \frac{k_2|\psi_{se}|^2 r^2 d\Omega}{k_1|\psi_{inc}|^2} \tag{24}$$

or

$$\frac{d\sigma}{d\Omega} = \frac{k_2}{k_1}\sum_{n_o} P_{n_o}|f_{n_o n_o}(\theta)|^2 \tag{25}$$

where Equation (25) is obtained from Equation (24) by averaging over distribution of the initial states $|n_o\rangle$ of the system.

Inelastic scattering cross-section can likewise be obtained from the scattering amplitude by

$$\frac{d^2\sigma}{d\Omega d\omega_2} = \frac{k_2}{k_1}\sum_{n,n_o} P_{n_o}|f_{nn_o}(\theta)|^2 \delta\left(\omega_1-\omega_2+\frac{E_{n_o}-E_n}{\hbar}\right) \tag{26}$$

where an additional average over the unknown final states $|n\rangle$ is made together with a delta function factor ensuring the energy conservation

$$\hbar\omega = \hbar\omega_1 - \hbar\omega_2 = E_n - E_{n_o} \tag{27}$$

Using expression of scattering amplitude [Equation (23)] in Equation (26) we then get (by writing $|i\rangle \equiv |n_o\rangle$, $|f\rangle \equiv |n\rangle$)

$$\frac{d^2\sigma}{d\Omega d\omega_2} = \frac{k_2}{k_1}\sum_{i,f} P_i \left|\langle f|\sum_{\ell=1}^{N} b_\ell e^{i\underline{Q}\cdot\underline{R}_\ell}|i\rangle\right|^2 \delta\left(\omega + \frac{E_i - E_f}{\hbar}\right) \tag{28}$$

Expression similar to Equation (1) is recovered except for a trivial kinematic factor k_2/k_1 when all nuclei are identical, i.e., $b_j = b$,

$$\frac{d^2\sigma}{d\Omega d\omega_2} = b^2\frac{k_2}{k_1}\sum_{i,f} P_i \left|\langle f|\sum_{\ell=1}^{N} e^{i\underline{Q}\cdot\underline{R}_\ell}|i\rangle\right|^2 \delta\left(\omega + \frac{E_i - E_f}{\hbar}\right) \tag{29}$$

We note that in this case $(d\sigma/d\Omega)_o = b^2$ which is the differential cross-section for a single bound nucleus.

Coherence and Incoherence

Cross-sections in Equations (25) and (28) contain a coherent superposition of phase factors from each nuclei $\exp(i\underline{Q}\cdot\underline{R}_\ell)$ only when all nuclei look identical to incoming neutrons. In practice there are two sources of incohence. First of all, even a chemically homogeneous system contains isotopes of the same element. Secondly, neutron-nuclear interaction is spin dependent, namely the scattering length b_ℓ is dependent on mutual orientations of neutron spin relative to nuclear spin. Therefore even for an unpolarized incident neutron beam, expression of Equation (28) has to be averaged over all possible nuclear isotopic and spin states. We denote this average by a bar. Consider then the factor

$$\overline{\sum_{i,f} P_i \left|\langle f|\sum_{\ell=1}^{N} b_\ell e^{i\underline{Q}\cdot\underline{R}_\ell}|i\rangle\right|^2 \delta\left(\omega + \frac{E_i - E_f}{\hbar}\right)} ,$$

which, by using an integral representation of delta function

$$\delta\left(\omega + \frac{E_i - E_f}{\hbar}\right) = \frac{1}{2\pi}\int_{-\infty}^{\infty} dt\ e^{-i\omega t}\ e^{\frac{i}{\hbar}(E_f - E_i)t},$$

can easily be transformed into a time-dependent form

$$\frac{1}{2\pi}\int_{-\infty}^{\infty} dt\ e^{-i\omega t}\sum_i P_i \langle i|\ \overline{\sum_{\ell,\ell'} b_\ell b_{\ell'} e^{-i\underline{Q}\cdot\underline{R}_\ell(o)}\ e^{iQ\cdot R_{\ell'}(t)}}\,|i\rangle .$$

If one now assumes that the isotopic states and spin states of each nucleus are completely uncorrelated with its position one can rewrite the above expression as

$$\frac{1}{2\pi}\int_{-\infty}^{\infty} dt\ e^{-i\omega t}\sum_i P_i \langle i|\sum_{\ell,\ell'} \overline{b_\ell b_{\ell'}}\ e^{-i\underline{Q}\cdot\underline{R}_\ell(o)}\ e^{iQ\cdot R_{\ell'}(t)}|i\rangle . \tag{30}$$

Consider now the average

$$\overline{b_\ell b_{\ell'}} = \overline{b_\ell^2} = \overline{b^2} \qquad \text{when } \ell = \ell'$$

$$= \overline{b_\ell b_{\ell'}} = (\bar{b})^2 \qquad \text{when } \ell \neq \ell'$$

or combined

$$\overline{b_\ell b_{\ell'}} = [\,\overline{b^2} - (\bar{b})^2\,]\,\delta_{\ell\ell'} + (\bar{b})^2 \tag{31}$$

$$\equiv b_{inc}^2\,\delta_{\ell\ell'} + b_{coh}^2 \tag{32}$$

Substituting Equation (32) back to Equation (30), we see that the dynamic structure factor (30) can always be decomposed into a single particle term (incoherent)

$$Nb_{inc}^2 S_s(\underline{Q}.\omega) = Nb_{inc}^2\int_{-\infty}^{\infty}\frac{dt}{2\pi}\, e^{-i\omega t}\sum_i P_i \langle i|e^{-i\underline{Q}\cdot\underline{R}(o)}\ e^{i\underline{Q}\cdot\underline{R}(t)}|i\rangle \tag{33}$$

and a pair term (coherent)

$$Nb_{coh}^{2}\, S(\underline{Q}, \omega) =$$

$$Nb_{coh}^{2} \int_{-\infty}^{\infty} \frac{dt}{2\pi}\, e^{-i\omega t} \sum_{i} P_i \langle i| \frac{1}{N} \sum_{\ell,\ell'} e^{-i\underline{Q}\cdot\underline{R}_{\ell}(o)}\, e^{i\underline{Q}\cdot\underline{R}_{\ell'}(t)} |i\rangle. \tag{34}$$

It is of interest to work out an example of spin incoherence for two isotopes of hydrogen. For a given nuclear spin I there are two possible spin states of neutrons as shown in the following:

Combined Spin States	Multiciplicity	Scattering Length
$I + 1/2$	$2I + 2$	a_+
$I - 1/2$	$2I$	a_-

Therefore

$$\bar{b} = \frac{2I + 2}{4I + 2} a_+ + \frac{2I}{4I + 2} a_- \tag{35}$$

$$\overline{b^2} = \frac{2I + 2}{4I + 2} |a_+|^2 + \frac{2I}{4I + 2} |a_-|^2 \tag{36}$$

For light hydrogen and heavy hydrogen we have

	I	$a_+(10^{-12}$cm)	$a_-(10^{-12}$cm)	$\bar{a}(10^{-12}$cm)	σ_{coh}(barn)	σ_{inc}(barn)
H^1	1/2	1.085	-4.750	-0.3742	1.77	79.9
D^2	1	0.950	0.098	0.6671	5.59	2.2

It should be noted that, because $a_- < 0$, in H^1 the coherent cross-section is much smaller than the incoherent cross-section. Since the average scattering length $\bar{a}$ of H^1 is negative, it is possible for example to mix a proportion of light water and heavy water so that the average scattering amplitude of the mixture changes continuously from a negative value to positive values. $\bar{b}$ for O^{16} is 0.5804×10^{-12}cm; therefore, $\bar{b}(H_2O) = -0.168 \times 10^{-12}$cm, and $\bar{b}(D_2O) = 1.915 \times 10^{-12}$cm. The mixture has an average coherent length per unite volume (in unit of 10^{-12}cm Å^{-3})

$$\langle b \rangle = [-0.00562 + 0.0697\, P_{D_2O}]$$

where P_{D_2O} is the percentage of D_2O in water. This possibility is the basis for the so-called contrast variation method.

3. X-RAY SCATTERING

Scattering of x-rays is usually treated classically in standard text books starting from Maxwell's equations.[6] This is usually satisfactory for understanding x-ray diffraction phenomena and also for propagation of x-rays in perfect crystals. However, besides these coherent elastic phenomena there are other incoherent inelastic phenomena such as Compton scattering and x-ray Raman scattering. In order to include these latter phenomena in a coherent way, a quantum mechanical treatment is more suitable. We shall therefore use a non-relativistic quantum electrodynamic treatment for the x-ray scattering.

The primary interaction of x-rays with a material system is through the photon-electron coupling. The standard Hamiltonian of a N-electron system in the presence of electro-magnetic field is

$$H = H_s + H_R + \sum_{j=1}^{N} -\frac{e}{mc}\vec{p}_j\cdot\vec{A}_j + \sum_{j=1}^{N}\frac{e^2}{2mc^2}A_j^2 \tag{37}$$

where the first term is the Hamiltonian of the electronic system in the absence of EM field, the second term is the Hamiltonian of a pure radiation field, the third term (called V_1) and the fourth term (called V_2) are interaction terms. $\vec{p}_j$ is the momentum operator of j-th electron (m is the electron mass), and $\vec{A}_j$ is the vector potential of the EM field at the position of the j-th electron, given by

$$\vec{A}_j = \sum_{k,\lambda}\sqrt{\frac{2\pi\hbar c}{kV}}\,\hat{\varepsilon}_{k\lambda}\left[a_{k\lambda}e^{i\underline{k}\cdot\underline{r}_j} + a^{+}_{k\lambda}\,e^{-i\underline{k}\cdot\underline{r}_j}\right]. \tag{38}$$

In Equation (38), $\hat{\varepsilon}_k$ is the polarization vector of photon state $|\underline{k}, \lambda\rangle$, $a_{k\lambda}$, $a^{+}_{k'\lambda}$, are annihilation and creation operators of the photons, and V is the volume of the box enclosing the EM radiation. In terms of the creation and annihilation operators, the Hamiltonian of the radiation field can be written as

$$H_R = \sum_{k,\lambda}\hbar\omega_k(a^{+}_{k\lambda}a_{k\lambda} + \tfrac{1}{2}). \tag{39}$$

Single Free Electron Initially at Rest

Since coupling of EM wave to an electron is weak (coupling

constant $\sim 1/137$) we can use first order time-dependent perturbation theory. In the free electron case $\vec{p} = 0$ and we have only to consider V_2 interaction. We want to calculate the transition rate from an initial state

$$|i\rangle = |A; 1_{k\lambda}, 0_{k'\lambda'}\rangle \tag{40}$$

to a final state

$$|f\rangle = |B; 0_{k\lambda}, 1_{k'\lambda'}\rangle \tag{41}$$

where A and B denote the initial and final state of the electon. Using the golden rule

$$W_{fi} = \frac{2\pi}{\hbar} |V_{fi}|^2 \rho(E_f) \tag{42}$$

and noting that

$$V_{fi} = \langle B; 0_{k\lambda}, 1_{k'\lambda'}| \frac{e^2}{2mc^2} A^2 | A; 1_{k\lambda}, 0_{k'\lambda'}\rangle$$

$$= \frac{e^2}{mc^2} \frac{2\pi\hbar c}{V} \frac{1}{\sqrt{kk'}} \left(\hat{\varepsilon}_{k\lambda}, \hat{\varepsilon}_{k'\lambda'}\right) e^{i(\underline{k}_1 - \underline{k}_2)\cdot \underline{r}} \delta_{AB} \tag{43}$$

and the

$$\rho(E_f) = \frac{V}{(2\pi)^3} \frac{\omega_2^2}{hc^3} \left(\frac{\omega_2}{\omega_1}\right) d\Omega \tag{44}$$

for the final density of states of a freely recoiling electron, we can express the differential cross-section as (write $\omega_1 = ck_1 \equiv ck$, $\omega_2 = ck_2 \equiv ck'$)

$$d\sigma = \frac{W_{fi}}{c/V} = \left(\frac{e^2}{mc^2}\right)\left(\hat{\varepsilon}_{k_1} \cdot \hat{\varepsilon}_{k_2}\right)^2 \left(\frac{\omega_2}{\omega_1}\right)^2 d\Omega \tag{45}$$

or

$$\left(\frac{d\sigma}{d\Omega}\right)_{pol} = r_o^2 \left(\hat{\varepsilon}_{k_1} \cdot \hat{\varepsilon}_{k_2}\right)^2 \left(\frac{\omega_2}{\omega_1}\right)^2 \tag{46}$$

where r_o is the classical radius of electron (e^2/mc^2). Equation (46) is nothing but the non-relativistic Thomson cross-section for an unbound electron. For unpolarized incident x-rays one averages Equation (46) over all possible initial polarizations to get

$$\left(\frac{d\sigma}{d\Omega}\right)_{unpol} = r_o^2\left(\frac{1+\cos^2\theta}{2}\right)\left(\frac{\omega_2}{\omega_1}\right)^2 . \quad (47)$$

Here θ is the scattering angle.

Bound Electrons

For bound electrons $\vec{p}_j \neq 0$ and, in principle, both V_1 and V_2 have to be included in the calculation of transition rate. However, it was shown by Eisenberger and Platzman[7] that when $\hbar\omega_1$ and $\hbar\omega_2$ are much larger than $E_f - E_i$, where E_i and E_f are initial and final state energies of the atom, the contribution to the transition rate from V_1 is small enough to be neglected. Typically $\hbar\omega_1 \sim \hbar\omega_2 \sim$ 20 KeV (Mo K_α -line, λ = 0.62Å) while $E_f - E_i$ varies from 20-100 eV for most light elements. This condition is thus met and one has only to retain V_2. A straightforward calculation such as the free electron case then gives

$$\left(\frac{d\sigma}{d\Omega}\right)_{pol} = r_o^2\left(\hat{\varepsilon}_{k_1}\cdot\hat{\varepsilon}_{k_2}\right)^2\left(\frac{\omega_2}{\omega_1}\right)\left|\langle f|\sum_{j=1}^{N} e^{i\underline{Q}\cdot\underline{r}_j}|i\rangle\right|^2 . \quad (48)$$

We have here only (ω_2/ω_1) factor because in the case of a bound electron, one should not include the recoil factor in the density of state factor. One uses instead

$$\rho(E_f) = \frac{V}{(2\pi)^3}\,\frac{\omega_2^2}{\hbar c^3}\,d\Omega . \quad (49)$$

The double differential cross-section is likewise put into a form similar to Equation (1)

$$d^2\sigma/d\Omega d\omega_2$$

$$= r_o^2\left(\hat{\varepsilon}_{k_1}\cdot\hat{\varepsilon}_{k_2}\right)^2\left(\frac{\omega_2}{\omega_1}\right)\sum_f\left|\langle f|\sum_{j=1}^{N} e^{i\underline{Q}\cdot\underline{r}_j}|i\rangle\right|^2\delta\left(\omega+\frac{E_i-E_f}{\hbar}\right)$$

$$= \left(\frac{d\sigma}{d\Omega}\right)_{Thomson}\left(\frac{\omega_1}{\omega_2}\right)\sum_f\left|\langle f|\sum_{j=1}^{N} e^{i\underline{Q}\cdot\underline{r}_j}|i\rangle\right|^2\delta\left(\omega+\frac{E_i-E_f}{\hbar}\right) . \quad (50)$$

Note that the kinematic factor in this case is $\omega_1/\omega_2 = k_1/k_2$ which is just the inverse of that in the neutron cross-section [Equation (28)].

Referring to the dynamic structure in Equation (50), we can distinguish between three cases for an atomic system:

A. Take $|i\rangle = |0\rangle$ to be the ground state of the atom. Then if $|f\rangle = |n\rangle$ is a discrete excited electronic state, the scattering is called "electronic Raman scattering" and is an incoherent scattering.

B. If $|f\rangle$ is in the continum states of the atom then the scattering is called "Compton scattering" which is also an incoherent scattering.

C. If $|i\rangle = |f\rangle = |0\rangle$, that is, when the scattering does not involve excitation of the electronic system (therefore elastic), then it is called "Rayleigh scattering" which is a coherent scattering. We shall be interested only in the Rayleigh scattering in the case of the x-ray diffraction experiment.

Eisenberger and Platzman[7] calculated relative magnitudes of contributions from each of the three processes for hydrogen atom as a function of parameter Qa (a = Bohr radius). They showed that for small Qa (<0.5) Rayleigh scattering is dominant, but for large Qa (>2.0) Compton scattering is a dominant factor. Raman scattering is always small and only appreciable around qa $\sim$ 1.0.

For unpolarized x-rays, the Rayleigh cross-section is then

$$\left(\frac{d\sigma}{d\Omega}\right)_R = r_o^2 \left(\frac{1+\cos^2\theta}{2}\right) \left| \langle 0| \sum_{j=1}^{N} e^{i\underline{Q}\cdot\underline{r}_j} |0\rangle \right|^2 \qquad (51)$$

where $|0\rangle$ is the ground state of the system. Consider the so-called scattering factor

$$f(\underline{Q}) = \langle 0| \sum_{j=1}^{N} e^{i\underline{Q}\cdot\underline{r}_j} |0\rangle \ . \qquad (52)$$

Introducing the average charge density at $\underline{r}$ by

$$\rho(\underline{r}) = \langle 0| \sum_{j=1}^{N} \delta(\underline{r} - \underline{r}_j) |0\rangle \ , \qquad (53)$$

we can rewrite the scattering factor as

$$f(\underline{Q}) = \int d^3r\rho(\underline{r})\, e^{i\underline{Q}.\underline{r}} \tag{54}$$

which is seen to be a Fourier transform of charge density distribution in the system. In condensed phases where electrons can be regarded as tightly bound to atoms with center position at $\underline{R}_\ell$. We can further write

$$f(\underline{Q}) = \langle 0|\sum_\ell\sum_{j=1}^{Z} e^{i\underline{Q}.(\underline{R}_\ell+\underline{r}_{\ell j})}|0\rangle$$

$$= \sum_\ell e^{i\underline{Q}.\underline{R}_\ell} \langle 0|\sum_{j=1}^{Z} e^{i\underline{Q}.\underline{r}_{\ell j}}|0\rangle$$

$$= \sum_\ell f_\ell(\underline{Q})\, e^{i\underline{Q}.\underline{R}_\ell} \tag{55}$$

where $f_\ell(\underline{Q})$ is the atomic form factor given by

$$f_\ell(\underline{Q}) = \int d\tau\, \overline{\psi}_o^* \sum_{j=1}^{Z} e^{i\underline{Q}.\underline{r}_j}\, \overline{\psi}_o \tag{56}$$

and $\overline{\psi}_o$ is the atomic ground state wave function. We notice $f_\ell(Q) \underset{Q\to 0}{=} Z$ (atomic number) so that x-ray scattering amplitude is very weak for light elements. In terms of the atomic form factor [Equation (51)] can be rewritten as

$$\left(\frac{d\sigma}{d\Omega}\right)_R = r_o^2\left(\frac{1+\cos^2\theta}{2}\right)\left|\sum_\ell f_\ell(Q)\, e^{i\underline{Q}.\underline{R}_\ell}\right|^2 . \tag{57}$$

For a monoatomic single crystal, $f_\ell(\underline{Q}) = f(\underline{Q})$, and the diffraction condition is satisfied only at reciprocal lattice points, i.e., only when $\underline{Q} = \underline{G}$. Hence, an absolute atomic structure factor measurement at various $\underline{G}$ can be used to determine

$$|f(\underline{G})|^2 = \left|\int d\tau\, \overline{\psi}_o^* \sum_{j=1}^{Z} e^{i\underline{Q}.\underline{r}_j}\, \overline{\psi}_o\right|^2 , \tag{58}$$

which can provide information on the ground state electronic wave function.

4. LIGHT SCATTERING

Quantum mechanical formulation of scattering of electromagnetic waves in the optical frequencies can be based on the same Hamiltonian [Equation (37)] which we used before for the x-ray case,

$$H = H_s + H_R + V_1 + V_2. \tag{59}$$

Since for optical frequencies $\hbar\omega_1 \sim \hbar\omega_2 \sim 2$eV are of the same order of magnitude as atomic energy levels, both interaction terms V_1 and V_2 contribute to the matrix element in the transition rate. In general it can be shown[8] that the sum of the interaction energies V_1 and V_2 can be transformed into an equivalent series of multipole interactions. This is especially convenient at low frequencies when ka << 1 or, in other words, when the wave length of light is large when compared to the atomic size. In this case we have only to retain the leading term in the multipole expansion, which is just the electric dipole term, i.e.,

$$V_1 + V_2 \longleftrightarrow -\sum_{\ell} \vec{\mu}_\ell \cdot \vec{E}_\ell \tag{60}$$

$\vec{\mu}_\ell$ is the dipole moment of the μ-th atom in the system, and $\vec{E}_\ell$ the electric field vector at the ℓ-th atomic site and is given by

$$\vec{E}_\ell = -\frac{1}{c}\frac{\partial \vec{A}_\ell}{\partial t}$$

$$= \sum_{k,\lambda} i\sqrt{\frac{2\pi\hbar\omega_k}{V}}\, \hat{\varepsilon}_{k\lambda}\left[a_{k\lambda} e^{\underline{ik}\cdot\underline{R}_\ell} - a^+_{k\lambda} e^{-\underline{ik}\cdot\underline{R}_\ell}\right]. \tag{61}$$

Since the interaction V is a linear combination of $a_{k\lambda}$ and $a^+_{k\lambda}$, the transition matrix element between an initial and final states of the form

$$|i\rangle = |A; 1_{k\lambda}, 0_{k'\lambda'}\rangle \; ; \quad E_i = E_A + \hbar\omega_k \tag{62}$$

$$|f\rangle = |B; 0_{k\lambda}, 1_{k'\lambda'}\rangle \; ; \quad E_f = E_B + \hbar\omega_{k'} \tag{63}$$

is of the second order type, i.e.,

$$\sum_I \frac{\langle f|V|I\rangle\langle I|V|i\rangle}{E_i - E_I} \quad .$$

Furthermore, there are two possible intermediate states

$$|I>_1 \;=\; |I;\, 0_{k\lambda},\, 0_{k'\lambda'}> \quad ; \quad E_{I_1} = E_I \tag{64}$$

$$|I>_2 \;=\; |I,\, 1_{k\lambda},\, 1_{k'\lambda'}> \quad ; \quad E_{I_2} = E_I + \hbar\omega_k + \hbar\omega_{k'} \tag{65}$$

Taking into account the energy conservation relation that $\hbar\omega_k + E_A = \hbar\omega_{k'} + E_B$, the transition matrix element works out to be a sum of two terms for each atom

$$M_\ell \;=\; -2\pi\hbar\sqrt{\omega_k\omega_{k'}}\; e^{i\underline{Q}.\underline{R}_\ell}\; x$$

$$\sum_I \left\{ \frac{<B|\mu^\ell_{k'}|I><I|\mu^\ell_k|A>}{E_A-E_I+\hbar\omega_k} + \frac{<B|\mu^\ell_k|I><I|\mu^\ell_{k'}|A>}{E_A-E_I-\hbar\omega_{k'}} \right\} \tag{66}$$

where abbreviations $\mu^\ell_k \equiv \hat{\varepsilon}_{k\lambda} \cdot \vec{\mu}_\ell$; $\mu^\ell_{k'} \equiv \hat{\varepsilon}_{k'\lambda'} \cdot \vec{\mu}_\ell$ are made.

Scattering Cross-section for an Atom

Using the "golden rule" we work out the differential cross-section to be

$$\left(\frac{d\sigma}{d\Omega}\right)_{Raman} = kk'^3 \left| \sum_I \left\{ \frac{<B|\mu_{k'}|I><I|\mu_k|A>}{E_A-E_I+\hbar\omega_k} + \frac{<B|\mu_k|I><I|\mu_{k'}|A>}{E_A-E_I-\hbar\omega_{k'}} \right\} \right|^2 \tag{67}$$

In the special case of elastic scattering (Rayleigh scattering) we set $\omega_k = \omega_{k'} = \omega$, $E_A = E_B = E_0$, $|A> = |B> = |0>$, $|I> = |n>$ and get

$$\left(\frac{d\sigma}{d\Omega}\right)_R = k^4 \left| \frac{1}{\hbar} \sum_n \left\{ \frac{<0|\mu_{k'}|n><n|\mu_{k'}|0>}{\omega_{no} - \omega} + \frac{<0|\mu_k|n><n|\mu_{k'}|0>}{\omega_{no} + \omega} \right\} \right|^2$$

$$= k^4 \left| \sum_{\alpha,\beta} \varepsilon'_\alpha\, P_{\alpha\beta}(\omega)\varepsilon_\beta \right|^2 . \tag{68}$$

In Equation (68) the quantity $P_{\alpha\beta}(\omega)$, given by

$$P_{\alpha\beta}(\omega) = \frac{1}{\hbar} \sum_n \left\{ \frac{<0|\mu_\alpha|n><n|\mu_\beta|0>}{\omega_{no}-\omega-i\eta} + \frac{<0|\mu_\beta|n><n|\mu_\alpha|0>}{\omega_{no}+\omega+i\eta} \right\} , \tag{69}$$

is the atomic polarizability tensor.

To show this latter point we consider an atomic depole subjected to a time-dependent EM field with an interaction energy of a form

$$\hat{A}F(t) = -\hat{\mu}\cdot\hat{\varepsilon}E_0 e^{-i\omega t+\eta t} \tag{70}$$

and compute the response of a system property $\hat{B}$ via a standard linear response formula[9]

$$<\hat{B}(t)> - B_o = \chi_{BA}(\omega)e^{-i\omega t+\eta t} \tag{71}$$

where the "admittance" function is given by

$$\chi_{BA}(\omega) = \frac{1}{i\hbar}\int_0^\infty d\tau <0|[\hat{B}_I(\tau), \hat{A}]|0> e^{i\omega\tau-\eta\tau} \tag{72}$$

We are interested in the case $\hat{A} \equiv -\vec{\mu}\cdot\hat{\varepsilon}E_0$ and $\hat{B} = \mu_\alpha$. Since in the absence of the field the atomic dipole is randomly oriented with its equilibrium value $B_o = 0$, we have

$$<\hat{\mu}_\alpha(t)> = \chi_{\alpha\beta}(\omega)\, e^{-i\omega t+\eta t}$$

where $\chi_{\alpha\beta}(\omega)$ may be worked out from Equation (72) to be

$$\chi_{\alpha\beta}(\omega) = \frac{1}{\hbar}\sum_n \left\{\frac{<0|\mu_\alpha|n><n|\mu_\beta|0>}{\omega_{no} - \omega - i\eta} + \frac{<0|\mu_\beta|n><n|\mu_\alpha|0>}{\omega_{no} + \omega + i\eta}\right\}$$

$$= P_{\alpha\beta}(\omega)E_0 . \tag{73}$$

Notice that, if the atomic charge distribution is spherically symmetric, $P_{\alpha\beta}(\omega)$ reduces to a scalar polarizability $\alpha(\omega)$, i.e.,

$$P_{\alpha\beta}(\omega) = \alpha(\omega)\delta_{\alpha\beta} \tag{74}$$

and the Rayleigh cross-section is simply

$$\left(\frac{d\sigma}{d\Omega}\right)_R = k^4|\hat{\varepsilon}_{k_1}\cdot\hat{\varepsilon}_{k_2}|^2\,|\alpha(\omega)|^2 . \tag{75}$$

Now, consider a system of N optically isotropic atoms in thermal equilibrium. Remember from Equation (66) that μ_ℓ contains a phase factor $e^{i\underline{Q}\cdot\underline{R}_\ell}$. We obtain the differential cross-section by summing the transition matrix element over ℓ before squaring:

$$\left(\frac{d\sigma}{d\Omega}\right)_R = k^4|\hat{\varepsilon}_{k_1}\cdot\hat{\varepsilon}_{k_2}|^2|\sum_\ell \alpha_\ell(\omega)\ e^{i\underline{Q}\cdot\underline{R}_\ell}|^2 \quad . \tag{76}$$

The double differential cross-section is likewise obtained by averaging Equation (76) over the initial external motional states of the thermal equilibrium system and then summing over the final motional states with an energy conserving factor:

$$\frac{d^2\sigma}{d\Omega d\omega} = k^4|\hat{\varepsilon}_{k_1}\cdot\hat{\varepsilon}_{k_2}|^2\sum_{i,f} P_i|\langle f|\sum_\ell \alpha_\ell e^{i\underline{Q}\cdot\underline{R}_\ell}|i\rangle|^2\,\delta\left(\omega+\frac{E_i-E_f}{h}\right)$$

$$= k^4|\alpha|^2|\hat{\varepsilon}_{k_1}\cdot\hat{\varepsilon}_{k_2}|^2 S(Q,\omega). \tag{77}$$

We see that, for a monoatomic system, the basic cross-section $(d\sigma/d\Omega)_o$ is nothing but the Rayleigh scattering cross-section for an isolated atom, and here $\alpha(\omega)$ plays the role of scattering length in the case of thermal neutron scattering.

5. DIGRESSION: CLASSICAL V.S. QUANTUM MECHANICAL DESCRIPTIONS OF SCATTERING

It is useful at this point to return to the question: What is the relative merit of using quantum mechanical versus classical descriptions of scattering of electromagnetic wave by a material medium? Since Maxwell's equations are a valid description of EM phenomena almost down to the atomic scale; in many ways it is nicer to start from the classical wave equation.

Consider Maxwell's equation in a non-magnetic medium without free charge and current density. Setting the time dependence of all fields to be $e^{-i\omega t}$, we have

$$\nabla\times\vec{E} - (i\omega/c)\vec{B} = 0 \qquad \text{(First Maxwell's equation)} \tag{78}$$

$$\nabla\times\vec{B} + (in^2\omega/c)\vec{E} = 0 \qquad \text{(Third Maxwell's equation)} \tag{79}$$

Eliminating $\vec{B}$ from Equations (78) and (79) we get

$$\nabla\times\nabla\times\vec{E} - (n^2\omega^2/c^2)\vec{E} = 0$$

or

$$\nabla\times\nabla\times\vec{E} - n^2k_o^2\vec{E} = 0 \tag{80}$$

where n is the index of refraction of the medium and k_o is the wave number in vacuum. Equation (80), with a subsidiary condition $\nabla\cdot\vec{E}=0$ (which follows from Maxwell's fourth equation $\nabla\cdot\vec{D}=0$), supports only propagation of a plane wave with a modified wave number $k \equiv nk_o$. In order to have scattering one must have an inhomogeneity of the index of refraction. Let the dielectric inhomogeneity be given by

$$n^2 \equiv \varepsilon = \varepsilon_o + \Delta\varepsilon(\underline{r}) \equiv n_o^2 + n^2(\underline{r}) \tag{81}$$

where the fluctuation $\Delta\varepsilon(\underline{r}) << \varepsilon_o$. Then Equation (80) can be rewritten,

$$\nabla x \nabla x \vec{E} = \nabla(\nabla\cdot\vec{E}) - \nabla^2\vec{E} = -\nabla^2\vec{E},$$

to get

$$\nabla^2\vec{E} + k^2\vec{E} = -k_o^2\ \Delta\varepsilon(\underline{r})\vec{E} \tag{82}$$

where $k = n_o k_o$ is the propagation wave number in the medium in the absence of fluctuations. This is a vector analogue of the scalar Schrodinger wave equation which we solved for the neutron scattering. It describes propagation of EM waves in a material with local fluctuations of dielectric constant. Quantum theory is only needed to calculate the material property in terms of atomic properties. Once this is done the solution of the scattering problem is entirely classical. The standard method, analogous to what we did in Section 2, can be used to establish a scattering solution

$$\vec{E}_s(\underline{r},t) = -\frac{e^{ikr}}{r}\frac{k^2}{4\pi}(\overline{\overline{I}}-\hat{r}\hat{r})\cdot\vec{E}_o e^{-i\omega t}\int_V d^3r'\Delta\varepsilon(\underline{r}')e^{i\underline{Q}\cdot\underline{r}'} \tag{83}$$

where $\overline{\overline{I}}$ is the unit dyad, $\hat{r}$ the unit vector in the scattered direction, and $\vec{E}_o$ the incident field vector.

In the case of Rayleigh scattering of x-rays $k = k_o$ (in vacuum), the dielectric constant fluctuation at the x-ray frequency is given by a free electron expression[10]

$$\Delta\varepsilon(\underline{r}) = -\frac{4\pi e^2 n(\underline{r})}{m\omega^2} \tag{84}$$

where $n(\underline{r})$ is the number density of electrons at r. Therefore

$$\vec{E}_s(\underline{r},t) = \frac{e^{ikr}}{r}(\bar{\bar{I}}-\hat{r}\hat{r})\cdot\vec{E}_o\left(\frac{e^2}{mc^2}\right)\int_V d^3r'\ n(\underline{r}')e^{i\underline{Q}\cdot\underline{r}'}. \tag{85}$$

We observe that

$$|(\bar{\bar{I}} - \hat{r}\hat{r})\cdot\vec{E}_o|^2 = \sin^{-2}\gamma \tag{86}$$

where γ is the angle between $\hat{E}_o$ and $\hat{r}$. The differential cross-section is just the scattering amplitude squared, with a unit amplitude incident wave. Thus

$$\left(\frac{d\sigma}{d\Omega}\right)_R = \left(\frac{e^2}{mc^2}\right)\sin^2\gamma|n(\underline{Q})|^2 \tag{87}$$

which is exactly the same as the quantum mechanical result obtained from Equation (48) by setting $|i\rangle = |f\rangle = |0\rangle$, i.e.,

$$\left(\frac{d\sigma}{d\Omega}\right)_R = \left(\frac{e^2}{mc^2}\right)^2(\hat{\varepsilon}_{k_1}\cdot\hat{\varepsilon}_{k_2})|\langle 0|\sum_{j=1}^{N} e^{i\underline{Q}\cdot\underline{r}_j}|0\rangle|^2, \tag{88}$$

noting $\sin^2\gamma = (\hat{\varepsilon}_{k1}\cdot\hat{\varepsilon}_{k2})^2$. The Fourier transform of the electron density in the classical expression is to be understood quantum mechanically as

$$n(\underline{Q}) = \langle 0|\sum_{j=1}^{N} e^{i\underline{Q}\cdot\underline{r}_j}|0\rangle. \tag{89}$$

In the case of light scattering in the commonly used VV geometry, the factor $(\bar{\bar{I}}-\hat{r}\hat{r})\cdot\vec{E}_o$ equals $\vec{E}_o$. In condensed phases the dielectric constant fluctuation may contain a slow time dependence caused by slowly varying density fluctuations of wave lengths of the same order of magnitude as the wave length of light. We therefore put

$$\Delta\varepsilon(\underline{Q},t) = \left(\frac{\partial\varepsilon}{\partial n}\right)_T n(\underline{Q},t) \tag{90}$$

in Equation (83) to get

$$\vec{E}_s(\underline{r},t) = -\frac{e^{ikr}}{r}\vec{E}_o e^{-i\omega t}\frac{k^2}{4\pi}\left(\frac{\partial\varepsilon}{\partial n}\right)_T n(\underline{Q},t). \tag{91}$$

Equation (91) can then be used to obtain the commonly used expression for the double differential cross-section[11]

$$\frac{d^2\sigma}{d\Omega d\omega_2} = \frac{k^4}{16\pi^2}\left(\frac{\partial\varepsilon}{\partial n}\right)^2 \frac{1}{2\pi}\int_{-\infty}^{\infty} dt\ e^{-i\omega t} < n(-\underline{Q},0)n(\underline{Q},t)>. \tag{92}$$

6. PHYSICAL INTERPRETATION OF DYNAMIC STRUCTURE FACTORS

In previous sections we have shown that all three scattering experiments, involving thermal neutrons, x-rays and light, lead to double differential cross-sections expressable in terms of a dynamic structure $S(\underline{Q},\omega)$ or its self part $S_s(\underline{Q}\omega)$. It is desirable to give some physical interpretation on the relation between dynamics as measured in $(\underline{Q}\omega)$ space and dynamics as pictured in real space time $(\underline{r}, t)$.

Define a thermal average of a physical quantity $\hat{O}$ by

$$<\hat{O}> = \sum_i P_i <i|\hat{O}|i>$$

In terms of this average the dynamic structure factors can be written as

$$S(\underline{Q},\omega) = \frac{1}{2\pi}\int_{-\infty}^{\infty} dt\ e^{-i\omega t}\ \frac{1}{N}\sum_{\ell,\ell'} <e^{-i\underline{Q}.\underline{R}_\ell(0)}\ e^{i\underline{Q}.\underline{R}_{\ell'}(t)}> \tag{93}$$

$$S_s(\underline{Q},\omega) = \frac{1}{2\pi}\int_{-\infty}^{\infty} dt\ e^{-i\omega t} <e^{-i\underline{Q}.\underline{R}(0)}\ e^{i\underline{Q}.\underline{R}(t)}>. \tag{94}$$

One often calls the factor in the integrand the intermediate scattering functions:

$$F(\underline{Q},t) = \frac{1}{N}\sum_{\ell,\ell'} <e^{-i\underline{Q}.\underline{R}_\ell(0)}\ e^{i\underline{Q}.\underline{R}_{\ell'}(t)}> \tag{95}$$

$$F_s(\underline{Q},t) = <e^{-i\underline{Q}.\underline{R}(0)}e^{i\underline{Q}.\underline{R}(t)}>. \tag{96}$$

A frequency domain experiment measures the dynamic structure factor, whereas a time domain experiment measures the intermediate scattering function. Among frequency domain techniques, we have, for neutrons, crystal spectrometry and time of flight measurements, and for photons, optical mixing spectroscopy used in light scattering experiment. Time domain techniques include the newly developed neutron "spin-echo" spectrometer, and the extensively used photon correlation spectroscopy.

In order to discuss dynamics in (Q,ω) space versus $(\underline{r}, t)$ space it is most instructive to use examples. Consider, first, incoherent scatterings. These situations arise in incoherent scattering of neutrons from hydrogen containing substances, or in light scattering from a system of independently moving particles. One is observing, in the case, a single particle motion in the system. The simplest model of thesingle particle motion is a random walk model. A particle moves in an arbitrary direction for a step length ℓ with a mean speed V_o (for a mean duration $\tau = \ell/V_o$), then turns around to a randomly selected direction and repeats the same motion. This is a Gaussian Markoff process, and we can take an exponentially decaying velocity auto-correlation function:

$$\langle V(0)V(t)\rangle = V_o^2 e^{-t/\tau} . \tag{97}$$

One can calculate the intermediate scattering function[12] by using Equation (97) to obtain

$$F_s(Q, t) = \text{Exp}\left[-Q^2\ell^2(t/\tau - 1 + e^{-t/\tau})\right] \tag{98}$$

where $F_s(Q, t)$ depends only on $|\vec{Q}|$ because the motion is isotropic. It is instructive to consider two limiting cases:

A. $\underline{Q\ell \ll 1 \text{ (or } Q^{-1} \gg \ell)}$.

For small values of $Q\ell$, the half width of $F_s(Q, t)$ occurs at large time $(t/\tau \gg 1)$ where we may neglect $\exp(-t/\tau)$ and 1 compared to t/τ in the second bracket of Equation (98). Hence

$$F_s(Q, t) \xrightarrow[Q\ell \ll 1]{} e^{-Q^2(\ell^2/\tau)t} = e^{-Q^2 Dt} . \tag{99}$$

We can interpret this result by noting that $R \equiv Q^{-1} \gg \ell$, meaning the spatial resolution R of the probe is much larger than the particle mean free path ℓ. In this case the probe is sampling the particle motion with such a coarse resolution that it sees many changes of directions characterized by a particle undergoing diffusion. This corresponds to an observation of the long time limit of $F_s(Q, t)$ which is an exponential decay with a diffusion constant $D = \ell^2/\tau$. Fourier transform of Equation (99) gives a Lorentzian spectral density function

$$S_s(Q, \omega) \xrightarrow[Q\ell<<1]{} \frac{1}{\pi} \frac{DQ^2}{\omega^2 + (DQ^2)^2}, \tag{100}$$

which has a spectral with DQ^2 inversely proportional to the decay time constant of $F_s(Q, t)$.

B. <u>$Q\ell>>1$ (or $Q^{-1}<<\ell$).</u>

In this opposite limit the half-width of $F_s(Q, t)$ occurs at $t/\tau<<1$ so that we are allowed to expand the exponential factor $e^{-t/\tau}$ to get

$$F_s(Q, t) \xrightarrow[Q\ell>>1]{} e^{-\frac{1}{2}Q^2\ell^2(\frac{t}{\tau})^2} = e^{-\frac{1}{2}Q^2V_o^2t^2} . \tag{101}$$

Interpretation of this limit can be made by saying that the spatial resolution of the probe is so high that it samples motions corresponding to small segments of the free moving paths. This limit is characterized by a Gaussian time dependence of $F_s(Q, t)$ with a decay time $t_{\frac{1}{2}} = \sqrt{2\ell n2}/QV_o$.

The corresponding dynamic structure factor

$$S_s(Q, \omega) = \frac{1}{\sqrt{2\pi}\, QV_o} e^{-(\omega^2/2Q^2V_o^2)} \tag{102}$$

also shows a well known Doppler broadened line shape with a half-width $\omega_{\frac{1}{2}} = \sqrt{2\ln 2}\cdot QV_o$ which is again inversely proportional to $t_{\frac{1}{2}}$. Figure 3 gives plots of the reduced decay constant $t_{\frac{1}{2}}QV_o$ and spectral width $\omega_{\frac{1}{2}}/QV_o$ as a function of the observational parameter $Q\ell$. An inverse relationship

$$(t_{\frac{1}{2}}QV_o)^{-1} \propto \frac{\omega_{\frac{1}{2}}}{QV_o} \tag{103}$$

follows from the fact that $F_s(Q, t)$ and $S_s(Q, \omega)$ are a Fourier transorm pair.

Figure 4 shows an example of $S_s(Q, \omega)$ as measured by an incoherent neutron quasi-elastic scattering from compressed hydrogen

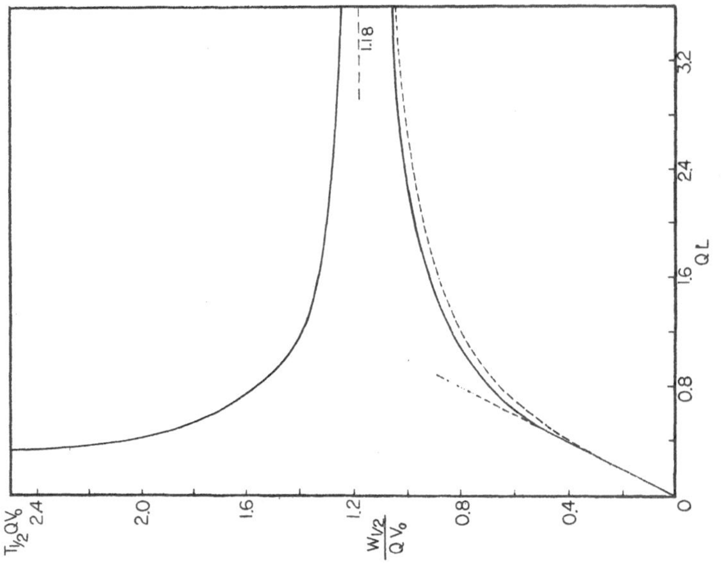

Figure 3. Plots of the reduced half-width, $t_{\frac{1}{2}}QV_0$, of $F_s(Q,t)$ given by Equation (98) and the reduced half-width, $\omega_{\frac{1}{2}}/QV_0$, of the corresponding $S_s(Q,\omega)$ as function of the parameter $Q\ell$. These waves illustrate nicely a continuous transition from the collision (i.e., change of direction) dominated diffusion behavior at small $Q\ell$ to the free particle-like behavior at large $Q\ell$ for particles undergoing independent random walks. The dashed line in the lower part of the figure refers to a different kind of random walks, namely, particles undergoing binary collisions as in the case of low density gases.

at 85°K.[13] Y is a parameter essentially equal to $(Q\ell)^{-1}$. This parameter can be changed by changing the scattering angle (thus changing Q) or by changing gas pressure to vary ℓ. Looking from the top to bottom curves in Figure 4 one can clearly see a gradual transition from a Doppler broadened Gaussian line at the top to collision dominated Lorentzian line at the bottom. A plot of reduced width $\omega_{\frac{1}{2}}/QV_0$ versus $Q\ell$ is given by a dash line in lower part of Figure 3. The experimental line widths do not quite agree with the random walk model (given by the solid line) because in a gas a molecule collides with another molecule, leaving with some degree of persistence in the original direction of the incidence. It is

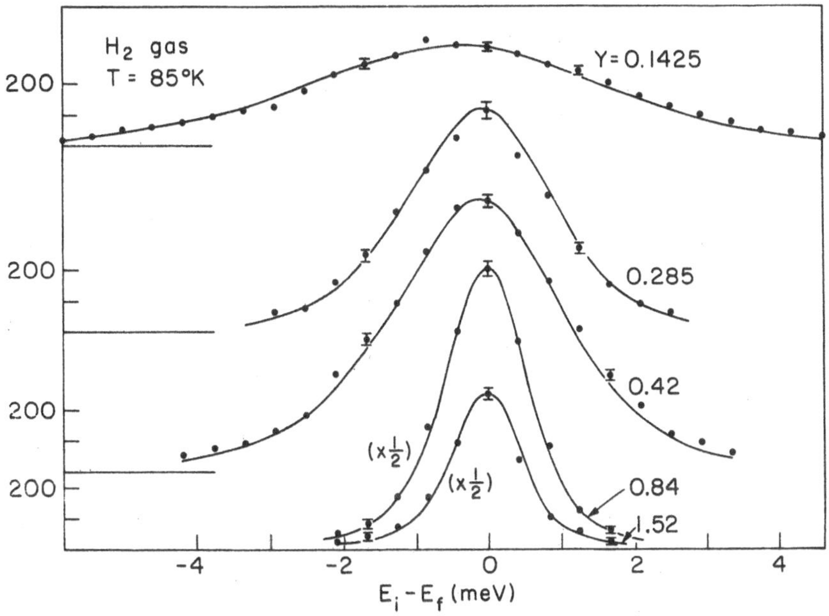

Figure 4. A series of the frequency domain function $S_s(q,\omega)$ as measured by an incoherent neutron quasi-elastic scattering from compressed hydrogen gas at 85°K.[13] The Abssisa is $\hbar\omega$ in unit of meV and the ordinate is proportional to $S_s(Q,\omega)$. The parameter $Y \equiv (Q\ell)^{-1}$. As the gas is compressed, the mean free path between collision decreases or equivalently Y parameter increases. Notice that as Y = 0.14 the line shape is more like a Gaussian but as Y increases to 1.5 the line shape changes to more like a Lorentzian.

therefore not accurate to model the collision with a random change of direction as in the random walk model.

Figure 5 shows an example of the time domain measurement.[14] It gives the reduced width $t_{\frac{1}{2}}QV_0$ of $F_s(Q,t)$ measured from a collection of independent, randomly moving, E. Coli bacteria in quasi-elastic light scattering. The half-width is plotted against scattering angle θ which is related to $Q = 2k \sin\theta/2$. The plot resembles the top curve of Figure 3 because, qualitatively speaking, bacteria moving in "motility buffer" undergo a kind of random walk as a result of chemotaxis.

Information contained in the full (or coherent) dynamic structure factor is of a different nature as compared with the single particle (or self) dynamic structure factor we discussed above. Note that the local density function, defined by

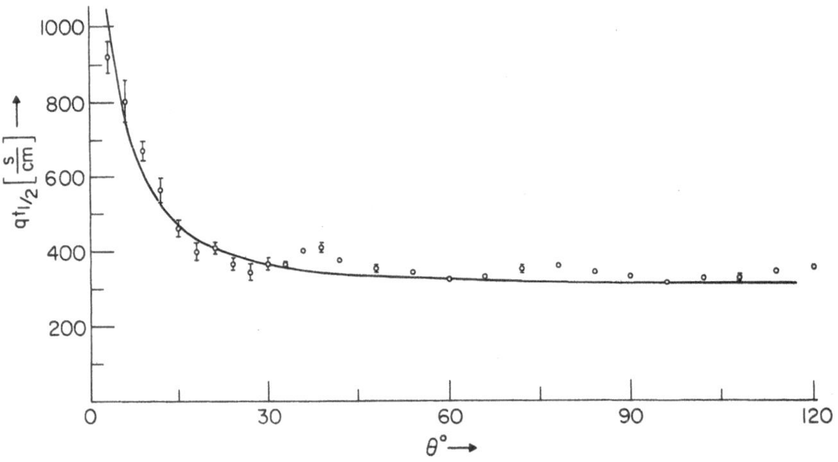

Figure 5. An example of the time-domain experiment done with photon correlation technique. This is a plot of reduced half-widths $t_{1/2}Q$ ($Q \equiv q$ in the figure) vs scattering angle θ, of $F_S(Q,t)$ measured from a collection of independently moving E. Coli bateria.[14] The bacteria are chemotactic to oxygen and are known to undergo "run-twiddle" motions. The twiddle can be regarded as a series of small step motions with $\ell \sim 0.1$-0.3 μm. This is essentially an experimental realization of the upper curve in Figure 3. At $\theta \lesssim 10°$ the condition $Q\ell < 1$ is satisfied and consequently the reduced half-widths start to rise rapidly.

$$n(\underline{r}, t) = \sum_{\ell} \delta[\underline{r} - \underline{R}_{\ell}(t)], \tag{104}$$

has a Fourier transform

$$n(\underline{q}, t) = \sum_{\ell} e^{i\underline{q}\cdot\underline{R}_{\ell}(t)} . \tag{105}$$

$n(\underline{q}, t)$ expresses a collective density oscillation of a wave vector $\underline{q}$ which exists in any condensed phase at finite temperatures. In an equilibrium system, $n(\underline{q}, t)$ is a stationary process so it is more natural to talk about the density correlation function

$$\frac{1}{N} \langle n(-\underline{q}, 0)n(\underline{q}, t)\rangle . \tag{106}$$

From Equation (95), one immediately realizes that a coherent scattering experiment measures, directly, such a density correlation function at a particular wave vector $\underline{Q}$ imposed by the probe, i.e.,

$$\underline{q} = \underline{Q}. \tag{107}$$

Thus, the half-width of $F(\underline{Q}, t)$ indicates the "decay time" of the density fluctuation of a wave vector $\underline{q} = \underline{Q}$. Similarly, the half-width of $S(\underline{Q}, \omega)$ reflects the corresponding "decay rate" of the density oscillation.

An interesting situation occurs when local order of a fluid is characterized by a correlation length ξ. A well known example is given by an one-component fluid near its liquid-gas critical point or a two-component fluid near its de-mixing critical point. In the limit of $Q\ell << 1$ the radiation is probing the system with a spatial resolution much coarser than the correlation range, and one is effectively seeing many correlated regions each diffusing independently like a Brownian particle with radius ξ. The density or the concentration fluctuation spectrums have line-widths Γ_Q of the form

$$\Gamma_Q \underset{Q\xi \to 0}{\frown} \frac{k_B T}{\eta \xi} Q^2 \sim \frac{Q^3}{Q\xi} . \tag{108}$$

On the other hand, in the limit $Q\xi > 1$, the spatial resolution of the probe is so good that it does not see the correlation range ξ any longer and one gets a ξ independent line width

$$\Gamma_Q \underset{Q\xi > 1}{\frown} Q^3 . \tag{109}$$

The theory of critical slowing down in fluids is by now well known[15], thus no attempt is made to elaborate it here except to show one example (Figure 6) from a two-component fluid of a 50-50 mixture of n-hexane and nitrobenzene near room temperature. This work, of C. C. Lai,[16] clearly shows that the Γ_Q/Q^3 versus Q plot has a region inversely proportional to $Q\xi$ when $Q\xi < 1.0$ and a region approaching a constant when $Q\xi > 1.0$.

ACKNOWLEDGEMENT

I appreciate the assistance from Paul Wang in the preparation of the figures. This work is supported by the National Science Foundation grant #NSF PCM7815844.

REFERENCES

1. J. S. Higgins, L. K. Nicholson, and J. B. Hayter, "Observation of single chain motion in a polymer melt," preprint (1980).
2. V. F. Sears, "Dynamic theory of neutron diffraction," Can. J. Phys. 56: 1262 (1978).
3. Bound scattering length refers to scattering length of a nucleus which is fixed in space. Measurement of scattering length

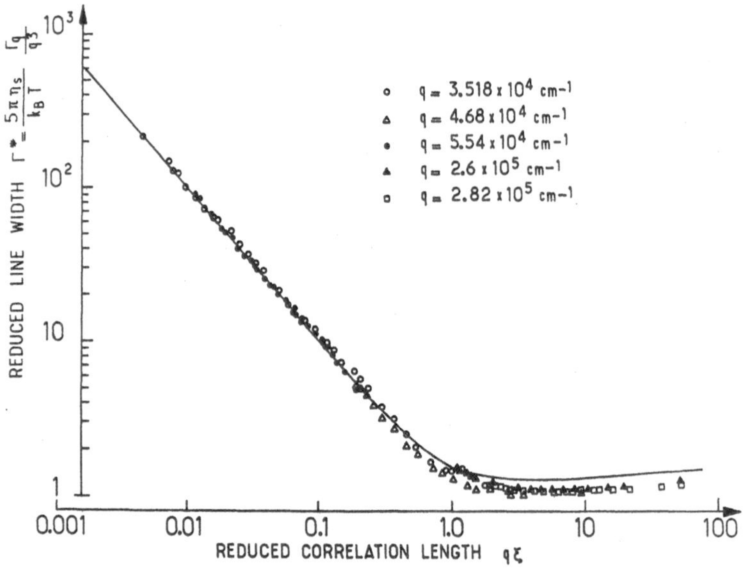

Figure 6. A reduced line-width of quasi-elastic light scattering spectra from a binary liquid mixture n-hexane and nitrobenzene near the critical point, plotted as a function of a parameter $Q\xi$ ($Q \equiv q$ in the figure) where ξ is the correlation length of the critical flunctuations.

is normally made in a situation where the nucleus is free to recoil. The measurement gives free scattering length a which is related to b by $b = [(A+1)/A]a$ where A is atomic weight of the nucleus. See G. E. Bacon, Neutron Diffraction, 2d ed., Oxford (1975) for tabulation of values of b.

4. H. Rauch, and D. Petrascheck, "Dynamical neutron diffraction and its application," in Neutron Diffraction, ed. by H. Dachs, Springer-Verlag, Berlin (1978).
5. The Fermi approximation is discussed clearly in R. G. Sachs, Nuclear Theory, Addison-Wesley, Reading, Mass. (1953). p. 91.
6. See for example, L.D. Landau, and E. M. Lifshitz, Electrodynamics of Continuous Media, Addison-Wesley, Reading, Mass. (1960). Chapter 15.
7. P. Eisenberger, and P. M. Platzman, "Compton scattering of X-rays from bound electrons," Phys. Rev. A2: 415 (1970).
8. A detailed discussion of this transformation can be found in E. A. Power, and T. Thirunamachandran, "On the nature of the Hamiltonian for the interaction of radiation with atoms and molecules," Am. J. Phys. 46: 370 (1978).
9. See for example, P. Nozieres, Theory of Interacting Fermi Systems, W. A. Benjamin, New York (1964).

10. C. Kittel, Quantum Theory of Solids, John Wiley, New York (1963).
11. See derivation of Equation (92) from Equation (91) in S. H. Chen, "Structure of liquids," in Physical Chemistry - An Advanced Treatise, ed. by H. Eyring, D. Henderson, and W. Jost, vol. VIIIA, Academic Press, New York (1971).
12. G. E. Uhlenbech, and L. S. Ornstein, "On the theory of Brownian motion," Phys. Rev. 36: 823 (1930).
13. S. H. Chen, Y. Lefevre, and S. Yip, "Kinetic theory of collision line narrowing in pressurized hydrogen gas," Phys. Rev. A8: 3163 (1973).
14. M. Holz, and S. H. Chen, "Quasielastic light scattering from migrating chemostatic bands of E. Coli," Biophys. J. 23: 15 (1978).
15. K, Kawasaki, "Mode coupling and critical dynamics," in Phase Transitions and Critical Phenomena, ed. by C. Domb, and M. S. Green, vol. Va, Academic Press, New York (1976).
16. S. H. Chen, C. C. Lai, and J. Rouch, "Experimental confirmation of renormalization-group prediction of critical concentration fluctuation rate in hydrodynamic limit," J. Chem. Phys. 68: 1994 (1978); and C. C. Lai, Ph.D. thesis, "Light Intensity Correlation Spectroscopy and Its Application to Study of Critical Phenomena and Biological Problems," MIT (1972).

NEUTRON SMALL ANGLE SCATTERING, INSTRUMENTS AND APPLICATIONS

J. Schelten

Institut für Festkörperforschung
der Kernforschungsanlage Jülich
D-517 Jülich, West Germany

CONTENTS

1. INTRODUCTION TO SMALL ANGLE NEUTRON SCATTERING (SANS)

In this lecture the basic features of small angle neutron scattering as a tool to investigate solids are described. In discussing the design and construction of the most powerful instruments it is explained why the instruments are up to 100 m long, why moderately monochromatized neutron beams are used, and why long wavelength neutrons are favorable. In three examples the strength of this method, in comparison to competitive methods, is demonstrated. The examples, if referred to by their results, are:

(i) a spiral spin-order in higher manganese silicides was discovered by small angle neutron scattering which was not seen in a conventional neutron diffractometer;

(ii) the predicted correlation hole in the pair correlation function of polymer centroids was experimentally verified;

(iii) by combining light and neutron results, a reliable new method was developed to determine the critical exponent η of the second order phase transformation of a binary liquid mixture.

Frequently one thinks that small angle X-ray and neutron scattering (SAXS and SANS) are methods for determining the radius of gyration R_g of a scatterer. By plotting logarithmically the scattered intensity versus the square of the scattering vector Q, where Q is defined by

$$Q = \frac{2\pi}{\lambda} 2 \sin \frac{\theta}{2} \tag{1}$$

with λ the neutron wavelength and θ the scattering angle, one usually gets a straight line whose slope is related to the radius of gyration. In a rather crude way R_g is a measure of the size of the scatterer, which may be a macromolecule in solution or a precipitate in a matrix. This geometrical aspect of SAS is discussed in Guinier's book on X-ray diffraction and one certainly is not impressed by the fact that one can use neutrons to measure rather precisely the mean diameter of a precipitate or a macromolecule. However, there is more to SAS than just this for a variety of reasons.

(i) The radius of gyration can be utilized to discriminate between various models. An example will be provided below.

(ii) Any SAS pattern has an intensity axis which can be interpreted in terms of the number of coherently scattering atoms if calibrated to absolute units. In this case the intensity has many important implications.

(iii) A quite natural extension of the SAS measurements, after having determined the forward scattering value and radius of gyration, is to a higher Q range. At intermediate Q, i.e. $QR_g > 1$ and $Q \cdot a < 1$ where a is the atomic distance, a few general theoretical aspects of the scattering function are known. For instance, the scattering from particles with sharp boundaries obeys a Q^{-4} law and the scattering from polymers with a gaussian conformation obeys a Q^{-2} law. Furthermore, the integral SAS given by

$$\int \underline{Q}^2 \frac{d\Sigma}{d\Omega} (\underline{Q})\, d_3\underline{Q} \tag{2}$$

where $(d\Sigma/d\Omega)(Q)$ is the differential cross section is related to fluctuations of the scattering length density.

In the following example it is illustrated how informative an SAS signal can be. By blending into a hydrogenous polystyrene sample 5% deuterated polystyrene polymer molecules of the same molecular weight, information on the polymer conformation in amorphous polystyrene can be deduced. Figure 1 shows the measured scattering pattern. The polymer molecules are neither stretched out nor collapsed into dense packed coils since curve 1 and 3 are inconsistent with the experimental scattering function. However, the conformation is a gaussian chain with dimensions as in θ solution (curve 2).

For an explanation of this result we follow de Gennes[1] and consider interactions between segments of chain molecules

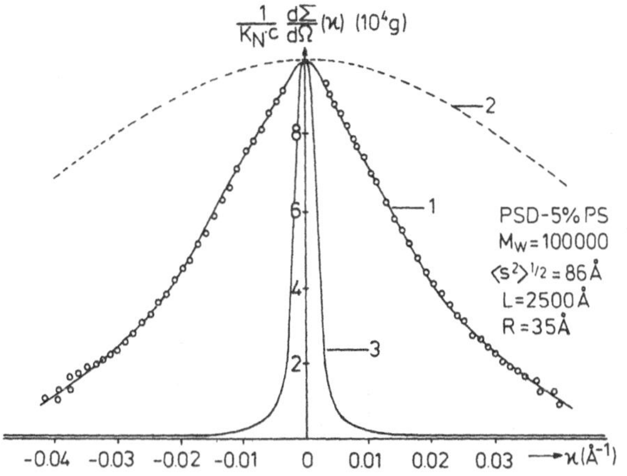

Figure 1. Small angle scattering from 5% hydrogenous polystyrene in deuterated polystyrene at room temperature.

$$V(r) = V_s(r) + V_d(r) \tag{3}$$

where V_s and V_d refer to interactions between segments belonging to the same chain and to different chains, respectively. For a homogeneous polymer, as holds for polystyrene in the amorphous and glassy state, the interactions are proportional to the segment densities $c_s(r)$ and $c_d(r)$. Thus the total interaction V(r) is proportional to $c_s(r) + c_d(r)$. Since the total segment density is constant irrespective of the polymer conformation, interactions need not be considered and a gaussian segment distribution is observed.

By referring to a rather incomplete list of results which were obtained from my own SAS studies, I further stress the point that there is more to SAS than merely measuring the radius of gyration.

(i) The existence of magnetic single-domain behavior was verified by comparison of the magnetic SAS from Co precipitates in copper at zero magnetic field and at saturation field.

(ii) The microscopic magnetic field distribution of flux lines in type II superconductors was determined from the neutron diffraction of flux line lattices.

(iii) In an unusual application of SANS the drift velocity of flux lines in type II superconductors due to a dc current perpendicular to the magnetic flux was measured.

(iv) The reaction time of H/D exchange for hemoglobin dissolved in D_2O was estimated from a real-time SANS experiment.

These few examples demonstrate that it is essential to obtain a sensitive signal from the solid on a correct scale. The information, however, which one deduces from the signal, depends strongly

Figure 2. View at the FRJ-2 and the SANS instrument.

on the imagination of the experimentalist and on his understanding of the scattering mechanism.

2. DESIGN AND CONSTRUCTION OF SANS INSTRUMENTS

The most powerful instrument is certainly the D11 at the HFR reactor of the ILL in Grenoble. Design and construction of D11 is similar to that of the FRJ-2 in Jülich (Figure 2). Both instruments are placed outside the reactor hall and use cold neutrons from a Cold Source. The total length of the instrument is 80 m in Grenoble and 40 m in Jülich. The neutron beams are monochromatized by mechanical velocity selectors, and multi-detector systems are used to detect the scattered neutrons.

In the following, the design of a small angle neutron scattering instrument will be discussed in detail and the questions why poor wavelength resolution can be tolerated, why the instrument must be so long and why long wavelength neutrons are useful will be answered.

The purpose of any SAS experiment is to obtain scattering data at small scattering vectors Q

$$Q = k \cdot \theta \ll a^{-1} \qquad (4)$$

This requires long wavelength ($2\pi/k$) and/or small scattering angles. An SAS experiment makes sense only if the uncertainty δQ of the

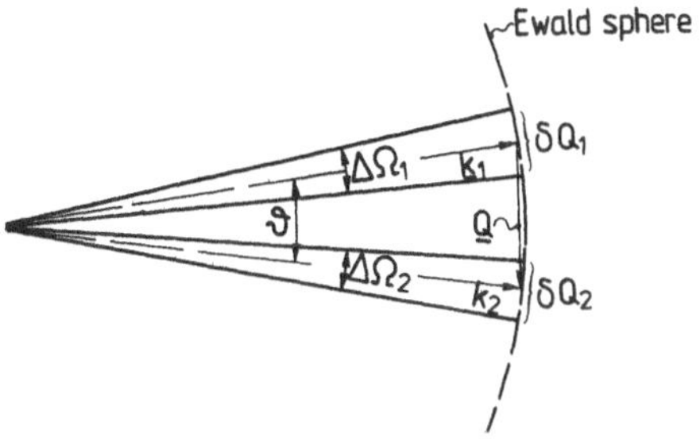

Figure 3. Resolution elements in k-space.

scattering vector is smaller than Q. The various contributions of δQ are due to finite apertures in the collimator, finite illuminated cross sections of sample and detector, and finite monochromatization. As can be seen from Figure 3, there are three contributions of δQ:

$$\begin{aligned} \delta Q_1 &= k\Delta\Omega_1^{\frac{1}{2}} \\ \delta Q_2 &= k\Delta\Omega_2^{\frac{1}{2}} \\ \delta Q_3 &= Q \cdot \frac{\Delta\lambda}{\lambda} \end{aligned} \tag{5}$$

where $\Delta\Omega_1$ is the solid angle of the collimation, $\Delta\Omega_2$ is the solid angle subtended by the cross sections of the illuminated sample and of a detector element, and $\Delta\lambda$ is the wavelength band width of the monochromator. By simply adding these three contributions one gets, with Equation (5),

$$\frac{\delta Q}{Q} = \frac{\Delta\Omega_1^{\frac{1}{2}}}{\theta} + \frac{\Delta\Omega_2^{\frac{1}{2}}}{\theta} + \frac{\Delta\lambda}{\lambda} \tag{6}$$

The criterion which determines these quantities and leads to an optimal instrument is the maximizing of the scattered intensity P under the constraint that δQ/Q is fixed. The scattered intensity is given by

$$P = L_o\Delta\Omega_1 \frac{\Delta\lambda}{\lambda} \Delta\Omega_2 \; F_S t_S \; e^{-\mu_S t_S} \cdot \frac{d\Sigma_S}{d\Omega}(Q) \tag{7}$$

Here, L_o is the luminosity of the neutron source per unit area, time, solid angle and relative wavelength band, F_S is the sample cross section, t_S is the sample thickness, μ_S is the sample attenuation coefficient, and $(d\Sigma/d\Omega)$ (Q) is the differential cross section of the sample per unit volume. The maximization of P leads to the condition that the three resolution contributions have the same magnitude

$$\frac{\Delta\Omega_1^{1/2}}{\theta} = \frac{\Delta\Omega_2^{1/2}}{\theta} = \frac{\Delta\lambda}{\lambda} \qquad (8)$$

A reasonable choice for each contribution is a value between 0.1 and 0.3. Then, sophisticated and time-consuming resolution correction procedures can be omitted since the resolution corrections to the measured scattering functions are only a few percent in almost all cases.

It is instructive to express the scattering power in terms of $\delta Q/Q$. Apart from a numerical value which is not of interest, P is given by

$$P = L_o \left(\frac{\delta Q}{Q}\right)^5 \left(\frac{Q}{k}\right)^4 F_S t_S e^{-\mu_S t_S} \frac{d\Sigma}{d\Omega}(Q) \qquad (9)$$

An ambitious resolution, let us say only three times better than mentioned above, implies a reduction of scattered intensity by a factor $3^5 = 250$. With such an unnecessary high resolution, most of the SANS experiments which were successfully performed would not have worked.

So far this optimization did not involve the illuminated sample cross section F_S. According to Equation (9), a large F_S is advised since P is proportional to F_S. The consequences are a long instrument, large area detecting systems, and big specimens. The increase of the scattered intensity by increasing the illuminated cross section of the specimen works only because the neutron sources, which are reactors, provide neutron fluxes over huge source areas. The characteristic features of neutron sources are that the luminosity is weak and the source area is large in comparison to light and X-ray sources.

One may ask how one should distribute the available total length of an instrument into collimation and flight path in front of and behind the sample. In a high resolution experiment one obtains maximum scattered intensity if one places the sample in the middle between the entrance slit of the collimator and the detector. Some figures will illustrate the need of long instruments. A length of 80 m is required for a scattering experiment with the extreme resolution $\delta Q = 3 \cdot 10^{-4}$ Å^{-1} if a sample of 20 mm ϕ should be illuminated and if a neutron wavelength of 10 Å is chosen. If one cuts down the 80 m long instrument to the convenient length of 2 m the diameter of the illuminated sample cross section can only be 0.5 mm to maintain the same high resolution. This reduction is on the expense of scattered intensity by a factor of $(20/0.5)^2 = 1600$. Such a reduction in intensity cannot be tolerated in most of the scattering experiments which are performed at present.

Finally, the neutron wavelength dependence of the scattered intensity is discussed in order to find its optimum value. In Equation (9), L_o, $(Q/k)^4$, and $e^{-\mu_S t_S} \cdot t_S$ are λ dependent. To a good approximation $L_o(\lambda)$ is given by

$$L_o(\lambda) = \frac{\phi_{th}}{2\pi} e^{-\left(\frac{\lambda_T}{\lambda}\right)^2} \left(\frac{\lambda_T}{\lambda}\right)^4 \tag{10}$$

with ϕ_{th} the thermal flux and λ_T a characteristic wavelength which is a function of the moderator temperature. The effect of moderation by the Cold Source is essentially an increase of λ_T from about 1.8 Å to about 4 Å. For $\lambda > \lambda_T$, $L_o(\lambda)$ is roughly proportional to λ^{-4}. This dependence is cancelled by the term $(Q/k)^4$ since it is proportional to λ^4 for given Q. Thus, if λ is chosen larger than λ_T, the remaining λ dependence of P in Equation (9) is the one of the term $t_S \cdot e^{-\mu_S t_S}$. The product is maximized with respect to a variation of t_S if $t_S = \mu_S^{-1}$ and its maximum value is e/μ_S. This value has no wavelength dependence if the attenuation is solely due to incoherent scattering, and is proportional to λ^{-1} if the attenuation is solely due to absorption. Because of this, the scattered intensity P does not strongly depend on the wavelength and the optimal wavelength is about 4 Å if a thermal neutron beam is used in a SANS instrument and about 8 Å if the subthermal neutron beam for a Cold Source is used. Apart from this consideration it should be mentioned that the gain from a Cold Source is about a factor of 15 with respect to the same reactor without the Cold Source.

A summary of this optimization consideration will be given by comparing the scattered intensities of a typical X-ray and neutron SAS experiment performed at the same resolution. The most powerful conventional X-ray source nowadays is a 100 kW rotating anode which has a luminosity of Cu K_α radiation, which is seven orders of magnitude larger than the luminosity of the most powerful neutron source (the HFR at Grenoble) for neutrons of 8 Å. However, this drawback in luminosity is more than compensated by several means which will be described by considering, for the two cases, the scattered intensity P. There is a factor of 700 in favor of neutrons because of the ratio of wavelength $(\lambda_n/\lambda_x)^4$. In order to obtain the same Q resolution, the X-ray radiation must be collimated much better before and behind the specimen, i.e. $\Omega_1^{\frac{1}{2}} \sim \lambda^{-1}$ and $\Omega_2^{\frac{1}{2}} \sim \lambda^{-1}$. Furthermore, it was discussed that a wavelength band of $(\Delta\lambda/\lambda) = 0.1$ is tolerable without affecting the total Q resolution $(\delta Q/Q)$ significantly. By monochromatizing the neutron beam with a mechanical velocity selector, such a relative broad wavelength distribution is utilized in SANS experiments. For X-rays using Cu K_α radiation, the wavelength width $(\delta\lambda/\lambda)$ is determined by the intrinsic width of the K_α line which is about 0.610^{-3}. The ratio of both widths yields another factor of $(0.1/0.6 \cdot 10^{-3}) = 150$ in favor of neutrons.

The neutron instruments are of such a length that, with specimens of at least 1 cm^2 cross section, the required resolution can be maintained. In the X-ray case the illuminated cross section is limited to 1 mm^2 since the source area is of about this cross section. This adds another factor of 100 in favor of the neutron case. Finally, the sample thickness must be considered according to Equation (9). Because of the strong absorption of X-ray scattering, the optimal sample thickness is usually much smaller than 1 mm in the X-ray case. For SANS the optimal sample thickness is in almost all cases between 1 mm and 10 cm. This leads to another factor in the average of 10 in favor of neutrons. The product of these four factors of $700*150*100*10 \cong 10^8$ does more than compensate for the drawback in luminosity, which was of seven orders of magnitude. The differential scattering cross section was not considered since on the average there is not very much difference between the neutron and X-ray case; however, in particular cases the difference is enormous, as, for instance, if Pb precipitates in Be (which is a favorite X-ray case) and deuterated polymers in hydrogenous polymers (which is a favorite neutron case) are considered. Nevertheless, the comparison has shown that, on the average, at present an SAS experiment with neutrons is more sensitive than one with X-rays.

3. EXAMPLES OF SYSTEMS STUDIED BY SANS

After this discussion of the design and construction of SANS instruments, the purpose and the results of three recent SANS experiments will be briefly discussed. The first one is an investigation of magnetic ordering of $MnSi_x$ which has, besides its interesting result, an instrumental aspect. The second experiment provides a deep understanding of the segment distribution of polymers in the melt. The third is a combination of light and neutron scattering for a precise determination of the critical exponent η. It fits quite well to this conference.

A. Investigation of Magnetic Ordering of $MnSi_x$

From the high field susceptibility measurements (Figure 4), it was suggested that below the Neél temperature of about 40 K, a magnetic ordering occurs in the higher manganese silicide. By neutron diffraction, attempts were made to verify this prediction. However, no magnetic Bragg peaks could be observed at zero external field below the Neél temperature. The difference powder patterns measured at $4^{o}K$ and $100^{o}K$ were flat in the entire Q range and did not indicate any weak magnetic peaks. Because of this result, any magnetic ordering in $MnSi_x$ was denied. However, what could not be observed with thermal neutrons in a diffractometer became observable with subthermal neutrons in a SANS instrument. At $4^{o}K$ two sharp peaks occur at small Q around the origin of reciprocal space (Figure 5). They are of magnetic origin since they disappear at temperatures above T_N and when a magnetic saturation field is applied. This SANS result leads to

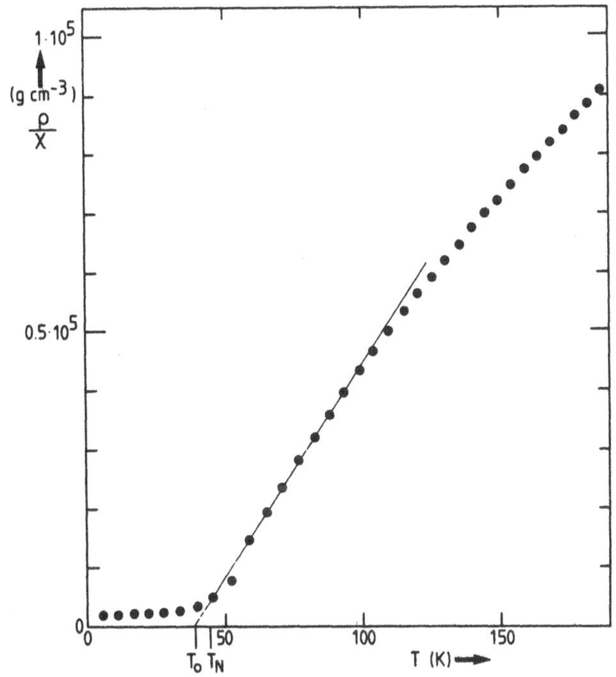

Figure 4. Inverse high field susceptibility of $MnSi_{1.7}$ versus temperature.

the interpretation that the ordered state of $MnSi_x$ is a spiral spin structure with a periodicity length of 160 Å. It is certain that nobody can do a complete structure analysis with only one measured Bragg peak. Nevertheless, the spiral spin structure with the spin direction $\underline{S}$ perpendicular to the spiral axis is the simplest explanation consistent with limited amount of data which is available in this case.

It remains to be explained why the magnetic reflection around the origin of reciprocal space could be measured while all other satellite reflections which exist around ordinary Bragg peaks could not be measured. There is simply an intensity and sensitivity argument in favor of SANS. The integrated reflectivity of a Bragg peak is for poly-crystalline specimens proportional to $\lambda^3/\sin\theta$. In the SANS case, where a satellite around the origin of reciprocal space was measured, the values of λ and $\sin\theta$ are 8 Å and 0.02 rad. In the diffractometer case the values are for satellites around ordinary Bragg peaks $\lambda = 2$ Å and $\sin\theta > 0.5$. With these numbers one calculates that the integrated intensity of the satellite observed by SANS is more than 40,000 times larger than all the other existing satellites which could not be observed by thermal neutron diffraction. In addition, the Q-resolution of conventional diffractometers is not high enough to separate the magnetic satellites from the ordinary Bragg

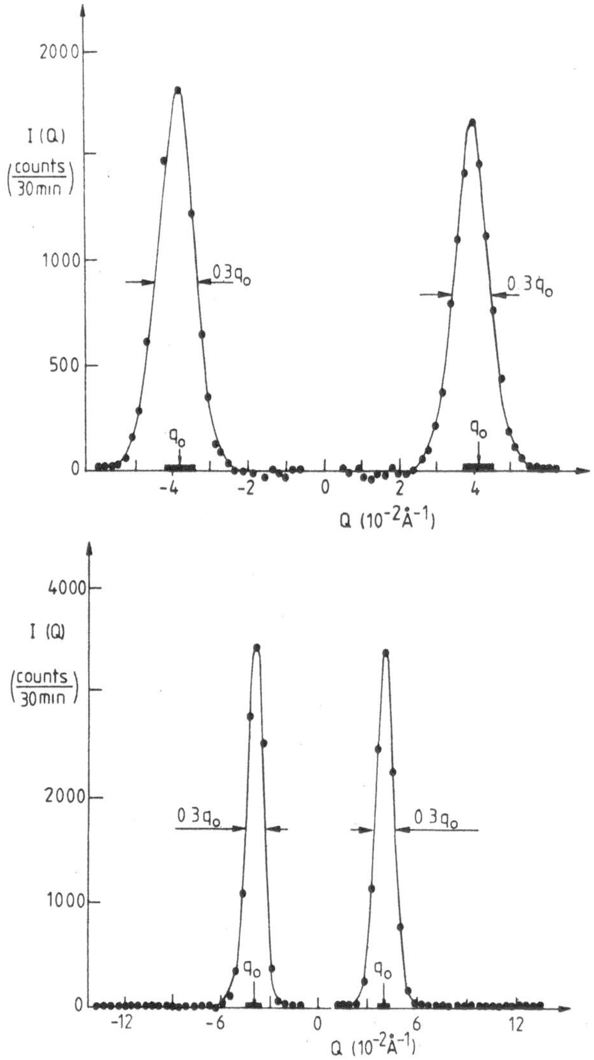

Figure 5. Magnetic satellites of $MnSi_{1.7}$ around the origin of reciprocal space measured at 4.2 K.

reflections in $MnSi_x$ in order that the weak magnetic diffraction appears to be superimposed on strong nuclear reflections. This is in contrast with the SANS experiment where the magnetic satellites were well resolved.

An interesting result of this SANS experiment with $MnSi_x$ was that the magnitude of the Mn spin as determined from the integrated intensity of the measured satellite was four times larger than the

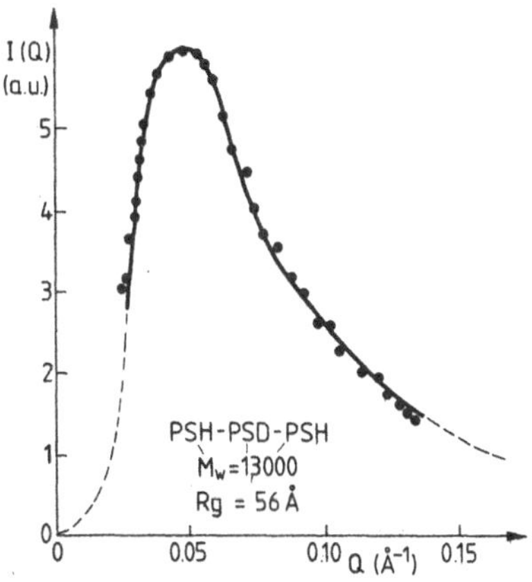

Figure 6. SANS from tri block polystyrene PSH-PSD-PSH with M_w of each block 13000.

Mn spin as determined from saturation measurements. This can be explained by longitudinal spin fluctuations as they are discussed in itinerant magnets. For further details one is referred to Nakajima and Schelten.[2]

B. Segment Distribution of Polymers in the Melt

In the next SANS experiment the organization of polymer segments in amorphous polystyrene is investigated. The result is based on a different kind of tagging. Instead of having a small fraction of polymer molecules predeuterated, all polymer molecules were deuterated at their middle part. All polymers are three-block polymers. The scattering pattern from such a specimen is a sharp peak at $Q \approx 0.05$ $Å^{-1}$ with asymmetric shape (Figure 6). The skillful preparation and the scattering experiments were done by the French group in Strasbourg, Saclay and Grenoble.[3] The interpretation of the scattering result is that there is a repulsion between the central parts of the polymers which can be understood according to de Gennes in the following way. If one focuses at a particular chain and at its center of mass, then the space around the center is partly filled by segments of that chain. Consequently, the number of segments of other chains is reduced in this region. In particular, this is the case for deuterated segments of other chains which label their center of mass. Thus, a correlation hole between the centers of chains is the result of this space filling and its range is typically of the order of the radius of gyration of the chain molecules. In a random phase approximation the correlation hole and the scattering function has been calculated by de Gennes[1] and numerically evaluated by Duplessix et al.[3] A fit to the experimental data is shown in Figure 6.

C. Combination of Light and Neutron Scattering for Determination of Critical Exponent η

In the third application a recent experiment will be described in which light and neutron scattering have been used in order to determine the critical exponent η. This exponent is defined by the asymptotic behavior of the correlation scaling function g(x)

$$g(x) \propto x^{-(2-\eta)} \tag{11}$$

where $x = Q \cdot \xi(\tau)$, with $\xi(\tau)$ the correlation length and τ the reduced temperature $(T - T_c)/T_c$. According to Equation (11), η is merely a small correction to an x^{-2} asymptotic behavior and is therefore very difficult to measure. Many attempts have been made in the past to determine its value, and one knows that a η determination is really a challenge. The idea of our experiment is to obtain, from light scattering, the critical concentration of the investigated system butyric acid/D_2O, its critical temperature T_c, and the temperature dependence of the correlation length ξ, and to measure with neutrons, at large values of Q which cannot be reached by light scattering, the critical opalescence. With both results a correlation scaling function can be constructed by utilizing the scaling hypothesis.

$$S(Q,\tau) = F(\tau) \cdot g(x) \; ; \; x = Q \cdot \xi(\tau) \tag{12}$$

The scattering functions $S(Q,\tau)$ measured by neutrons over a large Q range at various temperatures close to T_c had a considerable overlapping x region. By adjusting the temperature factors a continuous scaling correlation function $g_{exp}(x)$ was constructed. In order to determine η, data of $g_{exp}(x)$ with $x \geq 2$ as shown in Figure 7 were fitted by the Fischer Langer function

$$g_{FL}(x) = C_1 x^{-(2-\eta)} \left[1 + C_2 x^{-\frac{1-\alpha}{\nu}} + C_3 x^{-\frac{1}{\nu}} \right]$$

with theoretical values for C_2, C_3 and $\alpha = 0.1$, and data of $g_{exp}(x)$ with $x \geq 20$ as shown in Figure 8 were fitted by the asymptotic function

$$g_\infty = C_1 x^{-(2-\eta)}$$

The weighted least square fits were performed with only two adjustable parameters, C_1 and η. The results for C_1 and η were, for both theoretical functions within their uncertainties, the same.

The determined value for $\eta = 0.086 \pm 0.026$ is definitely larger

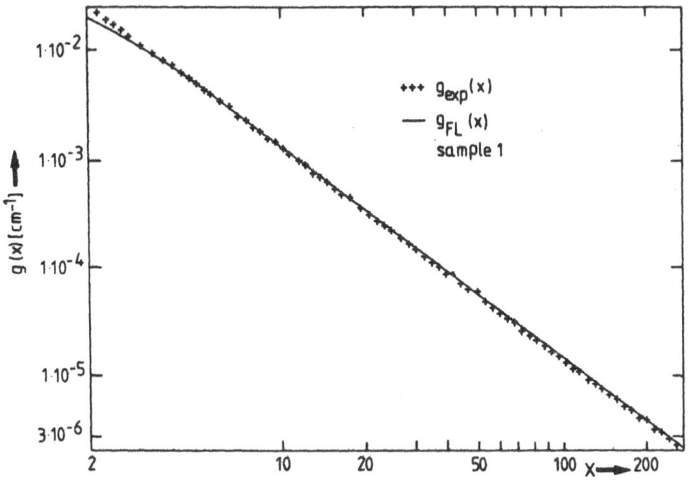

Figure 7. Experimental scaling correlation function for isobutyric acid/D_2O for $x \geq 2$.

than the theoretical value for an Ising model of d = 3 and n = 1. In this context it may be mentioned that the exponent $\eta^* = 2 - \eta$ which appears in all theoretical expressions has been determined with a relative accuracy of 1.5% and deviates from the theoretical value by only 2.5%. Further discussions are given by Schneider et al.[4]

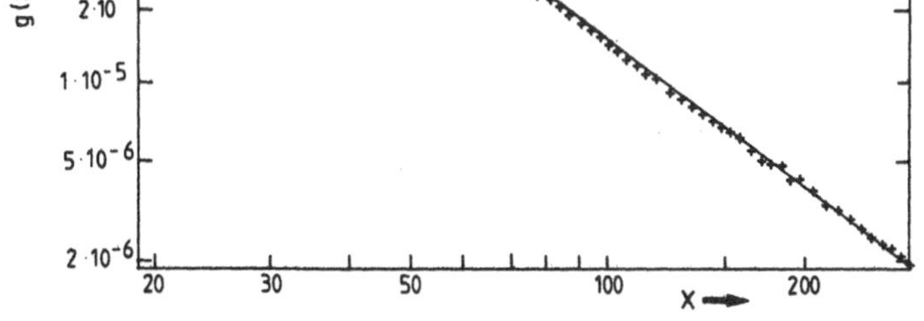

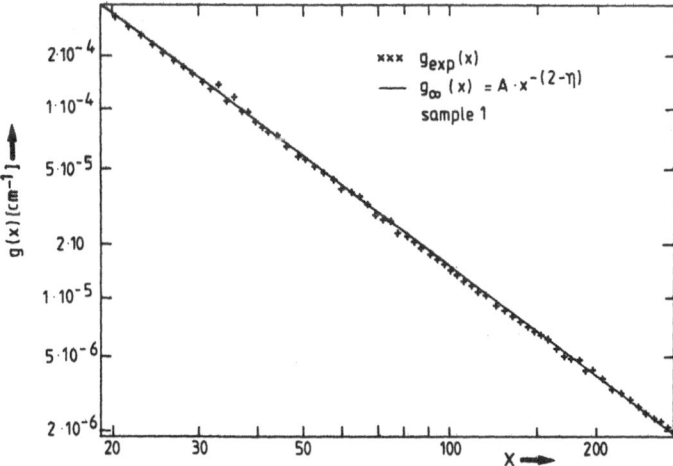

Figure 8. Experimental scaling correlation function for isobutyric acid/D_2O for $x \geq 20$.

REFERENCES

1. P. G. de Gennes, "Scaling Concepts in Polymer Physics", Cornell University Press, Ithaca, N. Y. (1979).
2. T. Nakajima and J. Schelten, J. of Magnetism and Magnetic Materials 21: 157 (1980).
3. R. Duplessix, J. P. Cotton, H. Benoit, and C. Picot, Polymer 20: 1181 (1979).
4. R. Schneider, L. Belkoura, J. Schelten, D. Woermann, and B. Chu, Phys. Rev. B 22: 5507 (1980).

QUASIELASTIC NEUTRON SPIN-ECHO SPECTROSCOPY

John B. Hayter

Institut Laue-Langevin, 156X

38042 Grenoble Cédex, France

CONTENTS

1. INTRODUCTION

In this lecture, I want to develop the way in which dynamic neutron small-angle scattering can complement light-scattering, by observing a different region of momentum space. In particular, I shall discuss in detail a new technique, neutron spin-echo[1], which opens the way to a neutron equivalent of photon correlation spectroscopy. I shall then give some examples of its application to studies of polymers, polyelectrolytes and micelles.

To situate the notation, I shall first recall a few standard results. Neutrons with de Broglie wavelength λ have momentum $\hbar\,\underline{k}$, where $\underline{k}$ is the neutron wave-vector ($|\underline{k}| = 2\pi/\lambda$). In general, when the neutron crosses an interface, the momentum changes from $\underline{k}_o$ to $\underline{k}_f$, and the refractive index for the interface is

$$n = |\underline{k}_f| / |\underline{k}_o| \qquad (1)$$

Neutrons undergo refraction in a way analogous to light. In a non-magnetic system, simple physical arguments show[2,3] that

$$n \simeq 1 - N\bar{b}\,\lambda^2/2\pi \qquad (2)$$

with

$$n\bar{b} = \sum_j N_j b_j \qquad (3)$$

where N_j is the number density of scattering centres of type j, with coherent scattering length b_j. (Recoil effects are absent in forward scattering, and the bound scattering length is involved, even for studies of liquids. Typical values of b for common isotopes are $-3.74(^1H_1)$, $+6.67$ $(^1H_2)$, $+6.65$ $(^6C_{12})$, $+5.80$ $(^8O_{16})$ $\times 10^{-15}$m). Since b_j may be positive, n may be less than unity, and neutron optics extend the possibilities available with light. The use of contrast matching is one important application.

Neutron small-angle scattering is basically neutron refraction at small but finite momentum transfer $\underline{Q} = \underline{k}_f - \underline{k}_o$. ($\underline{Q}$ is also variously referred to in the literature as $\underline{\kappa}$, $\underline{k}$ or $\underline{q}$). At finite Q, neutrons begin to probe the details of the atomic arrangement in the target, but for small Q (long wavelength) the scattering is sensitive only to rather gross details, averaged over distances of order 1/Q. For a correctly chosen Q-range, desired features of a complex system (such as the overall dimensions of a macromolecule) can therefore be studied without needing to model the fine detail. This means, for example, that the average (3) may still be used to calculate the scattering length density, in a region containing many atoms.

If ξ is a characteristic correlation length of the system under study, small-angle scattering takes place around $Q\xi = 1$. There is thus a fundamental scaling phenomenon; qualitatively similar physical systems will give similar scattered intensity distributions as a function of the product $Q\xi$. (An extreme example is given by the structures of the liquid alkali metals, which are effectively scaled versions of one another[4]). Because neutron wavelengths are typically 2-3 orders of magnitude smaller than for light, scattered neutron momentum transfers can be much larger than for scattered photons. Since the same physical phenomena appear in the same (dimensionless) range of $Q\xi$, neutrons complement light scattering by observing similar systems in different correlation length ranges.

These ideas have of course been employed in structural studies (using elastic scattering) for many years (see, e.g.[5,10]). Until recently, however, neutron spectroscopists have been unable to observe quasielastic scattering at small angles; that is, there has been no dynamic neutron equivalent of photon correlation spectroscopy (PCS). Conventional neutron scattering techniques measure in the frequency domain, while PCS measures in the time domain. Although the two methods are in principle equivalent (being related by a Fourier transform), they are not equivalent in practice, because a real experiment cannot collect data over a large enough dynamic range (in, say, frequency) to allow accurate Fourier transformation to the other (time) domain. To discuss the implications, I shall summarize some basic scattering theory, restricting the discussion to quasielastic coherent scattering, which will be our main interest.

The intensity of neutrons scattered with momentum change from $\underline{k}_o$ to $\underline{k}_f$ (energy change from E_o to E_f) is given by the double differential cross-section[11]

$$\frac{\partial^2 \sigma_{coh}}{\partial \Omega' \partial E_f} = \frac{k_f}{k_o} \cdot \frac{\sigma_{coh}}{4\pi} \cdot N \cdot S(\underline{Q}, \hbar\omega) \quad (4)$$

where Ω' denotes solid angle, $\sigma_{coh} = 4\pi\bar{b}^2$ is the cross section for coherent scattering and N is the number of scattering centres. The function $S(\underline{Q}, \hbar\omega)$ is variously known as the coherent scattering law, the scattering function or the dynamical structure factor. The energy transfer $\hbar\omega = E_f - E_o$ is usually expressed simply as the corresponding frequency, ω. In the case of isotropic liquids or solutions, only the magnitude of $\underline{Q}$ is important, and I shall denote the scattering law simply by $S(Q,\omega)$. For quasielastic scattering $k_f \simeq k_o$ and $Q \simeq (4\pi/\lambda) \sin\theta$ for scattering through angle 2θ.

$S(Q,\omega)$ is calculated via the correlation function

$$F(Q,t) = \frac{1}{N} \sum_{j,k} \langle \exp[-i\,\underline{Q}\cdot\underline{R}_k(0)]\exp[i\underline{Q}\cdot\underline{R}_j(t)] \rangle \quad (5)$$

where $\underline{R}_j(t)$ is the position vector of the j'th atom at time t, and the angular brackets denote a thermal average. The function $F(Q,t)$ is also known as the intermediate scattering law, since it is related to the scattering law $S(Q,\omega)$ by a temporal Fourier transform

$$S(Q,\omega) = \frac{1}{2\pi\hbar} \int F(Q,t) \exp(-i\omega t)\, dt \quad (6)$$

and to the distribution function

$$G(\underline{r},t) = \frac{1}{(2\pi)^3} \int F(Q,t) \exp(-i\underline{Q}\cdot\underline{r})\, d\underline{Q} \quad (7)$$

by a spatial Fourier transform. Classically (i.e., when $\underline{R}_k$ and $\underline{R}_j$ commute), $G(\underline{r},t)\,d\underline{r}$ is the probability that, given a particle at the origin at time t = 0, any particle is in volume $d\underline{r}$ at time t. From Equation (6) and Equation (7) it is seen that $S(\bar{Q},\omega)$ is the space-time Fourier transform of $G(\underline{r},t)$.

Setting t = 0 in Equation (5) gives the usual expression for the structure factor S(Q), so that S(Q) = F(Q,0). The normalized intermediate scattering law, which will be of central interest later, is thus

$$F_N(Q,t) = F(Q,t)/F(Q,0) = F(Q,t)/S(Q) \tag{8}$$

The normalized n'th moment of $S(Q,\omega)$ is defined by

$$\langle\omega^n\rangle = \int \omega^n S(Q,\omega)\, d\omega \Big/ \int S(Q,\omega)\, d\omega \tag{9}$$

where the denominator is easily seen to be the structure factor. Now, on the classical assumption that velocities of different particles taken at the same time are on average uncorrelated, $S(Q,\omega)$ is an even function of ω, so all odd moments are zero. The normalized second moment is given by[12]

$$\langle\omega^2\rangle = \hbar^2 Q^2/m\beta\, S(Q) \tag{10}$$

where m is the particle mass and $\beta = 1/k_B T$ is the inverse thermal energy. For quasielastic scattering from a classical liquid or solution, it is therefore not until the 4th moment that dynamic fluctuations from the average enter the shape of the scattering law, since the mean velocity $\langle v^2\rangle \sim 1/m\beta$. The familiar de Gennes narrowing given by Equation (10) merely reflects the average structure. This gives rise to Egelstaff's rule, that quasielastic scattering from different classical liquids generally looks the same.

The use of conventional neutron scattering to observe details of the dynamics of particles in solution therefore requires extremely accurate measurements of the wings of the frequency spectrum $S(Q,\omega)$, since these are most heavily weighted in calculating fourth (and higher) moments. The wings are precisely the region where data is statistically least accurate, and it is not surprising that even high resolution back-scattering spectrometry[13] has found the study of small-angle quasielastic scattering intractable.

The problem may be avoided, in principle, by Fourier inversion, which takes the wings of the frequency spectrum into the origin of the time spectrum; that is, by measuring F(Q,t) directly, as is done in PCS. Fourier transforming Equation (6) yields

$$F(Q,t) = \int S(Q,\omega) \exp(i\omega t)\, d(\hbar\omega) \tag{11}$$

From the condition of detailed balance[11]

$$S(Q,\omega) = \exp(\hbar\omega\beta)\, S(Q,-\omega) \qquad (12)$$

For quasielastic scattering, $\hbar\omega$ is small relative to the thermal energy, and the exponential factor is effectively unity. (In the systems I shall consider later, $\hbar\omega$ is typically 10^{-8} eV, or $<10^{-6}$ k_BT at room temperature). $S(Q,\omega)$ may therefore be taken as even, so that only the real part of Equation (11) is non-zero:

$$F(Q,t) = \int S(Q,\omega) \cdot \cos \omega t \cdot d(\hbar\omega) \qquad (13)$$

Many attempts have been made in the past to measure $F(Q,t)$ via Equation (13), using electro-mechanical modulation added to a conventional spectrometer.[14] In the context of macromolecular studies, two problems arise. The first is the difficulty of achieving perfect modulation; harmonics present in the modulation add spurious components to $F(Q,t)$. The second is one of time-scale, which is related to resolution in the frequency domain. Conventional neutron spectrometers achieve frequency resolution by preparing and comparing well-defined initial and final neutron momentum states $\underline{k}_o$ and $\underline{k}_f$. Phase-space considerations show that the definition of these states can only be improved by throwing away unwanted neutrons, so that higher resolution implies a corresponding, and drastic, reduction of intensity. (This is a practical consequence of Liouville's theorem). A frequency resolution of $\Delta\omega$ corresponds to a relaxation time $\tau \sim 2\pi/\Delta\omega$. In studies of macromolecules, τ is typically $\gtrsim 10^{-7}$s, so that $\hbar\Delta\omega \lesssim 4 \times 10^{-8}$ eV. Even at the Grenoble high flux reactor, such resolution cannot be achieved with conventional techniques, without reducing the scattered neutron intensity well below the useable practical limits which are set by the background.

Neutron spin-echo spectrometry[1] side-steps these difficulties by taking a fundamentally different approach. The key to the method lies in avoiding having to prepare well-defined *momentum* states $\underline{k}_o$ and $\underline{k}_f$ in order to achieve sufficient resolution. Instead, precessing *spin*-states before and after scattering are compared, in a way that measures the difference of states directly. There is thus one quantum mechanical measurement less in a neutron spin-echo experiment than in a conventional neutron measurement. The resolution of the method is not coupled to the definition of $\underline{k}_o$, provided the differential spin-state measurement can be made individually on each neutron. The beam used can therefore have a broad momentum spread, with correspondingly high intensity. I shall now show how this can be achieved.

2. THEORY OF NEUTRON SPIN-ECHO

The neutron carries a magnetic moment $\mu = -1.91304\ \mu_N$, associated with a spin $S = 1/2$. (μ_N is the nuclear magneton). In conventional neutron polarization analysis, these spins are aligned parallel (+)

or anti-parallel (-) to a magnetic guide field, which I shall take to define the z-direction. After scattering, the final polarization is analysed in the same direction (see, for example, Ref. 3). If the intensities analysed along +z and -z are called I^+ and I^-, respectively, the polarization of the beam is defined as

$$\langle P \rangle = (I^+ - I^-)/(I^+ + I^-) \tag{14}$$

where the average is over all neutrons in the beam. The polarization is inherently normalized, $-1 \leq \langle P \rangle \leq 1$, because of this form of intensity ratio.

The key to understanding neutron spin-echo lies in realizing that it is not necessary to consider only the beam-averaged polarization. Each neutron may be separately considered as polarized, in that it has spin components (in the axes defined by the field) which define a polarization vector

$$\underline{P} = (P_x, P_y, P_z) \tag{15}$$

even when $\langle P \rangle = 0$. If the x and y components are non-zero, the spin will precess about the field with the well-known Larmor frequency

$$\begin{aligned} \omega_L &= 2|\mu|\ H/\hbar \\ &= 1.8325 \times 10^8\ H\ s^{-1} \end{aligned} \tag{16}$$

where H is the induction of the magnetic guide field, which is assumed to vary direction in space slowly relative to ω_L.

The important feature of Equation (16) is that it is independent of the magnitudes of P_x and P_y, provided at least one x-y component is non-zero. All neutrons whose spin has been turned out of the z-direction will precess at the same frequency ω_L, regardless of the actual angle between $\underline{P}$ and z. The time spent by a neutron of wavelength λ in a constant field H of length ℓ is

$$t' = \ell\lambda m_n/h \tag{17}$$

where m_n is the neutron mass, so the precession angle accumulated in traversing the field will be

$$\begin{aligned} \phi\ (H\ell,\lambda) &= \omega_L\ t' \\ &= (4\pi|\mu|m_n/h^2)\ H\ell\lambda \end{aligned} \tag{18}$$

The angle $\phi(H\ell,\lambda)$ thus provides a direct measure of the neutron wavelength.

Suppose for the moment that it is possible to switch on spin-precession at a precise point during the neutron flight through a spectrometer, so that the spin of a neutron of wavelength λ will have precessed $\phi(H\ell,\lambda)$ after travelling a further distance ℓ in the field H. If at this point, the direction of precession is reversed, travelling the same distance again in the same field will subtract $\phi(H\ell,\lambda)$ from the precession angle, and the net angle accumulated at the point 2ℓ distant from the origin will be zero. Furthermore, it will be identically zero, *independent of the wavelength* λ. If all spins had the same orientation at the origin, they would therefore again all have the same orientation a distance 2ℓ later, independent of their velocity. This is neutron spin-echo focussing.

Now place a sample at the midpoint. After a quasielastic scattering event which changes the neutron energy by $\hbar\omega$, the wavelength will have changed to $\lambda + \Delta\lambda$, where (to first order in $\Delta\lambda$)

$$\Delta\lambda = (m_n \lambda^3/2\pi h)\omega \tag{19}$$

The net precession angle at distance 2ℓ from the origin will now be

$$\phi_{net} = \phi\ (H\ell,\lambda) - \phi\ (H\ell,\ \lambda + \Delta\lambda)$$

$$= (4\pi|\mu|m_n/h^2)\ H\ell.\Delta\lambda \tag{20}$$

so that ϕ_{net} is a linear measure of the wavelength *change*. For quasielastic scattering, Equation (20) may be written:

$$\phi_{net} = \omega t \tag{21}$$

where from Equation (19) the parameter t, which has the dimension of time, is

$$t = (2|\mu|m_n{}^2/h^3)\ H\ell.\lambda^3 \tag{22}$$

Measurement of the spin-precession phase-angle at the point 2ℓ therefore gives a direct measure of the quasielastic energy transfer $\hbar\omega$ which occurred on scattering. Note that it is unnecessary to measure the phase-angles (i.e.: wavelengths) separately before and after scattering, unlike the need to know $\underline{k}_o$ and $\underline{k}_f$ separately in conventional spectrometry.

Each neutron arriving at the focussing point thus carries the information about its own particular scattered energy transfer, encoded as a spin-precession angle. To show how to decode this information, I must now consider more explicitly the initial supposition, that spin-precession can be turned on (or, as will be seen, turned off) at a precise point in the beam path.

Precessing polarization is created by the use of a spin-turn coil.[1] This is a thin flat rectangular coil placed with its flat face perpendicular to the beam, so that neutrons experience a sudden (non-adiabatic) change in field direction on entering the coil. Provided the change of field direction is fast compared with ω_L, the neutron spin-direction will remain fixed with respect to the laboratory on entering the coil, and a spin initially aligned with the external guide field will enter the coil at an angle to the net internal field. The spin will thus precess about this net field for the time taken to cross the coil, before exiting, again non-adiabatically. The coil field and dimensions may be arranged, for a given wavelength, so that the final spin-direction on leaving the coil is at any desired angle to the guide field.

The two most commonly used configurations are shown schematically in Figure 1.

In Figure 1(a), $H_1 = H_o$ and $\underline{H}_1 \perp \underline{H}_o$, so that the net field in the coil makes an angle $\pi/4$ with the guide field. The coil thickness is chosen so that the neutron transit time corresponds to π in Larmor precession angle about the net field; on exit, the neutron spin has turned $\pi/2$ with respect to $\underline{H}_o$.

The action of a general spin-turn coil may be described by a wavelength-dependent transfer matrix, $\underline{\underline{T}}$, which relates the polari-

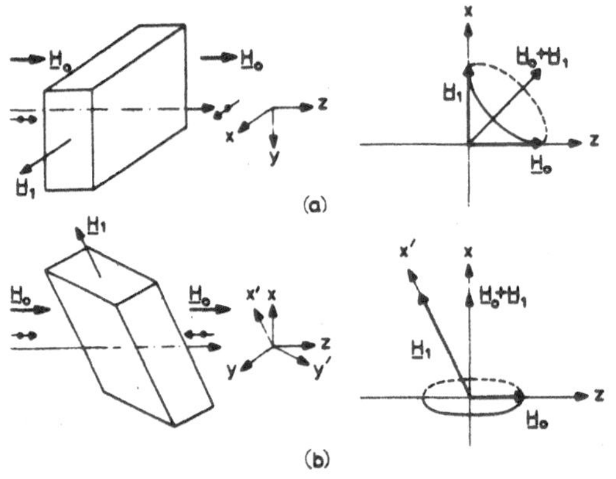

Figure 1. Spin-turn coil arrangements (a) $\pi/2$ turn. (b) π turn. H_o: Guide field. H_1: Coil field. Dotted arrows represent neutron spins.

zation on exit, $\underline{P}'$, to the initial polarization $\underline{P}$:

$$\underline{P}' = \underline{\underline{T}}\underline{P} \tag{23}$$

For a $\pi/2$ spin-turn coil set to work at wavelength λ_o, with $\underline{H}_o$ // z and $\underline{H}_1$ // x, the transfer matrix at any wavelength λ is[15]

$$\underline{\underline{C}} = 1/2 \begin{bmatrix} 1 + \cos\psi & -\sqrt{2}\sin\psi & 1 - \cos\psi \\ \sqrt{2}\sin\psi & 2\cos\psi & -2\sqrt{\sin}\psi \\ 1 - \cos\psi & \sqrt{2}\sin\psi & 1 + \cos\psi \end{bmatrix} \tag{24}$$

where $\psi = \lambda\pi/\lambda_o$. For an incident beam with not too large a wavelength spread, $\psi \simeq \pi$ and Equations (23) and (24) reduce to the transformation

$$\underline{P} = (P_x, P_y, P_z) \underset{\pi/2}{\longrightarrow} \underline{P}' = (P_z, -P_y, P_x) \tag{25}$$

The fundamental action of a $\pi/2$-turn coil is thus to interchange the z and x components of the polarization. A spin initially polarized in the z (field) direction will be turned into the x-direction, and will precess about the guide field on leaving the spin-turn coil.

Figure 1(b) demonstrates the geometry needed to create a π-turn; in this case, the net field is perpendicular to the guide field, and the transfer matrix is:

$$\underline{\underline{F}} = \begin{bmatrix} 1 & 0 & 0 \\ 0 & \cos\psi & -\sin\psi \\ 0 & \sin\psi & \cos\psi \end{bmatrix} \tag{26}$$

which for $\lambda \simeq \lambda_o$ gives

$$\underline{P} = (P_x, P_y, P_z) \underset{\pi}{\rightarrow} \underline{P}' = (P_x, -P_y, -P_z) \tag{27}$$

A π-turn coil thus has two fundamental actions:
(i) it acts as a spin-flipper, since it reverses P_z, and (ii) it reverses the direction of precessing polarization, since it also reverses one of the two components orthogonal to z.

I now have all the elements needed to create a spin-echo focussing configuration, and, as will be seen shortly, to decode the information stored in the phase-angle ϕ_{net} at the focussing point. Figure 2 shows an idealized typical arrangement.

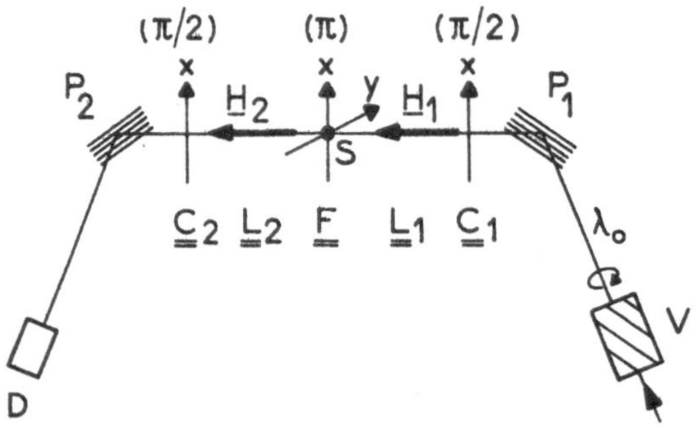

Figure 2. Schematic neutron spin-echo experiment.
V: Velocity selector. H_1, H_2: Main guide fields.
D: Detector. Other notation is defined in the text.

The neutron beam, of mean wavelength λ_0, is polarized by reflection from a magnetic mirror P_1, which aligns all spins in the z (guide-field) direction. A $\pi/2$ spin-turn coil C_1 starts precession, which is reversed in direction by a π-turn coil F placed next to the sample S. A second $\pi/2$ spin-turn coil C_2 is placed at the focussing point, where ϕ_{net} is to be measured. (C_1 and C_2 are in magnetically symmetric positions with respect to F; that is, the field integrals in the two precession regions are the same). It is the sequence $\pi/2 - \pi - \pi/2$ which gives neutron spin-echo its name.

Since precession is about z, the components of the polarization at C_2 are

$$P_x = \cos \phi_{net}$$
$$P_y = \sin \phi_{net} \tag{28}$$

where the x-direction is along the field $\underline{H}_1$ of the $\pi/2$ coil. Now the analysing mirror P_2 transmits the P_z (field direction) component of the polarization to the detector with a probability P, which is easily calculated to be[16]

$$P = 1/2\ (1 + P_z) \tag{29}$$

Reference to Equation (25) shows that the action of C_2 will be to stop precession by turning P_x into P_z, so that P_x becomes the analyzed component. The probability of detection is thus, from Equations (21), (28), and (29)

$$P = 1/2\ (1 + \cos \omega t) \tag{30}$$

In turn, the probability that there was an energy transfer $\hbar\omega$ on scattering is governed by $S(Q,\omega)$. Provided the analyzer makes no energy discrimination over the range of energy transfers involved, and the detector integrates over all energies, the final analyzed intensity is therefore

$$I^E = \frac{A}{2}\int_{-E_0}^{\infty} S(Q,\omega)\ (1 + \cos \omega t)\ d\ (\hbar\omega) \tag{31}$$

The constant A contains all experimental factors such as the initial intensity, sample size and cross-section, and detector efficiency, which is independent of ω for quasielastically scattered cold neutrons.

The last step is to reverse the field direction in the second $\pi/2$ spin-turn coil C_2 (by reversing the current), and to remeasure the intensity. The precessing polarization is now analyzed π radians out of phase with the previous measurement, and the analyzed intensity is

$$\overline{I^E} = \frac{A}{2}\int_{-E_0}^{\infty} S(Q,\omega)\ [1 + \cos\ (\omega t + \pi)]\ d\ (\hbar\omega) \tag{32}$$

The intensities I^E and $\overline{I^E}$ correspond to I^+ and I^- in conventional polarization analysis (π in precession corresponds to spin-flip in linear polarization), and by analogy with Equation (14) the spin-echo polarization is defined as

$$P_E = (I^E - \overline{I^E})\ /\ (I^E + \overline{I^E}) \tag{33}$$

Substituting Equations (31) and (32) in (33) then gives

$$P_E = \int_{-E_0}^{\infty} S(Q,\omega)\cos\omega t . d(\hbar\omega)\ \Big/ \int_{-E_0}^{\infty} S(Q,\omega) d(\hbar\omega) \tag{34}$$

The cut-off introduced by E_0 is of no consequence in quasielastic scattering ($|\hbar\omega| << E_0$), and the lower limits in Equation (34) may be formally extended to infinity, to yield finally via Equations (8) and (13)

$$P_E = F_N(Q,t) \tag{35}$$

which is the desired result.

I want to emphasize three important features. First, the modulation in Equation (34) is obtained by fixed-axis projection of a rotating vector, and is therefore exactly cosinusoidal. Second, the two intensities are measured under rigorously identical geometrical conditions (since they are related simply by current reversal in one coil), and the normalization is therefore absolute. Finally, the time variable t, given by Equation (22), may be varied in fixed geometry merely by varying the guide field.

3. PRACTICAL NEUTRON SPIN-ECHO

I shall briefly describe the only neutron spin-echo spectrometer in routine operation to date. This is the IN11 spectrometer, built by myself and F. Mezei at the I.L.L. Without entering into technical details, I shall use the description to point out the practical implications of the several implicit and explicit assumptions made in deriving Equation (35). The general arrangement used on IN11 to study quasielastic small-angle scattering is shown in Figure 3.

The cold neutron beam from the reactor is monochromatised by a helical velocity selector ($\lambda_o \simeq 8$ Å, $\Delta\lambda/\lambda_o \simeq 9\%$), and then polarized by mirror reflection. The main precession regions L_1 and L_2 are each 2m long solenoids 0.3m in diameter, generating up to 50 mT field; this produces up to 4×10^4 radians precession in each arm, and corresponds (using Equation (22)) to a maximum Fourier time $t \lesssim 10_{ns}$.

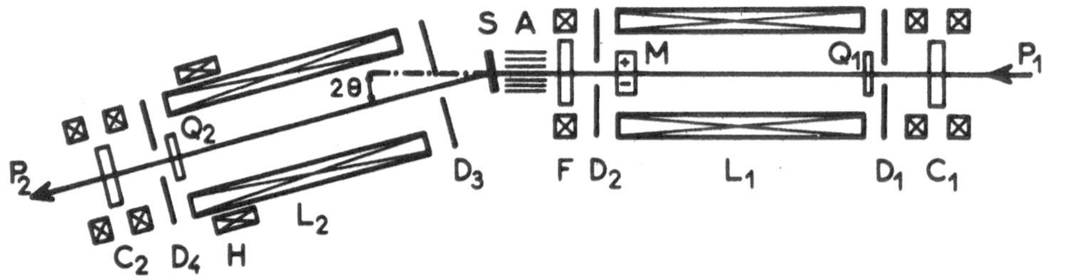

Figure 3. Spin-echo spectrometer layout. P_1: Incident polarized beam. C_1, C_2: $\pi/2$ spin-turn coils. F: π spin-turn coil. L_1, L_2: Larmor precession regions. S: sample. D_1-D_4: Collimating diaphragms. P_2: Final beam to polarization analyzer and detector. Other notation is defined in text.

Each spin-turn coil is made from 1mm enamelled Al wire wound on a 1.1mm pitch on a 100mm wide, 6mm thick Al plate, in which a 50 x 50 mm^2 hole is first cut to allow passage of the beam. The coil is 150mm long. Setting for a π or $\pi/2$ turn is straightforward, and is described in Reference 15.

As the main fields are varied to vary t, the stray fields perturb the spin-turn coils C_1, C_2 and F. A correction coil is therefore associated with each spin-turn coil to maintain constant efficiency. A further correction coil H is used to adjust the exact magnetic symmetry of the two arms. Angular collimation is controlled by circular collimating diaphragms D_1-D_4. Collimation of $\pm$ 0.25° is typically used, and a Soller collimator A is placed after the π-turn coil to eliminate small-angle scattering from the Al wire.

The ultimate performance of the spectrometer is limited by several sources of error which can violate assumptions of the theory; the polarizing mirrors are not 100% efficient, the main fields are inhomogeneous, and the beam is not monochromatic. The first point is straightforward to correct. If the efficiencies of polarizer and analyzer are P_1 and P_2, respectively ($P_1 \simeq P_2 \simeq 0.95$ on IN11), then Equation (35) is modified to [16]

$$P_E = P_1 P_2 . F_N(Q,t) \tag{36}$$

Inhomogeneities enter in an unusual way, compared with usual magnet design problems, since for spin-echo it is the homogeneity of the field-integral along the beam path which matters. On IN11, spiral coils Q_1 and Q_2 are used to compensate for the different field integrals experienced by off-axial neutrons. There is still some residual inhomogeneity, however, since the finite divergence leads to different path-lengths for different neutrons. The finite scattering angle also causes path-length differences across the sample, which are approximately compensated by a rectangular figure-8 coil, M. If the remaining inhomogeneity is small, it may be shown to reduce P_E by a multiplicative factor which depends on sample size.[16] The effect may thus be removed by normalizing to the signal from a purely elastic scatterer of the same geometry as the sample. (A 5% solid solution of perdeutero-polystyrene in polystyrene is used for this purpose). Using Equation (36) in the two cases gives

$$F_N(Q,t) = (P_E)_S/(P_E)_N \tag{37}$$

where (P_E) is calculated from Equation (33), and S and N refer to measurements on sample and normalizing standard, respectively. In this case the factor P_1P_2 cancels and need not be known.

The wavelength spread enters in three ways. The most obvious is in the Q-resolution; most macromolecular scattering has no fine structure in Q, and $\Delta\lambda/\lambda \sim 9\%$ is acceptable from this point of view. Less obvious are the effects on the spin-echo measurement. The first of these is that the spin-turn coils will not operate according to Equations (25) and (27), but must be described by the full $\underline{\underline{C}}$ and $\underline{\underline{F}}$ operators (24) and (26). A full matrix calculation of the spectrometer response[15] shows that a spread of 9% in $\Delta\lambda/\lambda$ has negligible effect on the total response, because the off-diagonal elements in each operator tend to cancel one another. If needed, however, the exact response can be calculated for a given wavelength spread.[15,17]

The second effect of $\Delta\lambda/\lambda$ is in the definition of the Fourier time t, which varies as λ^3 (Equation (22)). The situation is saved by the fact that Q and t enter $F_N(Q,t)$ in the form $Q^n t$ in most studies of macromolecular systems. For diffusion, for example, n = 2 and the final λ-dependence is only linear; this has little effect if there is no fine structure in $F_N(Q,t)$, but again may be calculated explicitly if needed. In the next section I shall consider an example where n = 3, in which case the λ-dependence is completely cancelled.

I have assumed that there is no inelastic contamination of the quasielastic scattering. Provided any other frequencies present are well above the range of interest, they will not interfere with the measurement. This is because large energy transfers completely depolarize the corresponding spin-echo signal, and enter as a background.

This is, in fact, the basis of a new use of neutron spin-echo, to remove thermal diffuse scattering from diffraction measurements.[18]

The final problem is spin-incoherent scattering from the sample, which contaminates the coherent signal. The analysis of this problem is outside the scope of this talk, and I shall merely state the result:[16] if Equation (33) is replaced by

$$P_E = f(I^E - \overline{I^E})/(I^+ - I^-) \qquad (38)$$

where f is the π-turn coil efficiency, then Equation (37) will be independent of spin-incoherent effects if the coherent and incoherent dynamics are assumed to be the same. (Note that the machine parameter f again cancels in Equation (37)). In the usual case of small incoherent contamination, this rather crude assumption about the dynamics has negligible effect on the results.

Finally, I remark without further comment that IN11 can measure its own wavelength spectrum using a different Fourier technique, *assymetric* spin-echo, which is fully described in Reference 19.

4. MACROMOLECULAR STUDIES USING NEUTRON SPIN-ECHO

In the past two years neutron spin-echo has been applied with some success to several problems in macromolecular chemistry. The first example I shall discuss is the confirmation of the Dubois-Violette-de Gennes formulation[20] of the Zimm model[21] for the dynamics of a single polymer chain in dilute solution.

The description of polymer relaxation in solution, on the scale of segmental diffusion (Flory radius[22] $R_o > r >$ segment length ℓ), is based on the use of hydrodynamic equations.[20,23] The hydrodynamic interaction between the monomers has to be included to obtain physically realistic predictions.[21] The excluded volume interaction has also been considered.[24,25] Much of the work has been concerned with scaling laws, and a useful review is given by de Gennes.[26] The influence of concentration is treated by introducing a correlation length ξ which corresponds roughly to the distance between entanglement points of different polymers; over distances $r < \xi$, essentially single-chain behaviour should still be observed. In good solvents, ξ scales with concentration as $c^{-3/4}$,[27] and the width of the relaxation spectrum is predicted to scale with Q^3.[23,25] Some progress was made in examining these scaling laws using neutron backscattering and time-of-flight spectrometry.[28-30] Scaling theories do not give information about the shape of the scattering law, however, and the neutron measurements were unable to obtain detailed information on the spectral shape, for the reasons discussed in the introduction.

A direct theoretical attack on the problem was undertaken by Dubois-Violette and de Gennes,[20] who found

$$F_N(Q,t) = (\Gamma t)^{2/3} \int_0^\infty du \cdot \exp\{-(\Gamma t)^{2/3} u[1 + h(u)]\} \qquad (39)$$

where

$$h(u) = \frac{4}{\pi} \int_0^\infty \frac{dy}{y^3} \cdot \cos y^2 \left[1 - \exp(-y^3 u^{-3/2})\right] \qquad (40)$$

The relaxation frequency Γ is given by

$$\Gamma = \frac{1}{6\pi\sqrt{2}} \cdot \frac{k_B T}{\eta} \cdot Q^3 \qquad (41)$$

where η is the solvent viscosity. This result is interesting because it predicts a universal function for all polymers, Γ scaling with Q^3 and depending only on the bulk solvent parameter T/η.

Neutron spin-echo measurements were undertaken by two groups to test Equations (39) - (41) on dilute solutions of polydimethylsiloxane (PDMS) and polymethylmethacrylate (PMMA) in C_6D_6,[31] and polystyrene (PS) and polytetrahydrofuran (PTHF) in both CS_2 and C_6D_6.[32] Typical spin-echo data[31] is shown in Figure 4.

The exponential decay function corresponds to $S(Q,\omega)$ being Lorentzian; this is clearly not the case. Equation (39) fits much better, although the deviations may be outside experimental error. Since the prefactor in Equation (41) is constant in a given experiment, a fit of all spectra from each sample was made with common variation of T/η. Results for two samples are shown in Figure 5, where $\Gamma/Q^3 \propto T/\eta$ is plotted against Q.[31] This clearly confirms the Q^3 law.

The absolute values of the prefactors, however, vary from sample to sample, and Equation (41) is apparently not universal. In this context it is worth noting that PDMS has a particularly flexible chain and may represent a limiting case.[33]

Results at higher concentration are shown in Figure 6, where Γ/Q^2 is plotted versus Q. While the results at 0.05 $g \cdot cm^{-3}$ follow a straight line, indicating a Q^3 law, deviations from this behaviour occur at small Q as the concentration increases. This is an experimental indication for the crossover from Q^3 to Q^2 predicted by de Gennes[25] for entangled systems. (Q is still too large here for the Q^2 behaviour to result from diffusion of the polymer as a whole).

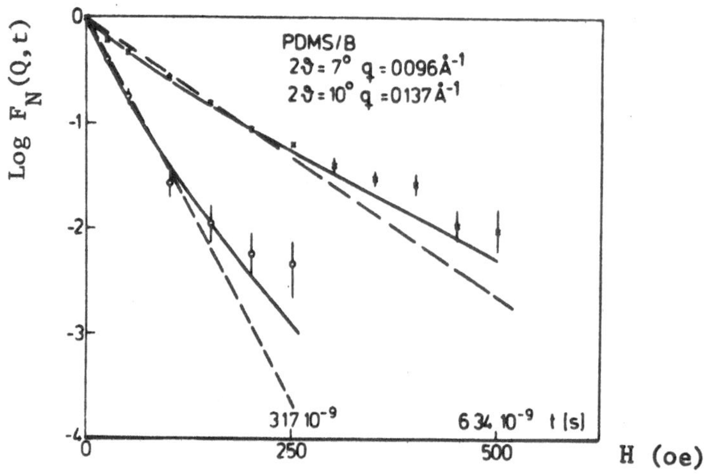

Figure 4. Logarithm of $F_N(Q,t)$ measured on 5% PDMS/C_6D_6 as a function of magnetic field H, corresponding to time t. —— Fit to Equation (32) --- Exponential decay function.

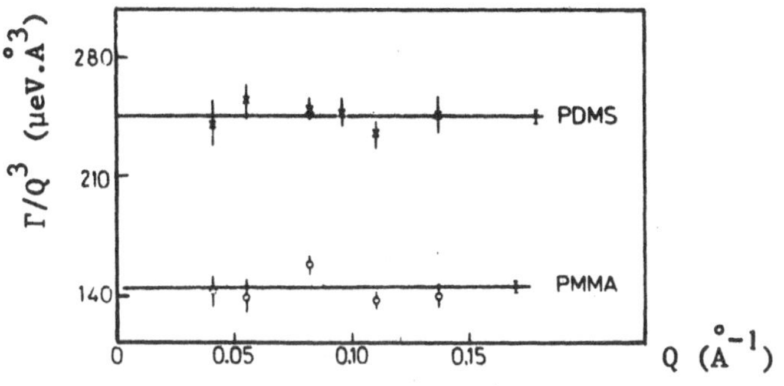

Figure 5. Variation of Γ/Q^3 with Q. —— Result of a fit with common parameter variation.

The position of the crossover point allows a dynamical estimate of the screening length ξ. The values given in Figure 6 were obtained assuming $Q^*\xi = 1$, where Q^* is the position of the crossover in momentum space.

Recent studies[34] have also shown a Q-dependent temperature crossover in the dynamics of dilute polymer solutions. This crossover is associated with the transition from theta to good solvent conditions.

Polymer melts have also been studied. The theory is based on the reptating chain model discussed by de Gennes[35,36] and by Doi and Edwards.[37] Polymer chains cannot pass through one another, so large scale motions are imagined to proceed by reptation of the polymer

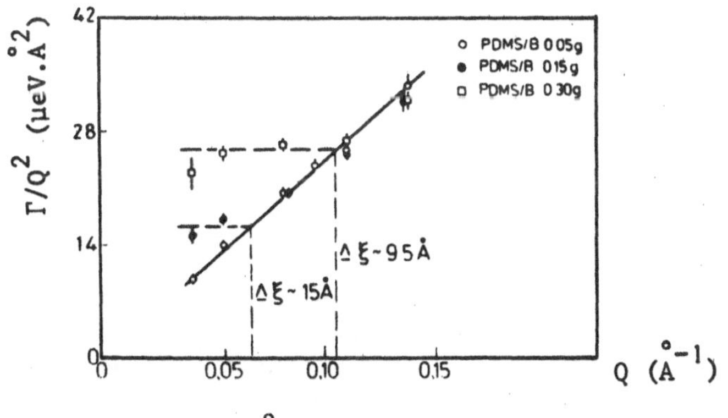

Figure 6. Variation of Γ/Q^2 with Q. —— Common variation of parameter fit to data at 0.05 $g \cdot cm^{-3}$. The dashed lines are guides for the eye.

chain, within a tube of radius a and length L such that aL is the mean square end-to-end distance of the chain. Measurements have been performed on 3% solutions of PTHF in perdeutero - PTHF (PTDF) at temperatures of 110°C and 150°C, using both neutron spin-echo and backscattering spectroscopy.[38] Results were analyzed in terms of a relaxation frequency Ω taken from the initial slope of $F_N(Q,t)$. This is given by Akcasu et al[39] as

$$\Omega = \partial \log F_N(Q,t)/\partial t \big|_{t\to o}$$

$$= \frac{1}{12} \cdot \frac{k_B T}{\xi_o} \cdot \sigma^2 Q^4 \qquad (42)$$

where ξ_o is a friction factor per step-length σ. The Q-dependence is thus stronger than for solution (cf. (41)). Results are shown in Figure 7.

Qualitatively the behaviour in the melt and in dilute solution are somewhat similar, with a faster Q-dependence of Ω in the melt. The lines shown on Figure 7 would give $\Omega \propto Q^n$ with $2.4 < n < 2.8$ in solution and $3.2 < n < 3.8$ in the melt. Such power laws are not very meaningful here, however, because the Q-range is transitional between $Qa < 1$ and $Qa > 1$. The sensitivity to Q is evident from the recent work of de Gennes[36] who predicts a Q^6 dependence of the coherent dynamics for large enough Q ($QR_o > 1$), with different scaling properties for the incoherent scattering. This leads to an important factor which may distort the apparent low-Q slope, namely the use of back scattering to extend the Q range. Neutron spin-echo provides polarization analysis, which may be used to separate coherent from incoherent scattering,[40] and the IN11 data points shown in Figure 7 are for coherent scattering. The backscattering data contains an unknown but certainly significant incoherent contribution, with somewhat different dynamics, and it may be fortuitous that a linear plot is obtained over the whole range. The question of polymer melt dynamics still needs further investigation.

Another field being studied by spin-echo is that of polyelectrolytes in solution, where strong Coulomb interactions are added to the inter- and intrapolymer forces. The static structure factor S(Q) shows a pronounced interaction peak at a Q value Q_m in these systems. A preliminary spin-echo study[41] of the solution dynamics shows that the Q-dependent diffusion coefficient D(Q) decreases sharply with Q below $Q = Q_m$ to a nearly constant value above Q_m. If sufficient salt is added to screen the electrostatic interactions, D(Q) becomes independent of Q. These features have been explained in terms of a new picture for the correlations, based on (i) the notion of a correlation hole, of diameter $\xi = \ell\phi^{-1/2}$, and (ii) locally rigid chains with a persistence length $b = \ell\phi^{-1}$, where ϕ is the

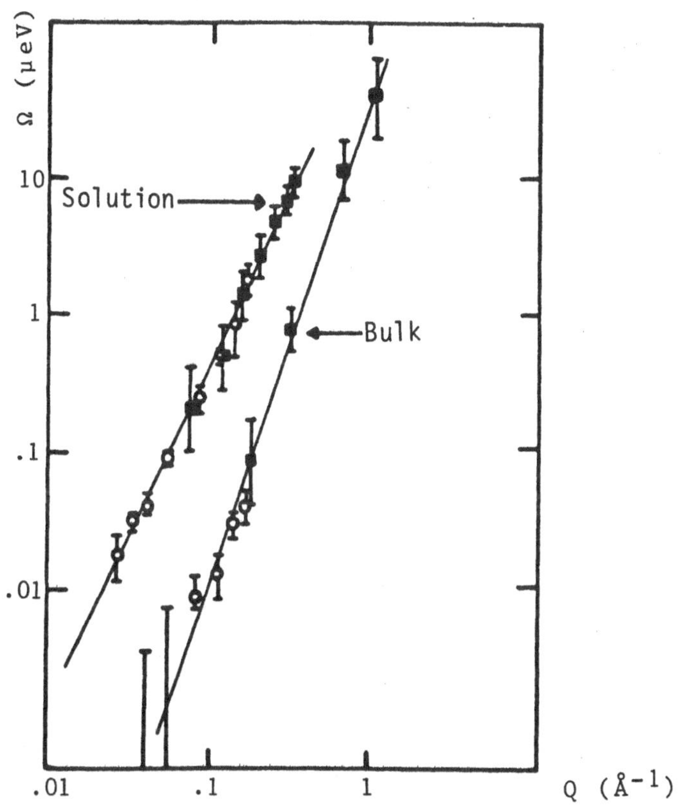

Figure 7. Variation of relaxation frequency Ω with Q for a dilute (3%) solution of PTDF/CS_2 (T = 30°C), and a melt sample of 3% PTHF/PTDF (T = 110°C). ■Backscattering data. O Spin-echo data. The lines are guides for the eye.

polymer volume fraction.

The final example I want to discuss also involves the dynamics of a strongly interacting system, namely charged spherical micelles in concentrated solution. The system chosen was sodium dodecyl sulphate (SDS) in aqueous solution, with and without added salt. Typical spin-echo data[42] is shown in Figure 8.

In the absence of interactions, the data of Figure 8 would be described by the usual result for diffusion (see, e.g. Reference 43)

$$F_N(Q,t) = \exp(-D_o Q^2 t) \tag{43}$$

where D_o is the diffusion coefficient. The decay in Figure 8 is

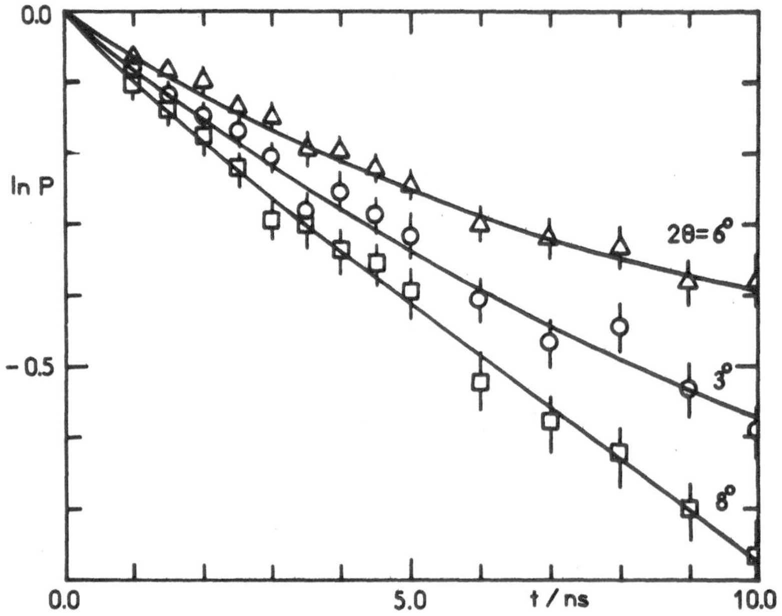

Figure 8. Typical variation of the logarithm of spin-echo polarization with time for a micelle solution (0.4M SDS). Scattering angles correspond to momentum transfers of 0.04 $Å^{-1}$ (3°), 0.079 $Å^{-1}$ (6°) and 0.106 $Å^{-1}$ (8°). Lines are fits to Equation (44) truncated at second order.

clearly not exponential, and the variation with Q is more complicated than predicted by Equation (43). It may be understood in terms of a generalized hydrodynamic theory (see e.g. Refs. 44,45), which leads to a cumulant expansion for the intermediate scattering law:

$$F_N(Q,t) = \exp\{-[c_1(Q)t + c_2(Q)t^2/2! + \ldots]\} \tag{44}$$

where

$$c_1(Q) = D_o Q^2 [1 + H(Q)]/S(Q) \tag{45}$$

with H(Q) determined by hydrodynamics. If the hydrodynamic interactions are ignored as a first approximation, H(Q) = 0 and Equations (44) and (45) yield

$$c_1(Q) = -\partial \log F_N(Q,t)/\partial t\Big|_{t\to o}$$

$$= D_o\, Q^2/S(Q)$$
$$= D_{eff}\, Q^2 \qquad (46)$$

where, by analogy with Equation (43), D_{eff} is a Q-dependent effective diffusion coefficient.

The static intermicellar structure factor S(Q) was measured using neutron small-angle scattering.[42] In the case of interacting spherical particles the scattered intensity is given by[10]

$$I(Q) \propto P(Q)\cdot S(Q) \qquad (47)$$

where P(Q) is the single-particle scattering function. Typical data is shown in Figure 9.

A closed analytic structure factor is now available for charged spheres in a screening medium,[46] and S(Q) was obtained by fitting the latter theory to the data using a model P(Q).[42] This fit yields the radius a of the micelles, from which D_o is calculated using the Stokes-Einstein relation

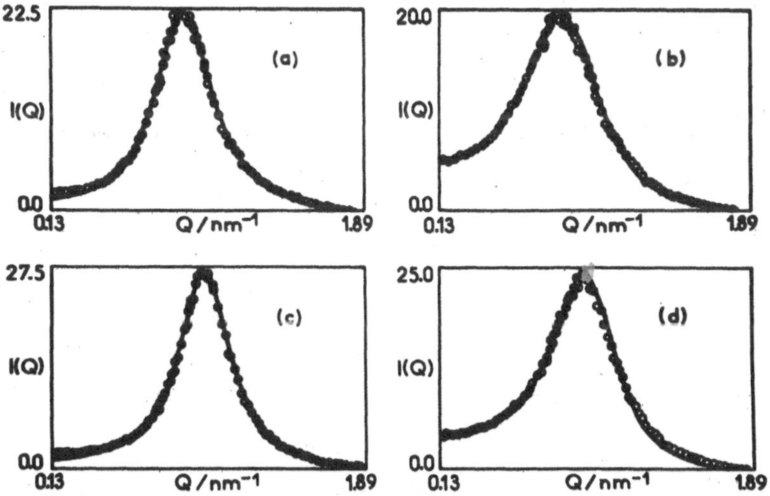

Figure 9. Variation of scattered intensity (arbitrary units) with Q for solutions of SDS/D_2O.
(a) 0.4M SDS; (b) 0.4M SDS + 0.1M NaCl; (c) 0.6M SDS; (d) 0.6M SDS + 0.1M NaCl
—— Fits to Equation (47) (see text).

$$D_o = (1 - \phi)\ k_BT/6\pi\eta a \tag{48}$$

where ϕ is the volume fraction; $(1 - \phi)$ is a reference-frame correction to volume-fixed conditions.

From Equation (46)

$$D_o/D_{eff} = S(Q) \tag{49}$$

and this comparison (due to Pusey[47]) is shown in Figure 10, where D_{eff} has been calculated from the initial slopes of the spin-echo data. Q has been multiplied by $\sigma = 2a$ to provide a dimensionless scale.

This procedure allows a fully self-consistent simultaneous analysis of static and dynamic data. It is important to note that the parameters compared in Figure 10 are absolutely scaled, and no fitting is involved in the comparison. (This is a consequence of the absolute normalization in the spin-echo polarization measurement).

At higher concentrations, the hydrodynamic interaction becomes non-negligible (Figure 11).

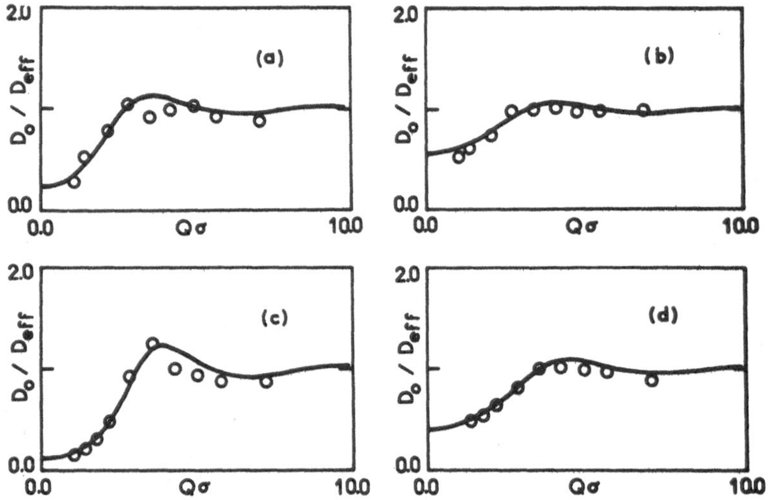

Figure 10. Dimensionless comparison of static and dynamic micelle data. O D_o/D_{eff} —— S(Q)
(a) 0.04M SDS; (b) 0.04M SDS + 0.1M NaCl; (c) 0.1M SDS; (d) 0.1M SDS + 0.1M NaCl.

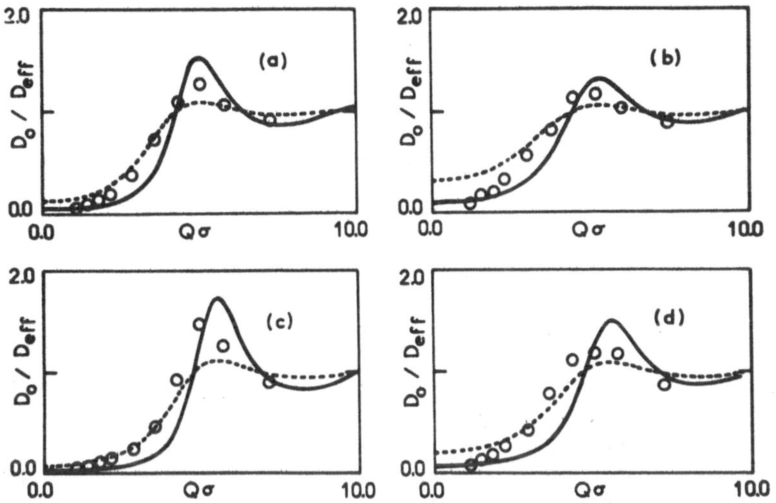

Figure 11. Dimensionless comparison of static and dynamic data for concentrations corresponding to Figure 9. O D_o/D_{eff} —— S(Q) --- S(Q)/{1 + 0.75[S(Q) - 1]}

The dashed lines represent a simplified inclusion of hydrodynamics based on a hard-sphere theory due to Ackerson,[48] which gives

$$H(Q) = 0.75\ [S(Q) - 1] \tag{50}$$

so that

$$D_o/D_{eff} = S(Q)/\{1 + 0.75\ [S(Q) - 1]\} \tag{51}$$

Equation (51) represents an improvement over Equation (49), but there is clearly still scope for a more detailed treatment of the hydrodynamics.

This combined use of neutron small-angle scattering and neutron spin-echo shows, for the first time, that small-angle scattering data from strongly interacting systems may be treated in a self-consistent manner. This opens the way to the study of many interesting colloidal systems, for example charged microemulsions and biological macromolecules. The same analytical technique can also be applied to PCS studies, the choice of neutrons or photons being an experimental convenience to situate the range of Q, according to the size of the particles to be studied.

REFERENCES

1. F. Mezei, Z. Phys. 255: 146 (1972).
2. G. E. Bacon, Neutron Diffraction, 3d ed., Clarendon Press, Oxford, England (1975).
3. J. B. Hayter, Polarised Neutrons, in Neutron Diffraction, ed. by H. Dachs, Springer, Berlin (1978).
4. N. W. Ashcroft and J. Lekner, Phys. Rev. 145: 83 (1966).
5. B. J. Berne and R. Pecora, Dynamic Light Scattering, John Wiley, New York (1976).
6. E. R. Pike and H. Z. Cummins, Photon Correlation and Light Beating Spectroscopy, Plenum, New York (1974).
7. A. Guinier and G. Fournet, Small-Angle Scattering of X-rays, John Wiley, New York (1955).
8. S. H. Chen and S. Yip, Spectroscopy in Biology and Chemistry, Academic Press, New York (1974).
9. B. Jacrot, Rep. Progr. Phys. 39: 911 (1976).
10. G. Kostorz (ed.), Neutron Scattering, in Treatise on Materials Science and Technology, vol. 15, Academic Press, New York (1979).
11. W. Marshall and S. W. Lovesey, Theory of Thermal Neutron Scattering, Clarendon Press, Oxford, England (1971).
12. P. G. de Gennes, Physica 25: 825 (1959).
13. M. Birr, A. Heidemann, and B. Alefeld, Nucl. Instr. Methods 95: 435 (1971).
14. G. Badurek and G. P. Westphal, Nucl. Instr. Methods 128: 315 (1975) and references cited therein.
15. J. B. Hayter, Z. Phys. B31: 117 (1978).
16. J. B. Hayter, I.L.L. Report 78HA50, Institut Laue-Langevin, Grenoble, France (1978).
17. J. B. Hayter, Theory of Neutron Spin-Echo Spectrometry, in Proc. International Workshop on Neutron Spin-Echo, ed. F. Mezei, vol. 128, Springer Lecture Note Series, Berlin (1980).
18. J. B. Hayter, M. S. Lehmann, F. Mezei, and C. M. E. Zeyen, Acta. Cryst. A35: 333 (1979).
19. J. B. Hayter and J. Penfold, Z. Phys. B35: 199 (1979).
20. E. Dubois-Violette and P. G. de Gennes, Physics 3: 181 (1967).
21. B. H. Zimm, J. Chem. Phys. 24: 269 (1956).
22. P. Flory, Principles of Polymer Chemistry, Cornell University Press, Ithaca, New York (1971).
23. A. Z. Akcasu and H. Gürol, J. Polym. Sci., Polym. Phys. Ed. 14: 1 (1976).
24. P. G. de Gennes, Macromolecules 9: 587 (1976).
25. P. G. de Gennes, Macromolecules 9: 594 (1976).
26. P. G. de Gennes, Nature 282: 367 (1979).
27. M. Daoud, J. P. Cotton, B. Farnoux, G. Jannink, G. Sarma, H. Benoit, R. Duplessix, C. Picot, and P. G. de Gennes, Macromolecules 8: 804 (1975).
28. G. Allen, R. E. Ghosh, J. S. Higgins, J. P. Cotton, B. Farnoux, G. Jannink, and G. Weill, Chem. Phys. Lett. 38: 557 (1976).

29. J. S. Higgins, G. Allen, R. E. Ghosh, W. S. Howells, and B. Farnoux, Chem. Phys. Lett. 49: 197 (1977).
30. A. Z. Akcasu and J. S. Higgins, J. Polym. Sci., Polym. Phys. Ed., 15: 1745 (1977).
31. D. Richter, J. B. Hayter, F. Mezei, and B. Ewen, Phys. Rev. Lett. 41: 1484 (1978).
32. L. K. Nicholson, J. S. Higgins, and J. B. Hayter, Proc. International Workshop on Neutron Spin-Echo, ed. F. Mezei, vol. 128, Springer Lecture Note Series, Berlin (1980).
33. J. S. Higgins, R. E. Ghosh, W. S. Howells, and G. Allen, J.C.S. Faraday II 73: 40 (1977).
34. D. Richter, B. Ewen, and J. B. Hayter, Phys. Rev. Lett. 45: 2121 (1980).
35. P. G. de Gennes, J. Chem. Phys. 55: 572 (1971).
36. P. G. de Gennes, J. Chem. Phys. 72: 4756 (1980).
37. M. Doi and S. F. Edwards, J.C.S. Faraday II 74: 1789, 1802, 1818 (1978).
38. J. S. Higgins, L. K. Nicholson, and J. B. Hayter, Polymer 22: 163 (1981).
39. A. Z. Akcasu, M. Benmouna, and C. C. Han, Polymer 21: 866 (1980).
40. R. M. Moon, T. Riste, and W. C. Koehler, Phys. Rev. 181: 920 (1969).
41. J. B. Hayter, G. Jannink, F. Brochard-Wyart, and P. G. de Gennes, J. de Physique 41: L451 (1980).
42. J. B. Hayter and J. Penfold, J.C.S. Faraday I, in press.
43. A. R. Altenberger and J. M. Deutch, J. Chem. Phys. 59: 894 (1973).
44. W. Hess and R. Klein, Physica 94A: 71 (1978).
45. W. Hess and R. Klein, Physica 99A: 463 (1979).
46. J. B. Hayter and J. Penfold, Mol. Phys. 42: 109 (1981).
47. P. N. Pusey, J. Phys. A 8: 1433 (1975).
48. B. J. Ackerson, J. Chem. Phys. 64: 242 (1976).

THE NATIONAL CENTER FOR SMALL-ANGLE SCATTERING RESEARCH*

W. C. Koehler, R. W. Hendricks, H. R. Child,
S. P. King, J. S. Lin, and G. D. Wignall

Oak Ridge National Laboratory
Oak Ridge, Tennessee

CONTENTS

1. INTRODUCTION

The National Center for Small-Angle Scattering Research (NCSASR) is a national user-oriented facility dedicated to studying the structure of matter on the scale of tens to hundreds of angstroms (10^{-8} cm). The center, sponsored by the National Science Foundation (NSF) through an interagency agreement with the Department of Energy (DOE), was established at the Oak Ridge National Laboratory (ORNL) in 1978. It makes state-of-the-art equipment for neutron and x-ray small-angle scattering experiments available to qualified investigators from university, industrial, and other government laboratories.

---

*Research sponsored by the National Science Foundation under interagency agreement #40-637-77 under Union Carbide Corporation contract W-7405-eng-26 with the U.S. Department of Energy.

NCSASR facilities include full-time use of a new NSF-funded, 30-m small-angle neutron scattering (SANS) instrument at the High Flux Isotope Reactor (HFIR) and a fraction of the beam time on four existing DOE-funded instruments. These are a 10-m x-ray scattering (SAXS) camera, a high intensity Kratky camera with a linear position-sensitive detector, a 5-m SANS facility at the Oak Ridge Research Reactor (ORR), and a high-resolution, double-perfect-crystal SANS instrument at the HFIR.

Beam time on the various instruments is assigned on the basis of the scientific merit of the proposed research. Proposals are reviewed by an external panel of experts who judge the quality of the research and determine the suitability of the NCSASR facilities for accomplishing it. Special procedures have been established for the proprietary and applied problems. At present, nine scientists representing the areas of biology, chemistry, materials, and polymer science, areas in which SANS and SAXS techniques are particularly fruitful, serve on this panel.

The NCSASR is jointly responsible to DOE, through ORNL management, and to the NSF. The performance of the center is reviewed by a policy committee consisting of six distinguished scientists appointed by NSF and ORNL. The policy committee is broadly representative of the user community and includes experts in the fields of biology, chemistry, materials science, polymer science, and diffraction physics. The committee also provides guidance on the policies of the center and reviews the professional accomplishments of the staff.

In addition to the program and policy committees, a formal users' organization has been established. This group, through its executive committee, serves in an advisory capacity and communicates general users' concerns to NCSASR and the policy committee.

Except for proprietary research investigations, there are no charges for beam time on the instruments, for staff time, or for the use of any equipment associated with the center. Extraordinary costs, however, are borne by the user.

Proposal forms, report forms, and a general guide to the NCSASR facilities may be obtained from: NCSASR Secretary, Solid State Division, Oak Ridge National Laboratory, P. O. Box X, Oak Ridge, TN 37830.

2. THE INSTRUMENTS

A. The 30-m Small-Angle Neutron Scattering Facility[1,2]

The 30-m SANS instrument is the realization of a joint NSF-

DOE-ORNL project, initiated in January 1978, to provide a national facility for carrying out small-angle neutron scattering experiments in the United States. Three years later, the design goals set for the first phase of construction have been achieved. Measurements of machine performance have shown that the instrument will serve the needs of the majority of scientists in the polymer, materials, and chemical sciences communities and that it will adequately meet the requirements of many users in the biological sciences. Planned future improvements will make it useful to an even broader spectrum of scientists.

The 30-m instrument utilizes pinhole geometry with collimating slits of 0.5-3.0 cm diameter separated by a distance of 10 m. The detector, with an active area of 64 by 64 cm^2 and resolution element dimensions of 0.5 by 0.5 cm^2, can be positioned at any distance from 1.5 to 18.5 m from the specimen by moving a motor-driven detector carrier along rails in the evacuated flight path. The standard incident wavelength, provided by a bank of pyrolitic graphite crystals, is 4.75 A, although this can be changed to 2.38 A by substituting graphite for the cold beryllium filter normally in position. The changeover time is less than five minutes and the procedure permits experiments to be performed with increased flux over a wider range of scattering angles.

The specimen chamber is designed to accommodate standard samples or, if necessary, it can be fitted with specialized ancillary equipment. Slit positioning and sample stage positioning are carried out with computer-controlled stepping motors.

The data acquisition and processing system has been adapted from a comprehensive package used successfully for six years on the ORNL 10-m SAXS camera. The software, representing more than five man years in development, permits the user to interact with the dedicated MODCOMP II computer in a simple and direct fashion.

Specifications for the 30-m SANS instrument are listed in Table 1. A schematic drawing of the instrument is given in Figure 1.

The first experiments by external users were carried out during the fall of 1980. By the end of the year some 50 research proposals had been approved for beam time. Experiments completed or scheduled include, among others, investigation of the mechanism of transformation in alloys, structures and conformation of polymers and polymer blends, stress induced cavitation in metals and alloys, size distribution of hydrocarbon filled voids in oil shales, "molecular weights" of coals in solution, interaction of nucleosomes with chromosomal proteins, and the structure of shark fin collagens.

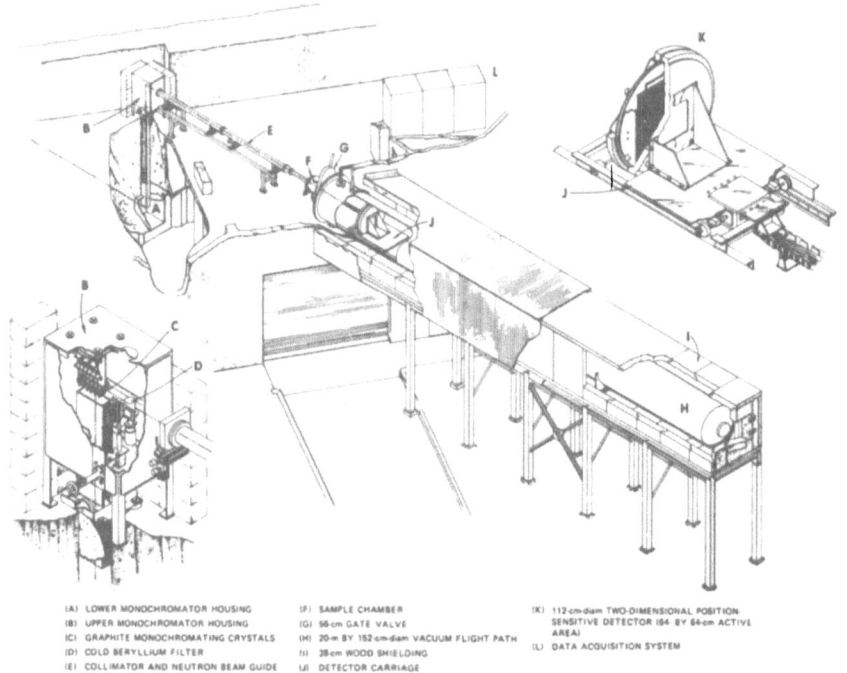

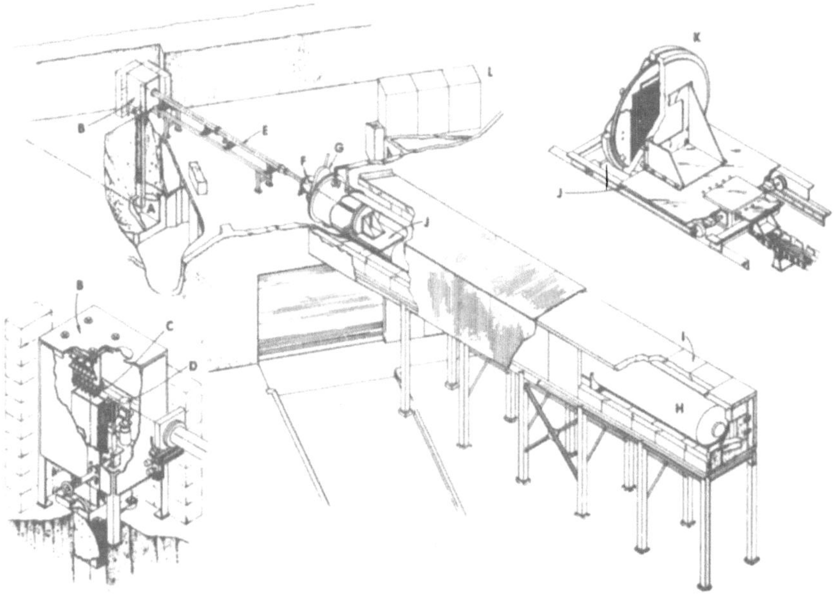

Figure 1. Schematic diagram of the 30-m SANS facility. The neutron beam used by the instrument is transported from the beam room to the SANS Facility level by Bragg reflection from pyrolitic graphite crystals scattering at 90°.

B. The 10-m SAXS Instrument

The 10-m SAXS camera has been in routine operation in the Metals and Ceramics Division of ORNL for over six years. It is a unique instrument in the United States and it represents an enormous improvement over previously existing SAXS facilities. It was designed and built originally by R. W. Hendricks as part of a program to investigate void size distribution in irradiated metals. It has since been applied to studies of the crystallization of polymers, voids in coal macerals, and collagen fibrils among many other problems.

This new scattering camera utilizes a rotating anode x-ray source, crystal monochromatization of the incident beam, pin-hole collimation, and a two-dimensional position-sensitive proportional counter. Because the sizes of the x-ray focal spot, the sample, and the resolution element of the detector are each approximately 1 x 1 mm^2, the camera was designed so that the focal spot-to-sample

Table 1. The 30-m SANS Instrument Specifications

Beam tube: HB-4B (High Flux Isotope Reactor)
Monochromator: six pairs of pyrolitic graphite crystals
Incident wavelength: 4.75 A or 2.38 A
Wavelength resolution: $\Delta\lambda/\lambda = 6\%$
Source-to-sample distance: 10 m
Beam size at specimen: 0.5-3.0 cm diam
Sample-to-detector distance: 1.5-18.5 m
K range: $5 \times 10^{-3} \leq K \leq 0.6\ A^{-1}$
Detector: 64 by 64 cm^2
Flux at specimen: $10^4 - 10^5$ neutrons $cm^2\ s^{-1}$ depending on slit sizes and wavelength

and sample-to-detector distance may each be varied in 0.5 m increments up to 5 m, to provide a system resolution in the range 0.5 to 4.0 mrad. A large, general purpose specimen chamber has been provided into which a wide variety of special purpose specimen holders can be mounted. The detector has an active area of 200 x 200 mm^2 and has up to 200 x 200 resolution elements. The data are recorded in the memory of a minicomputer by a high speed interface which uses a microprocessor to map the position of an incident photon into an absolute minicomputer memory address. With this interface, over 10^5 events/sec can be recorded. The data recorded in the computer memory can be processed on-line by a variety of programs designed to enhance the users[1] interaction with the experiment. Among these are routines for background and detector sensitivity correction, contour and perspective plotting of the two-dimensional data, a variety of averaging schemes for determining circular averages and/or one dimensional slices of the data, file management programs to handle the large quantity of data produced by a two-dimensional detector, and an interactive package for communications with a central computing facility via a dedicated hardwired link. At the highest angular resolution (0.4 mrad) the flux incident on the specimen is 1.0 x 10^6 photons/sec with the x-ray source operating at 45 kV and 100 mA. Since January 1, 1978, a fraction of the beam time on the instrument (about 30%) is available to external users as part of the resources of the NCSASR.

In addition to the ancillary equipment listed in Table 2 a deformation device for dynamic experiments is under development and nearing completion.[4] This is being designed for use, at first, on the 10-m SAXS camera and it will be available to qualified users of the NCSASR facilities.

This device, a specimen holder and control system, is intended primarily for the study of the dynamics of polymer deformation and relaxation, and for strain-induced crystallization by x-ray small-angle scattering. The system, with its associated electronics, provides for maintaining a polymer specimen in a He atmosphere in the temperature range from $^{-}150$ to 300 C. It will permit (i) preloading a specimen to a specified fixed elongation and then superimposing an oscillatory load with strain amplitude of up to 2 mm at frequencies between 0.001 and/or 100 Hz, or (ii) providing a one-

Table 2. The 10-m SAXS Instrument Specifications

Monochromator:	hot-pressed pyrolitic graphite
Incident wavelengths:	1.542 A(CuK_α) or 0.707 A(MoK_α)
Source-sample distances:	0.5, 1.0, 1.5, ..., 5.0 m
Beam size at specimen:	0.1 by 0.1 cm (fixed)
Sample-detector distances:	1, 1.5, 2.0, ..., 5 m
K range covered:	$3 \times 10^{-3} \leq K \leq 0.3$ A^{-1} (CuK_α)
	$6 \times 10^{-3} \leq K \leq 0.6$ A^{-1} (MoK_α)
Maximum flux at specimen:	10^6 photons per second on sample-irradiated area 0.1 by 0.1 cm
Detector:	two-dimensional position-sensitive proportional counter with 20- by 20-cm^2 active area and potentially 200 by 200 resolution elements (electronic resolution 0.1 by 0.1 cm^2)
Background on detector:	45 cps electronic noise distributed uniformly over detector and parasitic slit edge scattering typically 20-160 cps, depending on sample-to-detector distance, beam stop size, etc.
Ancillary equipment:	special sample chambers include furnace ($20 \leq T \leq 200^\circ C$) and liquid sample holder

shot high speed elongation of the specimen (>500% elongation in about 10 msec). The system should be valuable as well for studying the dynamics of deformation and relaxation in biological materials such as muscle.

Sophisticated computer interfaces and software have been developed in order to acquire and process large amounts of data rapidly and to permit interactive data analysis.

C. The Kratky SAXS Camera[5]

This instrument, and the two to be described subsequently, will for different reasons, probably not be in as heavy demand by users as the 30-m SANS and 10-m SAXS facilities. The Kratky camera has a conventional Kratky small-angle collimation system that has been modified to allow the use of a one-dimensional position-sensitive X-ray detector (Table 3). The detector was designed specifically

Table 3. The Kratky SAXS Camera Specifications

Monochromators: Ni-filter for CuK_α radiation, Zr-filter for MoK_α radiation, and energy discriminator on detector

Incident wavelengths: λ = 1.542 A (CuK_α)
λ = 0.707 A (MoK_α)

Source-to-sample distance: 29.8 cm
Sample-to-detector distance: 151.5 cm

K ranges covered: $2 \times 10^{-3} \leq K \leq 0.21$ A^{-1} (CuK_α)
$4 \times 10^{-3} \leq K \leq 0.44$ A^{-1} (MoK_α)

Maximum flux at specimen: 15×10^6 photons per second on sample-irradiated area 60 μm by 1.5 cm

Detector: One-dimensional position-sensitive proportional counter 7.6 cm in length with an electronic spatial resolution 0.04 cm

Background on detector: 17 cps scattered from a typical empty cell window

Ancillary equipment: Data display; thermostat for sample temperature control in the range $0 \leq T \leq 150°C$

Special features: ModComp II minicomputer for camera control: data desmearing programs available for use with a long-slit camera and has uniform sensitivity over the entire beam in the slit-length direction. The camera has relatively low background scattering because of its unique collimation geometry, and thus it is particularly suitable for SAXS studies of any systems that have inherently weak, isotropic scattering. The camera is not suitable for the study of systems with anisotropic scattering patterns because of the complicated smearing of the scattered intensity and the problems with anisotropic collimation corrections. Such systems are better studied on the 10-m SAXS camera.

D. The High-Resolution Double-Perfect-Crystal SANS Instrument[6,7,8]

This facility is characterized by its capability to analyze neutron scattering at very small angles with high resolution (Table 4). This is achieved by placing the sample between two perfect crystals of silicon, as shown in Figure 2. The first silicon crystal acts as a neutron monochromator by Bragg reflection of the HB-3 HFIR beam. The second silicon crystal is rotatable about a vertical axis, and serves to analyze the specimen-scattered neutrons

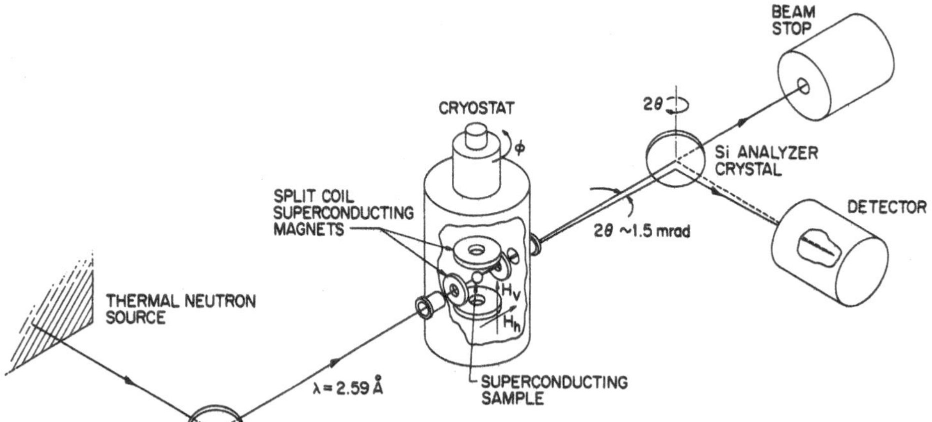

Figure 2. The high-resolution double-perfect-crystal SANS instrument.

Table 4. The Double-Crystal SANS Instrument Specifications

Beam tube: HFIR HB-3A

Monochromator: Silicon in <111> Bragg reflection

Incident Wavelength: 2.6 A

Wavelength resolution: $\Delta\lambda/\lambda \simeq 2\%$

Source-to-sample distances: 2.5 m (monochromator to sample) and 4.5 m (monochromator to source)

Beam size at specimen: maximum - 4 cm high by 2 cm wide

Angular resolution: $\Delta K \simeq 4 \times 10^{-5}$ A^{-1} (determined by FWHM of double-crystal Bragg peak)

Sample-to-detector distance: approximately 100 cm (measured through a Bragg reflection at analyzer crystal)

K ranges covered: $0.001\ A^{-1} \leq K \leq 0.02\ A^{-1}$

Maximum flux at specimen: approximately 10^4 neutrons cm^{-2} s^{-1}

Detector: BF_3 proportional counter (active area 20 cm^2)

Background: dependent on beam area and K value (varies approximately as K^{-1} for $0.001 \leq K \leq 0.006$ and saturates at 0.3 counts cm^{-2} s^{-1} for $K > 0.006\ A^{-1}$.

Auxiliary equipment: cryostat and peripherals for 1.5 K < T < 300 K: vertical magnetic field capability to ∿20 kG and horizontal magnetic field capability to ∿6 kG

---

by means of Bragg reflection into a fixed detector. Since the beam collimation of the machine is effectively a vertical slit, the facility is well suited for specialized applications where the scattering is isotropic or where the physical system possesses translational invariance in one dimension, as does, for example, a fibrous or filamentary material.

The principal advantage of the instrument is its capability to accommodate large samples possessing gross structure without a sacrifice in resolution ($\Delta K \simeq 4 \times 10^{-5}\ A^{-1}$). The disadvantages involve (a) the inability to measure scattered intensity at more than one point in reciprocal space at a given time (i.e., no possibility of area detection), and (b) intrinsically high angle-dependent background resulting from the double-crystal technique.

The principal applications of this instrument have been to investigations of the properties of flux-line lattices in type-II superconductors. The experimental arrangement shown in Figure 2 illustrates the disposition of the horizontal or vertical fields that can be applied to scattering specimens.

E. The 5-m SANS Instrument[9]

This instrument is a relatively low-flux medium resolution SANS facility located at the Oak Ridge Research Reactor. Time on this instrument was made available, on request, to external users since January 1, 1978. It has been used as well in DOE programmatic research on voids in irradiated materials, metallurgical phase transformations, magnetic critical scattering, spin glass systems, and amorphous magnetic materials for example. During the construction of the 30-m machine it was used by external users for feasibility experiments and for experiments involving strong scatterers. Outside users will probably use the 30-m machine rather than the 5-m instrument at least until such time as the former becomes saturated.

3. APPLIED TECHNOLOGY AND PROPRIETARY RESEARCH

The primary objective of the NCSASR is to advance basic science through the use of small-angle scattering techniques. In pursuing this general objective the center encourages the use of the facilities by industrial, government, and academic institutions. A modest amount of time has therefore been allocated to engineering, applied technology, and proprietary research, in addition to fundamental nonproprietary studies. Hence, empirical studies having potential for important technological applications but no likelihood of external publication may also be accommodated on the instruments. Similarly, proprietary research is encouraged by both NSF and ORNL, and general policies for the performance of such research have been developed.

REFERENCES

1. W. C. Koehler and R. W. Hendricks, "The United States national small-angle neutron scattering facility," J. Appl. Phys. 50: 1951 (1979).
2. W. C. Koehler and R. W. Hendricks, "Construction and operation of a national user-oriented small-angle neutron scattering facility at the Oak Ridge National Laboratory," Research Proposal to the National Science Foundation, April 1977.

3. R. W. Hendricks, "The ORNL 10-meter small-angle x-ray scattering camera," J. Appl. Crystallogr. 11: 15 (1978).
4. R. W. Hendricks, "Design and construction of a specimen holder and control system for dynamic x-ray and neutron small-angle scattering studies of the plastic deformation of polymers," Research Proposal to the National Science Foundation, February 1979.
5. T. P. Russell, R. S. Stein, M. K. Kopp, R. E. Zedler, R. W. Hendricks, and J. S. Lin, "The application of a one-dimensional position-sensitive detector to a Kratky small-angle x-ray camera," ORNL/TM-6678 (January 1979).
6. F. Tasset, D. Christen, S. Spooner, and H. A. Mook, "Small-angle double-silicon-crystal neutron diffraction studies of the fluxoid lattice in superconducting niobium," Proceedings of the Conference on Neutron Scattering, Gatlinburg, Tennessee, June 6-10, 1976, R. M. Moon, ed., p. 482.
7. D. K. Christen, S. Spooner, P. Thorel, and H. R. Kerchner, "Determination of low-field critical parameters of superconducting niobium by small-angle neutron diffraction," J. Appl. Crystallogr. 11: 650 (1978).
8. D. K. Christen, H. R. Kerchner, S. T. Sekula, and P. Thorel, "Equilibrium properties of the fluxoid lattice in single-crystal niobium. II. Small-angle neutron diffraction measurements," Phys. Rev. B21: 102 (1980).
9. H. R. Child and S. Spooner, "New small-angle neutron scattering (SANS) instrument at ORNL using a position sensitive area detector," J. Appl. Crystallogr. 13: 259 (1980).

RECENT DEVELOPMENTS IN PHOTON CORRELATION AND SPECTRUM ANALYSIS TECHNIQUES: I. INSTRUMENTATION FOR PHOTODETECTION SPECTROSCOPY

C. J. Oliver

Royal Signals and Radar Establishment

Malvern, Worcs, UK

CONTENTS

ABSTRACT

Initially the nature of the signal in photodetection spectroscopy is discussed and the implications for instrumentation examined. Since a major limitation of such spectroscopy is imperfections in source or detector these are discussed and their effect demonstrated. The spectral estimators in time-delay space (the autocorrelation function) and the frequency domain (the periodogram) are demonstrated to be equivalent so that choice of either depends on engineering considerations. Finally a selection of instruments operating in either space are discussed and their particular advantages or drawbacks indicated culminating in some conclusions on the choice of suitable instrumentation for photodetection spectroscopy.

1. INTRODUCTION

In this part we shall endeavour to establish the essential features of instrumentation capable of measuring the spectral properties of a detected optical signal in either the time-delay or frequency domain. In Section 2 we shall consider the nature of the signal available as defined by its moments and its higher-order correlation properties. The importance of the signal statistics in terms of the factorisation properties of the correlation functions and the effect of the Poisson photodetection process are indicated. Having considered the basic data, this set of photodetection events, we next consider data-sampling and processing schemes (Section 3). In particular the photodetection autocorrelation function and power spectrum estimators are defined and suitable architectures for implementing such processes illustrated. In order for instrumentation to measure these functions successfully the equipment must itself be free from distortion. In Section 4 we examine some of the imperfections encountered in sources and detectors and their effects. Examples of results in which theory and experiment are in close agreement are given to show that with sufficient attention to detail near-perfect performance can be achieved so that the measured functions can be reliable.

If we now assume that the recorded photodetection events are undistorted we next examine the time and frequency domain estimators in more detail (Section 5). We derive expressions for the mean and variance of the spectral estimates and examine them for bias and consistency. Section 5 effectively establishes the information-content limits on the spectral estimates obtained in photodetection spectroscopy. These limits will be discussed further in Part II. In addition to these fundamental limits there are the restrictions imposed by the use of non-ideal instruments. The practical limitations of various processing methods in the time and frequency domains are examined in Section 6. Finally various conclusions on the choice of instrumentation for photodetection spectroscopy are outlined in Section 7.

2. NATURE OF THE SIGNAL

It is essential to understand the nature of the signal from which we hope to extract information about the physical properties of a system under study. In Figure 1 we show a typical light-scattering experiment in which laser radiation is scattered by refractive index fluctuations in the scattering region into a detector. This type of experiment relies on the fact that if the source properties are known any changes will have been introduced by the scattering medium. Photon correlation spectroscopy or spectral analysis, for example, thus derives information about the physical behaviour of the scattering region from the time-dependent fluctuations of the detector output. The received field can be

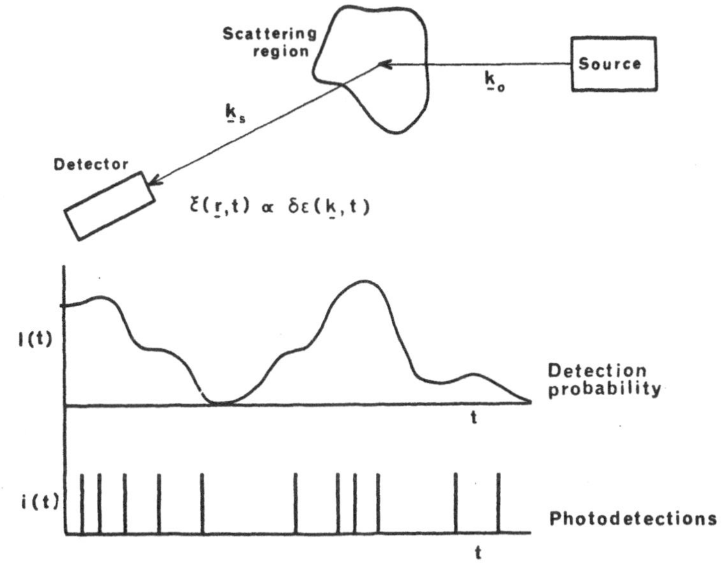

Figure 1. Block diagram for a typical laser scattering experiment. The detection probability, I(t) (i.e. the intensity), and the corresponding photodetection pulses, i(t), are shown.

expressed as the sum of positive and negative frequency components, $\mathcal{E}^+(t)$ and $\mathcal{E}^-(t)$ which are themselves defined by Fourier decomposition (Jakeman, 1974). Thus

$$\mathcal{E}(t) = \sum_{\omega>0} a^*_\omega \exp\left[-i\omega t\right] + \sum_{\omega\geq 0} a_\omega \exp\left[i\omega t\right]$$

$$= \mathcal{E}^+(t) + \mathcal{E}^-(t), \quad (1)$$

where

$$a_\omega = a^*_{-\omega} = \frac{1}{T}\int_{-T/2}^{+T/2} \mathcal{E}(t) \exp\left[-i\omega t\right] dt. \quad (2)$$

Complete information about the scattered field would be contained in knowledge of all the correlation coefficients and moments. The simplest requirement would be to know the field probability distribution, $P(\mathcal{E})$, or its moments, $\langle\mathcal{E}^r\rangle$, and the first-order temporal correlation coefficient

$$G^{(1)}(\tau) = \langle \mathcal{E}^{+}(0)\mathcal{E}^{-}(\tau)\rangle . \tag{3}$$

However, a photodetector does not respond to the field but to the square of the field envelope (Glauber, 1963), loosely termed the intensity, i.e.

$$I(t) = \mathcal{E}^{+}(t)\mathcal{E}^{-}(t) = |\mathcal{E}(t)|^{2}, \tag{4}$$

so that no direct information is available in the detector output regarding the field moments or first-order correlation function. Indeed the simplest measureable quantities are the intensity probability distribution, P(I), or the intensity moments, $\langle I^{r}\rangle$, and the second-order temporal correlation coefficient

$$G^{(2)}(\tau) = \langle I(0)I(\tau)\rangle . \tag{5}$$

These correlation coefficients are conveniently expressed in normalised forms such that

$$g^{(1)}(\tau) = G^{(1)}(\tau)/\langle I\rangle \tag{6}$$

and

$$g^{(2)}(\tau) = G^{(2)}(\tau)/\langle I\rangle^{2} . \tag{7}$$

In order to deduce information about the field from the measured intensity we require to know the field factorisation properties. Perhaps the simplest case, which corresponds to the ideal laser, is for a coherent source in which the intensity is constant. Here the normalised intensity moments are given by

$$n^{(r)} = \langle I^{r}\rangle/\langle I\rangle^{r} = 1 \tag{8}$$

and the normalised intensity autocorrelation function by

$$g^{(2)}(\tau) = 1. \tag{9}$$

Another common case is when the source or scattering region comprises many statistically independent contributions. Invoking the central limit theorem (Cramer, 1946) we find that the field will become gaussian distributed as the number of contributions increases without limit. Here

$$P(\mathcal{E}) = \frac{1}{\sqrt{2\pi\langle \mathcal{E}^{2}\rangle}} \exp\left[-\frac{\mathcal{E}^{2}}{2\langle \mathcal{E}^{2}\rangle}\right], \tag{10}$$

$$n^{(r)} = \langle I^r \rangle / \langle I \rangle^r = r! \tag{11}$$

and

$$g^{(2)}(\tau) = 1 + |g^{(1)}(\tau)|^2. \tag{12}$$

This last expression, due to Siegert (1943), relates the observed second-order autocorrelation function to the envelope-squared of the field autocorrelation function.

In general fields are not of this simple nature and the thrust of the last few years has led to study of a wide range of non-gaussian fields. In principle if the field factorisation properties are known measurement of the intensity autocorrelation function yields the field autocorrelation function since factorisation implies that all higher-order statistical properties are functions of the lower order ones. It should be noted at this point that it is possible to obtain direct measurement of the first-order, field, autocorrelation function in laser scattering experiments if some of the original laser source radiation is mixed with the scattered radiation (heterodyne detection). In this case the observed intensity autocorrelation function is given by

$$g^{(2)}(\tau) = 1 + \frac{2I_o I_s}{(I_o+I_s)^2} g_s^{(1)}(\tau) \cos \tau\Delta\omega + \frac{I_s^2}{(I_o+I_s)^2}\left(g_s^{(2)}(\tau)-1\right) \tag{13}$$

where I_o and I_s are the reference and scattered intensities, $g_s^{(1)}(\tau)$ and $g_s^{(2)}(\tau)$ represent the first and second-order scattered radiation autocorrelation functions and $\Delta\omega$ represents any frequency difference between the reference and scattered beams. Thus it is possible to make direct measurement of the first-order correlation coefficients by this means if necessary provided that $I_o > I_s$.

So far we have confined our attention to the detected intensity. However, photodetection itself is a Poisson random process in which the mean detection rate is proportional to the intensity. The probability of obtaining a photodetection within a period T is related to the integrated intensity by the expression due to Mandel (1959), i.e.

$$P(n,T) = \int_o \frac{(\eta E)^n}{n!} \exp[-\eta E] P(E)\, dE \tag{14}$$

where η is the detector quantum efficiency and E is given by

$$E = \int_{t-T/2}^{t+T/2} I(t)dt. \qquad (15)$$

It is this train of photodetection events which carries all the available information (see Figure 1).

3. DATA SAMPLING AND PROCESSING SCHEMES

In order to extract information from this train of photodetection events we need, in principle, to record the time of arrival of each event. With weak signals and low count rates this approach may be feasible but the more general approach is to count the events recorded during some sample time T chosen to be short compared with any signal of interest, i.e. such that $f_{max}T < 1$. Even the photomultiplier tube itself performs this operation since it effectively integrates the total charge occurring during the basic photodetection pulse width which may well include several actual events. In photocounting operation the detector output is first standardised and then individual detections are counted in a series of contiguous samples of duration T. The probability distribution of the number of counts per sample time is derived from Equation (14). For a constant intensity, e.g. an ideal laser source, this yields a Poisson photon-counting distribution.

$$P(n,T) = \frac{\bar{n}^{n}}{n!} \exp[-\bar{n}] \qquad (16)$$

where $\bar{n}$ is the mean count-rate per sample time T. Intensity-fluctuations, induced for example in scattering laser light, would result in a non-Poisson distribution. Thus measurement of the photodetection statistics can be used as a means of extracting spectral information (Jakeman et al, 1968a).

A more direct measure of spectral parameters can be obtained from measurement of the intensity autocorrelation function or the power spectrum. Suppose the photo-counting data consists of a string of N samples of duration T where the jth sample has contents n(j). The estimate of the autocorrelation function is then defined by (Oppenheim and Shafer, 1975; Oliver 1979)

$$\hat{G}^{(2)}(k) = \frac{1}{N} \sum_{j=0}^{N-1-k} n(j)n(j+k) \qquad (17)$$

where kT is the delay value. This function is symmetrical about $k = 0$ so it could be implemented using the architecture of Figure 2. Equation (17) shows that during the jth sample we need to perform the product of the present input, $n(j)$, with the contents of inputs, $n(j \pm k)$, shifted by $\pm kT$. The delayed signals are conveniently obtained by storing the data in a shift register of length K. Each sample time the data is shifted to the right and the K products computed. In addition to the delay element the architecture therefore requires a multiplier for each delay element. The products are then accumulated in K counters which record the individual autocorrelation function estimates, $\hat{G}^{(2)}(k)$, as a running summation of the data received. This architecture is attractive for digital implementation since, with the exception of the parallel multiply and accumulate operation, the structure is essentially naturally pipelined. This means that data is fed in serially without any necessity for storing a large number of samples such as would be required in a batch processing method.

In the frequency domain the power spectrum estimator for the data set could be defined by Oppenheim and Shafer (1975) and Oliver (1979)

$$\hat{P}(\omega) = \frac{1}{N}\left|\sum_{j=0}^{N-1} n(j) \exp\left[-i\Omega j\right]\right|^2 \tag{18}$$

where $\Omega = \omega T$. In this form the structure is definitely a batch

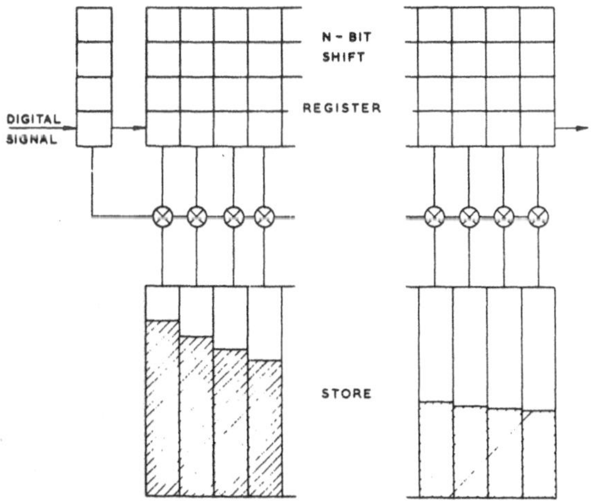

Figure 2. Block diagram of a many-bit digital autocorrelator

process which is not very conveniently implemented. If we consider the estimate for frequency $\omega = 2\pi\ell/NT$, where ℓ is an integer, then the above estimate corresponds to the periodogram of the data (Oppenheim and Schafer, 1975). Expressed in terms of discrete frequencies in this manner Equation (18) becomes

$$\hat{P}(\ell) = \frac{1}{N}\left|\sum_{j=0}^{N-1} n(j) \exp\left[-\frac{i2\pi\ell j}{N}\right]\right|^2 . \tag{19}$$

This discrete Fourier transform can be recast by making the substitution

$$-2\ell j = -\ell^2 - j^2 + (\ell - j)^2$$

into Equation (19) as proposed by Bluestein (1970). Hence

$$\hat{P}(\ell) = \frac{1}{N}\left|\exp[-i\pi\ell^2/N]\sum_{j=0}^{N-1} n(j) \exp[-i\pi j^2/N]\exp[i\pi(\ell-j)^2/N]\right|^2$$

$$= \frac{1}{N}\left|\sum_{j=0}^{N-1} n(j) \exp[-i\pi j^2/N] \exp[i\pi(\ell-j)^2/N]\right|^2 . \tag{20}$$

Equation (20) can be broken down into two processes: multiplication of the data, $n(j)$, by a negative-slope linear frequency-modulation (chirp), $\exp[-i\pi j^2/N]$, followed by convolution with a positive chirp of the same gradient, $\exp[i\pi(\ell - j)^2/N]$. This is known as the chirp-z transform because of the discrete time variable adopted. The choice of the power spectrum rather than the Fourier transform of the batch of data eliminates one chirp multiplication in Equation (20). The basic architecture for this implementation is shown in Figure 3. The incident signal samples are first multiplied by a down chirp and then convolved with the up-chirp of the same slope in a filter before being envelope detected to give the power spectrum. This architecture processes a batch of data with both input and output being serial in nature. A detailed discussion of different variants on this architecture, and on spectrum analysis in general, can be found in Roberts et al (1980). Where the input data is noisy, i.e. contains very few photodetection pulses, the average over many separate estimates of the periodogram can be obtained by summation in a many-channel digital integrator following digitisation in an analog-to-digital converter (ADC) (Alldritt et al, 1978).

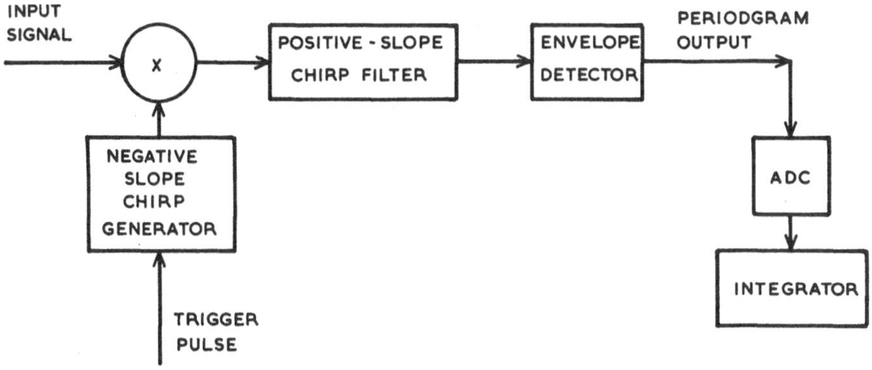

Figure 3. Block diagram of the chirp-Z transform architecture for obtaining the periodogram power spectrum estimate.

4. SOURCE AND DETECTOR IMPERFECTIONS

An ideal optical instrument to extract the required information about some medium from the scattered radiation should use every photodetection event, for highest efficiency, and must not introduce any distortion or extraneous signal. Obviously the more detailed the information that is needed the more extreme the fidelity requirement. As we shall see comparatively small errors will lead to major discrepancies where complex information retrieval is attempted. As an example of the sort of complicated information that one may wish to acquire consider Figure 4 which illustrates histogram treatments of the distribution of diffusion coefficient for a solution of polystyrene in trans-decalin as a function of temperature. This example is taken from a wide range of similar examples by Chu and fellow-workers (e.g. Chu and Nose, 1979). Without going into detail, it is obviously essential to be able to demonstrate that the variations in the calculated distributions of diffusion coefficient are not introduced by instrumental properties or the results of random noise fluctuations but are indeed entirely due to the process under study.

In this section we shall consider some of the problems introduced by real sources and detectors. We shall gear our thinking to a laser scattering experiment such as that represented in Figure 1. At this stage we shall not investigate effects due to unexpected components (e.g. dust) in the scattering region, which will be considered in the second lecture, but will assume that all fluctuations introduced in the scattering region may be considered to be signal and that only source and detector effects need concern us. Many of these source and detector defects and their implications have been discussed previously (Oliver, 1974) so that only a summary of the

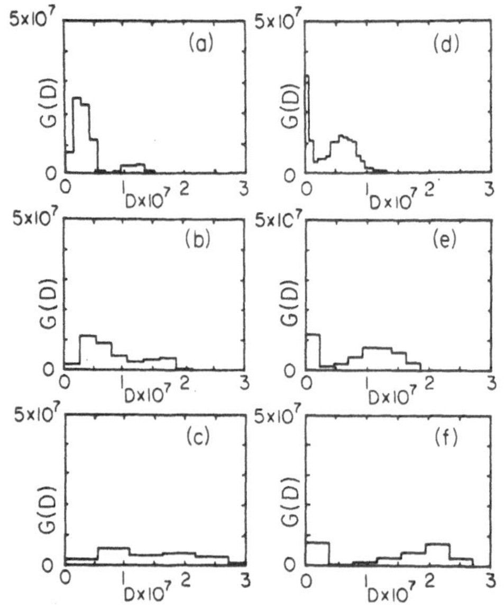

Figure 4. An example of the complexity of the information we would wish to extract from photodetection spectroscopy measurements. The plots show the dependence of the probability distribution, G(D), of the diffusion coefficient, D, on temperature for a solution of polystyrene in trans-decalin. (a), (b), (c) solutions 0.233 g/g measured at 90°; (a) 20° C, (b) 30° C, (c) 40° C; (d), (e), (f) solutions 0.346 g/g measured at 135°; (d) 20° C, (e) 30° C, (f) 40° C. Results taken from Chu and Nose, 1979.

salient features will be given here. The imperfections can be investigated via the observed photon-count statistics (Foord et al, 1969) or via the autocorrelation function (Oliver, 1974). As shown in Equation (16) for a coherent, constant-intensity source the photon-count statistics should be Poisson with factorial moments given by (Jakeman et al, 1968b)

$$N^{(r)} = \langle n(n-1)\ldots(n-r+1)\rangle/\langle n\rangle^{r} = 1,$$

and a normalised autocorrelation function [Equation (9)].

$$g^{(2)}(\tau) = 1.$$

In Table I we show a comparison of experiment with theory for the normalised factorial moments of the photon-counting distribution, for coherent (laser) and incoherent (scattered) radiation

Table 1. Normalised factorial moments for coherent and incoherent light

	Normalised Factorial Moment	10^7 Samples			
		Experiment		Theory	
COHERENT SOURCE	$n^{(1)}$	1.0000		1.0000	
	$n^{(2)}$	1.0002		1.0000 ± 0.0009	
	$n^{(3)}$	1.0001		1.0000 ± 0.0017	
	$n^{(4)}$	1.0006		1.0000 ± 0.0030	
	$n^{(5)}$	1.0030		1.0000 ± 0.0120	
	$n^{(6)}$	0.9940		1.0000 ± 0.0360	
		10^5 Samples		9×10^6 Samples	
		Experiment	Theory	Experiment	Theory
INCOHERENT SOURCE	$n^{(1)}$	1.000 ± 0.036	1.000 ± 0.038	1.000	1.000
	$n^{(2)}$	2.000 ± 0.150	2.000 ± 0.160	2.004	2.000 ± 0.017
	$n^{(3)}$	6.050 ± 0.850	6.000 ± 0.850	6.050	6.000 ± 0.090
	$n^{(4)}$	24.700 ± 5.900	24.000 ± 5.700	24.700	24.000 ± 0.600
	$n^{(5)}$	130.000 ±45.000	120.000 ±49.000	130.000	120.000 ± 5.000
	$n^{(6)}$	850.000 ± 461	720.000 ± 510	850.000	720.000 ±54.000

(Jakeman et al, 1968b). If we consider first the coherent source results we see that experiment and theory are identical within the theoretical errors. This implies that source, detector and autocorrelator are free from distortion. The measurement with an incoherent source, obtained by scattering light from particles undergoing diffusion, again shows agreement within the theoretical error between experiment and theory. These measurements vindicate the use of this equipment for autocorrelation function measurements. A further test of the detector and correlator combination is shown in Figure 5 which illustrates normalised autocorrelation functions obtained with a white-light, constant-intensity, source running from a large accumulator (Paul and Pusey, 1980). The measurements were taken over several hour period at various different count rates and a sample time of 5 μs. The autocorrelation functions show slight evidence of correlation over the first couple of delays (10 μs), probably due to the correlations within the pm tube, but are

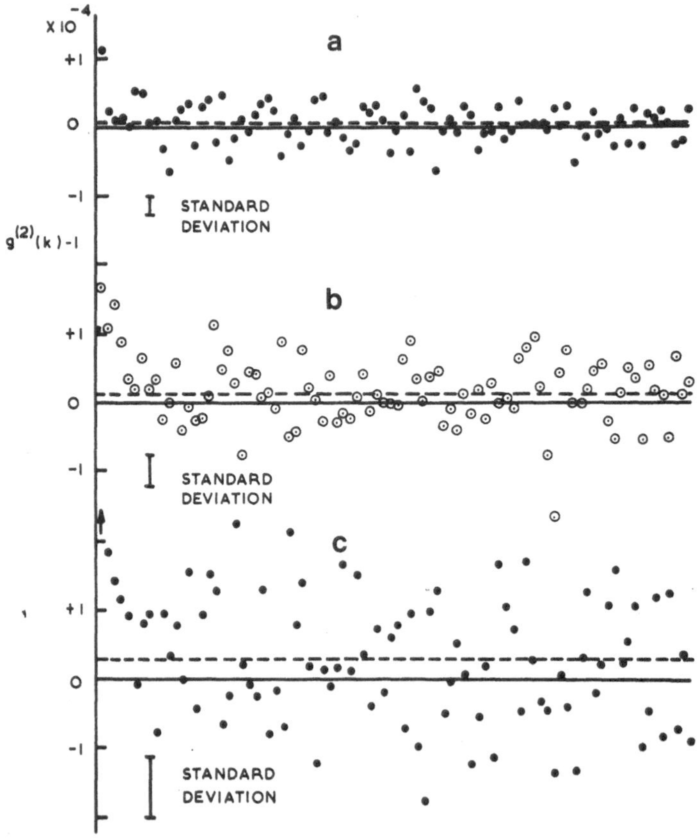

Figure 5. Measured autocorrelation functions for constant intensity illuminiation obtained after many hours integration. T = 5 μs. Results taken from Paul and Pusey, 1980.

essentially flat with a value near 1.0 over the remaining 85 coefficients. The correlation functions were actually constructed by averaging a large number, M, of independently-normalised measurements each obtained with N samples and a mean counts per sample time <n> where: (a) M = 37, N = 2 x 10^8, <n> = 1.15; (b) M = 26, N = 4 x 10^8, <n> = 0.49; (c) M = 26, N = 4 x 10^8, <n> = 0.205. The observed standard deviations of the coefficients, included in Figure 5, were (a) 2.7 x 10^{-5}, (b) 4.5 x 10^{-5}, (c) 9 x 10^{-5} which differ from the relevant theoretical predictions (Jakeman, 1974), (a) 1.0 x 10^{-5}, (b) 2.0 x 10^{-5}, (c) 3.9 x 10^{-5}, since that theory only applies in the small <n> limit. The average values of the 85 coefficients, shown as dashed lines, are approximately within the expected error derived from the observed standard deviation having values of (a) 1.000004 ± 0.000003, (b) 1.000012 ± 0.000005, (c) 1.000027 ± 0.000001. The observed scatter of the individual coefficients in Figure 5 reveals no evidence of synchronous noise in the autocorrelation function. This provides yet another test of the photon-correlation instrument demonstrating that it will indeed measure the real properties of the incident radiation on the detector. Some such check should be a standard procedure - particularly, if complicated correlation functions are to be analysed - since any departure from this ideal behaviour will result in distortion and misinterpretation of the data.

A. Source Imperfection

Ideally a laser should provide a coherent field of constant intensity. Departure from the ideal behaviour may take the form of plasma oscillations (Oliver, 1974) where gross effects may be observed or, more subtily, long-term amplitude drift associated with thermally-induced charges in laser cavity alignment and hence gain. Provided these latter effects take place on a time scale long compared with any other process under study the most obvious effect would be a misnormalisation of the autocorrelation function. Suppose we take N/2 samples with mean rate $\bar{n}$ and N/2 samples with mean rate $\bar{n}(1 + \beta)$. The normalised correlation function will then be given by (Oliver 1974)

$$g^{(2)}_{obs}(\tau) = G^{(2)}_{obs}(\tau)/n^{-2}_{obs} \simeq 1 + \frac{\beta^2}{4} . \tag{21}$$

Thus for a drift of 1% one might expect a misnormalisation of 2.5 x 10^{-5}. This may not appear significant when considering normal intensity-fluctuation spectroscopy (IFS) in which the intercept lies at $g^{(2)}(o) \approx 2$. Indeed, it has been shown (Oliver, 1974) that forcing a fit to such data in a single component delay would yield an apparent line-width, Γ_β, given by

$$\frac{\Gamma_\beta}{\Gamma} \simeq 1 - \frac{3\beta^2}{4} \tag{22}$$

where Γ is the true linewidth, i.e. an error of only 7.5 x 10^{-5} results from a long term drift of 1%. This effect is considerably magnified, however, where heterodyne detection is used since, as shown in Equation (13), the intercept is now reduced by the ratio I_s/I_o with an associated corresponding increase in the error in linewidth. Since we require to dominate the IFS term in Equation (13) by the cross term, values for I_s/I_o of 1% are not uncommon leading to an error of 0.75%. A 10% drift in laser power would increase these effects by a factor of 100 in which case the IFS error in linewidth would now be o.75% while the heterodyne error would be 75%.

B. Detector Imperfections

Broadly speaking, in photodection spectroscopy detector performance provides a more servere limitation than the source. Let us now consider the different detector imperfections and their effects. The photocathode emits single electrons which require multiplication by a factor of about 10^7 before they can be detected reliably in high-speed counting circuitry. In practice the most effective wide-band amplifier is the electron-multiplier tube. However, multiplication by secondary emission is a statistical process which introduces two new properties into the description of the photomultiplier operation: the total charge distribution (pulse-height) and the detector rise-time and dead-time (pulse shape).

Gain fluctuations: Since the gain is a statistical process the probability distribution of charge in the output pulse is no longer a delta function, corresponding to a single electron, but relates to the multiplier design characteristics. Analysis show (Oliver, 1974) that the excess noise introduced into the detector output is governed by the parameter $R_N = 1 + [Var(q)/\bar{q}^2]$, where q is the charge per pulse. This distribution of charge per pulse has been discussed by many authors, but particularly by Prescott (1966) who analysed the gain process assuming that secondary emission is a Poisson process with a variable secondary-emission ratio across the dynode or a reduced collection efficiency from stage to stage. While other mechanisms also contribute to the output charge distribution, in particular to the probability of small pulses being recorded (Coates, 1970; Oliver and Pike, 1970), this treatment still remains an elegant general approach representing an extension of the work of Lombard and Martin (1961), who assumed a pure Poisson multiplication process, to include the effects of fluctuation secondary-emission ratios suggested by Baldwin and Friedman (1965). In Prescott's treatment he suggested a statistical model to describe secondary emission in photomultipliers based on the negative binomial (Polya) distribution. Poisson and exponential charge distributions emerge as special cases. He shows that the relative variance after several stages is given by

$$\frac{\mathrm{Var}(q)}{\bar{q}^{2}} = \frac{b\mu + 1}{\mu - 1} \qquad (23)$$

where μ is the mean gain per stage and b is a parameter related to the dynode inhomogeneity. With b = 0 this corresponds to a pure Poisson while b = 1 yields an exponential probability distribution. An example of a measured pulse-height distribution for an EMI6256 photomultiplier is shown in Figure 6 together with theoretical predictions taken from Prescott. From Equation (23) we see that the excess noise parameter will lie in the range

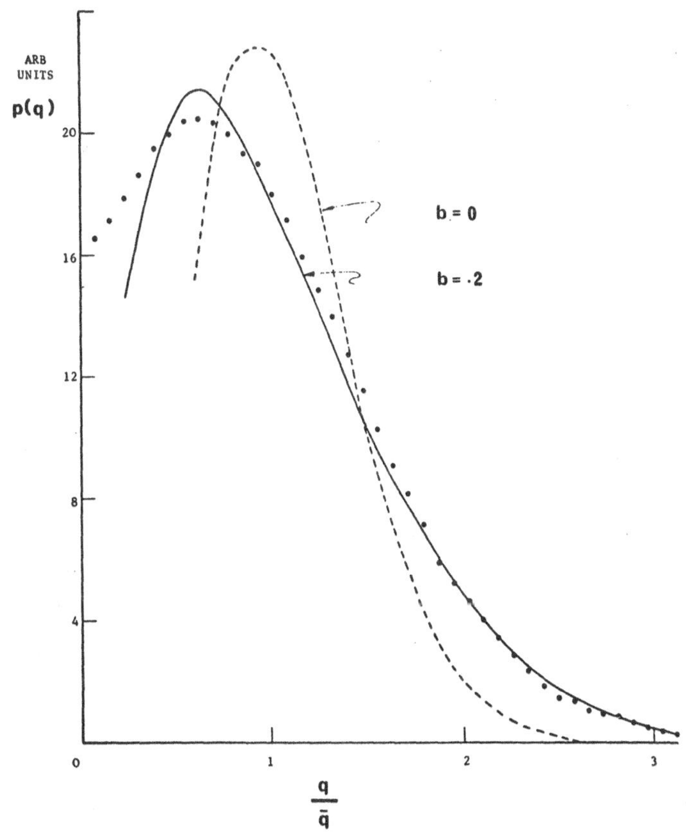

Figure 6. Measured pulse-height distribution for an EMI 6256 Photomultiplier used with a gain per stage of 6.5. The theoretical predictions for b = 0 (dashed) and b = 0.2 (continuous) are included for comparison.

$$\frac{\mu}{\mu - 1} < R_N < \frac{2\mu}{\mu - 1} . \tag{24}$$

For a typical value ($\mu = 5$) the excess noise introduced by the tube gain lies between 1.25 and 2.5. This excess noise may be removed by standardising the detector output in a discriminator, which is conventially termed photoncounting operation. This restores the essential digital nature of the signal photodetections at the expense of losing those pulses whose pulse-height lies below the discriminator threshold. From Figure 6 it is apparent that tubes with a value of b near zero lose fewer pulses due to this cause.

Pulse shape effects: The electron optics of the multiplier tube are not capable of delivering all the charge at the anode simultaneously. The time of flight of the secondary electrons between dynodes is not a fixed quantity nor, as we have seen, is the secondary emission ratio. Euling (1964) has analysed the time statistics of photodetection pulses and derived expressions for the random fluctuations in the apparent detection time. When using photomultipliers two particular time scales are important. Firstly in photo-counting operation a dead-time, τ_D, is associated with every pm tube which is the total width at the base-line of the pulse. This ensures that there is no correlation between observed pulses due to the pulse shape. Suitable dead-times are thus selected to meet the criterion of no correlation (Foord et al, 1969). As the incident photon arrival rate increases towards τ_D^{-1} there will be pulse pile-up at the multiplier anode introducing dead-time effects. Indeed in the limit as the intensity increases the tube output becomes a continuous current so that the discriminator saturates. Thus at high count rates the use of photon-counting can introduce distortion. However, it is well known from nucleonics practice that two photomultipliers used in delayed coincidence (or in cross-correlation) enable one to resolve events down to the scale of the multiplier rise time, τ_r. Use of a beam splitter and two photomultiplier tubes in this fashion will indeed enable events on this time scale to be measured (e.g. Morgan and Mandel, 1966) provided that the individual tube count rates are less than τ_D^{-1}. If we are prepared to sacrifice the digital precision of photon-counting where high rates are encountered then analog processing of the tube output allows us to extract information about processes on time scales of τ_r or greater for all intensities. This is important where we require quick measurements as in tracking (Oliver, 1980a).

Having examined the effects of the multiplication process in terms of gain fluctuation and pulse shape which occur with genuine photodetection events, let us next consider the introduction of spurious signals in the photomultiplier tube. There are two main causes for such unwanted signals - dark current and correlated after-pulses.

Dark current: The dark currents observed with a pm tube are mainly the result of thermionic emission from the cathode and as such can be removed by cooling. However other contributory factors such as radioactivity appear to lead to a residual dark count rate which is of order 0.5 per second with small photo-cathode tubes such as the ITT FW130 (Oliver and Pike, 1968). With some tubes this residual count rate shows evidence of correlation between output pulses with an average of up to 3.5 correlated pulses being observed spread over 10 μs by Rodman and Smith (1963). As we shall see such correlations have impact both on the correlation function and on the measurement of count rate.

Correlated after-pulses: There are various feedback mechanisms in photomultiplier tubes which can give rise to correlated after-pulses. These can be divided into internal processes, such as optical, X-ray or positive-ion feedback from the anode region induced by the presence of large pulses, or external processes associated with the HT supply or the output circuitry, such as finite recovery time following a large pulse. The most serious feedback effects are those which occur at some significant delay after the initial pulse. An example is shown in Figure 7 which demonstrates the presence of positive-ion feedback in an RCA PF1011 photomulti-

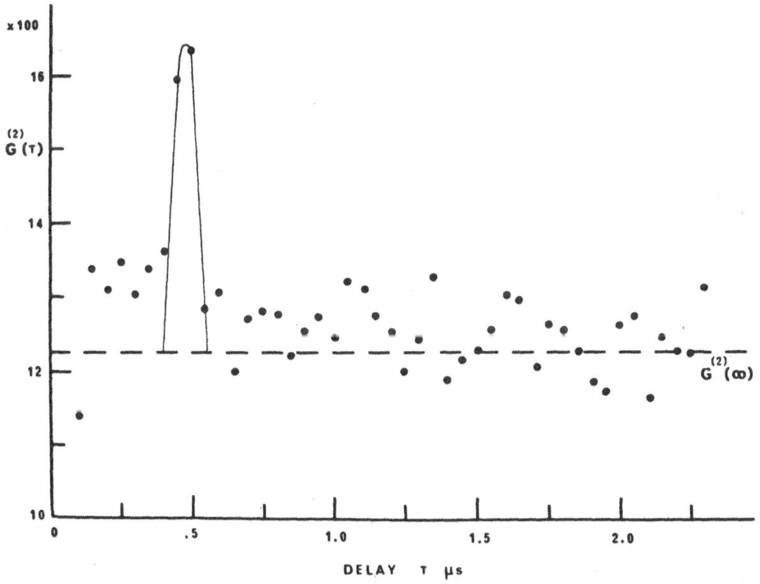

Figure 7. Measured autocorrelation function obtained with coherent illumination RCA PF1011 photomultiplier. T = 50 ns, <n> = 0.00035. The theoretical normalisation, $G^{(2)}(\infty)$, is shown as a dashed line; the continuous line indicates positive-ion after-pulsing at τ = 0.5 μs.

plier tube (Oliver, 1974). Optical or x-ray feedback is effectively instantaneous so that the correlated pulse would emerge with a delay equal to the tube transit time (30-40 ns). This may be avoided by choice of suitable discriminator dead-time when photon-counting. The time-of-flight of the ion back to the cathode in positive-ion feedback is a function of its mass and the EHT voltage but is typically in the region of 0.5 to 1.0 μs as shown in Figure 7.

In order to investigate the scale of the effect of such feedback let us assume that output pulses have a probability $\alpha(\tau)$ of being followed by a correlated after-pulse at delay τ. The observed mean count rate, $\bar{n}_{obs}$, will then be given by

$$\bar{n}_{obs} = \bar{n}_o(1 + \alpha) \tag{25}$$

where $\bar{n}_o$ is the true count rate and

$$\alpha = \int_0^\infty \alpha(\tau)\, d\tau \ .$$

The normalised correlation coefficient for constant intensity is modified by an amount which depends on the detail of the feedback mechanism. For small α the result is independent of the mechanism being given by (Oliver, 1974)

$$g^{(2)}_{obs}(\tau) \simeq 1 + \frac{\alpha(\tau)}{\bar{n}_o} \ . \tag{26}$$

Thus the effect is most marked with low count rates. In the example shown $\bar{n} = 0.00035$ and the after-pulse probability was calculated to be 2 per 10^4. We see therefore that very small after-pulse probabilities can be serious in terms of the distortion they introduce to the autocorrelation function. Great attention should be paid to the choice of suitable tubes which have been designed to minimise these feedback processes.

To conclude this section on source and detector imperfections we should note that choice of a suitable combination of source and detector which actually yields the theoretical performance is an essential prerequisite of any serious photodetection spectroscopy experiment. Effort spent obtaining a near-perfect instrument will save one being misled at some future point by instrumental artifacts.

5. SPECTRAL ESTIMATORS USED IN PHOTODETECTION SPECTROSCOPY

Assuming that we have now selected an ideal source and detector let us next consider the possible spectral estimators that can be used in investigating the properties of the received radiation and hence the scattering process. The forms of the photon-counting autocorrelation function and power spectrum have already been given in Equations (17) and (19) respectively. Since these refer to Fourier transform domains with respect to each other it is necessary to establish what relationships exist between the methods and whether either is preferable. Let us continue to consider what information can be extracted from a batch of N samples of photon-counts. In many applications using autocorrelation the total number of samples is much greater than the number of delay coefficients computed, i.e. N > k. In this case no simple relation between the estimators in time and frequency space can be derived. We shall consider this case later but in the meantime restrict our consideration to the short batch of N data samples. As we have seen the periodogram power spectral estimator is given by

$$\hat{P}\left(\frac{2\pi\ell}{NT}\right) = \frac{1}{N}\left|\sum_{j=0}^{N-1} n(j)\exp\left[-\frac{i2\pi\ell j}{N}\right]\right|^2$$

which can be expanded and rearranged (Oliver, 1979) to yield

$$\hat{P}\left(\frac{2\pi\ell}{NT}\right) = \frac{2}{N}\sum_{j=0}^{N-1}\sum_{j=0}^{N-1-k} n(j)\,n(j+k)\cos\left(\frac{2\pi\ell k}{N}\right) - \frac{1}{N}\sum_{j=0}^{N-1} n^2(j)$$

$$= \sum_{k=-N+1}^{N-1} \hat{G}^{(2)}(k)\exp\left[-\frac{i2\pi\ell k}{N}\right] . \qquad (27)$$

This identity between the periodogram and the autocorrelation function estimators can be regarded as the appropriate form of the Wiener-Khinchin relation for finite sets of discrete samples. Since the two estimators are related by a reversible process [Equation (27)] there is no reason to suppose that spectral parameters can be determined more accurately in one space than another. This would imply that one's choice of method should be determined either by engineering considerations or by virtue of special properties of the signal or backgound noise. In order to illustrate the equivalence of operation in the two domains, Figure 8 shows results obtained in some anemometry measurements using both a photon correlator and a surface acoustic wave (SAW) spectrum analyser (Alldritt et al, 1978). Here a Doppler difference measurement is being performed with a fringe spacing of about 10 μm and a flow speed of approximately 24 ms^{-1}. The mean count rate was about 2 MHz or 0.8 per Doppler cycle. A turbulent flow was obtained which was not

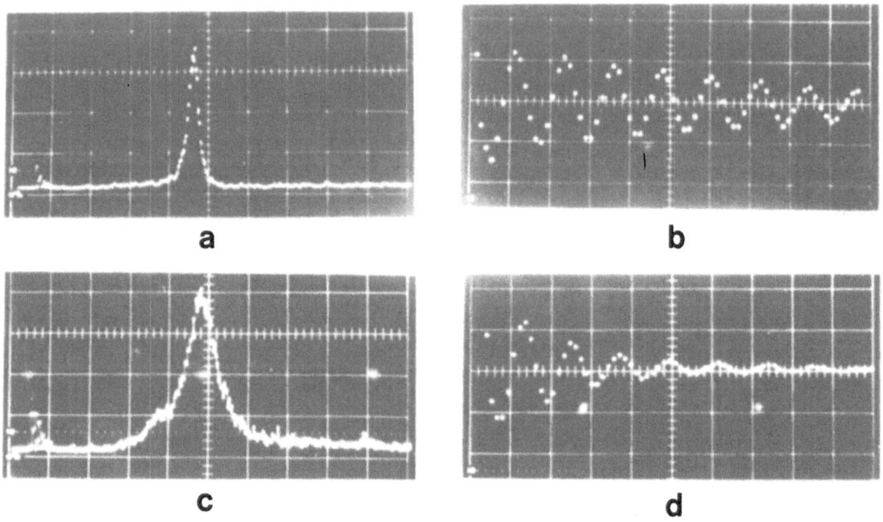

Figure 8. Comparison of photodetection spectroscopy measurements in time-delay and frequency domains. SAW spectra (a, c) and autocorrelation functions (b, d) from a visible laser velocimeter are shown: all 5s integration time, count rate about $2 \times 10^6\ s^{-1}$; spectra 0-5 MHz full horizontal scale, correlation functions 50 ns per channel. (a), (b) near laminar flow; (c), (d) turbulent flow.

Guassian by nature. Inverting the autocorrelation function produced a virtually identical asymmetric power spectrum to that obtained with direct spectrum analysis. Turbulence levels of 11.6% and 12% respectively were obtained for the two methods showing that equivalent information can indeed be obtained in the two domains.

While this comparison serves to show that the mean values of the estimators in the two domains are related via a straightforward Fourier inversion a much more detailed study is required to show how the accuracy with which the spectra or derived spectral estimators can be obtained in the two spaces depends on signal and noise properties. Such a detailed comparison has already been given (Oliver, 1979); only a synopsis of the important results need be given here. It is necessary when comparing the estimators in the different domains to establish whether there is any bias in the estimator, i.e. the extent to which the ensemble average of the estimator differs from its true value, and also to determine whether the estimators are consistent, i.e. whether the variance tends to zero as the number of samples increases without limit. We require, therefore, to derive the mean and variance of the spectral estimators. The value of the variance will also indicate the accuracy with which the spectral estimators are obtained.

If we consider first the autocorrelation function estimator defined in Equation (17) then the ensemble average value will be given by

$$\langle \hat{G}^{(2)}(k) \rangle = \frac{1}{N} \sum_{j=0}^{N-1-|k|} \langle n(j)\, n(j+k) \rangle = \left(1 - \frac{|k|}{N}\right) G^{(2)}(k) \quad (28)$$

so that it can be seen to be biassed by an amount proportional to k/N. When performing analysis in the time domain it is often convenient to use a modified version of the estimator which is unbiassed, namely

$$\hat{G}^{(2)}(k) = \frac{1}{N - |k|} \sum_{j=0}^{N-1-|k|} n(j)\, n(j+k), \quad (29)$$

though it should be noted that the Fourier transform of Equation (29) does not yield the periodogram estimator. The variance of this new estimator will be given, in general, by (Oliver, 1979)

$$\mathrm{Var}\ \hat{G}^{(2)}(k) = \frac{1}{(N-k)^2} \left\{ \sum_{j=0}^{N-1-k} \langle n^2(j) n^2(j+k) \rangle + 2 \sum_{j=0}^{N-1-2k} \langle n(j) n^2(j+k) n(j+2k) \rangle + \sum_{j=0}^{N-1-K} \sum_{\substack{\ell \neq j \\ \neq j-k \\ =0}}^{N-1-k} \langle n(j) n(j+k) n(\ell) n(\ell+k) \rangle \right\} - [G^{(2)}(k)]^2 \quad (30)$$

for $k < N/2$. For $k \geqslant N/2$ a simpler expression is obtained, namely

$$\mathrm{Var}\ \hat{G}^{(2)}(k) = \frac{1}{(N-k)^2} \left\{ \sum_{j=0}^{N-1-k} \langle n^2(j) n^2(j+k) \rangle \right.$$

$$+ \sum_{j=0}^{N-1-k} \sum_{\substack{\ell \neq j \\ =0}}^{N-1-k} <n(j)n(j+k)n(\ell)n(\ell+k)> \Bigg\} - \left(G^{(2)}(k)\right)^2 \quad (31)$$

As these expressions stand one cannot make further deductions about the variance without introducing knowledge of the properties of the photoncounts, n(j). The simplest case is to assume that the n's are uncorrelated, which would be true either for wideband signals ($T > \tau_c$) or for weak ones ($\bar{n} < 1$). Hence

$$<\hat{G}^{(2)}(k)> \quad = \quad G^{(2)}(k) \quad = \quad <n>^2 \qquad (32)$$

for the mean value and

$$\text{Var } \hat{G}^{(2)}(k) \quad = \quad \frac{1}{N-k} \left\{ <n^2>^2 - <n>^4 \right\} \qquad (33)$$

when k > N/2 with an additional term

$$+ \ 2 \ \frac{(N - 2k)}{(N - k)^2} <n>^2 \text{ Var } n$$

where k < N/2 for the variance. From Equation (32) it can be seen that the estimator is unbiassed; this was already chosen to be the case for all statistics. Equation (33) shows in addition that the estimator is consistent. In general, when it is not possible to factorise Equations (30) and (31) because the photoncounts in adjacent samples are correlated, this autocorrelation function estimator will not be consistent.

The equivalent analysis for the periodogram spectral estimator defined in Equation (19) shows that

$$\hat{P}(\ell) \quad = \quad \frac{1}{N} \sum_{j=0}^{N-1} \sum_{k=0}^{N-1} <n(j)n(k)> \exp\left[- \frac{i2\pi\ell}{N} (j-k) \right] \qquad (34)$$

which is an unbiassed estimator. Similarly the variance will be given by (Oliver, 1979)

$$\text{Var } \hat{P}(\ell) = \frac{1}{N^2} \left\{ \sum_{j=0}^{N-1} <n^4(j)> + \sum_{j=0}^{N-1} \sum_{\substack{k \neq j \\ =0}}^{N-1} \left[4<n^3(j)n(k)> \exp\left[- \frac{i2\pi\ell}{N}(j-k)\right] \right. \right.$$

$$+ 2\langle n^2(j)n^2(k)\rangle + \langle n^2(j)n^2(k)\rangle \exp[-\frac{i4\pi\ell}{N}(j-k)]\Big]$$

$$+\sum_{j=0}^{N-1}\sum_{\substack{k\neq j\\=0}}^{N-1}\sum_{\substack{p\neq k\\ \neq j\\=0}}^{N-1}\Big[4\langle n^2(j)n(k)n(p)\rangle \exp[-\frac{i2\pi\ell}{N}(k-p)]$$

$$+ 2\langle n^2(j)n(k)n(p)\rangle \exp[-\frac{i2\pi\ell}{N}(2j-k-p)]\Big]$$

$$+\sum_{j=0}^{N-1}\sum_{\substack{k\neq j\\=0}}^{N-1}\sum_{\substack{p\neq k\\ \neq j\\=0}}^{N-1}\sum_{\substack{q\neq p\\ \neq k\\ \neq j\\=0}}^{N-1}\langle n(j)n(k)n(p)n(q)\rangle \exp[-\frac{i2\pi\ell}{N}(j-k+p-q)]\Bigg\}$$

$$- [P(\ell)]^2. \tag{35}$$

As with the autocorrelation function we can only proceed by using the factorisation properties of the photoncounts. Assuming uncorrelated counts as before we obtain

$$\langle\hat{P}(\ell) = \text{Var } n \tag{36}$$

and

$$\text{Var } \hat{P}(\ell) = (\text{Var } n)^2 + \frac{1}{N}\Big\{\langle n^4\rangle - 4\langle n\rangle\langle n^3\rangle - 3\langle n^2\rangle^2 + 12\langle n\rangle^2\langle n^2\rangle - 6\langle n\rangle^4\Big\} . \tag{37}$$

From Equation (36) we see that the periodogram is an unbiassed estimator. Equation (37) shows that it is not a consistent estimator even for uncorrelated data samples since the dominant term in the variance is independent of N. This should not be taken to imply, however, that there is any difference between the spectral information contained in the periodogram (which is not consistent) and the autocorrelation function estimator (which is consistent) since they are related by a linear process (the Fourier transform).

6. INSTRUMENTS FOR SPECTRAL ANALYSIS

In the previous section we examined how we could extract as

much information as possible from a batch of N photon-counting samples. This can be regarded as establishing information-content limits on spectral measurements. In the present section we shall consider the limitations introduced by real, as opposed to ideal, instrumentation. In Section 6.A we shall consider a variety of instruments that operate in the time domain measuring the autocorrelation function or delayed-coincidence probability. Section 6.B concentrates on instruments which operate in the frequency domain.

The detailed comparison of the instruments will be concerned with various aspects of their performance:

- The efficiency with which all the incoming data is used. This will obviously be greater with parallel rather than single-channel instruments.

- Whether the processor uses the amplitude information present in the signal or introduces distortions due to hard-limiting (clipping). One way to achieve such many-bit processing is to use analog methods.

- Analog methods tend to introduce performance problems related to linearity due to thermal drifts etc. causing difficulty in terms of reproducibility. This is particularly serious with single-channel methods.

- The upper-frequency limit may be imposed either by the pm tube rise time (a minimum of about 1 ns) or by the instrument itself. The tube upper frequency limit will thus be given as 1 GHz. Smaller upper limits will be basically instrumental.

A. Time-Domain Instruments

In this section six different methods will be considered. These are summarised, together with salient features, in Table 2. A previous discussion of many of these features will be found in Oliver (1974).

Delayed coincidence: The simplest method of extracting time-domain information is to use the delayed coincidence method (Morgan and Mandel, 1966; Rebka and Pound, 1967; Scarl, 1968). Only one delay probability is measured at a time so it is inherently an inefficient method. It is also fundamentally a single-bit method only yielding the undistorted delayed-coincidence result when the time resolution is made sufficiently short for there to be a negligible probability of detecting more than one photon per sample time. This effect serves to increase still further the inefficiency of such a method. It is, however, valuable when very short time delays of order 1 ns are involved.

Table 2. Summary of the properties of instrumentation for photodetection spectrocopy in the time domain

Method	Parallel-Channel Efficiency?	Many-bit?	Linearity and Distortion	Upper Frequency Limitation
Delayed coincidence	No	No	No distortion for short resolving times or low count-rates	Pm tube rise time ~ 1 GHz
Start/stop	Partial (only single stop)	No	Requires very low count-rates to avoid distortion	Pm tube rise time ~ 1 GHz
Multistop	Partial	No	Less distortion than start/stop	Pm tube rise time ~ 1 GHz
Multiscale	Partial	Yes	Digital processing	Instrumental limit ~ 100 MHz
Clipped autocorrelation	Yes	No	Digital processing. No distortion for Gaussian signals or low count-rates	Instrumental limit ~ 100 MHz
Random-clipped autocorrelation	Yes (with reduced efficiency)	Yes	Digital processing. Must choose clipping range correctly	Instrumental limit ~ 100 MHz
Many-bit autocorrelation	Yes	Yes	Digital processing	Instrumental limit ~ 100 MHz

Start/stop methods: The next level of sophistication in such instruments is to use a start/stop method in which the delay from a start pulse to the next stop pulse is measured, usually with a time-to-voltage converter and multichannel pulse-height analyser storage. This technique accepts stops from all delays within the range with a resolution similar to the delayed coincidence method. However any pulse following the first stop is lost resulting in distortion of the conditional probability. The observed conditional probability for detection of photons at time 0 and time kT will then be given by (Coates, 1968)

$$P_{obs}(0, kT) \approx P_c(0, kT) \left(1 - \sum_{j=1}^{k-1} P(jT)\right) \tag{38}$$

where $P_c(0, kT)$ is the true conditional probability and $P(jT)$ is the probability of detecting an event in the jth delay interval. The sum therefore represents the probability of a pulse occurring during the interval between delays 0 and T. For a constant intensity a distortion of 1% would be observed with a 100 channel instrument when the count rate per sample (resolution) time is 10^{-4}. This distortion may be overcome by providing a correction (Coates, 1968) or by using an inhibit technique (Davis and King, 1970) such that only events in which one stop pulse was recorded over the entire delay range are accepted. However, the correction method is inaccurate at higher count rates and the second becomes progressively less efficient.

Multistop techniques: An obvious extension of the single-stop methods is to use a many-stop method in which the time of arrival of several pulses is recorded following the start. In general this method has the same advantages as the two previous methods but allows higher count rates without appreciable distortion. A useful variant, which can be described as triggered multichannel scaling, allows a very large number of pulses to be recorded following the initial start (Chen and Polonsky-Ostrowsky, 1969). This technique is limited by the multichannel scaling rate and is certainly not capable of operating at frequencies higher than about 100 MHz.

While the delayed coincidence method is only a single-channel method both the start/stop and multistop methods offer a parallel-channel efficiency advantage in that stop pulses at any delay can be accepted. This advantage is still only partial in that it ignores the contributions due to considering stop pulses as starts in a further set of measurements. In principle the time relationships between every pulse should be used. This is achieved when autocorrelation techniques are applied.

Clipped autocorrelation: One of the commonest forms of autocorrelation instruments is based on a single-bit delay register so that one takes the cross-correlation between a single-bit, clipped, signal and the present full signal. This has been described and analysed many times in the literature (Jakeman and Pike, 1969; Foord et al, 1970; Jakeman, 1974; Oliver, 1974). A block diagram of the architecture is shown in Figure 10. It offers the engineering advantage of simplicity in that the many-bit delay (shown in Figure 2) is replaced by a single-bit delay and the parallel set of multipliers can be replaced by "and" gates. Clock rates of up to 100 MHz are possible with current instruments of this type. However this instrument achieves this simplicity at the expense of flexibility. For light having Gaussian statistics the normalised single-clipped autocorrelation function has the form described by Jakeman (1974)

$$g^{(2)}_{clip}(k) = 1 + C|g^{(1)}(k)|^2 \tag{39}$$

which is obviously similar to the Siegert relation [Equation (12)]. The constant C in Equation (39) depends on a complicated combination of the effects of sample time, clip level, count rate and detector area (Jakeman, 1970; Jakeman et al, 1971; Koppel, 1971). The distortion due to clipping is not important in this case and it can readily be shown that the loss in accuracy is very small provided that clipping is performed near the mean count rate (Hughes et al, 1973). However if non-Gaussian statistics are encountered Equation

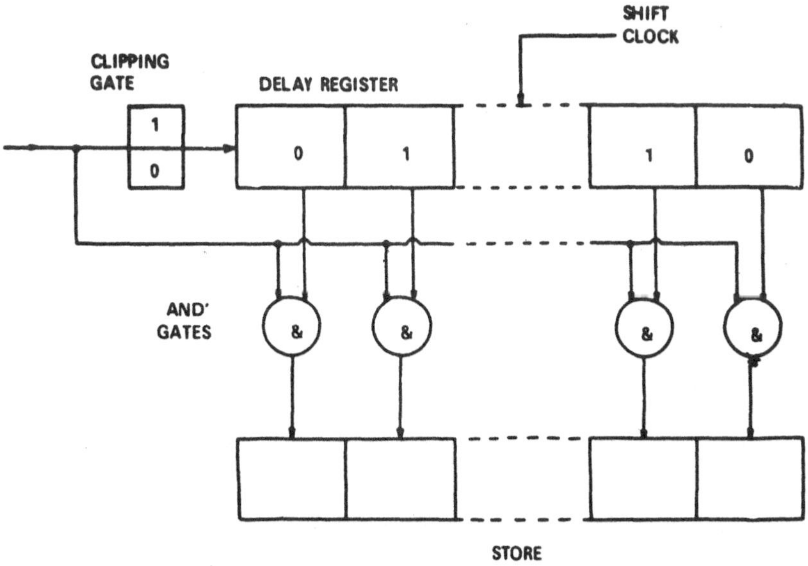

Figure 9. Block Diagram of a single-clipped digital autocorrelator.

(39) no longer applies so that the observed, clipped, autocorrelation function is now distorted with respect to the true many-bit function.

Randomly-clipped autocorrelation: It can be shown that one-bit delay methods can be used with non-Gaussian signals provided that a uniform random selection of clip levels, spanning the whole amplitude range of the signal, is used (Jakeman et al, 1972). The commonest method for achieving this is scaling (Pusey and Goldberg, 1971; Shaefer and Berne, 1972; Jakeman et al, 1972). This method allows a uniform near-random selection of clip levels to be made at the expense of efficiency due to clip levels at the extremes of the amplitude range passing nearly all 1's or all 0's. Measurements with Gaussian sources (Jakeman et al, 1972) have shown that this loss in accuracy is not very significant compared with full correlation at least with low count rates $\bar{n} \leqslant 1$. For non-Gaussian statistics the loss in accuracy would be expected to become more pronounced.

Many-bit autocorrelation: As the price of integrated circuits is reduced while their complexity is increased it is becoming progressively more practical to achieve a full autocorrelator structure as shown in Figure 2 with at least 4 bits in the delay register. This offers the full flexibility in terms of signal statistics at the expense of a reduction in the upper frequency limit from about 100 MHz for single-bit instruments to about 10 MHz. One would expect that the progress of time should allow increases in speed without attendent corresponding increases in price and power consumption.

B. Frequency-Domain Instruments

In this section we shall consider four different methods of achieving spectral analysis in the frequency domain. The salient features of these techniques are summarised in Table 3. The processing method is either digital or analog. In the latter case the usual analog disadvantages apply whereas digital processing is exact.

Scanned filter: The simplest instrument one can envisage operating in frequency space is the conventional tuned-filter scanned spectrum analyser. This consists of a single tuned filter which is slowly swept in frequency over the range of interest at a rate determined by the filter bandwidth. It is apparent that each small range of frequencies inside the filter pass-band is only analysed once during each sweep. This imposes a major limitation with non-stationary signals since we need to average over a very large number of sweeps to ensure that a reasonable ensemble average is obtained at each frequency otherwise excess fluctuation will be

Table 3. Summary of the properties of instrumentation for photodetection spectroscopy in the frequency domain

Method	Parallel-channel efficiency?	Many-bit?	Processing method	Upper frequency limitation
Scanned filter	No	Yes	Analog	Pm tube rise time 1 GHz
Bank of filters	Yes	Yes	Analog	Pm tube rise time 1 GHz
Fourier transform (Array processor)	Yes	Yes	Digital	Instrumental limit 80 kHz in real time; 100 MHz in batch mode
SAW spectrum analyser	Yes	Yes	Analog	Instrumental limit 100 MHz

obtained in the spectral estimate. Thus not only is this method inefficient in that it is a single-channel technique but it can result in misleading spectral measurements with non-stationary signals. In terms of the upper frequency limit this will be determined by the photodetector rise time since the instrument is basically analog. A maximum frequency of order of 1 GHz is thus possible with fast pm tubes. The fact that the system is analog in nature means that problems such as thermal drift need careful attention.

Bank of filters: A much more efficient means of measuring the spectrum is to use a bank of filters spanning the range of interest. This will offer the full theoretical accuracy. However the practical difficulty of maintaining tuning and gain stability of such an (assumed analog) instrument over, say, 100 filters at specified frequencies and bandwidths is likely to provide an over-riding limitation.

Discrete Fourier Transform: Another method offering the full processing efficiency, capable of operating with many-bit inputs but with the additional advantages of digital operation, is the Discrete Fourier Transform implemented on a computer. Once the data is initially digitised this system is stable and linear. However this technique suffers from the limited throughput rate of general purpose computers and so is only of real value with low-frequency signals or with data arriving in batches. For example a 1024 point real Fourier Transform would take typically about 12 ms in an Array Processor (Floating Point Systems Ltd, AP 120-B) corresponding to a real-time processing at a sampling rate of 80 kHz. Where batch mode operation is acceptable the primarily limitation is likely to be the analog-to-digital converter at the input restricting the maximum frequency to about 100 MHz.

Chirp-Z Transform, Surface Acoustic Wave (SAW) spectrum analyser: The implementation of the Chirp-Z algorithm has been discussed in Section 2 while the architecture is illustrated in Figure 3. A convenient means of implementing this structure is to use surface acoustic wave components to provide the initial down chirp and the second up-chirp convolving filter. A description of the mode of operation of such an instrument will be found in Alldritt et al, 1978, and Oliver, 1978. To summarise, this technique uses a SAW spectrum analyser to obtain an estimate of the periodogram of a 25 μs block of data every 50 μs. Typically the incident signal is Taylor-weighted to reduce the sidelobes introduced by truncation of the data into short batches. This process is by no means essential but serves to simplify the observed spectrum at the expense of degrading the resolution by about 50%. Each estimate of the spectrum, which may be very noise-like, is digitised by an analog-to-digital converter before being summed in a digital integrator as shown in Figure 3 (Alldritt et al, 1978). The integrator allows estimates to be accumulated until an adequate signal-to-noise ratio on the spectrum is obtained. Since the components of the spectrum analyser are initially analog in nature the expected linearity will be restricted in the usual fashion. In addition the filter itself introduces a non-flat background (± 0.5 dB being realistic) which, while it can be calibrated out, complicates the analysis. The final limitation is the inflexibility of the resolution, batch duration and frequency range which are initial fixed design parameters. This limitation is not fundamental since use of an initial bandwidth expander (or compressor) buffer will allow other frequency ranges to be processed. Perhaps the main advantage of this instrument lies in its ability to perform tracking operations in anemometry very simple (e.g. Figure 10) (Alldritt et al, 1978; Watson, 1980). At present an instrument with a sampling rate of up to about 20 MHz is available; in the near future an extension to 100 MHz will be possible.

7. CONCLUSIONS

Having started by examining the fundamental nature of the signal in Section 2 followed by the introduction of the photon-counting data sampling and the processing methods in the time and frequency domain in Section 3, we examined the main source and detector defects in Section 4. The most serious defects approved to be those in the detector. However it is possible with care to achieve correlation measurements with a constant-intensity source in which the

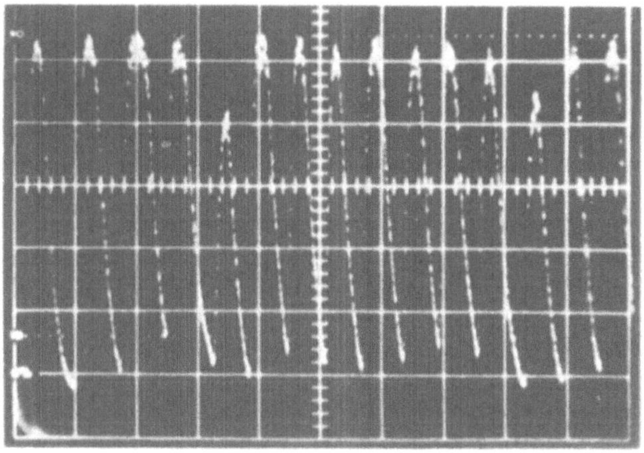

Figure 10. Example of velocity tracking in laser velocimetry. The variation of the peak frequency with time is shown measured at 500 Hz rate; horizontal scale 1 s/division, vertical scale 1 MHz/division.

observed autocorrelation function is extremely close to the theoretical value as shown in Figure 5 (Paul and Pusey, 1980). Only when this is the case is it possible to perform any detailed analysis of autocorrelation functions. The more sophisticated the analysis technique the more important it is that the source, detector and analyser combination is performing correctly.

From the discussion of the instrumental limitations in Section 6 we see that no single instrument covers all options. While some are free of the inefficiency problems arising from single-channel operation others restrict the type of signal that can be studied or introduce linearity and stability problems which make difficult the measurement of monotonically decaying functions, such as are encountered in diffusion experiments. Thus one needs to select that instrument that most closely matches requirements for the specific experiment. Three particular areas seen to offer more clear-cut choices than others:

A. Where low-frequency information is required for ultimate accuracy one should use photon-counting operation with all-digital many-bit processing either in a computed Fourier Transform or an autocorrelator. The advantage of these digital methods is that they offer accurately flat autocorrelation functions or spectra for constant intensity. Thus complicated functions can be measured with confidence. In particular monotonic functions such as these encountered in diffusion measurements are best treated in this manner.

B. Where the signal count-rate is low one requires longer integration time and the flatness of the background becomes still more important. Thus photon-counting together with digital processing is mandatory. Provided the count-rate is low enough one can even use single-bit delay instruments for arbitrary signals since clipping does not introduce serious distortion.

C. Where high-speed tracking of velocity is required it is important to use many-bit processing to avoid the distortion and loss of accuracy introduced by clipping techniques. With the Doppler difference method often used (Alldritt et al, 1978) the SAW spectrum analyser plus integrator offers a useful technique. A typical transit in which about 50 photodetections were recorded gave rise to the spectrum shown in Figure 11 (Abbiss et al, 1980). It is obvious that detecting the peak frequency component can be readily achieved. It has been shown (Oliver, 1980b) that about this number of detections would be required per Doppler burst in order to achieve adequate tracking performance. An alternative to the SAW spectrum analysis would be to use a clipped digital correlator followed by a fast Cosine Fourier Transform (Brown et al, 1979). This offers advantages in linearity but introduces distortion at the higher

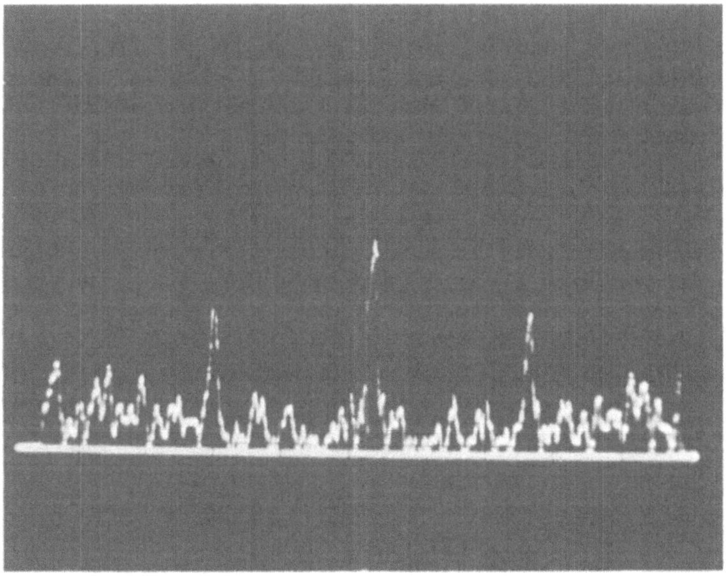

Figure 11. Measurement of Doppler difference frequency from a single-scatterer transit in laser velocimetry. Periodogram over the range -3MHz to +3MHz is obtained with a SAW spectrum analyser. The dc term (at center) and predominant frequency (1.5 MHz) are clearly visible. About 50 photodetections gave rise to this signal; the fluctuations are due to photon noise

intensities one requires in tracking operation. Both instruments have the same information-content performance limits in photon-counting operation.

Other experimental conditions impose combinations of these factors. A suitable balance should be obtained in each situation. A more critical approach to the choice of instrumentation at an early stage would generally improve the reliability of the final conclusions in any experiment.

REFERENCES

Abbiss, J., R. Jones, C. J. Oliver, P. R. Sharpe, and J. M. Vaughan, 1981, to be published.

Alldritt, M., R. Jones, C. J. Oliver, and J. M. Vaughan, J. Phys. E-11: 116 (1978).

Baldwin, G. C., and S. I. Friedman, Rev. Sci. Instr. 36: 16 (1965).

Bluestein, L. I., IEEE Trans. AU-18: 451 (1970).

Brown, R. G. W., R. D. Callan, W. Jenkins, R. Jones, I. Miller, C. J. Oliver, E. R. Pike, D. J. Watson, and M. Wedd, Physica Scripta 19: 365 (1979).

Chen, S. H., and N. Polonsky-Ostrowsky, Opt. Comm. 1: 64 (1969).

Chu, B., and T. Nose, Macromolecules 12: 599 (1979).

Coates, P. B., J. Phys. E-1: 878 (1968).

Coates, P. B., J. Phys. D-3: 1290 (1970).

Cramer, H., Mathematical Methods of Statistics, (New Jersey: Princeton University Press, 1946).

Davis, C. C., and T. A. King, J. Phys. A-3: 101 (1970).

Euling, R., J. Appl. Phys. 35: 1391 (1964).

Foord, R., E. Jakeman, C. J. Oliver, E. R. Pike, R. J. Blagrove, E. Wood, and A. R. Peacocke, Nature 227: 242 (1970).

Foord, R., R. Jones, C. J. Oliver, and E. R. Pike, Appl. Opts. 8: 1975 (1969).

Glauber, R. J., Phys. Rev. Lett. 10: 84 (1963).

Hughes, A. J., E. Jakeman, C. J. Oliver, and E. R. Pike, J. Phys. A-6: 1327 (1973).

Jakeman, E., J. Phys. A-3: 201 (1970).

Jakeman, E., in Photon-Correlation and Light-Beating Spectroscopy, edited by H. Z. Cummins and E. R. Pike, (New York: Plenum, 1974). p. 75.

Jakeman, E., C. J. Oliver, and E. R. Pike, J. Phys. A-1: 406 (1968a).

Jakeman, E., C. J. Oliver, and E. R. Pike, J. Phys. A-1: 497 (1968b).

Jakeman, E., C. J. Oliver, and E. R. Pike, J. Phys. A-4: 827 (1971).

Jakeman, E., C. J. Oliver, E. R. Pike, and P. N. Pusey, J. Phys. A-5: L93 (1972).

Jakeman, E., and E. R. Pike, J. Phys. A-2: 411 (1969).

Koppel, D. E., J. Appl. Phys. 42: 3216 (1971).

Lombard, F. J., and F. Martin, Rev. Sci. Instr. 32: 200 (1961).

Mandel, L., Proc. Phys. Soc. 74: 233 (1959).

Morgan, B. L., and L. Mandel, Phys. Rev. Lett. 16: 1012 (1966).
Oliver, C. J., in Photon-Correlation and Light-Beating Spectroscopy, edited by H. Z. Cummins and E. R. Pike, (New York: Plenum, 1974). p. 151.
Oliver, C. J., J. Phys. D-11: 2499 (1978).
Oliver, C. J., J. Phys. A-12: 591 (1979).
Oliver, C. J., J. Phys. D-13: 1145, 1577 (1980a,b)
Oliver, C. J., and E. R. Pike, J. Phys. D-1: 1459 (1968).
Oliver, C. J., and E. R. Pike, J. Phys. D-3: L73 (1970).
Oppenheim, A. V., and R. W. Schafer, Digital Signal Processing, (Englewood, NJ: Prentice-Hall, 1975).
Paul, G. L., and P. N. Pusey, to be published (1980).
Prescott, J. R., Nucl. Instr. Meths. 39: 173 (1966).
Pusey, P. N., and W. I. Goldberg, Phys. Rev. A3: 766 (1971).
Rebka, G. A., and R. V. Pound, Nature 180: 1035 (1957).
Roberts, J. B. G., G. L. Moule, and G. Parry, IEEE Proc. 127: 76 (1980).
Rodman, P. J., and H. J. Smith, Appl. Opts. 2: 181 (1963).
Scarl, D. B., Phys. Rev. 175: 1661 (1968).
Schaefer, D. W., and B. J. Berne, Phys. Rev. Lett. 28: 475 (1972).
Siegert, A. J. E., MIT Rad. Lab. Report No. 465 (1943).
Watson, D., to be published (1981).

RECENT DEVELOPMENT IN PHOTON CORRELATION AND SPECTRUM ANALYSIS TECHNIQUES: II. INFORMATION FROM PHOTODETECTION SPECTROSCOPY

C. J. Oliver

Royal Signals and Radar Establishment

Malvern, Worcs, UK

CONTENTS

ABSTRACT

The extraction of information from the measured correlation function or power spectrum is considered. Of particular importance in this context is the selection of an appropriate model for the physical process involved. We shall discuss the information-content limit on the model complexity and the extent to which any model can be justified. In this context a discussion of experimental artifacts which distort the expected data and can thus invalidate the model will be given. Having chosen some model the accuracy with which the parameters of that model can be established from measurements will be discussed with particular reference to Gaussian-Lorentzian light. We conclude with a set of guidelines to govern the choice of experimental parameters in photodetection spectroscopy experiments.

1. INTRODUCTION

In this Part we shall consider the way in which we can extract information from photodetection spectroscopy measurements. We shall assume that ideal instrumentation is available satisfying the requirements outlined in Part I. Having obtained an autocorrelation function or periodogram for the process under study we next discuss the interpretation of this data. There are four main questions than can be identified in such analysis:

A. How is the physics of the process introduced into the interpretation? (Section 2).

B. What fundamental limit is there on the number of physical parameters that can be determined from a given experiment? (Section 3).

C. What effect can experimental artifacts introduce which distort the result and make interpretation in the light of the model adopted under A. inappropriate? (Section 4).

D. For a given model under well-behaved conditions what factors govern the accuracy with which the parameter values can be determined? (Section 5).

The discussion is not intended to provide universal answers so much as to ask questions which should be considered in any series of measurements where reliability and accuracy are required.

2. THE PHYSICAL MODEL

Having measured an autocorrelation function or periodogram there are analysis steps which can be taken immediately which make no a priori assumptions about the physical process giving rise to the data (i.e. the physical model). One can, for example, perform data transformation between time and frequency space if required. Such analysis involving no physical model is of very limited applicability. Further progress requires us to represent the physical process taking place by some model which may be of varying levels of complexity. By fitting the observed function we obtain values of the parameters of the model which characterise the process under study. Let us discuss the relationship between the physical insight into the nature of the process and the model used to interpret it in terms of an example from the field of laser Doppler velocimetry.

In Figure 1 we show measured power spectra and autocorrelation functions obtained in anemometry measurements using the Doppler difference technique (Alldritt et al, 1978). Since the experiment is to measure a Doppler difference frequency the simplest model

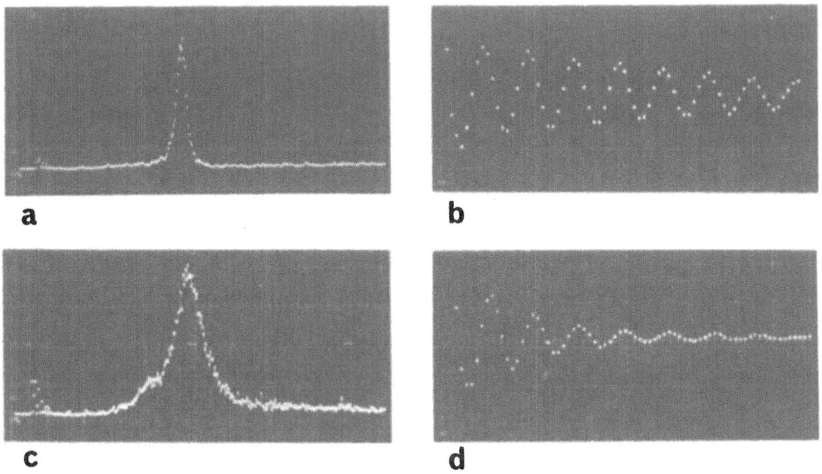

Figure 1. Comparison of photodetection spectroscopy measurements in time-delay and frequency domains (taken from Alldritt et al, 1978). SAW spectra (a,c) and autocorrelation functions (b,d) for a visible laser velocimeter are shown: all 5s integration time, count rate about 2 x 10^6 s^{-1}; spectra 0-5 MHz full horizontal scale, correlation functions 50 μs per delay channel. (a), (b) represent near laminar flow, (c), (d) turbulent flow.

would be to assume that the result corresponded to a single velocity and hence frequency. In this case the peak frequency (a), (c) or period (b), (d) would yield a value for the velocity parameter. A study of the spectra or correlation functions reveals that such a model is inadequate. The spectral broadening and damping of the correlation function show that other processes than a single linear velocity are occurring. We can proceed in a model-independent fashion by introducing other parameters such as spectral width and even skewness and curtosis etc. In principle we can increase the number of terms towards the number of points measured. However such a polynomial fit would have no physical significance. Indeed, most of the coefficients would be correlated and would be determined by the statistical fluctuations of the data points. Physical significance arises in the assignment of the values of the parameters of the spectrum to the properties of the velocity distribution. For example, we might assume that the linewidth is caused by turbulence. In the limit one could regard the power spectrum as representing the probability distribution of velocity since each velocity component would give rise to a different Doppler shift contribution to the spectrum. However we know that such a model would not be valid since effects such as the beam profile and other instrumental functions have also contributed to the observed spectrum. Thus a model-free approach, while it demonstrates general features of what is taking place, has its limitations in terms of giving physical information.

In principle we would like our model to be capable of verification. This implies that the data should be such that all competing models, relating to other possible physical processes, can be demonstrated to fit the observation significantly less well. Where that happens we can have confidence both in the model and in the parameter values which characterise that model. In order to achieve this we need to establish:

A. What are the fundamental limitations on the amount of information that can be extracted from the data.

B. Whether these limitations allow us to verify a particular model as against other contenders.

C. What effect experimental artifacts will have, through distorting the observations, on the interpretation in terms of some model.

These points are discussed respectively in Sections 3, 4.A and 4.B. Subsequently the accuracy with which the parameters of a particular model can be estimated will be examined in Section 5.

3. INFORMATION-CONTENT LIMIT ON THE MODEL COMPLEXITY

An eigenvalue theory has been developed (McWhirter and Pike, 1978; McWhirter, 1980) which relates the number of degrees of freedom one can choose in interpreting data to the accuracy of that data. Another paper in this volume (Pike, 1980) will be applying this approach to the analysis of polydisperse scattering data. In this section we shall consider only a grossly simplified approach to the problem which predicts the same form of logarithmic spacing of parameters as the detailed theory. Suppose we have a batch of data of duration T, then the minimum frequency that can be measured is given by the Nyquist limit, i.e.

$$f_{min} \approx \frac{1}{T} \tag{1}$$

This is also the fundamental resolution (i.e. $\Delta f = f_{min}$) of the measured information within the batch of data. For a diffusion process we might expect the effective batch length to be related to the coherence time of the radiation, τ_c (= $1/\Gamma$ where Γ is the linewidth), so that

$$\frac{\delta\Gamma}{\Gamma} \approx \frac{\alpha}{\Gamma\tau_c} = \alpha \tag{2}$$

where

$$\tau_c = \frac{1}{D_T K^2}, \qquad (3)$$

D_T is the translational diffusion coefficient, K is the scattering vector defined by

$$K = \frac{4\pi}{\lambda} \sin\frac{\theta}{2}$$

where λ is the radiation wavelength in the medium and θ the scattering angle. α is a coefficient whose value depends on the standard deviation of the data (McWhirter, 1980). Similarly, in velocimetry the basic batch duration is determined by the transit time of a scatterer through the laser beam. Hence

$$\Delta f \approx \frac{1}{\tau_o} \qquad (4)$$

where

$$\tau_o = \frac{W_o}{v}, \qquad (5)$$

W_o is the beam radius and v is the scatterer velocity. But,

$$f = \frac{2\pi v}{\lambda} \qquad (6)$$

so that

$$\frac{\Delta f}{f} = \alpha \cdot \frac{2\pi}{\lambda_o W_o} \;. \qquad (7)$$

Thus the frequency spacing is again logarithmic with a value of α determined by the noise in the data.

Let us consider some of the implications of these findings to the analysis of the diffusion of polydisperse scatterers. It has been suggested by Chu and his fellow workers (Chu, Gulari and Gulari, 1979; Gulari, Gulari, Tsunashima and Chu, 1979; Chu and Nose, 1979) that the observed field autocorrelation function can be calculated by summing over the contributions from scatterers within each of a narrowband of linewidths. Thus

$$|g^{(1)}(k)| = \sum_{j=1}^{n} a_j \int_{\Gamma_j-\Delta\Gamma/2}^{\Gamma_j+\Delta\Gamma/2} \exp(-\Gamma\ kT)d\Gamma \tag{8}$$

where a_j is the light intensity contributed by the system exhibiting linewidths between $\Gamma_j-\Delta\Gamma/2$ and $\Gamma_j+\Delta\Gamma/2$, n is the number of elements in the histogram, $\Delta\Gamma(=\Gamma_{max} - \Gamma_{min/2})$ is the width of each step and Γ_{max} and Γ_{min} represent the maximum and minimum linewidths in the distribution for Γ (and hence the diffusion coefficient D_T). The observed intensity correlation function,

$$g^{(2)}(k) - 1 = |g^{(1)}(k)|^2,$$

can be computed and the values of a_j obtained by non-linear least squares fitting. Hence one can calculate a histogram denoting the probability of scatterers having a certain linewidth, Γ, or diffusion coefficient. The choice of the number of elements in the histogram is of great importance as is the distribution of step lengths. Equation (2) makes two main contributions to the form this histogram analysis should take:

o The spacing, $\Delta\Gamma$, should be logarithmic rather than uniform for optimum performance.

o High statistical accuracy data is required to enable n to take values greater than one.

When applying these conclusions in the context of velocimetry a significant difference may arise from the fact that, in general, only one scatterer is present at a time. Provided that its velocity does not change appreciably within its transit of the beam the spectrum for a single transit is close to a single-frequency, broadened by the beam profile and instrument function. Individual measurements can be performed, therefore, for each transit with only one frequency parameter. This imposes much weaker requirements on thestatistical accuracy of the original correlation function than if several parameters needed estimation simultaneously. There is thus a threshold effect that if the available scattered signal is too weak to determine the principal frequency (and hence velocity) within a single transit data must be summed over a much larger time and interpreted with a sophisticated model. This introduces the type of discussion in terms of logarithmic separation of frequency and the number of components given previously for diffusion processes.

In general there is an advantage from the point of view of the total signal energy requirement in performing "tracking" measurements on a single, slowly-changing parameter rather than deconvolving the effect of many variables. Where possible experiments should be designed to exploit this feature.

4. DATA INTERPRETATION

A. Model Verification

Let us now investigate how measured autocorrelation functions and periodograms can be interpreted in order to give useful information about the process under study. We shall illustrate this discussion with some results of Paul and Pusey (1980) shown in Figure 2. Two features are immediately apparent from a study of the data. Firstly, the scatter of points is very small. This statistical accuracy arises because the measured result is made up of many individual measurements of duration 10^3s. These are then independently normalised and summed so that the total data represents many hours of integration. The photodetection count rate was around $10^5 s^{-1}$ so that the final function was derived from between 10^9 and 10^{10} photodetection events. Following Jakeman (1974) the expected standard deviation in the coefficients is thus of order 2×10^{-5}. Secondly, there is no evidence of misnormalisation in the data. That is to say that this autocorrelation function, obtained using intensity-fluctuation (self-beat) spectroscopy, is normalised against the square of the mean-value of the counts per sample time. The fact that this normalisation appears reasonable is a strong indication that no unknown experimental artifacts are corrupting the data.

The physical process under investigation in this example is the "long-time tail" in Brownian motion. Early computer experiments (Alder and Wainwright, 1967, 1970) predicted a $t^{-3/2}$ asymptotic decay in the velocity autocorrelation function of an atom in a simple dense fluid. This should contribute a $t^{-\frac{1}{2}}$ term to the autocorrelation function. Unfortunately earlier experimental work in this field (Boon and Bouiller, 1976; and Bouiller, Boon and Deguent, 1978) led to statistical errors of the same order as the effect under study so that detailed comparison between models is not possible. The simplest model to represent a diffusion process giving the correlation function shown in Figure 2(a) is the linear fit to a logarithmic plot of the correlation function taken from Paul and Pusey (1980), i.e.

$$\frac{1}{2} \ell n \left[g^{(2)}(k) - 1 \right] = \ell n c^{\frac{1}{2}} - K^2 a kT \tag{9}$$

where $g^{(2)}(k)$ is the observed correlation function at delay kT, T is the sample time, c is the correlation function intercept [$= g^{(2)}(0) - 1$], a is a linear coefficient and K is the magnitude

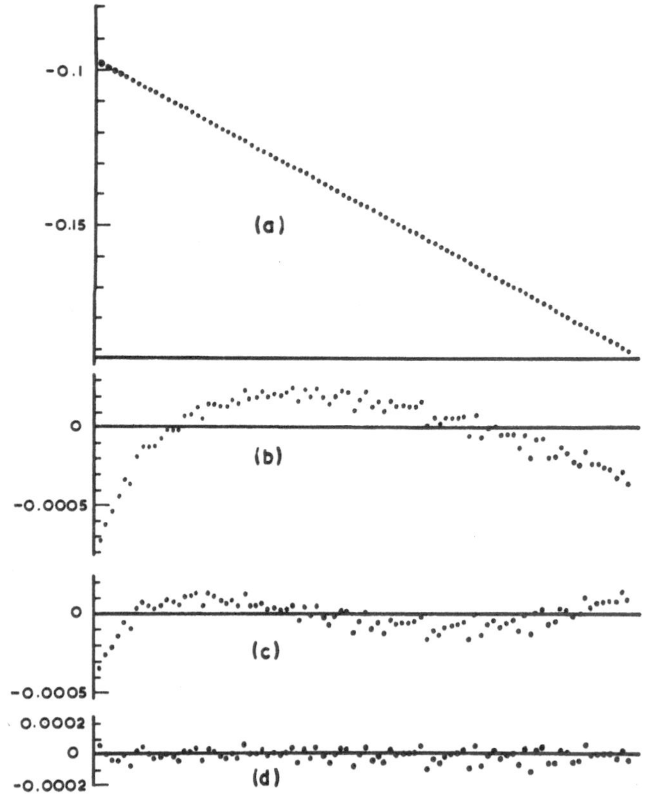

Figure 2. Correlation function measurements on light scattering from large polystyrene spheres demonstrating the presence of a long-time tail (taken from Paul and Pusey, 1980). Results are integrated over many hours. The scattering angle was 118° and the sample time, T, 5 μs. The abscissae corresponds to delay in steps of T while the ordinate corresponds to: (a) $\frac{1}{2}\ell n\ [g^{(2)}(kT)-1]$; (b), (c), (d) the residuals of the difference between the measured correlation function and the fit for (b) linear (c) quadratic (d) long-time tail models.

of the scattering vector. The long-time tail model predicts a form

$$\frac{1}{2}\ \ell n\left[g^{(2)}(k) - 1\right] = \ell n c^{\frac{1}{2}} - K^{2}[akT - b(kT)^{\frac{1}{2}}] \qquad (10)$$

where b is another coefficient.

As we have already seen the physical insight into the process is contained in the model. It is obviously essential to verify that this model does in fact represent the true situation. In Figure 3 we outline a series of steps that are always required when interpreting the measured data in order to justify the choice of model. Initially we have the measured function together with a proposed model. This model is characterised by parameters, values of which are obtained by fitting to the data. It is desirable to establish whether such a fit is unique - that is to determine if there is another model which could also fit this data. It is not actually possible to perform this test rigorously but with simple models it is certainly possible to choose between a few possibilities. Let us illustrate such a process from the results of Paul and Pusey.

Initially we adopt the simple model, Equation (9) and plot the residuals, i.e. the difference between the original data and the best fit result. These are plotted in Figure 2(b) and clearly demonstrate downwards curvature to the correlation function. Models based on polydispersity, rotational diffusion or including

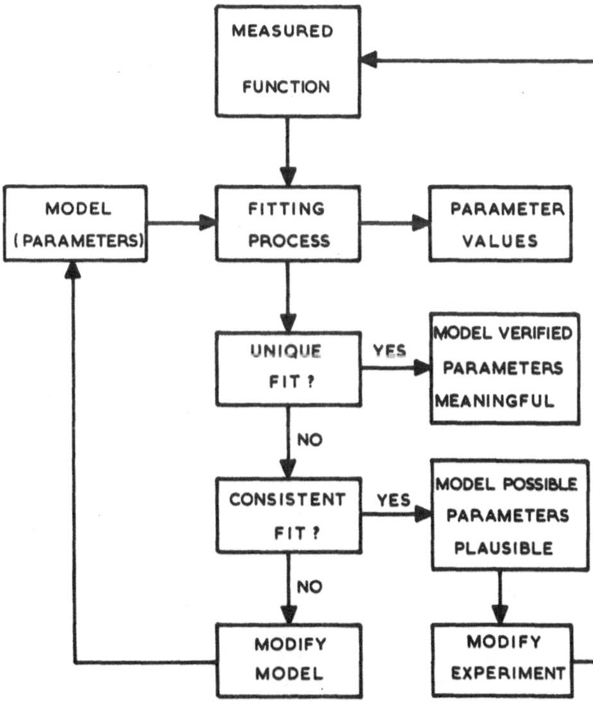

Figure 3. A flow chart for model verification in photodetection spectroscopy.

experimental artifacts such as dust or flare, discussed in Section 4.B, predict upward curvature and so are eliminated. Only bulk motion through the laser beam or the existence of the long-time tail would predict the correct sense for the curvature. If, however, we adopt the model-independent cumulants approach we next expand in terms of a quadratic delay term, i.e.

$$\frac{1}{2}\, \ell n \left[g^{(2)}(k) - 1 \right] = \ell n\, c^{\frac{1}{2}} - K^2 [a\, kT - d(kT)^2] \tag{11}$$

where d is a coefficient. While the residuals in this case, shown in Figure 2(c), are less than for a linear fit significant systematic effects are still apparent. In contrast the residuals for the long-time tail model, Equation (10), shown in Figure 2(d) have only a random scatter. Provided that the possibility of bulk motion can be eliminated (see Section 4.B) the model is thus verified. Of course it is never possible to eliminate <u>all</u> other models so that a fit could never be described as unique unless the original data was noiseless. However, for simple models such as single exponentials or single frequencies one can probably achieve the required demonstration.

The fitting process may not satisfy the test for uniqueness (Figure 3) due to two causes. Firstly, the model may actually be incorrect for a variety of reasons [as in Figures 2(b), 2(c)], or secondly, the statistical accuracy of the data may not be adequate to distinguish between models. One then proceeds to a weaker test which is for consistency. This test distinguishes between the situation where an incorrect model is proposed and that in which the statistical errors are too great. In the latter case one would find that the model chosen gave a resultant residual plot that was perfectly consistent with the data but that it could also be equally well interpreted in the light of a different model. For example, if the measured autocorrelation function of Figure 1 had a statistical spread of about 1 part in 10^3 instead of 2 parts in 10^5 the plot of the residuals in Figure 2(b) would be indistinguishable from that in 2(c) or 2(d) so that all models would be equally valid. Under these circumstances the parameter values obtained are only plausible. One might apply the principle of Occam's razor and select the simplest but there is no internal evidence within the measured data to justify this approach.

At this point one can adopt two possible routes for progress. Either one can use external evidence about the nature of the model to resolve the ambiquity or one has to modify the experiment, perhaps merely by integrating for longer to reduce the statistical uncertainty. In a large class of experiments one is interested in monitoring the behaviour of a well-understood system. In this case it is adequate to establish that the data is consistent with the model. As an example we show in Figure 4 a measured autocorrelation function for adenovirus (Oliver et al, 1976). Here electron microscopy had shown that the virus was icosahedral in shape so that one would expect the diffusion to be dominated by the translational component. Indeed the measured correlation function shows no evidence of departure from a single exponential within the statistical limits, thus demonstrating consistency. Since the long-time tail model would also fit this data it should be recognised that the model is not unique. In Figure 5 we show the histogram probability distribution of the observed single coherence time, τ_c, obtained from a set of 192 measurements. The predicted Gaussian distribution of the correct standard deviation is included for comparison. Thus we can demonstrate experimental consistency in that the measured data shows no sign of unexpected fluctuations. Combination of an a priori knowledge of the virus structure and the consistent experimental results allow one to have confidence in the meaningfulness of the calculated parameters (in this case the translational diffusion coefficient) within the experimental error.

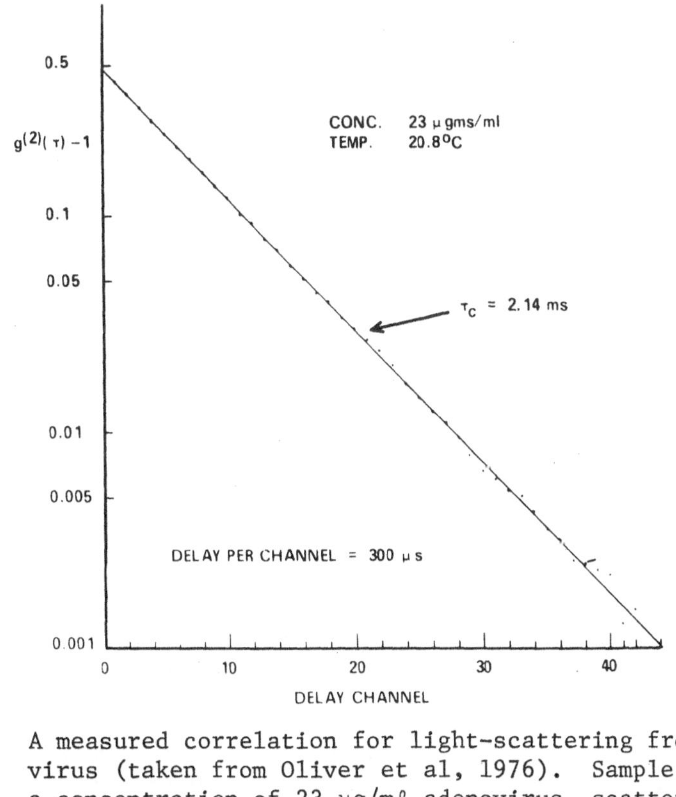

Figure 4. A measured correlation for light-scattering from adenovirus (taken from Oliver et al, 1976). Sample contained a concentration of 23 μg/mℓ adenovirus, scattering angle 35° 22', sample time 300 μs. The measurement took 150 s and determined D_T with an accuracy of 1%.

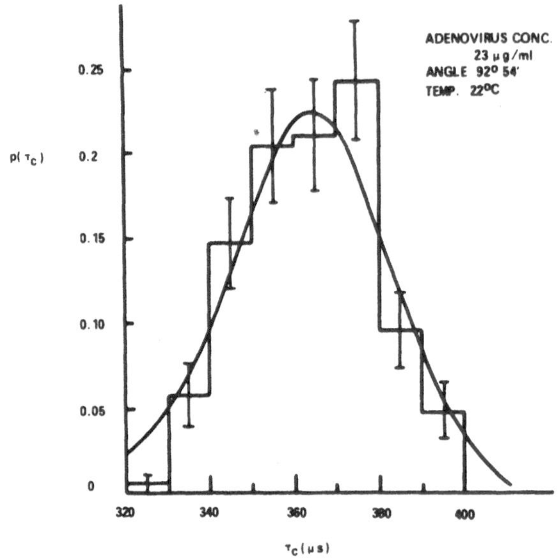

Figure 5. The probability of the estimated coherence time, τ_c, obtained by light scattering at an angle of 92° 54' on the sample used in Figure 4 (taken from Oliver et al, 1976). The histogram of the results from 192 measurements, each taking 4 s and determining τ_c with an accuracy of 5.5%, are compared with the predicted Gaussian distribution with the theoretical fractional standard deviation of 5.1%.

If the fit to the measured data (Figure 3) is demonstrated to be inconsistent, as in Figure 2(b), then one requires a new model. The analysis loop is then repeated in terms of this new model until a satisfactory model is found [Figure 2(d)].

In Section 3 we demonstrated that it is good practice to minimise the complexity of the model for the system under study by suitable choice of experimental technique. Since the number of independent parameters that can be evaluated is a function of the accuracy of the original data a less sophisticated model places less stringent demands on that data. In the present section the discussion on model verification has shown that this too is more readily achieved if the physical process can be represented by a simple model. Let us now see how this applies to the analysis of polydispersity. In the first place it might appear preferable to represent the polydispersity situation by the simple model in which the mean diffusion coefficient is obtained from the linear term in a cumulants expansion while the width is obtained from the quadratic term. It is certainly correct that such a two parameter model will result in accurate values of the parameters. However there is no information as to the detailed form of the distribution and the results can be actually misleading, as in the example with a long-time tail. Until the actual details of the system behaviour are understoond one needs to use analysis methods which are as free as possible from assumptions about the nature of the signal. It is only when the physics of the process is established that one can afford to choose the simplest consistent model and analyse the data to obtain values of the appropriate parameters. Continual consistency checks on the data are required in order to demonstrate that the process itself is still well-represented. Changes may occur due to unforeseen experimental conditions.

B. The Effect of Experimental Artifacts on the Model

So far we have assumed that all the information contained within the measured autocorrelation function or periodogram results from a well-behaved sample. In practice there is a wide variety of effects which render such a well-characterised situation unlikely. These experimental difficulties can be divided into six basic categories:

(1) Electronic problems;
(2) Sample contaminants;
(3) Optical problems;
(4) Bulk motion in the sample;
(5) Interparticle effects; and
(6) Number fluctuations.

In this section we shall examine the causes of these problems, identify their effect on the observed data and suggest approaches which overcome or minimise the problem.

First let us consider how we should handle the measured autocorrelation function. Similar discussions can be outlined for operation in frequency space. However, precision measurements of diffusion-like processes will, as indicated in Part I, be only possible with digital techniques. At present the digital autocorrelator is the only practical instrument for most applications of this type. In Part I, Equation (12), we showed that the normalised autocorrelation function should take the form

$$g^{(2)}(\tau) = G^{(2)}(\tau)/\langle n\rangle^2 = 1 + |g^{(1)}(\tau)|^2 \tag{12}$$

which is the Siegert relation (1943). Thus the value of the unmormalised autocorrelation function in the long-delay limit should be equal to $\langle n\rangle^2$ provided that there is no correlation between events on the time scale KT, where KT is the longest delay value computed. If the model for the physical process predicts that there should be no such correlation then comparison of the theoretical background with the observed value is a first, extremely important, means of verifying the validity of that model. Indeed it is essential where sophisticated interpretation is to be attempted. In order for the analysis of Figure 2 to be performed it had to be demonstrated that the normalisation was accurate to at least 1 part in 10^5 (Paul and Pusey, 1980). The commonly-adopted method of normalising the data against some long-delay coefficient may be misleading since it disguises long-term effects which cause misnormalisation with resultant curvature of the correlation function. In order not to produce distortion these long-delay coefficients must be calculated at a delay of at least 12 coherence times for the slowest process under study. The method can only work if this slowest timescale is still very much faster than any experimental artifact. Thus if the data can be normalised against the theoretical value $\langle n \rangle^2$, this is valuable evidence in confirming the adequacy of the experiment. Other approaches represent dangerous compromises.

Let us now discuss the problems summarised at the beginning of this section and see how they affect the observed data:

(1) Electronic problems. These may arise in the laser source, the photodetector or the processor as discussed in Part I. It is instructive to note that the maximum permissible misnormalisation of 1 part in 10^5 would arise from an intensity drift of only 0.6% over the course of the experiment in intensity-fluctuation spectroscopy [Equation (20) of Part I] or about two orders of magnitude less drift in heterodyne spectroscopy. This highlights a major difficulty in investigating complicated processes with heterodyne spectroscopy. Intensity-fluctuation spectroscopy would seem preferable when possible. Indeed Paul and Pusey have demonstrated that achieving the requisite normalisation is possible in intensity-fluctuation spectroscopy when sufficient attention is paid to the detail of the experimental technique. Thus with care it should be possible to reduce all electronic distortion to an acceptable level.

(2) Sample contaminant. When preparing the samples it is obviously essential to ensure that they only contain what is required. Contaminants, such as dust, will contribute significantly to the observed function making the result inconsistent with the expected model The presence of dust in the sample will have two main effects since such large particles are likely to be slow-moving, strong, scatterers. First, they will serve as local oscillator sources and interfere with the other scattered radiation to give rise to a heterodyne term in the observed correlation function. Secondly, since there are relatively few such scatterers, one will observe intensity-fluctuations due to number fluctuations in the probe volume. The observed autocorrelation function with dust particles acting as a local oscillator source can be approximately represented by (Oliver, 1974)

$$g^{(2)}(k) \approx 1 + \frac{2\bar{n}_s\bar{n}_d}{(\bar{n}_s+\bar{n}_d)^2} g_s^{(1)}(k) \cos kT\Delta\omega + \frac{\bar{n}_s^2}{(\bar{n}_s+\bar{n}_d)^2} [g_s^{(2)}(k) - 1] + \frac{\bar{n}_d^2}{(\bar{n}_s+\bar{n}_d)^2} [g_d^{(2)}(k) - 1], \tag{13}$$

where $\bar{n}_s$, $\bar{n}_d$ are the mean signal and dust counts per sample time, $g_s^{(1)}(k)$, $g_s^{(2)}(k)$ are the signal correlation functions, $\Delta\omega$ represents any Doppler shift frequency due to movement of the dust and $g_d^{(2)}(k)$ is the intensity autocorrelation function for scattering by the dust. The last term is a comparatively flat addition to the theoretical background which could be removed by measurement of $G^{(2)}(k)$ at long delay ($kT \approx 12\tau_c$). Even allowing that the dust term <u>is</u> flat over the appropriate region there will still be an error in the fitted autocorrelation function due to the present of the second term. If the Doppler shift is assumed to be zero then the error in the apparent linewidth can be deduced from Figure 6 (Oliver, 1974). This plots the dependence of the ratio of the intercepts of the autocorrelation function, $g^{(2)}(0)-1$, and the ratio of the apparent (single) coherence time to the true one as a function of the ratio of the local oscillator to the signal mean count ($\bar{n}_o/\bar{n}_s$). Assuming that $\bar{n}_o = \bar{n}_d$, we can deduce the effect of dust on the apparent linewidth when we force-fit to a single exponential. For small fractions of spurious scattering the apparent linewidth obtained by assuming a single exponential fit would be in error by

$$\frac{\Delta\Gamma}{\Gamma} \approx -\frac{\bar{n}_d}{\bar{n}_s} . \tag{14}$$

The only way to overcome this problem is to introduce a local oscillator deliberately and perform heterodyne detection subject to the reservation noted in Section 4.B.(1) above.

It must be realised that the separation of contaminants from genuine scatterers is somewhat arbitrary. In the first place various other mechanisms can give rise to a positively-curved logarithmic plot (such as rotational diffusion or polydispersity). Indeed dust is only an extreme case of polydispersity. The scatterers themselves may tend to aggregate and thus introduce self contamination. If we obtain agreement between the long-time correlation coefficients

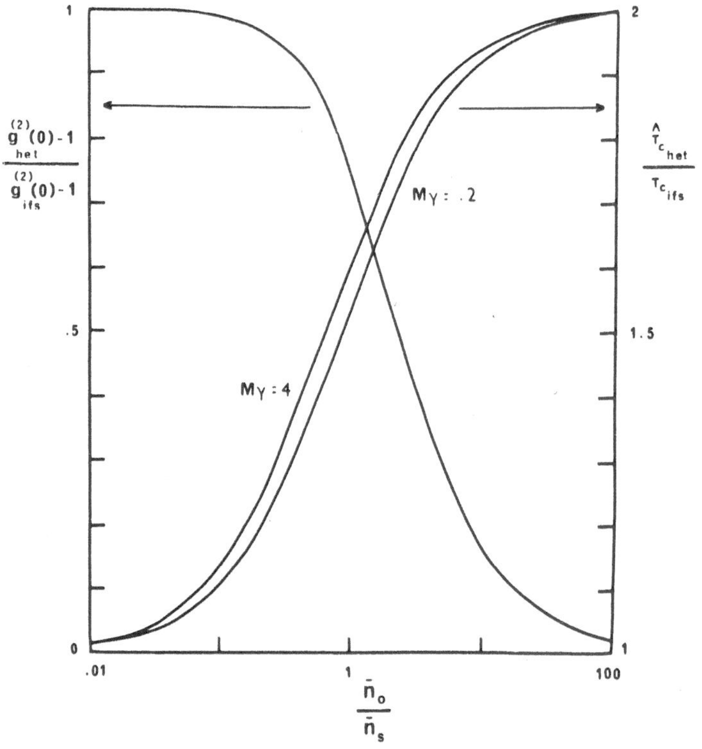

Figure 6. The dependence of the apparent coherence time, $\hat{\tau}_{c_{het}}$, obtained by force-fitting mixed scattered and local oscillator fields to a single exponential, on the ratio of the local oscillator to scattered intensities, $\bar{n}_o/\bar{n}_s$ (taken from Oliver, 1974). The ratio of the apparent coherence time $\hat{\tau}_{c_{het}}$, to the true IFS one, $\tau_{c_{IFS}}$, is given. Also shown is the dependence of the ratio of the intercepts, $g^{(2)}(0)-1$, for the mixed (het) case to the pure IFS case.

and the theoretical normalisation then the time scales of the processes within the sample are similar and it is no longer appropraite to try to interpret part of the result as arising from a "dust" contribution. A more sophisticated model needs to be introduced involving polydispersity, perhaps, expressed in terms of the cumulants method or the histogram approach. A comparison of some possible methods is given in Chu et al (1979).

Where contamination really consists of very large "dust" particles it is possible to use experimental methods to overcome the distortion. Provided that such particles cross the beam comparatively seldom their passage will be marked by a significant increase in count rate. Thus one could construct an integrated autocorrelation function by summing a large number of comparatively short ones and rejecting those in which an excess intensity fluctuation was noted. With suitably prepared samples such a fluctuation should only occur on a timescale of about one per hour.

(3) Optical problems. The primary optical problem that may be encountered in laser scattering experiments is flare from components in the system entering the photodetector. As demonstrated in Figure 6 flare will increase the apparent coherence time due to the scatterers. Indeed it results in an upwards-curved correlation function due to the presence of the two components, the IFS term and the heterodyne cross-term. Force-fitting to a single exponential with a flare contribution of 10% would increase the apparent coherence time by 12%. Thus for 1% accuracy in the linewidth flare must be reduced below about 1%. The most significant effect of flare, however, lies in its contribution to the second cumulant due to the curvature of the correlation function. This has great significance when investigating polydispersity, as demonstrated by Chu et al (1979) in discussion of the effects of flare. Here the histogram model fails to account for small fractions of stray light (flare) in an IFS (self-beat) experiment giving rise to a heterodyne cross-term in the observed correlation function. The model interprets all deviation from a linear logarithmic plot as being due to polydispersity in the sample. Figure 7 shows computed histogram fits, taken from Chu et al (1979), in which different fractions of flare are introduced. It is apparent that as small a contribution as 0.5% has a marked effect on the width of the distribution while a 2% contribution even makes the distribution appear bimodal. From Figure 7 the mean value of Γ and the total width $2\Delta\Gamma$ is found to be $\Gamma = 1278$, $2\Delta\Gamma = 194$ (no flare), $\Gamma = 1278$, $2\Delta\Gamma = 378$ (0.5% flare) and $\Gamma = 1298$, $2\Delta\Gamma = 480$ (2% flare). The effect of 2% flare on the mean is to give an increase of about 2% in the apparent value of linewidth, Γ. Force-fitting to a single exponential as an alternative model would yield a 2% reduction in the linewidth. Thus choice of model determines how the effect of stray light influences the estimated value of a given parameter. Much more significant than the effect on the linewidth is the apparent increase of 100% in the measured polydispersity of the sample due to only 0.5% of stray light. This would suggest that flare must be reduced to 1 part in 2000 for a 10% increase in the apparent polydispersity.

This extreme sensitivity to the effects of flare makes it imperative that the optical design should be such that the acceptance angle of the system is well-defined and does not include any trace of flare. Some receiver arrangements suffer from diffraction when used with small apertures which allows flare from cell walls to be detected. Since only very small flare contributions can be accepted it is difficult to demonstrate that no local oscillator is entering the detector in any particular measurement. Measure-

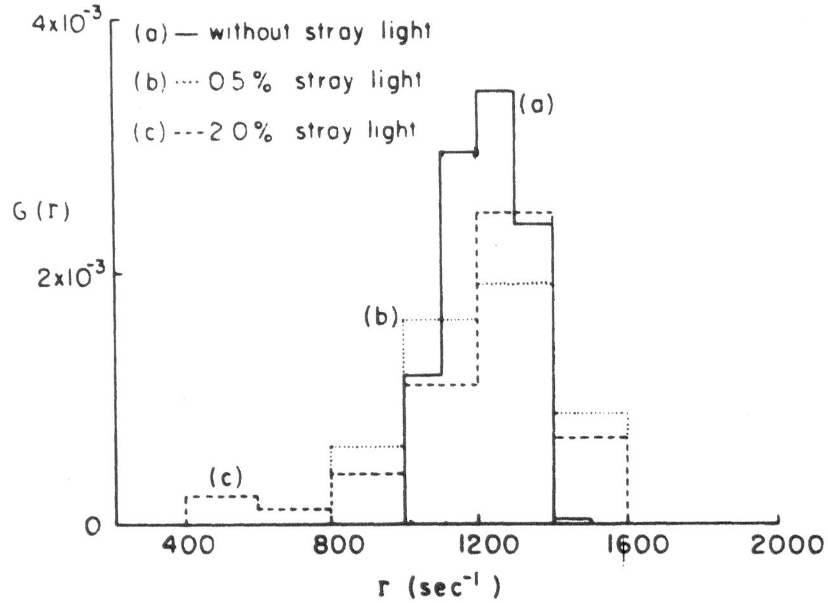

Figure 7. The computed distribution of linewidths, G(Γ), for scattering off latex spheres as stray (heterodyne) light is introduced (taken from Chu et al, 1979). The same sample is used (a) without stray light, (b) with o.5% stray light and (c) with 2% stray light.

ment of the dependence of linewidth on angle can sometimes reveal the presence of flare since this is often strongly angular dependent. When in doubt local-oscillator signal should be deliberately introduced and the nominal IFS and heterodyne linewidths compared. From Figure 6 we should also note that it is important in heterodyne detection to use enough local oscillator, i.e. $\bar{n}_o \gg \bar{n}_s$, so that one is actually in the heterodyne limit. Study of the linewidth distortion in Figure 6 suggests that $\bar{n}_o \geqslant 100\ \bar{n}_s$. However, the analysis will be just as sensitive to curvature due to small IFS contributions as it was to small heterodyne (flare) contributions. This would suggest an order of magnitude increase in the local oscillator is necessary. Such an increase would render the reservations noted previously about the use of heterodyne detection with complex models even more important.

A second optical problem is the question of multiple scattering. Supposing a large fraction γ of the incident radiation is transmitted through the sample then a fraction $1-\gamma$ has undergone a single scattering and a fraction $(1-\gamma)^2$ has undergone double scattering. Thus the fraction of the total scattering due to multiply scattering is $\sim 1-\gamma$ which may be a few per cent. Obviously the properties of multiply-scattered radiation will differ from singly-scattered (Bertolotti, 1977) introducing a distortion into the correlation function. This may be reduced by a reduction in the sample concentration and hence an increase in γ. This in turn means that a longer experiment will be needed to achieve the same statistical accuracy.

A third optical problem relates to back-scatter. Where the incident radiation is normal to the sample container an appreciable fraction will be reflected back along the same light path and may be scattered through the complementary scattering angle $(180-\theta)$. Provided that the cell dimension is less than the coherence length of the laser constructive interference can occur. This effect can only be minimised by index-matching as far as possible.

(4) Bulk motion in the sample. Two principal modes of bulk motion within the scattering sample may be evident. Any sample may undergo convection. This problem may be minimised by use of a small sample within a large, well-stirred, water bath in which adequate time has been allowed for the sample to stabilise. Since convection results in the scatterers being swept across the laser beam there is a changing population within the beam giving a contribution to the correlation function of the form $\exp[-\tau^2/\tau_c^2]$ where $\tau_c = W_o/v_c$, W_o is the beam radius and v_c is the convection velocity across the beam. A second, similar, problem, evident in solutions or suspensions of large particles, is that the process of sedimentation will contribute a vertical coherent-motion to the scatterers. The scattering volume will thus contain a changing population due to this motion. This will contribute a term to the correlation function of the form $\exp[-\tau^2/\tau_s^2]$ where $\tau_s = W_o/v_{sed}$ and v_{sed} is the sedimentation velocity. This contribution also gives a downward curvature to the correlation function. However, in general, these effects are insignificant except in extreme cases.

(5) Sample effects. Another effect that could occur within the sample is interaction between particles. This is obviously critically dependent on the concentration. In the analysis we regard each particle as moving within the suspending medium completely independent of all other particles. Provided particles are well separated by, say, at least 10 diameters it seems unlikely that interactions will occur. However, in some cases one can obtain aggregation of scatterers with attendant distortion. These problems require careful attention when samples are prepared.

(6) Number fluctuations. In order to achieve a good separation of scatterers we require them to be separated by at least 10 diameters (say). For large particles this could well mean that the total number, m, within the scattering volume is sufficiently small

authors (Benedek, 1968; Haus, 1969; Cummins and Swinney, 1970; Jakeman et al, 1970, 1971; Degiorgio and Lastovka, 1971; Kelly, 1971; Jakeman, 1972; Saleh and Cardoso, 1973; Hughes et al, 1973; Jakeman, 1974; Oliver, 1974, 1978, 1979). Both experiment, simulation and theory are considered in the past five references. With the exception of Jakeman 1972, and Oliver 1978 and 1979, this work concentrates on IFS (self-beat) measurements; the exceptions include analysis of heterodyne detection. The model for this process assumes that the field can be represented by a stochastic first-order Markoff process characterised by a linewidth $\Gamma(= 1/\tau_c)$. The field is complex so that both real and imaginary components at time kT will take the form (Hughes et al, 1973; Oliver, 1978)

$$\mathcal{E}(kT) = \exp\left[-\Gamma T(k-j)\right]\mathcal{E}(jT) + \alpha \tag{17}$$

where k, j are integers and α is a Gaussianly-distributed random variable. Simulation of any combination of experimental conditions can thus be performed starting from Equation (17). In general the performance is a complicated function of all the operating conditions which cannot be factorised into separate contributions. In various limits, however, the effects may be treated individually and then combined. For simplicity we shall adopt this approach in our discussion.

A second model comes from laser Doppler velocimetry (Figure 1). Here the light from the laser is split and then recombined at a region in some air flow, for example. Seed particles passing through this region give rise to sinusoidal intensity fluctuation weighted by the beam profile (assumed Gaussian) of the form (Oliver, 1980):

$$I(t) = I_o \exp\left[-\frac{2t^2}{\tau_o^2}\right]\left[1 + m\cos(2\pi f t + \phi)\right], \tag{18}$$

where t is measured from the time the particle crosses the beam axis, τ_o is the beam transmit time, m is the fringe contrast, f is the Doppler difference frequency, I_o is the peak intensity and ϕ is an unknown phase. The observed correlation function is also a Gaussian-weighted sinusoidal function of the form (Oliver, 1980):

$$G^{(2)}(\tau) = I_o^2 \frac{\tau_o}{T}\frac{\sqrt{\pi}}{2}\exp\left[-\frac{\tau^2}{\tau_o^2}\right]\left(1 + \frac{m^2}{2}\cos 2\pi f\tau\right). \tag{19}$$

Error analysis has only been performed for particle transits along the axis and does not account for many of the experimental variables treated in the Gaussian-Lorentzian case.

for the Poisson number fluctuation

$$\frac{\delta m}{m} = \frac{1}{\sqrt{m}} \tag{15}$$

to be appreciable. These number fluctuations contribute an additional flat component to the expected correlation function of the form

$$\delta g^{(2)}(k) = \frac{<\delta m^2>}{<m>^2} [g^{(2)}(0) - 1], \tag{16}$$

which may well be a few parts per thousand. Thus there will be an apparent misnormalisation of the correlation function leading to interpretation problems as already noted.

To summarise this section, therefore, we note that, in essence, all the effects listed above render interpretation of the observed data in the light of the expected model inappropriate. Any valid model must include all physical processes whether these are basic or represent experimental artifacts. In view of the limited number of parameters which can be extracted from a given measurement good experimental technique consists in reducing the number of parameters expected in a given experiment to a minimum. In particular all distorting processes should be eliminated or minimised. The example, taken from Chu et al (1979), showed how significant experimental distortion of the correlation function could be. It reveals that meticulous experimental technique of a high calibre is required before confidence can be placed in results where any but simple processes are involved. It is worth noting that the example taken from Paul and Pusey (1980) used only a two parameter model but was based on data which showed no departure from this model outside the statistical errors of about 2 parts in 10^5. For more complex models involving more parameters the statistical accuracy of the data should obviously be even better.

5. ACCURACY

Having established the nature of the received field and selected some appropriate model to represent the process under study the next stage is to consider the accuracy with which the parameters of that model can be evaluated from the measured data. To date effort has been concentrated mainly on two simple models which represent a large number of physical situations. One model relates to the diffusion of monodisperse particles in which the scattered radiation has Gaussian statistics and a Lorentzian spectrum. Most early experimental work on diffusion fitted this model. Theoretical discussion of the accuracy with which the linewidth for the diffusion process could be measured has been attempted by many

A. Dependence on Normalisation Method

In any photodetection spectroscopy measurement we are presented with a set of unnormalised data. It has been demonstrated that there are advantages in normalising the autocorrelation function or periodogram in that some methods of so doing reduce the statistical fluctuations in the estimators (Jakeman et al, 1971; Hughes et al, 1973; Oliver, 1974, 1978, 1979). The theoretical value at large delay or high frequency are the respective normalisation quantities. For the autocorrelation function, as we have seen,

$$G^{(2)}(k \to \infty) = \langle n \rangle^2 \tag{20}$$

while for the periodogram in the uncorrelated limit [Equation (35) of Part I] yields

$$\hat{P}(f \to \infty) = \text{Var } n = \sigma_n^2 . \tag{21}$$

The normalised functions are thus given by

$$g^{(2)}(k) = G^{(2)}(k)/\langle n \rangle^2 \tag{22}$$

and

$$P(\ell) = P(\ell)/\sigma_n^2 . \tag{23}$$

In any experiment of finite length, we obtain estimates of the true quantities which we denote $\hat{G}^{(2)}(k)$, $\hat{n}$, $\hat{P}(\ell)$ and $\hat{\sigma}_n^2$. It is important to establish the most useful combinations for normalisation. Three principal means of normalisation have been analysed (Oliver, 1978, 1979) which make different use of these estimator values.

(1) "True" normalisation: In this method the normalised estimator for the autocorrelation function for a given experiment is derived by dividing the unnormalised estimator, $\hat{G}^{(2)}(k)$, by the long-term average (i.e. true mean value) of n, so that

$$\hat{g}_T^{(2)}(k) = \hat{G}^{(2)}(k)/\langle n \rangle^2 . \tag{24}$$

For the periodogram the unnormalised estimator, $\hat{P}(\ell)$, is divided by the long-term average of the variance so that

$$\hat{P}_T^{(2)}(\ell) = \hat{P}(\ell)/\sigma_n^2 . \tag{25}$$

(2) "Self" normalisation: In this method the unnormalised spectral estimators, $\hat{G}^{(2)}(k)$ and $\hat{P}(\ell)$ are divided by the measured estimators for the background obtained during the same experiment, $\hat{n}$ and $\hat{\sigma}_n^2$ respectively. Thus

$$\hat{g}_s^{(2)}(k) = \hat{G}^{(2)}(k)/(\hat{n})^2 \tag{26}$$

and

$$\hat{p}_s(\ell) = \hat{P}(\ell)/\hat{\sigma}_n^2 \,. \tag{27}$$

(3) "Far-point" normalisation: In practice many workers use normalisation based on some long-delay or high-frequency channel. An equivalent approach to the previous two methods then defines far-point estimators as

$$\hat{g}_F^{(2)}(k) = 1 + \frac{\hat{G}^{(2)}(k) - \hat{G}^{(2)}(M-1)}{\langle n\rangle^2} \tag{28}$$

and

$$\hat{p}_F(\ell) = 1 + \frac{\hat{P}(\ell) - \hat{P}(M-1)}{\sigma_n^2} \tag{29}$$

where M represents the largest-delay or highest-frequency channel respectively.

A detailed discussion about the bias, consistency and relative variance of these estimators for short data batches of length N samples have been given (Oliver, 1979). Long data sets in autocorrelation function analysis have been discussed in Oliver (1978). The results are summarised in Table 1.

From Table 1 we see that the greatest statistical accuracy is achieved with self-normalisation while the poorest occurs with far-point normalisation. Self-normalisation introduces a bias of order 1/N which may be significant when complicated model comparison is intended, as in Section 4. Provided such a bias is not an overriding limitation, we can conclude that it is advantageous to use the self-normalised estimator in either the time-delay or frequency domain. If such a bias is a limitation then true normalisation (or no normalisation) should be adopted. In no situation does far-point normalisation offer a statistical advantage though it may well be required to overcome experimental problems. As we have already pointed out the fact that the estimator is consistent or not has

Table 1. A comparison of the properties of the differently-normalised spectral estimators. The ratio of the theoretical variances are mainly for short data batches ($N = 128$, $\langle n \rangle = 10$, $k = 1$; Oliver, 1979) though the results in brackets correspond to long-data sets ($N = 10240$, $\langle n \rangle = 10$; Oliver, 1978).

Estimator	Normalisation Method	Ratio of Variance	Bias	Consistency
Autocorrelation function	True	1 (1)	Unbiased	Consistent
	Self	0.058 (0.92)	Order (1/N)	Consistent
	Far-Point	60 (1.84)	Unbiased	Inconsistent
Periodogram	True	1	Unbiased	Inconsistent
	Self	0.96	Order (1/N)	Inconsistent
	Far-Point	2	Unbiased	Inconsistent

no importance in determining spectral parameters. By the same token the ratio of variances of the estimators may be a misleading criterion. As a demonstration of the different methods we shown in Table 2 the mean and standard deviations of the spectral parameters for coherence time (τ_c) and period (τ_p) in simulation of heterodyne photodetection of Gaussian-Lorentzian light (Oliver, 1979). The parameters used were: 200 sets of 16 batches of length $N = 128$, (local oscillator) $\bar{n}_o = 10$, (signal) $\bar{n}_s = 1$, $\tau_c = 10T$, $\tau_p = 10T$. Least squares fitting was performed and the means and standard deviations calculated. The most significant difference appears to be that use of far-point normalisation in autocorrelation function measurements is considerably worse in determining the coherence time than the other methods though the frequency is essentially measured with the same accuracy. This results from effective misnormalisation of the data due to the larger statistical fluctuations in the background estimator with this method which are particularly apparent where short data batches are used as in this example. Similar results apply with long data sets (Oliver, 1978) as shown in Table 3. With direct detection far-point normalisation is found to be the poorest since the relative statistical fluctuations in the autocorrelation function are greatest. With homodyne (heterodyne at zero frequency shift) detection true normalisation gives

Table 2. A comparison of the errors arising from the differently-normalised estimators with heterodyne photodetection data analysed in short batches. The results correspond to simulation with parameter values: 200 sets of 16 batches of length N = 128, $\langle n_o \rangle$ = 10, $\langle n_s \rangle$ = 1, τ_c = 10T, τ_p = 10T.

Spectral Estimator	Normalisation Method	$\frac{\delta\Gamma}{\Gamma}$ %	$\frac{\delta f}{f}$ %
Autocorrelation function	True	19	2.2
	Self	21	2.1
	Far-point	71	2.3
Periodogram	True	23	2.4
	Self	22	2.4
	Far-point	25	2.5

the worst performance since the fluctuation in the estimated background is the over-riding factor. Overall therefore we may deduce that self-normalisation gives the best statistical accuracy with either short or long data sets.

Table 3. A comparison of the errors arising from the differently-normalised autocorrelation function estimators with long data sets. The results correspond to simulation with parameter values: M = 32, N = 10240, $\langle n_o \rangle$ = 10, $\langle n_s \rangle$ = 0.05, τ_c = 10T, τ_p = 10T.

Normalisation Method	Direct Detection	Homodyne Detection	Heterodyne	Detection
	$\frac{\delta\Gamma}{\Gamma}$ %	$\frac{\delta\Gamma}{\Gamma}$ %	$\frac{\delta\Gamma}{\Gamma}$ %	$\frac{\delta f}{f}$ %
True	43	151	18	2.4
Self	39	21	18	2.4
Far-point	71	32	21	2.2

B. Dependence on the Data-Set Length

Generally, autocorrelation functions are constructed from long data sets from which M delay coefficients can be calculated. Periodograms, however, are constructed from short data sets of duration M, say. There is no reason why autocorrelation should not be used in the batch-processing mode and continuous spectrum analysis based on the sliding chirp-Z transform is also possible (Broderson et al, 1976). Let us therefore consider the result of using either a single long data set of N samples, from which M coefficients are calculated, or by summing over N/M individual batches of length M.

Simulation and theory for the self-normalised functions (Oliver, 1979) indicate that in the weak-signal limit, where the performance is dominated by the Poisson photodetection statistics, the use of the long data set is marginally preferable. In the strong-signal limit ($\bar{n}_s \gtrsim 1$), such that the intensity fluctuation statistics are dominant, simulation indicates an advantage of about 40% in the accuracy with which spectral parameters can be obtained where short data batches are summed. Statistically speaking there is, therefore, little real advantage in either method. Since the self-normalised estimator is biased to order 1/M use of short data batches will tend to distort the data. This distortion due to data batch length is not merely restricted to the self-normalised estimator. Though both the periodogram and autocorrelation function estimators are statistically unbiased when used with true or far-point normalisation the fact that the estimator is made up of a truncated batch of data means that the mean value of the spectral estimators will differ from the long-term true spectrum or autocorrelation function. This effect is clearly visible in Figure 8, which shows the autocorrelation function and power spectrum estimators derived from simulation of heterodyne photodetection of Gaussian-Lorentzian light (Oliver, 1979). The effect is clearly demonstrated by the periodogram where the theoretically predicted form of the estimator for the short data batch is noticeably asymmetric and differs from the true spectrum. However, the average value of the true and far-point normalised periodogram is equal to this theoretical value, confirming that they are indeed unbiased estimators, while the value of the self-normalised estimator shows evidence of biasing. The effect is disguised in the autocorrelation function estimators by the large error in the normalisation value with far-point normalisation. The essential difference between this distortion and the statistical bias introduced by use of the self-normalised estimator is that this effect is analytic. Provided that the data is fitted to the correct form, which includes the effects of truncation, the true spectral parameter values will be obtained. Choice of the length of the data set and the form of normalisation will thus depend largely on what information we require to extract.

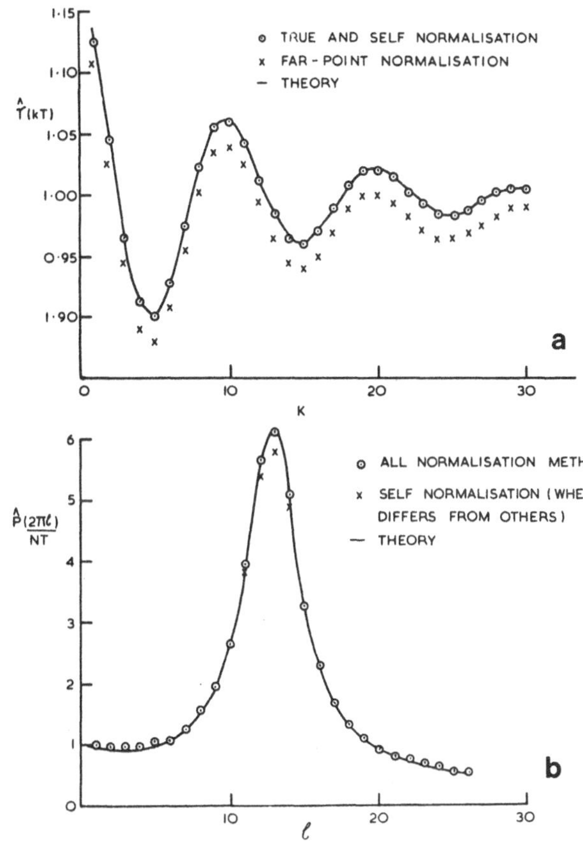

Figure 8. A comparison of the different normalisation techniques for (a) the autocorrelation function estimator and (b) the periodogram. The results are for heterodyne photodetection of narrow-band Gaussian-Lorentzian light under the following conditions: 3200 batches, M = 128, $\langle n_o \rangle = 10$, $\langle n_s \rangle = 1$, $\tau_c/T = 10$, $\tau_p/T = 10$. The full curves represent theory.

C. Dependence on Operating Conditions

The fractional error in linewidth Γ, or frequency, f, obtained by fitting the observed data to theory for either Gaussian-Lorentzian light or laser Doppler velocimetry can be derived under certain restricted conditions (Jakeman et al, 1971; Jakeman, 1972, 1974; Saleh and Cardoso, 1973; Oliver, 1978, 1980). For Gaussian-Lorentzian light the restriction is that only with weak signals can the effects of the experimental variables be factorised. For the velocimetry example the form of the correlation function given in Equation (19) is only valid for scatterers passing through the center of the beam-crossing region. The model itself is incorrect under all other conditions. In each case the fundamental limitation will be imposed by photodetection noise. For Gaussian-Lorentzian light the error in linewidth Γ is then given by

$$\frac{\delta\Gamma}{\Gamma} = \frac{4.6}{r_s}\sqrt{\frac{\Gamma}{\mathbf{T}}} \qquad (30)$$

for IFS (self-beat) measurements or

$$\frac{\delta\Gamma}{\Gamma} = \frac{1.4}{r_s}\sqrt{\frac{\Gamma}{\mathbf{T}}} \qquad (31)$$

for heterodyne detection measurements (Jakeman, 1972; Oliver, 1978) where r_s is the signal photodetection rate and $\mathbf{T}$ is the total experiment duration. The frequency on the other hand has an error given by (Oliver, 1978)

$$\frac{\delta f}{f} = \frac{0.20}{r_s f(MT)^{3/2}\mathbf{T}^{1/2}} \qquad (32)$$

where M is the number of correlation coefficients calculated.

All three results depend inversely on the count rate, r_s, and on the square root of the experiment duration $\mathbf{T}$. In Figure 9 we show a comparison of the errors in linewidth for direct and heterodyne detection as a function of count rate (Oliver, 1978). The full curves represent the IFS theory of Jakeman et al (1970, 1971) and the higher, dashed, curve, showing better agreement, is the equivalent theory of Saleh and Cardoso (1973), including the effect of correlation between samples. At low count rates ($\bar{n}_s < 0.01$) the ratio of fractional errors in IFS and heterodyne spectroscopy are as shown in Equation (30) and (31) with the predicted inverse dependence on $\bar{n}_s$ ($= r_sT$). As the count rate increases the performance is no longer limited by shot-noise so that the error reaches an asymptotic minimum, the same for both methods, in which the error is determined by the intensity fluctuations. Figure 10 illustrates the dependence of linewidth and frequency errors on count rate in heterodyne detection. Both curves show the predicted initial slope [Equations (21) and (22)] and the same saturation effect at high $\bar{n}_s$. The ratio of the errors does not follow Equations (31) and (32) because of interactions between parameters (Oliver, 1978).

One can conclude from the study of the effect of count rate with Gaussian-Lorentzian sources that:

- Heterodyne detection is only better than direct detection

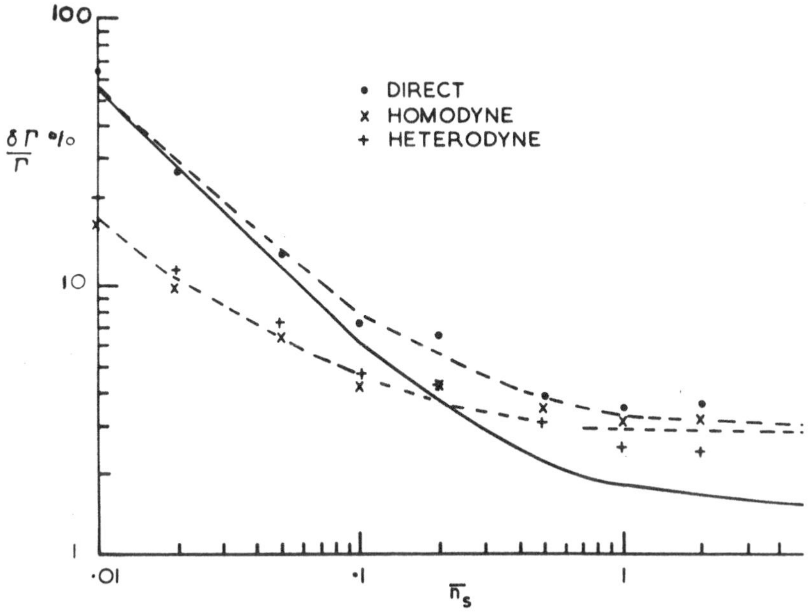

Figure 9. The dependence of the linewidth error, $\delta\Gamma/\Gamma$, on the signal count rate, $\bar{n}_s$, with direct (IFS, self-beat), homodyne (heterodyne at zero frequency shift) and heterodyne detection (taken from Oliver, 1978). $M = 32$, $\bar{n}_o = 10$, $\Gamma T = 0.1$, $fT = 0.1$, normalised to $N = 10^5$. The full curve is the direct detection theory of Jakeman et al, 1971; the higher dashed curve is the equivalent theory of Saleh and Cardose, 1973; the lower dashed line is a smooth curve through the homodyne and heterodyne data.

(IFS) with weak signals since it serves to reduce the effects of photon noise.

• In the strong-signal limit photon-noise is negligible and the accuracy is determined by the intensity-fluctuation noise. There is little to be gained in increasing the signal beyond $\bar{n}_s \approx 1$.

With velocimetry data the error in the Doppler difference frequency, f, from a single particle transit is given by (Oliver, 1980)

$$\frac{\delta f}{f} = \frac{0.90}{m^2 \tau_o \langle N_s \rangle f} \tag{33}$$

where N_s is the number of detections per transit, m is the fringe

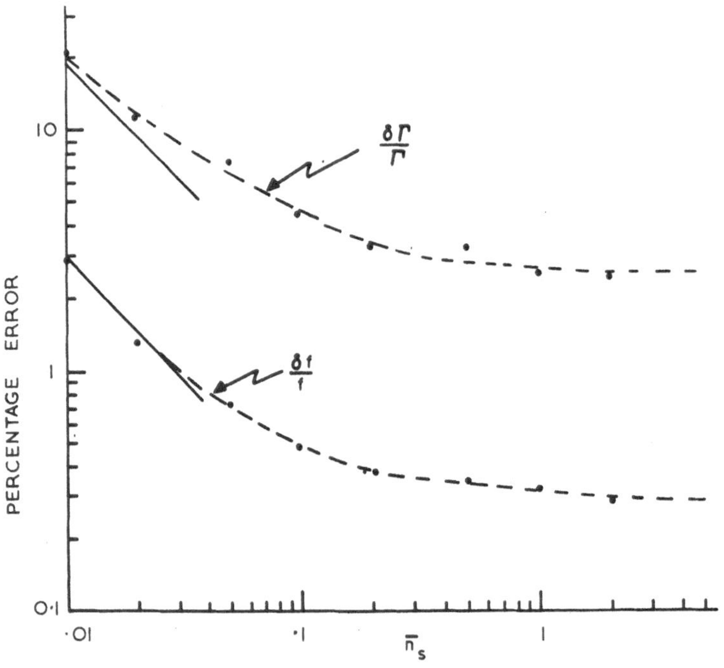

Figure 10. A comparison of the frequency shift, $\delta f/f$, and linewidth, $\delta\Gamma/\Gamma$, errors as a function of count rate, $\bar{n}_s$, in heterodyne detection (taken from Oliver, 1978), under the same conditions as Figure 9. The full lines represent the weak-signal theory with corrections for the effects of Γ, f and T, and dashed lines are smooth curves through the data.

visibility and τ_0 the radius transit time (= W_0/v). Within the limitations of the model this expression holds for both weak and strong signals since the form of the signal is deterministic. This error depends on signal strength in the same way as for Gaussian-Lorentzian signals in the weak limit. It also depends inversely on the square of the depth of modulation of the signal (the fringe visibility). Finally it depends inversely on the product $f\tau_0$ where

$$f = v/d, \tag{34}$$

v is the velocity across the virtual fringes and d is the fringe separation. Thus the error depends inversely on the number of fringes per beam radius (W_0/d). This is discussed in detail in Oliver (1980), and represents the extent of the analysis of errors in Doppler velocimetry at present.

Let us now return to the Gaussian-Lorentzian case with its fuller analysis and examine how the other experimental parameters affect the errors. Experimental variables fall into two principal categories:

- Variables relating to the signal to be processed;
- Variables relating to the processing technique.

Let us examine these variables and how the achieved accuracy is influenced by them:

(1) Incident signal variables: (a). The scattering region under study will be illuminated over some radius, W_o, determined by the optics of the incident laser radiation. Reducing W_o will increase the source brightness. However, it should be noted that this may also result in number fluctuations arising due to there only being a small number of scatterers within the illuminated volume. (b). The scattered radiation will be collected over a receiver of radius, W_R. For this combination of W_o and W_R a coherence rate, A_c, can be defined over which radiation from the scattering region will interfere constructively at the receiver. A coherent receiver radius, W_{coh}, can be defined in terms of the illuminated radius, W_o, such that

$$W_{coh} = \frac{1.92\,\lambda\,D}{\pi\,W_o} \tag{35}$$

for IFS (self-beat) measurements (Scarl, 1968; Jakeman et at, 1970a) or

$$W_{coh} = \frac{\lambda\,D}{\pi W_o} \tag{36}$$

for heterodyne measurements (Jakeman et al, 1975), where D is the separation of scattering region and receiver. In order to increase the effective received radiation we need to increase the coherent-mixing radius W_{coh}. Increased source brightness by reducing W_o achieves this. Increasing D has no effect on the amount of collected radiation provided that the receiver is also scaled correctly. The radiation scattered will give rise to some photodetection rate R (measured in counts per coherence area per coherence time) which will be inversely proportional to W_o^2 when all other parameters are unchanged. In general the performance can be shown to depend on the ratio $A/A_c(=W_R^2/W_{coh}^2)$ and the value of R. In Figure 11, we show results of measurements of the error in linewidth, $\delta\Gamma/\Gamma$, as a function of A/A_c and R (Hughes et al, 1973). Firstly, suppose we consider the dependence of $\delta\Gamma/\Gamma$ on R for fixed A/A_c (= 0.2 say) and fixed sample time (T = 0.1). For weak signals $\delta\Gamma/\Gamma$ is proportional to R. However as R increases the error

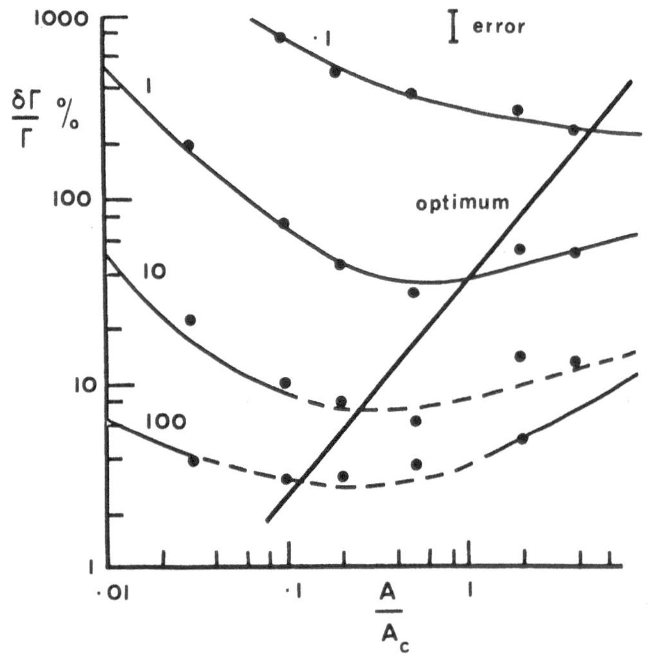

Figure 11. The dependence of the linewidth error on detector area (A/A_c) for various count-rates, R, per coherence time per coherence area (taken from Oliver, 1974).

tends to some asymptotic value at about R = 100 at which point the mean counts per sample time will be about $\bar{n}_s = 2$. Little is gained by increasing R further. Secondly, it is advantageous with weak signals, on the left of Figure 11, to increase the receiver area; though this will tend to average out the intensity fluctuations, particularly when $A/A_c > 1$, it also increases the mean counts per sample time and so reduces the photon-noise fractional error in the measured coefficients. The balance between these two effects will determine the overall influence of the varying receiver area. From the right hand side of Figure 11 we see that too great an increase in A will reduce accuracy. With all except very weak signals use of a detector such that $A/A_c \sim 1$ appears to give optimum results. When heterodyne detection is used with the receiver aperture defined by the local oscillator beam this condition is automatically satisfied (Jakeman et al, 1975). (c). As mentioned above the scattered radiation gives rise to a certain count rate R per coherence area per coherence time. R can be increased therefore, by increasing the coherence time. This is a function of angle, being reduced as the scattering angle is reduced in scattering from diffusion. There may well be a statistical advantage, therefore, in using forward-angle scattering. However this will introduce problems of flare and dust scattering which may well nullify the gain. (d). If heterodyne detection is used so that both frequency and linewidth are to be measured then there will be some mutual dependence. In Figure 12, the dependence of the error in the period, $\delta\tau_p/\tau_p$, on the coherence time, τ_c, is illustrated in the lower traces. These show that increasing the coherence time for a fixed period, so that the number of cycles of the frequency per coherence time is increased, gives reduced errors. The upper traces show that the accuracy with which the coherence time can be obtained is not strongly dependent on the value of the coherence time. Similarly we see from Figure 13 that the dependence of the error in the linewidth is not strongly dependent on the value of frequency.

(2) Processing technique variables. Being presented with some detector output train of photon arrival pulses it is necessary

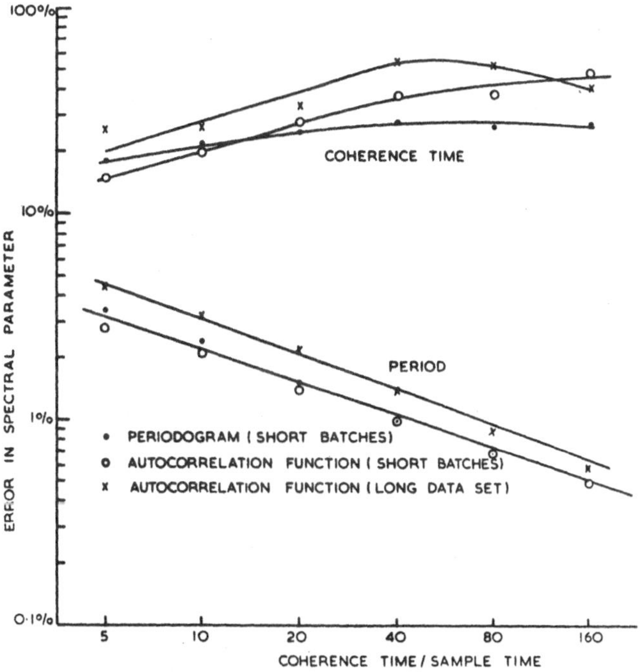

Figure 12. The dependence of the errors in coherence time, $\delta\tau_c/\tau_c$, and period, $\delta\tau_p/\tau_p$, on the number of samples per coherence time, T/τ_c, (taken from Oliver, 1979). Smooth curves are drawn through simulation results obtained by summing 3200 short batches of M=128 samples of heterodyne detection of marrowband Gaussian-Lorentzian light. $\bar{n}_o = 10$, $\bar{n}_s = 1$, $\tau_p/T = 10$.

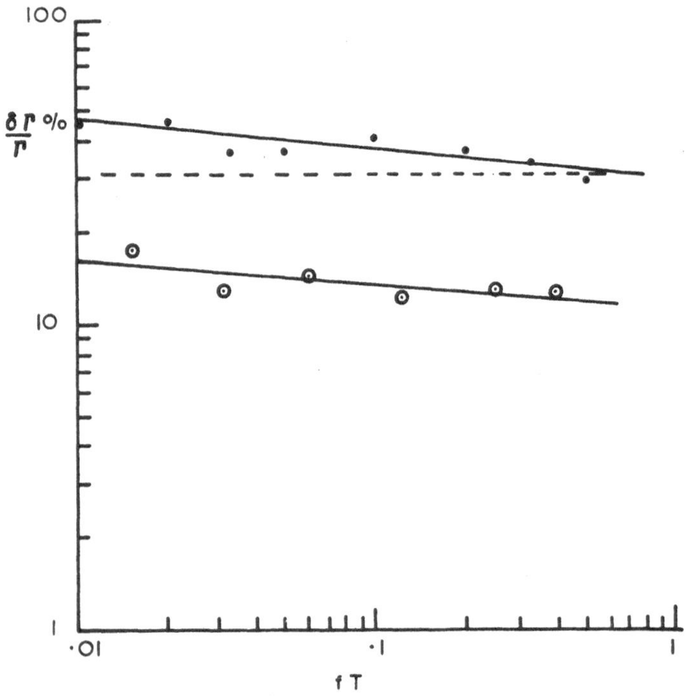

Figure 13. The dependence of $\delta\Gamma/\Gamma$ on f with heterodyne detection of both weak (•) and strong (⊙) signals: for weak signal M = 32, $\bar{n}_o$ = 10, $\bar{n}_s$ = 0.02, ΓT = 0.1, N = 10240, for strong signals ΓT = 0.0078. Full lines are drawn through the simulated data (taken from Oliver, 1978); the result for homodyne detection under the same (weak signal) conditions is shown as a dashed line.

to consider how best to process the data in order to extract the desired information with the greatest accuracy. (a). We showed in Part I that processing in either frequency or time space is effectively equivalent proving that full processing is possible. At the highest speeds it is still necessary to resort to single delay-bit methods in autocorrelation such as clipping or scaling. It can be shown that choice of a clip level equal to the mean counts per sample time gives the highest accuracy (Hughes et al, 1973) which is insignificantly worse than for full processing with Gaussian signals. (b). A second limitation in most processing instruments is that only a restricted number (M) of discrete frequencies or time-delays can be processed simultaneously. It is necessary to select the processor sampling time T so as to optimise the determination of τ_c (the coherence time) and τ_p (the period) from, for example, a correlation function representing a maximum delay improving the accuracy with which individual correlation coefficients can be determined resulting from longer sample times will be overcome by the averaging out of the intensity fluctuations. (c). The final experimental variable is the number of sample times, N, from which the spectral estimator is constructed. Whatever the source of uncertainty, either photodetection or intensity-fluctuation noise, the relative error in either linewidth or frequency shift is proportional to the square root of the experiment duration **T** (= NT) for other variables fixed (Oliver, 1978). The same result follows from the weak-signal theory of Equations (30) and (32).

In practice it is impossible to separate all these effects as we have done above. The errors in Γ and f are complicated functions of the variables R, A, A_c, W_o, W_R, τ_c (= $1/\Gamma$), T, τ_p (= 1/f), $\bar{n}_s$, $\bar{n}_o$, clip level (if applicable) M and N. Good experimental technique consists in choosing the most suitable combination of all these variables for the particular experiment in hand. The relationship between M, f, τ_c and the choice of sample time, T, in heterodyne detection has been analysed in some detail (Oliver, 1978) leading to the set of conclusions summarised in Table 4 describing how to achieve the minimum fractional error in the parameters. The quantity minimised refers to which parameter the most attention is paid to. In some cases either the linewidth and/or the frequency shift can be varied, as noted under the column labelled "conditions". In addition it is sometimes possible to use a shifted local oscillator so that the observed shift in now given by

$$f = f_o - f_1 + f_s$$

where f_o is the original laser frequency, f_1 is a shifted local oscillator frequency and f_s is the shift introduced in scattering. We wish to determine f_s as accurately as possible and so wish to minimise $\delta f/f_s$. It is then sometimes advantageous to use $f_o \neq f_1$, which is the general situation. The results may apply to either weak or strong signals or to both (third column). In the fourth column we note the conditions that the previous three columns imply for the selection of the sample time T. In addition the optimum choice of linewidth, Γ, or frequency shift, f, is noted in columns 5 and 6 respectively where these are variable. For example, if we wish to minimise both $\delta\Gamma/\Gamma$ and $\delta f/f$, where Γ and f are both fixed, then for both weak and strong signals the sample time should be selected to lie between T = 1/2f and T = $2/M\Gamma$ with the former condition dominant. If we allow Γ to vary (see next row) then we select T $\sim$ 1/2f and adjust Γ so that $\Gamma \approx 0.7$ f/M. This table can be applied to all heterodyne situations. The IFS (self-beat) conditions are found by ignoring the dependence on f and minimising $\delta\Gamma/\Gamma$, i.e. the first row such that T $\sim 2/M\Gamma$ as already noted.

of MT. In Figure 14, the dependence of the linewidth error on the ratio ΓT ($= T/\tau_c$) is shown for a fixed correlation function length of M = 32 delays. The best accuracy in both direct and heterodyne detection with weak signals is obtained at $T/\tau_c \approx 0.08$ so that the record subtends about 2.5 coherence times. A similar result is apparent in Figure 13 in which 128 frequency and delay values were computed. The optimum performance for determining coherence time was at about $MT \sim 2\tau_c$. The data is normalised to total samples rather than total number of coherence times which explains the difference between this and Figure 14. In general this does indeed represent the most suitable choice of sample time (Hughes et al, 1973; Oliver, 1974). The strong-signal results in Figure 14 indicate that the accuracy is not degraded by only taking small fractions of a coherence time for the maximum delay by reducing the sample time. This follows because the errors are still determined by intensity-fluctuations rather than photon-noise. Figure 14 also shows that where all possible delays are computed ($M = \infty$) the accuracu remains at an asymptotic value once the sample time is reduced so that the coherence time spans about 12 samples. It is important therefore not to use too long a sample time otherwise the effect of

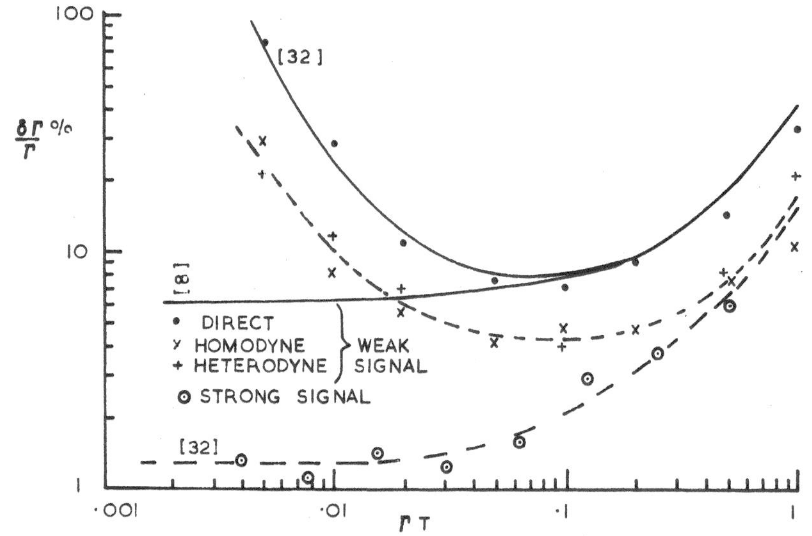

Figure 14. The dependence of $\delta\Gamma/\Gamma$ on ΓT with both weak and strong signals in direct, homodyne and heterodyne detection (taken from Oliver, 1978). $M = 32$, $\bar{n}_o = 10$, $\bar{n}_s\tau_c/T$ (counts per coherence time) = 1, $fT = 0.1$, normalised to $NT/\tau_c = 10^4$. The full lines represent the direct detection theory for [M] channels; dashed lines are drawn through the rest of the simulated data.

Table 4. Summary of the selection conditions for T, Γ and f to minimise the errors.

Quantity Minimised	Conditions	Signal	T	Γ	f
$\frac{\delta\Gamma}{\Gamma}$	Fixed Γ, f	Both	$T \sim \frac{2}{M\Gamma}$	-	-
	Fixed f, variable Γ	Both	$T \sim \frac{2}{M\Gamma}$	$r_s/\Gamma \sim 1$	-
	Fixed Γ, variable f	Both	$T \sim \frac{2}{M\Gamma}$	-	$f \sim \frac{M\Gamma}{4}$
$\frac{\delta f}{f}$	Fixed Γ, f	Both	$T \sim \frac{1}{2f}$	-	-
	Fixed f, variable Γ	Both	$T \sim \frac{1}{2f}$	$\Gamma << \frac{4f}{M}$	-
	Fixed Γ, variable f	Weak	$T \sim \frac{1}{2f}$	-	$f \sim \frac{M\Gamma}{6.4}$
	Fixed Γ, variable f	Strong	$T \sim \frac{1}{2f}$	-	$f >> \Gamma$
$\frac{\delta f}{f_s}$	Fixed Γ, variable f	Weak	$T \sim \frac{1}{2f}$	-	$f \sim \Gamma$
	Fixed Γ, variable f	Strong	$T \sim \frac{1}{2f}$	-	$f >> \Gamma$
$\frac{\delta\Gamma}{\Gamma}, \frac{\delta f}{f}$	Fixed Γ, f	Both	$\frac{1}{2f} \leq T < \frac{2}{M\Gamma}$	-	-
	Fixed f, Variable Γ	Both	$T \sim \frac{1}{2f}$	$\Gamma \sim \frac{0.7f}{M}$	-
	Fixed Γ, Variable f	Weak	$T \sim \frac{1}{2f}$	-	$f \sim \frac{M\Gamma}{6.4}$
	Fixed Γ, Variabe f	Strong	$T \sim \frac{1}{2f}$	-	$f \sim \frac{M\Gamma}{4} (M>>4)$
$\frac{\delta\Gamma}{\Gamma}, \frac{\delta f}{f_s}$	Fixed Γ, Variable f	Weak	$T \sim \frac{1}{2f}$	-	$f \sim \frac{M\Gamma}{16}$
	Fixed Γ, Variable f	Strong	$T \sim \frac{1}{2f}$	-	$f \sim \frac{M\Gamma}{4} (M>>4)$

6. CONCLUSIONS

Let us now summarise the points discussed in the form of a set of suggested experimental procedures which, if followed, should lead to reliable and meaningful results. These conclusions relate basically to the physical model and to experimental technique.

A. Physical Model

(1). Initially choose the simplest plausible model to represent the region under study. The more parameters the more accuracy is required and the harder the model is to test.

(2). Perform experiments of adequate length to achieve the necessary statistical accuracy.

(3). Verify the model chosen by comparison with other possibilities including:

- different physical processes in the system under study;
- extraneous experimental artifacts.

(4). Modify model or experiment as required and repeat.

(5). Once the physical model is understood it may be advantageous to simplify the model to reduce the number of parameters, e.g. well behaved polydisperse samples could be treated in terms of second cumulant rather than needing the histogram method.

B. Experimental Technique

(1) Test for experimental distortions due to:

- electronics; usually evident as extraneous correlations and/or misnormalisation of the correlation function;
- sample contaminants; e.g. dust, results in upwards curvature of correlation function and, possibly, misnormalisation.
- optical problems; e.g. flare, multiple scattering or backscatter, usually resulting in upwards curvature of the correlation function;
- bulk motion of the scatterers; e.g. convection and sedimentation, usually resulting in downwards curvature (Gaussian shape) of the correlation function;
- inter-particle effects; visible as lattice-like behaviour of the sample in terms of the angular variation of the scattered intensity;
- number flutuations; resulting in apparent misnormalisation due to constant background term.

Electronics faults can be identified by measuring direct laser illumination. The effects of sample contaminants, some optical problems, especially flare, and inter-particle effects can be identified by the angular dependence of the scattered light. The effects of bulk motion and number fluctuations can be identified by varying the illuminating beam radius.

(2). Select conditions for greatest accuracy by:

- increasing source brightness by reducing the illuminated radius;
- choosing receiver area to subtend about one coherence area;
- increasing the coherence time of the scattered light by varying the scattering angle;
- perhaps introducing a local oscillator for heterodyne detection to improve statistical accuracy with weak signals. It will also help to cover the effects of flare and dust but demands still greater stability than IFS (self-beat) operation;
- choosing the sample time so that $T \sim 1/2f$ when one wishes to measure frequency alone;
- choosing the sample time so that $T \sim 2/M\Gamma$, i.e. the correlation function subtends about two coherence times, where one wishes to measure linewidth alone;
- choosing $T \lesssim 1/2f$, i.e. the Nyquist condition, where both quantities are required since the effect on frequency determination is greater;
- maintaining many cycles per coherence time, i.e. $\Gamma < f$, if Γ or f can be varied;
- as an exception to the above condition, where one is able to use a shifted local oscillator in heterodyne detection and is interested in minimising the error in the measurement of the original frequency shift, selecting the final frequency shift such that f $(= f_o - f_1 + f_s) \sim \Gamma$;
- having selected the sample time, setting the correct clipping (k) or scaling (s) level where appropriate such that $k \approx \bar{n}$ or $p(n \geqslant s) \to 0$ respectively.

Following these rules-of-thumb more reliable information should be achieved in photodetection spectroscopy. It cannot be stressed too often that poor experimental technique can never really be rectified by subsequent sophisticated data extraction and will not only reduce the accuracy of the final result but may render it meaningless.

REFERENCES

Alder, B.J., and T. E. Wainwright, Phys. Rev. Lett. 18: 988 (1967).
Alder, B.J., and T. E. Wainwright, Phys. Rev. A-1: 18 (1970).
Alldritt, M., R. Jones, C. J. Oliver, and J. M. Vaughan, J. Phys. E-11: 116 (1978).
Benedek, G. B., Polarisation, Matiere et Rayonnment, (Paris: Presses Universitaries de France, 1968).
Bertolotti, M., in Photon Correlation Spectroscopy and Velocimetry, edited by H. Z. Cummins, and E. R. Pike (New York: Plenum, 1977).

p. 22.
Boon, J. P., and A. Bouiller, Phys. Lett. A-55: 391 (1976).
Bouiller, A., J. P. Boon, and P. Deguent, J. de Phys. 39: 151 (1978).
Broderson, R. W., C. Hewes, and D. D. Buss, IEEE Trans. ED-23: 143 (1976).
Chu, B., E. Gulari, and E. Gulari, Physica Scripta 19: 476 (1979).
Chu, B., and T. Nose, Macromolecules 12: 599 (1979).
Cummins, H. Z., and H. L. Swinney, Progress in Optics, edited by E. Wolf, (Amsterdam: North-Holland, 1970). Vol. 8.
Degiorgio, V., and J. B. Lastovka, Phys. Rev. A-4: 2033 (1971).
Gulari, E., E. Gulari, Y. Tsunashima, and B. Chu, Polymer 20: 347 (1979).
Haus, H. A., Proc. Int. School of Physics 'Enrico Fermi', (Course 42), (New York: Academic Press, 1969). p. 111.
Hughes, A. J., E. Jakeman, C. J. Oliver, and E. R. Pike, J. Phys. A-6: 1327 (1973).
Jakeman, E., J. Phys. A-5: 249 (1972).
Jakeman, E., in Photon-Correlation and Light-Beating Spectroscopy, edited by H. Z. Cummins, and E. R. Pike (New York: Plenum, 1974). p. 75.
Jakeman, E., C. J. Oliver, and E. R. Pike, Adv. Phys. 24: 349 (1975).
Jakeman, E., E. R. Pike, and S. Swain, J. Phys. A-3: 255 (1970).
Jakeman, E., E. R. Pike, and S. Swain, J. Phys. A-4: 517 (1971).
Kelly, H. C., J. Quant. Electron (IEEE) 7: 180 (1971).
McWhirter, J. G., Optica Acta 27: 83 (1980).
McWhirter, J. G., and E. R. Pike, J. Phys. A-11: 1729 (1978).
Oliver, C. J., in Photon-Correlation and Light-Beating Spectroscopy, edited by H. Z. Cummins and E. R. Pike (New York: Plenum, 1974). p. 151.
Oliver, C. J., Adv. Phys. 27: 387 (1978).
Oliver, C. J., J. Phys. A-12: 591 (1979).
Oliver, C. J., J. Phys. D-13: 1145 (1980).
Oliver, C. J., K. F. Shortridge, and G. Belyavin, Biochem. Biophys. Acta 437: 589 (1976).
Paul, G. L., and Pusey, P. N., 1980, to be published.
Pike, E. R., in Scattering Techniques Applied to Supra-Molecular and Non-Equilibrium Systems, (1980) NATO Advanced Study Institute, Wellesley, USA.
Saleh, B. E. A., and M. F. Cardoso, J. Phys. A-6: 1897 (1973).
Scarl, D. B., Phys. Rev. 175: 1661 (1968).
Siegert, A. J. E., MIT Rad. Lab. Report No. 465 (1943).

POLARIZED DYNAMIC LIGHT SCATTERING AS A PROBE OF MACROMOLECULAR INTRAMOLECULAR MOTIONS

R. Pecora

Department of Chemistry
Stanford University
Stanford, California

CONTENTS

ABSTRACT

Dynamic polarized light scattering is routinely used to study translational diffusion of macromolecules in solution. For macromolecules comparable in size to the wavelength of light, it may also be used to probe macromolecular rotations and long range intramolecular motions. The basic theory of dynamic light scattering from these "large" macromolecules as well as techniques of data fitting are discussed. A brief survey of experimental results for both synthetic and biological macromolecules is given.

1. INTRODUCTION

Light scattering is an interference phenomenon. Light scattered from each of the scatterers in a scattering medium into a given angle arrives at the detector with a different phase. The phase of light scattered from each scatterer depends, among other things, upon the scatterer position. The intensity at a detector resulting from the interference of these light waves will therefore depend upon the relative positions of the scatterers. Changes in these positions due to thermal motions will give rise to fluctuations in the resultant scattered intensity. Thus, measurement of the intensity fluctuations can be used to probe the thermal motions of the scatterers. Dynamic light scattering (DLS) is the name given to light scattering experiments which measure these intensity fluctuations (or quantities related to them) and extract information about molecular dynamics from them.

The quantity characterizing thermal motions most often measured by polarized DLS experiments on dilute solutions of macromolecules is the macromolecular translational diffusion coefficient. DLS is now probably the best method for rapid determination of macromolecular diffusion coefficients.

However, in cases where the macromolecular size approaches that of light, macromolecular intramolecular motions may also affect the intensity fluctuations probed by a polarized DLS experiment. This case is very useful if one wishes to study intramolecular motions, although in cases in which one is interested only in translational diffusion times it may present obstacles to quick measurement of the diffusion coefficient.

It is the purpose of this article to give a brief review of the theoretical background and of some of the work and problems in polarized DLS from these very large molecules. Only dilute solutions are discussed. The large amount of current work on semi-dilute and concentrated solutions is not discussed at all, except peripherally in the case of DNA's where it is not yet clear whether or not "slow modes" observed in these systems can be explained solely on the basis of single chain approximations. No attempt is made to be comprehensive; only selected works which illustrate important points are discussed.

2. THEORETICAL BACKGROUND

Consider a collection of N identical scatterers each with isotropic, constant, polarizability α. The ith scatterer is at position $\vec{r}_i(t)$ at time t. The scattered electric field at scattering angle θ is then proportional to

$$E_S(t) = \alpha \sum_{i=1}^{N} e^{i\vec{K}\cdot\vec{r}_i(t)} \tag{1}$$

where $\vec{K}$ is the scattering vector, whose length is dependent on λ, the wavelength of light (measured in the scattering medium), and θ,

$$K = \frac{4\pi}{\lambda} \sin(\theta/2)$$

The quantities measured in DLS experiments may usually be related to the time correlation function of E_S,

$$S(t) \equiv \langle E_S(t) E_S^*(0) \rangle$$

where the angular brackets denote an ensemble average. From Equation (1) and the definition of S(t),

$$S(t) = \alpha^2 \langle \sum_i \sum_j e^{i\vec{K}\cdot(\vec{r}_i(t) - \vec{r}_j(0))} \rangle \tag{2}$$

The quantity in the exponential in Equation (2) represents the difference in phase between the light scattered from particle i at time t and that scattered from particle j at time 0. This phase difference depends, of course, upon $\vec{K}$.

For a dilute solution of macromolecules the scatterers i and j may be considered to be "segments" on the same molecule. In dilute solution, scatterers on different molecules are uncorrelated so that intermolecular interference is negligible and may be ignored in the sum in Equation (2).

Let $\vec{R}$ be the position of a point within the molecule (e.g. center of resistance) and let $\vec{b}_j$ locate segment j relative to $\vec{R}$,

$$\vec{r}_j(t) = \vec{R}(t) + \vec{b}_j(t).$$

Thus,

$$(\vec{r}_j(t) - \vec{r}_i(0)) = (\vec{R}(t) - \vec{R}(0)) + (\vec{b}_j(t) - \vec{b}_i(0))$$

and the time correlation function in Equation (2) becomes

$$S(t) = \alpha_M^2 \langle e^{i\vec{K}\cdot(\vec{R}(t)-\vec{R}(0))} \sum_{i,j} \frac{1}{N^2} e^{i\vec{K}\cdot(\vec{b}_j(t)-\vec{b}_i(0))} \rangle \tag{3}$$

where $\alpha_M = N\alpha$ is the macromolecular polarizability.

The first exponential on the right-hand-side of Equation (3) depends on the translational motion of the macromolecule as a whole while the second term depends upon the relative motion of the segments. If the molecule is small, that is, if its characteristic dimension ℓ is small enough so that $K\ell \ll 1$, then the "intramolecular" exponential terms contribute a constant factor of 1 and (omitting the polarizability factor),

$$S(t) = \langle e^{i\vec{K}\cdot(\vec{R}(t) - \vec{R}(0))}\rangle. \quad (4)$$

If the mechanism for translation is isotropic translational diffusion with diffusion coefficient D, it is a simple matter to show that[1]

$$S(t) = e^{-K^2Dt}. \quad (5)$$

Thus, if $K\ell << 1$, this technique will not give any direct information about intramolecular flexing motions or molecular rotation. It could, however, be used indirectly to probe these motions since in many cases, intramolecular motions affect the diffusion coefficient D. Note that even if the characteristic dimension is large, K can be made small by performing the experiment at low angles.

3. APPLICATION TO RIGID MACROMOLECULES

The simplest case to treat in which the "intramolecular" terms become important is that of a rigid rod of length L undergoing independent rotational and translational diffusion. In this case it is not difficult to show that the time correlation function in Equation (2) becomes

$$S(t) = e^{-K^2Dt}(S_o(KL) + \sum_{n=1}^{\infty} S_n(KL)e^{-2n(2n+1)D_Rt}) \quad (6)$$

where D_R is the rotational diffusion coefficient of the long axis of the rod and the coefficients S_n are well-known functions[2] of the dimensionless parameter KL.

For $KL \to 0$, $S_o(KL) \to 1$ and all other $S_n(KL) \to 0$. Thus, in the limit of small KL we recover the result that only translational motion of the rod is directly observable by polarized DLS.

For $KL \leq 8$, Equation (6) is well approximated by only the first two terms,

$$S(t) = e^{-K^2Dt}(S_o(KL) + S_1(KL)e^{-6D_Rt} + \ldots) \quad (7)$$

When $KL \leq 3$ the term containing $S_1(KL)$ is also negligible. Thus, for a given rod shaped molecule, experiments at low angle such that $KL \leq 3$ may be used to obtain the translational diffusion coefficient. Then experiments performed in the range $3 \leq KL \leq 8$ may be used in conjunction with the value of D obtained at low angles and the theoretical values of S_o and S_1 to obtain D_R.

Several authors have reported polarized DLS experiments on dilute solutions of tobacco mosaic virus (TMV), a thin rod about 3000 Å in length, and have extracted both translational and rotational diffusion coefficients from the data using the above procedure or variants of it.[3-7]

This procedure, however, becomes difficult in practice for molecules that are much longer than TMV. To attain the required low values of KL, experiments would have to be done at low scattering angles, a region where such irritants as dust, stray light and long-averaging times require great effort on the part of the observer to obtain good data.

Loh and co-workers[7,8] have discussed some of the difficulties that arise in the interpretation of polarized DLS experiments on very long-thin rods:

(1) As discussed above terms of higher order than $S_1(KL)$ become important in the time-correlation function, Equation (6). This renders fitting the data rather difficult.

(2) Long thin rods are expected to have a more complex correlation function than that given by Equation (6) because of the large anisotropy of the molecular translational diffusion coefficient.[5,9,10]

(3) The longer the rod, the more likely that it might exhibit enough flexibility to affect the scattered field correlation function.

(4) As in all experiments on macromolecular systems, polydispersity might give significant distortions to the theoretical forms predicted for monodisperse systems.

Loh and co-workers[7,8] have performed polarized DLS experiments on long filamentous viruses up to 20,000 Å in length. Factor one above renders extraction of a rotational diffusion coefficient from the data difficult. They conclude that factor two is important for the longer viruses and that factor three may indirectly influence the decay rate of the terms in the correlation function containing the rotational diffusion constants.

For uniform rigid spherical macromolecules, there is, of course, no contribution to the scattered field time correlation function due to molecular rotation. A general theory of DLS (including both polarized and depolarized scattering) from rigid molecules exhibiting cylindrical symmetry has been given by Aragon and Pecora.[11] Applications of this theory to the calculation of the correlation functions for lollipop shaped structures (a model for T-Even Bacteriophages) have been performed by Wilson and Bloomfield[12] and by Koopmans and co-workers.[13]

4. FLEXIBLE COILS

Polarized DLS experiments on dilute solutions of flexible coils of radius of gyration R_G exhibit scattered field correlation functions of the form given in Equation (5) when $KR_G \ll 1$. However, when $KR_G > 1$ contributions from intramolecular modes of motion strongly influence the correlation functions and the coils no longer appear to be point scatterers.

For the free draining version of the Rouse-Zimm bead-spring dynamical model of the flexible coil, Pecora[14] has shown the scattered field time correlation function is given by

$$S(t) = e^{-K^2Dt}(S_o(KR_G) + S_1(KR_G) \times e^{-2/\tau_1 t} + \ldots) \quad (8)$$

in the regime $KR_G \simeq 1$. In Equation (8), τ_1 is the longest intramolecular relaxation time of the chain. S_o and S_1 are functions given by Pecora.[14] As $KR_G \to 0$, $S_o \to 1$ and $S_1 \to 0$ as is expected on physical grounds. Modifications of the expressions given by Pecora for the S_o, S_1 and τ_1 appearing in Equation (8) occur when hydrodynamic interactions between the chain segments are taken into account. Perico et al. have studied this case in detail.[15]

Several experiments have been performed on polystyrenes in solution in order to extract values of D and τ_1 using Equation (8).[16-22]

As is the case with rod-shaped molecules, polydispersity also complicates the interpretation of the results for coils. In addition, for flexible coils there is only a small region of KR_G for which the correlation function includes contributions from only one internal mode.

To circumvent these difficulties many authors have proposed using cumulant methods to characterize dynamic scattering from large flexible molecules.[23-29]

In this method the first cumulant (the negative of the initial slope of the normalized time-correlation function),

$$\Omega(K) \equiv -\lim_{t\to 0} \frac{dS(K,t)}{dt}$$

is used to characterize the data.

For a monodisperse system with $KR_G \ll 1$, the first cumulant is easily seen from Equation (5) to be

$$\Omega = K^2 D.$$

Büldt[23] has shown that the first cumulant for the free draining Rouse-Zimm model is given by

$$\Omega = \frac{K^2 D}{P(KR_G)}$$

where $P(KR_G)$ is the coil structure factor.[14]

A major advantage of using the first cumulant to characterize the data is that it can often be calculated for given polymer models as a function of temperature and concentration for all values of KR_G, even though the full correlation function S(t) cannot be calculated. Akcasu and co-workers[24,29] and Burchard and co-workers[25-28] have performed such calculations for various polymer models by using the projection operator techniques that were first introduced into polymer science by Zwanzig and Bixon.[30,31]

Akcasu and co-workers and Burchard and co-workers have shown that the first cumulant contains information about chain parameters such as the effective segment length, the friction coefficient per segment, the radius of gyration and polydispersity. We should note also that the formulas for the first cumulant at large K may be applied to the interpretation of quasi-elastic neutron scattering experiments from polymers.[24,29,32]

Expressions for the full time correlation function for infinitely long Rouse-Zimm chains in the $KR_G >> 1$ limit have been derived by DuBois-Violette and de Gennes. Adam and Delsanti[34] find good agreement between the shape of their experimentally determined correlation functions for high molecular weight polystyrenes and the shape predicted by DuBois-Violette and de Gennes.[33] Huang and Frederick[34] find essentially the same dependence of decay constant of the correlation function on K as Adam and Delsanti.

Akcasu <u>et al</u>. have applied a rather complicated data analysis procedure utilizing the first cumulant to DLS experiments on solutions of very high molecular weight polystyrene (MW = 4.8 x 10^6).[29] Because of the difficulty of measuring $\Omega(K)$ directly for macromolecules at fairly large values of KR_G, these authors calculate a "shape function" which is defined as ln S(t). This shape function depends, among other things, upon the dimensionless parameters Ωt and KR_G. By comparing the theoretical shape function with the experimental ln S(t), values of both $\Omega(K)$ and R_G may be simultaneously extracted from the data. These authors find "satisfactory agreement" between their experimental results and their theoretical predictions.

5. SEMI-FLEXIBLE MACROMOLECULES

Many macromolecules, especially those of biological importance, are semi-flexible and require more complex dynamical theories to describe their intramolecular motions than do either rigid rods or flexible coils. Semi-flexible molecules whose intramolecular motions have been studied by polarized DLS include F-actin,[35-38] plasmodium actin,[39] bacterial flagella,[40] DNA,[41-53] and synthetic polynucleotides.[54-56]

Since there is no rigorous theory for the dynamics of these molecules, there is no generally accepted theoretical form for fitting the data. Consequently, there has been much controversy over the methods of data fitting and the physical significance of the numbers obtained from a given fitting procedure.

Pecora has formulated a theory for once-broken rods.[57] Fujime and co-workers[58-60] have used the Harris-Hearst[61] model of the dynamics of semi-flexible polymers to compute polarized DLS correlation functions and power spectra. They have, in addition, performed extensive DLS experiments using a spectrum analyzer on the muscle protein F-actin and its complexes with myosin fragments.[35-36] They developed methods of data fitting which allowed extraction of intramolecular relaxation times for actin and its complexes. Carlson and Fraser[37-38] also studied F-actin and its complexes with myosin subfragments using intensity autocorrelation techniques. These authors failed to find evidence for relaxation times due to molecular flexing. The reasons for these discrepancies is not yet clear.

The ribonucleic acids, especially the viral DNA's, are a class of large semi-flexible macromolecules which have been studied by several research groups using polarized dynamic light scattering. The major problem is again fitting raw data. Most experiments on DNA's in the region where $KR_G \gg 1$ exhibit more than 1 exponential in the scattered field time correlation function. Most authors find that 2 exponentials are usually adequate to fit the data although it is usually clear that the correlation function is probably more complicated than this. The experiments usually (although not always) consist in extracting the relaxation times from the two exponential fits and then observing the effects of variations in solution conditions on these times (e.g. temperature, ionic strength, addition of an intercalating dye, binding of proteins, etc.). The variations in times are then interpreted in terms of molecular conformations and conformational changes.

A major controversy in the interpretation of these spectra is the existence (or nonexistence) of a very slow component in the polarized time correlation functions of calf thymus, N1 and λ-phage DNA's.[48,62] This slow mode apparently disappears at high salt concentrations[62] (1 M NaCl).

The theory leading to Equation (8) predicts that at infinite dilution, the slowest time in the correlation function is that containing the translational diffusion coefficient. However, the observed slow mode relaxation time is much too slow to be attributed to translational diffusion.

Lee _et al._[62] have proposed a theory of this slow mode which starts with the observation that the instantaneous shape of a relatively flexible polymer is non-spherical. The crowded solution conditions under which DNA's are usually studied then result in a severe anisotropy of the translational motion of a given DNA molecule. The DNA molecule can translate in the direction of its long molecular axis but because of the blocking action of its neighbors cannot move perpendicular to it. The intramolecular motions tending to reorient the long axis will then be strongly coupled to the rotational motion. In fact, an intramolecular motion which relaxes the extension along the long-axis and distends it somewhere else will result in an effective rotation of the coil. The "effective" rotational diffusion time is then comparable to the longest intramolecular relaxation time. Lee _et al._[62] find qualitative agreement of this model with the published DNA data. Mathiez _et al._[56] have also observed very slow modes in the scattering from synthetic polynucleotides in water and also in polystyrenes dissolved in ethyl acetate. These authors contend that these very slow modes are a nonequilibrium effect related to the translational diffusion of knots formed by entanglements of different polymer chains.

Studies of homogenized calf-thymus DNA by Chen _et al._[48], however, show no evidence of a slow mode, although the homogenized DNA is of lower molecular weight than the native form and the studies were conducted at relatively high ionic strength. Recent attempts by Wang and Pecora[63] to detect the previously observed slow mode in λ-phage DNA[44] have failed. The slow mode is observed only in samples into which small amounts of dust have been introduced.[63]

Lin and Schurr[52] have proposed a method of fitting the "fast" part of the DNA correlation function which is based on numerical calculations of the correlation functions for Rouse-Zimm (bead-spring) chains. Basically they fit this part of the correlation function to a single-exponential plus a baseline. They perform a series of measurements as a function of K and plot the reciprocal of the resulting relaxation time as K^2. The models predict two linear regions in the $1/\tau$ vs. K^2 plots. The slope of the linear region at low K is proportional to the DNA translational diffusion coefficient while that at high K is proportional to that of a "segmental" diffusion coefficient. The midpoint (K_m) of the transition region between the low and high K limits is related to the mean squared distance between beads b^2,

$$b^2 = 8/K_m^2 .$$

Combining these parameters with the theoretical relations of Lin and Schurr allows calculation of three parameters entering the Rouse-Zimm model, namely, the segmental friction coefficient, the mean-squared distance-between beads (b^2) and the number of beads in the chain. Lin and Schurr have extracted values of these parameters for calf-thymus DNA in 1 M NaCl (where the very slow mode is no longer observed). Parthasarathy et al.[53] have used this model to interpret data on calf-thymus DNA in 5.5 M LiCl. They conclude that the DNA in a solution with this very high salt concentration undergoes a tertiary collapse. In going from 0.4 M NH_4Ac to 5.5 M LiCl, with no change in molecular weight, they estimate that the calf thymus DNA undergoes an eight-fold contraction of its Stokes hydrodynamic volume.

The Lin-Schurr technique for analyzing polarized DLS experiments of DNA's in solution is very intriguing. More theory and experiments are needed to show whether or not it is applicable to DNA's under a wide variety of conditions and to clarify the meaning of the parameters obtained. The relation between the Lin-Schurr theory and the projection operator theory of Akcasu et al.[29] is as yet unclear.

6. CONCLUSIONS

Polarized DLS has shown itself to be a powerful technique for studying the dynamics of very large molecules in dilute solution. However, as the brief survey above shows, there remain serious difficulties in data characterization and of relating the data to molecular parameters. The major advance in the past few years, at least for solutions of very flexible molecules, has been the use of projection operator techniques.[24-32] Application of these theoretical techniques to the interpretation of experiments on semi-flexible molecules coupled with improvements in experimental techniques should help resolve some of the present controversies in this field.

REFERENCES

1. R. Pecora, J. Chem. Phys. 40:1604 (1964).
2. R. Pecora, J. Chem. Phys. 48:4126 (1968).
3. H. Z. Cummins, F. D. Carlson, T. J. Herbert, and G. Woods,

Biophys. J. 9:518 (1969).
4. S. Fujime, J. Phys. Soc. Japan 29:416 (1970).
5. D. W. Schaefer, G. B. Benedek, P. Schofield, and E. Bradford, J. Chem. Phys. 55:3884 (1971).
6. T. A. King, A. Knox, and J. D. G. McAdam, Biopolymers 12:1917 (1973).
7. E. Loh, E. Ralston, and V. N. Schumaker, Biopolymers 18:2549 (1979).
8. E. Loh, Biopolymers 18:2569 (1979).
9. H. Maeda and N. Saito, J. Phys. Soc. Japan 27:984 (1969).
10. H. Maeda and N. Saito, Polymer 4:309 (1973).
11. S. R. Aragon and R. Pecora, J. Chem. Phys. 66:2506 (1977).
12. R. R. Wilson and V. A. Bloomfield, Biopolymers 18:1543 (1979).
13. G. Koopmans, B. J. Van der Meer, P. C. Hopman, and J. Greve, Biopolymers 18:1533 (1979).
14. R. Pecora, J. Chem. Phys. 43:1562 (1965); 49:1032 (1968).
15. A. Perico, P. Piaggio, and C. Cuniberti, J. Chem. Phys. 62: 2690 (1975).
16. T. F. Reed and J. F. Frederick, Macromolecules 4:72 (1971).
17. O. Kramer and J. E. Frederick, Macromolecules 5:69 (1972).
18. W. Huang and J. E. Frederick, Macromolecules 7:34 (1974).
19. T. A. King, A. Knox, and J. D. G. McAdam, Chem. Phys. Lett. 19:351 (1973).
20. T. A. King, A. Knox, and J. D. G. McAdam, J. Polym. Sci. Polym. Symp. 44:195 (1974).
21. J. D. G. McAdam and T. A. King, Chem. Phys. 6:109 (1974).
22. B. E. A. Saleh and J. Hendrix, Chem. Phys. 12:25 (1976).
23. G. Büldt, Macromolecules 10:919 (1977).
24. Z. Akcasu and H. Gurol, J. Polym. Sci. Polym. Phys. Ed. 14:1 (1976).
25. W. Burchard, Polymer 20:577 (1979).
26. W. Burchard, Macromolecules 11:455 (1978).
27. M. Schmidt and W. Burchard, Macromolecules 11:460 (1978).
28. W. Burchard, M. Schmidt, and W. H. Stockmayer, Macromolecules 13:580 (1980).
29. A. Z. Akcasu, M. Benmouna, and C. C. Han, Polymer 21: 866 (1980).
30. M. Bixon, J. Chem. Phys. 58:1459 (1973).
31. R. Zwanzig, J. Chem. Phys. 60:2717 (1974).
32. A. Z. Akcasu and J. S. Higgins, J. Polym. Sci. Polym. Phys. Ed. 15:1745 (1977).
33. E. DuBois-Violette and P. G. de Gennes, Physics 3:181 (1967).
34. W. N. Huang and J. E. Frederick, J. Chem. Phys. 58:4022 (1973).
35. S. Ishiwata and S. Fujime, J. Phys. Soc. Japan 30:302 (1970); 31:1601 (1971).
36. S. Ishiwata and S. Fujime, J. Mol. Biol. 68:511 (1972).
37. F. D. Carlson and A. B. Fraser, in "Photon Correlation and Light Beating Spectroscopy," H. Z. Cummins and E. R. Pike, eds., Plenum, New York (1974).
38. F. D. Carlson and A. B. Fraser, J. Mol. Biol. 89:283 (1974).
39. S. Fujime and S. Hatano, J. Mechanochem. Cell Motility 1:81

(1972).
40. S. Fujime, M. Maruyama, and S. Asakura, J. Mol. Biol. 68:347 (1972).
41. S. B. Dubin, J. H. Lunacek, and G. B. Benedek, Proc. Nat. Acad. Sci. U.S.A. 57:1164 (1967).
42. K. S. Schmitz and J. M. Schurr, Biopolymers 12:1543 (1973).
43. R. L. Schmidt, Biopolymers 14:521 (1973).
44. K. S. Schmitz and R. Pecora, Biopolymers 14:521 (1975).
45. K. L. Wun and W. Prins, Biopolymers 14:111 (1975).
46. D. Jolly and H. Eisenberg, Biopolymers 15:61 (1976).
47. J. M. Schurr, Q. Rev. Biophys. 9:109 (1976).
48. F. C.-Chen, A. Yeh, and B. Chu, J. Chem. Phys. 66:1290 (1977).
49. R. L. Schmidt, J. A. Boyle, and J. A. Mayo, Biopolymers 16: 317 (1977).
50. R. L. Schmidt, M. A. Whitehorn, and J. A. Mayo, Biopolymers 16:327 (1977).
51. M. Caloin, B. Wilhelm, and M. Daune, Biopolymers 16:2091 (1977).
52. S. C. Lin and J. M. Schurr, Biopolymers 17:425 (1978).
53. N. Parthasarathy, K. S. Schmitz, and M. K. Cowman, Biopolymers 19:1137 (1980).
54. P. Mathiez, G. Weisbuch, and C. Mouttet, J. Physique Lett. 39:L139 (1978).
55. P. Mathiez, C. Mouttet, and G. Weisbuch, Biopolymers 18:1465 (1979).
56. P. Mathiez, C. Mouttet, and G. Weisbuch, J. Physique Lett. 41:519 (1980).
57. R. Pecora, Macromolecules 2:31 (1969).
58. S. Fujime, J. Phys. Soc. Japan 29:751 (1970).
59. S. Fujime and S. Ishiwata, J. Phys. Soc. Japan 29:1651 (1970).
60. S. Fujime and M. Maruyama, Macromolecules 6:237 (1973).
61. R. A. Harris and J. E. Hearst, J. Chem. Phys. 44:2595 (1966).
62. W. I. Lee, K. S. Schmitz, S. C. Lin, and J. M. Schurr, Biopolymers 16:583 (1977).
63. C. C. Wang and R. Pecora, unpulished results.

DYNAMIC DEPOLARIZED LIGHT SCATTERING FROM MACROMOLECULAR SYSTEMS

R. Pecora

Department of Chemistry
Stanford University
Stanford, California

CONTENTS

ABSTRACT

Dynamic depolarized light scattering may be used to study rotational diffusion of rigid macromolecules and both local and long range intramolecular motions of flexible macromolecules. It may also be used to probe the rotational motion of small molecules dispersed in polymeric systems. Both the basic theory of dynamic depolarized light scattering and examples of its applications to macromolecular systems are discussed.

1. INTRODUCTION

Dynamic depolarized light scattering (DDLS) has been extensively utilized to study rotational motion of small molecules in both neat liquids and solutions.[1] These studies have contributed a great deal to our understanding of such topics as the relationship between single-molecule and collective reorientation times, reorientation time viscosity dependences, the anisotropy of reorientational motion and static and dynamic pair correlations of molecular orientation.

Relatively few depolarized light scattering experiments have, however, been performed on macromolecular systems. DDLS experiments on macromolecular systems are somewhat more difficult than those on systems composed of small optically anisotropic molecules because the signals are usually weaker and the time scales slower. Such experiments can, however, be readily performed if state of the art equipment is utilized and careful attention is given to certain experimental pitfalls.

2. RIGID MOLECULES

Rigid molecules which are optically anisotropic give rise to depolarized light scattering. In addition to translational motions, rotational motions change the fluid polarizability (in a laboratory fixed frame of reference) and affect the spectral distribution of the scattered light. If molecules in solution are cylindrically symmetric with polarizability anisotropy β, they give rise to a Lorentzian depolarized spectrum,[2]

$$I_{VH} = A\beta^2 \frac{(q^2D + 6\Theta)}{\omega^2 + (q^2D + 6\Theta)^2} \tag{1}$$

where A is a constant, q the scattering vector, ω the frequency difference of the scattered light frequency from that of the laser and D and Θ are, respectively, the translational and rotational diffusion coefficients.

If the spectral half-widths are $\gtrsim 10^6$ Hz, the spectrum can be measured directly using a Fabry-Perot interferometer as the monochromator. At such large frequency changes the q^2D term in Equation (1) is usually negligible and the rotational diffusion coefficient may be obtained directly from the half-width of the scattered spectrum. Experiments of this type have been performed by several groups on globular proteins with molecular weights less than 50,000.[3,4]

If the spectral half-widths are smaller than 10^6 Hz, the preferred method of data analysis is photon correlation. The time correlation function of the scattered depolarized intensity is given by

$$\langle I_{VH}(0)I_{VH}(t)\rangle = B\beta^2 e^{-2(q^2D + 6\Theta)t} \tag{2}$$

where B is a constant.

For the large molecules which have rotational motions slow enough to easily measure with photon correlation techniques, the q^2D term is often not negligible. In order to circumvent the difficulty of extracting Θ from the depolarized time correlation function when D contributes significantly, one may perform the measurements at very low or even zero scattering angle ("forward depolarized light scattering") where q is small or zero and hence q^2D may be neglected relative to 6Θ.[5] Performing the experiment at very small q also avoids complications due to molecular structure factors which can have the effect of changing the single-exponential correlation function in Equation (2) into a sum of exponentials.

A frequent procedure for characterizing rigid macromolecules in solution (especially globular proteins) is to combine polarized light scattering measurements of D with depolarized light scattering measurements of Θ. The rotational and translational diffusion coefficients obtained can then be used along with theoretical relations between these quantities to obtain the solution dimensions. This procedure (utilizing Fabry-Perot interferometry) has been used for instance in studies of lysozyme,[3,4] gramicidin[6] and bovine pancreatic trypsin inhibitor.[7]

Another application of DDLS is to use the rotational diffusion coefficient obtained from depolarized light scattering in combination with relaxation times obtained from NMR experiments. The combination can be used to separate overall molecular rotational motion from local motions of groups within a molecule.[4]

The concentration dependence of Θ is also amenable to study by DDLS. Of particular interest is the behavior of Θ in the semi-dilute region for long-thin rod molecules. For a rod of length L and cross section diameter d, this region lies roughly at number concentrations c such that

$$\frac{1}{L^3} \ll c \ll \frac{1}{dL^2}$$

Doi[8] has predicted that Θ in this region is related to its infinite dilution value Θ^o by

$$\Theta = \beta \frac{\Theta^o}{c^2L^6}$$

where β is a constant of order 1. Experiments performed in my laboratory on light meromyosin[9] and poly-γ-benzyl-L-glutamate[10] indicate that the inverse c^2L^6 dependence of Θ is approximately correct although β is several orders of magnitude larger than 1.

It should be pointed out that for rigid molecules of arbitrary symmetry five relaxation times should contribute to the depolarized spectrum. Thus, for instance, five Lorentzians should appear in an interferometry experiment on such a system.[2] In principle, the number of Lorentzians appearing in the spectrum is an indication of the symmetry of the molecule. In practice, however, it is extremely difficult to distinguish Lorentzians (or exponentials) which have similar decay constants. To my knowledge no one has as yet reported any experiments which clearly detect the expected number of Lorentzians (or exponentials) for solutions of asymmetric rigid molecules.

3. NON-RIGID MOLECULES

The experimental techniques used for DDLS on non-rigid molecules are the same as those used for rigid molecules. An added complication for these systems is that intramolecular relaxation processes can contribute to the spectra (or correlation functions). The mechanism for their appearance is different from that for dynamic polarized spectra where the structural fluctuations must be of the order 1/q in order to be observable. In the depolarized case, a local change in the polarizability due to tumbling or torsion of some optically anisotropic group could affect the depolarized spectrum. The depolarized spectrum is then capable of giving information about "local" as well as long-range motions.

Bauer, Brauman and Pecora have observed the depolarized spectra of polystyrenes in dilute solution.[11] The spectra exhibit at least two components--a slow component whose relaxation time is proportional to molecular weight and a fast component whose relaxation time is independent of molecular weight. For dilute polystyrenes in CCl_4, the fast relaxation time is 4.5 ± 1.0 nsec. The slow relaxation time in these systems is identified with half the relaxation time of the longest Rouse-Zimm mode for the chain, and the fast, molecular-weight-independent time is assigned to a local, correlated motion of phenyl groups about the main chain backbone. Monte Carlo simulations of the dynamics of chains on a tetrahedral lattice with optically anisotropic side groups give depolarized spectra that are qualitatively similar to those observed in polystyrenes.[12] Moro and Pecora[13] have attempted to explain the dominance of the longest chain mode in the low frequency part of the depolarized spectrum by introducing local transverse stiffness into the chain. Experiments from my laboratory on polyamic acid and DNA's in solution indicate that local motions contribute to the depolarized spectra of these polymers.

The observation of depolarized scattering spectra in semidilute solutions should be feasible. To my knowledge no experiments of this type have as yet been reported.

4. BULK POLYMERS AND SMALL MOLECULES IN AMORPHOUS POLYMERS

Few depolarized light scattering experiments have been performed on bulk polymers, although we expect this to become an active field in the next few years. Patterson et al.,[14] for instance, have studied depolarized scattering from polystyrene near the glass-rubber transition using photon-correlation spectroscopy. They find highly non-exponential correlation functions that are difficult to characterize and interpret.

Motions of small optically anisotropic molecules dissolved in polymers that exhibit relatively small optical anisotropies may also be studied by depolarized light scattering. Ouano and Pecora[15] have, for instance, studied the rotational motion of chlorobenzene in amorphous poly(methyl methacrylate). They find two distinct relaxation regions. A fast region with relaxation times in the picosecond range and a slow region with multiple relaxation times in the range from milliseconds to nanoseconds. These authors interpret their results in terms of a theory of restricted rotational diffusion.[16]

REFERENCES

1. D. R. Bauer, J. I. Brauman, and R. Pecora, Ann. Revs. Phys. Chem. 27:443 (1976).
2. R. Pecora, J. Chem. Phys. 49:1036 (1968).
3. S. B. Dubin, N. A. Clark, and G. Benedek, J. Chem. Phys. 54: 5158 (1971).
4. D. R. Bauer, S. J. Opella, D. J. Nelson, and R. Pecora, J. Am. Chem. Soc. 97:2580 (1975).
5. A. Wada, N. Suda, T. Tsuda, and K. Soda, J. Chem. Phys. 50:31 (1969).
6. S. Michielsen and R. Pecora, Proceedings of the Conference on Photon Correlation Techniques in Fluid Mechanics, Stanford University Press (1980).
7. A. Flamberg, C.-C. Wang, and R. Pecora, unpublished results.
8. M. Doi, J. de Physique 36:607 (1975).
9. C.-C. Wang and R. Pecora, unpublished results.
10. K. M. Zero and R. Pecora, Polymer Preprints, 22: 98 (1981).
11. D. R. Bauer, J. I. Brauman, and R. Pecora, Macromolecules 8: 433 (1975).
12. C. W. Cornelius, Ph.D. Thesis, Stanford University (1979).
13. K. Moro and R. Pecora, J. Chem. Phys. 69:3254 (1978); 72:4958 (1980).
14. G. D. Patterson, C. P. Lindsey, and J. R. Stevens, J. Chem. Phys. 70:643 (1979).
15. A. C. Ouano and R. Pecora, Macromolecules, 13: 1167, 1173 (1980).
16. C.-C. Wang and R. Pecora, J. Chem. Phys. 72:5333 (1980).

THE ANALYSIS OF POLYDISPERSE SCATTERING DATA

E. R. Pike

Royal Signals and Radar Establishment

Malvern, UK

CONTENTS

1. INTRODUCTION

There will be no need at this school for a lengthy introduction to the problem outlined in my title. It is well-known that light scattering from a suspension of macromolecules undergoing Brownian motion can be used to gain information about their size and shape.[1] In particular, for spherical monodisperse particles the normalised "self-beat" photon correlation function, $g^{(2)}(\tau)$, of the scattered light has the simple exponential form

$$g^{(2)}(\tau) = 1 + C|g^{(1)}(\tau)|^2 = 1 + Ce^{-2D_T K^2 \tau} \tag{1}$$

where K is the scattering wave vector, C is an experimental constant, $g^{(1)}(\tau)$ is the first-order correlation function (the Fourier transform of the optical spectrum) and D_T is the linear translational diffusion coefficient, related to the particle radius, a, by the Stokes-Einstein relation

$$D = \frac{kT}{6\pi\eta a} \tag{2}$$

η is the kinematic viscosity of the medium, T the temperature and k, Boltzmann's constant.

If a reference beam is used of sufficient strength the photon correlation function becomes

$$g^{(2)}(\tau) = 1 + Cg^{(1)}(\tau) = 1 + Ce^{-D_T K^2 \tau} \tag{3}$$

When a polydisperse suspension is present the distribution of radii gives rise to a range of values of D_T and hence the correlation function is made up of a mixture of exponentials. It is convenient to introduce the variable

$$\Gamma(a) = D_T K^2 \tag{4}$$

representing the decay time of $g^{(1)}(\tau)$ in the monodisperse case. In the more general situation the first-order correlation function, which may be determined experimentally by subtracting unity and taking the square root of the self-beat function, or directly from the heterodyne result, may be expressed as

$$g^{(1)}(\tau) = \int_0^\infty e^{-\Gamma\tau} C(\Gamma)p(\Gamma)d\Gamma \tag{5}$$

which we may write

$$g^{(1)}(\tau) = \int_0^\infty e^{-\Gamma\tau} f(\Gamma)d\Gamma \tag{6}$$

where

$$f(\Gamma) = C(\Gamma)p(\Gamma) \tag{7}$$

The Γ-dependent "constant", $C(\Gamma)$, takes into account the variation of absolute scattering power as a function of particle size given, for example, for spherical particles, by the Mie scattering formulae.

Equation (6) has the form of a Laplace transform, so the first problem in deducing information about polydispersity is that of Laplace inversion of the data.

The Laplace transform is a member of the general class of Fredholm equations of the first kind

$$g(x) = \int_a^b K(x,y)f(y)dy \qquad c \leqslant x \leqslant d \tag{8}$$

where the kernel K(x,y) is continuous. We may note that many other problems in the interpretation of photon correlation functions have this form with different kernels. For example, the extraction of a velocity probability distribution, p(v), in laser anemometry[2] is a problem of this type with kernel

$$K(v,\tau) = \left[1 + m^2 \cos\left(\frac{2\pi v\tau}{s}\right)\right] e^{-v^2\tau^2/r^2}, \tag{9}$$

s is the fringe spacing, m^2 is a visibility factor and r is the laser beam radius.

These Fredholm equations are well-known also in other branches of physics and an exceptionally well studied example has the kernel

$$K(x,y) = \frac{\sin c(x - y)}{\pi(x - y)} \tag{10}$$

which arises in the theory of diffraction-limited imaging and in the theory of electrical communication. In this particular case a theory of "information" has been developed to form a framework for the discussion of the inversion problem which quantifies an intrinsic difficulty that inversion is found to be unstable, that is to say that a minute change in the data can give rise to a large change in the solution.[3] Information theory postulates a number of "degrees of freedom" for the solution (Shannon number or number of resolution elements) which gives an upper limit to the number of independent parameters available from the data.

This instability under inversion or so called ill-posedness is generic to the whole class of Fredholm equations of the first kind and manifests itself also, therefore, in the Laplace inversion required for the analysis of polydispersity. We shall see that the effect can be studied by a generalization of information theory, and a generalized Shannon number will tell us how many parameters in a distribution of radii or diffusion coefficients can be determined.

2. SINGULAR-VALUE ANALYSIS AND CONDITION NUMBER

The difficulty of inversion of Equation (8) can be easily illustrated. Let us assume that there exists a unique solution $f_o(x)$ corresponding to given noiseless data $\bar{g}_o(x)$. Then we may add to that solution a function

$$f^{(n)}(x) = C \sin (nx) \tag{11}$$

where C and n are constants. Now the Riemann-Lebesgue theorem tells us that

$$\lim_{n \to \infty} \int_a^b K(x,y) \sin (ny)dy = 0 \tag{12}$$

so that $\bar{g}_o$ corresponds arbitrarily closely to a set of widely different functions $f = f_o + f^{(n)}$. Small departures of g from $\bar{g}_o$ can thus perturb wildly the result of the inversion.

Many scientists in the past, coming upon this problem in their own fields for the first time without being aware of its ill-posed nature, have tried to solve the inversion numerically by converting the integral equation into a linear algebraic system by sampling the data and solution uniformly in a finite number of points. Suppose that g is given in M points and f by an N point quadrature where $N \leq M$; the integral equation becomes

$$Af = g \tag{13}$$

where A is an M x N matrix of components

$$A_{mn} = A(x_m, y_n)w_n \tag{14}$$

where the weights w_n depend on the form of quadrature, and f and g are vectors $\{f(y_n)\}^n$ and $\{g(x_m)\}$ in N- and M- dimensional Euclidean domains F and $\bar{G}$ respectively. In the presence of errors in g the data may lie in a larger domain G and the solution f may depart from f_o. Let us investigate the possible extent of the departure. We know that

$$Af_o = \bar{g}_o \tag{15}$$

and for any inversion method in the presence of noise a procedure exists which must project g into $\bar{g}$ s.t.

$$Af = \bar{g} \tag{16}$$

the vector $\bar{g}$ corresponding to g could be found if necessary by the forward transformation of the solution obtained. We thus have

$$A(f - f_o) = \bar{g} - \bar{g}_o \tag{17}$$

and if the error in g is bounded such that the vector $\bar{g} - \bar{g}_o$ has "length" ε, by which we mean that its Euclidean norm or scalar product with itself obeys

$$||\bar{g} - \bar{g}_o||_M = (\bar{g} - \bar{g}_o, \bar{g} - \bar{g}_o)_M \leq \varepsilon^2 \quad (18)$$

then we have

$$(A(f - f_o), A(f - f_o)) \leq \varepsilon^2 \quad (19)$$

which may be rewritten in the form

$$(A^*A(f - f_o), (f - f_o)) \leq \varepsilon^2 \quad (20)$$

where A* is the N x M Hermitian conjugate of A. A*A is an N x N, symmetric, non-negative matrix so that it can be diagonalised. We denote its eigenvalues in decreasing order by λ_n^2. $\lambda_n(>0)$ is called a singular value of A. Equation (20) defines the interior of an N-dimensional "ellipsoid" in the solution space F centred on f_o, with axes along the eigenvectors, u_n of A*A and with semi-axial lengths ε/λ_n. The ratio λ_1/λ_N of the longest to the shortest axis is called the condition number of the matrix A. For most problems of interest A will have a sequence of eigenvalues decreasing to zero so that the finer the sampling the larger the condition number and hence the longer the largest dimension of the "error ellipsoid" and the greater the possibility for a large change in the solution arising from a small change in the data. The situation is shown schematically in Figure 1. Points in the M-dimensional ball of radius ε about $\bar{g}_o$ in the data space, $\bar{G}$, give inversions inside the N-dimensional ellipsoid centred on $\bar{f}$ in the solution space F.

This argument however may be turned on its head. Since we have postulated a 1:1 mapping $A:F \rightarrow \bar{G}$, the inverse mapping $A^{-1}:\bar{G} \rightarrow F$ must exist so that

$$A^{-1}\bar{g} = f \quad (21)$$

$$A^{-1}(\bar{g} - \bar{g}_o) = (f - f_o) \quad (22)$$

and the whole argument can be repeated to show that a ball of radius $\varepsilon\lambda_1$, centred on f_o maps into an ellipsoid centred on $\bar{g}_o$ and must therefore have maximum axial length ε. These regions are schematically shown shaded in Figure 1. Thus the instability with respect to noise is greatest in the unshaded region around $\bar{g}_o$ which lies in those direction u_n where the information about the object is most strongly attenuated by the operator A.

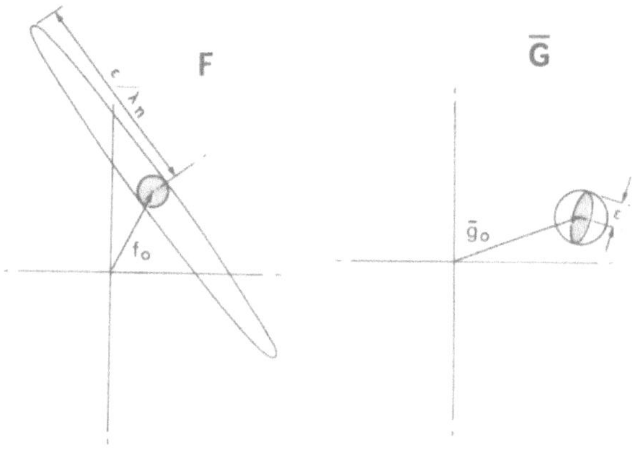

Figure 1. Corresponding domains in solution space F and data space $\bar{G}$.

More complication arises if we consider the projection of real data g into the space $\bar{G}$. There is no guarantee that any particular noise vector $\bar{g} - \bar{g}_0$ lies in the space $\bar{G}$. On the contrary, for the Shannon problem we know that any vector in $\bar{G}$ has no frequency components outside a given band limit. Thus high frequency noise will require adding some extra dimensions to $\bar{G}$ for a geometrical description. We illustrate this in Figure 2. A new dimension out of the paper has been added to Figure 1 for $\bar{G}$ to describe the data space G. The larger ellipsoid illustrated in G will represent the error-vector domain, say a certain level of white noise. Note that we have not taken a sphere about $\bar{g}_0$ this time for the error-vector domain in G since white noise would only lie in a "spherical" domain in frequency space. This sphere will map into some more general shape of domain in G and its intersection with $\bar{G}$ may not depart too much from the shaded region in Figure 1, since the expansion in eigenfunctions of band-limited noise will be similar to that of a band-limited object. What we are postulating, therefore, is that ill-conditioning is the result of the projection of "out-of-band" noise onto $\bar{G}$ which could

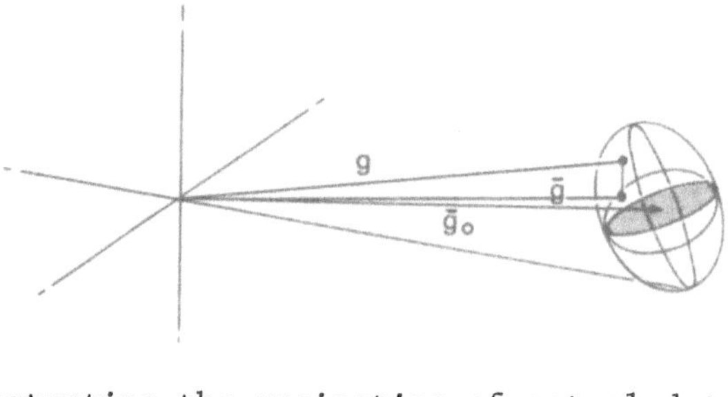

Figure 2. Illustrating the projection of actual data onto data which could have arisen from possible solutions.

well lie in the dangerous unshaded region of Figure 1. In the conventional Shannon problem the operator A* is used to perform this projection and we have no way of avoiding the resulting ill-conditioning as the sampling of the object gets more finely divided. Similar considerations apply to the analysis of polydisperse data. If inversion were attempted by equally spaced samples in $F(\Gamma)$ and $g(\tau)$ the eigenvalue analysis of McWhirter and Pike[4] shows that in the continum limit the eigenvalues are

$$|\lambda_n^{\pm}|^2 = \frac{\pi}{\cosh \pi n} \tag{23}$$

which decays are $\pi e^{-\pi n}$ for large n giving rise to exponentially increasing condition numbers.

Recent work by Bertero and Pike (to be published) shows that in the diffraction-limited imaging problem, in the absence of truncation of the image, the out-of-band noise lies in the null space of a known operator A* in G. The error ellipsoid of Figure 2 can thus be projected into the shaded region of $\bar{G}$, in principle therefore, removing the instability. It is too soon to assess the performance of such an inversion procedure or whether these results can be adapted to the polydispersity problem. These new ideas, however, give promise of much better inversion procedures for the future with condition numbers reducing as the length of correlation function used increases.

3. MODELLING AND THE USE OF PRIOR KNOWLEDGE

If we disregard, for the present, any new techniques along the lines of the previous paragraph we are faced with the result

$$A^*Au_n = \lambda_n^2 u_n \tag{24}$$

of the singular-value analysis of the previous section giving rise to large condition numbers when too many resolution elements are attempted. Let us introduce the vectors[3]

$$v_n = \frac{1}{\lambda_n} Au_n \tag{25}$$

Operating on both sides of this equation with A* we find that

$$A^*v_n = \lambda_n u_n \tag{26}$$

also from Equation (25)

$$Au_n = \lambda_n v_n \tag{27}$$

The $\{u_n\}$ and $\{v_n\}$ form bases in F and $\bar{G}$ respectively. We have seen that solutions f given by

$$||Af - \bar{g}||_{\bar{G}} \leq \varepsilon \tag{28}$$

lie in the full ellipsoid of Figure 1 with axes along the u_n and without further a priori information any of these solutions is as good as any other. We must, therefore, investigate how to use various facts which may be assumed about the solution to restrict the possibilities. We may call this procedure modelling, that is assuming the solution to conform to some a priori model form. In the strongest case this would demand a solution in terms of a given finite set of parameters while weaker models would impose global constraints only. Both features could also be combined in a more composite model.

Let us consider first the imposition of constraints. In addition to the inequality (Equation (28)) above we may require that it obey the constraint

$$||Bf||_F \leq E \tag{29}$$

where B is chosen according to available prior knowledge. The simplest example for B is the identity operator which in the diffraction problem bounds the energy radiated and in the polydispersity problem would essentially bound the total scattering mass. The technical requirements on B as a constraint operator are that it is densely defined in F and has a continuous inverse. A can be the full integral operator

$$(Af)(x) = \int_a^b K(x,y)f(y)dy \tag{30}$$

Our problem[3] is now to find that function $\tilde{f}$ which minimises $||A\tilde{f} - \bar{g}||$ under the constraint

$$||B\tilde{f}||_F \leq E \tag{31}$$

Such a function will also minimise the functional

$$Q(f) = ||Af - \bar{g}||^2_{\bar{G}} + \left(\frac{\varepsilon}{E}\right)^2 ||Bf||^2_F$$

$$= ||Af||^2_{\bar{G}} - 2(\bar{g},Af)_{\bar{G}} + ||\bar{g}||^2_{\bar{G}} + \left(\frac{\varepsilon}{E}\right)^2 ||Bf||^2_F \tag{32}$$

$$= ||Af||^2_{\bar{G}} + \left(\frac{\varepsilon}{E}\right)^2 ||Bf||^2_F - 2(\bar{g},Af)_{\bar{G}} + ||\bar{g}||^2_{\bar{G}}$$

which may be written

$$Q(f) = (Cf,f)_F - 2(\bar{g},Af)_{\bar{G}} + ||g||^2_{\bar{G}} \qquad (33)$$

where we have introduced the operator

$$C = A^*A + \left(\frac{\varepsilon}{E}\right)^2 B^*B \qquad (34)$$

It can be seen by inspection, remembering that $(\bar{g},Af)_{\bar{G}} = (A^*\bar{g},f)_F$, that the minimising solution obeys

$$C\tilde{f} = A^*\bar{g} \qquad (35)$$

In the simplest case where B = I, the identity operator, it may easily be deduced that

$$\tilde{f} = \sum_{n=0}^{\infty} \frac{\lambda_n}{\lambda_n^2 + (\varepsilon/E)^2} g_n u_n \qquad (36)$$

where

$$g_n = (\bar{g},v_n)_{\bar{G}} \qquad (37)$$

As long as

$$Q(\tilde{f}) \leq 2\varepsilon^2 \qquad (38)$$

we may take $\tilde{f}$ as an estimate of the solution. This is said to be regularised following Tikhonov[5] and Miller[6]. Graphically the procedure, in this simple case, can be visualised in Figure 3 where the domain of previously possible values is now restricted to lie inside the large sphere.

Use of Equation (36) damps down the possible increase in axial lengths of the error domain as N increases.

A simpler model is obtained by explicitly neglecting the higher

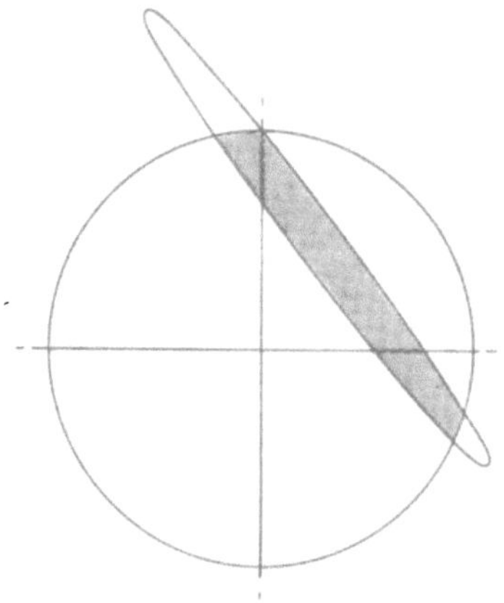

Figure 3. Restriction of possible solutions by regularisation.

terms of the series (Equation (36)) beyond a point where the coefficients drop below a certain value related to ε/E. It is usual to cut them off without regularisation to obtain an estimate (within the error ellipsoid):

$$f = \sum_{n=0}^{N} \frac{1}{\lambda_n} g_n u_n \tag{39}$$

This procedure of truncation of the eigenfunction series is sometimes called numerical filtering.[7]

In 1978 McWhirter and Pike[4] discovered the eigenfunctions, u_n of the Laplace transform. These functions are

$$u_n(\tau) = \frac{[\Gamma(\frac{1}{2} + in)]^{\frac{1}{2}} \tau^{-\frac{1}{2}-in}}{[\pi|\Gamma(\frac{1}{2} + in)|]^{\frac{1}{2}}} \tag{40}$$

where Γ is the gamma function and, of course, one may use both the real and imaginary parts separately. The singular value spectrum was given above in Equation (23). Equipped with these functions McWhirter and Pike were able to perform accurate numerical inversions of the Laplace transform. An analytic form for $f(\Gamma)$ was chosen as input

$$f(\Gamma) = \Gamma e^{-\Gamma} \tag{41}$$

which allowed the forward transform to be performed analytically

$$g(\tau) = (1 + \tau)^{-2} \tag{42}$$

Equation (39) was then used to recover the input function in the presence of increasing amounts of added noise. The features described above were demonstrated, the input being recovered at low noise levels but instability setting in to prevent a satisfactory result when a sufficient number of terms for reconstruction could not be used.

Unlike the Shannon problem where the fall off of the eigenfunction spectrum is very sharp and the number of terms of Equation (39) which can be used (the Shannon number) is insensitive therefore to the actual noise level, we now have to tailor the number of terms used to the actual noise level present. Thus a generalised information theory can be applied with a noise-dependent number of "degrees of freedom".

4. EXPONENTIAL SAMPLING

The eigenfunction truncation method gives rise to a model for $f(\Gamma)$, using real and imaginary parts of the u_n, of the form[8]

$$\approx f(\Gamma) = \int_{-\omega_{max}}^{\omega_{max}} \beta(\omega)\Gamma^{-\frac{1}{2} + i\omega} d\omega \tag{43}$$

expressing the series (Equation (39)) as an integral. (Γ here denotes the slope of the correlation function, not the gamma function of Equation (4)!). By making the substitution

$$\Gamma = e^x \tag{44}$$

this equation becomes

$$e^{x/2} \approx f(e^x) = \int_{-\omega_{max}}^{\omega_{max}} \beta(\omega) e^{i\omega x} d\omega \tag{45}$$

McWhirter[8] has pursued an argument familiar in communication theory to show that, due to the band limitation at ω_{max}, the eigenfunction truncation solution has the alternative expression.

$$\approx f(\Gamma) = \sum_{n=-\infty}^{\infty} \approx f(\Gamma_n) \left(\frac{\Gamma}{\Gamma_n}\right)^{-\frac{1}{2}} \frac{\sin\left[\omega_{max}(\log\Gamma - \log\Gamma_n)\right]}{\omega_{max}(\log\Gamma - \log\Gamma_n)} \tag{46}$$

where

$$\Gamma_n = e^{\frac{n\pi}{\omega_{max}}} \tag{47}$$

We thus have a "point-spread function" of logarithmic argument which defines a criterion, analogous to that of Rayleigh in the diffraction-limited imaging problem, whereby points of $f(\Gamma)$ cannot be resolved between Γ_i and Γ_j unless

$$\Gamma_j \geq \Gamma_i e^{\frac{\pi}{\omega_{max}}} \tag{48}$$

Following this theme McWhirter introduced a simple model consisting of an exponentially spaced histogram for $f(\Gamma)$ and has used this successfully to recover a double gaussian analytical input function. Over a limited range away from the origin the exponential histogram approximates the linear histogram which has been used extensively by Chu[9] with additional measures of constraint to relieve the ill-conditioning problem. Work by McWhirter and Parker (unpublished) has shown that in the absence of additional constraints the exponential histogram always gives more independent parameters than the linear one before instability sets in.

5. INTERPOLATION

Rather than utilising McWhirter's exponential histogram model Ostrowsky and Pike (to be published) have pointed out that a δ-function model of exponentially spaced sample values is simple to calculate and gives the possibility of two different interpolation schemes. Suppose that the experimental input were band-limited in the general sense of the previous section with p degrees of freedom. The "information content" of the distribution $f(\Gamma)$ would then be completely described by p exponentially spaced sample values. Just as in the analogous case in communication theory where the information content of a Fourier transform of a band-limited function is entirely contained in the coefficients of the fast Fourier transform, these points can be interpolated to give the complete function either by successively phase shifting the transform coefficients and calculating the output at p new places each time or by using the sin x/x interpolation formula. In our problem this would be achieved by use of Equation (46). In either case the complete distribution is recovered if p coefficients are retrievable.

We conjecture that out-of-band noise will be more successfully eliminated the more restricted the *a priori* range of $f(\Gamma)$ and the larger the range of the correlogram $g(\tau)$ analysed, as we have outlined in section 2 and Figure 2. More work, however, is urgently necessary to verify these ideas.

6. NUMERICAL RESULTS

We have explained above that the "resolution" which can be achieved in the recovery of a distribution of radii is a function of the noise present on the data. Simulated results can be obtained in which analytically related $p(\Gamma)$ and $g(\tau)$ are numerically inverted which are subject only to computer noise arising from the finite arithmetic length. Real data, however, will always be much less accurate than this and we need to investigate the resolution or number of exponentially spaced samples of $p(\Gamma)$ we may recover in the presence of various amounts of experimental noise for given input distribution functions.

To give something more than academic interest to our initial calculations we have used as an input a distribution of radii, measured from a real sample by equilibrium sedimentation, of silver halide particles used by Kodak Ltd for the preparation of a photographic emulsion. The aim of the work is to evaluate, in collaboration with colleagues at Kodak, the possible use of photon correlation spectroscopy for this type of measurement in the photographic industry.

We first digitised the distribution, taken from a chart recording as $p(1/a^2)$ and rescaled the abscissa to $1/a$. A first exercise, in which different scattering factors were neglected, involved determining the effects of computer noise only by taking M sample points of the resulting distribution $p(\Gamma)$ at exponentially spaced abscissa values Γ_i. A correlation function

$$G^{(2)}(mT) = \sum_{i=1}^{M} p(\Gamma_i)\, e^{-\Gamma_i mT} \qquad m = 1, 2 \ldots 100 \qquad (49)$$

was then constructed over 100 equally spaced delay times, T, chosen so that the mean value $\bar{\Gamma}$ obeyed approximately the relation $\bar{\Gamma}T = 0.1$. This ensured that the 100th channel was well into the tail of the correlation function. M values of $p(\Gamma)$ were then fitted by the computer program, using a stable least squares procedure, on the same abscissa values. It was found that perfect inversions resulted with very few samples but that by the time M reached 11 the computer errors began to affect the inversion significantly. The results for M = 10, 12, and 13 are shown in Figures 4a, b, and c respectively. Slightly more realistic inversions were then performed in which 50 values of $f(\Gamma)$ weighted by a suitable Mie factor, were used as input to construct $G^{(2)}(\tau)$. Various values of M were then tried for the number

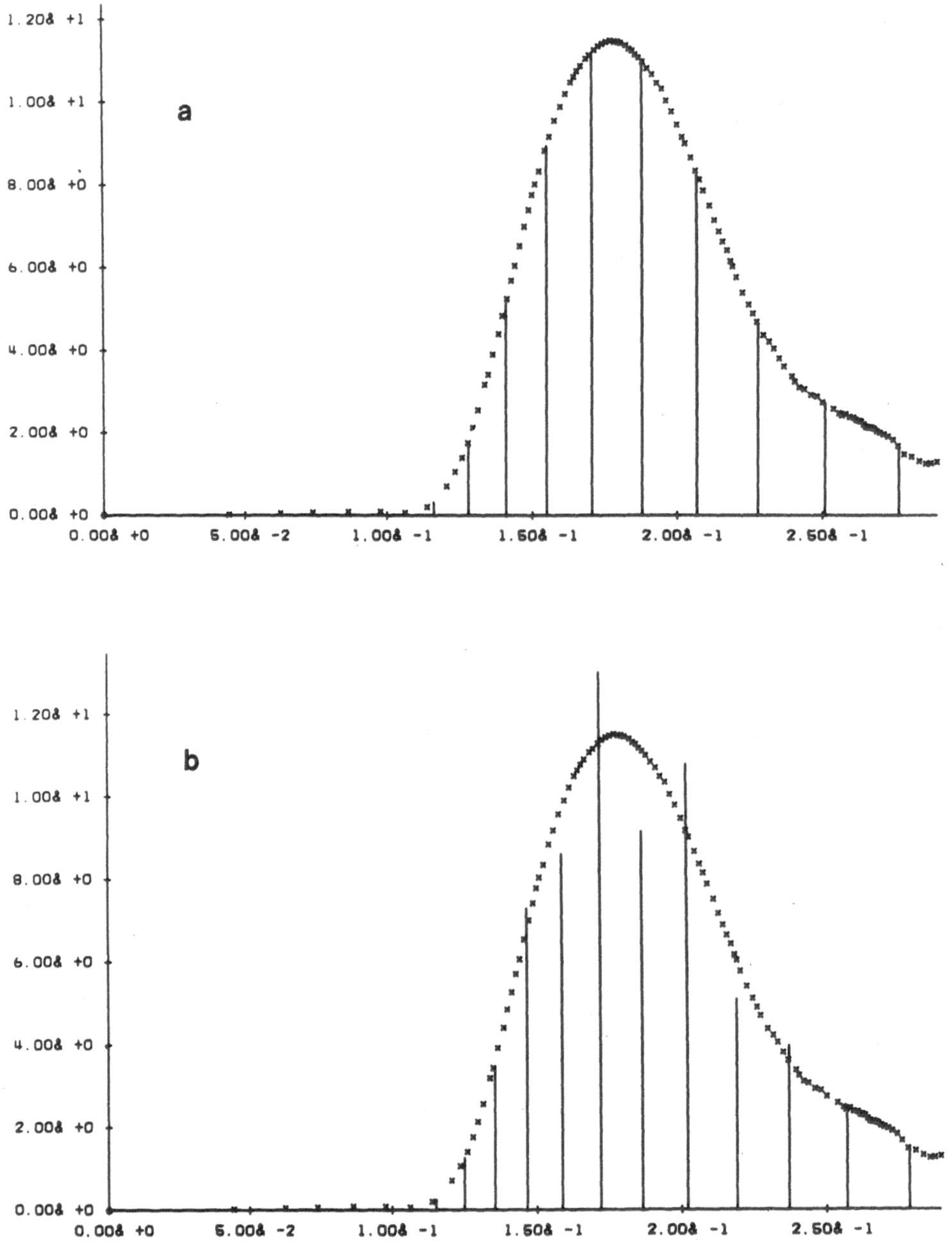

Figure 4. (a) 10 parameter inversion from 10 input exponentials; (b) 12 parameter inversion from 12 input exponentials

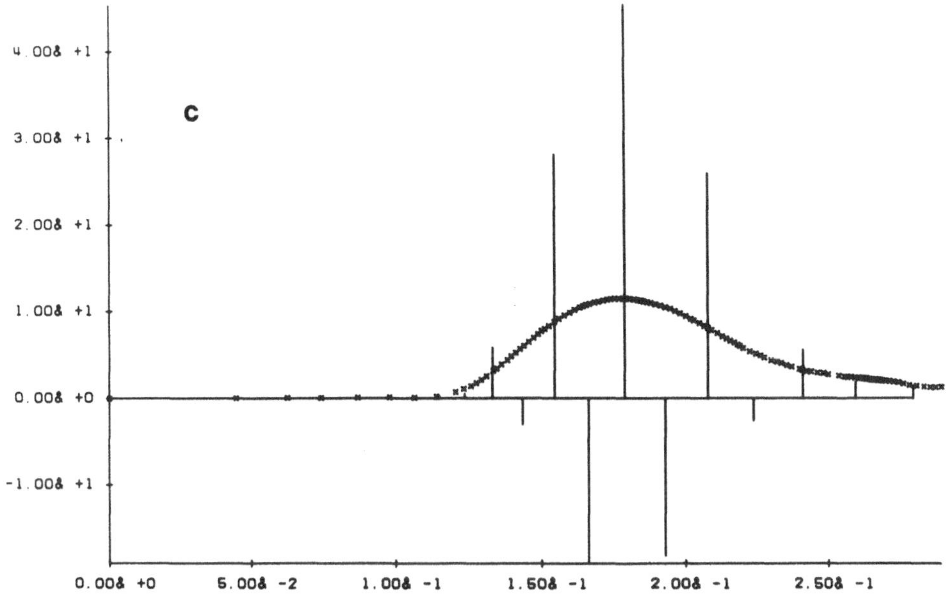

Figure 4. (c) 13 parameter inversion from 13 input exponentials

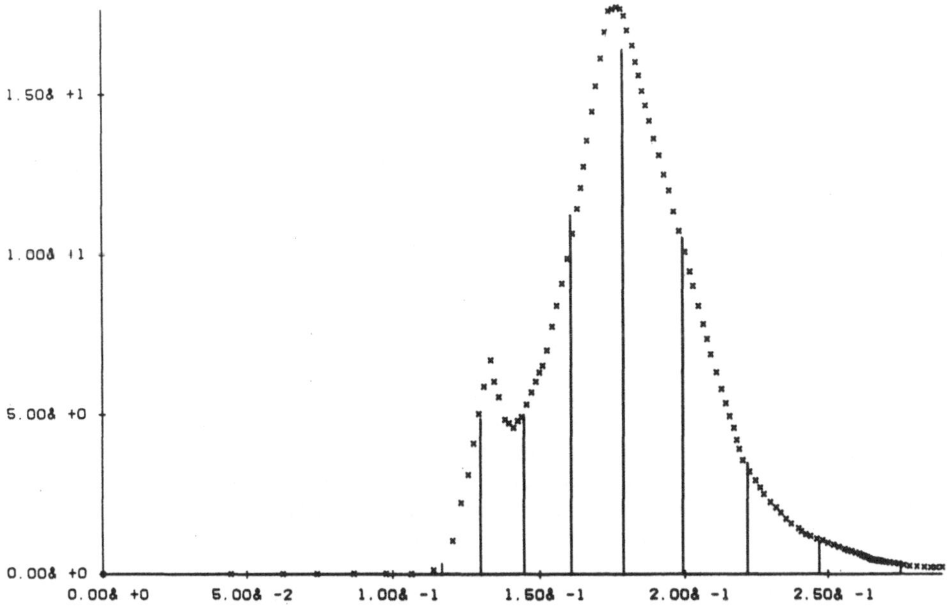

Figure 5. 9 parameter inversion from 50 input exponentials

of output ordinate samples. The results were somewhat worse than using equal numbers of input and output samples but it was found that up to nine samples could be obtained before the output became badly distorted. Figure 5 shows an inversion with "nine degrees of freedom" from this series.

We now required to add noise to simulate a real photon correlation experiment. To do this we used the result of Jakeman, Pike and Swain[10] that the noise in a "self-beat" experiment to determine a single diffusion constant is given by

$$\mathrm{Var}\left[g^{(2)}(\tau)\right] = \frac{1}{N\gamma}\left[\frac{1}{2}\left\{1 + 8e^{-x} - e^{-2x}(5 + 2x)\right\} + \frac{2}{\Gamma}(1 + e^{-x})^2 + \frac{1 + e^{-x}}{r^2\gamma}\right] \tag{50}$$

Here

$$\gamma = \Gamma T \tag{51}$$

$$x = 2\gamma\tau \tag{52}$$

and

$$r = \bar{n}/\gamma \quad , \tag{53}$$

$\bar{n}$ is the mean number of photons detected per sample time and N the total number of samples of duration T. The fact that Equation (50) applies to the monodisperse rather than the polydisperse situation will make a difference which we assume will not be important unless very large polydispersities are present and even in this case some indication of the quantity of noise in an experiment of duration N samples would be expected.

To make quite sure that formula (50) is correct we ran sets of experiments with a carefully set up spectrometer using a monodisperse polystyrene solution and computed the normalised correlation coefficient variances. We were able to verify that the theoretical and experimental values were in agreement to within a few tens of percent. Noise was thus added to the constructed correlation function coefficients according to this prescription and characterised by various values of N from 10^7 upwards with fixed values of $\gamma = 0.1$ and $r = 10$.

Fifty values of $f(\Gamma)$ were input to construct the correlation function as before, including the Mie factor weighting. At $N = 10^7$ it was found that three or four output values only could be used.

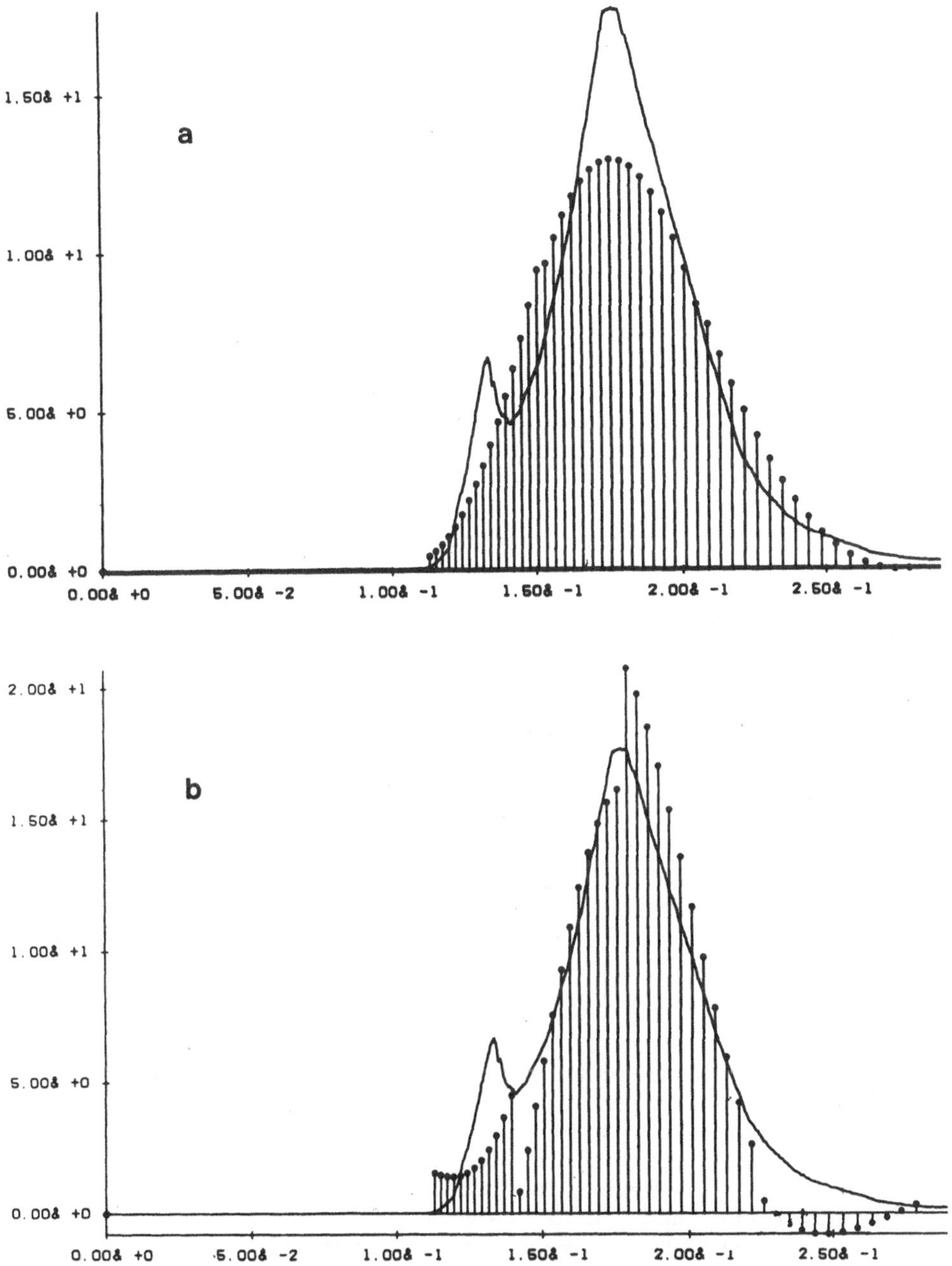

Figure 6. (a) 3 parameter inversion from 50 input exponentials with interpolation; (b) 4 parameter inversion from 50 input exponentials with interpolation.

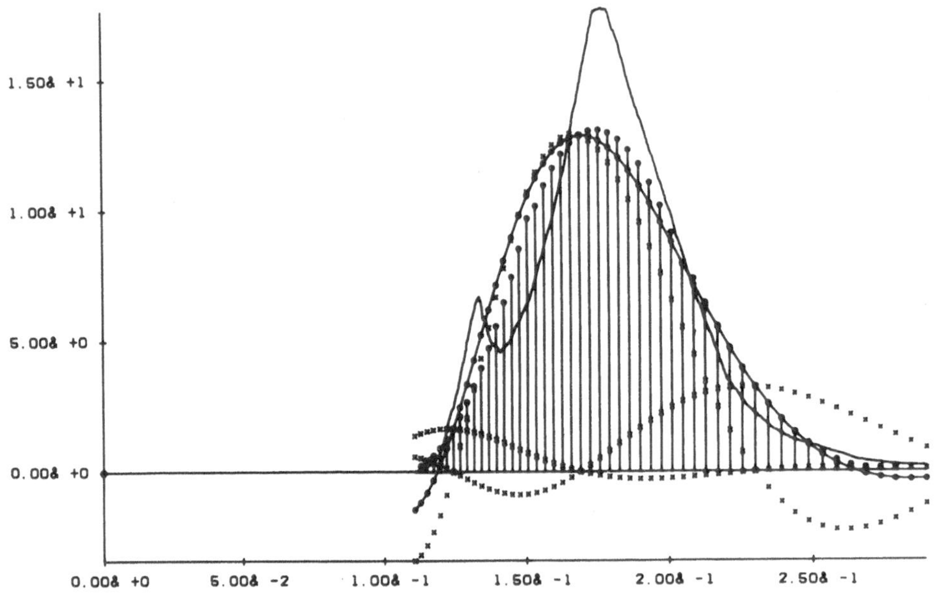

Figure 7. Figure 6a with functional interpolation of three output parameters

This is of the same order as has been previously found using the histogram method on experimental data in this laboratory and by Ostrowsky using both δ-fn sampling and the histogram method in Nice (private communication). There seems to be no way to avoid these low "information throughputs" in this type of experiment without increasing the constraints and in the absence of such extra a priori information input (positivity being one which clearly should be incorporated as soon as possible) we have used the interpolation methods described in § 5 to improve the interpretation of the output. In Figures 6a and 6b we show results for $M = 3$ and $M = 4$ respectively, shifting the sample points and in Figure 7 an interpolation by Equation (46) for the case $M = 3$. It is clear that at this level of noise the structure in $f(\Gamma)$ produced by the Mie factor is an embarrassment requiring a higher resolution than is available to reconstruct satisfactorily. The effects of this on the recovery of the input distribution of radii is shown in Figures 8a and 8b which depict the final output, after removal of the Mie factor, overlaid on the input for the same cases $M = 3$ and 4, respectively, as were illustrated in Figure 6. To give some idea of the fits obtained, in Figure 9 the noisy correlogram is shown overlaid with all the three-sample fits of Figure 6a.

A final numerical experiment consisted of attempting to reconstruct a δ-function as the input for $p(a)$. In the presence of computer noise only the output for $M = 13$ without interpolation is shown in Figure 10.

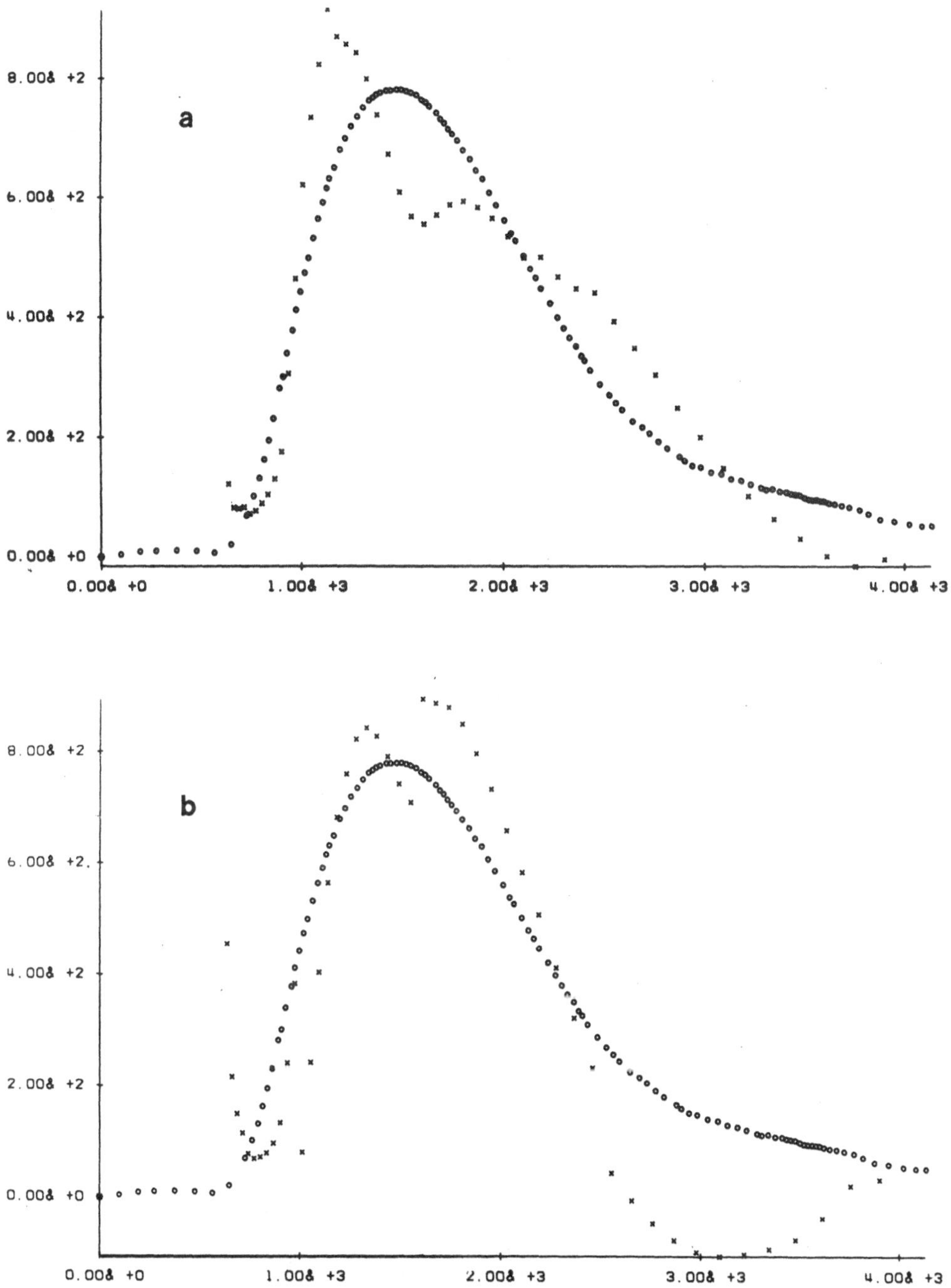

Figure 8. Final output compared with input after removal of Mie factor for (a) 3 outputs; (b) 4 outputs

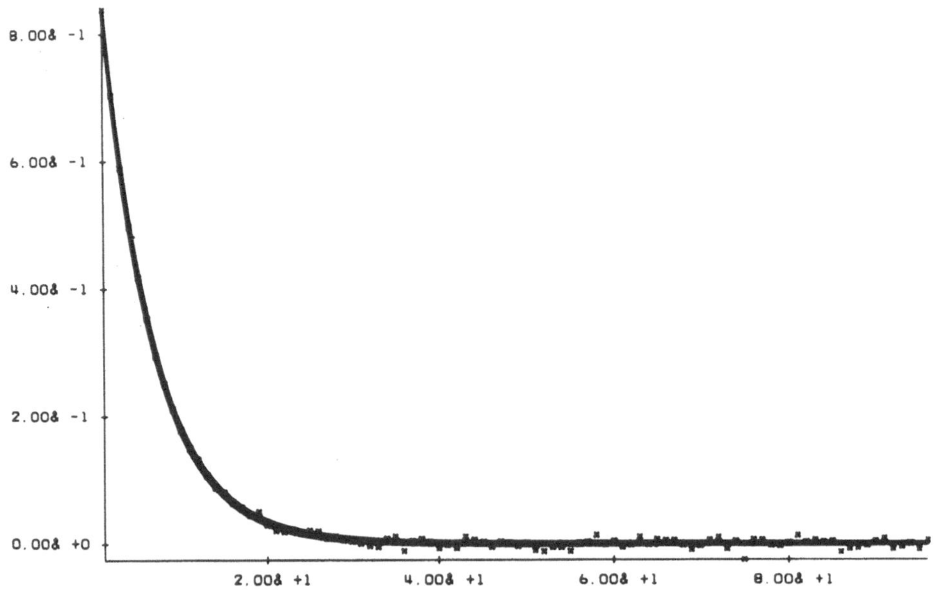

Figure 9. Correlogram with 10^7 samples overlaid with fits of Figure 6a

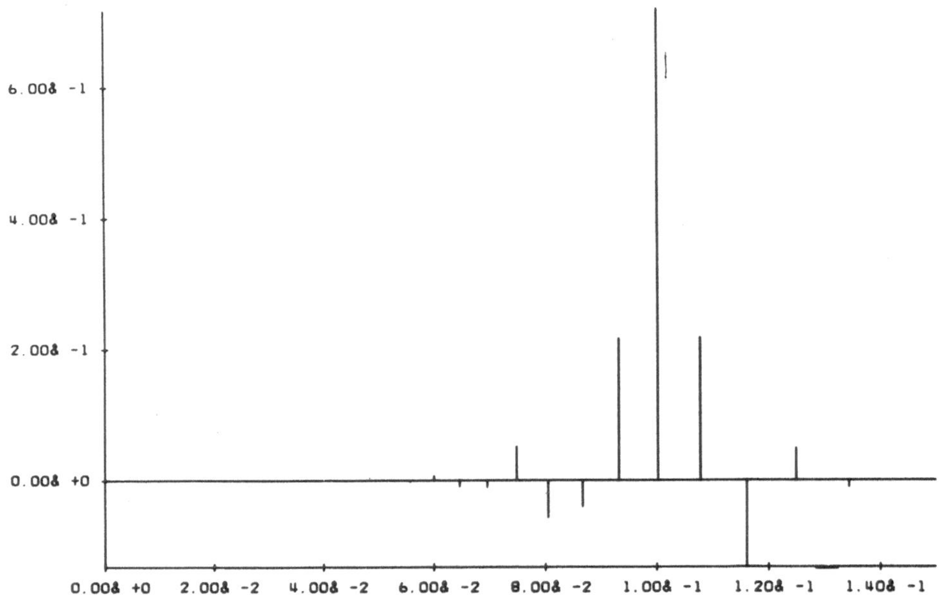

Figure 10. "Impulse response" using 13 output parameters from a single exponential input

This can be regarded as the "impulse response" of the inversion procedure at this level of noise with the particular limits on Γ and τ used, and has the general form of the point-spread function of Equation (46).

To summarise these numerical calculations, we have developed a method which can be used to invert polydisperse distributions of particles which do not display fine structures in $f(\Gamma)$ either due to the distributions themselves or to the effects of Mie weighting. This would be the case with smooth distributions of particles in the Rayleigh scattering range.

To cope with structure not within these limits we will carry out further work along the lines indicated above, by linear programming using Simplex or non-negative least square methods, for example, to ensure positivity or perhaps more easily by the addition of a measure of regularization, and we are also investigating further possible error-free projection methods using larger lengths of the correlogram record, as mentioned at the end of § 2 above.

ACKNOWLEDGEMENTS

My first grateful acknowledgement must be to Mrs. Pat Parker of RSRE who has been responsible for the numerical work of Section 6 and has in addition given help in the preparation of the manuscript. Elizabeth Whiting did the verification of errors described in Section 6 and Dr. John McWhirter has joined in profitable discussions. Mr. E. M. L. Parkin of Kodak Ltd, Harrow, provided the raw data on the silver halide distribution. Finally, Drs. Mario Bertero of the University of Genoa and Nicole Ostrowsky of the University of Nice have been in close touch with this work and I have derived great benefit from counting them both as colleagues.

REFERENCES

1. Photon Correlation and Light Beating Spectroscopy, H. Z. Cummins and E. R. Pike, eds. (Plenum, New York, 1974).
2. Photon Correlation Spectrometry and Velocimetry, H. Z. Cummins and E. R. Pike, eds. (Plenum, New York, 1977).
3. M. Bertero, C. de Mol, and G. A. Viano, Inverse Scattering Problems in Optics, H. P. Baltes, ed. (Springer, Berlin, 1980).
4. J. G. McWhirter and E. R. Pike, J. Phys. A: Math. Gen. 11: 1729 (1978).
5. A Tikhonov and V. Arsenine, Methodes de Resolution des Problèmes Mal Posés (Mir, Moscow, 1976).
6. G. F. Miller, Numerical Solution of Integral Equations, L. M. Delves and J. Walsh, eds. (Clarendon Press, Oxford, 1974).

7. S. Twomey, J. Franklin Inst. 279: 95 (1965).
8. J. G. McWhirter, Optica Acta 27: 83 (1980).
9. B. Chu, Es Gulari, and Er Gulari, Physica Scripta 19: 476 (1979).
10. E. Jakeman, E. R. Pike, and S. Swain, J. Phys. A: Gen. Phys. 4: 517 (1971).

A REVIEW OF TECHNIQUES FOR TIME-RESOLVED X-RAY STUDIES ON MUSCLE

A. R. Faruqi and H. E. Huxley

MRC Laboratory of Molecular Biology
Hills Road
Cambridge CB2 2QH, England

CONTENTS

1. INTRODUCTION

The molecular structure of muscle is of great importance in modern biology for a variety of reasons but mainly because it allows an insight into the mechanism of conversion of chemical energy into mechanical work. This is a fundamental and general problem encountered not only in muscle but also in other situations, e.g. cell motility, cytoplasmic streaming or flagellar movement. The use of structural methods, viz electron microscopy and X-ray diffraction, has complemented biochemical and physiological techniques in elucidating the structure and function of different types of muscle[1,2,3]. We will restrict our discussion to vertebrate striated muscles in this paper as they have been most extensively used in structural studies.

Muscle is a dynamic system, the changes in internal structure resulting in external changes, the most important of which is the generation of force used for performing mechanical work; it is obviously important to find out the nature of these basic structural changes on a molecular level which are responsible for the generation of force. Time-resolved X-ray diffraction has proven to be a very important technique in studying such structural changes during contraction because live muscles can be studied in as close to 'natural' conditions as possible without any structural modifications (such as heavy metal staining as in electron microscopy). Provided changes in the X-ray pattern can be recorded sufficiently rapidly during contraction, a tool is available, in principle, for finding out the nature of structural changes.

The main theme of the paper is concerned with the application of several new techniques (discussed in detail in Section 4) to time-resolved studies; these include the application of synchrotron radiation, position sensitive detectors and electronic routing systems along with semi-automatic software for data analysis. A brief introduction to muscle structure and X-ray diagram is given in Section 2 followed by a discussion on the typical counting rates in the X-ray pattern and the technical requirements that this poses in Section 3. Finally we show some recent results illustrating the technique in Section 5.

2. MUSCLE STRUCTURE AND X-RAY DIAGRAM

The aim of this brief introduction to the molecular structure of vertebrate striated muscle is to give some background information about those parts of the regular structure which contribute strongly to the low angle X-ray pattern and to show how changes in this structure might contribute to changes in the pattern.

The contractile apparatus of the muscle is contained within long, thin myofibrils, 1 to 2 μm in diameter and several centimetres long. The myofibrils are packed together to form fibres which have a diameter of ∿ 100 μm. The main protein components are contained in two partially overlapping filaments, 'thick' filaments which are composed mainly of myosin and 'thin' filaments consisting mainly of actin (shown in Figure 1). To give some rough dimensions, the myosin filaments are ∿ 1.6 μm in length and 100-120 Å in diameter and the actin filaments ∿ 1 μm in length and 50-70 Å in diameter. The actin filaments are attached to a structure known as the Z-line, which also delineates a standard unit of the muscle cell known as a "sarcomere" which is typically ∿ 2.5 μm in resting muscle. Actin filaments on either side of the Z-line have opposite polarity, a fact that is used in force generation by cross-bridges projecting from the myosin filaments. Cross-bridges are thought to attach to binding sites on actin, go through a 'tilting' motion[4] producing a relative sliding movement between the filaments, and then become detached; they go repetitively through this cycle in an asynchronous manner, i.e. different cross-bridges perform the cycle at different times.

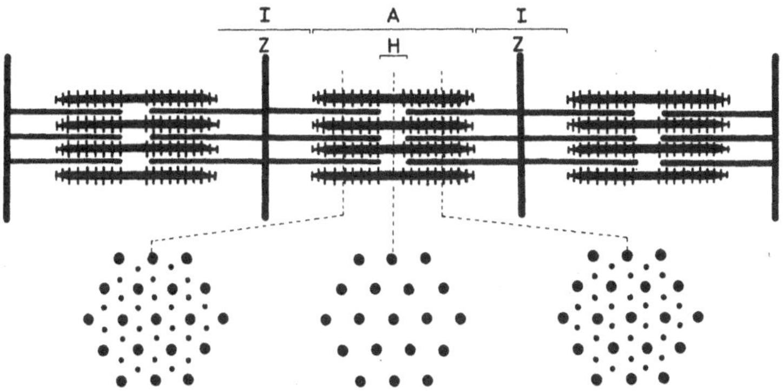

Figure 1. Schematic diagram showing the arrangement of thick and thin filaments in striated muscle. Myosin cross-bridges, extending out of the thick filament, are shown at 90° to the filament for simplicity. The basic sarcomere, referred to in text, is the region between the Z-bands. (From Huxley[8]).

As the 'polarity' of actin filaments and of the cross-bridges is reversed on either side of the sarcomere the net effect is a shortening of the sarcomere and consequent force generation.

A considerable effort has been spent in trying to understand the low angle pattern from muscle[5,6], and we will summarize the relevant parts of this pattern. Figure 2 shows a typical low angle X-ray pattern recorded on a synchrotron camera at EMBL, Hamburg.[7] The most striking feature of the pattern is a series of layer lines, known to arise from the myosin filament periodicity with a spacing of 429 Å and strong sampling on the meridian at 1/3 of this, i.e. at 143 Å. Actin layer lines are comparatively weak and are not visible in Figure 2; the layer lines at 59 Å and 51 Å can be clearly seen in the layer line plots shown in Figure 11.

The intensity distribution on the myosin layer lines was explained[5] on a proposed model with a pair of cross-bridges every 143 Å, successive pairs of cross-bridges being related by a 3-fold screw axis, i.e. arranged on a 6/2 helix, provided suitable dimensions could be assumed (compatible woth electron microscopic evidence) for the cross-bridges. More recent models[2] require a different value for the number of cross-bridges per 143 Å but the basic helix pitch of 429 Å is not affected.

The X-ray pattern along the equator (inset, Figure 2) can be explained[8] on the basis of the transverse (i.e. perpendicular to the filament direction) arrangement of the thick and thin filaments as shown in Figure 3(a) and (b). The myosin filaments lie on a crystal-

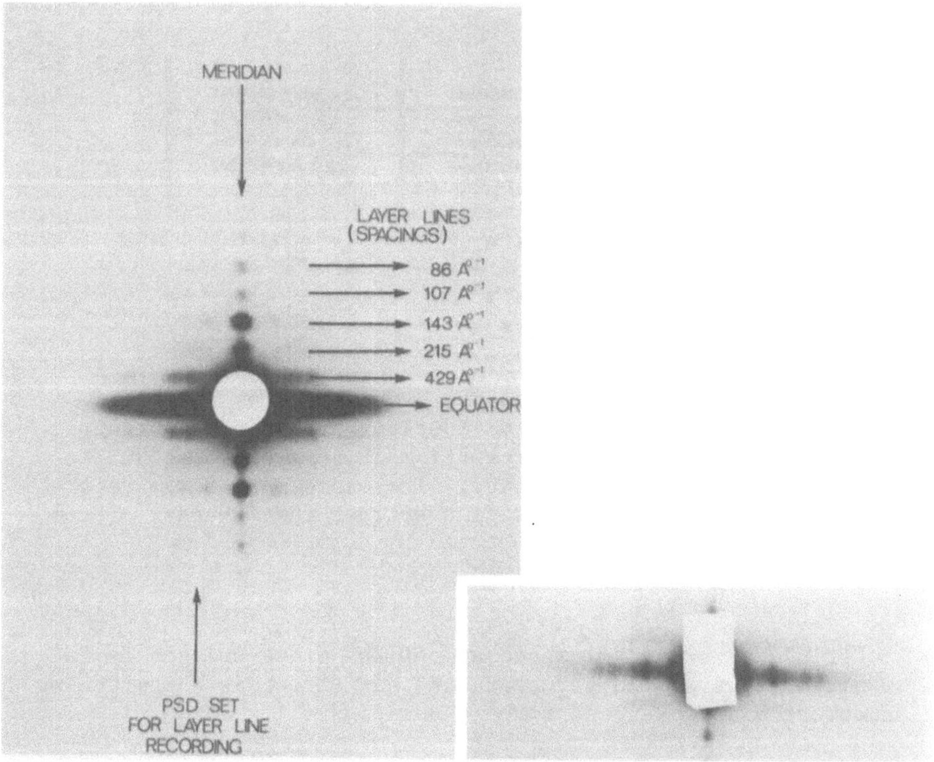

Figure 2. Small angle X-ray pattern from resting frog sartorius muscle showing the layer lines and meridian. The equatorial pattern is shown more clearly in the inset. (From Huxley et al).[7]

lographic plane which can be indexed as (1,0) which gives rise to the strong inner reflection while the (1,1) planes, containing both myosin and actin, give rise to the outer reflection. The equatorial pattern shown in Figure 4, taken with a position sensitive detector, helps to clarify this detail.

Three main changes in the X-ray pattern were observed during contraction[5,6,7]:

(i) the off-meridional layer lines reduced in intensity considerably; this was interpreted as due to an increase in disorder in the regular helical arrangement of cross-bridges on the myosin filament during contraction; due to the asynchronous nature of the cross-bridge cycle, the cross-bridges are distributed over a range of axial positions, i.e. with increased disorder.

(ii) the meridional reflection at 143 Å sometimes reduced also but not to the same extent as the layer lines.

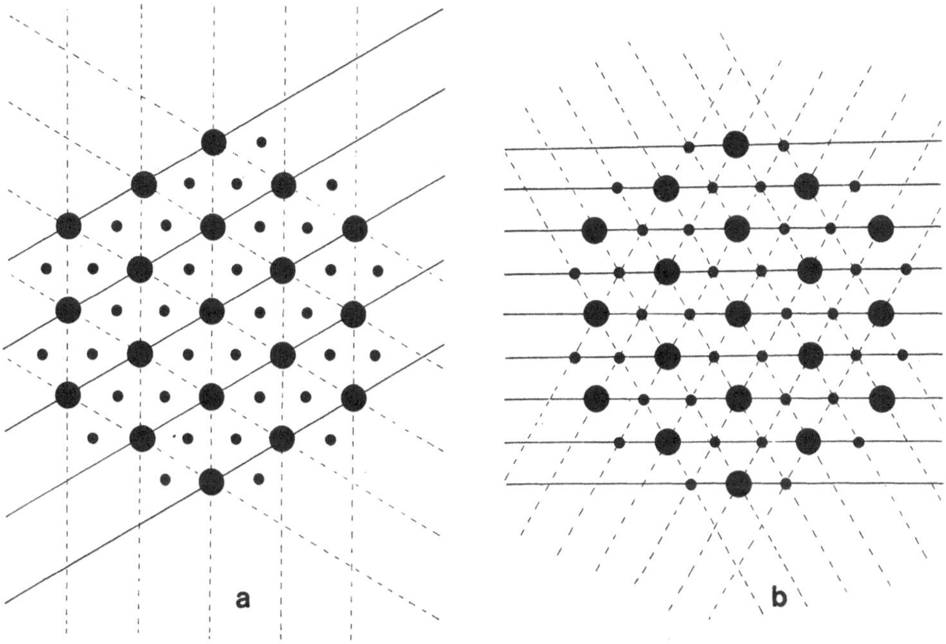

Figure 3. Transverse view of the thick and thin filaments in the overlap region (region A in Figure 1) of striated muscle. The thick filaments are arranged on a hexagonal lattice with the thin filaments occupying the trigonal points. The (1,0) and (1,1) scattering planes are shown in (a) and (b) respectively. During contraction when there is a transfer of mass from the thick to thin filaments, the (1,0) decreases in intensity as mass moves out of the scattering plane but the (1,1) increases in intensity as it contains both the thick and thin filaments and actin-attached bridges. (From Huxley).[8]

(iii) the intensity of the (1,0) equatorial reflection went down while the (1,1) went up; this was interpreted as a transfer of mass from the myosin filaments to the actin filaments, i.e. a movement of cross-bridges from near the myosin to the actin filaments.

We next consider the nature of questions that can be posed, given the existence of the new technology, and the type of experiments which could help to answer some of these questions. This is a brief summary from a more detailed discussion given by Huxley et al.[9]

The main question regarding the cross-bridge cycle relates to whether unequivocal evidence can be obtained that during tension generation cross-bridges attach to actin and go through either a

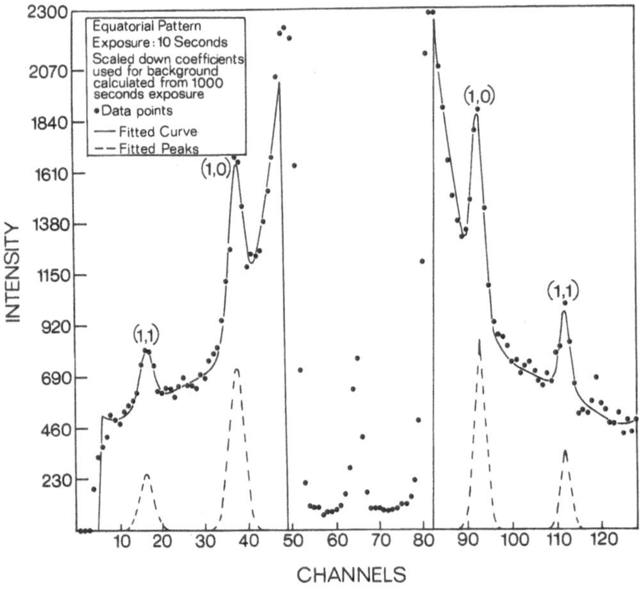

Figure 4. Equatorial pattern from resting frog sartorius muscle recorded with an RC-Delay Line PSD. Solid line is χ^2-minimised fit to data assuming a fourth order polynomial background and gaussian peaks.

tilting motion or some other structural change resulting in the shortening and hence the force generation.[4] A number of secondary questions relate to the measurement of rate constants for various steps in the cross-bridge cycle, viz for attachment and detachment and their correlation with analogous mechanical measurements made by physiologists with good time resolution.[10,11]

The main evidence for the attachment of cross-bridges comes from changes observed in the equatorial pattern and the layer lines during contraction; some of this evidence is discussed in Section 5. More direct evidence would consist of the observation of a rigor-like pattern, i.e. myosin layer lines 'labelled' with the characteristic symmetry of the actin helix during contraction. The change in the angle of attached cross-bridges during the force generating step could be looked for by studying the 143 Å meridional reflection during a quick release, of the type used by A. F. Huxley and Simmons[10] to study tension transients in single muscle fibres.

3. COUNTING RATES AND EXPERIMENTAL REQUIREMENTS

To appreciate the technical requirements for recording time-resolved X-ray patterns from contracting muscle and obtain meaningful results it is essential to find out what counting rates one would expect in the complete X-ray pattern, in selected regions of the pattern, and in individual reflections.[9] The main aim of such a discussion is to make clear the experimental requirements in terms of high intensity X-rays (viz synchrotron radiation), fast electronic detectors for converting the X-ray pattern into digital information and other special electronic circuits for storing information.

The diffraction pattern from muscle, arising from the ordered part of the structure, is superimposed on a relatively high background produced by a small angle scattering from the non-ordered part of the structure and the aqueous medium within the muscle. The background is particularly high for the region corresponding to smaller angles, i.e. corresponding to large spacing reflections. For example, the signal-to-background ratio, i.e. the integrated peak area divided by integrated background area[12], for the (1,0) equatorial reflection, which has a spacing of ∿ 350 Å at resting length, is only ∿ 0.5. The first myosin layer line at 429 Å (as sampled by the detector, see Figure 11) has a signal-to-background ratio of < 0.1. The total counting rates can be calculated from the fact that the fraction of incident X-rays scattered into the 'X-ray pattern' from a typical specimen is roughly ∿ 10^{-4}, the most intense scattering taking place along the equator, the meridian being less intense and the rest of the scattered intensity being along layer lines and in the 'background'.

The flux obtainable from a conventional rotating anode X-ray generator (Elliott GX-13) using a single focusing monochromator (producing a line focus) is ∿ 2×10^8 photons/second[9]; with synchrotron radiation even for double focusing cameras the flux increases by a factor of between 200 - 1000.[7] The counting rates in various parts of the muscle pattern can be calculated for various cameras by scaling the following values given for an incident beam flux of 10^{11} photons/second, a fairly typical figure obtainable for DORIS operation in a dedicated mode for synchrotron radiation experiments at a suitable energy.

Equatorial Pattern (One-dimensional slice)	∿ 4×10^6 counts/sec
(1,0) Equatorial Reflection	∿ 2×10^5 counts/sec
143 Å Meridional Reflection	∿ 4×10^4 counts/sec
429 Å Myosin Layer Line (sampled in one quadrant only)	∿ 4×10^3 counts/sec

Three main hardware requirements of the X-ray detection and Electronic Routing system can now be formulated.

A. X-ray Detector

If a linear position sensitive detector is used to record the equatorial pattern it has to be capable of counting at 2 - 5 MHz;

an area detector would need to count at speeds in excess of 10 MHz. For the weaker parts of the pattern, for example parallel to the meridian across the layer lines the counting rates are much lower.

Good spatial resolution required of laboratory based linear detectors[12] is not so important for synchrotron radiation cameras which usually have a larger focus ($\sim$ 1 mm). Provided it is sufficient to sample the peak profile at about five points, a spatial resolution of $\sim$ 200 μms is adequate.

B. Electronic Routing

Data from the position sensitive detector is normally stored (in the case of a static pattern) in a computer memory; for time-resolved studies where a contraction cycle is broken up into a number of segments, it is necessary to store patterns from successive 'time-frames' in successive portions of the memory, each time frame showing a 'snapshot' of the pattern corresponding to the time when it was taken. A typical contraction may last several hundred milliseconds and the time course of the complete cycle can be studied with relatively poor time resolution, say 10 or 50 milliseconds and the region of particular interest at 1 millisecond. The short time frames impose a stringent requirement on both the detector and the routing electronics. For example, in the case of the 143 Å reflection, $\sim$ 40 counts per time frame (of 1 ms) are expected; the detector has to record the pattern for 1 millisecond then read and store it in the appropriate part of the memory without adding noise or significant dead-time.

C. Data Summation

Even with powerful synchrotron sources it is not yet possible to get statistically significant data from a single contraction on the 1 millisecond time scale. It is possible for the muscle to go through several hundred twitches given adequate 'resting' period between twitches without an unreasonable fall-off in tension generated; the electronics is required to sum the data from successive twitches in the appropriate time frames of the memory.

The essential features of collecting time-resolved X-ray patterns are shown in a schematic form in Figure 5.[13] The basic time synchronisation between the muscle contraction and the data storage is provided by a Time Synchronisation Generator which produces a datum time (t=0) which controls the stimulus of the muscle and the start of data storage in the computer memory. Any other external changes, e.g. length steps, are also synchronised with the datum pulse. The cycle repeats itself automatically at preset intervals corresponding to the muscle resting period.

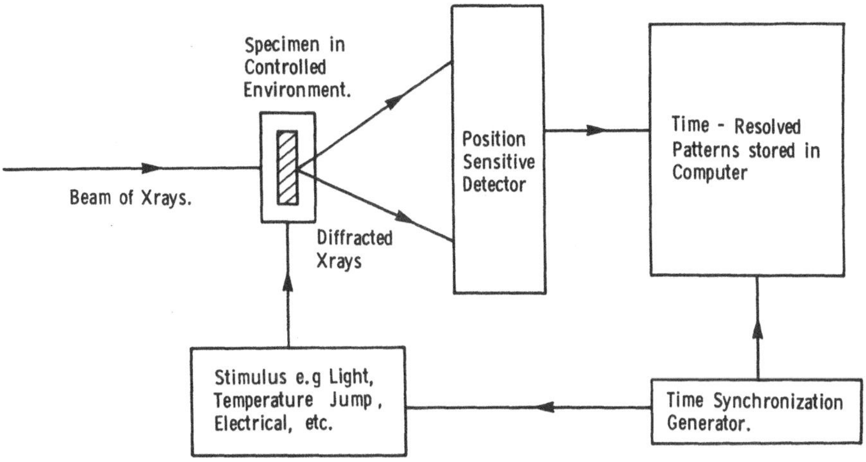

Figure 5. Schematic diagram of data collection system for recording time resolved diffraction patterns.

4. TECHNIQUES

Two relatively new techniques have contributed enormously to the success of muscle time-resolved X-ray measurements: high intensity X-ray sources based on synchrotron radiation and positive sensitive detectors. As mentioned earlier, it is possible to obtain X-ray fluxes from a synchrotron radiation camera which are higher by about a factor of 1000 over the most powerful conventional X-ray generators. The increase in recording speed of X-ray patterns due to the introduction of position sensitive detectors over film methods depends on the type of reflections measured, but is at least a factor of 10 and is even more for the weaker and more diffuse reflections. Combined together, this represents an increase in speed of $\sim 10^4 - 10^5$, a formidable gain.

We have divided techniques into three sub-sections:

(a) Synchrotron radiation. The design of double-focusing cameras, particularly for muscle studies, is described.

(b) Position sensitive detectors. The two types of detectors which have been used for most muscle experiments are described in some detail. A newer type of parallel readout detector is also described as ot shows considerable promise for being able to cope with high rates.

(c) Data analysis. The large amount of data generated requires a considerable amount of interactive computing and curve fitting for semi-automated data analysis; some examples are given to show how this problem has been dealt with.

A. Synchrotron Radiation

As mentioned above the main interest in using synchrotron radiation for muscle X-ray work stems from the fact that X-ray intensities are several orders of magnitudes greater than are obtainable from conventional X-ray generators, allowing exposure times for experiments to be correspondingly reduced.

Some other special features of synchrotron radiation, viz tunability of wavelengths, polarisation of the beam in the plane of the electron orbit or the time-structure due to electron bunching in the machine, have not been made use of in muscle work, as yet.

A number of recent review papers describing the general properties and characteristics of synchrotron radiation and its applications to small angle scattering have been published[14-17]. There are a number of fundamental differences between the source properties of synchrotron radiation as compared to conventional X-ray generators and some of these are reflected in the difference in design of small-angle cameras which might be constructed for doing similar experiments. The main difference between a synchrotron and a conventional source is that, in the former, the emission of X-rays is a continuous spectrum as a function of wavelength whereas in the latter, it is generally confined to the inner electron shell transitions of the target metal used in the X-ray generator (for copper targets the wavelength selected is usually the $K\alpha$ at 1.54 Å). The first essential feature of the synchrotron camera, therefore, is to provide a means of selecting a narrow wavelength band from the incident radiation, before it can be used as a scattering probe. The emission of synchrotron radiation from an electron in orbit is confined to very small angles ($\lesssim 10^{-4}$ radian) relative to the forward direction which implies a high degree of parallelism in the beam, whereas the emission of X-rays from X-ray generators is virtually isotropic. To increase the photon flux it is necessary to maximise the aperture of the synchrotron camera, i.e. to collect as much of the incident beam as possible and after focussing to bring it to as small a focus as possible in the plane of the detector or film. Finally, because of the high levels of background radiation near synchrotron cameras it is necessary to be able to remotely control the focusing and collimating elements. To summarize then, the general requirements for a synchrotron camera are:

(i) it should be capable of selecting a narrow band-pass of wavelength from the essentially 'white' incident beam.

(ii) it should have as large an aperture as possible to maximise the photon flux falling on the specimen.

(iii) it should be capable of focussing in two planes to improve camera resolution and to make it more compact.

The general features of a small angle double-focussing camera for use with synchrotron radiation are shown in Figure 6 which has been taken from Haselgrove et al[18]; cameras based on similar design prin-

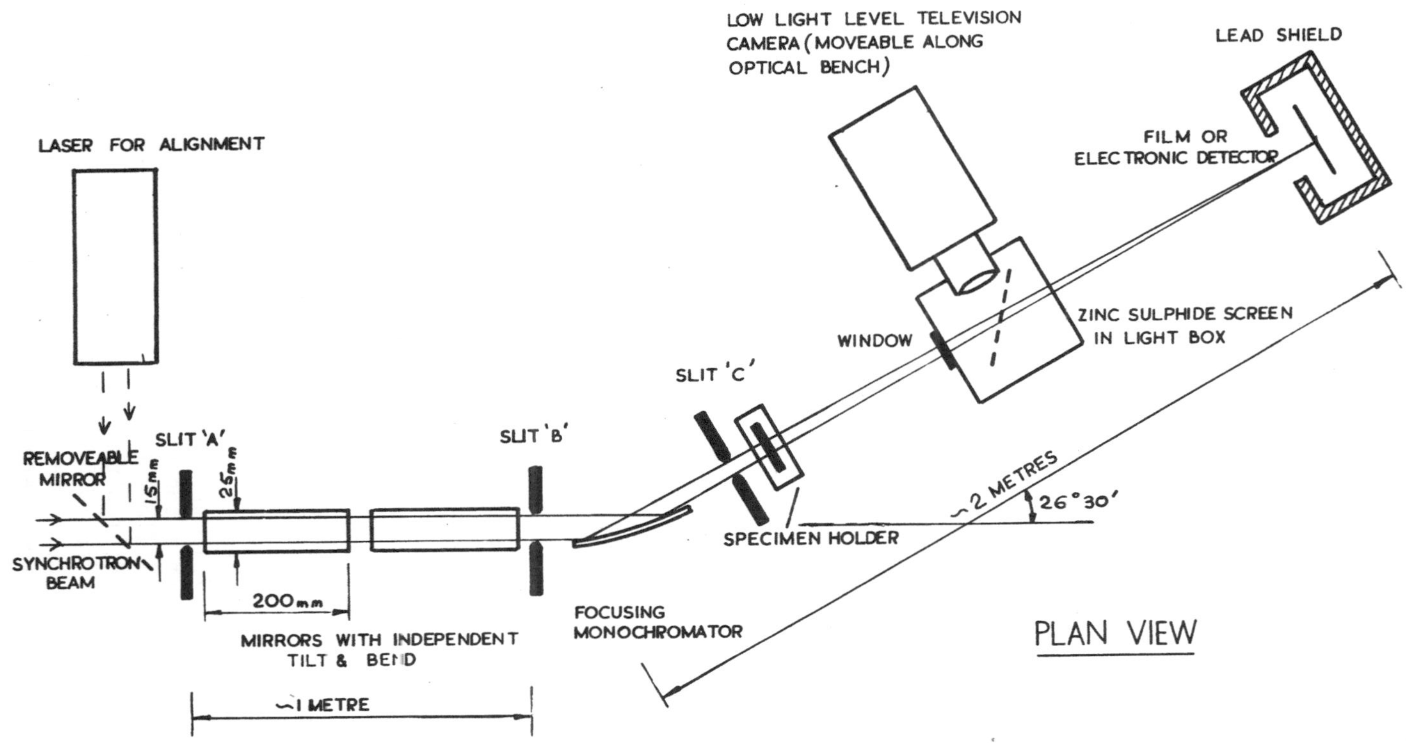

Figure 6. Schematic diagram of a small angle synchrotron radiation camera (From Haselgrove et al).[18]

ciples have been built at Stanford[19] and Hamburg[20,21]. Focussing in the vertical plane is achieved by using total reflection of the beam from two mirrors which are bent to a radius of ∿ 1 km and typical angles of incidence are ∿ 3.5 milli-radian. Under these conditions there is total reflection of wavelengths $\gtrsim$ 1.5 Å, i.e. longer wavelengths than 1.5 Å are transmitted but shorter wavelengths are cut out.[22] The total vertical aperture with two 20 cm mirrors is only 1.4 mm which does not cover the full vertical distribution of the beam; later cameras[21] have used eight sections of mirrors (total length 160 cms) to cover most of the vertical aperture (but without bending). A different approach, adopted by Webb <u>et al</u>[19] is to use a single sheet of float glass (120 cms x 30 cms x 0.5 cms) bent into an asymmetric elliptic profile with the help of eight adjustable shims.

Focussing in the horizontal plane is done by an asymmetrically cut crystal monochromator bent to the required radius of curvature in a crystal bender. Normally crystals of quartz ($10\bar{1}0$ plane) or germanium (111 plane) are used. The Stanford monochromator[19] uses a silicon crystal bent into a logarithmic spiral while a triangular crystal of germanium, bent into the arc of a circle, is used at Hamburg.[21]

Collimation is essential to reduce unwanted scattered radiation; there are three pairs of slits, each consisting of a horizontal and a vertical slit, at strategic points in the camera.

Haselgrove et al reported a flux of 3.6 x 10^8 photons/sec with the synchrotron NINA running at 5 Gev and 15 mA of circulating beam current while in Hamburg[21], a flux of ∿ 5 x 10^{11} photons/sec is reported with the storage ring (DORIS) operating at 4.6 Gev and 20 mA. The increase in flux is partly due to geometrical effects (e.g. increase in vertical aperture, smaller source-to-camera distance) but in the main comes from the basic operating difference between a synchrotron and a storage ring; the former has a poor duty cycle and X-ray intensities are critically dependent upon the mode of electron extraction.[23]

B. Position Sensitive Detectors

The most commonly used position sensitive detector for recording muscle X-ray diffraction patterns until a few years ago (∿ 1972) was simple X-ray sensitive film which can be densitometered to obtain fairly accurate values of integrated intensities. There are three main advantages in using film recording: (i) it is simple and does not require sophisticated electronics (densitometry is another matter, (ii) it can record data at very high rates without saturation effects and (iii) the spatial resolution is excellent, being typically $\lesssim$ 10 µm. However these are offset by some fairly serious disadvantages:

(i) films have to be developed, fixed and dried before any quantitative information can be extracted about the peak intensities, etc. Even for initial camera setting-up purposes, which require minimal information, the 'film processing' delay is most time consuming.

(ii) there is generally an inherent fog level corresponding to O. D. $\sim$ 0.1 on the film and this represents an unavoidable background which makes the measurement of weak or diffuse reflections more difficult and less accurate.

(iii) time-resolved measurements on a millisecond time scale are difficult to envisage with film recording without elaborate 'cine-film' type set-ups.

(iv) the linearity of film response[24] is good only within the range O. D. $\sim$ 0.05 - 1.0 and with special corrections can be pushed up to O. D. $\sim$ 2.0. This corresponds to an intensity dynamic range of only 100:1 which is somewhat restricted and means recording on several films to cover an adequate dynamic range.

Position Sensitive Detectors (PSDs for short) based on electronic methods get around most of the film disadvantages mentioned above, though they cannot normally count at the maximum rates possible with film and the spatial resolution is usually worse. Due to technical limitations in the development of PSDs[47], most of the time-resolved muscle work has been done with linear detectors rather than area detectors; two similar types of position encoding mechanisms have been used and these are discussed in some detail below. A third type of linear detector, based on parallel readout and potentially capable of high counting rates is also described. A television camera-based area detector[25] has been used for some low time resolution work[7] but a discussion of that detector is beyond the scope of this paper.

(1) PSDs based on RC-delay lines. The earliest position sensitive detector used in recording muscle diffraction patterns had RC-delay line position encoding. The principle of this type of detector, originally proposed by Borkowski and Kopp[26] for nuclear physics applications, was introduced to biological work by Gabriel and Dupont[27] and an improved spatial resolution version was built for muscle work by Faruqi.[28]

The principle of operation of the detector relies on substituting a highly resistive anode wire in a 'normal' proportional counter. The high resistance of the anode, in conjunction with the anode-cathode capacitance, forms an RC delay line along the length of the detector. The position of an incident photon is determined in the following way. If a charge Q is injected at a point x along the anode, it results in a current flow to both ends of the anode wire where a voltage is generated across a suitable terminating resistance. The position information is contained in the risetimes of the two anode pulses travelling in opposite directions which, after processing in a suitable time constant filter amplifier and double differentiation produce bipolar pulses. The zero crossing point of the bipolar pulses are proportional to the risetime and can be determined with zero crossing discriminators. The time difference in the two channels is directly proportional to the incident photon's position and it can be digitised with a Time to Amplitude converter and stored in a Multichannel Analyzer or computer.

The great advantage of this type of detector, particularly for normal laboratory based experiments is the excellent spatial resolution that can be obtained, provided a xenon-based gas mixture is used at several atmospheres pressure.[28] Due to the fine focii obtainable with fine-focus X-ray tubes (∿ 100 μm), it is necessary for optimum signal-to-noise ratio to have the detector resolution about the same. For example, it has been shown[12] that if the detector resolution were reduced from 100 μm (fwhm) to 750 μm at a specimen-to-detector distance of about 60 cms, the signal-to-background ratio of the muscle equatorial reflection (1,0) is reduced from 0.5 to about 0.1. With the detector running at normal, i.e. 100 μm (fwhm), resolution, it gives very good patterns. Figure 4 shows a typical equatorial pattern with the prominent (1,0) and (1,1) reflections as discussed in Section 2. The fitted curve is a 4th order polynomial with gaussian peaks using χ^2-minimisation.[12]

The main drawback in using this detector lies in the radiation damage to the high resistance anode wire and consequent loss of resolution and linearity. The usual material used for the high resistance anode wire is carbon-coated quartz fibre; the diameter of the quartz fibre is typically 25 μm and the carbon coating thickness is ∿ 200 Å. It has been reported that the wire becomes damaged during normal use[12,29], probably by etching of the carbon coating on the fibre. This makes the detector very unsuitable for high flux measurements.

(2) PSDs based on LC-delay lines. There are two significant advantages in using LC delay lines PSDs as compared to the RC delay line detectors. First, because the anode wire is a 'normal' metal wire, e.g. gold-plated tungsten wire rather than a special high resistance fibre, the problems of radiation damage are significantly reduced. Second, due to the lower optimum time constants of delay lines it is possible to count at higher rates than in RC detectors.

The general principle of the method is to break up the cathode into segments (or separate wires) and couple these directly or capacitatively to a delay line with suitable pulse transmission properties.[30,32] The induced charge on the cathode is transferred to the delay line and picked up at either end by charge sensitive preamplifiers and amplified with suitable time constants. The difference in arrival time of the signal at the two ends of the delay line is directly proportional to the distance to the point of origin of the charge, i.e. to the position of the incident photon.

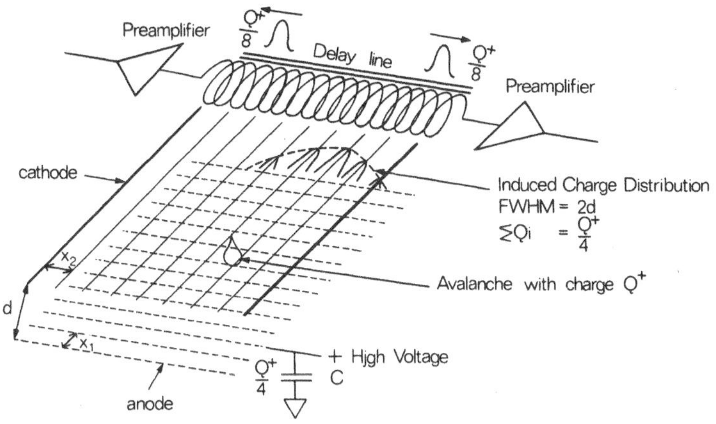

Figure 7. Schematic diagram explaining the generation of induced pulses on the cathode plane and the measurement of the centroid of the avalanche with a delay line (From Lacey and Lindsay).[31]

The mechanism of charge induction on the cathode and its couplin to the delay line is illustrated with the help of Figure 7 from Lacey and Lindsay.[31] The situation shown in the figure represents the charge at about 100 ns after the start of the avalanche, when the total charge in the avalanche is ∿ 10% of the total charge liberated. For normal anode-cathode spacings the image charge splits up into equal amounts on anode and cathode. If all the charge from the cathode can be transferred to the delay line, approximately one-eighth of total charge travels in either direction of the delay line.

For simple imaging of charges on planes, if a single charge is imaged on a plane electrode the charge distribution is expected to be given by

$$\delta(x) \propto \frac{1}{d^2 + x^2}$$

where d is the distance between charge and imaging plane, x is the distance from centroid in the imaging plane, and $\delta(x)$ is the charge per unit area at x. Measurements[31] confirm that the distribution of image charge does indeed follow this distribution. The induced charge from the various cathode wires is transferred to the delay line which converts the spatial charge distribution into a time distribution of charge; this gives an accurate measurement of the charge centroid with the help of suitable charge sensitive amplifiers and constant fraction discriminators. The spacing of the cathode wires is not critical for spatial resolution as they are used merely as charge-sensing devices, several of which contribute to the production of the final delay line pulse. It is quite common to have a spatial resolution ∿ 1/10 of the cathode wire spacing.

A number of linear detectors based on this principle have been used in small angle experiments.[33-35] An example of a muscle resting layer line pattern is shown with the Data Analysis in Figure 11(a) and a contracting layer line pattern in Figure 11(b); in both these patterns a five-point smoothing has been performed.

(3) <u>Parallel readout multiwire linear detectors</u>. To increase the counting capacity of the linear detector one approach is to break up the detector into its individual resolution elements and read them out in parallel.[12,36] For both types of detectors described above viz, the RC and LC-types, each detected photon effectively disables the complete detector until its position has been extracted and digitised, a time period of the order of a few microseconds. For reasonable dead-times this restricts the count rate over the complete detector to ∿ 200 kHz, a figure which is well below the required rates for measurements on the most intense part of the muscle pattern synchrotron cameras as discussed in an earlier section.

We have built a system for rapid acquisition of time-resolved data, based on a 'conventional' Multiwire Proportional Chamber equipped with an amplifier, discriminator and scaler on each wire.[37-39] The system has facilities for recording time-resolved patterns with a time resolution down to 1 millisecond (this can be reduced even further with some additional hardware); the system is shown in schematic form[37] in Figure 8. The timing for the data collection is synchronised by a Master Timer which can be pre-programmed (by software) to set the time slots, number of time slots, stimulus delay, etc.

An important feature of the system is the rapid transfer of data between the scalers and a random access memory. The transfer is made with the help of a special Camac module which acts like an auxiliary controller. This special module is pre-programmed with the Camac addresses of both the scalers and the memory; on receiving the appropriate trigger it goes through a read/write cycle in which it reads the contents of the first scaler and writes it in the memory and goes repetitively through this cycle until all the scaler data has been stored. The scalers have special circuits which enable the data to be latched in an output gate very fast, so that they can continue counting while the data transfer is taking place. This method reduces the dead time between time slots to only a few microseconds. The data acquisition system can be looked upon as a form of 'multi-channel-multiscaler' which has potentially more general applications in experimental situations where a large number of time-varying functions have to be measured simultaneously.

Tests carried out on the system show that it can cope with peak

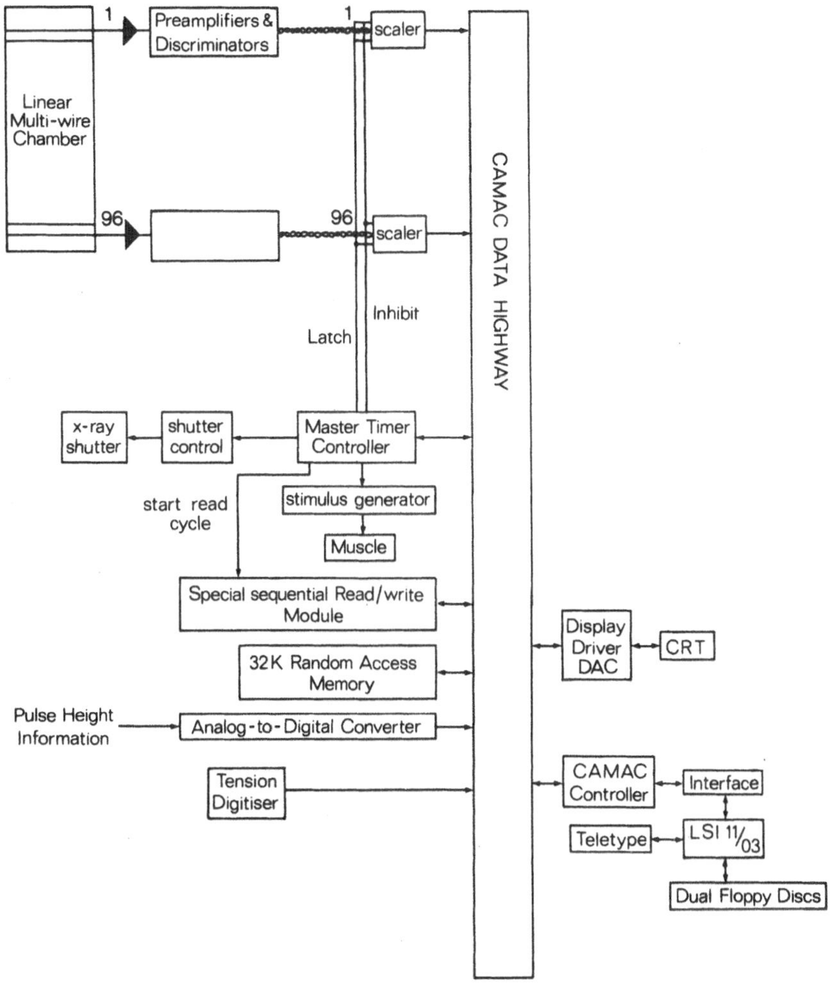

Figure 8. A schematic diagram of the data acquisition system for the multiwire linear detector.

rates of 4 MHz in each counting channel, i.e. with 400 MHz over 100 parallel counting channels, storing the time-resolved data into 100 (or more) separate time bins. The inclusion of the X-ray detector in the test system reduces these rates considerably because of the rate limitations imposed by the physics of the detection process (this figure is normally $\sim 2.10^4$ photons/mm^2/sec). As an example of the typical counting rates possible in the detector, a high voltage setting curve for a single wire is shown in Figure 9. The source of X-rays for this measurement was a (15 mm x 1) beam produced by an asymmetrically cut and bent quartz crystal. It is clear that the plateau is quite a reasonable one and in a separate experiment, using

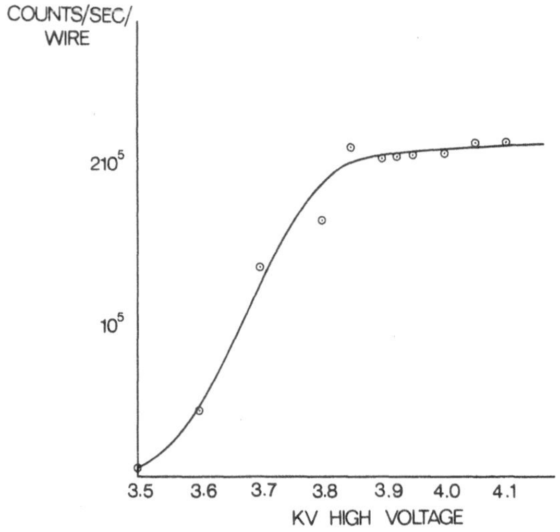

Figure 9. High voltage plateau for multiwire linear detector for 8 kev X-ray photons; gas mixture consists of argon 70% and isobutane 30%.

successively higher attenuation filters, it was confirmed that the inefficiency at the plateau was very small, $\lesssim$ 5%. With the given line-focus geometry, then, each wire can cope with ∿ 200 kHz, i.e. over the complete detector consisting of 100 wires, one can expect to count at ∿ 20 MHz with small losses; this figure is related to the beam geometry and for a 'smaller' beam parallel to the anode wires will be correspondingly reduced due to saturation effects in the detector mentioned above.

As an example of a static pattern recorded on this system, Figure 10 shows a small angle pattern from collagen which has a strong periodicity at ∿ 670 Å. The detector used to record this pattern has anode wire spacing of 1 mm, i.e. the spatial resolution is restricted to 1 mm. In order to improve the spatial resolution we are planning to use smaller spacing wire chambers ($\lesssim$ 0.5 mm) which are coming into use in High Energy Physics.[40]

C. Data Analysis

A common feature of the X-ray patterns, whether recorded along the equator, the meridian or across the layer lines, is a steeply rising background towards the 'centre' of the pattern, i.e. towards low angles; superimposed on the background are the reflections or layer lines which have to be measured, as discussed in Section 2. The main object of the analysis program is to separate the background from the peaks by drawing in a suitable 'background' curve

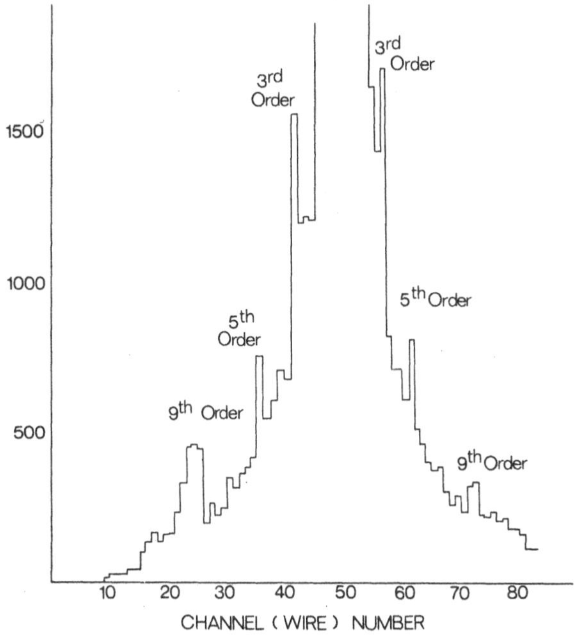

Figure 10. Collagen pattern recorded with the multiwire linear detector.

and integrating the counts within the peak. This has to be done for all the time frames during the twitch and this gives the variation in the peak areas during the contraction cycle. A secondary aim in some of the runs is to record parameters relating to the peak, viz centroid, width, etc. to obtain further information.

The large quantity of data to be analyzed requires some measure of automation in the analysis. To give some idea of the amount of data for processing (we restrict this discussion to data collected on the storage ring DORIS at EMBL, Hamburg)[7,41], each time frame consists of 256 channels of 16-bit data; the number of time frames in a data set varied between experiments from a few to 256. The total number of data sets were ∿ several hundred. Data was stored in binary files and, for reasons of time and memory economy, programs used Direct Access read methods for reading selected portions (i.e. individual time frames) of the data set file.

The computer system used in the analysis, a PDP 11/10 allows a fairly rapid point-mode display of spectra in a Tetronix Storage Oscilloscope (Model 611) and this was used as a basis for a degree of interactive computing as described below. The basic feature of the analysis program is the choice of certain data channels as 'background' through which a polynomial, usually second or third (only occasionally fourth) order was fitted using a standard least-squares fitting procedure. There are several problems in this approach. The background is not constant throughout a data set (i.e. over all the frames representing a contraction cycle) so a perfectly reasonable background curve fitted to the control patterns may not appear valid for some parts of the contracting pattern. This difficulty can be partially overcome for slower changes with improved S/N of individual frames by taking the background averaged from several frames (five was found adequate), two from either side of the frame in question and the frame itself, and smoothing the result. It is not desirable to include more frames than five in situations where the background itself may be going through time-dependent changes; and if the changes are rapid the background has to be calculated for individual frames. The 'signal', i.e. the primary pattern, is not affected by the background computations; the latter is subtracted from the 'signal' to get the 'signal-background' and the integrated peak intensities. It is then relatively simple to calculate the width or centroid of the 'stripped' peak.

We can illustrate the use of this program by an example taken from a data set consisting of the variations in the myosin-related layer line intensities during an isometric twitch.[7] Figure 11(a) shows the 'resting' pattern which is used as a control and Figure 11(b) shows the pattern at the maximum of contraction, i.e. when the peaks are at a minimum. Along with the five layer-line peaks, the two reflections at 59 Å and 51 Å from the actin helix can also be seen and these do not change significantly during the twitch. The intensities of the reflections are integrated during the primary 'analysis' run and stored on disk for further analysis and plotting.

There are two ways of plotting the time course of the recorded intensities. The first method shown in Figure 13(a) shows the changes in the 429 Å layer line (the strongest of the myosin related layer lines) during a twitch in a frog sartorius muscle at 10°C. To correlate with physiological activity the relative tension generated by the muscle is drawn on the same time scale; the arrow indicates the time when a single stimulus was applied. The second way of plotting the intensities is shown in Figure 13(b) which shows the same data as Figure 13(a) but the 429 Å intensity is now plotted on a relative scale to show more clearly the correlation between intensity changes and tension generated.

Only the 429 Å layer line plots are shown to illustrate the method; but any of the other reflections can also be plotted in the same manner. To improve statistics data can be averaged from several data sets, either at the primary spectrum level or at the integrated peak intensity level. Provision is also made in the program for determining the weighted mean of the peak (i.e. peak centre) and its width.

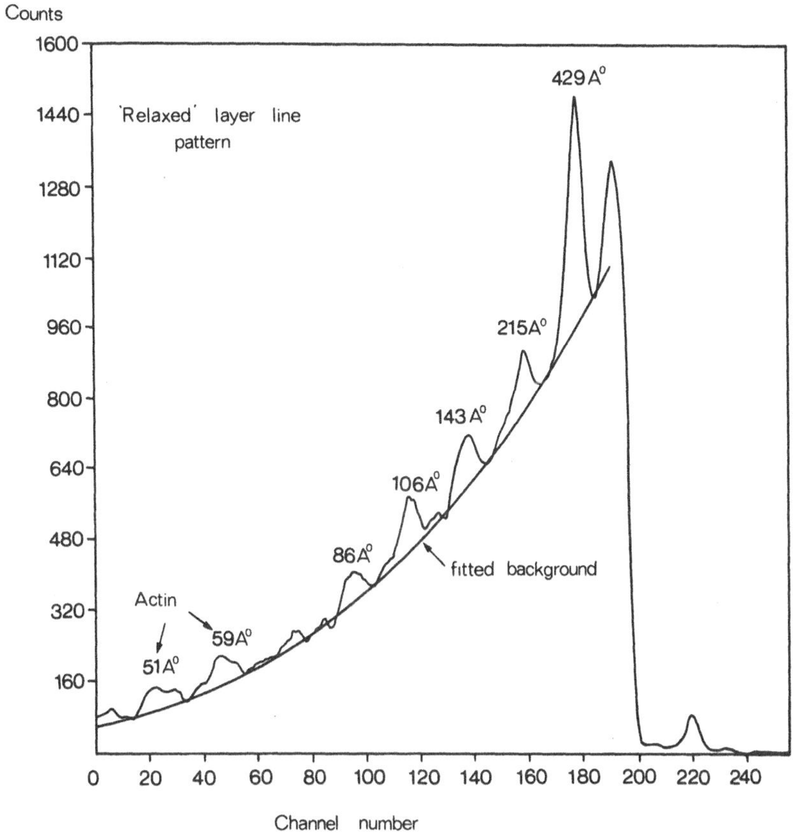

Figure 11(a). Layer line pattern from resting frog sartorius muscle; the five visible myosin layer lines are labelled with the appropriate spacing along with the two actin layer lines. A program-generated background fit is also shown.

A different approach to data analysis, based on χ^2-minimisation[41], has been used previously on equatorial patterns collected on a laboratory based camera.[12] Two main assumptions have to be made:

(i) that it is possible to describe the background by a polynomial function of order n, and

(ii) that the peaks can be described by a well-defined function, such as a gaussian.

Having described the X-ray pattern as a sum of background and peaks in terms of a number of parameters, viz the polynomial coefficients and the parameters describing the gaussian peak, the program proceeds to minimise the differences between the (program) predicted

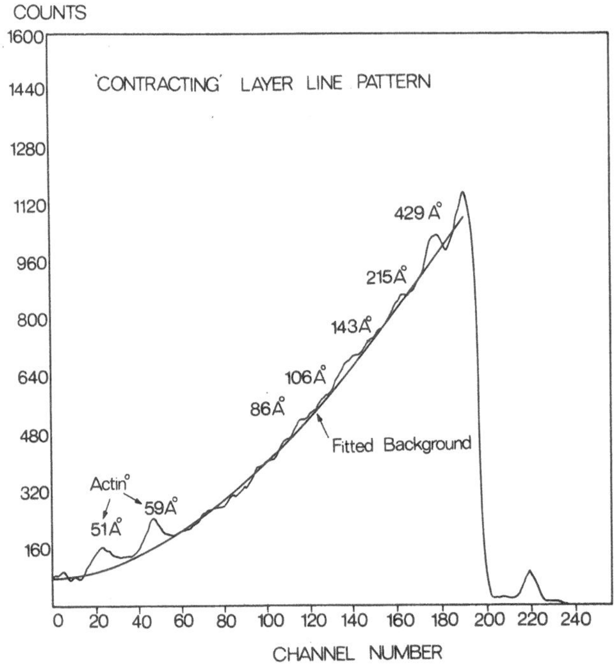

Figure 11(b). Same as in (a) but this time at maximum of contraction.

values for the intensity and the experimental data. The iterative procedure stops when it finds a suitable minimum.

An example of a χ^2-minimised fit to a muscle equatorial pattern is shown in Figure 4; the polynomial coefficients used in this case are scaled-down values from those obtained in a longer exposure to improve the signal-to-noise ratio. The background was chosen to be a fourth order polynomial and the peaks were specified as gaussians. The broken line at the bottom of the plot shows the 'stripped' peaks.

5. EXAMPLES

It is not possible to give even a short summary of all the muscle results obtained with time-resolved X-ray diffraction and we can only refer to some of them in original literature.[42-45] We have chosen just three examples to illustrate the general nature of the findings which hopefully give clues to the mechanism of force generation and answer some of the questions posed in Section 2.

The first example[46] shows the changes in the equatorial pattern (Figure 12) from a frog sartorius muscle during an isometric twitch at 2°C. The changes in both the (1,0) and the (1,1) reflections are

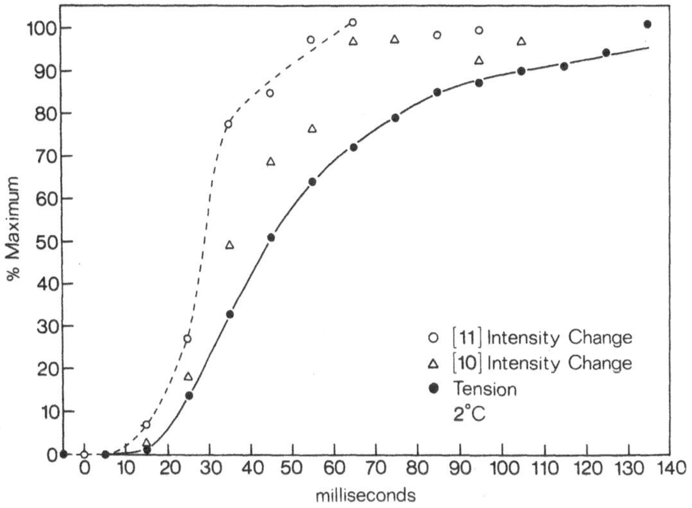

Figure 12. Relative changes in the equatorial reflections (1,0) and (1,1) during a twitch at 2°C along with the tension. (From Huxley).[46]

plotted as relative changes, which for convenience are shown with the same sign; in fact, the (1,0) decreases and the (1,1) increases in intensity during a twitch. This diagram shows that the structural change precedes the tension generation by 10-20 ms, which possibly represents a rate constant limiting the force generation after the cross-bridges have attached to actin.

The second example is taken from time-resolved study of myosin layer lines with the help of synchrotron radiation.[7] Figure 13(a) shows the variation in the strongest layer line (429 Å) during a twitch at 10°C. It is more convenient to plot the intensity changes (as for the equatorials) on a relative scale along with the tension; these changes are shown in Figure 13(b). The time course of the intensity changes appear to be well correlated with the tension generated and might be just slightly ahead in time. The interpretation of such a change is that during the onset of contraction, cross-bridges rapidly go through their cycle of attachment and detachment with consequent loss of the myosin helical pattern. At the end of the twitch the cross-bridges resume their helical positions and the pattern quickly returns to the 'resting' state.

The third example is also taken from Huxley et al[7]; it concerns the time course of the 143 Å meridional reflection during an isometric twitch at 10°C as shown in Figure 14. The intensity, which also decreases during the twitch, is again plotted on a relative scale to show more clearly the correlation with tension generation. An

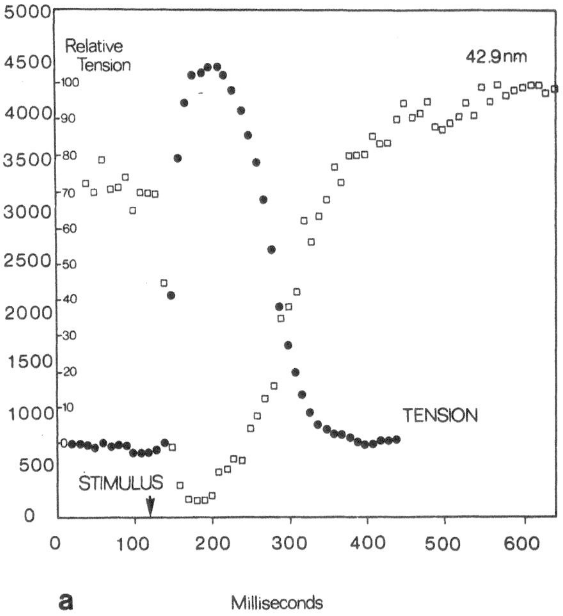

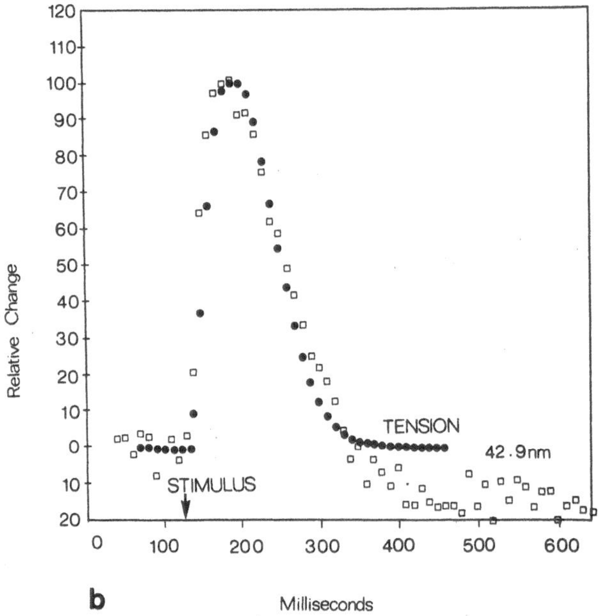

Figure 13. (a) Changes in the 429 Å myosin layer line from a frog sartorius muscle during a twitch at 10°C. (b) Relative changes in the 429 Å intensity during a twitch. There is good correlation between intensity and tension changes.

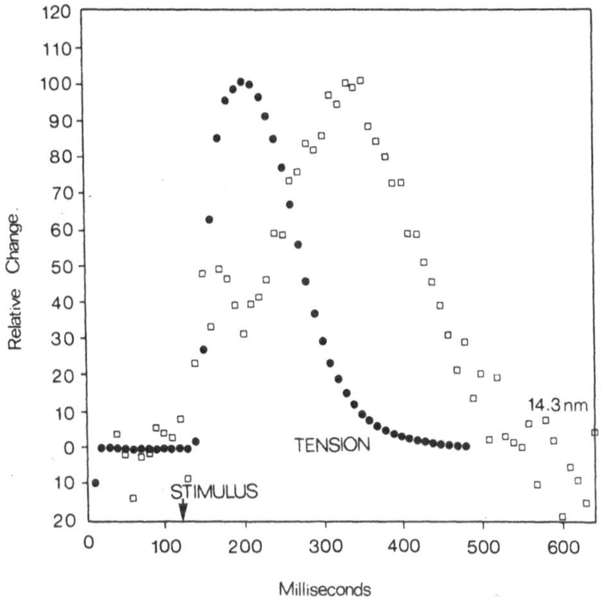

Figure 14. Relative changes in the 143 Å meridional intensity from a frog sartorius muscle during a twitch at 10°C.

interesting effect seen in this plot is that although the initial decrease in the 143 Å intensity is similar to that of the layer lines, it makes a partial recovery with a later large decrease in intensity after the tension has returned to its normal resting value. This 'bi-phasic' behavior has been interpreted as being possibly due to disordering in the myosin filaments during the relaxation phase; this would affect the 143 intensity far more than the 429 layer line intensity.

ACKNOWLEDGEMENTS

We are grateful for discussions and suggestions regarding the software for synchrotron data analysis to Drs. T. S. Horsnell and M. H. J. Koch.

REFERENCES

1. H. E. Huxley, in "The Structure and Function of Muscle", (Academic Press, New York, 1972). pp. 301-387.
2. J. M. Squire, Ann. Rev. of Biophys. and Bioeng. 4: 137 (1975).
3. Cold Spring Harbor Laboratory, The mechanism of muscle contraction, Cold Spring Harbor Symposia on Quantitative Biology 32: (1973).
4. H. E. Huxley, Science 164: 1356 (1969).

5. H. E. Huxley and W. J. Brown, J. Mol. Biol. 30: 383 (1967).
6. H. E. Huxley, J. Mol. Biol. 37: 507 (1968).
7. H. E. Huxley, A. R. Faruqi, J. Bordas, M. H. J. Koch, and J. R. Milch, Nature 284: 140 (1980).
8. H. E. Huxley, Proc. Roy. Soc. London Ser. B. 178: 131 (1971).
9. H. E. Huxley, A. R. Faruqi, J. Milch, and J. Bordas, "Proceedings of Study Weekend 26-28 Jan (1979) Daresbury Laboratory", R. B. Cundall and I. Munro, eds.
10. A. F. Huxley and R. M. Simmons, Nature 233: 533 (1971).
11. A. F. Huxley and R. M. Simmons, Cold Spring Harbor Symposia on Quantitative Biology 37: 669 (1972).
12. A. R. Faruqi and H. E. Huxley, J. Appl. Cryst. 11: 449 (1978).
13. A. R. Faruqi, Nucl. Instr. Meth. 156: 19 (1978).
14. H. B. Stuhrmann, Q. Rev. Biophys. 11: 71 (1978).
15. K. R. Lea, Phys. Letters (Section C) 43: 337 (1978).
16. H. Winick and A. Bienenstock, Ann. Rev. Nucl. Part. Sci. 28: 559 (1978).
17. G. N. Kulipanov and A. N. Skrinskii, Sov. Phys. Uspekhi 20(7): 559 (1978).
18. J. C. Haselgrove, A. R. Faruqi, H. E. Huxley, and U. W. Arndt, J. Phys. E. (Sci. Instr.) 10: 1035 (1977).
19. N. G. Webb, S. Samson, R. M. Stroud, R. C. Gamble, and J. D. Baldeschwieler, J. Appl. Cryst. : 104 (1977).
20. J. Barrington-Leigh and G. Rosenbaum, J. Appl. Cryst. 7: 117 (1974).
21. J. Hendrix, M. H. J. Koch, and J. Bordas, J. Appl. Cryst. 12: 467 (1979).
22. A. Franks and P. Breakwell, J. Appl. Cryst. 7: 122 (1974).
23. A. R. Faruqi and J. C. Haselgrove, J. Phys. E. (Sci. Instr.) 10: 1042 (Appendix) (1977).
24. H. Morimoto and R. Uyeda, Acta Cryst. 16: 1107 (1963).
25. G. T. Reynolds, J. R. Milch, and S. M. Gruner, Rev. Sci. Instr. 49: 1241 (1978).
26. C. J. Borkowski and M. K. Kopp, Rev. Sci. Instr. 39: 1515 (1968).
27. A. Gabriel and Y. Dupont, Rev. Sci. Instr. 43: 1600 (1972).
28. A. R. Faruqi, J. Phys. E. (Sci. Instr.) 8: 633 (1975).
29. A. R. Faruqi, IEEE Trans. Nucl. Sci. NS-22: 2066 (1975).
30. A. Rindi, V. Perez-Mendez, and R. J. Wallace, Nucl. Instr. Meth. 77: 325 (1970).
31. J. L. Lacey and R. S. Lindsay, Nucl. Instr. Meth. 119: 483 (1974).
32. P. Lecomte, V. Perez-Mendez, and G. Stoker, Nucl. Instr. Meth. 153: 543 (1978).
33. H. Hashizume, K. Mase, Y. Amemiya, and K. Kohra, Nucl. Instr. Meth. 152: 199 (1978).
34. A. Gabriel, Rev. Sci. Instr. 48: 1303 (1977).
35. D. Ortendahl, V. Perez-Mendez, J. Stoker, and W. Beyermann, Nucl. Instr. Meth. 156: 53 (1978).
36. R. Kahn and R. Fourme, Abstract 14.2-2, Acta Cryst. : 5330 (1978).

37. A. R. Faruqi and H. E. Huxley, Proceedings of the 4th Taniguchi Symposium on Biophysics : 367 (1978).
38. A. R. Faruqi, IEEE Trans. Nucl. Sci. NS-27: 644 (1980).
39. A. R. Faruqi and C. C. Bond, Nucl. Instr. Meth. (in press).
40. "Proceedings of the Wire Chamber Conference", W. Bartl and M. Regler, eds., (North Holland, Amsterdam, 1978). A similar volume for the 1980 Proceedings has also been published.
41. F. James and M. Ross (1975), CERN DD/75/20.
42. I. Matsubara, I. Yagi, and H. Yagi, J. Physiol., Lond. 278: 297 (1978).
43. R. J. Podolsky, R. St. Onge, L. Yu, and R. W. Lymn, Proc. Nat. Acad. Sci. 73: 813.
44. N. Yagi, M. H. Ito, H. Nakajima, T. Izumi, and I. Matsubara, Science 197: 685 (1977).
45. H. Sugi, Y. Amemiya, and H. Hashizume, Proc. Japan Acad. 54(B): 559 (1978).
46. H. E. Huxley, in "The Molecular Basis of Force Development in Muscle", N. B. Ingels, ed., (Palo Alto Medical Research Foundation, 1979).
47. A. R. Faruqi, in "The Rotation Method in Crystallography", U. W. Arndt and A. J. Wonacott, eds., (North Holland, Amsterdam, 1977). pp. 227-243.

PART II

Polymer And Colloidal Systems

STATIC AND DYNAMIC PROPERTIES OF POLYMER SOLUTIONS*

B. Chu

Chemistry Department
State University of New York at Stony Brook
Long Island, New York 11794

CONTENTS

ABSTRACT

Applications of photon correlation spectroscopy and angular distribution of absolute scattered intensity to the studies of dynamic and static properties of polymer solutions in terms of large scale motions, such as translational motion of the polymer coil, pseudo-gel motion and internal motion, are reviewed. In particular, the asymptotic nature of the scaling concept is emphasized with examples based on measurements of the polystyrene-trans-decalin system.

1. INTRODUCTION

Our understanding of the statistical properties of flexible linear polymers in solution has been advanced significantly in recent years. These advances are made possible partly because of the availability of modern scattering techniques in small angle

*Work supported by the U.S. Army Research Office and the National Science Foundation.

neutron scatterings (SANS), small angle x-ray scattering (SAXS) and light-scattering spectroscopy - the main topic of this NATO Advanced Study Institute (ASI); and partly because of the discovery by de Gennes of an important relationship between polymer statistics and critical phenomena, leading to many simple asymptotic scaling laws.

The main theme of my two talks deals with studies of the static and dynamic properties of flexible linear polymers in solution by means of light-scattering spectroscopy.

As the wavelength of visible light in a scattering medium is a few hundred nanometers which can probe local concentration fluctuations in a polymer solution at spatial scales of order K^{-1} (ranging from about 1000 Å to a few microns), the angular distribution of scattered light measures mainly large-scale averaged spatial correlations, such as the radius of gyration, r_g. The scattering wave vector $K(=(4\pi/\lambda)\sin(\theta/2)$ with λ and θ being the wavelength of incident radiation in the scattering medium and the scattering angle, respectively) for SANS and SAXS is larger: for most experiments K^{-1} ranges from less than 1000 Å to tens of Angstroms. Therefore, SAXS and SANS probe smaller scale spatial correlations of local concentration fluctuations.

The dynamical motions of flexible polymers in solution are complex. These could include translational motions of the entire polymer coil ($Kr_g << 1$), internal modes of large polymer coils ($Kr_g >> 1$), motions of entangled polymer coils ($C > C^*$, with C^* being the overlap concentration) and local segmental motions involving conformational changes related to only a small number of monomers. For large scale motions where the whole polymer chain is approximated as beads connected by springs, the characteristic times ranging from $1 - 10^{-6}$ sec can be measured by means of photon correlation spectroscopy. For local segmental motions where atomic structure begins to play a significant role, the frequencies involved are in the GHz range and require the use of Fabry-Perot interferometry or inelastic neutron scattering. With the recent spin-echo method which will be discussed at this NATO ASI, frequencies down to $2 \times 10^8 \text{sec}^{-1}$ have been achieved. I shall limit my discussions to large scale motions of linear coiling macromolecules in solution. Rotational motions of rigid particles involving depolarized Rayleigh scattering will also be reviewed at this ASI.

In polymer solution dynamics, a variety of new theoretical methods, such as the renormalization group approach, and experimental techniques, such as photon correlation spectroscopy, are routinely used. These disciplines are unfamiliar to many of the polymer scientists. Therefore, my talks aim to provide a general introduction of our present understanding of static and dynamic properties of linear flexible polymers in solution with emphasis on the application of the histogram method[1-4] to studies of the structure and dynamics of entangled polymer coils near theta condition in semidilute solutions.[5-10] The book on scaling concepts in polymer physics by De Gennes[11] is highly recommended and requires very careful reading. It is particularly simple in appearance and presents the scaling concepts in phenomenological terms that are understandable. I shall make no attempt to recite all the references but shall mention only pertinent ones which include many articles representing the joint activities at Strasbourg (CRM), Saclay, and College de France (STRASACOL).[12] In semidilute solutions the readers should refer to the initial papers[13,14] as well as more recent developments[15,16] by the French groups; and in dilute solutions, the paper by Azcasu and Han[17] represents recommended readings.

2. PHASE DIAGRAMS

Phase diagrams of polymer solutions have been classified into two types of cloud-point curves in the temperature-concentration diagram as shown in Fig. 1. The upper critical mixing point (UCP) moves toward lower concentrations and higher temperatures with increasing molecular weight of the polymer and coincides with the Flory theta (Ⓗ) temperature at infinite molecular weight and zero concentration. The other type of cloud-point curve lies at temperatures in the neighborhood of the gas-liquid critical temperature of the solvent. Immiscibility occurs if the temperature of the polymer solution is raised above the lower cloud-point curve (LCP) or lowered below the upper cloud-point curve. The Flory theta temperature is referred to as the limit of the upper critical solution temperature at infinite molecular weight. Another "theta" temperature may be related to the lower critical solution temperature at infinite molecular weight. Therefore, although the solvent quality becomes better as we increase the temperature of the polymer solution from $\tau = 0$, there is a limit in the goodness of the solvent quality. As we move closer to the LCP, its influence becomes more dominant with further increases in temperature. Then, the solvent quality becomes worse again as shown in Fig. 1. The region of complete miscibility for some polymer/solvent systems may be quite limited. Consequently, it is important to note that good solvent quality may never be reached by simply increasing the temperature of polymer solutions. In some cases, the upper critical solution temperature (UCST) and the lower critical solution temperature (LCST) are merged to give an "hour glass" shape in the polymer-poor solvent system, such as polystyrene-acetone.[18] Values of pairs of UCST and LCTS for polymer/solvent systems have been reported.[19-23] The existence of UCST and LCST pairs is a very widespread phenomenon in polymer solutions and has been discussed in terms of the Patterson-Delmas theory[24] of corresponding states.

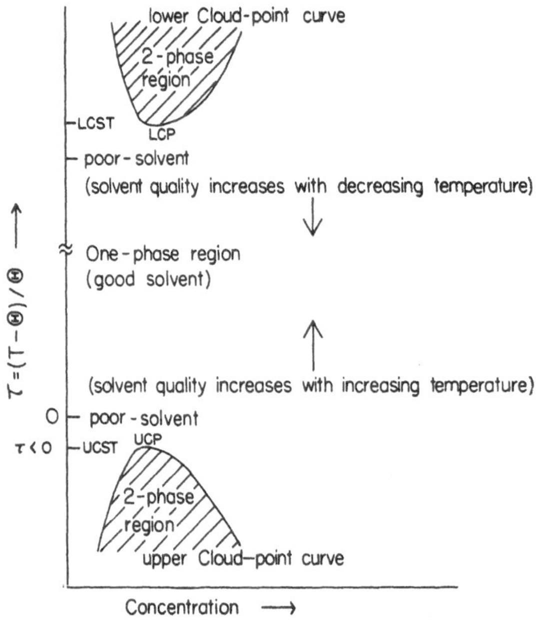

Figure 1. Schematic phase diagram of a monodisperse polymer solution with upper and lower critical solution temperatures. UCST and LCST denote the upper and the lower critical solution temperatures, respectively. UCP and LCP denote the upper and the lower critical mixing points, respectively. The reduced temperature $\tau = 0$ at the theta Ⓗ temperature. The solvent quality becomes better as we increase the temperature from $\tau = 0$. However, the influence of the LCP becomes more dominant as we move closer to LCP by further increases in temperature.

Figure 2 shows a schematic temperature concentration diagram of Daoud and Jannink. In interpreting this figure, I shall provide a few comments which may prove to be helpful in our understanding of the cross-over in polymer solutions.

1. Figure 2 represents a schematic temperature-concentration diagram far away from the lower critical mixing point. It assumes that we can change a theta solvent to a good solvent by increasing only the temperature.

2. Regions I' and III represent near theta condition while regions I and II denote good solvent condition. However, the change of solvent quality from poor (theta) to fair and then good is not sharp. In some cases, good solvent quality with dominant excluded volume effects may not be attainable.

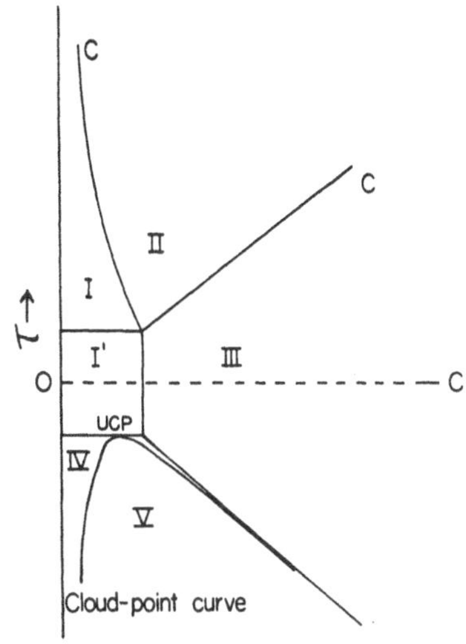

Figure 2. Schematic temperature-concentration diagram of a monodisperse polymer solution[25] far away from the LCP. $\tau = (T-\Theta)/\Theta$. Regions I, I' and IV: dilute solution range. Regions II and III: semidilute solution range. Region V: 2-phase region. I' and III are near theta condition. In the vicinity of UCP and along the cloud-point (coexistence) curve, critical fluctuations occur. Polymer coils expand with an increasingly more positive second virial coefficient A_2 as we increase the temperature above $\tau = 0$.

3. Although polymer statistics of entangled systems in the semidilute region is amenable to scaling laws[11,14] these laws are valid asymptotically for long chains, in very good solvents, and in a semidilute regime where the chains do overlap, but where the concentration is still low enough, so that the friction processes remain dominated by the solvent viscosity.[26]

4. The overlap concentration C^* [$=M/N_A\rho_s r_g^3$ with M, N_A and ρ_s being the polymer molecular weight, the Avogadro number and the solvent density, respectively] is defined arbitrarily as polymer coils begin to overlap at concentrations much lower than C^*.

5. The crossovers from dilute to semidilute solutions, especially near theta conditions, are not sharp. Therefore, except for the cloud-point curve in Fig. 2, the regions (I-IV) do not have sharp boundaries for polymers of finite molecular weights. Nevertheless, the temperature concentration diagram is extremely

useful in providing us with a semiquantitative description of well defined regions where polymer solutions behave distinctly. We shall follow Fig. 2 in discussing the static and dynamic properties of polymers in solution.

3. PAIR CORRELATION FUNCTION AND STATIC STRUCTURE FACTOR

The pair correlation function for a set of N beads (or segments), representing the monomers of a single polymer, each located at a position $\vec{r}_i$ $(i = 1,\ldots,N)$ has the form[15]

$$g(\vec{r}) = \frac{1}{N^2} \sum_{ij}^{N} \langle \delta(\vec{r} - \vec{r}_i - \vec{r}_j) \rangle \tag{3.1}$$

where < > denotes the average over all configurations with

$$\langle r_g^2 \rangle = \int r^2 g(r)\, d\vec{r} \Big/ \int g(r)\, d\vec{r} \tag{3.2}$$

The Fourier transform of $g(r)$

$$S(\vec{K}) = \frac{1}{2\pi} \sum_{ij}^{N-1} \langle e^{i(\vec{K}\cdot(\vec{r}_i - \vec{r}_j)} \rangle \tag{3.3}$$

is directly measurable by means of SAXS, SANS, and light scattering. Alternatively, for a single polymer chain of N segments, Eq. (3.3) can be written as

$$S(\vec{K}) = \sum_{n=0}^{N-1} (N-n) \langle \cos \vec{K} \cdot \vec{r}_n \rangle \tag{3.4}$$

where the running index n measures the distance between two beads, (i,j) along the polymer chain. The parameter n plays an important role in the crossover behavior of $S(\vec{K})$ as the K dependence of $S(\vec{K})$ depends on the scaling of $\langle r_n^2 \rangle$ with n. The simple blob concept assumes that there exists a particular n, called n_τ, such that the scaling below n_τ is different from the scaling above n_τ. Then,

$$S(\vec{K}) = S^{I}(\vec{K}) + S^{II}(\vec{K}) \tag{3.5}$$

where

$$S^{I}(\vec{K}) = \sum_{n=0}^{n_\tau - 1} (N-n) \langle \cos \vec{K} \cdot \vec{r}_n \rangle \tag{3.6}$$

$$S^{II}(\vec{K}) = \sum_{n=n_\tau}^{N-1} (N-n) \langle \cos \vec{K} \cdot \vec{r}_n \rangle \tag{3.7}$$

At $K > (\langle r_n^2 \rangle)^{-1/2} = K^*$, $S^{I}(\vec{K}) > S^{II}(\vec{K})$. Conversely, $S^{II}(\vec{K})$ is dominant in the range $K < K^*$. In the simple blob theory, we have assumed that the pair correlation function jumps from Gaussian to excluded volume statistics at $n = n_\tau$. In reality, the crossover from Gaussian to excluded volume statistics is not sharp and a modified blob model has been proposed.[27]

4. STATIC PROPERTIES OF POLYMER SOLUTIONS

In discussing the static and dynamic properties of polymers in dilute solutions, we are particularly interested to examine the scaling behavior in good and theta solvents as well as the crossover behaviors, e.g., the temperature and the molecular weight dependence of those properties according to regions I, I', and IV of Fig. 2; yet at all times we must remember that the scaling laws are asymptotic formulas valid for polymers at infinite molecular weight.

4.1 End-to-End Distance and Thermal Blob Size in Dilute Solution

At infinite dilution, we deal with an isolated polymer chain of N beads (or monomers). According to the virial theory, the excluded volume v has the form:

$$v = \int (1 - e^{-V(r)/kT}) d\vec{r} \tag{4.1}$$

where V(r) is the interaction energy of a pair of dissolved monomers separated by a distance r. $v = 0$ at the theta (Θ) temperature signifying the disappearance of the effective two-body interaction between monomers. v is positive above the Θ temperature and negative below the Θ temperature. For T near Θ, v has the approximate form

$$v \sim a^3 (T - \Theta)/\Theta = a^3 \tau \tag{4.2}$$

where a^3 is the monomer volume. However, the excluded volume v swells the coil only if $v/a^3 \gg N^{-1/2}$,[28] or

$$\tau \sim (v/a^3) \gg N^{-1/2} \tag{4.3}$$

Eq. (4.3) provides us with a crossover condition for the blob concept, i.e., there is a characteristic n_τ with $n_\tau = \tau^{-2}$ such that all distances $\langle r_n^2 \rangle (n \geq n_\tau)$ are swollen and all distances $\langle r_n^2 \rangle (n \leq n_\tau)$ are unperturbed obeying Gaussian statistics. For an ideal chain and for an excluded volume chain with $n_\tau = 1$, the mean end-to-end distance of the polymer coil is given by

$$<R_{1N}^2> = N^{2\nu}a^2 \tag{4.4}$$

where ν = 1/2 and 3/5 for ideal (theta) and good solvents, respectively. However, as $n_\tau = \tau^{-2}$, we have blobs, each with size of

$$\xi_\tau^2 = <r_{n_\tau}^2> = n_\tau/a^2 \tag{4.5}$$

and the mean end-to-end distance of the polymer coil is

$$<R_{1N}^2> = (N/n_\tau)^{2\nu} <r_{n_\tau}^2> \tag{4.6}$$

where (N/n_τ) is the number of thermal blobs per chain as shown in Fig. 3(a). By substituting $n_\tau = \tau^{-2}$ and Eq. (4.5) into Eq. (4.6), we get

$$<R_{1N}^2> = N^{6/5}\tau^{2/5}a^2 \tag{4.7}$$

for ν = 3/5.

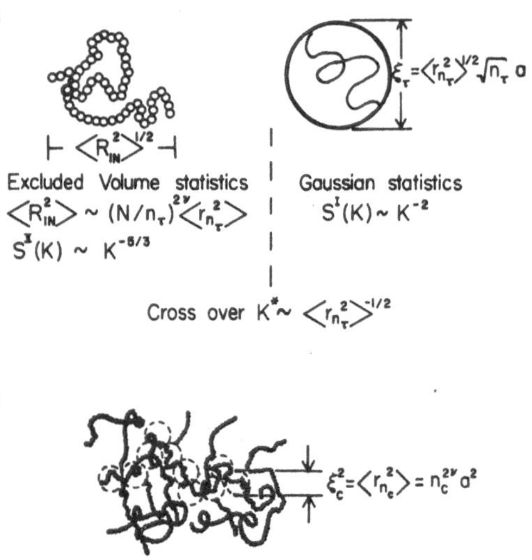

Figure 3a. An isolated chain of (N/n_τ) thermal blobs. Thermal blob size ξ_τ: $<r_{n_\tau}^2> = n\ a^2H$

Figure 3b. A particular chain in the semidilute regime with mean distance between interchain contacts ξ_c: $\xi_c^2 = <r_{n_c}^2> = n_c^{2\nu}a^2$

4.2 End-to-End Distance and Mesh Size in Semidilute Solution

As we increase the concentration, we reach an overlap concentration C*, arbitrarily defined as $M/N_A\rho_s r_g^3$, where the coils begin to entangle, as shown schematically for the three concentration ranges in Fig. 4. In the semidilute regime, the excluded volume parameter v is screened beyond a characteristic value n_c by the finite density of coil segments. Then, the chain behaves ideally. Figure 3b shows one particular chain in the semidilute regime. The mesh size or the main distance between interchain contacts ξ_c decreases with increasing concentration. Inside one blob (or for distances smaller than ξ_c and larger than ξ_τ), the chain does not interact with other chains, the pair correlation function obeys the excluded volume statistics and the distances $<r_n^2>^{1/2}$ for $n \leq n_c$ are swollen. For $n > n_c$, the distances $<r_n^2>^{1/2}$ remain unperturbed obeying Gaussian statistics in semidilute solutions.

The mesh size ξ_c can be defined as

$$\xi_c^2 = n_c^{2\nu}a^2 \tag{4.8}$$

and the mean end-to-end distance of the particular chain in the semidilute regime has the form

$$<R_{1N}^2> = (\frac{N}{n_c})\xi_c^2 \tag{4.9}$$

As $n_c \sim c^{-5/4}$, we get

$$<R_{1N}^2> \sim Nc^{-1/4}a^2 \tag{4.10}$$

where c is the monomer concentration.

In semidilute good-solvent solutions, there are two characteristic correlation lengths ξ_τ (Eq. (4.5)) and ξ_c (Eq. (4.8)) as shown schematically by Fig. 3. As n_c is normally greater than n_τ, $\xi_c(=n^\nu{}_c a)$ is greater than $\zeta_\tau(=n_\tau{}^{1/2}a)$. For real systems, we can have a concentration $\tilde{c}$ such that $\xi_c = \xi_\tau$. As $c^* \sim N/<R_{1N}^2>^{3/2}$ with $<R_{1N}^2>$ obeying Eq. (4.7) in a good solvent, $\xi_c = a(a^3)^{-3/4}\tau^{-1/4}$. Then, with $\xi_\tau = a/\tau$, we have

$$\tilde{c} = \tau/a^3 \tag{4.11}$$

In a good solvent, this condition can be met only at fairly high concentrations, i.e., $\tilde{c} > c^*$, since $c \sim 1/a^3$ is the crossover to the concentrated solution and $\tau \sim 1$ is in the good solvent

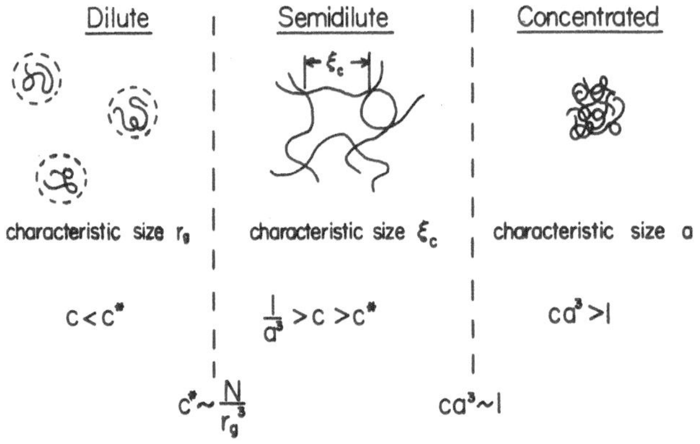

Figure 4. Schematic presentation of dilute, semidilute and concentrated solution. C = number of monomers over cm^3. The regions are not sharp. More generally, we should use the statistical segment length b instead of a.

region. In the neighborhood of $\tilde{c}$, there are two correlation lengths of comparable magnitude. Therefore, the simple scaling laws should break down. For $c > \tilde{c}$, the two blobs, as shown in Figs. 3a and 3b, have merged and the polymer chain should have a near ideal behavior. Schaefer, Joanny and Pincus[29] have referred the regime $\tilde{c} < c < a^{-3}$ as a marginal semidilute region with solvent quality exhibiting $\tilde{c} \sim c^*$. There the excluded volume interactions are weak and can be treated as a perturbation to a Gaussian coil. Results for the theta solvent can be derived in a similar manner with $\nu = 1/2$. The concentration, molecular weight and temperature dependence of the radius of gyration and of the correlation length (mesh size) ξ_c are listed in Table 1. A few of the exponents in the scaling laws such as $r_g^2 \sim N^{6/5}$ and N in dilute-good and dilute-theta regimes have been tested.

4.3 Light Scattering Intensity Studies

We can obtain the excess Rayleigh ratio $R_c(0)$ due to concentration fluctuations from absolute excess scattered intensity extrapolated to zero scattering angle, the refractive index increment $(\partial n/\partial C^V)_{T,P}$ by means of a differential refractometer or a Rayleigh interferometer and the refractive index n (not the running index n as in Eq. (3.4)) by means of a refractometer and compute the osmotic compressibility

$$(\partial\pi/\partial C^V)_{T,P} = H^*(\partial n/\partial C^V)^2_{T,P}C^V RT/R_c(0) = HC^V RT/R_c(0) \qquad (4.12)$$

TABLE 1

Concentration, Molecular Weight and Temperature Dependence of the Radius of Gyration r_g and of the Correlation Length ξ_c Based on the Scaling Concept.

Region[(a)]	Designation (Fig. 1)	r_g^2	ξ_c
Dilute - good	I	$N^{6/5}\tau^{2/5}$	-
Dilute - theta	I'	N	-
Semidilute - good	II	$Nc^{-1/4}\tau^{1/4}$	$c^{-3/4}\tau^{-1/4}$
Semidilute - marginal	III ($\tilde{c} > c^*$)	N	$c^{-1/2}\tau^{-1/2}$
Semidilute - theta	III ($\tilde{c} > c^*$)	N	c^{-1}
Dilute - chain collapse	IV	(ref. 30, 31)	-
Critical - coexistence curve	V	critical-point behavior	

(a) There are some variations in stating the regions:

Region	Alternate Definition
Theta	Tricritical
Good (I and II)	Excluded volume; Critical (which is not region V)

where $H^* = 4\pi^2 n^2/(N_A \lambda_0^4)$ with C^V, N_A, T, P and R being the polymer concentration in g/cm³, the Avogadro number, the absolute temperature, the pressure, and the gas constant, respectively.

In dilute solution, the virial expansion has the form

$$(\partial\pi/\partial C^V)_{T,P} = RT/M_w + 2A_2RTC^V + \ldots \qquad (4.13)$$

where A_2 is the second virial coefficient and M_w, the weight average molecular weight. At infinite dilution, $(\partial\pi/\partial C^V)_{T,P} = RT/M_w$. The Ⓗ temperature corresponds to the temperature at which $A_2 = 0$ when M is infinite. At finite scattering angles and finite concentrations

$$\frac{HC^v}{R_c(C^v,K)} = \left(\frac{1}{M_w} + 2A_2C^v\right)\left(1 + (1 + 2A_2C^vM_w)^{-1}\frac{\langle r_g^2(C^v)\rangle_z}{3}K^2\right) \quad (4.14)$$

where we have defined an apparent radius of gyration

$$\langle r_g^2(C^v)\rangle_z^* = r_g^{*2} = r_g^2(C^v)/(1 + 2A_2C^vM_w) \quad (4.15)$$

with $r_g^2(C^v) \equiv \langle r_g^2(C^v)\rangle_z$. According to the scaling concept, $(\partial\pi/\partial C^v)_{T,P} \sim C^{\sim 0}$, $C^{5/4}$, C^2 in dilute, semidilute, and concentrated solutions, respectively.

In the dilute region, Figure 5 shows (a) plots of $(\partial\pi/\partial C^v)_{T,P}$ versus concentration C^v at 20°, 30° and 40°C for the NBS 705 standard polystyrene in trans-decalin[7] and (b) plots of $(\partial\pi/\partial C^v)_{T,P}/T$ versus concentration C(g/g of solution) for the TSK-1500 polystyrene in trans-decalin at different temperatures[9] where the superscript v in $(\partial\pi/\partial C^v)_{T,P}$ has been dropped. For the NBS 705 polystyrene with $M_w = 1.79 \times 10^5$) Eq. (4.13) is valid over a concentration range of a few percent while for the TSK-1500 polystyrene with $M_w \sim 1.3 \times 10^7$ Eq. (4.13) is valid at concentrations $C < 1 \times 10^{-4}$ g/g. Figure 6 shows plots of the second virial coefficient A_2 versus temperature. For both polystyrene samples, $A_2 = 0$ at 20.5 C which is the Ⓗ temperature for the polystyrene-trans-decalin system. Figure 7 shows (a) a plot of $\langle r_g^2\rangle_z^{1/2}$ versus temperature and (b) a log-log plot of $\langle r_g^2\rangle_z$ versus (T - Ⓗ) [= τⒽ]. The radius of gyration varied from 99 nm at 17.2°C to 175 nm at 40.0°C. The collapse of polymer coils has been studied by de Gennes[30] and Akcasu.[31] Experiments[32] on the coil-globule transition for 2.6×10^7 molecular weight polystyrene in cyclohexane have shown a sharp decrease in the hydrodynamic radius down to 50 nm.

By defining a linear expansion factor

$$\alpha_s = \langle r_g^2\rangle_{z,T}^{1/2} / \langle r_g^2\rangle_{z,T=\Theta}, \quad (4.16)$$

Akcasu and Han[17] have plotted α_s as a function of N/N_τ, as shown in Figure 1 of ref 20 where $N_\tau (\equiv n_\tau)$ is a temperature-dependent cutoff to separate Gaussian and excluded volume regimes as shown in Fig. 3a; N becomes the equivalent number of links in a statistical chain. Each link has local rigidity and has a persistence length b. For polystyrene, b/a = 20 and with a monomer molecular weight of 104, we get

$$\frac{N}{N_\tau} = \frac{M}{416}\tau^{*2} \quad (4.17)$$

where $\tau^* = (T - \Theta)/T$ and the experimental points are adjusted

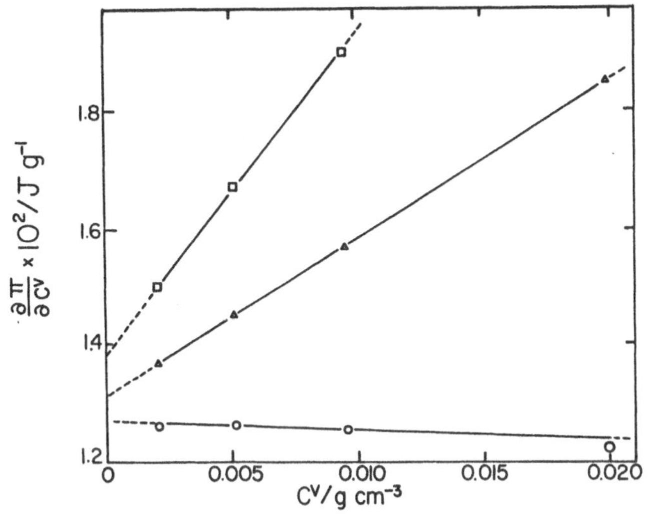

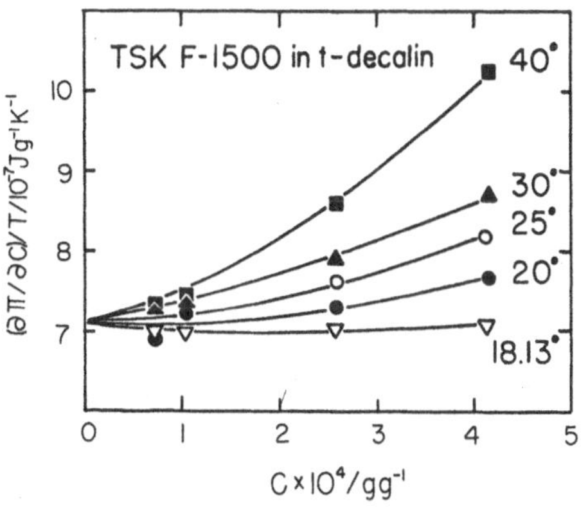

Figure 5a. Plots of $\partial\pi/\partial C^V$ vs. C^V at 40°C (hollow squares), 30°C (hollow triangles), and 20°C (hollow circles) for polystyrene (NBS 705 standard) in trans-decalin. (See refs. 7 and 9.)

Figure 5b. Plots of $(\partial\pi/\partial C)_{T,P}/T$ vs. concentration C(g/g of solution) at different temperatures in the dilute solution region. Note: the curvature of the isotherms signifies that the expression $(\partial\pi/\partial C)_{T,P}/T = R/M_w + 2A_2RC$ is valid only over very limited ranges in the concentration where C is expressed in g/cm^3 except as noted. (See refs. 7 and 9.)

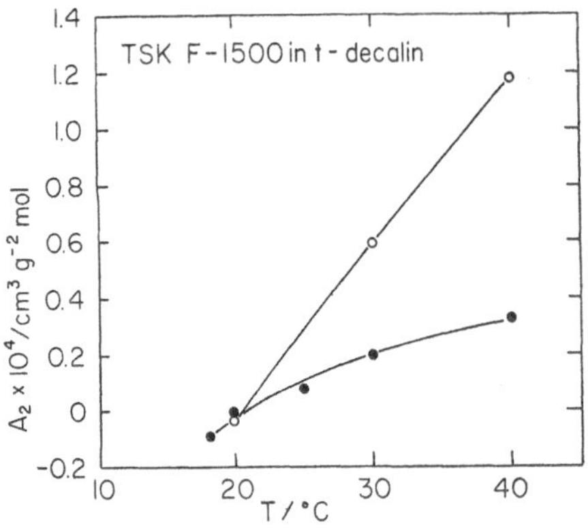

Figure 6. Plots of the second virial coefficient A_2 vs. temperature in the dilute solution region: NBS 705 standard (M_w = 1.79 x 10^5),[16] open circles; and TSK F-1500 ($M_w \sim 1.4 \times 10^7$), solid circles. (See ref. 9.)

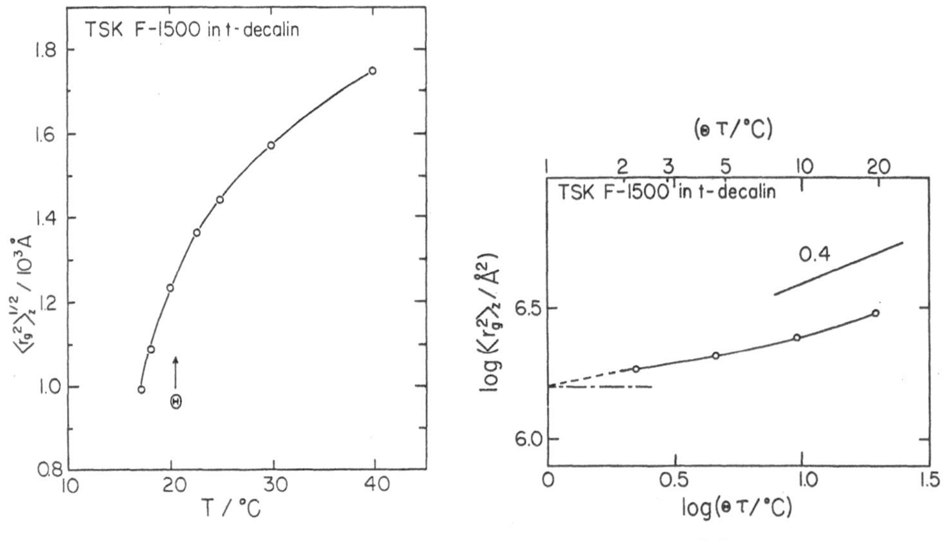

(a) (b)

Figure 7. (a, left) plot of $(r_g^2)z^{1/2}$ vs. temperature for TSK F-1500 in transdecalin, θ = 20.5°C. (b, right) Log-log plot of $(r_g^2)_z$ vs Ⓗτ. The dot-dash line indicated log $(r_g^2)_z$ at τ=0. In the Gaussian regime, the measured exponent at τⒽ=20° is expected to be smaller than 0.4. τⒽ $\sim r_g^{0.4}$ in the asymptotic limit; τ=(T-Ⓗ)/Ⓗ. (See ref. 9) to fit the blob theory by setting the proportionality constant to 0.2. The results are summarized in Fig. 8. α_s exhibits a slower approach to the asymptotic behavior at small values of N/N_τ than the simple blob theory.

Figure 9 shows a schematic diagram of experimental ranges of TSK F-1500[9,10] and NBS 705[6,7] standard polystyrene in transdecalin in the $\tau M^{1/2}$- C/C^* plot [10]. The broken line represents an estimated coexistence curve of the TSK F-1500 polystyrene in transdecalin based on earlier measurements of lower molecular weight polystyrene samples by Kuwahara and his coworkers[33]. The scaling concept is applicable only over very limited ranges as we approach the theta temperature and decrease the molecular weight.

Figure 10 shows plots of osmotic compressibility as a function of concentration for the NBS 705 standard polystyrene ($M_w = 1.79 \times 10^5$) in transdecalin. Figure 11 shows (a) log-log plots of $(\partial\pi/\partial C)_{T,P}$ versus C(g/g) for polystyrene in transdecalin and (b) log-log plots of $(\partial\pi/\partial C)_{T,P}/T$ versus C/C^* for the TSK F-1500

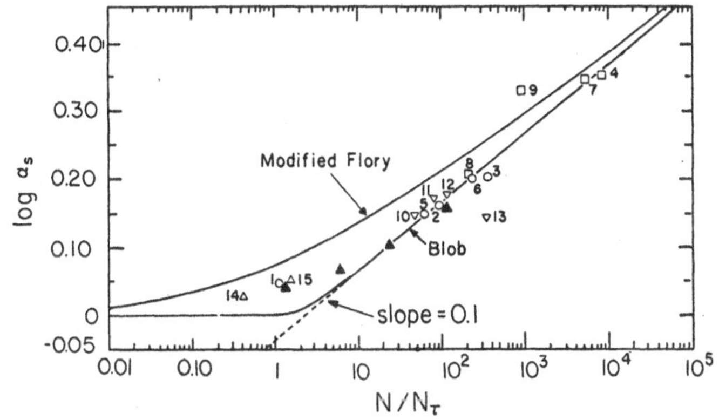

Figure 8. The variation of the expansion factor α_s for radius of gyration with $N/N_\tau = (M\ \tau^{*2}/416)$. For filled triangles, we used $\Theta = 20.5°C$ with trans-decalin as solvent, $M_w \sim 1.17 \times 10^7$, the number of monomers in a statistical length n = 20, and the molecular weight per monomer A = 104: open triangles, polystyrene ($M_w = 1.79 \times 10^5$) in trans-decalin; open circles, cyclohexane; inverted open triangles, toluene; open squares, benzene. The labeled number corresponds to Table I of ref. 7. The values of numbers 9 and 13 are not consistent with the others. (See ref. 9.)

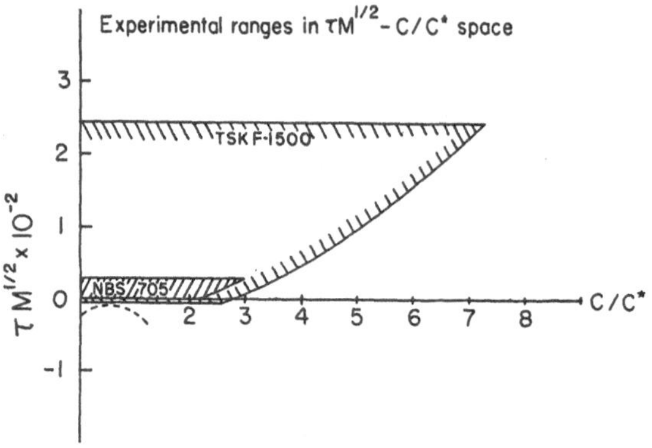

Figure 9. Schematic diagram of experimental ranges of TSK F-1500 and NBS 705 Standard[1,2] polystyrene samples in the $\tau M^{1/2}$-C/C* plane. (See ref. 10.)

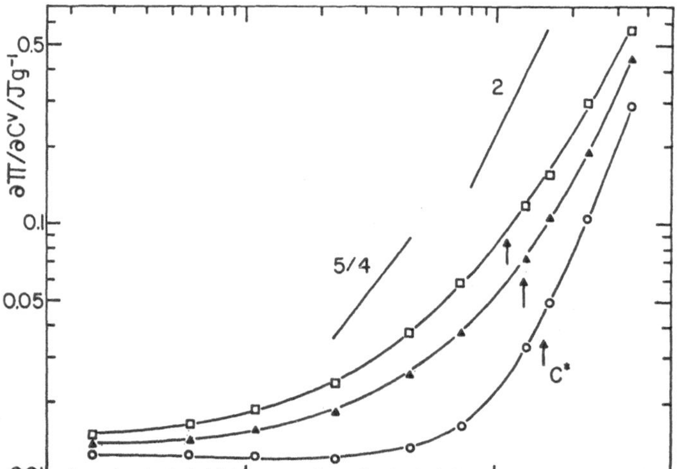

Figure 10. Plots of osmotic compressibility as a function of concentration at 40°C (hollow squares), 30°C (hollow triangles),[6] and 20°C (hollow circles) NBS 705 Standard in trans-decalin. (See ref. 7.)

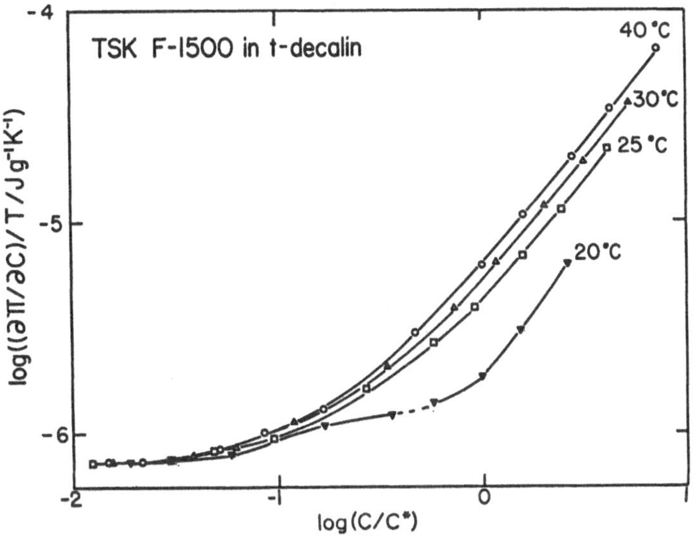

Figure 11. (a) Log-log plots of $(\partial\pi/\partial C)_{T,P}$ vs. C(g/g) for polystyrene (TSK F-1500) in trans-decalin. At $T \geq 25°C$ and at $C > C^*$, the slopes are very close to 5/4, predicted for the semidilute region. The depressions at 19 and 20°C are the results of critical effects. (b) Log-log plots of $(\partial\pi/\partial C)_{T,P}/T$ vs. C/C^* at 20,25,30, and 40°C. $(\partial\pi/\partial C)_{T,P}/T \propto (1/N)(C/C^*)^m$. Critical effects can easily be observed at 20°C before the overlap concentration is reached. (See ref. 10.)

polystyrene in trans-decalin. The transition for the osmotic compressibility from the dilute to the semidilute regime is represented by gradual upward increases without sharp identifiable slope changes. For the polystyrene-trans-decalin system, the overlap concentration $C^* \sim 10$ wt % and 0.5 wt % for molecular weights of 1.79×10^5 and 1.3×10^7, respectively. We have taken $(\partial n/\partial C)_{T,P}$ to be independent of concentration. Therefore, there is a measurable error in the absolute magnitude of $(\partial\pi/\partial C)_{T,P}$ at concentrations $C \gtrsim C^* \sim 10$ wt %. However, the qualitative shape of the curves should remain unchanged for Fig. 10. The assumption for a concentration independent $(\partial n/\partial C)_{T,P}$ is valid for Fig. 11. In Fig. 11, we can observe regions with $(\partial\pi/\partial C)_{T,P}$ very nearly proportional to $C^{5/4}$ beyond the overlap concentration. For $C > C^*$,

$$(\partial\pi/\partial C)_{T/P} \propto (C/C^*)^m \text{ or } \tau^n C^m \qquad (4.18)$$

where $m = 1/(3\nu - 1)$ and $n = 3(2\nu - 1)/(3\nu - 1)$. Thus, if $\nu = 3/5$, $m = 5/4$ as shown in Fig. 11, and $n = 3/4$, as shown in Fig. 12.

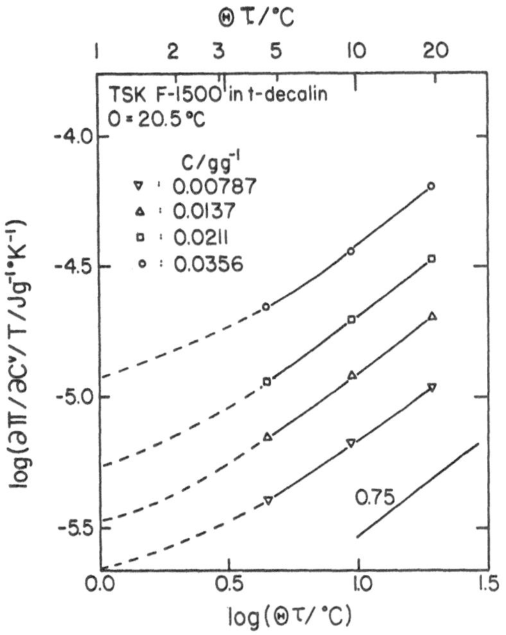

Figure 12. Log-log plots of $(\partial\pi/\partial C)_{T,p}/T$ vs. Ⓗτ. $n = 3r/(3\nu - 1) = 0.75$ for $\nu = 0.6$ and $r = 0.2$. (See ref. 10.)

For the high molecular weight sample at Ⓗ $\tau \sim 20°$ the radius of gyration as shown in Fig. 7 did not reach the asymptotic value of 0.2 for r in a good solvent. Yet in terms of $(\partial\pi/\partial C)_{T,p}/T$ we have evidently reached a value of 5/4 for m in a good solvent, as predicted by the scaling concept. It appears that the range of the asymptotic region where simple power laws hold depends upon the nature of the parameters of interest. We knew that the rate of convergence to the asymptotic limit ($N \to \infty$) is faster for static properties than for dynamic properties.[34] Under our experimental conditions, we have now shown that among the static properties, $(\partial\pi/\partial C)_{T,p}/T$ approaches $\tau^{0.75}$ earlier than $\langle r_g^2\rangle_z$ approaches $\tau^{0.4}$. The deviation could be attributed to N or solvent quality. At the critical mixing point, $(\partial\pi/\partial C)_{T,P} = 0$. Figure 13 shows plots of $(\partial\pi/\partial C)_{T,P}^{-1}$ versus C(g/g) at different temperatures.

We define a characteristic length L based on the relation:

$$R_{vv}(\theta) = R_{vv}(0)/(1 + \frac{L^2K^2}{3}) \qquad (4.19)$$

In dilute solutions, L is similar to $\langle r_g^2\rangle_z^{*1/2}$ of Eq. (4.15) which increases with increasing temperature as $L \propto \tau^r$ with $r = 0.2$ in the dilute-good region. In the transition from dilute to semi-dilute solutions, L is an apparent parameter which is a composite

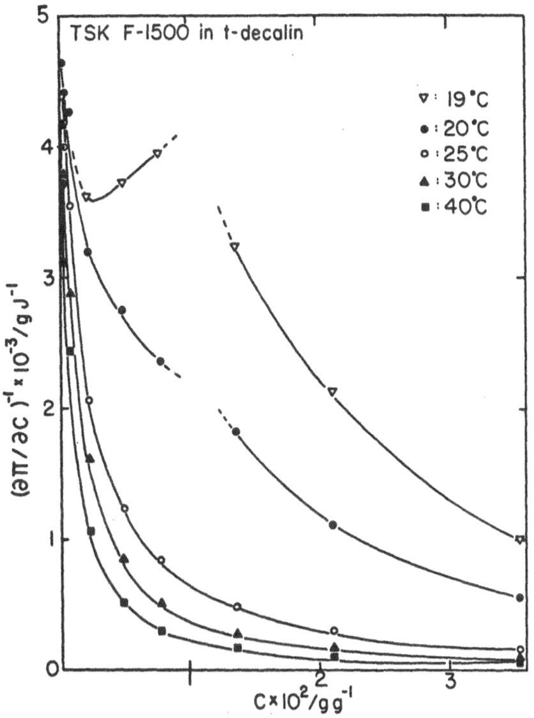

Figure 13. Plots of $(\partial\pi/\partial C)_{T,P}^{-1}$ vs. C(g/g) at different temperatures. For a binary fluid mixture, $(\partial\pi/\partial C)_{T,P} = 0$ at the critical mixing point. C* = 0.0136 g/g of solution at 20°C. (See ref. 10.)

of polymer coil size and distance between adjacent entanglement points. As $\xi_c \propto c^{-3/4}\tau^{-1/4}$ and $c^{-1/2}\tau^{-1/2}$ in semidilute-good and semidilute-marginal solutions, we see that at (H) $\tau \sim 20°$, the polystyrene-trans-decalin system may be more in the semidilute-marginal regime than the semidilute-good regime, as shown in Figs. 14 and 15. The critical-point behavior for L is shown in Fig. 16.

5. DYNAMIC PROPERTIES OF POLYMER SOLUTIONS

5.1 $Kr_g \ll 1$.

In dilute solutions, we have access to the translational diffusion coefficient of the entire polymer coil at $Kr_g \ll 1$, whereas at $Kr_g \gg 1$, the dynamic structure factor is related to internal motions of the polymer chain. Figure 17 shows a schematic diagram of experimental ranges of TSK F-1500[9,10] and NBS 705 standard[6,7] polystyrene in trans-decalin in the Kr_g - C/C* plane.

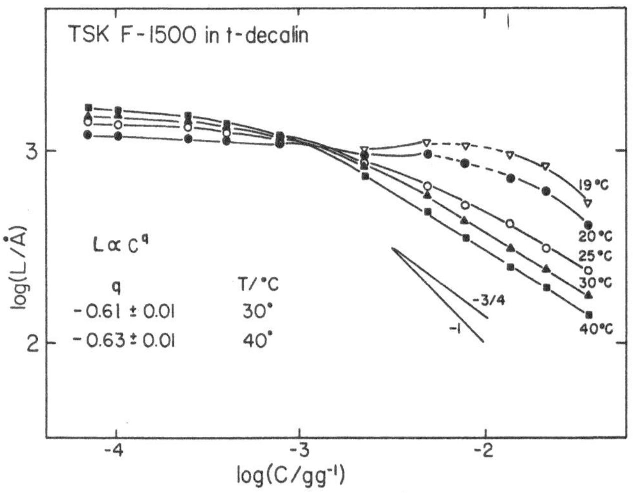

Figure 14. Log-log plots of L vs. C(g/g) at different temperatures. In the semidilute regime, $L \propto C^q$, with $q = \nu/(1-3\nu) = -3/4$ for $\nu = 0.6$. (See ref. 10).

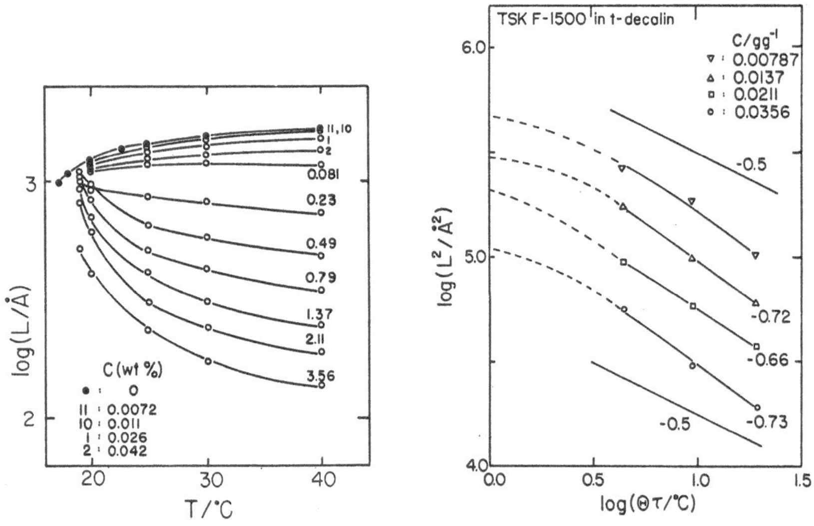

Figure 15. Log-log plots of L^2 vs. $\Theta\tau$ in the semidilute region. (See ref. 10.)

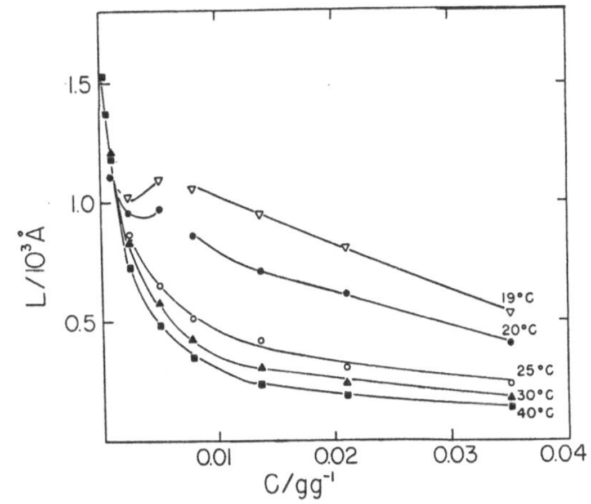

Figure 16. Plots of isotherms of L vs. C(g/g). The anomaly occurs near the critical mixing point where the critical solution temperature is estimated to be about 19°C. (See ref. 10.)

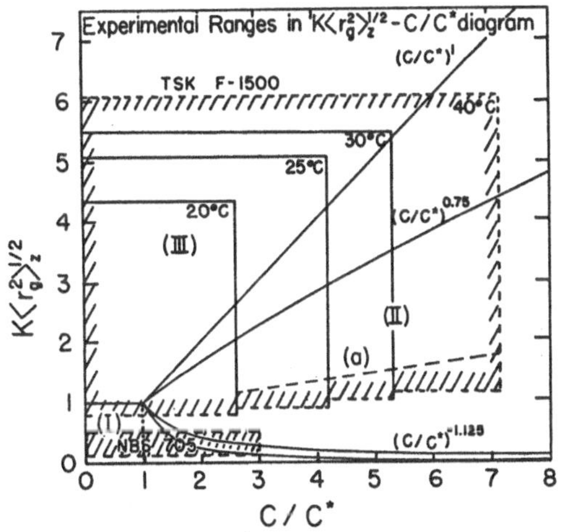

Figure 17. Schematic diagram of experimental ranges of TSK F-1500 and NBS 705 Standard,[6,7] polystyrene samples in the $K(r_g^2)_z^{1/2}$- C/C^* plane: I, $KL \ll 1$, dilute solutions; II, pseudogel domain; and III, $KL \gg 1$, internal motions dominate. The curve below $K_{min}L = (C/C^*)^{-1.125}$ corresponds to $KL = (C/C^*)^{-2}$. Curve a represents the changes in the $K(r_g^2)_z^{1/2}$- C/C^* plane for Figure 14 at $C = 0.0356$ g/g of solution, $\theta = 30°$, and temperatures from 20 to 40°C. (See ref. 10.)

At $Kr_g << 1$,

$$D = D_o(1 + k_D C^v + \ldots) \tag{5.1}$$

where $D_o(=(k_BT/f_o)$ is the translational diffusion coefficient at infinite dilution, with k_B and f_0 denoting the Boltzmann constant and the frictional coefficient at infinite dilution. The hydrodynamic and thermodynamic factors are combined in k_D as

$$k_D = 2A_2M - k_f - N_AV_1/M \tag{5.2}$$

where k_f is the first-order friction coefficient and V_1 is the polymer molecular volume. At all concentrations, the mutual diffusion coefficient has the form:

$$D = \frac{C^v}{f}\left(\frac{\partial \mu}{\partial C^v}\right)_{T,P} \tag{5.3}$$

where the chemical potential $(\partial\mu/\partial C^v)_{T,P}$ can be related to the osmotic compressibility by the relation

$$C^v(\partial\mu/\partial C^v)_{T,P} = \frac{(1 - \bar{v}C^v)M}{N_A}(\partial\pi/\partial C^v)_{T,P} \tag{5.4}$$

where $\bar{v}(= N_AV_1/M)$ is the partial specific volume of the polymer. By substituting Eq. (5.4) into Eq. (5.3), we get

$$f = \frac{M(1 - \bar{v}C^v)}{DN_A}(\partial\pi/\partial C^v)_{T,P} \tag{5.5}$$

All the properties on the right side of Eq. (5.5) are accessible experimentally. However, if we try to determine the translational (mutual) diffusion coefficient by means of light-scattering experiments using the relation $D = \Gamma/K$, the physical interpretation becomes more complex in semidilute solutions because, in addition to the translational motion of the entire polymer coil, there can be faster pseudo-gel motions.

In dilute solutions, the diffusion coefficient is dominated by backflow properties[35] which predicts

$$D = \frac{k_BT}{6\pi\eta_o r_h} \tag{5.6}$$

where η_o and r_h are the solvent viscosity and the effective hydrodynamic radius of the polymer coil, respectively. This means that[34]

$$\frac{1}{r_h} \sim \int \frac{1}{r} g(r)\, d\vec{r} \Big/ \int g(r)\, d\vec{r} \tag{5.7}$$

which implies that the dynamic size is more sensitive to the short range portion of g(r). For N not too large, r_h is closer to the Gaussian chain value ($\sim N^{1/2}$). Ultimately, for N very large, $r_h \sim r_g \sim N^{3/5}$ in the dilute-good regime.

5.2 $Kr_g >> 1$

In the intermediate K-region defined by $Kr_g >> 1$ and $Kb << 1$ and for an isolated Gaussian polymer coil of infinite size with backflow hydrodynamic interactions in a Ⓗ -solvent. Dubois-Violette and de Gennes[36] derived an expression for the net time correlation function

$$|g^{(1)}(K.t)| \left(\equiv \frac{S(K,t)}{S(K,0)}\right) = \begin{cases} (t/\tau_K)^{2/3} \int_0^\infty du \exp[-(t/\tau_K)^{2/3} u(1+h(u))] & \text{when } t \neq 0 \\ 1 & \text{when } t = 0 \end{cases} \tag{5.8}$$

where $h(u) = \frac{4}{\pi}\int_0^\infty \frac{dy \cos y^2}{y^3} [1 - \exp(- y^3 u^{-3/2})]$ with the inverse characteristic time τ_K defined by

$$\frac{1}{\tau_K} = \left(\frac{1}{6\pi}/\sqrt{2}\right) \left(k_B T/\eta_o\right) K^3 \tag{5.9}$$

in the double limit of $Kb \to 0$ and $Kr_g \to \infty$.

In actual experiments, the conditions $Kb << 1$ and $Kr_g >> 1$ are difficult to satisfy for polymers of finite molecular weights, especially if the solution is near theta condition. Furthermore, in the large-K range, local molecular segmental motions appear while in the smaller-K range, translational motion of the whole polymer coil becomes increasingly important. There is also a cross-over time from Rouse-like behavior to Zimm-like (long-time) behavior.

Benmouna and Akcasu[37] proposed an alternative method by introducing a characteristic frequency Ω as the initial slope of log S(K,t), i.e.,

$$\Omega = -\lim_{t\to 0} \frac{d}{dt} \log S(K,t) \quad (5.10)$$

The initial slope $\Omega(K)$ can be computed for all values of K as a function of concentration and temperature. Therefore, Eq. (5.10), which is different from Eq. (5.8), is not restricted to the $Kr_g >> 1$ condition and dilute solutions near the theta temperature. We should, however, like to remark here that Eq. (5.8) appears to be applicable over a much broader concentration and solvent quality range than is appreciated in the literature.

In semidilute solutions, there can be at least five distinct dynamic processes:

1. Cooperative diffusion D_g is related to the hydrodynamic motion of a portion of the polymer chain between entanglement points and has a characteristic time faster than the motion of the entire polymer coil. According to Table I, $\xi_c \propto r_g(C^*/C)^{3/4}$ in a good solvent. Then,

$$D_g = k_B T/(6\pi\eta_o \xi_h) \sim C^{3/4} \quad (5.11)$$

with $\xi_h \sim \xi_c$. In the semidilute domain of Fig. 17, Adam and Delsanti[38] obtained $D_g \sim C^{0.67\pm0.02}$. The lower exponent value simply implies that the asymptotic limit was not reached.

2. Individual diffusion D_{self} is related to the Brownian motion of one chain immersed in a solution of other chains. D_{self} is small and is much slower than D_g. For the reptation model, D_{self} is proportional to $C^{-7/4}M^{-2}$. The effect can be measured using IR spectroscopy[39] and forced Rayleigh scattering.[40] There is a discrepancy of the scaling prediction with mechanical data.[16]

3. In the decay of local concentration fluctuations, motions related to entire polymer coils are slower than the cooperative diffusive motions of (1). The cross-over from dilute to semidilute solutions is not sharp, and for $C \sim C^*$, $|g^{(1)}(\tau)|$ measures both (1) and (3).

4. Local segmental motions involving only a few neighboring monomers can be measured by means of depolarized Rayleigh scattering for anisotropic segments. These motions are much faster and are usually not accessible using photon correlation spectroscopy.

5. The short time behavior of $S(K,t)$ is influenced by all the internal modes collectively in the intermediate $Kr_g >> 1$, $Kb << 1$ range. For $Kr_g > 1$ and $C \sim C^*$, both internal dynamics and pseudogel motions are present. Thus, as shown in Fig. 17, all changes from (I) to (II), (I) to (III), and (II) to (III) involve fairly

broad transitions. $|g^{(1)}(\tau)|$ contains information related to (1), (3) and (5) and should not be analyzed by fitting a single exponential function of time to the dynamic structure factor.

5.3 Photon Correlation Measurements[7-10]

The measured single-clipped photocount autocorrelation function has the form:

$$G_k^{(2)}(\tau) = A(1 + \beta|g^{(1)}(\tau)|^2 \tag{5.12}$$

where k is the clipping level, A is the background, and β is an unknown parameter in the data-fitting procedure. $\tau(=I\Delta\tau)$ is the delay time with I and $\Delta\tau$ being the channel number and the increment delay time, respectively.

For a polydisperse example,

$$|g^{(1)}(\tau)| = \int_0^\infty G(\Gamma) \exp(-\Gamma\tau)d\Gamma \tag{5.13}$$

where $G(\Gamma)$ is the normalized distribution of decay rates.

Two methods of data analysis were used: (1) cumulants method and (2) histogram method.

1. Cumulants Method.[41]

The average linewidth $\bar{\Gamma}$ is defined as

$$\bar{\Gamma} = \int G(\Gamma)\Gamma d\Gamma \tag{5.14}$$

$$\mu_i = \int (\Gamma - \bar{\Gamma})^i G(\Gamma)d\Gamma \tag{5.15}$$

with $\mu_2/\bar{\Gamma}^2$ as the variance.

2. Histogram Method.[1-4]

$$|g^{(1)}(\tau)| = \sum_{j=1}^{m} G(\Gamma_j) \int_{\Gamma_j-\Delta\tau/2}^{\Gamma_j+\Delta\Gamma/2} \exp(-\Gamma I\Delta\tau)d\Gamma \tag{5.16}$$

where $G(\Gamma_j)$ is the integrated scattered intensity representing the amplitude of the Γ_j increment from $\Gamma_j - \Delta\Gamma/2$ to $\Gamma_j + \Delta\Gamma/2$, m is the number of steps in the histogram, and $\Delta\Gamma$ is the width at each step.

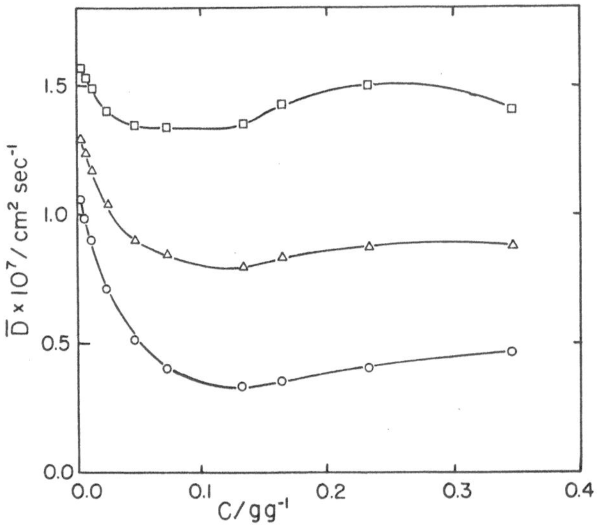

Figure 18. Plots of $\overline{D}$ vs. C at 40°C (hollow squares); 30°C (hollow triangles), and 20°C (hollow circles). NBS 705 Standard polystyrene in trans-decalin. (See ref. 7.)

Although the apparent mutual diffusion coefficient $\overline{D}$ for NBS 705 standard polystyrene in trans-decalin is in agreement with results determined by classical techniques[42], as shown in Fig. 18, a more detailed analysis of $|g^{(1)}(\tau)|$ in the semidilute regime clearly shows that the interpretation for $\overline{D}$ is complex. Figure 19 shows plots of $\mu_2/\overline{\Gamma}^2$ versus concentration for the NBS 705 standard polystyrene in trans-decalin. At low concentrations, $\mu_2/\overline{\Gamma}^2$ represents mainly the molecular polydispersity effect as shown in Figs. 20 and 21. At higher concentrations, $\mu_2/\overline{\Gamma}^2$ shows such high values which cannot be attributed to the molecular weight distribution function. Figure 22 shows plots of G(D) versus D for NBS 705 standard polystyrene in trans-decalin at different concentrations. By identifying and observing the changes of the translational diffusive motion of the polymer coil from dilute (Fig. 22 (a) and (b) to semidilute solutions (Fig. 22 (c) - (f)), we see the emergence of a faster gel-like mode. The histogram method also permits us to approximate the intensity contributions of the two motions whereby we can reconstruct $R_{VV}(K)$ as a function of concentration as shown in Fig. 23. The concentration C_g represents an upper limit of the semidilute region where the slow mode remains detectable. C_g decreases with increasing temperature.

In dilute solution, at $Kr_g > 1$, the characteristic time $1/\tau_K$ does not measure the translational diffusive motion of the entire polymer coil. The diffusion coefficient can be determined by

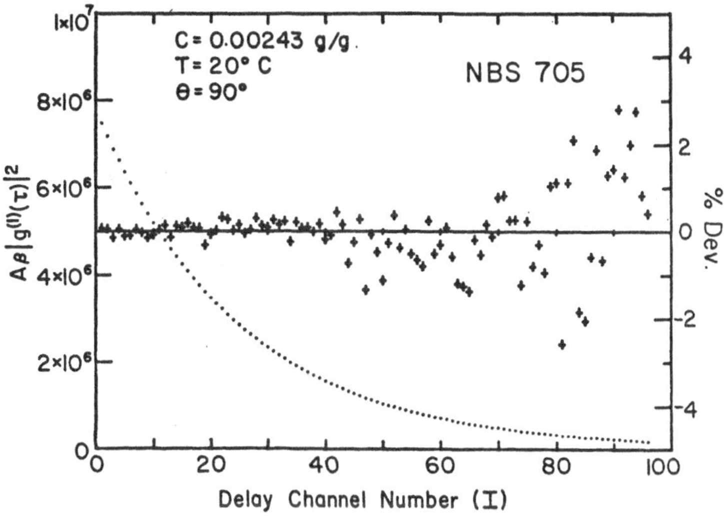

Figure 19. Plots of $\mu_2/\overline{\Gamma}^2$ vs. concentration (g/g of solution) for (NBS 705 standard) polystyrene in trans-decalin at 20°C (circles), 30°C (triangles), and 40°C (squares); θ = 90°. Solid symbols denote values from the cumulants fit, hollow symbols denote values from the histogram method; crosses denote computed values based on $\mu_2 = A_f A_s (\overline{\Gamma}_f - \overline{\Gamma}_s)^2 + A_s \mu_{2s} + A_f \mu_{2f}$. (See ref. 8.)

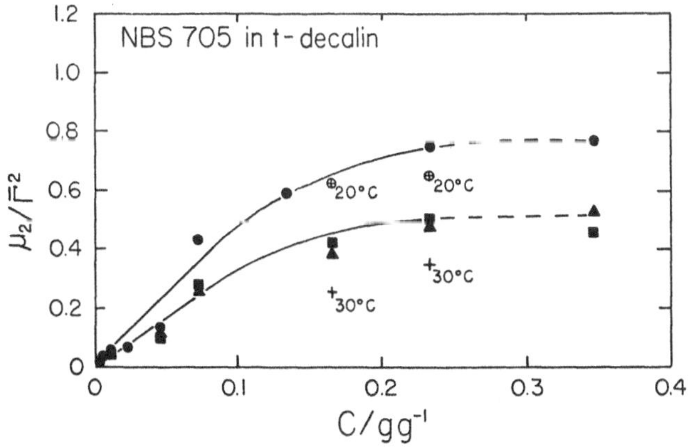

Figure 20. Plot of $A\beta|g^{(1)}(r)|^2$ and % dev. vs. delay channel number I. $\Delta\tau$ = 2.65 μs at θ = 90° and 20°C; C = 0.00243 g/g of solution. (See ref. 7.)

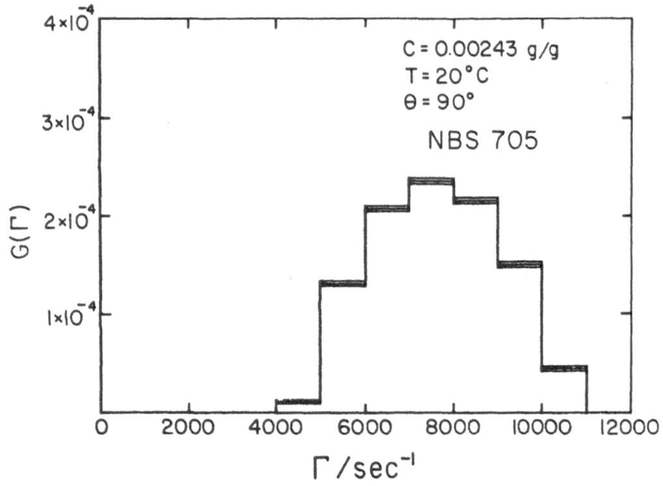

Figure 21. Plot of G(Γ) vs. Γ from a dilute polymer solution based on data shown in Fig. 20. (See ref. 7.)

scaling of Ω/K^2 versus Kr_g using the Benmouna and Akcasu theory[37(a)], by extrapolation, and by measurements at small scattering angles where $Kr_g \ll 1$. Figure 24 shows the variation of the expansion factor α_h for the hydrodynamic radius with N/N_τ ($=M_w\tau^{*2}/416$). Our results agree remarkably well with the blob theory[17] in the high N/N_τ range. Variation of the hydrodynamic radius r_h as a function of temperature is shown in Fig. 25. At temperatures greater than the Ⓗ temperature, r_g increases faster than r_h, while below the Ⓗ temperature the rates are comparable. The shape of the curve in Fig. 25 represents a middle section of the curve in Fig. 1 of ref. 32. However, in ref. 32 at high temperatures, r_h($\sim$125 nm) is too large for an accurate determination by the cumulants method using $\theta = 23°$ and $\lambda_o = 514.5$ nm. Therefore, Fig. 1 of ref. 32 has a substantial distortion at high temperatures.

In the semidilute regime and with $Kr_g > 1$, data analysis becomes difficult for lack of theory. We have used a simplified approximate expression:

$$|g^{(1)}(\tau)| = A_f \exp(-\bar{\Gamma}_f\tau + \mu_{2f}\tau^2/2) + A_s\exp(-\bar{\Gamma}_s\tau) \qquad (5.17)$$

to estimate the qualitative contributions of the slow and the fast modes. In other words, we have taken

$$G(\Gamma) = A_s\delta(\Gamma - \bar{\Gamma}_s) + A_f\exp[-(\Gamma - \bar{\Gamma}_f)^2/2\mu_{2f}] \qquad (5.18)$$

to be our linewidth distribution function even though we know that the dynamics of polymers in the intermediate Kr_g range and $C \sim C^*$

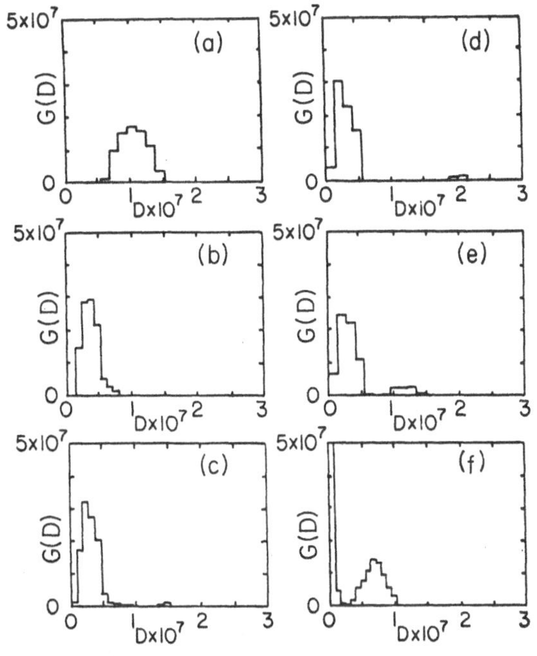

Figure 22. Plots of G(D) vs. D(cm^2/s) for polystyrene (NBS 705 standard) in trans-decalin at different concentrations, t = 20°C, θ = 90°: (a) 0.00243 g/g of solution; (b) 0.0731 g/g of solution; (c) 0.134 g/g of solution; (d) 0.165 g/g of solution; (e) 0.233 g/g of solution; and (f) 0.346 g/g of solution. (See ref. 8.)

contains several (more than two) modes of motion.

Figure 26 shows plots of $\overline{\Gamma}_f$, $\overline{\Gamma}_s$ and $\overline{\Gamma}$ versus concentration. At low concentrations, $\overline{\Gamma} = \overline{D}_s K^2$ yielding a hydrodynamic radius $\overline{r}_h$ of 86 nm at 30°C, in agreement with the results shown in Fig. 25. At higher concentrations, $\overline{\Gamma}_s$ decreases with increasing concentration because the polymer coils are entangled in the semidilute region. $\overline{\Gamma}_f$ represents the time behavior of large-scale internal motions at low concentrations ($C/C^* \lesssim 1$) and $Kr_g \sim 1$. As we increase the concentration, $C/C^* \sim 1$, $\overline{\Gamma}_f$ contains information related to the pseudogel motions as well as internal motions. A clear separation of the characteristic times has not yet been achieved, especially when the transition from region III to region II, as shown in Fig. 17, is very broad.

Fig. 27 shows a log-log plot of $T/\overline{\Gamma}_f\eta_o$ versus L. We take $\xi_h \propto T/\eta_o\overline{\Gamma}_f$ by assuming $\overline{\Gamma}_f \propto D_g$ which is not strictly valid and $L(\neq\xi_c)$ as a composite characteristic length. At 40°C, $\xi_h \propto \xi_c^{0.68}$.

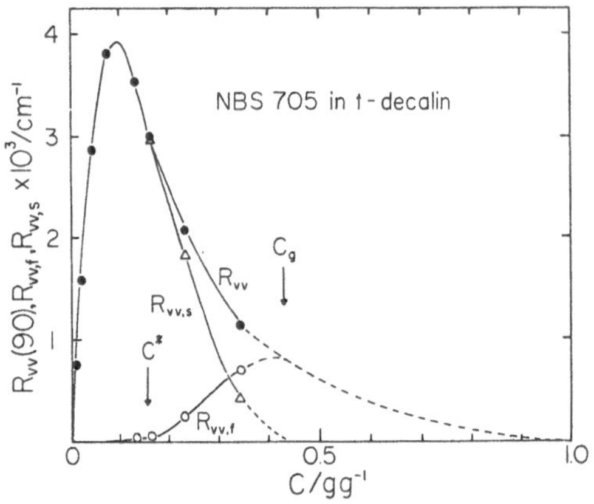

Figure 23. Plots of R_{VV} (solid circles), $R_{VV,f}$ (hollow circles), and $R_{VV,s}$ (hollow triangles) vs. concentration at $\theta = 90°$ and 20°C. $R_{VV} = R_{VV,f} + R_{VV,s}$; $R_{VV,f} = A_f R_{VV}$; $R_{VV,s} = A_s R_{VV}$. $C^* = M_W/(N_A \rho_s r_g{}^3)$. Schematically, we have neglected the density fluctuations of the polymer and of the solvent and have considered only the excess scattered intensity due to concentration fluctuations. (See ref. 8.)

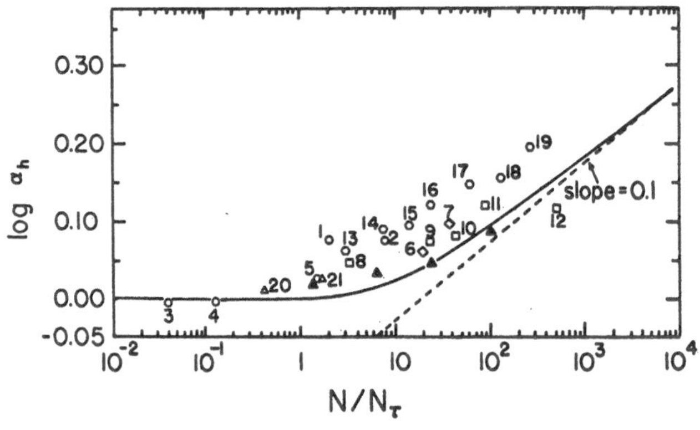

Figure 24. The variation of the expansion factor α_h for the hydrodynamic radius with $N/N_\tau = (M_W \tau^{*2}/416)$. The triangles represent our data in trans-decalin open triangles, $M_W = 1.79 \times 10^5$; filled triangles; $M_W = 1.17 \times 10^7$; open circles, cyclohexane; open diamonds, toluene; open squares, benzene; open hexagons, THF. The labeled number corresponds to Table II of ref. 17. (See ref. 9.)

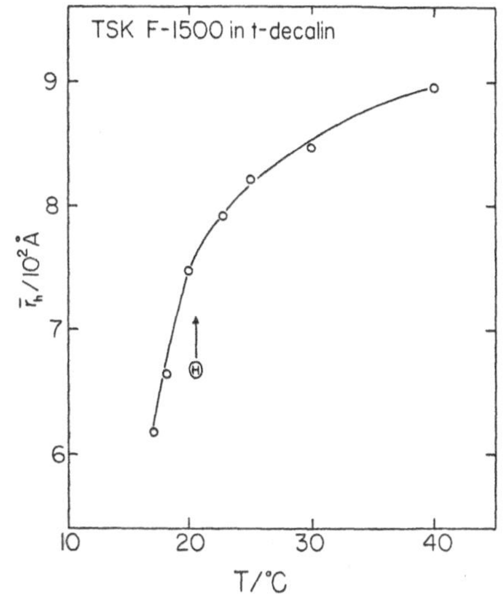

Figure 25. Variation of the hydrodynamic radius $\bar{r}_h$ as a function of temperature above and below the Ⓗ temperature. (See ref. 9.)

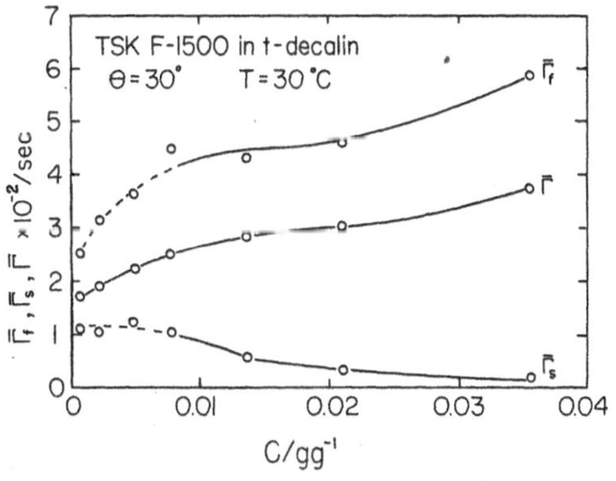

Figure 26. Plot of $\overline{\Gamma_f}$, $\overline{\Gamma_s}$, and Γ vs. concentration (g/g) at 30°C and θ = 30° for TSK F-1500 in trans-decalin. (See ref. 10.)

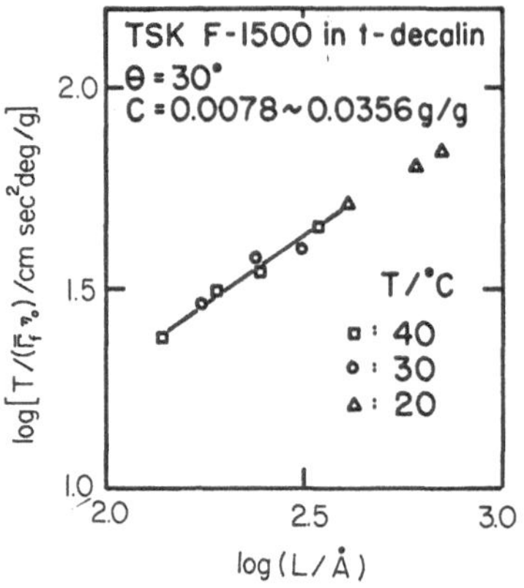

Figure 27. Log-log plot of $T/\overline{\Gamma}_f\eta_o$ vs. L at 20, 30, and 40°C, using $\overline{\Gamma}_f$ values at θ = 30°C and C from 0.00787 to 0.0356 g/g. The assumption $\overline{\Gamma}_f \sim D_g$ is not valid. neither is $L \equiv \xi$ due to pseudogel motion only; slope (at 40°C)$\sim$ 0.68. (See ref. 10.)

According to the scaling concept, $\xi_c \propto c^{-0.75}$ for ν = 3/5 and $\xi_h \propto c^{-0.75}$ for ν = 3/5. Therefore, $\xi_h \propto \xi_c$ in good solvents. However, as $\xi_h \propto c^{-0.43}$ and $\xi_c \propto c^{-0.63}$ experimentally, $\xi_h \propto \xi_c^{0.43/0.63} \sim \xi^{0.68}$ which is consistent with the results of Fig. 27. According to Table I, if $\xi_c \propto c^{-3/4}$ (semidilute-good) and $\xi_h \propto c^{-1/2}$ (semidilute-marginal), $\xi_h \propto \xi_c^{(1/2)(3/4)} \sim \xi_c^{2/3} \sim \xi_c^{0.67}$. The agreement, however, may be fortuitous because of the composite nature of $\overline{\Gamma}_f$ as well as L by light-scattering measurements.

Curve a of Fig. 17 represents temperature changes between 20° and 40°C at θ = 30° and C = 0.0356g/g of solution. As we increase the temperature, the radius of gyration increases resulting in increases of both C/C* and $K \langle r_g^2\rangle_z^{1/2}$. In fact, C/C* changes from 2.6 to 7.3 while $K \langle r_g^2\rangle_z^{1/2}$ increases from 1.21 to 1.71. As we enter into the semidilute regime, the slower motion of entire polymer coil translation becomes less important. It should be emphasized that laser light scattering was able to observe all modes of motions; i.e., translational, internal, pseudogel motions. Unfortunately, quantitative separation of three modes of motion is not yet feasible.

6. CONCLUSIONS

The scaling concept is a powerful tool for studying static and dynamic properties of polymer solutions. However, the asymptotic limit is quite difficult to reach under many actual experimental conditions. Complete separation of local internal motions and translational motion of the polymer coil as well as pseudogel motions are often difficult to achieve because the time correlation function measures all these modes of motion, especially when $Kr_g \sim 1$ and $C/C^* \sim 1$. The single exponential function has never been observed for polymer solutions in the semidilute regime because the boundaries specified for various domains are not sharp. At this time, definitive quantitative answers are available only in the asymptotic limit.

REFERENCES

1. B. Chu, Esin Gulari, and Erdogan Gulari, Phys. Scripta, 19, 476 (1979).
2. Esin Gulari, Erdogan Gulari, Y. Tsunashima, and B. Chu, J. Chem. Phys., 70, 3965 (1979).
3. Erdogan Gulari, Esin Gulari, Y. Tsunashima, and B. Chu, Polymer, 20, 347 (1979).
4. B. Chu and Esin Gulari, Macromolecules, 12, 445 (1979).
5. T. Nose and B. Chu, J. Chem. Phys., 70, 5332 (1979).
6. B. Chu and T. Nose, Macromolecules, 12, 347 (1979).
7. T. Nose and B. Chu, Macromolecules, 12, 590 (1979).
8. B. Chu and T. Nose, Macromolecules, 12, 599 (1979).
9. T. Nose and B. Chu, Macromolecules, 12, 1122 (1979).
10. B. Chu and T. Nose, Macromolecules, 13, 122 (1980).
11. P. G. De Gennes, Scaling Concept in Polymer Physics, Cornell University Press, New York, 324 (1979).
12. P. G. De Gennes, in "Les Houches, Session XXXI, 1978 - La matiere mal condensee/Ill-condensed matter," (R. Balian et al., eds.), North-Holland Publishing Company, 479-482 (1979).
13. M. Daoud, J. P. Cotton, B. Farnoux, G. Jannink, G. Sarmia, H. Benoit, R. Duplessix, C. Picot, and P. G. De Gennes, Macromolecules, 8, 804 (1975).
14. P. G. De Gennes, Macromolecules, 9, 587, 594 (1976).
15. B. Farnoux, F. Boue, J. P. Cotton, M. Daoud, G. Jannink, M. Nierlich and P. G. De Gennes, J. Phys. (Paris), 39, 77 (1978).
16. P. G. De Gennes, Nature, 282, 367 (1979).
17. A. Z. Akcasu and C. C. Han, Macromolecules, 12, 276 (1979).
18. K. S. Siow, G. Delmas, and D. Patterson, Macromolecules, 5, 29 (1972).
19. A. R. Shultz and P. J. Flory, J. Amer. Chem. Soc., 74, 4760 (1972).
20. G. Delmas and D. Patterson, Polymer, I, 513 (1966).
21. J. M. G. Cowie, A. Maconnachie, and R. J. Ranson, Macromolecules, 4, 57 (1971).
22. A. H. Liddell and F. L. Swinton, Discuss., Faraday Soc., 49, 115 (1970).
23. S. Saeki, N. Kuwahara, S. Konno and M. Kaneko, Macromolecules, 6, 246 (1973).

24. G. Delmas and D. Patterson, Int. Symp. Macromol. Chem. Toronto (1968); D. Patterson, J. Polym. Sci., Part C., 16, 3379 (1968).
25. M. Daoud and G. Jannink, J. Phys., (Paris), 37, 973 (1976).
26. P. G. De Gennes, Makrowol. Chem. Suppl., 3, 195 (1979).
27. J. Francois, T. Schwartz, and G. Weill, J. Phys. Lett. (orsay, Fr. P, 41, 69 (1980); Macromolecules, 13, 564 (1980).
28. P. J. Flory, Principles of Polymer Chemistry, Cornell University Press, New York, 6th ed. (1966).
29. D. W. Schaefer, J. F. Joanny and P. Pincus, "Dynamics of Semiflexible Polymers in Solution," reprint.
30. P. G. De Gennes, J. Phys. Lett. (Paris), 36, L-55 (1975); ibid, 39, L-299 (1978).
31. Z. Akcasu, private communication.
32. G. Swislow, S. T. Sun, I. Nishio, and T. Tanaka, Phys. Rev. Lett., 44, 796 (1980).
33. M. Nakata, S. Higashida, N. Kuwahara, S. Saeki, and M. Kaneko, J. Chem. Phys., 64, 1022 (1976).
34. G. Weill and J. Des Cloiseaux, J. Phys., 40, 99 (1979).
35. J. Kirkwood and J. Risemann, J. Chem. Phys., 16, 565 (1948).
36. E. Dubois-Violette and P. G. De Gennes, Physics, 3, 181 (1967).
37. (a) M. Benmouna and Z. Akcasu, Macromolecules, 11, 1187 (1978). (b) Z. Akcasu and M. Benmouna, ibid., 11, 1193 (1978).
38. M. Adam and M. Delsanti, J. Phys. Lett., 38, L271 (1977); Maceomolecules, 10, 1229 (1977).
39. J. Klein, Nature, 271, 143 (1978).
40. H. Hervet, L. Leger and F. Rondelez, Phys. Rev. Lett., 42, 1681 (1979).
41. D. E. Koppel, J. Chem. Phys., 57, 4814 (1972).
42. G. Rehage, O. Ernst, and J. Fehrmann, Discuss., Faraday Soc., 49, 208 (1970).

SMALL ANGLE NEUTRON SCATTERING BY POLYMER SOLUTIONS

B. Farnoux and G. Jannink

Laboratoire Léon Brillouin, CEN-SACLAY
BP n°2
91190 Gif sur Yvette, France

CONTENTS

1. INTRODUCTION

The small angle scattering (close to the incident direction[1]) reveals correlations at large distances inside the sample, as typically exhibited in polymer solutions. Earlier, small angle scattering experiments on dilute solutions using electromagnetic radiations were performed for the purpose of polymer characterization.[2] Small angle neutron scattering (SANS) is an experimental technique introduced about ten years ago for observing polymer conformations in all concentration ranges from dilute solution to the melt.[3-6]

In what follows, we will mainly be concerned with conditions necessary for the observation of small angle scattering.[8] After a brief review (Section 2) of the elementary relations between scattering amplitude, index of refraction, and scattered intensity, we will discuss in detail two concepts related to this last quantity, the contrast (Section 3) and the pair correlation function (Section 4).

Experimental results obtained by SANS have been extensively reviewed in many general articles.[9] We have limited our review to atactic polystyrene[6,10,11] (and sulfonated polystyrene) in solution. This polymer is taken as a prototype of the flexible coil. As an example of a problem studied by SANS, we consider polymer coil shape as a function of polymer concentration. Using light or X-rays, experimentation could only observe the coil configuration in dilute solutions. But with use of neutron scattering, investigation of the coil configuration in other environments such as semi-dilute and concentrated solutions becomes possible. We briefly recall the findings of the SANS by semi-dilute polymer solutions.

(1) Neutron, light scattering, and osmotic pressure data could all be interpreted in terms of a single length ξ[16];

(2) Characteristic domains of variation were found in the temperature concentration diagram.[17]
Results are mostly expressed in terms of power laws.[14,15] A major experimental problem is the determination of the characteristic exponents[13] and the spatial range over which a given power law holds true (cross over).[53]

Another typical problem investigated by SANS is the nature of strong polyelectrolye solutions. The scattering pattern is similar to the one associated with a long range ordered structure. However, experimenters are more and more convinced that the correct interpretation leads to a disordered structure.

In all these examples the polymer chain is in the equilibrium state. This state has been extensively studied for the last ten years. A new problem now is the study of polymer conformations in non-equilibrium states such as the stretched polymer chain. Neutron scattering in real experimental conditions are difficult because the scattered intensity is very low. These studies require new experimental approaches. One is described in Section 4.

2. ELEMENTARY RELATIONS

The experimental analysis of SANS is based on elementary relations.

A. The Scattering Amplitude by the Elementary Particle[8]

In the case of (high) polymers, the elementary particle is taken to be the monomer. The scattering amplitude by an isolated monomer is, for neutrons,

$$a = \lim_{K_o \to 0} \frac{m_n}{2\pi\hbar^2} \langle K_o \hat{r} | T | \vec{K}_o \rangle \qquad (1)$$

Here $\vec{K}_o$ is the incident wave vector, $\hat{r}$ is the unit vector pointing in the scattering direction, m_n is the neutron mass and T is the operator associated with the scattering.[18] The scattering amplitude by electromagnetic radiation[19] will either be defined by

$$a = K_o^2 \alpha \qquad (2)$$

where α is the polarizability (Raleigh scattering), or by

$$a = r_e Z \qquad (3)$$

where r_e is the classical radius of the electron and Z the effective number of scattering electrons (Thomson scattering, X-ray).[20]

The differential scattering cross section for the single collision of the radiation with the elementary particle is then

$$\Sigma = |a|^2.$$

B. Scattering Amplitude and Index of Refraction

The index of refraction and the index of refraction increment at constant volume, or at constant pressure, are observables from which the scattering amplitude can be determined.

Refraction is a result of multiple scattering. For a system of N elementary particles the index of refraction n is[19]

$$n^2 - 1 = -\frac{4\pi}{K_o^2} \frac{N}{V} a \qquad (4)$$

If the sample is made of $\alpha = 1, \ldots, m$ components with N_α elementary particles, we get

$$a_\alpha = -\frac{K_o^2}{2\pi} nV \left(\frac{\partial n}{\partial N_\alpha}\right)_{\{N\},V} \qquad (5)$$

where the notation[8] for the subscript indicates that the quantity n has to be taken as a function of the set {N} and the volume V.

This formula holds true for neutron scattering as well as light scattering. To be more precise we should consider the states of polarization of the incident and outgoing beams. Although such considerations are very important, we shall ignore them here.

C. Scattering Amplitudes and Scattered Intensity

The result of a small angle scattering experiment with any radiation of wavelength λ can be described by the scattering vector $\vec{q} = \vec{K} - \vec{K}_o$, where $\vec{K}_o$ and $\vec{K}$ are respectively the incident and scattered wave vectors. In small angle scattering we are only interested in elastic coherent scattering processes, for which

$$|\vec{q}| = \frac{4\pi}{\lambda} \sin \frac{\theta}{2} \tag{6}$$

where θ is the scattering angle.

Neutron wavelengths available from cold sources in nuclear reactors range from 2 Å to about 15 Å.[21] The associated values of the scattering vector q bridge the gap between those of light scattering and classical X-ray scattering, and are matched to the molecular dimensions encountered in polymer physics (from some Angströms to some thousand Angströms). Thus, as in light or X-ray scattering experiments[20], neutron scattering techniques can be used to measure the size of a polymer coil in the so-called "Guinier" range,

$$qR_G < 1, \tag{7}$$

where R_G is the radius of gyration of the coil. Moreover, neutrons can also be used to determine the monomer-monomer correlation function in the so-called "intermediate" range,

$$R_G^{-1} < q < \ell^{-1}, \tag{8}$$

where ℓ is the step length.

However, the striking advantage of this technique arises from the differences between interaction parameters a (Equation (1)) of various nuclei belonging to the molecules in the sample. For electromagnetic radiation, the interaction parameter a (Equations (2) and (3)) also varies with molecular species, but less strongly than in the case of neutrons. For instance, with neutron radiation the parameter a can be negative as well as positive. This yields a considerable change, which is the basis of SANS experiments for the investigation of polymer solutions and melts.

A particular feature of neutron interactions with nuclei is the dependence of the parameter a on the spin states of the neutron and the nuclei. This gives rise to an incoherent isotropic scattering, which is a background on which the coherent scattering is superimposed.

The scattering amplitude by a sample of N identical particles is, in the single collision approximation,

$$A(\vec{q}) = a \sum_{j=1}^{N} e^{i\vec{q}\cdot\vec{r}_j} \tag{9}$$

where $\vec{r}_j$ is the position of the jth particle. The coherent elastic scattering cross section is

$$\Sigma(q) = |A(\vec{q})|^2. \tag{10}$$

Thus, the scattered "intensity" measured by the detector is

$$I(q) = C[\Sigma(q) + \Sigma_i] \tag{11}$$

where C is a constant which depends on the diffractometer specifications. When C is a known quantity, the measurement is "absolute". Σ_i is the background, mainly due to the incoherent scattering.

If we are able to subtract from Equation (11) the incoherent background, then the result contains all the physical information about the structure of the sample. It is written as

$$\Sigma(q) = b^2 S(q) \tag{12}$$

where S(q) is the Fourier transform of the pair correlation function of the scattering particles. b^2 is the contrast, and does not depend on the scattering vector. It measures the ability of the radiation to "see" the "solute" against the "solvent" background. Clearly the larger the value, the more intense the scattering.

3. THE CONTRAST

A. Contrast Length

A polymer sample used in small angle scattering experiments may be a mixture of several components. How one particular component can be discerned among the others is given by the contrast between components. The contrast is described by a parameter, the contrast length b, whose value depends on the radiation used.[22]

Part of the interest in using neutron beams in polymer physics is based on this term. This is due to the nuclear nature of the neutron-matter interaction. As pointed out in the above section, this

Table 1. Coherent scattering length and incoherent cross section for some atoms.[23]

atom	isotope	$a_{coh}(10^{-12} cm)$	$\sigma_{inc.}$
Hydrogen	^{1}H	- 0.374	79.7
	^{2}D	0.667	2.2
Carbon	^{12}C	0.665	∿ 0
Nitrogen	^{14}N	0.940	0.3
Oxygen	^{16}O	0.580	∿ 0
Sulphur	^{35}S	0.284	∿ 0

interaction is measured by the coherent scattering length a, which has been measured for all elements.[23] Values for some chemical elements are listed in Table 1, together with the incoherent cross section.

There is a very large difference between the scattering length of hydrogen ($a_H = -0.374 \times 10^{-12}$ cm) and deuterium ($a_D = 0.667 \times 10^{-12}$ cm). Therefore, the replacement of a hydrogen atom by deuterium is a major modification to the scattering, but does not markedly influence the thermodynamic properties of the molecule. It is possible to modify the contrast of a given system by dividing the homogeneous part into labeled (i.e. deuterated) and non-labeled parts. The homogeneous part can be the solvent, the solute, or both. This is the basis of all experiments using the contrast variation method (see Section 3D).

The interaction of electromagnetic radiations (light or X-ray) does not exhibit such dramatic variations, and as we shall see below, these experiments are less fruitful.

Contrast lengths are essential quantities in the interpretation of small angle scattering experiments on multicomponent systems. The determination of these lengths depends essentially on the definition of the scattering experiment. The following is not intended to give any detailed derivation of these quantities but, rather, to introduce the basic equations and concepts which are used in SANS.[8]

B. Simple Model

We assume that the scattering is produced by a fixed number of elementary particles filling a fixed volume. For a one-component system, this model yields zero scattering; for a multiple component system, it yields a finite scattering cross section. We will follow the derivation given in reference 24.

In order to make calculations easier the following assumptions will be made:

(i) the scattering medium is incompressible, and

(ii) the volume V does not depend on the composition of the mixture.

Such a system can be viewed as a lattice, each lattice point being occupied either by a solvent molecule or by a monomer of the polymer molecule. To be more general, we can say that each lattice point is occupied by one molecule of species α ($\alpha = 1, \ldots, m$). The number N_α of α molecules is fixed and the number of sites is $\Sigma_\alpha n_\alpha N_\alpha$ where n_α is the number of monomers for one α molecule. The volume of one monomer is v_α.

The total concentration of α monomers is given by $n_\alpha N_\alpha/V$ and the local concentration is

$$c_\alpha(\vec{r}) = \sum_{a\varepsilon\alpha}^{N_\alpha} \sum_{j\varepsilon a}^{n_\alpha} \delta(\vec{r} - \vec{r}_{ja})$$

where $\vec{r}_{ja}$ is the position of the jth monomer of a molecule a belonging to the species α. Condition (i) is written as

$$\sum_{\alpha=1}^{m} v_\alpha c_\alpha(\vec{r}) = 1 \tag{13}$$

Thus, if the Fourier transform of the local concentration is expressed as

$$\tilde{c}_\alpha(q) = \int_V d^3r \, e^{i\vec{q}\cdot\vec{r}} \, c_\alpha(\vec{r}),$$

from Equation (13) we get

$$\sum_{\alpha=1}^{m} v_\alpha \tilde{c}_\alpha(0) = V \tag{14}$$

and, with the condition $q^3V >> 1$,

$$\sum_{\alpha=1}^{m} v_\alpha \tilde{c}_\alpha(q) = 0. \tag{15}$$

In a scattering experiment, the scattering amplitude, Equation (9), is written for a multicomponent system as

$$A(\vec{q}) = \sum_{\alpha=1}^{m} \sum_{a\varepsilon\alpha}^{N_\alpha} \sum_{j\varepsilon a}^{n_\alpha} a_{j\alpha} e^{i\vec{q}\cdot\vec{r}_{ja}} \tag{16}$$

where $a_{j\alpha}$ is the scattering length of the jth monomer of an α molecule. Generally the mixture contains one solvent. We indicate this solvent by setting $\alpha = 1$ and, with the assumption of equal volume between monomers and solvent molecules, the scattering amplitude becomes

$$A(\vec{q}) = \sum_{\alpha=2}^{m} \sum_{a\varepsilon\alpha}^{N_\alpha} \sum_{j\varepsilon a}^{n_\alpha} a_{j\alpha} e^{i\vec{q}\cdot\vec{r}_{ja}} + a_1\tilde{c}_1(q).$$

Using condition (i) (Equation (15)) we can write

$$A(\vec{q}) = \sum_{\alpha=2}^{m} \sum_{a\varepsilon\alpha}^{N_\alpha} \sum_{j\varepsilon a}^{n_\alpha} b_{ja}^{(1)} e^{i\vec{q}\cdot\vec{r}_{ja}} \tag{17}$$

where

$$b_{j\alpha}^{(1)} = a_{j\alpha} - v_\alpha \frac{a_1}{v_1} \tag{18}$$

is the contrast length with respect to the solvent molecule, for the jth monomer of the αth molecule. It should be noted that this result is coherent with the constant volume assumption only if the volume v_1 of the solvent molecule is equal to the volume v_α of the monomer. Equation (18) is applicable to any mixture as long as assumption (i) is fulfilled. For polymeric systems this contrast notion holds true for scattering vector values such that

$$q << \ell^{-1} \tag{19}$$

where ℓ is the monomer length. In that case the scattering length of a monomer α is the sum of the coherent scattering length a_i of each atom i (see Table I) of this monomer

$$a_\alpha = \Sigma_i a_i. \tag{20}$$

For example the scattering length for the normal polystyrene monomer (C_8H_8) is $a_H = 2.32 \times 10^{-12}$ cm and for the deuterated one (C_8D_8), $a_D = 10.70 \times 10^{-12}$ cm. These values clearly show that the contrast length (Equation (18)) can have very different values for normal and deuterated polymers.

C. Exact Model - Forward Scattering Calculation

We now consider an experiment in which the scattering is produced by $\alpha = 1, \ldots, m$ elementary particles fluctuating in a sample holder at fixed pressure P. The incident radiation intercepts a fixed volume fraction V of the sample, and the number N_a of particles in V is allowed to fluctuate. With this assumption a general formula for forward scattering

$$\lim_{q \to 0} \Sigma(q)$$

was derived by J. des Cloizeaux and G. Jannink[26]

$$\lim_{q \to 0} \Sigma(q) = \beta^{-1}\left\{\langle a|N\rangle^2 \frac{\chi_T}{V} + \langle b'|G_E^{-1}|b'\rangle\right\}. \tag{21}$$

In this expression $\beta = k_BT$ where k_B is the Boltzmann constant and T the temperature. $\langle a|$ is the vector in the m dimensional space, $a = a_\alpha (\alpha = 1, \ldots, m)$. a_α is the coherent scattering length (Equation (20)). $\langle N|$ is the vector $N_1, \ldots, N_m$. χ_T is the isothermal compressibility of the system. The vector $\langle b'|$ is

$$\langle b'| = |b\rangle - |t\rangle \frac{\langle t|b\rangle}{\langle t|t\rangle}$$

where (cf. Equation (18))

$$|b\rangle = |a\rangle - |v\rangle \frac{\langle N|a\rangle}{\langle N|v\rangle}$$

and $|t\rangle$ is an arbitrary vector (but not perpendicular to $|N\rangle$). $|v\rangle$ is the vector $v_1, \ldots, v_m$ and G_E is the projection of the operator

$$G_{\alpha\beta} = \left(\frac{\partial \mu_\alpha}{\partial N_\beta}\right)_{\{N\},P} \tag{22}$$

on the plane perpendicular to $|t\rangle$. $(\partial\mu_\alpha/\partial N_\beta)_{P,\{N\}}$ is the partial derivative of the chemical potential μ_α[36] considered as a function of pressure and the number of particles.

With the choice $|t\rangle = 1,0,0, \ldots, 0$ we get the result obtained a long time ago[25] (the solvent index is 1):

$$\lim_{q\to 0} \Sigma(q) = \beta^{-1}\left\{(\Sigma_\alpha a_\alpha N_\alpha)^2 \frac{\chi_T}{V} + \sum_{\alpha,\beta\neq 1} b'_\alpha b'_\beta \left(\frac{\partial \mu_\alpha}{\partial N_\beta}\right)^{-1}_{\{N\},P}\right\} \tag{23}$$

where

$$b'_\alpha = a_\alpha - v_\alpha \sum_{\gamma=1}^{m} a_\gamma \frac{N_\gamma}{V}. \tag{24}$$

According to Equation (5) we notice that in this case

$$b'_\alpha = K_o^2 \frac{n}{2\pi} V \left(\frac{\partial n}{\partial N_\alpha}\right)_{\{N\},P}, \tag{25}$$

a well-known result of light scattering. The forward scattering cross section (Equation (23)) is then the sum of two terms related to

(1) the scattering length a_α, for the term proportional to the compressibility, and

(2) the contrast length b'_α (Equation (24)) for the term proportional to the derivative of the chemical potential at constant pressure. Equation (23) can be used to determine the thermodynamic quantities of the scattering system viz the compressibility, the osmotic pressure, and the partial volumes v_α.

D. Values of Contrast Length

The monomer contrast length defined by Equation (18) is used in all neutron scattering experiments. The contrast length b'_α (Equation (24)) is expressed as a sum of these elementary lengths. Using the expression of the concentration $c_\gamma = N_\gamma/V$ of the γ particles, Equation (24) becomes

$$b'_\alpha = a_\alpha - v_\alpha \sum_{\gamma=1}^{m} a_\gamma c_\gamma$$

and, with the condition of Equation (13),

$$b'_\alpha = \sum_{\gamma\neq\alpha} c_\gamma v_\gamma \left(a_\alpha - v_\alpha \frac{a_\gamma}{v_\gamma}\right). \tag{26}$$

Thus, by setting

$$b_{\alpha 1} = \sum_{j=1}^{n_\alpha} b_{j\alpha}^{(1)}$$

and by returning to the more general case, Equation (18) my be written as

$$b_{\alpha\gamma} = a_\alpha - v_\alpha \frac{a_\gamma}{v_\gamma} \tag{27}$$

so that, from Equations (26) and (27), we get

$$b'_\alpha = \Sigma_{\gamma \neq \alpha} C_\gamma v_\gamma b_{\alpha\gamma} . \tag{28}$$

It is recalled that v_α is the molecular volume and c_α the concentration of the scattering units. Molecular volumes are related to partial specific volumes w_α by the relation

$$v_\alpha = \frac{w_\alpha M_\alpha}{A}$$

where M_α is the molecular weight and A the Avodadro number. The quantities w_α are determined by densitometric methods.[27,28]

For a given experiment, the components of the mixture must be chosen in order to give a contrast length [Equation (27)] as large as possible. Let us consider the contrast length per unit volume, given as

$$B_{\alpha\beta} = \frac{b_{\alpha\beta}}{v_\alpha} = \frac{a_\alpha}{v_\alpha} - \frac{a_\beta}{v_\beta} = A_\alpha - A_\beta , \tag{29}$$

where A_α is the coherent scattering lengths per unit volume

$$A_\alpha = \frac{a_\alpha}{v_\alpha} .$$

Values of a_α, A_α, and the incoherent scattering cross section are given in Table 2 for various solvents and polymers. By Equation (29), values of $B_{\alpha\beta}$ for a given mixture are obtained as the differences between the corresponding scattering lengths A_α. This difference can be visualized by the diagram of Figure 1:[29] the values of A_α are displayed on the same scale for solvent molecules and monomers and the contrast length $B_{\alpha\beta}$ for a given solvent β and a given polymer α is then directly estimated.

The main advantage of the neutron scattering technique when comparing with the light scattering technique is indicated as follows. In the latter case the expression of the contrast length is somewhat different. As noted in Equation (25), it is proportional

Table 2. Values of the coherent scattering length, coherent scattering length per unit volume and incoherent scattering cross section for some solvents and polymers (see Reference 29)

			$a(10^{-12} cm)$	$A(10^{+10} cm\ cm^{-3})$	σ_i(barn)
1. Solvents					
	Carbon disulphide	CS_2	1.23	1.23	0.01
	Benzene	C_6H_6	1.75	1.19	480
		C_6D_6	7.99	5.45	13.2
	Cyclohexane	C_6H_{12}	- 0.498	- 0.278	960
		C_6D_{12}	11.99	6.69	26.5
	Toluene	C_7H_8	1.66	0.94	640
		C_7D_8	10	5.68	17.7
	Water	H_2O	- 0.168	- 0.56	160
		D_2O	1.914	6.40	4.4
2. Polymers					
	Ethylene	$-CH_2-$	- 0.08	- 0.31	160
		$-CD_2-$	2	7.44	4.4
	Isoprene	$-C_5H_8-$	0.33	0.27	640
		$-C_5D_8-$	8.67	5.12	17.6
	Styrene	$-C_8H_8-$	2.32	1.41	640
		$-C_8D_8-$	10.7	6.61	17.6

to a readily measurable quantity, the increment of the refractive index. If this quantity is expressed as function of the concentration (in g cm^{-3}), the elementary constrast length is written as

$$b_{\alpha\beta} = \frac{K_o^2}{2\pi} n_o \left(\frac{\partial n}{\partial c_\alpha}\right)_{\{c\},P} \left(\frac{M_\alpha}{A}\right)$$

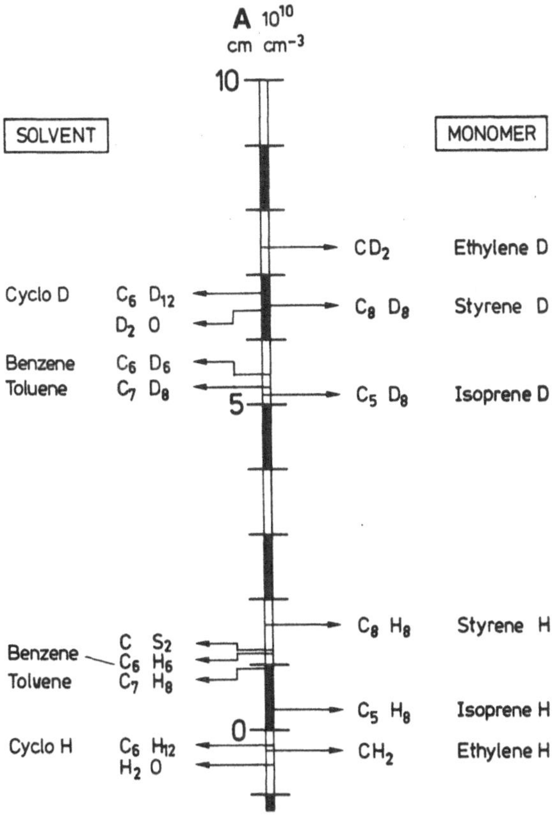

Figure 1. Scattering length by unit molecular volume for different solvents and polymers. The contrast length for a polymer-solvent mixture is given by the difference between these scattering lengths. (from Reference 29).

where n_0 is the solvent index. Using values given in Reference 30, the values of $B_{\alpha\beta}$ for some polymer solvent systems are given in Table 3.

A comparison with values obtained from Figure 1 shows that neutron contrast lengths are greater than these values by one or two orders of magnitude. This effect can be enhanced by using labeled and unlabeled molecules and by changing the composition of one or more species of the mixture. For example, this can be done by mixing X deuterated and 1-X normal solvent molecules. Then the scattering length $a_1(x)$ of one solvent molecule (subscript 1) is

$$a_1(x) = x\, a_D + (1-x)\, a_H \tag{30}$$

Table 3. Index increment, solvent index n_o[30] and contrast length per unit volume for a wavelength radiation λ_o = 5460 Å for polystyrene solutions

Solutions	$\partial n/\partial c(cm^3 g^{-1})$	n_o	$B_{\alpha\beta}(10^8 cm\ cm^{-3})$
Polystyrene/Benzene	0.1034	1.498	0.34
-Cyclohexane	0.1682	1.424	5.30
-Toluene	0.109	1.494	3.60
$-CS_2$	~ 0	1.630	~ 0

where a_D and a_H are, respectively, the scattering length of the deuterated and hydrogeneous molecules. The contrast length [Equation (27)] becomes

$$b_{\alpha 1}(x) = a_\alpha - v_\alpha \frac{a_1(x)}{v_1} \tag{31}$$

It is also possible to modify the scattering length of the solute by substitution of a fraction Y of deuterated molecules to the solute molecules.

By using selective labeling and contrast variations more information can be obtained from neutron scattering experiments than with other radiations. This point will be discussed in Section 4.

E. Experimental examples

Some examples of intensity scattered by polymer samples are shown in Figures 2, 3 and 4. Two curves are displayed on each figure. The upper one is the scattered intensity by the polymer mixture (polymer-solvent, polymer-polymer-solvent or polymer-polymer). It is the intensity given by Equation (11). The lower one corresponds to scattering by the "solvent" and is called the background. The excess scattering, obtained by substraction of the two curves, contains all the information about the structure of the scattering medium, as given by Equation (12).

Figure 2 shows the scattered intensity for two solutions of deuterated polystyrene in carbon disulphide. The concentrations are respectively 0.005 g cm^{-3} and 0.04 g cm^{-3}. In that case the background is the pure solvent CS_2, and the contrast is maximum as indicated by Figure 1.

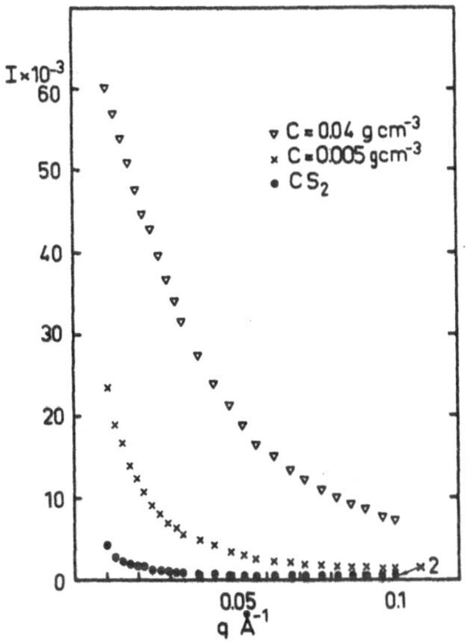

Figure 2. Scattered intensity as a function of the scattering vector q for solutions of deuterated polystyrene ($M_W = 5 \times 10^5$) in good solvent (CS_2) for two concentrations. The scattered intensity by pure solvent (closed circles) has been multiplied by a factor of 2.

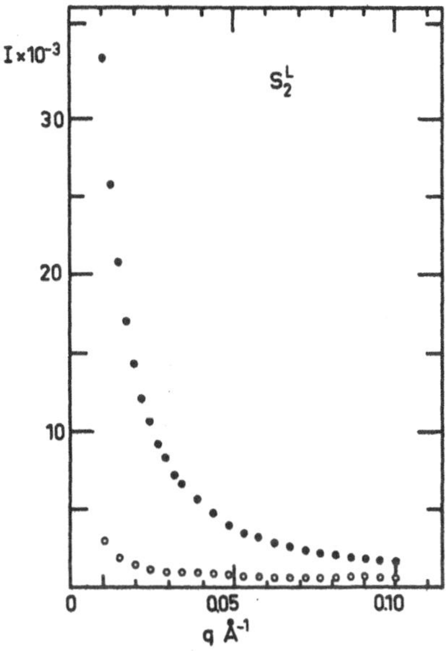

Figure 3. Plot of the scattered intensity versus scattering vector for a solution of protonated polystyrene ($M_W = 5x10^5$) in a good solvent (CS_2). This solution contains a concentration of $C_D = 0.005$ g cm^{-3} of deuterated polystyrene ($M_W = 5x10^5$). The total concentration is $C_T = 0.04$ g cm^{-3}. The lower curve is the intensity scattered by solution of protonated polystyrene ($C_H = 0.035$ g cm^{-3}).

The scattered intensity displayed on Figure 3 is recorded for a solution of hydrogeneous polystyrene in carbon disulphide. This solution contains a concentration $C_D = 0.005$ g cm^{-3} of deuterated polystyrene of the same molecular weight. The total concentration is $C_T = 0.04$ g cm^{-3}. In that case the background is a solution of protonated polystyrene with a concentration $C_T = 0.035$ g cm^{-3}. The data displayed in this figure emphasize the great contrast difference between the two polymer samples (see Figure 1).

Typical scattered intensities recorded for bulk samples are

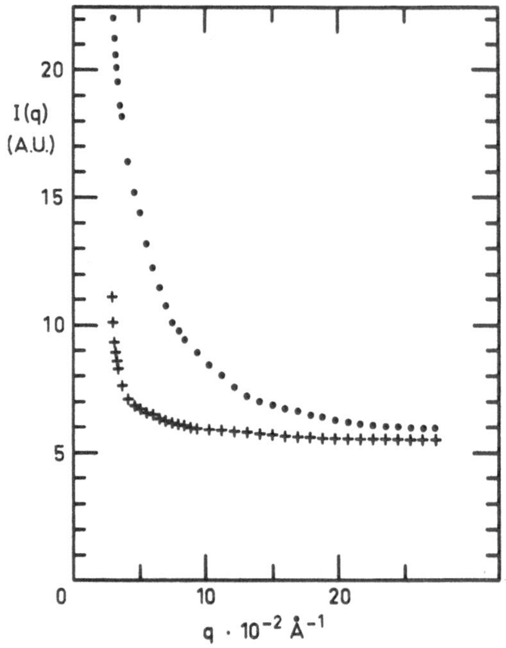

Figure 4. Scattered intensity as a function of the scattering vector for bulk polystyrene with a concentration of $C_D = 0.01$ g cm^{-3} of deuterated polystyrene ($M_W = 2.1 \times 10^4$). The lower curve is the intensity scattered by the protonated matrix ($M_W = 2 \times 10^4$) (see Figure 5 of Reference 10).

shown in Figure 4. The upper curve corresponds to the scattering by deuterated polystyrene dispersed in a matrix of undeuterated polystyrene. The concentration is $C_D = 0.01$ g cm^{-3}. The lower curve corresponds to the scattering by undeuterated polystyrene bulk material. Ideally this curve should show no dependence on the scattering vector. The increase at low angles is due to the divergence of the incident beam, since even without sample such an effect is observed.

We next consider two examples which are applications of Equation (23) for incompressible mixtures.

First we consider the case where the number of components m=2. For a simple mixture with two components, where the solute concentration is c_2, Equation (23), modified by a standard thermodynamic derivation,[31] becomes:

$$\lim_{q \to 0} \Sigma(q) = \beta^{-1} V\, b_2^2\, c_2\, (\partial\pi/\partial c_2)^{-1} \tag{32}$$

Then, measurements of the scattering cross section at the long wavelength limit lead to the determination of the osmotic compressibility $\chi_{TO} = \partial c_2/\partial\pi$. Precise determination of this quantity requires a so-called "absolute" measurement. Presently, this raises technical difficulties in neutron-scattering experiments which, however, do not appear in light scattering experiments. Typical values obtained from light scattering are displayed on Figure 5. Two sets of data are shown, both recorded for polystyrene solutions in the semi-dilute range. For the first one (open circle) the molecular weight is 7 x 10^6 and the solvent, benzene;[32] the second one is obtained from a solution in transdecalin at 40°C, the molecular weight being 1.3 x 10^7.[33] These data are consistent with the scaling law prediction:[7]

$$\chi_{TO} \sim c^{-5/4}$$

as displayed by the line with the slope 1/4.

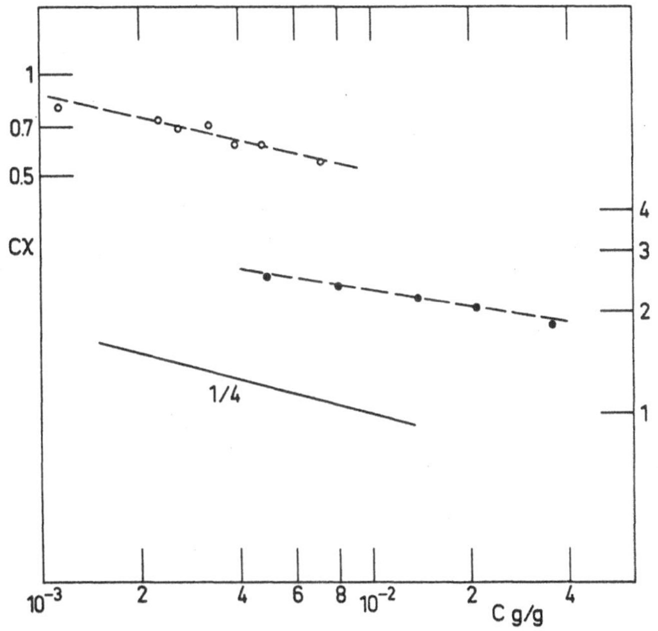

Figure 5. Log-log representation of the osmotic compressibility χ (multiplied by concentration C) versus concentration for polystyrene solution. Open circles: molecular weight M_W = 7 x 10^6, solution in benzene (data taken from Reference 32). Closed circles: molecular weight M_W = 1.3 x 10^7, solution in trans-decalin at 40°C (data taken from Reference 33). The line with slope 1/4 corresponds to the scaling law prediction.

When the number m of components is greater than two, the relation between $\Sigma(q)$ and thermodynamic quantities is less straightforward. When $m = 2$, relation of Equation (23) leads to the determination of the unique unknown term. For $m > 2$ we need some supplementary relations which can be obtained by varying the contrast lengths b_α. This method can be applied to cases where there are mutual interactions between components, described by chemical affinities. This is the case of polyelectrolyte solutions in water with added salt. If the water is denoted by 1, the salt by 2 and the polyelectrolyte by 3, the scattering cross section is then written as[37]

$$\lim_{q\to 0} \sum(q) = \beta^{-1}\left\{\frac{N_3}{V}\frac{(b_3+\zeta b_2)^2}{1-(N_3/V)(v_3+\zeta v_2)}\left(\frac{\partial\pi}{\partial N_3}\right)_D^{-1} + \frac{b_2^2}{G_{22}}\right\} \tag{33}$$

The index D indicates that the osmotic pressure gradient is to be measured at the dialysis equilibrium. The coefficient ζ, given as

$$\zeta = - G_{23}/G_{22},$$

is related to the preferential interaction coefficient between the salt and the polyelectrolyte.[34] The quantities $G_{\alpha\beta}$ are given by Equation (22).

Although relation of Equation (33) has been extensively used in light scattering experiments, neutron scattering techniques lead to more complete information because the contrast length can be varied. Following Equations (28) and (30) we get

$$b_\alpha(x) = a_\alpha - v_\alpha\{[x\, a_D + (1-x)\, a_H]\, c_1 + a_2c_2 + a_3c_3\}$$

where a_H and a_D are the scattering length for protonated and deuterated solvent. Let us suppose that there is a fraction x_1 for which $b_2(x_1) = 0$. In that case the scattering cross section, Equation (33), is directly proportional to the osmotic pressure gradient and does not depends on the solvent composition x. Moreover, if there exists another solvent composition x_o such that

$$b_3(x_o) + \zeta b_2(x_o) = 0, \tag{34}$$

the preferential interaction coefficient can be determined as long as the second term of the right hand of Equation (33) can be treated as a background. An experimental test of Equation (34) is now being performed with a solution of deuterated sulfonated polystyrene in a mixture of light and heavy water, with and without added salt.[35] Preliminary results are reported in Fig. 6. The square root of the recorded intensity at very small q value is plotted versus the content of

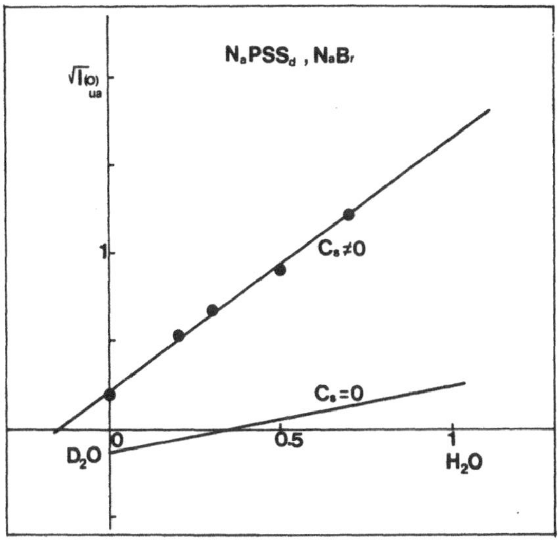

Figure 6. Variation of the square root of the scattered intensity at low scattering vector, as a function of the composition X of the mixture light water - heavy water, for a solution of polyelectrolyte (sulfonated polystyrene $M_W = 3 \times 10^4$) with ($C_S = 0$) and without ($C_S = 0$) salt. Polymer concentration $C_P = 4.76 \times 10^{-2}$ g cm^{-3}. Concentration of added salt (NaBr) $C_S = 1.47 \times 10^{-3}$ g cm^{-3}. The value X_o of the volume fraction $1-X_o$ of D_2O in water for which Equation (34) holds true is a function of added salt.[35]

light water and the value of the coefficient ζ is calculated from Equation (34). It can be seen that the value x_o of the volume fraction of D_2O in water for which Equation (34) holds true is a function of the added salt. The value of x_o for a solution without salt has been determined previously.[48]

4. THE SCATTERING FUNCTION

A. The Form Factor and the Structure Factor

For an incompressible solution of N polymers in a so vent, the coherent elastic cross section [Equation (10)] is written as

$$\Sigma(q) = \langle \sum_{\substack{ij \\ \alpha\beta}} b_{i\alpha}\, b_{j\beta}\, e^{i\vec{q}\cdot(\vec{r}_{i\alpha} - \vec{r}_{j\beta})} \rangle \tag{35}$$

with i and j being the indices of monomers belonging to polymers α and β, respectively. $r_{i\alpha}$ and $b_{i\alpha}$ are the positions and contrast length [Equation (18)] of the i-th monomer in molecule α. The sum in Equation (35) can be split into two terms. If, for reasons of simplicity, let us assume that the polymers are identical in all respects ($b_{i\alpha} \equiv b$), including molecular weight, we get

$$\frac{\Sigma(q)}{b^2} = \langle \sum_{\substack{j\,i \\ \alpha=\beta}} e^{i\vec{q}\cdot(\vec{r}_i-\vec{r}_j)} \rangle + \langle \sum_{\substack{i\,j \\ \alpha,\beta\neq\alpha}} e^{i\vec{q}\cdot(\vec{r}_{i\alpha}-\vec{r}_{j\beta})} \rangle$$

which can be written

$$\frac{\Sigma(q)}{b^2} = N\ S_1(q) + N(N-1)\ S_2(q) \tag{36}$$

where $S_1(q)$ is the polymer form factor which describes the conformation of one chain, and $S_2(q)$ is the structure factor of the solution,

$$S_1(q) = \langle \left| \sum_{j\in\alpha} e^{i\vec{q}\cdot\vec{r}_{j\alpha}} \right|^2 \rangle$$

$$S_2(q) = \langle \sum_{\substack{jk \\ \alpha,\beta\neq\alpha}} e^{i\vec{q}\cdot(\vec{r}_{j\alpha}-\vec{r}_{k\beta})} \rangle .$$

These two terms describe the mean configuration of polymers in solution and are functions of the concentration. The main problem in a scattering experiment is the determination of these two unknown functions. This can be achieved by the following ways:

i) Usually, the first term is determined by measuring the coherent scattering cross section for different concentrations. The extrapolation to the zero concentration limit gives the form factor $S_1(q, 0)$. This method can be used in all scattering experiments but in neutron scattering it is easier to choose a good contrast between polymer and solvent.

ii) The second method involves selective labeling of the polymer and has been used in some X-ray scattering experiment,[38] but is more powerful in neutron scattering. The solution is a mixture of labeled and normal polymers. The total concentration being fixed, the extrapolation is now done on the concentration of labeled molecules. This gives the form factor $S_1(q, c)$ at a given concentration c.

iii) A more useful scheme results in the determination of the two terms for any concentration by using both selective labeling and contrast variation. This last method generalizes the two others, so we will describe it in detail.[8]

Let X be the number fraction of labeled solvent molecules with coherent scattering length a_{L1}. The scattering length of a solvent molecule is, according relation, Equation (30),

$$a_1(x) = x\, a_{L1} + (1-x)\, a_1$$

and the contrast length of a monomer then is

$$b_1(x) = a - v\,\frac{a_1(x)}{v_1(x)} \tag{37}$$

where

$$v_1(x) = x\, v_1 + (1-x)\, v_{L1}.$$

Here, a is the scattering length and v the molecular volume of the monomer, and v_1 and v_{L1} the molecular volumes of the normal and deuterated solvent. If the solution contains a number fraction Y of labeled polymer molecules, the contrast length of a labeled monomer with coherent scattering length a_L and molecular volume v_L is

$$b_{L1}(x) = a_L - v_L\, a_1(x)/v_1(x). \tag{38}$$

Returning to the general case, a straightforward calculation reduces the coherent scattering [Equation (35)] to

$$\Sigma(q,c,x,y) = N\langle b^2(y)\rangle\, S_1(q,c) + N(N-1)\,\langle b(y)\rangle^2\, S_2(q,c) \tag{39}$$

where

$$\left.\begin{aligned} \langle b(y)\rangle &= b_1(x) + \bar{y}[b_{L1}(x) - b_1(x)] \\ &\text{and} \\ \langle b^2(y)\rangle &= b_1^2(x) + y[b_{L1}^2(x) - b_1^2(x)] \end{aligned}\right\} \tag{40}$$

Equation (39) provides a way to determine the two unknown quantities $S_1(q, c)$ and $S_2(q, c)$, by doing two experiments with two different contrast lengths. If we use the total scattering function $S_T(q, c)$ defined as

$$S_T(q, c) = N\, S_1(q, c) + N(N-1)\, S_2(q, c)$$

Equation (39) then can be written as[39]

$$\Sigma(q,c,x,y) = N\, Y(1-Y)[b(x)_{L1}-b(x)_1]^2 S_1(q,c) + \langle b(y)\rangle^2 S_T(q,c). \tag{41}$$

There are three experimental situations to be considered

i) Solution of identical labeled polymers. In that case Y=1, so Equation (41) reduces to

$$\Sigma(q,c,x,1) = b^2_{L1}(x)\; S_T(q,c) \tag{42}$$

the experiment yields a measure of the correlation between all monomers. The contrast variation is not generally used.

ii) In the bulk. The system is assumed to be incompressible so the total scattering functions $S_T(q)$ should vanish, as has recently been verified experimentally.[40] The scattering experiment provides a measurement of the single chain form factor $S_1(q,c_B)$, where c_B is the bulk concentration

$$\Sigma(q,c_B,Y) = N\, Y(1-Y)\; (a_L-a)^2\; S_1(q,c_B). \tag{43}$$

Furthermore, this expression indicates that $S_1(q,1)$ can be obtained even with a high concentration of labeled polymer, as already emphasized in an earlier experiment on bulk polystyrene.[10] The scattered intensity will reach the maximum for Y = 1/2.

iii) We can now discuss the use of the contrast variation defined by Equation (30). Let us suppose that there is a number fraction x_o of deuterated solvent molecules such that

$$b_1(x_o) = 0. \tag{44}$$

Then, Equation (39) takes the simple form

$$\Sigma(q,c,x_o,y) = Nb^2_{L1}(x_o)YS_1(q,c) + N(N-1)b^2_{L1}(x_o)Y^2S_2(q,c). \tag{45}$$

Now, if we take another number fraction Y of deuterated polymer, for example Y = 1, we get

$$\Sigma(q,c,x_o,1) = Nb^2_{L1}(x_o)S_1(q,c) + N(N-1)b^2_{L1}(x_o)S_2(q,c). \tag{46}$$

By combining these two equations, which contain two unknown quantities, it is possible to determine the form factor $S_1(q,c)$ and the structure factor $S_2(q,c)$ with only two experiments, viz.,

$$Nb_{L1}^2(x_o)Y(1-Y)S_1(q,c) = \Sigma(q,c,x_o,Y) - Y^2\Sigma(q,c,x_o,1) \tag{47}$$

and

$$N(N-1)b_{L1}^2(x_o)Y(1-Y)S_2(q,c) = \Sigma(q,c,x_o,Y) - Y\Sigma(q,c,x_o,1). \tag{48}$$

This method requires that Equation (44) is fulfilled. Values given on Figure 1 indicate that this fraction x_o exists in nearly all practical cases. For example, $x_o = 0.10$ for normal polystyrene in toluene and $x_o = 0.25$ for normal polystyrene in cyclohexane. If it is not possible to satisfy Equation (44), Equation (39) can be used in exactly the same way as Equation (45).

B. Scattering Vector Ranges and Physical Information.

The scattered intensity is seen to depend considerably upon the scattering vector q and this dependence is due to the existence of typical correlation ranges. In the "Guinier range" [Equation (7)] the form factor $S_1(q,c)$ is expressed as[20]

$$S_1(q,c) = 1 - \frac{q^2R_G^2}{3} \tag{49}$$

thus providing a measure of the radius of gyration. In the intermediate range [Equation (8)] details of the polymer conformation is observed. The scattering function is described by a power law of the scattering vector. For flexible polymer coils in dilute solution, the form factor can be written as[6]

$$S_1(q,\ c\rightarrow 0) \sim q^{-1/\nu} \tag{50}$$

where ν is the excluded volume exponent. In semi-dilute solutions the scattering function is expressed by a Lorentzian form[11]

$$S_T(q,c) \sim \frac{1}{q^2 + \xi(c)^{-2}} \tag{51}$$

from which the value of the screening length can be determined.

C. Experimental Examples

In order to illustrate the use of SANS we have consider the

determination of polymer conformations in the bulk and in semi-dilute solutions. These studies require measurement of the form factor $S_1(q,c)$, extracted from the scattering cross section by methods described in Sections 4A(i) and 4A(ii).

i) Bulk polymer. Experiments have been carried out in the Guinier range [Equation (7)] with samples of bulk amorphous polystyrene. These samples were a mixture of protonated polystyrene containing deuterated polystyrene with concentrations ranging from 0.5 to 2 x 10^{-2} g cm^{-3}. Different samples were made with different molecular weight M_W.

A typical example of the intensity scattered by such a sample is shown in Figure 4. The determination of the form factor $S_1(q,0)$ is carried out by the method described in Section 4A(i) and the value of the radius of gyration is obtained by using Equation (49). The result is displayed in Figure 7.

It was found that the coil configuration behaves in a way similar to the random walk, i.e.,

$$R_G \sim M_W^{0.5}.$$

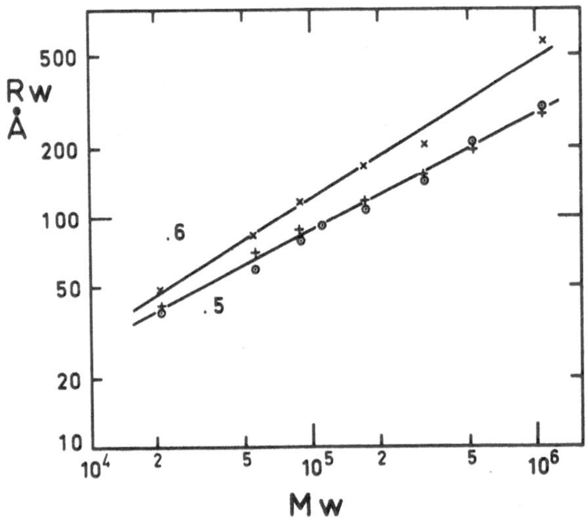

Figure 7. Radius of gyration plotted versus molecular weight in a log-log scale. The experimental data are obtained in different environments: Θ in the bulk, + in a theta solvent (cyclohexane), X in a good solvent (CS_2). (Reproduction of Figure 8 of Reference 10).

This relation was predicted[12] a long time ago but only could be tested recently when neutron scattering was made available. The same relationship holds true for the polymer coil in a poor solvent, i.e., a solution at a temperature close to the theta temperature, Θ. The polymer coil in a good solvent (i.e., in a solution at a temperature far away from Θ), exhibits a different behaviour. It is found that $R_G \sim M_W^{\nu}$, where $\nu \simeq .6$. Comparison between these three situations clearly supports the theoretical prediction: the polymer conformation is the same in a poor solvent and in the bulk.

This can also be proved by studying the scattering vector dependence of the scattering law in the intermediate range [Equation (8)], for three dilute solutions: in good solvent, in poor solvent, and in the bulk. The plot of the inverse of the scattered intensity as a function of q is shown in Figure 8 in a log-log scale. According to Equation (50) this is another way of determining the value of the excluded volume exponent. Again, the value of ν obtained in the bulk indicates the absence of an effective excluded volume interaction.

ii) Concentration effects. The results obtained in the bulk state raise the question of the polymer configuration variation

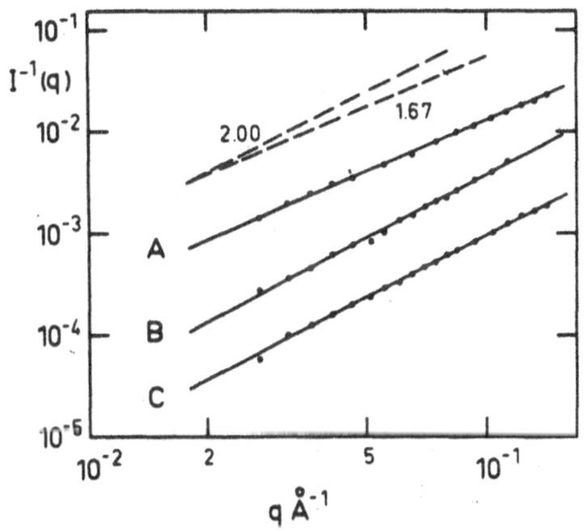

Figure 8. Inverse of the scattered intensity plotted versus scattering vector in a log-log scale for polymer in three dilute solutions: A) solution of deuterated polystyrene ($M_W = 1.2 \times 10^6$) in good solvent (CS_2) - concentration $C = 2.5 \times 10^{-3}$ g cm^{-3}. B) solution of protonated polystyrene ($M_W = 3.8 \times 10^6$) in theta solvent (deuterated cyclohexane, theta temperature 40°C) - concentration $C = 5 \times 10^{-3}$ g cm^{-3}. C) solution of deuterated polystyrene ($M_W = 1.1 \times 10^6$) in a matrix of hydrogeneous polystyrene of same molecular weight - concentration $C = 2.05 \times 10^{-2}$ g cm^{-3}. (Reproduction of Figure 1 of Ref. 6).

with concentration. The situation was clear in the two extreme concentrations. For dilute concentrations the polymer chain is expanded and the exponent in the power law dependence of the radius of gyration on the molecular weight is 0.6 (see Figure 7), at the bulk limit this exponent if 0.5.

Classical theory[12] predicted a decrease of the polymer dimension up to the overlap concentration c* (at that concentration the polymer chains begin to interpenetrate each other), where the polymer chain is in unperturbed state. In the scaling law description[7] the radius of gyration behaves as $c^{-0.25}$ for $c > c^*$.

This concentration dependence was tested in the following scattering experiment. The samples were a mixture of deuterated and protonated polystyrene dissolved in a good solvent. The form factor $S_1(q,c)$ is obtained by using the extrapolation described in Section 4A(ii). An example of such an extrapolation is displayed on Figure 9 for three total concentrations. The values of R_G as a

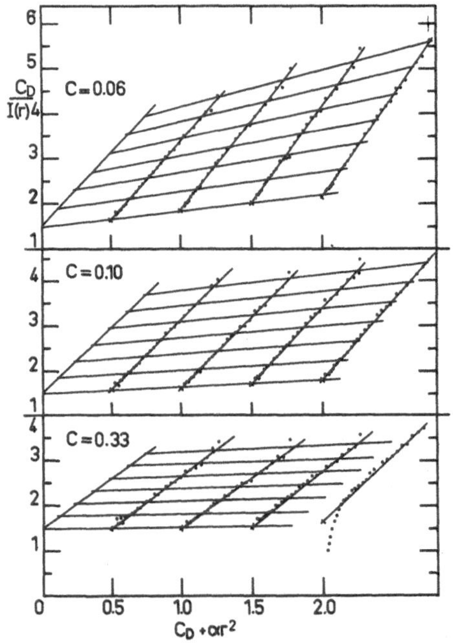

Figure 9. Inverse of the intensity scattered by solutions of hydrogeneous polystyrene (M_w=1.14x10^5) containing few deuterated polystyrene (same molecular weight), as a function of the square of the scattering vector (here noted r^2). C_D is the concentration of labeled polymer (C_D=0.02; 0.015; 0.01 and 0.005 g cm^{-3}). C is the total concentration $C = C_D + C_H$. Extrapolation to C_D=0 gives the form factor $S_1(q, c \to 0)$. (Reproduction of Figure 5 of Reference 11).

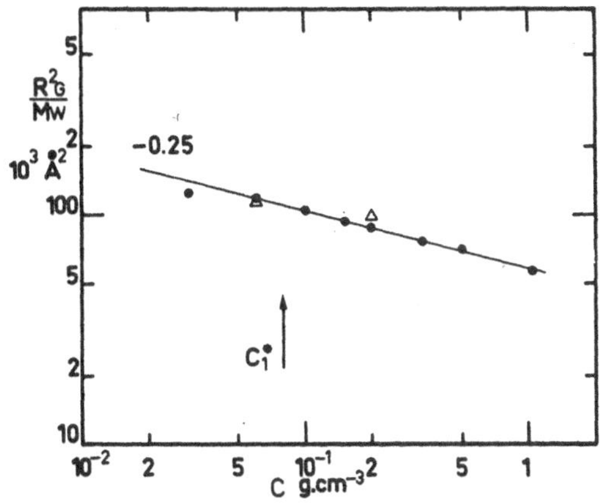

Figure 10. Log-log plot of the square of the radius of gyration divided by the molecular weights as a function of the concentration. Data are obtained for two different molecular weights: closed circles $M_W = 1.4 \times 10^5$, open triangles $M_W = 5 \times 10^5$. The slope determined by this plot is in excellent agreement with the scaling law prediction (Reproduction of Figure 7 of Reference 11).

function of the concentration are given in Figure 10 with a log-log scale. The variation is in agreement with the prediction of the scaling law theory, expressing the fact that, when the concentration increases, the intramolecular interactions become screened due to intermolecular interactions between segments on different chains. This effect is described by the screening length ξ.[16] For distances larger than ξ the portions of the same polymer chain have "forgotten" the presence of segments inside the length ξ, hence leading to an absence of excluded volume interaction; within the length ξ excluded volume still plays a role.

The screening length varies with the concentration according to the relation,[7] $\xi \sim c^{-3/4}$. It can be measured in a scattering experiment with a semi-dilute solution of identical polymers. According to the relation of Equation (42) this experiment leads to the determination of the total scattering function $S_T(q,c)$ which, for semi-dilute solutions, is given by Equation (51). A typical example of the intensity scattered by such a sample is given in Figure 2. The inverse of the scattering cross section is plotted on Figure 11 as a function of the square of the scattering vector. The lower curve corresponds to data recorded for dilute solution, and the variation is given by Equation (50) where $\nu = 0.6$.

The values of ξ, given by the abscissa intercept in Figure 11, are displayed in Figure 12 as a function of concentration in a log-

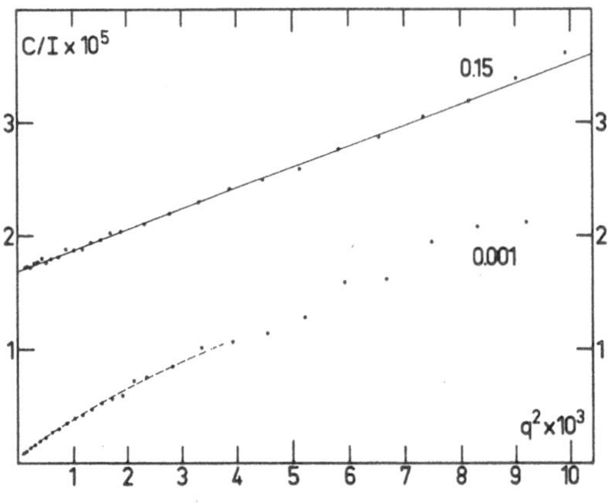

Figure 11. Inverse of the scattered intensity as a function of the square of the scattering vector q. Samples are solutions of deuterated polystyrene in good solvent (CS_2) at two concentrations. Upper curve $M_W = 5 \times 10^5$, C = 0.15 g cm^{-3} (semi dilute range), while lower curve $M_W = 1.1 \times 10^6$, C = 0.001 g cm^{-3} (dilute range). Note the difference in behavior. Value of the screening length is given by the extrapolation of the semi dilute solution curve.

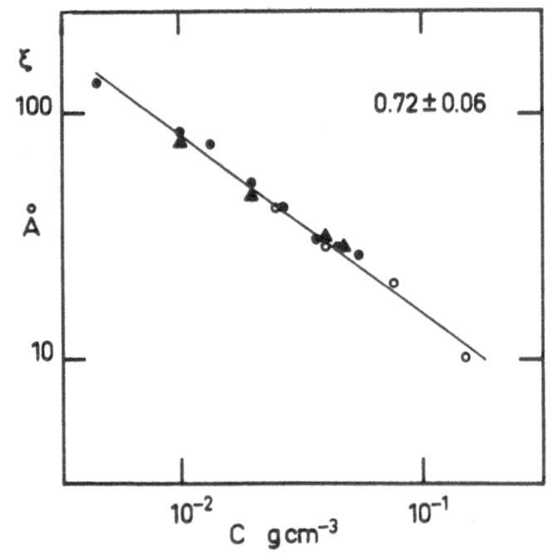

Figure 12. Log-log plot of the screening length as a function of concentration. The screening length is determined from the Lorentzian broadening [Equation (51)] of the scattered curves measured in the semi dilute range. Data are recorded for three different molecular weights: Closed circles $M_W = 2.1 \times 10^6$; closed triangles $M_W = 6.5 \times 10^5$; open circles $M_W = 5 \times 10^5$. (Reproduction of Figure 12 of Reference 11).

log scale. The value of the slope (0.72) is in good agreement with theoretical predictions.

The above examples are an illustration of the use of neutron scattering in the field of polymer physics. Since these early experiments, many other studies of the influence of temperature or concentration on polymer conformation have been done. It is not possible to give an exhaustive review of all these works.[50] They concern amorphous polymers,[41-44] gels and networks[45] and crystalline polymers.[46,47]

Finally, as pointed out recently,[48,49] the use of the method described in Section 4A(iii) enables measurement of the form factor $S_1(q,c)$:

i) Equation (45) provides a useful graphical method giving $S_1(q,c)$ and $S_2(q,c)$ from a plot of $\Sigma(q,c,x_o,y)/y$ versus y+q. This method has been applied[51] to data recorded in the experiment described in Section 4C(i). The result is displayed in Figure 13 for two total concentrations, one (c_T = 0.1 g cm^{-3}) above and the other (c_T = 0.03 g cm^{-3}) below the overlap concentration c*. The extra-

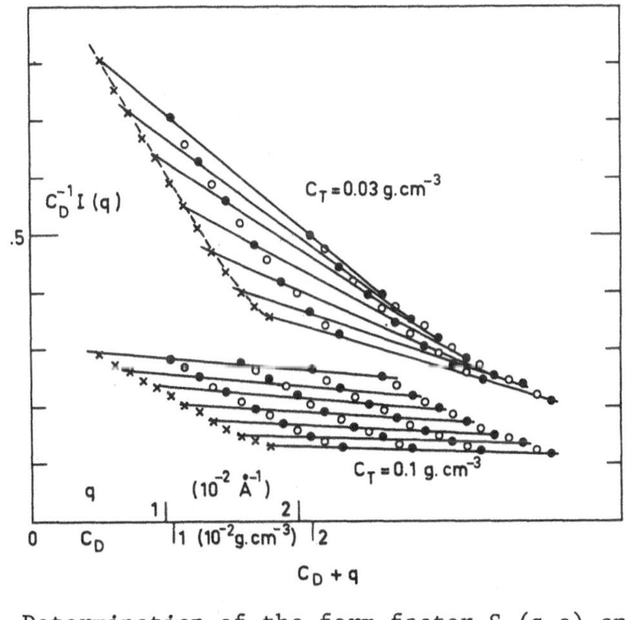

Figure 13. Determination of the form factor $S_1(q,c)$ and the structure factor $S_2(q,c)$ by using Equation (45): Samples are a mixture of deuterated and hydrogeneous polystyrene in solution in good solvent. The data are the same as in Figure 9.

polation to zero concentration in labeled polymer gives the form factor $S_1(q,c)$. The structure factor $S_2(q,c)$ is obtained from the slope of these extrapolation lines. The data for the higher concentrations clearly show the weakness of the interaction.

ii) This method is of particular interest when the q variation of the scattering cross section present a maximum. It is the case for scattered intensity by polyelectrolyte solution in pure water. A typical intensity distribution is shown in Figure 14 for different polymer concentrations.[52]

The form factor $S_1(q,c)$ and the structure factor $S_2(q,c)$ have been extracted using procedures described in Section 4A(iii). The result is displayed in Figure 15. It shows that the broad peak appears exclusively in $S_2(q,c)$ and consequently is due to an intra-

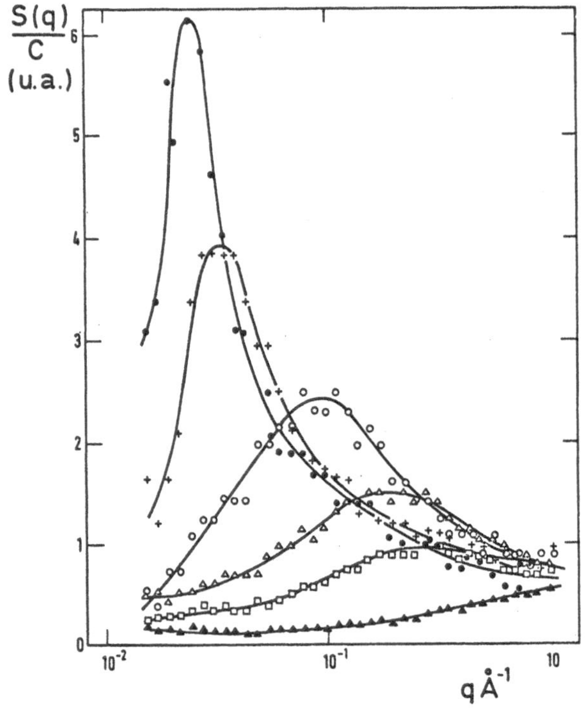

Figure 14. Scattered intensity as a function of the scattering vector for a solution of deuterated sulfonated polystyrene in pure water for different concentrations (in g cm^{-3})

•	10^{-2}	Δ	9.09×10^{-2}
+	1.96×10^{-2}	□	13.04×10^{-2}
o	4.76×10^{-2}	▲	23.00×10^{-2}

(Reproduction of Figure 3 of Reference 52).

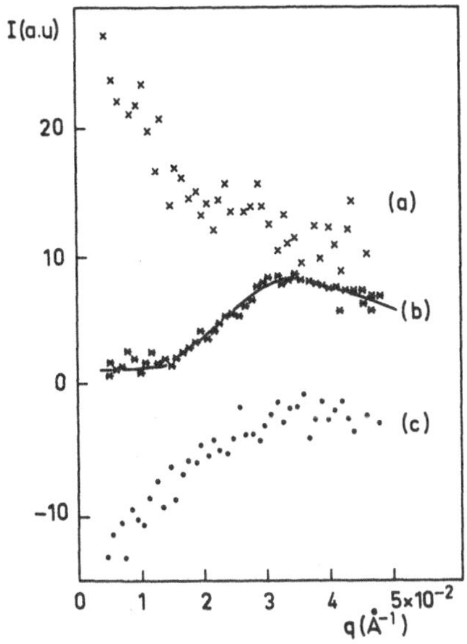

Figure 15. Decomposition of the measured scattering function (curve b) in form factor (curve a) and structure factor (curve c). Data are recorded for a solution of sulfonated polystyrene ($M_W = 3 \times 10^4$) in solution in pure water ($C = 2 \times 10^{-2}$ g cm^{-3}). (Figure 1 of Reference 48).

chain interaction. Furthermore, the negative sign corresponds to a repulsive interaction. This result is obtained in the intermediate scattering vector range [Equation (12)]. The same treatment can be applied to data recorded in the Guinier range [Equation (7)], allowing the determination of the radius of gyration.[48]

iii) The last example is an illustration of Equation (43). Data, recorded for bulk polystyrene samples with increasing concentration of deuterated polymer of same molecular weight, are displayed on Figure 16. The inverse of the scattered intensity multiplied by the factor Y(1-Y) is plotted versus the square of the scattering vector. It shows that the form factor $S_1(q,c_B)$ can be determined even for high concentration of labeled polymers.[40]

This method can be applied for the study of polymers chain out of equilibrium, in the stretched state for example. The large amount of deuterated chains will compensate for the low scattered intensity.

D. New Trends

SANS is a technique used to determine static correlation functions, characteristic of the sample structure. Recently, however,

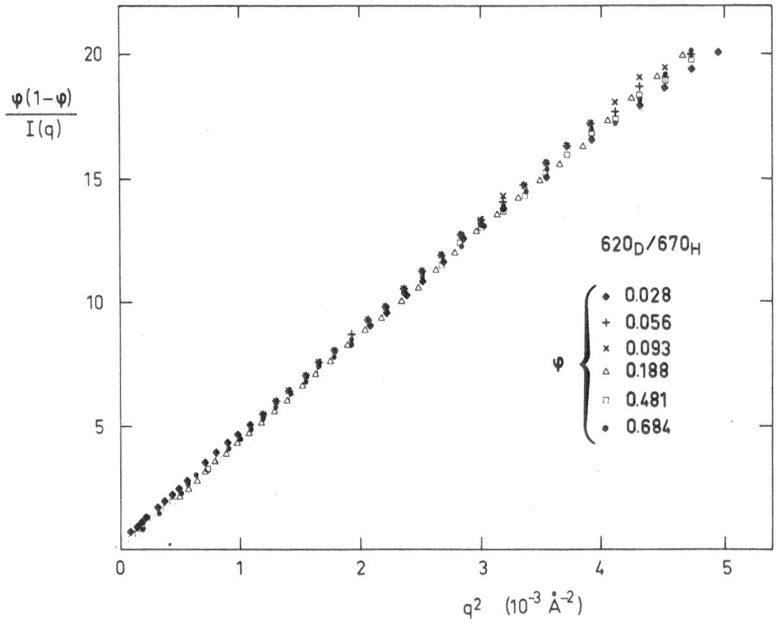

Figure 16. Inverse of the scattered intensity multiplied by $\varphi(1-\varphi)$ (φ the number fraction of labeled polymer) plotted versus the square of the scattering vector, for bulk protonated polystyrene ($M_W = 6.7 \times 10^5$) with increasing content of deuterated polymer ($M_W = 6.2 \times 10^5$). All these curves are nearly identical showing that the form factor can be obtained from Equation (43).

it has also become possible to derive static correlation functions from quasi-elastic scattering. The functions obtained from the interpretation of such experiments are not weighted in the same manner as those obtained from SANS and the comparison between elastic and quasi-elastic scattering therefore yields valuable information on corss-over problems.[54,55] Results depend on the performance of the spectrometer.

We also mention the application of SANS to the study of phenomena far from equilibrium (such as stress relaxation in rubbers). In this problem the experimental technique should be entirely reexamined. Use of spallation source could be of same advantage.[56]

5. CONCLUSION

We have derived the conditions necessary for the observation of neutron forward scattering by composition fluctuations in a mixture. These conditions are in great part based on the deuteration of organic substances. They are safe in so far as labeling by deuteration does not perturb the system, when it is not close to a critical demixtion point.

If the deuteration technique is developed in such a way as to label specific sites of molecules, the forward scattering experiment will give more detailed information concerning the spatial distribution. It is hoped that the experiment may one day observe an "elementary" pair correlation function in molecular systems.

ACKNOWLEDGEMENTS

We are greatly indebted for the collaboration of our colleagues J. P. Cotton, J. des Cloizeaux, M. Daoud, and M. Nierlich in writing this report.

REFERENCES

1. A. Guinier, and G. Fournet, Small Angle Scattering of X-rays, John Wiley, New York (1955).
2. See for example: D. McIntyre, and F. Cornick, eds., Light Scattering from Dilute Polymer Solutions, Gordon and Breach, New York (1964).
3. J. P. Cotton, B. Farnoux, G. Jannink, J. Mons, and C. Picot, C.R. Acad. Sci. C275: 175 (1972).
4. R. G. Kirste, W. A. Kruse, and J. Schelten, Makromol. Chem. 162: 299 (1973).
5. D. G. Ballard, J. Schelten, and G. D. Wignall, Eur. Polym. J. 9: 965 (1973).
6. J. P. Cotton et al, Phys. Rev. Lett. 32: 109 (1974).
7. P. G. de Gennes, Scaling Concepts in Polymers Physics, Cornell University Press, Ithaca, New York (1979).
8. A complete description will be available in a forthcoming book by J. des Cloizeaux and G. Jannink.
9. A. Maconnachie, and R. W. Richards, Polymer 19: 739 (1978).
10. J. P. Cotton et al, Macromolecules 7: 863 (1974).
11. M. Daoud et al, Macromolecules 8: 804 (1975).
12. P. J. Flory, Principles of Polymer Chemistry, Cornell University Press, Ithaca, New York (1967).
13. J. C. Le Guillou, and J. Zinn-Justin, Phys. Rev. Lett. 39: 95 (1977).
14. P. G. de Gennes, Phys. Lett. 38A: 339 (1972).
15. J. des Cloizeaux, J. Phys. (Paris) 36: 281 (1975).
16. S. F. Edwards, Proc. Phys. Soc. 88: 265 (1966).
17. M. Daoud, and G. Jannink, J. Phys. (Paris) 37: 973 (1976).
18. A. A. Davydov, Quantum Mechanics, 2d ed., Pergamon Press, New York (1964).
19. M. Born, and E. Wolf, Principles of Optics, 3d ed., Pergamon Press, New York (1965).

20. A. Guinier, Theorie et Technique de la Radiocristallographie, Dunod (1964).
21. See for example Figure 1 of Reference 10.
22. J. P. Cotton, CEA-N-1743 (1974).
23. G. E. Bacon, "Coherent neutron scattering amplitude," Acta Cryst. A28: 357 (1972).
24. J. P. Cotton, and H. Benoit, J. Phys. (Paris) 36: 905 (1975).
25. H. C. Brinkmann, and J. J. Hermans, J. Chem. Phys. 17: 574 (1949).
26. J. des Cloizeaux, and G. Jannink, Physica 102A: 120 (1980).
27. J. François, F. Candau, and H. Benoit, Polymer 15: 618 (1974).
28. J. François, and F. Candau, Eur. Polym. J. 9: 1355 (1973).
29. J. Ionescu, Thesis, Strasbourg (1979).
30. J. Brandrup, and E. H. Immergut, Polymer Handbook, part VII, J. Wiley, New York (1975).
31. P. A. Egelstaff, An Introduction to the Liquid State, Academic Press, London and New York (1967).
32. H. Benoit, and Picot, Pure Appl. Chem. 12: 545 (1966).
33. B. Chu, and T. Nose, Macromolecules 13: 122 (1980).
34. H. Eisenberg, Biological Macromolecules and Polyelectrolytes in Solution, Clarendon Press, Oxford (1976).
35. M. Nierlich, to be published.
36. C. Tanford, Physical Chemistry of Macromolecules, John Wiley, New York (1967).
37. G. Jannink, M. Nierlich, and C. Williams, C. R. Acad. Sci. (Paris) B290: 83 (1980).
38. H. Hayashi, F. Hamada, and A. Nakajima, Macromolecules 9: 543 (1976).
39. A. Z. Akcasu et al, J. Polym. Sci. 18: 863 (1980).
40. F. Boué, M. Nierlich, and L. Lieber, to appear [LLB-BP.2, 91190 Gif (France)].
41. M. Nierlich, J. P. Cotton, and B. Farnoux, J. Chem. Phys. 69: 1379 (1978).
42. R. W. Richards, A. Maconnachie, and G. Allen, Polymer 19: 266 (1978).
43. J. P. Cotton et al, J. Chem. Phys. 65: 1101 (1976).
44. B. Farnoux et al, J. Phys. (Paris) 39: 77 (1978).
45. R. Duplessix, Thesis, Strasbourg (1975). See also H. Benoit et al, Polym. Sci. A2, 14: 2119 (1976).
46. G. Lieser, E. W. Fisher, and K. Ibel, J. Polym. Sci. 13: 39 (1975); R. G. Kirste, B. R. Lehnen, and G. V. Schultz, Makrom. Chem. 177: 1137 (1976).
47. J. M. Guenet, and C. Picot, Polymer 20: 1483 (1979); D. M. Sadler, and A. Keller, Macromolecules 10: 1128 (1977).
48. C. E. Williams et al, J. Polym. Sci. Polym. Lett. 17: 379 (1979).
49. G. C. Summerfield, J. S. King, and R. Ullman, J. Appl. Cryst. 11: 548 (1978).

50. J. S. Higgins, and R. S. Stein, J. Appl. Cryst. 11: 346 (1978). Some recent works are published in J. Appl. Cryst. 11: 295-674 (1978).
51. J. P. Cotton, private communication.
52. M. Nierlich et al, J. Phys. (Paris) 40: 701 (1979).
53. B. Farnoux et al, J. Phys. (Paris) 39: 77 (1978).
54. M. Adam, and M. Delsanti, Macromolecules 10: 1229 (1977).
55. J. Hayter, G. Jannink, F. Brochard-Wyart, P. G. de Gennes, J. Phys. Lett. (Paris) 41: L451 (1980).
56. Y. Ishikawa, private communication.

QUASIELASTIC LIGHT SCATTERING FROM POLYMER GELS

Ralph Nossal

Physical Sciences Laboratory, DCRT
National Institutes of Health
Bethesda, Maryland

CONTENTS

ABSTRACT

Various schemes employing inelastic light scattering to measure rigidity parameters of polymer gels are reviewed. These include techniques which are understood to probe the longitudinal elastic modulus of polymer lattices excited by thermal fluctuations and, also, procedures which measure the bulk shear modulus of lattice and solvent together responding to weak mechanical perturbations. Theories which relate elasticity parameters to experimental measurements are outlined, and several applications are discussed.

1. INTRODUCTION

It seems to be difficult to provide a precise definition to the term "gel". In popular usage, the work refers to any semi-rigid 'jelly-like' or 'glue-like' material formed by the coagulation of a colloidal dispersion. Ferry[1] describes a gel as "a substantially diluted system which exhibits no steady-state flow". We restrict our attention, here, to polymer gels containing long polymer chains linked together to form an extended network which can resist stresses. The network is considered to be surrounded by fluid, which oftentimes is the solvent for the monomer from which the polymer chains are formed. Although polymer scientists are interested also in the properties of dry networks - i.e , where solvent has been drawn out of a gel - such materials cannot be studied by light scattering techniques.

The networks frequently are composed of linear polymers containing occasional interactions between the chains which give rise to a continuous three-dimensional matrix.[2] These interactions either can be fixed polyfunctional crosslinking units incorporated into the chains or labile entanglement points where the chains are sterically immobilized, the latter usually resulting in much weaker gels than do chemical crosslinks. Also, gels of intermediate strength may form when a network is stabilized by weak hydrogen bonding.

Quasielastic light scattering has been used to study a wide variety of gel-like substances. Examples are chemically crosslinked polyacrylamide[3,4] and polystyrene[5] gels, concentrated solutions of polystyrene,[6,7] hydrogen-bonded polysaccharide networks[8,9] and, among biological samples, fibrin clots stabilized both by hydrogen bonding and enzymatically induced covalent crosslinks,[10] cervical mucus,[11] and networks composed of actin and other contractile proteins.[12,13] The fraction of such gels which consists of polymer varies widely, from only tenths of a percent weight to volume in fibrin[10] and agarose[8,14] gels, to over 15 percent in samples containing polyacrylamide.[4,15]

The primary physical charactistic which distinguishes a gel from a viscoelastic fluid is the strength of the elastic response which such a material exhibits to an imposed mechanical stimulus. If the solvent and network are isotropic, the gel response is similar to that of a solid characterized by two elastic moduli.[16] (However, more complex representations are necessary if, e.g., the solvent exhibits liquid crystalline properties or the gel is inhomogeneously formed.)[17,18] Inelastic light scattering can be used to determine these moduli and, also, parameters relating to the dissipation of elastic energy either by friction between the polymer lattice and surrounding fluid or by internal rearrangements of the gel structure. If a sample is easily perturbed by an applied stress, the non-invasive character of light scattering makes such methodology preferable to conventional rheological techniques in which finite mechanical disturbances need to be imposed. Moreover, conventional rheological techniques provide values of bulk moduli, whereas certain light scattering techniques can be used to probe localized regions of a sample.

In the following we review several schemes which have involved inelastic light scattering to measure rigidity parameters of gels. They fall into two classes: i) where microscopic motions due to thermal fluctuations are probed, and ii) where bulk motions arising from weak mechanical disturbances of an entire sample are monitored. Some published applications also are discussed, mainly to illustrate

the possible uses of the techniques.

2. RAYLEIGH SCATTERING BY THERMAL FLUCTUATIONS

Consider a gel which is optically homogeneous on a length scale comparable to the wavelength of light. We assume that the irradiated region of the sample is free of macroscopic impurities and that it does not contain imperfections such as domain boundaries or inhomogeneous inclusions. It follows that the interference pattern at the detector varies because of local changes in density. We now briefly review a theory developed some years ago by Tanaka, Hocker, and Benedek[3] for the autocorrelation of photons scattered by thermally excited displacement waves in such materials.

Tanaka et al.[3] considered the polymer lattice to be an elastic continuum whose motion is stimulated by thermal fluctuations and damped by frictional interactions with surrounding solvent. Lattice motions were characterized by the displacement vector $\underset{\sim}{U}(\underset{\sim}{r},t)$ of a differential volume element whose equilibrium position is $\underset{\sim}{r}$. The quantity $\underset{\sim}{U}(\underset{\sim}{r},t)$ is a measure of changes in shape and volume, and indicates differences in locations of points within a body before and after a distortion (uniform translations or rotations of the entire sample are of no consequence). Fluctuations of the density $\rho(\underset{\sim}{r},t)$ are related to the displacement vector through the continuity equation from which, upon taking spatial Fourier transforms, one finds

$$\delta\rho(\underset{\sim}{Q},t) \simeq -i\rho_o\underset{\sim}{Q}\cdot\underset{\sim}{U}(\underset{\sim}{Q},t) \quad (1)$$

where ρ_o is the equilibrium density. The autocorrelation of photons is proportional to the field autocorrelation function $F_E^{(4)}(t) = \langle E_{sc}^*(Q,t)E_{sc}(Q,t)E_{sc}^*(Q,t+\tau)E_{sc}(Q,t+\tau)\rangle$ and, as the scattered field can be shown[20] to be proportional to the density fluctuation $\delta\rho$, it thus is seen from Equation 1 that the photon autocorrelation function depends on autocorrelation functions of $\underset{\sim}{U}(\underset{\sim}{Q},t)$. When the statistics of the scattered field are Gaussian, $F_E^{(4)}(t)$ factors as[19] $F_E^{(4)}(t) = |E_{sc}^*E_{sc}|^2\{1 + \langle E_{sc}^*(t)E_{sc}(t+\tau)\rangle^2/|E_{sc}^*E_{sc}|^2\}$ and the photon autocorrelation function depends upon autocorrelations of the displacement vector of the form $\langle U_i(Q,t)U_i(Q,t+\tau)\rangle$.

The kinetic equation for $\underset{\sim}{U}(\underset{\sim}{r},t)$ is[3]

$$\rho\frac{\partial^2 U}{\partial t^2} = \mu\nabla^2\underset{\sim}{U} + (K + \tfrac{1}{3}\mu)\underset{\sim}{\nabla}(\underset{\sim}{\nabla}\cdot\underset{\sim}{U}) - f(\partial\underset{\sim}{U}/\partial t) \quad (2)$$

where $\underset{\sim}{U}(\underset{\sim}{r},t)$ here is the displacement vector of the fiber network alone, ρ is the volume density of polymer, K and μ are the compressiblity and shear modulus of the lattice, and $\underline{f}$ is a friction coef-

ficient. The latter is assumed to be proportional to the viscosity of the solvent. Infinite spatial boundaries are assumed and Equation 2 then solved by taking Fourier-Laplace transforms. Nontrivial solutions of the form $U \sim \exp(i\omega t)$ are found if ω and Q satisfy

$$i\omega = -\frac{f}{2\rho}\left(1 \pm \sqrt{1 - 4C^2Q^2\rho^2/f^2}\right) \tag{3}$$

where, for waves propagating in the direction of $\underset{\sim}{Q}$, C is the longitudinal sound speed

$$C_\ell = [(K + \tfrac{4}{3}\mu)/\rho]^{1/2} \tag{4a}$$

or, for components propagating perpendicular to $\underset{\sim}{Q}$, C is the transverse sound speed

$$C_{tr} = (\mu/\rho)^{1/2} \tag{4b}$$

The character of the resulting waves depends on the values of K, μ, ρ, and f. When the product $2CQ\rho/f$ is less than 1, the excitations are non-propagating. Tanaka et al. observed that, in the limit $2CQ\rho/f << 1$, Equation 3 yields

$$-i\omega = \begin{cases} f/\rho \equiv (\tau_f)^{-1} & (5a) \\ \rho C^2Q^2/f \equiv (\tau_s)^{-1} & (5b) \end{cases}$$

and that the displacement autocorrelation function is given by

$$<U_j^*(Q,0)U_j(Q,t)> = <\overline{U_j(Q)^2}>[A_s e^{-t/\tau_s} + A_f e^{-t/\tau_f}] \tag{6}$$

They then showed that the amplitudes A_s and A_f are in the ratio $A_s/A_f = \tau_f/\tau_s = \rho^2C^2Q^2/f^2$ and, since for relatively stiff gels this term generally is very small (for polyacrylamide ca. 10^{-10}), that the first term within the square brackets in Equation 6 can be ignored.

The terms $\{<\overline{U_j(Q)^2}>\}$ were evaluated by thermodynamic arguments. Expressions for the field autocorrelation functions of both polarized and depolarized light were derived.[3,20] The resulting expression for polarized scattering is

$$<E^*_{pol}(Q,0)E_{pol}(Q,t)>$$

$$= \frac{I_o}{C}\left(\frac{\omega}{C}\right)^4 \frac{\sin^2\phi}{4\pi R^2}\left(\frac{\partial\varepsilon}{\partial\rho}\right)_T^2 \rho^2 \frac{LkT}{[K+\frac{4}{3}\mu]} \exp\left(-\frac{[K+\frac{4}{3}\mu]Q^2 t}{f}\right) \tag{7}$$

and for depolarized scattering is

$$<E^*_{dep}(Q,0)E_{dep}(Q,t)>$$

$$= \frac{I_o}{C}\left(\frac{\omega_o}{C}\right)^4 \frac{\sin^2\phi}{4\pi R^2}\left(\frac{\partial\varepsilon_D}{\partial U_{xy}}\right)_T^2 \frac{LkT}{\mu} \exp\left(-\frac{\mu Q^2 t}{f}\right) \tag{8}$$

where I_o is the intensity of incident light, ϕ is the angle between the polarization of the incident light and the scattering direction, c is the speed of light, c/ω_o is the wavelength of the incident radiation, R is the distance between the sample and the detector, L is the length of the illuminated sample region, k is Boltzmann's constant, and T is the absolute temperature. ε and ε_D are the diagonal and off-diagonal terms of the dielectric tensor. The quantity $(\partial\varepsilon/\partial\rho)_T$ is relatively easy to obtain by measuring the index of refraction of the sample; however, the term $(\partial\varepsilon_D/\partial U_{xy})$, which involves applying a shear deformation to the gel and measuring the ratio of depolarized to polarized intensity of light passed perpendicularly to the deformation, is more difficult to determine.

An example of an autocorrelation function of polarized light measured for polyacrylamide is shown in Figure 1a. The decay is seen to be well represented by a single exponential, with a decay rate proportional to Q^2 (see Figure 1b). The basic behavior predicted in Equation 7 indeed is observed. The data shown in Figure 1 are taken from the paper of Tanaka et al.,[3] but several laboratories (see e.g., Refs. 4, 5) have corroborated those results with similar materials. In contrast, measurements of μ which rely on Equation 8 have not yet been forthcoming, mainly because of the reduced intensity of depolarized scattering. (In the next section we describe another light scattering scheme which has been used to measure the shear modulus of polyacrylamide and other relatively soft gels).

Gels often contain what appears to be 'dust' or other particulate inhomogeneities. Also, macroscopic variations in refractive index (e.g., striations) can be frozen at the time of polymerization. These all can act as local heterodyning sources which cause deviations from the exponential behavior of the autocorrelation function, in which case measured photon autocorrelation functions would be predicted according to Equation 7 to vary as

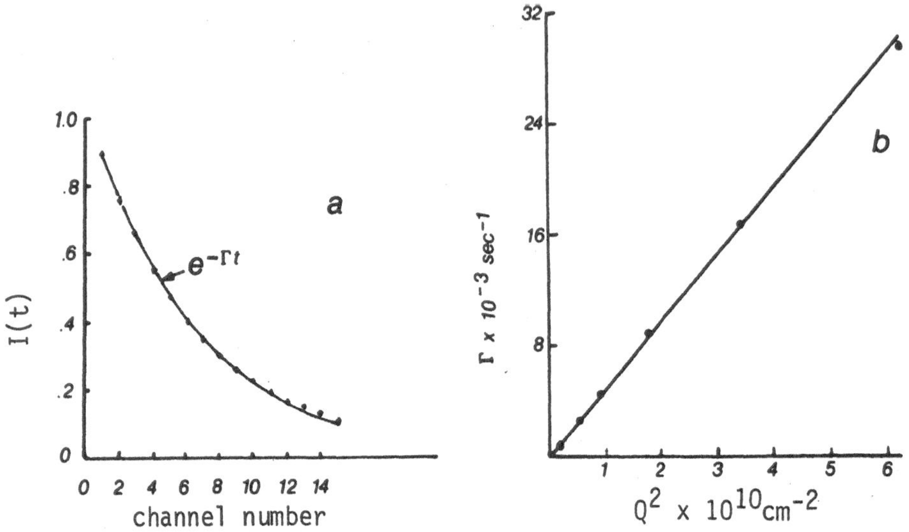

Figure 1. Inelastic light scattering from a 2.5% w/vol polyacrylamide gel (after Tanaka, et al.[3]). a) Exponentially decaying portion of the photon autocorrelation function. b) Variation of the decay constant $\Gamma = ([K + 4/3\mu]/f)Q^2$ with change in scattering angle, plotted as a function of the square of the magnitude of the Bragg scattering vector $Q = 4\pi n\lambda^{-1}\sin\theta/2$.

$$\langle n(0)n(\tau)\rangle \sim \langle E_{sc}^*(t)E_{sc}(t)E_{sc}^*(t+\tau)E_{sc}(t+\tau)\rangle$$

$$\sim \{b^2 + 2b(1-b)e^{-DQ^2t} + b^2e^{-2DQ^2t}\} \qquad (9)$$

where b is a constant. The signal would have the form of a double exponential, where the amplitudes and decay rates of the contributing terms would be related as above. Several authors[11,12,21] however, discerned deviations from "exponentiality" in their data which cannot be explained in terms of Equation 9, and have adopted other models.

If the terms which appear before the exponential in Equation 7 can be determined, values of the elastic moduli and friction coefficient can be separately evaluated. Tanaka, Ishimoto, and Ishiwata[22] combined quasielastic photon scattering with total intensity measurements to accomplish this task, but their method is somewhat tedious because each of the many factors which affect the intensity is individually measured. Hecht and Geissler[4] took advantage of the fact that their gel samples contained inclusions and imperfections which acted as local oscillators for heterodyning, and they expressed the time-varying part of the autocorrelation function as (cf. Equation 7)

$$S(t) = \frac{CI_0(N\Delta\tau)}{[K + \frac{4}{3}\mu]} \left(\frac{\partial \varepsilon}{\partial \rho}\right)^2 \rho^2 \exp\left(- \frac{[K + \frac{4}{3}\mu]Q^2 t}{f}\right) \tag{10}$$

where N is that total number of light pulses generated by the heterodyne reference source and $\Delta\tau$ is the clock period of the correlator.[6] The geometrical terms of Equation 7 now are absorbed in the constant C, which also includes such system parameters as the quantum efficiency of the detector and the ratio of the gate opening time of the correlator to the clock period. The factor C can be calculated by calibrating the spectrometer with concentrated polystyrene solutions,[6] for which the modulus $E = K + 4/3\mu$ is known from other sources. Typical data generated by this procedure, for various polyacrylamide samples, are shown in Figures 2a and 2b. (Decay rates were found to scale with Q^2).

The data given in Figure 2 indicate that the variation of the longitudinal modulus E with polymer concentration depends on the nature of the solvent. When the samples were polymerized in water and the temperature was sufficiently high, the modulus E is found (Figure 2a) to scale as $E \sim c^{2.35}$; the exponent is very close to the value 2.25 which has been predicted[17] for 'good' solvents (i.e., where the solvent is very much like the polymer, so the interaction energy of polymer with solvent is similar to solvent-solvent and polymer-polymer interaction energies). On the other hand, when a mixture of water and methanol is used as solvent, the modulus is found to scale (Figure 2b) approximately as c^3, which has been predicted[23] for 'poor' solvents (where the polymer chains tend to segregate and collapse). This good agreement between experimental results and theory tends to confirm the appropriateness of the physical model originally adopted by Tanaka et al.

A different procedure was employed by Munch et al.[5] in which inelastic light scattering measurements were combined with mechanical measurements of the modulus (unidirectional compression) in order to separately determine the friction coefficient. One problem with this scheme, however, is that low frequency elastic moduli determined by mechanical means are not necessarily equal to the high frequency moduli which enter into the inelastic light scattering measurements.[4] A separate theory was developed by these authors which resulted in expressions similar to those derived by Tanaka et al.[3] The theory relates fluctuations in network fibre density to concomitant changes in osmotic pressure and produces expressions for the decay of exponential autocorrelations in terms of molecular parameters.[5]

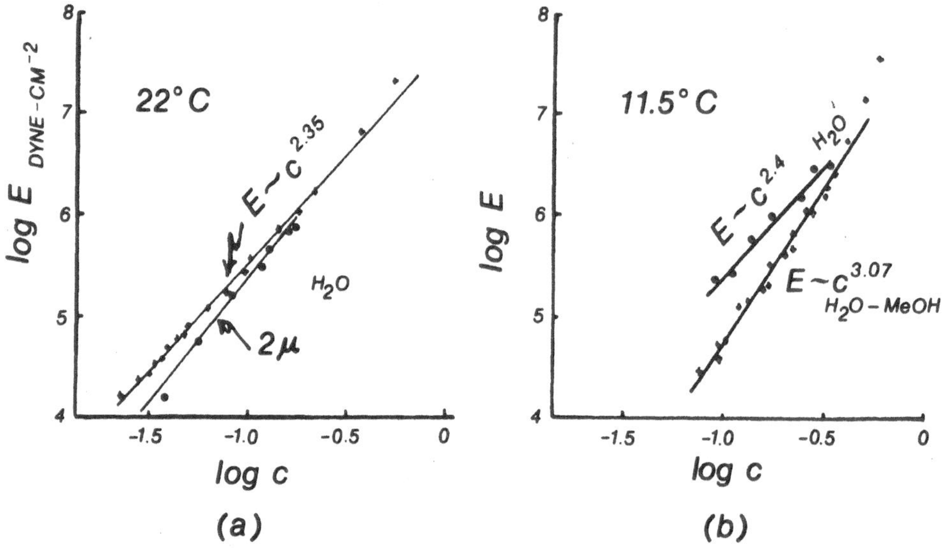

Figure 2. Concentration dependence of elastic moduli for polyacrylamide gels (after Geissler and Hecht[4]). a) Longitudinal elastic modulus $E = K + 4/3\mu$ of polyacrylamide-H_2O gels at 22°C (good solvent) determined by the light scattering technique described in Reference 6. Values of the shear modulus μ were determined by uniaxial compression. b) Longitudinal elastic modulus E at 11.5°C for polyacrylamide-MeOH/H_2O (poor solvent) and polyacrylamide-H_2O (good solvent) samples.

Another application of these techniques is to monitor phase changes in gels. Tanaka, Ishiwata, and Ishimoto[22] studied samples of 2.5% polycrylamide gels as they cooled, and discovered that both the intensity and correlation time increases as the temperature is lowered to the spinodal temperature T_s. Using a thermodynamic model which relates the compressibility modulus of a swollen network to the osmotic pressure and which employs expressions for the shear modulus derived from the theory of rubber elasticity, the authors predicted that the modulus should vary as $E = K + 4/3\mu = a(T - T_s)$, which indeed is observed in their data when it is interpreted in terms of Equation 7. From intensity data and the correlation times, the friction coefficient was determined and was found to decrease as the temperature decreases to T_s, a fact which was ascribed to the collapse of the lattice so that the "effective pore size becomes larger as the temperature decreases".[22]

Before completing this section, we note that when complex gels made of biological materials are investigated, one frequently finds that the physical model discussed above (where scattered light is related to the kinetics of lattice movements via $U(\underset{\sim}{r},t)$) cannot explain several experimental details. Carlson and Fraser[12] investigated samples of F-actin complexes and found that the relative amplitude of the time-varying part of the autocorrelation function decreased as gel was formed from actin polymers and myosin fragments. Also, the autocorrelation functions were found to have an angle dependence which was different from the Q^2 dependence discussed above. They thus formulated a model of harmonically bound Brownian particles subject to friction, and adopted a theory of Ornstein and Uhlenbeck[24] to obtain the following expression for the normalized intensity autocorrelation function $g^{(2)}(\tau)$[12]

$$g^{(2)}(\tau) = 1 + \exp\{-2DQ^2(1 - e^{-\gamma\tau})/\gamma\} - \exp\{-2DQ^2/\gamma\} \qquad (11)$$

where γ is defined as $\gamma \equiv k/f$, k being the stiffness coefficient and f the frictional factor, and D is a translational diffusion coefficient. Note that as the gel stiffens the relative amplitude of the time dependent term here decreases; also, as $\gamma \to 0$, the expression for freely diffusing particles $g^{(2)}(\tau) = 1 - \exp(-2DQ^2\tau)$ is recovered. Carlson and Fraser indicated that their simplified model was not meant to be a comprehensive description of molecular motions within a gel, but merely was proposed as an initial attempt to explain changes in light scattering spectra which were noted as F-actin gels were formed with crosslinking fragments of myosin. Subsequently, Wun and Carlson[15] examined polyacrylamide gels and obtained autocorrelation functions which could be fit by the expression

$$g^{(2)}(\tau) = 1+2\beta(1-\beta)[e^{-\Gamma_o\tau}-e^{-\Gamma_o}]+(1-\beta)^2[e^{-2\Gamma_o\tau}-e^{-2\Gamma_o}] \qquad (12)$$

where $\Gamma_o \equiv DQ^2/\gamma$ (cf. Equation 11) and β is the fraction of the total scattering intensity which is due to static scattering associated with microscopic inhomogeneities. Wun and Carlson studied the angle dependence of the amplitude $g^{(2)}(0)$ and, by comparing results with the observed angle dependence of Rayleigh ratios (normalized total integrated scattering intensities), construed that the static terms in the normalized photon autocorrelation functions cannot be explained simply as arising from internal heterocdyne sources considered independently of other aspects of gel structure (cf. Equation 22).

We note, also, a study of cervical mucus,[11] which is a complicated biological material containing a high concentration of complex glycoproteins in the form of an entangled random coli network. The observed correlation functions can be fitted by a double exponential, the fast decay showing a marked dependence on Q^2 (but not extrapolating to zero at small scattering angles), and the other having only a weak dependence on Q and being an order of magnitude slower. The absence of fixed crosslinks in this system seems to be the cause of the deviations form the predictions outlined above, and the theory may need modification in order to account for the labile nature of the interchain contacts.

3. MECHANICALLY EXCITED GELS

A different ligth scattering technique[25,26] can be used to determine the bulk shear modulus, G. The method involves using a weak mechanical field to excite standing displacement waves in a sample and then relating the frequencies of the excitations to the transverse sound speed C_{tr}. A major advantage of the procedure is that a sample does not need to be optically or mechanically homogeneous (in fact, the technique is somewhat easier to implement if a modest amound of gross optical heterogeneity is present). However, one complication is that excitation frequencies depend on the macroscopic dimensions of a specimen. The modulus G which is obtained is that of the compound gel composed of solvent and polymer lattice and, because the shear resistance of the solvent by itself is very low, G closely approximates the shear modulus μ of the lattice alone. Another parameter which in principle can be obtained is the gel viscosity η. This coefficient characterizes dissipation of elastic energy resulting from rearrangements of the interior gel structure such as changes in relative positions of entangled polymer chains or modifications of the interactions between lattice and solvent.

We again recall that the autocorrelation function of scattered photons $\langle n(t)n(t+\tau)\rangle$ is related to autocorrelations of the intensities of the scattered field as[19]

$$\langle n(t)n(t+\tau)\rangle \sim \langle E_{sc}^{*}(t)E_{sc}(t)E_{sc}^{*}(t+\tau)E_{sc}(t+\tau)\rangle \qquad (13)$$

To show how this function depends on the macroscopic gel displacements, we consider a model where the polymer lattice contains discrete dipoles located at positions $\{r_i\}$. In this case E_{sc} can be represented as[20,26]

$$E_{sc}(t) = \varepsilon(t)\int_{\substack{\text{scattering}\\ \text{volume}}} \sum_i \alpha_i \delta(\underset{\sim}{r} - \underset{\sim}{r}_i(t))e^{i\underset{\sim}{Q}\cdot\underset{\sim}{r}}\, d^3r \qquad (14)$$

where α_i is the "excess strength" of the ith dipole (α_i is proportional to the difference between the lattice and solvent polarizabilities) and $\varepsilon(t)$ is related to the incident field and the spectrometer geometry. Because $\underset{\sim}{r}_i(t)$ can be written as $\underset{\sim}{r}_i(t) = \underset{\sim}{r}_i^{0} +$

$\Delta r_i(t)$, where r_i^o is the location of the ith scattering center when the lattice is in its unstressed state, the Dirac delta function can be expanded as

$$\delta(r - r_i(t)) = \delta(r-r_i^o) + \frac{\partial}{\partial r_i}\delta(r-r_i)\Big|_{r_i=r_i^o}\cdot \Delta r_i(t)+\ldots \quad (15)$$

$$\approx \delta(r-r_i^o) - \frac{\partial}{\partial r}\delta(r-r_i)\Big|_{r_i=r_i^o}\cdot U(r,t)+\ldots \quad (16)$$

To obtain Equation 16 from Equation 15 we observe that deviations in lattice positions $\Delta r_i(t)$ can be represented to first order by the displacement vector $U(r,t)$, here pertaining to movements of differential volume elements of the entire macroscopic sample (not the lattice).

Thus, from Equations 14 and 16 we find

$$E_{sc}(t) = E_{sc}^o-[\varepsilon\int_V U(r,t)\cdot(\frac{\partial}{\partial r}\sum_i \alpha_i\delta(r-r_i^o))e^{iQ\cdot r}\, d^3r] \quad (17)$$

$$\approx E_{sc}^o + i\varepsilon Q\cdot\int_V U(r,t)\sum_i \alpha_i\delta(r-r_i^o)e^{iQ\cdot r}\, d^3r \quad (18)$$

$$\approx E_{sc}^o(1 + iQ\cdot U(\bar{r},t)) \quad (19)$$

where $\bar{r}$ is the center of the scattering volume. (To get the expression given in Equation 19, we noted that $U(r,t)$ varies on a length scale of the order of cm, and that dimensions of the scattering volume are much smaller). Many imprecise manipulations have been invoked to derive Equation 19, which therefore must be considered to be only a rather gross approximation. However, using Equations 13 and 19 provides the following expression for the photon autocorrelation function

$$\langle n(t)n(t+\tau)\rangle \approx \langle n\rangle_o^2(1+2[Q\cdot U(\bar{r},0)]^2+2\mathrm{Re}\{[Q\cdot U(\bar{r},0)]^*[Q\cdot U(\bar{r},\tau)]\})$$

$$= (\mathrm{const})_1+(\mathrm{const})_2\mathrm{Re}\{[Q\cdot U(\bar{r},0)]^*[Q\cdot U(\bar{r},\tau)]\} \quad (20)$$

where, for small displacements, the second term on the right hand side of Equation 20 is much smaller than the first. The term $\langle n\rangle_o^2$ is proportional to the scattering intensity from the unstressed sample, which can be quite large if the sample contains inclusions or fixed inhomogeneous regions.

where $\dot{U} \equiv \partial \underset{\sim}{U}/\partial t$. The coefficients χ and η can be interpreted as viscosity coefficients, and arise from general considerations of dissipative processes in solid-like materials.[16]

When Equation 23 is placed in Equation 21, the following is obtained:

$$\rho \frac{\partial^2 U}{\partial t^2} = (\mathcal{O}_\lambda + \mathcal{O}_\mu) \text{ grad div } \underset{\sim}{U} + \mathcal{O}_\mu \nabla^2 \underset{\sim}{U} + \underset{\sim}{F} \tag{24}$$

where the operators $\mathcal{O}_\lambda$ and $\mathcal{O}_\mu$ are given as $\mathcal{O}_\lambda = \lambda + \chi\, \partial/\partial t$ and $\mathcal{O}_\mu = \mu + \eta\, \partial/\partial t$. In cartesian coordinates Equation 24 thus becomes

$$\rho \frac{\partial^2 U_{xi}}{\partial t^2} = \mathcal{O}_\mu \nabla^2 U_{xi} + (\mathcal{O}_\lambda + \mathcal{O}_\mu) \frac{\partial}{\partial x_i} (\underset{\sim}{\nabla} \cdot \underset{\sim}{U}) + F_{xi} \tag{25}$$

for each component of the displacement vector $\underset{\sim}{U} = (U_x, U_y, U_z)$. Equation 25 pertains when gels are formed in cuvettes having rectangular cross-section.

Boundary conditions need to be specified before we can proceed. Two idealized cases result in easily solvable situations, both giving excitation frequencies which are either proportional to the longitudinal sound speed $C_\ell = [(K + 4/3\ G)/\rho]^{1/2}$ or the transverse sound speed $C_{tr} = [G/\rho]^{1/2}$, depending on the character of the mechanical excitation $\underset{\sim}{F}$ which is applied to the sample.[26] For reasons which are discussed at length later, shear waves, but not longitudinal waves, seem to correspond to the measure excitations.

The first set of boundary conditions mimics the case where a gel can slip easily along the sides of the cuvette. It involves setting the lateral stresses at the walls of the cuvette equal to zero, while asserting that the perpendicular detachment of the sample is restricted. The second case, which for plausible excitation schemes results in normal mode frequencies which are proportional to the transverse sound speed, mimics gels which adhere to the walls of the container.[26] These latter conditions are, for a rectangular cuvette,

$$U_y = U_z = 0,\ \partial U_x/\partial x = 0 \qquad \text{at } x = 0,\ a \tag{26a}$$

$$U_x = U_z = 0,\ \partial U_y/\partial y = 0 \qquad \text{at } y = 0,\ b \tag{26b}$$

$$U_x = U_y = 0,\ \partial U_z/\partial z = 0 \qquad \text{at } z = 0 \tag{26c}$$

and, if the top surface of the gel is free,

Expressions for the gel displacements $\underset{\sim}{U}(\underset{\sim}{r},t)$ are obtained by solving the elasticity equations. We again assume that the samples are spatially isotropic, in which case we have

$$\rho \frac{\partial^2 \underset{\sim}{U}}{\partial t^2} = \underset{\sim}{\nabla} \cdot \underset{\approx}{\sigma} + \underset{\sim}{F} \tag{21}$$

where $\underset{\sim}{F}$ is the vector of applied forces, and ρ now is the mass density of the <u>entire</u> gel. $\underset{\approx}{\sigma}$ is a generalized stress tensor. For small mechanical disturbances, the elements of the stress tensor are given in terms of the components of the displacement vector as[1,27]

$$\sigma_{ij} = \lambda \underset{\sim}{\nabla} \cdot \underset{\sim}{U} \delta_{ij}^{kr} + \mu \left(\frac{\partial U_i}{\partial x_j} + \frac{\partial U_j}{\partial x_i} \right) + \delta_{ij}^{kr} \int_{-\infty}^{t} \mathcal{L}(t-t') \nabla \cdot \dot{U}(t') dt'$$

$$+ \int_{-\infty}^{t} \mathcal{B}(t-t') \left(\frac{\partial \dot{U}_i(t')}{\partial x_j} + \frac{\partial \dot{U}_j(t')}{\partial x_i} \right) dt' \tag{22}$$

where δ_{ij} is the Kronecker delta. The constitutive equation given by Equation 22 describes relations between stress and strain in a cartesian coordinate system. For convenience we have chosen to use the Lamé coefficients λ and μ,[16] but Equation 22 equivalently can be written in terms of the bulk compressibility and shear moduli K and G, which are related to the Lamé coefficients as[16] $K = \lambda + 2/3\mu$, $G = \mu$. The kernels $\mathcal{L}(t)$ and $\mathcal{B}(t)$ relax to zero at long times.

A physical interpretation of Equation 22 is that the stress resulting from an impressed strain at first is maximal, but decreases as the interior structure of the sample gradually rearranges itself (i.e., soon after a strain is imposed the response given by Equation 22 is

$$\sigma_{ij} \approx [\lambda + \mathcal{L}(0)] \underset{\sim}{\nabla} \cdot \underset{\sim}{U} \delta_{ij}^{kr} + [\mu + \mathcal{B}(0)](\partial U_i/\partial x_j + \partial U_j/\partial x_i)$$

but is a lesser value thereafter). If the structural reorganization of the sample occurs rapidly on the timescale of experimental observation, as when studying low frequency standing waves of U(t), the kernels can be replaced by Dirac delta functions as $\mathcal{B}(t) \approx \eta\delta(t)$ and $\mathcal{L}(t) \approx \chi\delta(t)$, and Equation 22 becomes

$$\sigma_{ij} = \lambda \underset{\sim}{\nabla} \cdot \underset{\sim}{U} \delta_{ij}^{kr} + \mu \left(\frac{\partial U_i}{\partial x_j} + \frac{\partial U_j}{\partial x_i} \right) + \chi \underset{\sim}{\nabla} \cdot \dot{\underset{\sim}{U}} \delta_{ij}^{kr} + \eta \left(\frac{\partial \dot{U}_i}{\partial x_j} + \frac{\partial \dot{U}_j}{\partial x_i} \right) \tag{23}$$

$$\int_0^a \int_0^b (\sigma_{zx} U_x + \sigma_{zy} U_y + \sigma_{zz} U_z)\, dxdy = 0 \quad \text{at } z = c \tag{26d}$$

If, instead, the top surface is in contact with a cap (i.e., the gel is fully enclosed), a condition similar to Equation 26c is imposed. The conditions given by Equations 26a-26c essentially state that the sample cannot slip along the sides of the cuvette, but that the stress perpendicular to those surfaces is zero; the condition given by Equation 26d indicates that no net work is done in distorting a free surface.

When external forces and internal dissipation of elastic energy are neglected (i.e., take $\eta = \chi = 0$), the solutions of Equations 25 and 26 have the form[26]

$$\begin{aligned} U_x(r,t) &= \sum_{\alpha}\sum_{\beta}\sum_{\gamma} A_{\alpha\beta\gamma}(t)\cos\left(\frac{\alpha\pi x}{a}\right)\sin\left(\frac{\beta\pi y}{b}\right)\sin\left(\frac{\gamma\pi z}{c}\right) \\ U_y(r,t) &= \sum_{\alpha}\sum_{\beta}\sum_{\gamma} B_{\alpha\beta\gamma}(t)\sin\left(\frac{\alpha\pi x}{a}\right)\cos\left(\frac{\beta\pi y}{b}\right)\sin\left(\frac{\gamma\pi z}{c}\right) \\ U_z(r,t) &= \sum_{\alpha}\sum_{\beta}\sum_{\gamma} C_{\alpha\beta\gamma}(t)\sin\left(\frac{\alpha\pi x}{a}\right)\sin\left(\frac{\beta\pi y}{b}\right)\cos\left(\frac{\gamma\pi z}{c}\right) \end{aligned} \tag{27}$$

The time dependence of the coefficients $A_{\alpha\beta\gamma}$, $B_{\alpha\beta\gamma}$, and $C_{\alpha\beta\gamma}$ is given as $\exp(i\omega_{\alpha\beta\gamma}t)$, where $\omega_{\alpha\beta\gamma}$ is related to the wavenumbers α, β, γ as

$$\omega_{\alpha\beta\gamma} = \pi C\left[\left(\frac{\alpha}{a}\right)^2 + \left(\frac{\beta}{b}\right)^2 + \left(\frac{\gamma}{c}\right)^2\right]^{1/2} \tag{28}$$

C being either the transverse or longitudinal sound speeds. α and β are integers; γ is an integer if the cuvette is capped, or half integer if the top surface of the gel is free. The time dependence of the coefficients is found to be much more complicated when external mechanical forces and dissipation are included, and the amplitudes of the coefficients are proportional to the strengths of the force components. By considering simplified cases, e.g., where the forces are applied perpendicular to a cuvette face in only one direction, one finds that resonances in light scattering spectra occur at frequencies given by Equation 28, and have widths which depend on the viscosity coefficients.[25,26]

It is difficult to visualize the precise waveform of the excitations which occur in cuvettes having rectangular cross-section. However, mathematical complexities are lessened when cylindrical geometry is considered. If the external mechanical field has no angular structure - e.g., when a cuvette is subjected to a twist about its long axis - then, instead of Equation 25, one finds

$$\rho\frac{\partial^2 U_r}{\partial t^2} = [(\mathcal{O}_\lambda+2\mathcal{O}_\mu)(\frac{\partial^2 U_r}{\partial r^2}+\frac{1}{r}\frac{\partial U_r}{\partial r}-\frac{U_r}{r^2})+(\mathcal{O}_\lambda+\mathcal{O}_\mu)\frac{\partial^2 U_z}{\partial r\partial z}+\mathcal{O}_\mu\frac{\partial^2 U_r}{\partial z^2}]+F_r(r,z;t) \quad (29)$$

$$\rho\frac{\partial^2 U_\theta}{\partial t^2} = \mathcal{O}_\mu(\frac{\partial^2 U_\theta}{\partial r^2}+\frac{1}{r}\frac{\partial U_\theta}{\partial r}-\frac{U_\theta}{r^2}+\frac{\partial^2 U_\theta}{\partial z^2}) + F_\theta(r,z;t) \quad (30)$$

$$\rho\frac{\partial^2 U_z}{\partial t^2} = [(\mathcal{O}_\lambda+\mathcal{O}_\mu)(\frac{1}{r}\frac{\partial^2 (rU_r)}{\partial r\partial z})+(\mathcal{O}_\lambda+2\mathcal{O}_\mu)\frac{\partial^2 U_z}{\partial z^2}+\mathcal{O}_\mu(\frac{\partial^2 U_z}{\partial r^2}+\frac{1}{r}\frac{\partial U_z}{\partial r})]+F_z(r,z;t) \quad (31)$$

where U_r, U_θ, and U_z are the gel displacements along the radius, angular direction, and long axis of the sample.
These equations indicate that the angular displacments U_θ are associated only with shear waves (the operator $\mathcal{O}_\lambda$ does not appear in Equation 30).

If we assume that the gel sticks strongly to the walls of the cuvette, the boundary conditions on U_θ are

$$U_\theta(R,z;t) = 0 \quad (32a)$$

and, because for simplicity we here assume that the cuvette has a cap which is in contact with the top surface of the gel,

$$U_\theta(r,z;t) = 0 \qquad \text{at } z = 0,\ c \quad (32b)$$

External forces are transmitted to the sample where it interacts with the walls, and shear waves arise within the specimen because of the inertia of the interior. If the time dependence of the applied twist is given as $g_\theta(t)$ then, for small amplitudes and low twisting speeds, we may presume that the resultant internal body force is given as $\xi\rho r/R\ dg_\theta(t)/dt$. [The force on the gel must be proportional to a time derivative of $g_\theta(t)$ because, if the boundary of the sample is twisted very slowly, the interior follows immediately to its new equilibrium, position. However, if the time derivative were higher than first order, distortions due to rotations of steady velocity could not be accounted for. The factor r/R is a geometrical term, which implies that the body force varies as the tangential velocity. The coefficient ξ is unknown, but it probably varies with the viscosity η]. Equation 30 thus becomes

$$\rho\ \frac{\partial^2 U_\theta}{\partial t^2} = \mathcal{O}_\mu(\frac{\partial^2 U_\theta}{\partial r^2} + \frac{1}{r}\frac{\partial U_\theta}{\partial r} - \frac{U_\theta}{r^2} + \frac{\partial^2 U_\theta}{\partial z^2}) + \xi\rho\frac{r}{R}\ \dot{g}_\theta(t) \quad (33)$$

When dissipation and forcing terms are neglected, Equation 33 has

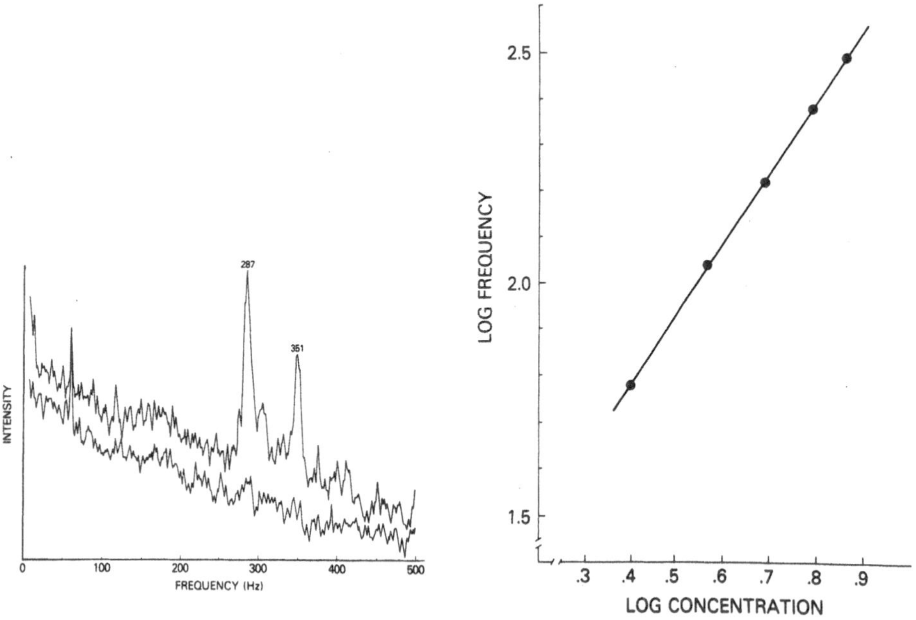

Figure 3. Examples of data obtained when an external mechanical field is used to induce standing waves in a gel contained in a light scattering cuvette (taken from Ref. 25). a) Spectral response at 22°C of a 7.5% polyacrylamide-H_2O gel formed in a 1 x 1 cm rectangular cross-section cuvette. The bottom curve is observed when no external mechanical excitation is applied. In contrast, the top curve is observed when a sound field, obtained from an audio-frequency sweep generator, is used to excite the sample (see Ref. 25). b) Resonant frequency (lowest order prominent peak) as a function of polymer concentration, from which one determines that μ varies approximately as the third power of the polymer concentration.

solutions

$$U_{\ell,n}(r,z;t) = J_1\left(\frac{\nu_\ell r}{R}\right)\sin\frac{n\pi z}{c}\, e^{i\omega_{\ell,n}t} \tag{34}$$

where $J_1(\cdot)$ is a first order Bessel function of the first kind[28] having zeros at $\{\nu_\ell\}$, and the frequencies $\omega_{\ell,n}$ are given as

$$\omega_{\ell,n} = [(\nu_\ell/R)^2 + n^2\pi^2/c^2]^{1/2}C_{tr} \equiv k_{\ell,n}C_{tr} \tag{35}$$

The complete Equation 33 is solved by noting that the force term $\xi\rho r/R\ \dot{g}_\theta(t)$ can be expanded in terms of the basis functions given in Equation 34, i.e.,

$$\xi\rho\frac{r}{R}\dot{g}_\theta(t) = \frac{4\xi\rho\dot{g}_\theta(t)}{\pi}\sum_{\ell=1}^{\infty}\sum_{n=1}^{\infty}\frac{[1-(-1)^n]}{n\nu_\ell J_2(V_\ell)} J_1(\frac{\nu_\ell r}{R})\sin\frac{n\pi z}{C} \tag{36}$$

where $J_2(\cdot)$ is the second order Bessel function, and we express $U_\theta(r,z;t)$ as

$$U_\theta(r,z;t) = \sum_{\ell=1}^{\infty}\sum_{n=1}^{\infty} a_{\ell,n}(t)J_1(\frac{\nu_\ell r}{R})\sin\frac{n\pi z}{C} \tag{37}$$

When Equations 36 and 37 are placed in Equation 33, the following equation for $a_{\ell,n}(t)$ is obtained

$$\frac{d^2a_{\ell,n}}{dt^2} = -\frac{1}{\rho}k^2_{\ell,n}\mathcal{O}_\mu(a_{\ell,n}) + \frac{4\xi[1-(-1)^n]}{\pi\ n\nu_\ell J_2(\nu_\ell)}\dot{g}_\theta(t) \tag{38}$$

which can be solved, for example, by Laplace transforms. If $g_\theta(t)$ is taken as $g_\theta(t) \equiv g_o\sin wt$, then $a_{\ell,n}(t)$ is found to be

$$a_{\ell,n}(t) = \frac{4wg_o[1-(-1)^n]}{n\pi\nu_\ell J_2(\nu_\ell)}\xi[\frac{1}{(w^2-k^2_{\ell,n}C^2_{tr})^2+k^4_{\ell,n}\eta^2w^2/\rho^2}]^{\frac{1}{2}}\sin(wt+\phi-\pi) \tag{39}$$

plus decaying initial value terms which have been neglected. The phase-shift ϕ which appears in this expression is related to the various gel parameters as

$$\tan\phi = \frac{k^2_{\ell,n}C^2_{tr} - w^2}{k^2_{\ell,n}\eta/\rho w} \tag{40}$$

Thus, from Equations 20, 37, and 39, we see that the amplitude of the oscillating photon autocorrelation function or, by the Weiner-Khintchine theorem, the amplitude of the spectral response near a resonance, is proportional to $w^2/\{(w^2-k^2_{\ell,n}C^2_{tr})^2 + k^4_{\ell,n}\eta^2w^2/\rho^2\}$. This function has a maximum at $w = k_{\ell,n}C_{tr}$, and the shear modulus thus easily can be determined.[10,25] A value of the dissipation parameter η can be obtained from the width of the resonance. However, if phase detection methods can be devised, it may be that such information can be more readily obtained from Equation 40.

An example of the spectral response exhibited by a moderately stiff gel when subjected to an applied sound field of varying frequency shown in Figure 3a. The location of the resonance maximum changes when the dimensions of the cuvette change or as the gel concentration or sample temperature is varied (Figure 3b). This technique recently has been used by Geissler and Hecht[29] to determine shear moduli of polyacrylamide-methanol gels at low temperature (poor solvent conditions), and previously was applied to polyacrylamide-water systems.[25] The techniques also have been used[10] to study the stiffening of fibrin gels when undergoing crosslinking by blood coagulation factor XIII (transglutaminase).

REFERENCES

1. J. D. Ferry, "Viscoelastic Properties of Polymers", 2nd ed., (Wiley, New York, 1970).
2. P. J. Flory, "Principles of Polymer Chemistry", (Cornell University Press, Ithaca, N. Y., 1953).
3. T. Tanaka, L. O. Hocker, and G. B. Benedek, J. Chem. Phys. 59: 5151 (1973).
4. A. M. Hecht and E. Geissler, J. Phys. (Paris), 39:631 (1978); E. Geissler and A. M. Hecht, J. Phys. (Paris), 39:955 (1978).
5. J. P. Munch, S. Candau, R. Duplessix, C. Picot, J. Herz, and H. Benoit, J. Polym. Sci. Polym. Phys. Ed. 14:1097 (1976).
6. E. Geissler and A. M. Hecht, J. Chem. Phys. 65:103 (1976).
7. J.P. Munch, S. Candau, J. Herz, and A. Hild, J. Phys. (Paris) 38:971 (1977).
8. K. L. Wun, G. T. Feke, and W. Prins, Faraday Discuss. Chem. Soc. 57:146 (1974).
9. W. Mackie, D. B. Sellen, and J. Sutcliffe, J. Polym. Sci., Polym. Symp. 61:191 (1977).
10. R. A. Gelman, J. Gladner, and R. Nossal, Biopolymers 19:1259 (1980).
11. W. J. Lee, P. Verdugo, R. J. Blandau, and P. Gaddum-Rosse, Gynecol. Invest. 8:254 (1977).
12. F. D. Carlson and A. B. Fraser, J. Mol. Biol. 89:273 (1974); J. Mol. Biol. 95:139 (1975).
13. A. B. Fraser, E. Eisenberg, W. W. Kielley, and F. D. Carlson, Biochemistry 14:2207 (1975).
14. S. L. Brenner, R. A. Gelman, and R. Nossal, Macromolecules 11:202 (1978).
15. K. L. Wun and F. D. Carlson, Macromolecules 8:190 (1975).
16. L. D. Landau and E. M. Lifshitz, "Theory of Elasticity", (Addison-Wesley, Reading, Mass., 1959).
17. P. G. deGennes, "Scaling Concepts in Polymer Physics", (Cornell University Press, Ithaca, N. Y., 1979).
18. F. Brochard, J. Phys. (Paris) 40:1049 (1979).
19. H. Z. Cummins and H. L. Swinney, in "Progress in Optics", E.

Wolf, ed., (North-Holland, Amsterdam, 1970) v. VIII, p. 133.
20. G. B. Benedek and T. Greytak, Proc. IEEE 53:1623 (1965).
21. J. D. G. McAdam, T. A. King , and A. Knox, Chem. Phys. Lett. 26:64 (1974).
22. T. Tanaka, S. Ishiwata, and C. Ishimoto, Phys. Rev. Lett. 38: 771 (1977).
23. M. Daoud and J. Janninck, J. Phys. (Paris) 37:973 (1976).
24. L. S. Ornstein and G. E. Uhlenbeck, in "Selected Papers on Noise and Stochastic Processes", N. Wax, ed., (Dover, New York, 1954). p. 93.
25. R. A. Gelman and R. Nossal, Macromolecules 12:311 (1979).
26. R. Nossal, J. Appl. Phys. 50:3105 (1979).
27. B. D. Coleman and W. Noll, Rev. Mod. Phys. 33:239 (1961).
28. M. Abramowitz and I. A. Stegun, "Handbook of Mathematical Functions", AMS 55, National Bureau of Standards (U.S. Gov't. Printing Office, Washington, D. C., 1964).
29. E. Geissler and A. M. Hecht, Macromolecules 13: 1276 (1980).

PHASE TRANSITION IN GELS

Toyoichi Tanaka, Shao-Tang Sun and Izumi Nishio

Department of Physics
Massachusetts Institute of Technology
Cambridge, MA 02139, USA

CONTENTS

1. INTRODUCTION

A gel is a crosslinked network of polymers immersed in a fluid. Gels play important roles in various aspects of our everyday life.

In the human body, most of the connective tissues are in a gel state. The vitreous humor of the eye is a gel occupying the space between the lens and the retina. The synovial fluid which lubricates the joints of bones is also a gel. In these biological examples, the fluid provides a medium through which substances utilized in metabolism, respiration and nutrition such as hormones, oxygen molecules and proteins are transported. The polymer network provides a structural frame that prevents the fluid from flowing away. "Jello" is probably the most familiar gel that we ordinarily encounter. In the typical dessert "jello" only 3% of its volume is the network of gelatin polymers -- the rest is sweetened flavored water. In this case the roles of the network and the fluid are obvious.

In chemistry and biochemistry, gels are used extensively as

matrices for chromatography and electrophoresis -- analytical methods that separate molecules according to their molecular weights and charges. In these techniques, the pore size of the crosslinked polymer network plays an essential role in its sieving effects. Gels are important intermediate products in polymer products such as rubbers, plastics, glues and membranes. Because of their scientific and technological importance, the physical chemistry of various gels has been studied extensively since the 1940's.[1]

In 1973, a new technique of light scattering spectroscopy was first introduced to gel studies.[2] It was demonstrated that by measuring the intensity and the time dependence of fluctuations of laser light scattered from a gel, it is possible to determine the viscoelastic properties of the gel, that is, the elasticity of the polymer network and the viscous interaction between the network and the gel fluid. Recently, with the help of this powerful technique, we have found critical phenomena in permanently crosslinked gels: as the temperature is lowered, the polymer network becomes increasingly compressible, and at a certain temperature it becomes infinitely compressible. At the same time, the effective pore size of the network increases and diverges. It is also observed that the gel volume changes reversibly by a factor as large as several hundred in response to an infinitesimal change in external conditions such as solvent composition, temperature, pH, or ionic strength, or when an electric field is applied across the gel.

In this lecture we will discuss phase transition phenomena of covalently crosslinked polyacrylamide gels. We will show that ionization of the gel network plays an essential role in the phase transition.

2. POLYACRYLAMIDE GELS

Let us first briefly describe the chemical structure of the acrylamide gels we have studied. As we shall see later, the chemical structure of the network plays an essential role in determining the collapse and other equilibrium behavior of gels. Acrylamide gel consists of linear polymer chains of acrylamide molecules crosslinked by bisacrylamide molecules, which consists of two acrylamide molecules connected together (Figure 1). The interstitial space of the polymer network is filled with water. We prepare a gel by dissolving acrylamide monomers (5 grams) and bisacrylamide monomers (0.133 grams) in 100 ml of water. Ammonium persulfate $(NH_4)_2S_2O_8$ (40 mg) and TEMED (tetramethyl-ethylene-diamine) then are added to the solution which is subsequently transferred to test tubes. These are the initiators that start the chain-reaction of the polymerization. First, a reaction between ammonium persulfate and TEMED occurs which produces free radical electrons on the TEMED molecules. These molecules of TEMED with free radicals are the nuclei of the polymerization. The free radical on the TEMED molecule attacks and opens one of the

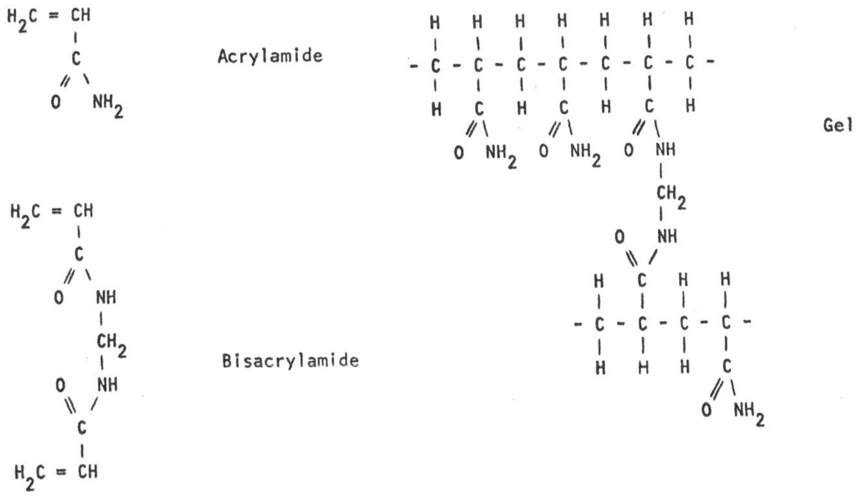

Figure 1. Chemical structure of polyacrylamide gel.

doubly-bonded carbon atoms of acrylamide and bisacrylamide monomers. One electron of the acrylamide or bisacrylamide double bond pairs with the odd electron of the free radical to form a bond between the free radical and this carbon atom; the remaining electron of the double bond shifts to the other carbon atom, which then becomes a free radical. In this way, the active center shifts uniquely to the newly added monomer, which then becomes capable of adding another monomer. The acrylamide monomers, having one double bond, are therefore polymerized into a linear chain. The bisacrylamide molecule, consisting of two connected acrylamide molecules, have two double bonds and serve as crosslinks. This chain reaction continues until a network consisting of practically an infinite number of bonded monomers is formed. Both polymerization and crosslinking begin almost instantaneously after the first free radical appears. The infinite polymer network is thus formed at this instant.

The gels are then taken out of the tubes. With syringe and needle, we carefully separate the gel from the tube wall by forcing water between the gel and the wall. Each gel sample is then soaked in water so that all the residual acrylamide, bisacrylamide monomers and the initiators are washed away. The gels are then immersed in a basic solution of NaOH or TEMED (pH 12) for zero to 60 days. During this immersion period the acrylamide groups of the network, $-CONH_2$, are hydrolyzed into carboxyl groups, $-COOH$, a quarter of which are ionized into carboxyl ions, $-COO^-$, and hydrogen ions, H^+. The polymer network becomes negatively charged having positive hydrogen ions, H^+, in the interstitial space. The longer the immersion time, the more charged the polymer network becomes. After the hydrolysis the gels are soaked in water to be washed. At this final stage of sample preparation the gel is largely swollen in water. A fully hydrolyzed gel, for example, swells 30 times from the original volume.

3. SWELLING EQUILIBRIUM

The washed gels were placed in acetone-water mixture of various compositions. After the gels had reached equilibrium, the diameter of each was measured using a dissecting microscope with calibrated eyepiece to determine its equilibrium volume. Figure 2 shows the relationship of acetone concentration to the swelling ratio ϕ/ϕ^*. The quantity ϕ/ϕ^* represents the ratio of final network concentration to initial network concentration for gels hydrolyzed in the TEMED solution for various lengths of time. The ratio is given by $(d/d^*)^3$ where d* and d are the initial and final equilibrium diameters of the gel, respectively. For swollen gels $\phi/\phi^* < 1$, whereas for shrunken gels $\phi/\phi^* > 1$.

For short hydrolysis times of 0 to 1 day, the final equilibrium network concentration changes continuously with the acetone concentration. If the gels are hydrolyzed in the TEMED solution for 2 days, the curve has a zero slope inflection point. For longer hydrolysis times discrete transitions are observed. Volume change at the transition becomes larger, and the acetone concentration required at the transition becomes higher. The last plot of the series in Figure 2 is for gels hydrolyzed in a 4% (volume) TEMED solution for 60 days. This gel exhibits a volume collapse of 350-fold.

So far, the collapse of the gels was caused by changing the acetone concentration in the mixture at room temperature. The collapse was also observed when, for the fixed acetone concentration, the temperature was varied. Figure 3 shows the swelling ratio curve of the gel (hydrolyzed for 8 days) immersed in the mixture of 42% acetone concentration. At temperatures higher than room temperature the network swells, and below it, it shrinks. There is a discontinuous volume change at room temperature.

4. EQUATION OF STATE AND PHASE DIAGRAM

In order to understand these phenomena we present a theory of phase transitions in ionic gels. We start with the Flory-Huggins formula for the osmotic pressure, π, of an ionic gel,[1,7]

$$\pi = \frac{NkT}{v}\left[\phi + \ln(1-\phi) + \frac{1}{2}\frac{\Delta F}{kT}\phi^2\right] + \nu kT\left[\frac{1}{2}\frac{\phi}{\phi_0} - \left(\frac{\phi}{\phi_0}\right)^{1/3}\right] + \nu f kT\left(\frac{\phi}{\phi_o}\right) \tag{1}$$

where N is Avagadro's number, k is the Boltzman constant, T is the

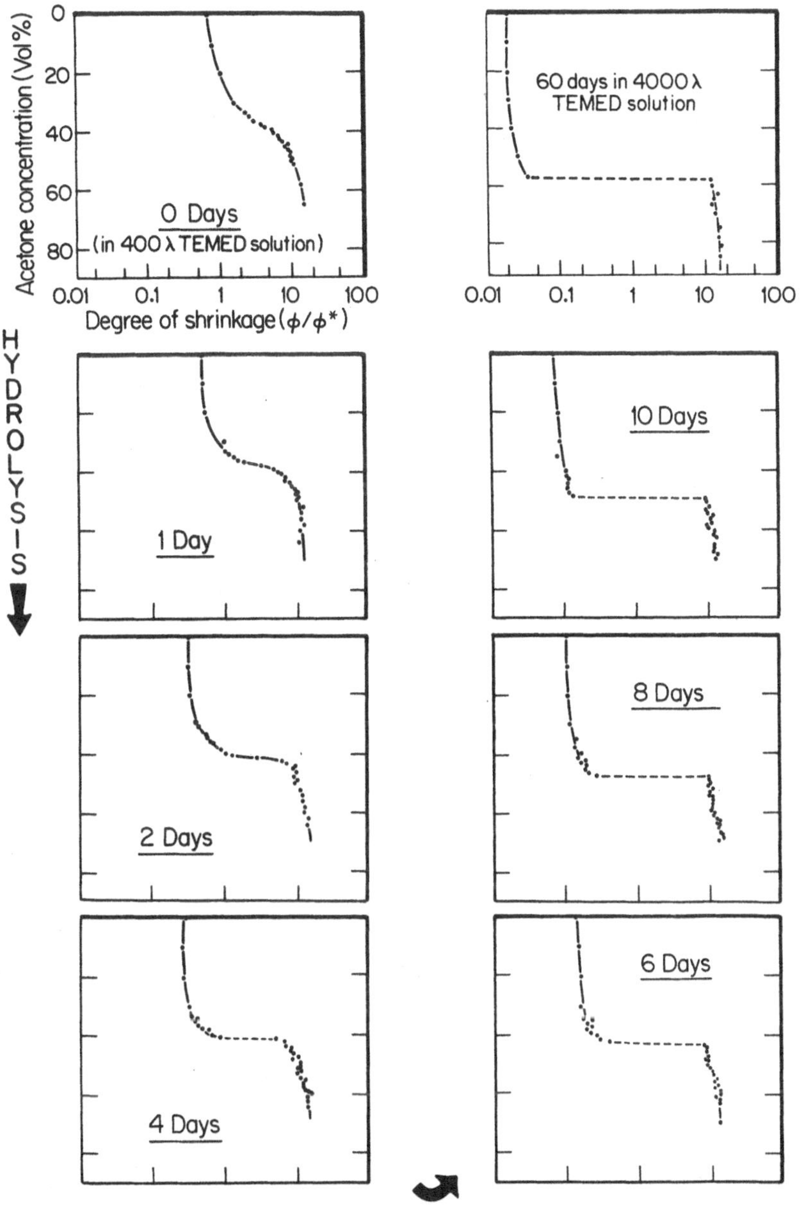

Figure 2. The acetone concentration dependence of the swelling ratio, ϕ/ϕ^*, for increasing hydrolyzation of the gel network. All curves are for gels immersed in 0.4% (volume) TEMED solutions for the times indicated, except for the upper right hand plot, which is for gels immersed in a 4% (volume) TEMED solution for 60 days.

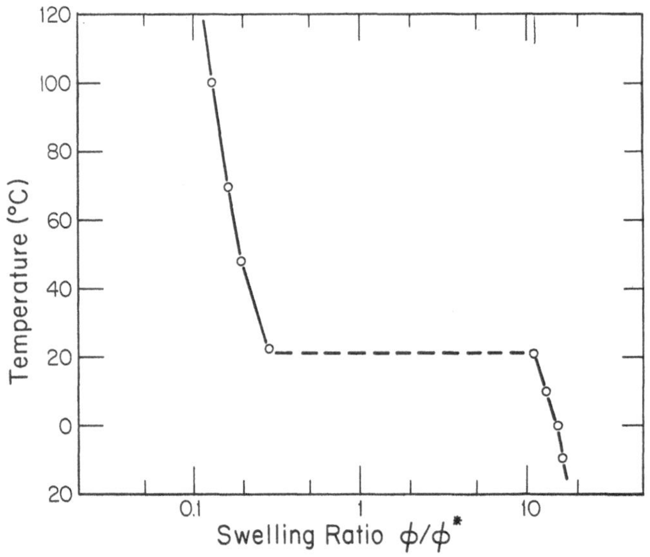

Figure 3. The temperature dependence of the swelling ratio, ϕ/ϕ^*, for a gel hydrolyzed for 8 days. The acetone composition in this swelling experiment was 42% by volume.

temperature, v is the molar volume of the solvent, ϕ is the volume fraction of the network, ΔF is the free energy decrease associated with the formation of a contact between polymer segments, ϕ_0 is the volume fraction of the network at the condition that the constituent polymer chains have random walk configurations, ν is the number of constituent chains per unit volume at $\phi = \phi_0$, and f is the number of dissociated hydrogen ions per effective chain.

The osmotic pressure of a gel must be zero for the gel to be in equilibrium with the surrounding solvent. Zero osmotic pressure is also necessary for the free energy of the gel, F, to be minimized, since $\pi = -\partial F/\partial V$, where V is the volume of the gel and $V \propto \phi_0/\phi$. From Equation (1), this condition may be expressed as

$$\tau \equiv 1 - \frac{\Delta F}{kT} = \frac{\nu v}{N\phi^2}\{(2f+1)[\frac{\phi}{\phi_0}]-2[\frac{\phi}{\phi_0}]^{1/3}\} + 1 + \frac{2}{\phi} + \frac{2\ \ln(1-\phi)}{\phi^2} \qquad (2)$$

The left side of Equation (2) corresponds to the reduced temperature, τ, which changes with temperature and solvent composition. The equation then determines the equilibrium network concentration as a function of the reduced temperature. For certain values of the reduced temperature, however, Equation (2) is satisfied by three values of ϕ, corresponding to two minima and one maximum of the free energy. The value of ϕ corresponding to the lower minimum

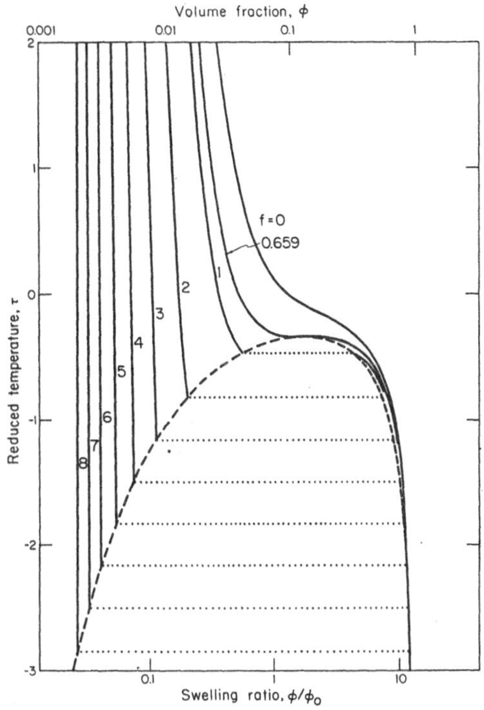

Figure 4. The states at which the osmotic pressure of the gel network is zero, for various values of f, the degree of ionization of the network, calculated using Equation (2), with the parameters $S_o \equiv \nu v/N\phi_0^3 = 10$ and $\phi_0 = 0.05$.

represents the equilibrium value. A discrete volume transition occurs when the two free energy minima have the same value.

In Figure 4 we plot the calculated equilibrium value of the swelling ratio, ϕ/ϕ_0, as a function of the reduced temperature for various degrees of hydrolysis, f. The theoretical curves are qualitatively in agreement with experimental results shown in Figure 2. As the graph shows, the hydrolysis of a small amount of acrylamide groups (for a change of f from 0 to 6) causes a drastic change in the collapse size at the transition. The situation may be understood if the term $\ln(1-\phi)$ in Equation (2) is expanded, retaining terms to order ϕ^3. This yields

$$t = S(\rho^{-5/3} - \rho^{-1}/2) - \rho/3 \quad (3)$$

where

$$t \equiv (1 - \Delta F/kT)(2f + 1)^{3/2}/2\phi_0, \quad (4)$$

$$S \equiv (v\nu/N\phi_0^3)(2f + 1)^4 \equiv S_0(2f + 1)^4 \quad (5)$$

and

$$\rho \equiv (\phi/\phi_0)\ (2f + 1)^{3/2} \quad (6)$$

In this approximation, the relationship between the renormalized reduced temperature, t, and the renormalized concentration, ρ, is determined by a single parameter S. In other words, the ratio of the two concentrations at the transition, which represents the size of the collapse, is a function only of S. Equation (5) shows that the S value changes very rapidly with f, the number of ionized groups per constituent chain. As f changes from 0 to 6, S changes by a factor of 3×10^4. The strong dependence of S on ionization accounts for the drastic change in the collapse with hydrolysis. In Figure 5 both S and f are plotted as functions of hydrolysis time.

5. pH AND SALT EFFECT

The role of ionization in the phase transitions of the gels is most conspicuous when the pH of the solvent is lowered. When the pH is changed - with the temperature and solvent composition fixed - the degree of ionization of the network is modified, altering the S value. Both continuous and discrete transitions may be induced by changing the pH. As Figure 4 shows, if the reduced

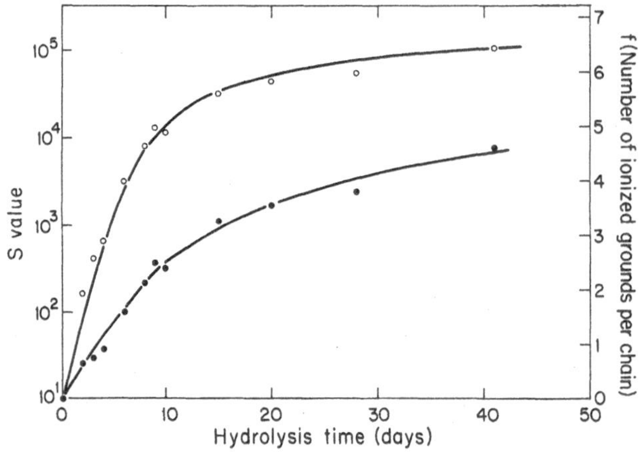

Figure 5. The value of f (open circles), the degree of ionization of the network, and $S - S_0(2f+1)^4$ (solid circles), the collapse parameter, as a function of hydrolysis time in 0.4% (volume) TEMED solutions. The values for f are determined using the experimental swelling curves and the theoretical curves of Figure 2. Solid curves are to guide the eye.

temperature is greater than -0.338, the equilibrium value of ϕ/ϕ_o increases continuously with f. However, below -0.338 the value of ϕ/ϕ_o changes discretely. Both types of transitions were observed. Water was used as a solvent in the case of the continuous transitions; a 50% acetone-water mixture was used in the discrete transition. The results are shown in Figure 6.

The effective ionization can also be modified by adding salts such as NaCl and $MgCl_2$, since the positive ions shield the negative charges of the polymer network. The transition was indeed affected when these salts were added. We observed that the transition concentration for the divalent ion, Mg^{++}, is 4000 times smaller than that for the monovalent ion, Na^+ (Figure 7). It is interesting to note that the divalent ions plays an essential role in muscle contraction.

6. CRITICAL BEHAVIOUR OF GELS

The critical divergence and slowing-down of concentration fluctuations in the polyacrylamide gels[3] are now described.

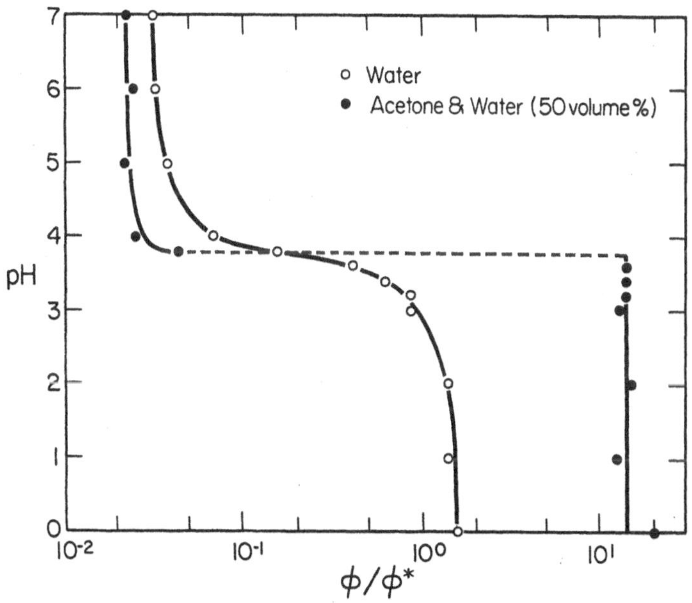

Figure 6. The pH dependence of the swelling ratio for gels hydrolyzed in a 4% (volume) TEMED solution for 60 days. The smooth curve is for gels immersed in water. The discontinuous curve is for gels in a 50% acetone-water mixture.

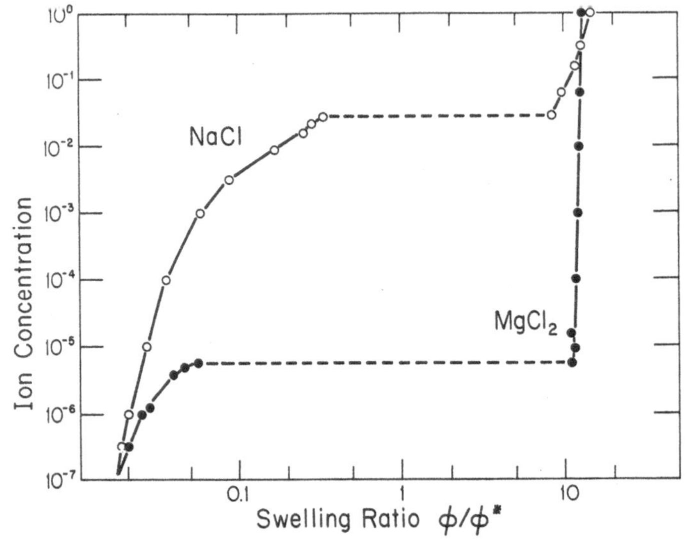

Figure 7. The dependence of swelling ratio, ϕ/ϕ^*, on salt concentration for highly hydrolyzed gels. The acetone composition in this swelling experiment was 50% by volume.

Tanaka, Hocker and Benedek[2] presented a theory which describes the concentration fluctuations in a gel polymer network and showed that the elastic modulus of the network can be determined from the intensity of light scattered from the fluctuations. They also showed that the ratio of the elastic modulus and the frictional coefficient of the network in the fluid medium can be determined from the correlation time of the scattered light. The normal modes of the concentration fluctuations in the gel network can be described as overdamped phonons. The phonons in the network are overdamped because of the large frictional drag of the fluid medium which is proportional to the velocity of the network relative to the fluid medium. The light incident on the gel is scattered by such overdamped phonons and the correlation function of the scattered light field is expressed in the following way:

$$\langle E(\vec{q},t)E(\vec{q},0)\rangle = C(\phi\partial n^2/\partial\phi)^2 \cdot \frac{kT}{K} \exp(-Kq^2t/\gamma) \equiv I \exp(-\Gamma t) \qquad (7)$$

Here E is the amplitude of the scattered light field, $\vec{q}$ is the scattering vector, n is the refractive index of the whole gel, K is the elastic modulus of the network in the fluid medium, and C is a constant depending only on the optical geometry. The frictional coefficient, γ, multiplied by the velocity of the network relative to the fluid medium gives the frictional force per unit volume of the network (Figure 8).

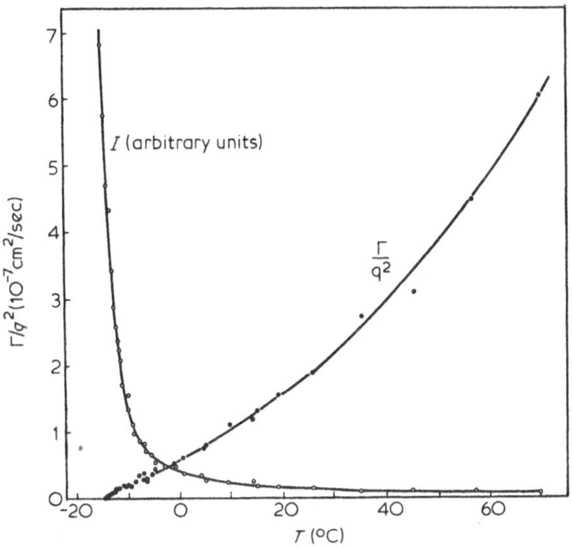

Figure 8. Intensity, I and the reciprocal correlation time (decay rate), Γ, of laser light scattered by a 2.5% polyacryla mide gel. Γ is divided by the square of the scattering vector, $|\vec{q}|^2$

Equation (7) has been confirmed experimentally by Tanaka, Hocker and Benedek,[2] and Munch et al.[4] In Figure 8, we show the temperature dependence of the intensity and the correlation time of laser light scattered by the 2.5% polyacrylamide gel.[3] Both the intensity and the correlation time of the scattered light increase by a factor of more than 200 as temperature decreases and appear to diverge at a certain temperature. This temperature corresponds to the spinodal temperature at which the elastic modulus of the network becomes zero. In other words, the network has infinite compressibility.[3] The elastic modulus K can be calculated using the equation of state (Equation 1):

$$K = \phi \partial \pi / \partial \phi \tag{8}$$

$$= \frac{NkT}{v} \{ \frac{\phi^2}{1-\phi} - \frac{\Delta F}{kT} \phi^2 + \frac{\nu v}{N} [(f + \frac{1}{2}) \frac{\phi}{\phi_0} - \frac{1}{3} (\frac{\phi}{\phi_0})^{1/3}] \} \tag{9}$$

$$= A \ (T - T_S) \tag{10}$$

where

$$A \equiv \frac{Nk}{v} \{ \frac{\phi^2}{1-\phi} + \frac{\nu v}{N} [(f + \frac{1}{2}) \frac{\phi}{\phi_0} - \frac{1}{3} (\frac{\phi}{\phi_0})^{1/3}] \} \tag{11}$$

$$T_S \equiv \frac{N\Delta F}{Av} \phi^2 \qquad (12)$$

T_S corresponds to the spinodal temperature. Equations (7) and (10) show that the scattered light intensity I, varies with temperature as:

$$1/I \propto 1/T_S - 1/T. \qquad (13)$$

This relation is confirmed[3] in Figure 9. Having understood the critical behaviour of the elastic modulus, let us examine the friction coefficient, Γ, between the polymer network and water. Since it is not easy to calculate directly, we calculated the decay rate, Γ, of concentration fluctuations using mode-coupling theory.[5] The result in the hydrodynamic regime ($q\xi < 1$) is given by:

$$\Gamma = (kT/6\pi\eta\xi)q^2 \qquad (14)$$

$$K = N_o(b/\xi)^2 kT \propto 1/\xi^2 \qquad (15)$$

$$\gamma = N_o(b/\xi)^2 6\pi\eta\xi \propto 1/\xi \qquad (16)$$

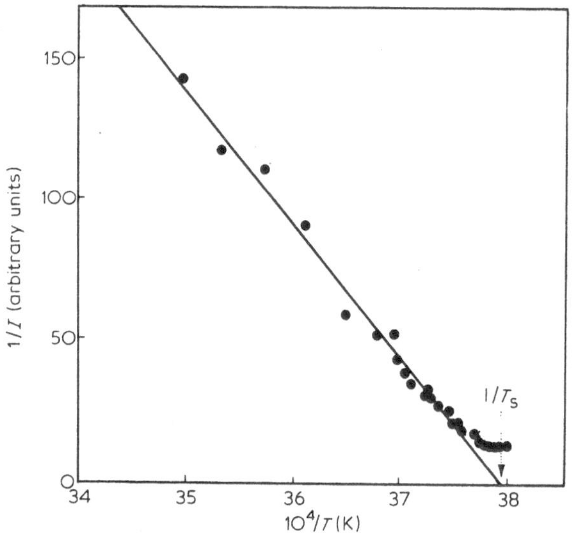

Figure 9. Inverse of the laser light intensity scattered by a 2.5% polyacrylamide gel as a function of reciprocal absolute temperature.

where N_o is a constant, b represents the length of segments constituting the network polymers, η is the viscosity of water and

$$\xi = b[T/(T - T_S)]^{\frac{1}{2}} \tag{17}$$

represents the correlation length over which the movements of the polymer segments are correlated. This correlation length diverges as the temperature approaches the spinodal temperature, T_S. In Figure 10, we compare the calculated curve of the decay rate, Γ, and the data.[3] They show excellent agreement.

These results indicated that as we approach the spinodal temperature the network becomes increasingly compressible ($K \to 0$) and porous ($\gamma \to 0$). How can the permanently crosslinked gel change its pore size with temperature? When the network becomes more compressible large concentration fluctuations appear in the network which lead to the formation of large regions of swollen or compressed regions in the network. The large regions of swollen network determine the effective pore size of network, which becomes increasingly larger as the spinodal temperature is approached.

We have determined the entire spinodal line using scattered light intensity measurements.[6] Gels made with 100 μl TEMED in 25 ml gel were hydrolyzed for 12 days. They were cut into pieces 1 cm long and dried at 30°C. Each dried gel was then put into an acetone-water mixture having a composition ratio of 44:56 by volume. After a certain amount of the mixture had soaked into the network, the gel was removed and placed in an ampoule. The ampoule was immediately sealed to prevent any loss of the solvent. Gels of different concentrations were prepared in this way. The network concentrations

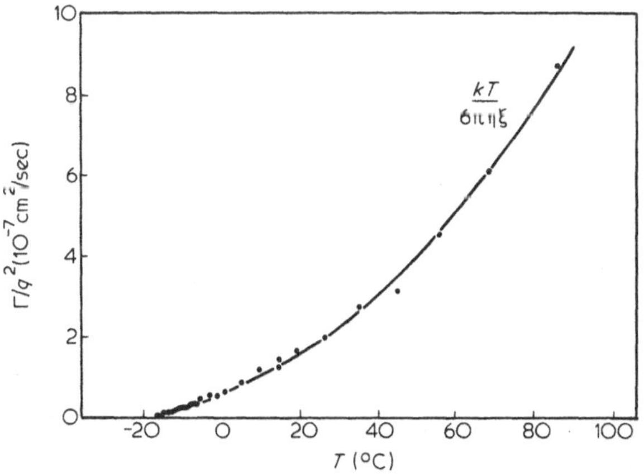

Figure 10. Decay rate calculated using a mode-coupling theory (———). •, measured data.

of each gel was determined by measuring the weights of dry and swollen gel.

The scattered light intensity at a 90° angle was determined for each sample as a function of temperature. It has been shown above that 1/I vs 1/T is linear. The experimental points are fitted to a straight line and the spinodal temperature is obtained from the intercept of the straight line with 1/T axis which corresponds to infinite scattered light intensity. The deviation of the data points from the straight line as the spinodal temperature is approached probably results from multiple scattering of the light. The spinodal temperature is plotted[6] as a function of gel concentration in Figure 11 (○). We also determined the swelling equilibrium concentration of the gel which was immersed in a large volume of the acetone-water mixture at various temperatures. They are also plotted in the Figure 11 (•).

7. COLLAPSE UNDER ELECTRIC FIELD

Finally we can induce the transition by applying an electric field across the gel. As shown in Figure 12a, we put a gel in a 50% acetone-water mixture and hold it between two electrodes. When an electric field (~ 0.5 V/cm) is applied across the electrodes, the negative charges on the polymer network are pulled toward the positive electrode. This produces a pressure gradient along the electric field. The pressure is strongest on the gel end at the

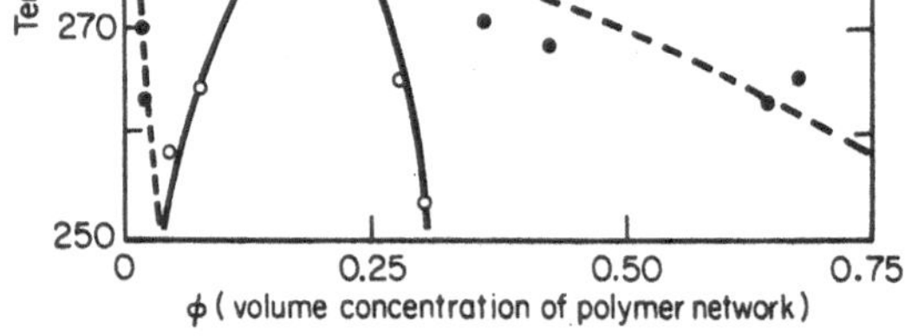

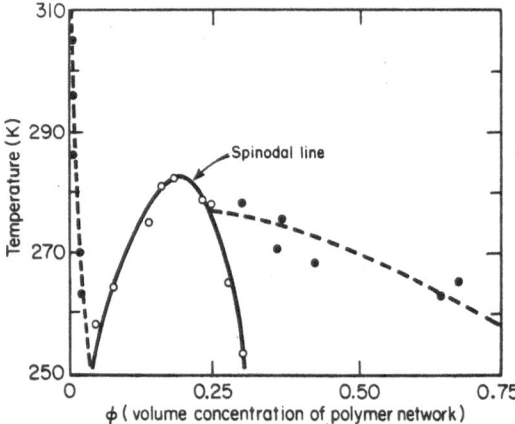

Figure 11. Spinodal line (○) determined by light scattering measurements. •, denote the equilibrium concentrations of the acrylamide gel which is immersed in a 44% acetone-water mixture at different temperature.

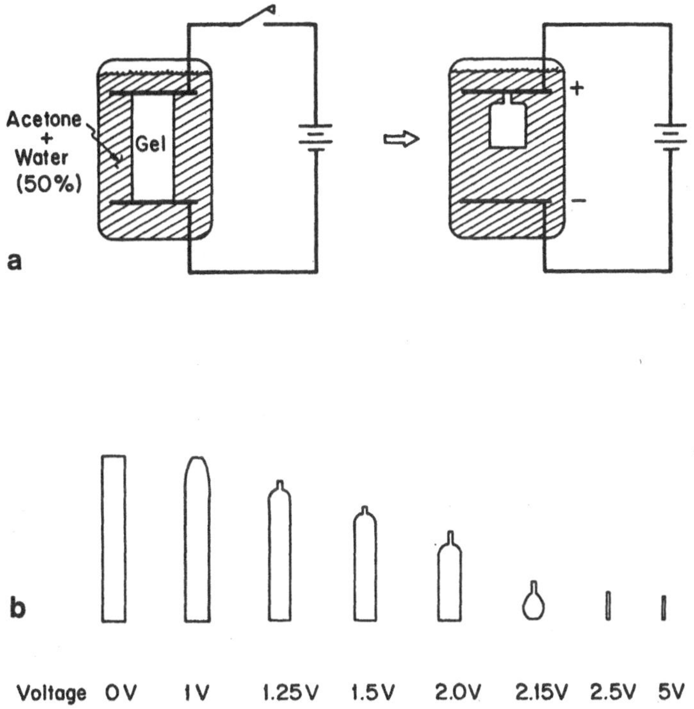

Figure 12. (a) Schematic diagram for the experimental arrangement. (b) Gel collapse as a function of the applied voltage.

positive electrode and zero at the other end of the gel. There is a critical pressure above which the gel collapses and below which it is swollen. The gel, therefore, has a bottle shape as shown in Figure 12b. The swollen portion becomes shorter as the electric field becomes stronger (Figure 12b). The collapsed part swells again when the electric field is turned off.

8. CONCLUSIONS

The drastic volume change upon infinitesimal variation of external conditions such as temperature, solvent, and an electric field across the gel may have a variety of applications. Various chemical engines may be designed using gels. A gel having such mechano-chemical amplification may be used as an artificial muscle. One may ask, then, what is the speed of contraction and relaxation of the gel? We observed that the time taken for swelling or shrinking is proportional to the square of a linear size of a gel. This is because the network motions obey the diffusion equation as was shown in Equation (7). For example, for a cylindrical gel of 1 cm diameter, it takes several days to reach a new equilibrium volume.

For cylindrical gel of 1 μm in diameter, which is the size of muscle filament, it would take only a few thousandths of a second. This is quick enough to be used as an artificial muscle. The discrete change in volume of a gel may be used as a mechano-chemical memory or a switch. The continuous control of the effective pore size seems to be important in understanding and improving the techniques of electrophoresis and chromatography.

In the eye, retinal detachment is caused by a collapse of the vitreous gel, which allows the retina to pull off the back of the eye. The opacification of the cornea gel is another common cause of human blindness. It is believed to be the result of separation of fluid domains in the cornea. The physics of the phase transitions in gels would shed a new light on these diseases. The concept of phase separation and the technique of laser scattering spectroscopy appear to be very useful in understanding and characterizing the sieve effects of the gels used in electrophoresis and chromatography. More generally, these studies can provide an improved perspective on the interdisciplinary field where polymer science and the physics of critical phenomena merge.

ACKNOWLEDGEMENTS

This lecture is merely a review of the work done in our laboratory at M.I.T. The people who participated in the work are: Dr. S. Ishiwata, Dr. C. Ishimoto, Dr. I. Ohmine, Dr. A. Hochberg, Dr. D. Nicoli, Mr. G. Swislow, Mr. D. J. Filmore, and Ms. A. Shah, in addition to the authors. It is our pleasure to thank Prof. G. B. Benedek for his helpful discussions and advice.

This work was supported by NSF CHE 77-26924 and NSF DMR 78-24185.

REFERENCES

1. P. J. Flory, Principles of Polymer Chemistry, (Cornell University Press, Ithaca, New York, 1953).
2. T. Tanaka, L. O. Hocker, and G. B. Benedek, J. Chem. Phys. 59: 5151 (1973).
3. T. Tanaka, S. Ishiwata, and C. Ishimoto, Phys. Rev. Lett. 38: 771 (1977).
4. J. P. Munch, S. Candau, R. Duplessix, C. Picot, J. Herz, and H. Benoit, J. Polym. Sci. 14: 1097 (1976).
5. T. Tanaka, Phys. Rev. A17: 763 (1978).
6. A. Hochberg, T. Tanaka, and D. Nicoli, Phys. Rev. Lett. 43: 217 (1979).
7. T. Tanaka, D. J. Fillmore, S.-T. Sun, I. Nishio, G. Swislow, and A. Shah, Phys. Rev. Lett. 45: 1636 (1980).

LIGHT SCATTERING STUDY OF GANGLIOSIDE MICELLES AND MIXED MICELLES WITH A NONIONIC AMPHIPHILE

Mario Corti and Vittorio Degiorgio[+]

CISE S.p.A.
P. O. Box 12081
Milano 20100, Italy

Sandro Sonnino and Riccardo Ghidoni

Department of Biological Chemistry
The Medical School, University of Milano
Milano, Italy

ABSTRACT

Aqueous solutions containing a mixture of two amphiphiles, a commercial nonionic surfactant, Triton X-100, and a biological lipid, the ganglioside GM1, are investigated by static and dynamic light scattering. It is found that mixed micelles are formed in the whole range of investigated molar ratios. The theory of light scattering from solutions of homogeneous micelles is generalized to the case of mixed micelles. The final formula is used to derive from the experimental data the aggregation number of the mixed micelle. A simple phenomenological law is proposed to describe the dependence of the aggregation number on the molar ratio between the two amphiphiles. The biochemical implications of the obtained results are shortly discussed.

1. INTRODUCTION

Several recent papers[1-3] have discussed the micellar properties of gangliosides which are anionic amphiphilic glycolipids occurring in neuronal plasma membranes.[4] Gangliosides appear to be involved in a variety of cell surface phenomena, and their interactions with

---

+ Researcher from the Italian National Research Council (CNR)

many biologically important molecules seem to be considerably influenced by their state of aggregation. Light scattering measurements[1] have shown that two of the most abundant brain gangliosides, GM1 and GD1a, form in aqueous solution disk-like micelles of large molecular weight (400000 - 500000) already at a concentration of 10^{-6} M. Moreover, the ganglioside concentration at which micellization takes place - the critical micelle concentration (cmc) - was reported[2,3] to be in the range 10^{-10} - 10^{-8} M.

In this paper we describe a light scattering experiment performed on aqueous solutions of the ganglioside GM1 plus a commercial nonionic amphiphile, Triton X-100. As it will be shown, the experimental data prove that mixed micelles are formed, and allow to derive such micellar parameters as the aggregation number and the hydrodynamic radius. The physicochemical information about GM1-Triton X-100 solutions is particularly relevant when correlated with the biochemical information which can be obtained by measuring the activity of an enzyme, galactose oxidase, which is known to catalyze the oxidation to aldehyde of the primary alcoholic group carried by the galactose residue terminally located in the hydrophilic portion of GM1. It is found indeed that the enzyme action is strongly influenced by the type of micelle in which GM1 is inserted.[5,6]

The present article is mainly focussed on the light scattering results. A more thorough discussion of the problem, particularly for the biochemical aspects, can be found in a previous work[5] (see also Ref. 6 for other biochemical and physicochemical data on the same system). We have treated with some detail the interpretation of scattered intensity data from solutions of mixed micelles, because this is a case of rather general interest which is not discussed, to our knowledge, in the available literature.

2. LIGHT SCATTERING FROM MIXED MICELLE SOLUTIONS

The total intensity I_s of the light scattered from the micellar solution is made up of two contributions, the first I_w due to scattering from the solvent alone and the second due to the presence of micelles. The expression relating the micellar weight M to $I_s - I_w$ is simply written for solutions containing a single micellar species. The experiments described in this paper concern however solutions of two amphiphiles, so that in principle more than one type of micelles may be found in solution. We will consider the two extreme cases:

a) the two amphiphiles micellize separately
b) the two amphiphiles form mixed micelles at all the investigated molar ratios.

For the case a) the relative normalized scattered intensity $I_r = \frac{I_s - I_w}{I_w}$ is simply given by the sum of the independent contributions from the two micellar species, that is

$$I_r = A\left[\left(\frac{dn}{dc}\right)_T^2 M_T(c_T - c_{OT}) + \left(\frac{dn}{dc}\right)_G^2 M_G(c_G - c_{OG})\right] \quad (1)$$

where c is the concentration and c_o is the critical micelle concentration, A is a calibration constant, $\frac{dn}{dc}$ is the refractive index increment and M is the micelle molecular weight. Subscripts T and G refer to Triton X-100 and GM1, respectively.

For a single species of homogeneous micelles of molecular weight M in solution the scattered intensity is proportional to:

$$N\left(\frac{dn}{dc} M\right)^2 \quad (2)$$

where N is the number of micelles in the scattering volume and the quantity in parenthesis is proportional to the electric field amplitude scattered by the single micelle. Considering now case b), we can say that the scattered amplitude from the individual mixed micelle is due to the sum of the contributions of each amphiphile present in the micelle weighted by its optical contrast represented by the refractive index increment. Consequently expression (2) becomes:

$$N\left[\left(\frac{dn}{dc}\right)_T n_T m_T + \left(\frac{dn}{dc}\right)_G n_G m_G\right]^2 \quad (3)$$

where m_T and m_G are the molecular weights of Triton and GM1 monomers, and n_T and n_G are the numbers of Triton and GM1 monomers in the mixed micelle. The number N of mixed micelles is readily calculated from the total amphiphile concentration which goes into micelles and the mixed micelle molecular weight M as:

$$N = N_{AV} \frac{c - c_m}{M} \quad (4)$$

where N_{AV} is Avogadro's number, c is the total amphiphile concentration and c_m is the total monomer concentration. Combining expression (3) and (4), after some rearrangements, we obtain

$$I_r = gA\left(\frac{dn}{dc}\right)_G^2 M(c - c_m) \quad (5)$$

where the dimensionless factor g is given by

$$g = \left[\frac{1 + \beta\alpha X}{1 + \alpha X}\right]^2 \tag{6}$$

with $\alpha = m_T/m_G$, $X = n_T/n_G$, and $\beta = (dn/dc)_T \,/\, (dn/dc)_G$. It should be noted that g becomes equal to one, independently of X, when the two amphiphiles have the same refractive index increment, that is, when $\beta = 1$. Equation 5 indicates that M can be derived from the measured I_r only if g and c_m are known. The factor g depends on X which is the molar ratio in the micelle and which does not generally coincide with the molar ratio in the solution because of the presence of free monomers. We recall that in the case of solutions containing a single amphiphile the concentration of free monomers can be taken equal to the cmc. This amounts to say that all the amphiphile added to the solution in excess of the cmc goes into micelles. For the two-amphiphile system the situation is more complicated. We will refer here to Clint's work[7] which treats the mixed micelle as an ideal mixture of its pure components. A more general approach was presented by Rubingh[8] by introducing activity coefficients. The cmc of the mixed micelle C_{cm} (we denote by capital C the molar concentrations) is connected to the cmc's of the single components by the relation[7,9]

$$\frac{1}{C_{cm}} = \frac{\alpha}{C_{cT}} + \frac{(1 - \alpha)}{C_{cG}} \tag{7}$$

where α is the molar fraction of Triton X-100. The treatment of Clint can also be used to calculate the monomer concentrations in the mixed micellar solution as a function of total amphiphile concentration.

Equation 12 of Ref. 7 gives the free monomer concentration of the amphiphile with the higher cmc as a function of the total concentration, and of the amphiphile mole fraction. For the case of the experiments described in this paper the above equation may be simplified because $C_{cG} \cong 10^{-9}$ M is much smaller than $C_{cT} = 0.32$ mM[10]. Neglecting c_{cG}, the Triton X-100 free monomer concentration c_{OT} is given by

$$C_{OT} = \frac{(C + C_{cT}) - \sqrt{(C + C_{cT})^2 - 4\alpha C C_{cT}}}{2} \tag{8}$$

Figure 1 reports the behavior of C_{OT} as a function of the Triton X-100 concentration with fixed (0.8 mM) GM1 concentration. An interesting result is that C_{OT} levels off at a much larger Triton concentration than cmc value for the pure Triton solution. We have also replotted in Figure 1 the surface tension data obtained by Masserini et al.[6] on Triton - 0.8 mM GM1 aqueous solutions with buf-

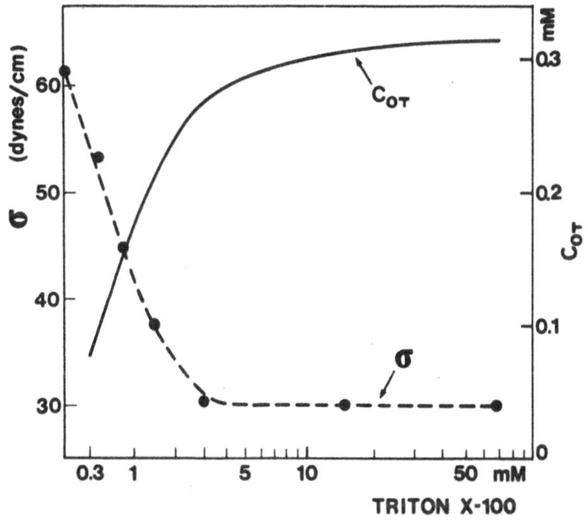

Figure 1. The concentration of free Triton X-100 monomers C_{OT} plotted as function of the total Triton X-100 concentration in an aqueous solution containing 0.8 mM GM1. The curve is calculated according to Equation 8. The experimental surface tension σ for the same solution is reported from Ref. 6. As explained in the text, the change of σ upon addition of Triton X-100 represents a measure of Triton monomer concentration.

fer 25 mM sodium phosphate/5 mM disodium EDTA (pH 7.0). We see that the surface tension σ starts from the value appropriate for GM1 aqueous solutions (≅ 60 dyne/cm)[11] and drops asymptotically to the value appropriate for Triton X-100 solutions (≅ 30 dyne/cm) when C_T becomes much larger than Triton X-100 cmc. The difference between the surface tension at zero Triton concentration and the surface tension at the Triton concentration C_T represents a measure of the free Triton monomer concentration at C_T. Figure 1 shows a good qualitative agreement between experiment and theory.

3. EXPERIMENTAL RESULTS AND DISCUSSION

The light scattering apparatus was equipped with an argon ion laser operating on the green 514.5 nm line and with a scattering cell temperature controlled within a few millidegrees. The average intensity of the scattered light was measured at $\theta = 90°$. The intensity correlation function was determined on the light scattered at 90° by a 108-channel digital correlator. Further details on the apparatus may be found in Ref. 12.

A description of the chemical composition and of the preparation procedure of the used ganglioside solutions is given in Ref. 5. We only recall here that the ganglioside GM1 was isolated from calf brain according to the method of Tettamanti et al.[13] and purified following the indications of Sonnino et al.[14] The Triton X-100 was gas-liquid chromatographic grade material from Merck (Darmstadt, West Germany).

The measured values of the relative scattered intensity I_r for 0.8 mM GM1 in 25 mM sodium phosphate/5 mM disodium EDTA buffer (pH 7.0) at 15°C are reported in Figure 2 as a function of the added concentration of Triton X-100. The intensity I_r at first decreases upon the addition of the nonionic surfactant. By further increasing the Triton concentration, I_r goes through a minimum and starts increasing.

The behavior of I_r as a function of C_T constitutes clear evidence for the formation of mixed micelles having a molecular weight smaller than the weight of GM1 micelles. In fact the hypothesis of unmixed micelles would lead to a scattered intensity increasing monotonically with C_T, as shown by the full line in Figure 2 which was calculated from Equation (1) with A = 5.86 cm^3 (see Ref. 5),

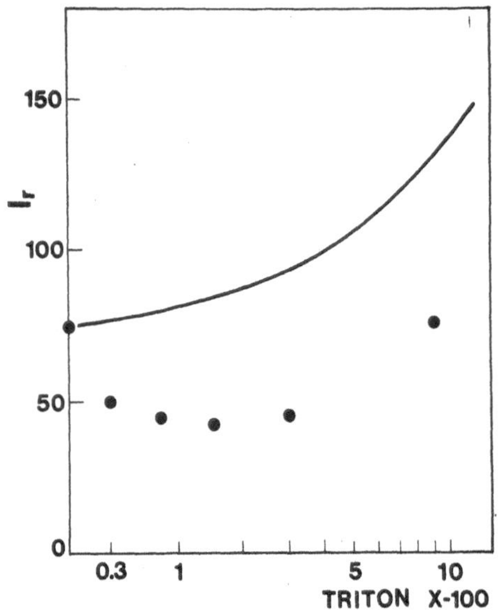

Figure 2. Behavior of the scattered light intensity I_r from mixed GM1-Triton X-100 micelles in 25 mM sodium phosphate/5 mM disodium EDTA (pH 7.0) at 15°C as a function of Triton X-100 molar concentration with fixed 0.8 mM GM1 concentration. The full line represents the scattered light intensity expected for unmixed micelles.

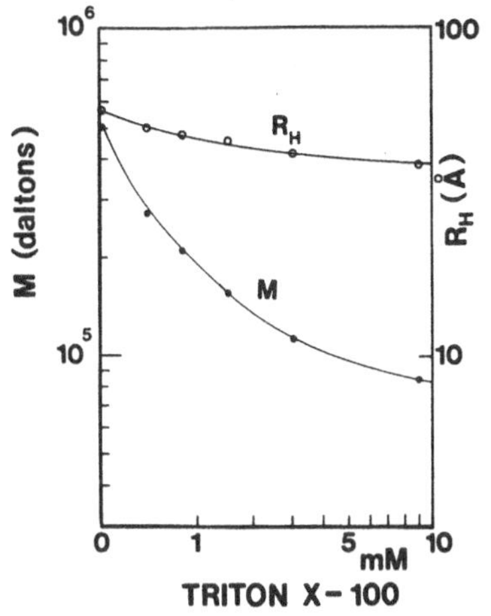

Figure 3. The molecular weight M and the hydrodynamic radius R_H of mixed GM1-Triton X-100 micelles in 25 mM sodium phosphate/5 mM disodium EDTA (pH 7.0) at 15°C as a function of Triton X-100 molar concentration with fixed 0.8 mM GM1 concentration.

$(dn/dc)_T$ = 0.150 cm^3/g (see Ref. 10), M_T = 78000, $(dn/dc)_G$ = 0.143 cm^3/g (see Ref. 1), and M_G = 495000.

By making the assumption that all the micelles are mixed micelles with the same proportion of amphiphiles, we calculated the molecular weight of the mixed micelle by means of Equation (5). As discussed in the previous Section, in order to apply Equation (5) one should know separately the free monomer concentration c_m as a function of the total concentration c. It can be easily shown that the concentration of free GM1 monomers is always negligible, since C_{cG} is much smaller than the used GM1 concentration. Therefore c_m practically coincides with the Triton monomer concentration c_{OT} given by Equation 8 and plotted in Figure 1. Furthermore, we have made the simple assumption that $(dn/dc)_T \cong (dn/dc)_G$, so that we can put g = 1 in the calculation of M. The plot of Figure 3 shows that the molecular weight M decreases from the value corresponding to the GM1 micelle to a value close to that found in a separate measurement for the Triton X-100 micelle, M_T = 78000.

The time-dependent part G(τ) of the intensity correlation function is an exponential for a solution of monodisperse micelles. The mass diffusion coefficient D is simply derived by fitting to the logarithm of the experimental data the expression

$$\ln G(\tau) = x_o - 2k^2 D\tau \qquad (9)$$

where x_o is also a fit parameter, and $k = (4\pi n/\lambda) \sin \theta/2$ is the modulus of the scattering vector. If the solution is polydisperse, the linear fit expressed by Equation (9) becomes inadequate because $G(\tau)$ is not a single exponential. However, the z-average diffusion coefficient can be still derived by taking the limit for $\tau \to 0$ of the slope of $\ln G(\tau)$.[12]

The hydrodynamic radius R_H is calculated from D by the Stokes-Einstein relation $D = k_B T/(6\pi\eta R_H)$, where k_B is the Boltzmann constant, T the absolute temperature, and η the viscosity of the solvent. We recall that $R_H = fR_e$, where R_e is the radius of the sphere having the equivalent volume, and $f \geq 1$ is a form factor. For a spherical particle, $f = 1$.

It should be stressed that the quantities M, as derived from Equation (5), and R_H, as derived from the Stokes-Einstein relation, may be considered as individual micellar properties only when the micelles behave as independent Brownian particles in solution. This may not be the case even at small amphiphile concentrations, as it will be discussed later on.

The time-dependent part $G(\tau)$ of the intensity correlation functions measured on 0.8 mM GM1 solutions (without Triton X-100) at 15°C and 30°C is a single exponential as shown by the semi-logarithmic plots reported in Figure 4, with an excellent reproduci-

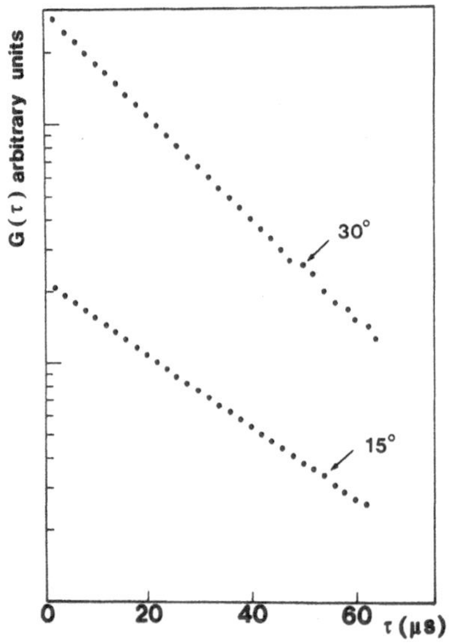

Figure 4. The time dependent part G (τ) of the intensity correlation function of laser light scattered from 0.8 mM GM1 in the same buffer as for Figures 2 and 3 at 15°C and 30°C.

bility from sample to sample. These results indicate that the new technique of preparation and purification of gangliosides, which leads to the formation of pure sodium salts, eliminates completely the fluctuations in the shape of G(τ) described in Reference 1, and therefore ensures a more precise characterization of micellar properties.

We have reported in Figure 3 the behavior of the hydrodynamic radius R_H as a function of the Triton X-100 concentration. The trend followed by R_H is qualitatively similar to that shown by the molecular weight M.

Table I summarizes the numerical values of the quantities discussed so far. It is evident that the mixed micelle critical concentration C_{cM}, as calculated from Equation (7) is very small for all the investigated Triton concentrations, so that neglecting the GM1 free monomer concentration is always a good approximation. The Triton free monomer concentration c_{OT} calculated from Equation (8), is not negligible with respect to the Triton concentration in the solution. Nevertheless its effect on the M values is of the

Table I. The critical micelle concentration C_{cM} of the mixed micelle, the free Triton-monomer concentration C_{OT}, the numbers of Triton and GM1 monomers n_T and n_G respectively, the molecular weight M and the hydrodynamic radius R_H reported for various GM1-Triton X-100 aqueous solutions at 15°C.

C_G(mM)	C_T(mM)	C_{cM}(mM)	C_{OT}(mM)	n_T	n_G	10^{-3}M	R_H(Å)
0.8	0.0	10^{-6}		0	328	495	55.9
0.8	0.4	1.5×10^{-6}	0.09	63	161	282	50.2
0.8	0.8	2.0×10^{-6}	0.14	89	109	220	47.5
0.8	1.5	2.9×10^{-6}	0.20	106	65	164	45.1
0.8	3.0	5.8×10^{-6}	0.25	111	32	118	41.4
0.8	9.0	12×10^{-6}	0.29	113	10	86	38.4
0.0	9.0	0.32	0.32	126	0	78	38.0

same order as the experimental uncertainties coming from light-intensity and refractive-index-increment measurements.

The number of monomers of each component in the micelle can be determined from M considering that the monomer ratio in the micelle should coincide with the molar concentration ratio in the solution once the concentration of free monomers is subtracted from the total concentration. The number of monomers n_T and n_G are found from the system of two equations:

$$M = n_T m_T + n_G m_G$$
$$n_T/n_G = (C_T - C_{OT})/C_G \qquad (10)$$

where C_{OG} was neglected with respect to C_G, $m_T = 620$, $m_G = 1510$, and, for our set of measurements, $C_G = 0.8$ mM. The obtained values are reported in Table I. With increasing C_T, n_T becomes of course larger, but the total number of monomers $n_T + n_G$ becomes smaller.

We are not aware of any theory which allows to predict the total number of monomers in the mixed micelle as a function of the molar ratio between the two components. It is very interesting to note that the experimental results fit well the relation

$$n_T/126 + n_G/328 = 1 \qquad (11)$$

for all values of C_T. The numbers 126 and 328 represent the aggregation numbers for pure Triton X-100 and pure GM1 micelles respectively. Relation (11) can be explained with the following geometrical picture in terms of the curvature of the micelle surface. In 4π steradians Triton X-100 alone accomodates 126 monomers and GM1 alone 328 monomers. Therefore the solid angle per monomer is $4\pi/126$ steradians for Triton X-100 and $4\pi/328$ steradians for GM1. If we make the assumption that in the mixed micelle each monomer takes the same solid angle taken in the homogeneous micelles, we obtain $4\pi = n_T(4\pi/126) + n_G(4\pi/328)$ which coincides with Equation (11). Our results suggest therefore that Triton X-100 and GM1 monomers contribute on the average the same amount of "curvature" regardless whether they are in the mixed micelle or not. This is true for all the investigated molar ratios. Another interesting example of mixed micelle formation is that of the system Triton X-100-sphingomyelin.[15] Sphingomyelin is a lipid which differs from GM1 in the fact that most of the molecule length is due to the hydrophobic portion, so that nearly parallel aggregation is favoured (liposomes). Therefore Triton X-100 molecules alone must accomodate the 4π steradians which is subtended by the mixed micelle surface, regardless of its shape. It is found indeed that the Triton X-100 monomer number in the mixed micelle does not change appreciably with Triton X-100 molar ratio.[15] This result is in agreement with Equation (11) if we consider the particular case when the solid angle relative to the lipidic monomers goes to zero.

We note finally that Equation (11) can be rearranged to express the molecular weight M of the mixed micelle in terms of the ratio $X = n_T/n_G$ and of the aggregation numbers n_{T0} and n_{G0} of the pure Triton X-100 and GM1 micelle respectively:

$$M = m_G n_{G0} \frac{1 + X\, m_T/m_G}{1 + X\, n_{G0}/n_{T0}} \tag{12}$$

4. CONCLUSIONS

We have described a light scattering study of an aqueous solution containing a mixture of two amphiphiles, a commercial nonionic surfactant, Triton X-100, and a biological lipid, the ganglioside GM1. Mixed micelles are formed in the whole range of investigated molar ratios. We have derived the formula which relates the scattered intensity I_r to the molar weight of the mixed micelle M. Such a formula constitutes a generalization of the well known expression for the single-amphiphile case, and shows that independent information about the free-monomer concentration may be needed to obtain M from I_r.

The aggregation number of the mixed micelle, measured as a function of the molar ratio between the two amphiphiles, is found to follow a simple phenomenological law expressed by Equation (11) or (12). It is possible that this law is of rather general applicability to mixed micellar solutions.

The interpretation of the light scattering results is particularly simple if the micellar solution behaves as an ideal solution, that is, micellar interactions are negligible. We have chosen for our discussion the data obtained at 15°C, instead of the 37°C temperature used in Reference 1, because the GM1-Triton X-100 solution presents a lower consolution point at 39°C, and critical effects are already very relevant at 37°C since the critical region of micellar solutions is much larger than that of usual binary mixtures.[10,16]

As we mentioned in the Introduction, the present work has a biochemical motivation. The activity of the enzyme galactose oxidase on GM1 is found to be very low when the solution contains homogeneous GM1 micelles. By adding growing amounts of Triton X-100 to the 0.8 mM GM1 solution, the enzyme activity increases at first very slowly, and becomes considerable when the Triton X-100 concentration exceeds 3 mM. The micellar parameters do not show any discontinuity at this concentration. It is possible that the enzyme starts to act on the GM1 inserted into the mixed micelle only when the Triton concentration is large enough to avoid contiguity among the GM1 hydrophilic heads on the micellar surface. Indeed the break point in the enzymatic activity corresponds to a ratio n_T/n_G about 3 which is just the minimum ratio needed to keep GM1 heads away one from the other on the micellar surface, as sketched in Figure 5. It should be noted that, by changing the GM1-Triton X-100 molar ratio, also the surface electric-charge distribution and the surface curvature are changed. This may influence the enzyme activity.

ACKNOWLEDGEMENTS

The experiments described in this paper are inserted in a research program we carry on in collaboration with the group directed by Professor G. Tettamanti at the Department of Biological Chemistry, Faculty of Medicine of the University of Milan, Milan, Italy.

This work was supported by CNR/CISE contract n.79.00764.02.

REFERENCES

1. M. Corti, V. Degiorgio, R. Ghidoni, S. Sonnino, and G. Tettamanti, Chem. Phys. Lipids 26: 225 (1980).
2. S. Formisano, M. L. Johnson, G. Lee, S. M. Aloj, and E. Edelhoch, Biochemistry 18: 1119 (1979).
3. W. Mraz, G. Schwarzmann, J. Sattler, T. Momoi, E. Seeman, and H. Wiegandt, Z. Physiol. Chem. 36: 177 (1980).
4. H. Wiegandt, Adv. Lipid Res. 9: 249 (1971).
5. M. Corti, V. Degiorgio, S. Sonnino, R. Ghidoni, M. Masserini, and G. Tettamanti, Chem. Phys. Lipids (to be published).

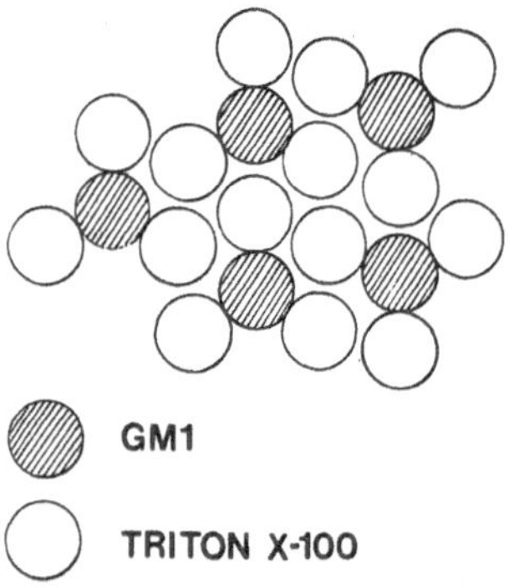

Figure 5. Sketch of the surface distribution of mixed Triton X-100-GM1 micelles which gives a monomer ratio $n_T/n_G = 3$.

6. M. Masserini, S. Sonnino, R. Ghidoni, and G. Tettamanti, Biochim. Biophys. Acta 601: 282 (1980).
7. J. Clint, Jour. Chem. Soc. 71: 1327 (1975).
8. D. N. Rubingh, in "Solution Chemistry of Surfactants," K. L. Mittal, ed., Plenum Press, New York (1979). vol. 1, p. 337.
9. H. Lange, and K. H. Beck, Kolloid Z. - Z. Polym. 251: 424 (1973).
10. M. Corti, and V. Degiorgio, Optic. Commun. 14: 358 (1975).
11. D. Gammack, Biochem. Jour. 88: 373 (1963).
12. M. Corti, and V. Degiorgio, Ann. Physique (Paris) 3: 303 (1978).
13. G. Tettamanti, F. Bonali, S. Marchesini, and V. Zambotti, Biochim. Biophys. Acta 296: 160 (1973).
14. S. Sonnino, R. Ghidoni, G. Galli, and G. Tettamanti, Jour. Neurochem. 31: 947 (1978).
15. S. Yedgar, Y. Barenholz, and V. G. Cooper, Biochim. Biophys. Acta 363: 98 (1974).
16. M. Corti, and V. Degiorgio, Phys. Rev. Letters 45: 1045 (1980).

KINETICS OF GROWTH OF PHOSPHOLIPID VESICLES

Didier Sornette and Nicole Ostrowsky

Laboratoire de Physique de la Matière Condensée
Associé au C. N. R. S. (LA 190)
Université de Nice, Parc Valrose
06034 Nice Cedex, France

CONTENTS

1. INTRODUCTION

Phospholipids are amphiphilic molecules. Such molecules consist of two parts - one easily soluble in water (the so-called polar head) and the other highly insoluble, composed of two aliphatic chains (see Figure 1a). When put into water, these molecules organize themselves so as to minimize their free energy, building up structures such as a spherical bilayer, called a vesicle (see Figure 1b). There has been considerable interest in studying such structures, in part because they represent a somewhat idealized model of a biological membrane. In addition, and from a more practical point of view, they are considered as potential drugs carriers which could eventually deliver their content to special body regions where membrane permeability would be altered, by a slight change in temperature, for example.

We also mention another line of current interest in the vesicle problem. In the well-known fundamental biological experiment where a number of simple elements are submitted to a very large electric discharge, some simple phospholipids are formed. Certain biologists thus infer that the resulting spontaneous vesicle forma-

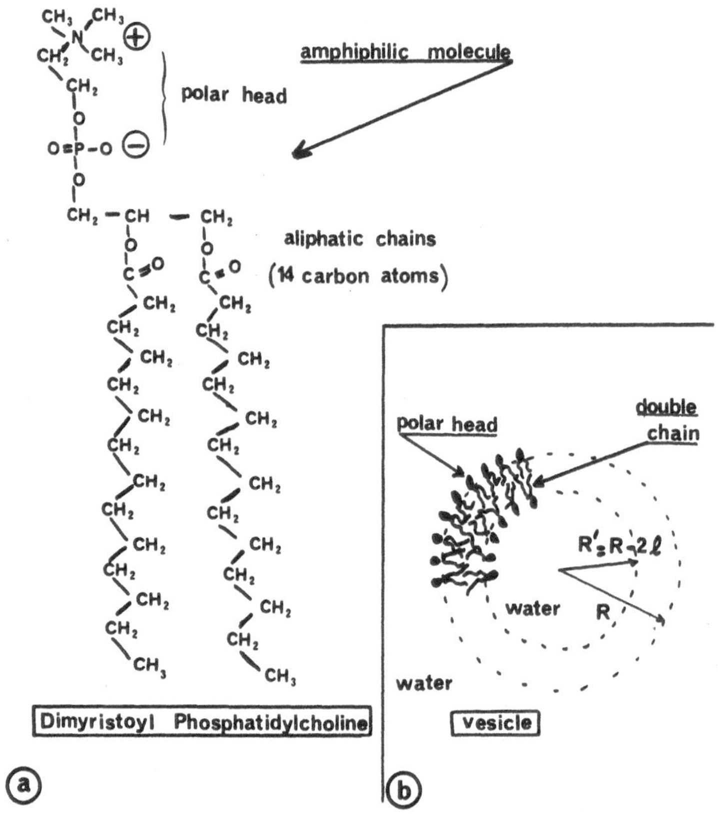

Figure 1. (a) Chemical structure of a particular phospholipid: Dimirystoyl phosphatidylcholine; (b) schematic representation of a vesicle.

tion could be a determining step in achieving the high concentration of various chemical agents necessary for the synthesis of more elaborate macromolecules.

The work we shall present here concerns the formation of such phospholipids vesicles and a study of their size as they pass through the phase transition of their aliphatic chains. We shall first briefly recall the basic theory for the formation of vesicles (Section 2), then describe the experimental results on the vesicles instability in the phase transition region (Section 3), and finally discuss different models to account for the kinetics of the observed vesicle growth (Section 4). An original method which we have derived to numerically reconstruct the very large size distribution function from the measured correlation function is briefly described in the other article by us in this volume.

2. FORMATION OF VESICLES

When an amphiphilic molecule leaves an aqueous surrounding to become part of a planar bilayer, there will be a net decrease $\Delta\mu$ of its free energy.[1] In addition to several constant terms related to the nature and size of the polar head and aliphatic chains, $\Delta\mu$ will be strongly dependent on the packing of the amphiphiles on the bilayer, characterized by the free surface a per polar head.

In a phenomenological way, this contribution can be described by two conflicting terms: 1) the attractive hydrophobic effect, which tends to minimize the contact surface between the aliphatic chains and the water and is of the form γa with $\gamma \simeq 50$ erg/cm^2 for the phosphatidylcholine polar head which we have studied; and, 2) the electrostatic repulsive term, mostly between the polar heads, which can be roughly approximated by a term inversely proportional to a, viz., C/a.

Minimizing $\Delta\mu = \gamma a + C/a$ leads to the concept of an optimal surface $a_o = \sqrt{C/\gamma}$, in terms of which we can rewrite $\Delta\mu$ as

$$\Delta\mu = -2\gamma a_o + \frac{\gamma}{a}(a - a_o)^2. \tag{1}$$

If we now transpose these results to a spherical bilayer, we must include a curvature term plus some geometrical constraints which yields an additional relation between the free surface a occupied by the polar heads and the vesicle's radius R,

$$\frac{v}{a} = \ell\left[1 - \frac{\ell}{R} + \frac{\ell^2}{3R^2}\right], \tag{2}$$

where v is the volume occupied by the two aliphatic chains, of length ℓ, which constitute the hydrophobic part of one amphiphilic molecule. As a result, $\Delta\mu$ becomes R-dependent [$\Delta\mu_R$] and, furthermore, one can show that there exists a critical radius R_c above which all polar heads may occupy their optimal surface area a_o.

It is now easy to derive the thermodynamic equilibrium size distribution of vesicles. The percentage of amphiphilic molecules participating in a vesicle of given radius R can be shown[1] to be the product of an energy term $\exp[-\Delta\mu_R/kT]$ which reaches a maximum and constant value for $R \geq R_c$, and an entropy term which favours the smallest vesicles' sizes. As a result, the equilibrium size distribution presents a maximum (at a value somewhat smaller than R_c because of the curvature term) with a fairly small dispersion, as we have experimentally checked with phospholipids of different aliphatic chain lengths.

3. PHASE TRANSITION - EFFECT ON THE VESICLES' STABILITY

A. Phase Transition

The above theory and experimental results were obtained for vesicles having their aliphatic chains in a pseudo-liquid form, i.e., for chains as flexible and mobile as in a liquid except at their anchoring point - the polar head, which remains on the surface of the bilayer. As the temperature is lowered, the chains undergo a first-order phase transition[2] at T_ϕ to a crystal-like phase where practically all the C-C bounds are in the "trans" conformation. The chains thus have an increased rigidity and a greater length ℓ'. This is accompanied by a tighter packing of the polar heads, which amounts to saying that their optimal free surface takes on a smaller value a_o'.

X-ray diffraction experiments[3] have determined the various geometrical parameters of a bilayer on both sides of the transition and the data are summarized below for dimyristoyl phosphatidylcholine, a phospholipid whose phase transition occurs around 23°C for the planar bilayer and spreads out down to 18°C for highly curved bilayers:

$$\ell \simeq 16\ \mathring{A} \qquad a_o \simeq 57\ \mathring{A}^2 \qquad \text{above } T_\phi$$

$$\ell' \simeq 19\ \mathring{A} \qquad a_o' \simeq 47\ \mathring{A}^2 \qquad \text{below } T_\phi\ . \qquad (3)$$

In addition to an increase in membrane permeability at the phase transition, a noticeable change in the vesicles' sizes has been qualitatively observed.[4] This can be readily understood by the following argument.[5] As a vesicle goes through the phase transition, the change in the packing of the spherical bilayer results in a surface tension which favours the opening of unstable pores, thus leading to the fusion of two vesicles.

B. Experimental Results

Using a standard 90° heterodyne scattering technique, we have studied the kinetics of the vesicles' growth.

The vesicles were prepared from Sigma synthetic dimyristoyl phophatidylcholine. The sonication and double centrifugation (at ≈ 100,000g) were performed at 40°C, that is, well above the phase transition temperature. Samples were then cooled down to a fixed temperature T below T_ϕ. The time evolution of the size distribution was subsequently measured.

The results obtained for T = 19°C are represented on Figure 2a, the radii being plotted in a logarithmic scale. For the sake of comparison, the time evolution of the size distribution for a

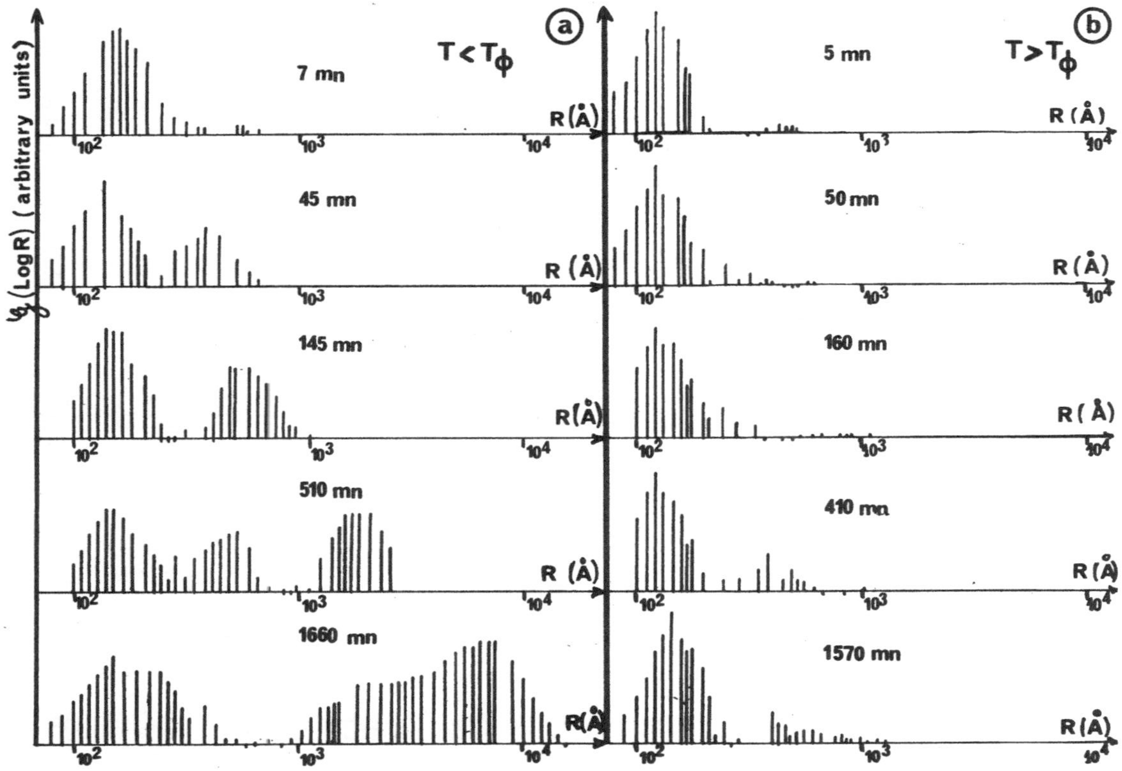

Figure 2. Time evolution of the size distribution of vesicles cooled down below T_ϕ (a) and kept above T_ϕ (b).

temperature T = 32°C, i.e., well above T_ϕ, is shown on Figure 2b. The ordinate $\mathcal{G}(\log R)$ represents the <u>intensity</u> contribution to the spectrum from vesicles having a radius R within the interval d(log R)

$$g(\tau) = A\int_0^\infty d(\log R)\, \mathcal{G}(\log R)\, e^{-Dq^2\tau} + BG, \tag{4}$$

where $g(\tau)$ denotes the experimentally measured correlation function. A and BG are experimental parameters depending on the heterodyne ratio and the optics of the experiment, q is the scattering wave vector, and D the translational diffusion coefficient, $D=kT/(6\pi\eta R)$.

The numerical procedure by which we deduced the function $\mathcal{G}(\log R)$ from $g(\tau)$ is summarized in the poster "Numerical Inversion of the Laplace Transform".[6] Figure 2a clearly shows the persistence of a large majority of small vesicles, and, as time progresses, the appearance of several minima in $\mathcal{G}(\log R)$ which will be interpreted in the next section.

4. KINETICS OF THE VESICLES' GROWTH

In order to interpret experimental data such as those presented in Figure 2a, we transposed to the vesicle problem the Smoluchowski Equation[7] first established to study the phenomenon of coagulation exhibited by colloidal particles.

Suppose that initially all the N_o vesicles have the same size. We shall say that they are in the monomer state, whose concentration at time t = 0 is therefore $n_1(t=0) = N_o$. The association of i monomers will yield the so-called i-mer state, whose concentration at time t is represented by $n_i(t)$. Assuming that the time between collisions of an i-mer and a j-mer is independent of i and j and is equal to t_c, and setting W_{ij} equal to the probability that an i-mer and a j-mer which have undergone a collision do associate to form an (i+j)-mer, we have

$$\frac{dn_i}{dt} = \frac{1}{N_o t_c}\left[\frac{1}{2}\sum_{j=1}^{i-1} W_{j,i-j}\, n_j(t)\, n_{i-j}(t) - n_i(t)\sum_{j=1}^{\infty} W_{ij}\, n_j(t)\right]. \tag{5}$$

Note that, for the usual concentrations ($N_o \approx 3\text{x}10^{13}$/cc), $t_c \simeq 5$ms.

As we measure the i-mer distribution by means of a light scattering method, we can readily distinguish between the two extreme types of associations, <u>aggregation</u> or <u>fusion</u>, which for a given i-mer correspond to drastically different radii and therefore different scattered intensities I(R) (see Figure 3).

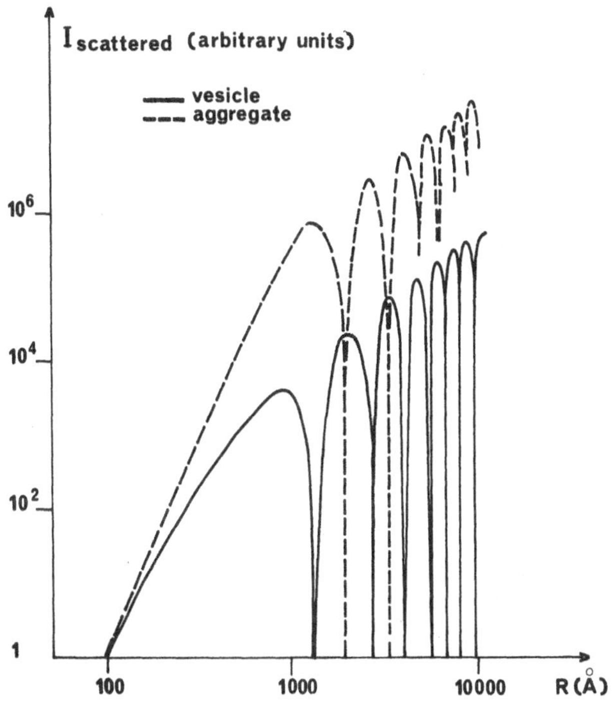

Figure 3. R-dependence of the intensity scattered by one vesicle (solid line) and one aggregate (dashed line).

We have solved Equation (5) for two different association models, $W_{ij} = W$ (normal fusion or aggregation), which yields

$$n_i(t) = \frac{(t/t_f)^{i-1}}{(1 + t/t_f)^{i+1}} \qquad \text{with } t_f = t_c/W, \tag{6}$$

and $W_{ij} = ij\,\tilde{W}$ (accelerated fusion or aggregation), which yields

$$n_i(t) = \frac{(2i\ t/\tilde{t}_f)^{i-1}}{i\ i!} \qquad e^{-2i\ t/\tilde{t}_f} \quad \text{with } \tilde{t}_f = t_c/\tilde{W}\ . \tag{7}$$

The results of the simulated size distribution functions as seen in a light scattering experiment are plotted on Figures 4 and 5. $\mathcal{G}(\log R)\ d(\log R)$ represents the scattered intensity from all particles of radius R within the interval d(log R) and is easily computed once we know the radius R(i) of an i-mer. If R_o is the radius of the monomer, we have:

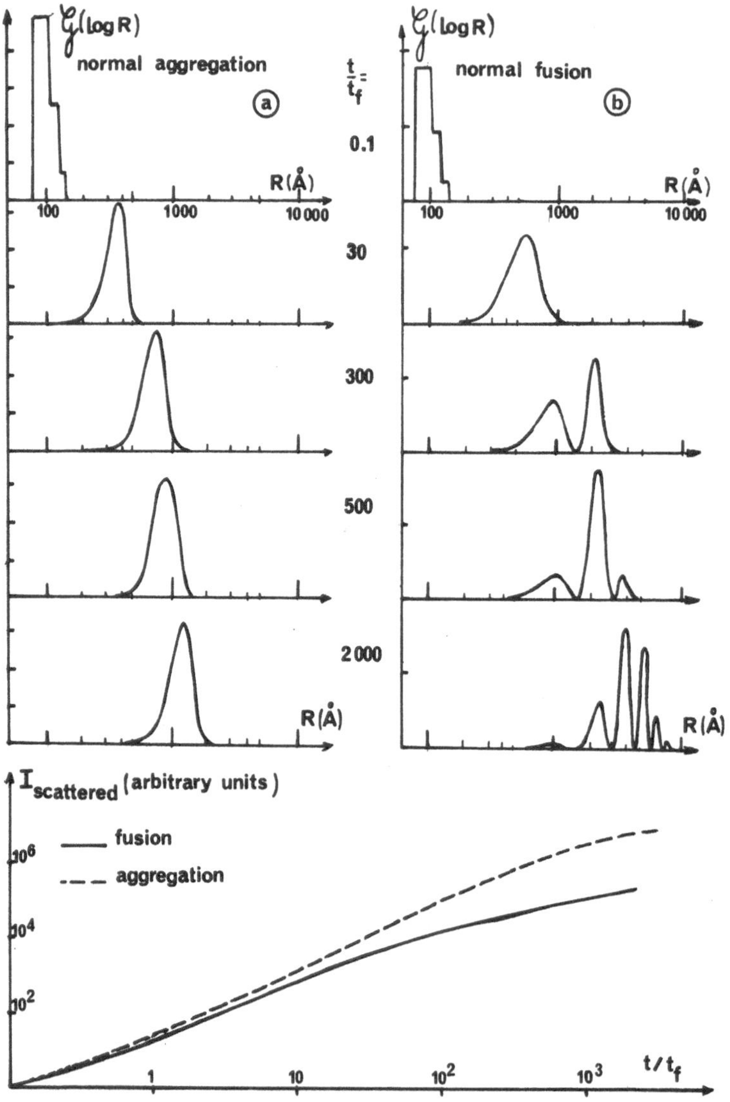

Figure 4. Simulated normalized size distributions for normal (W_{ij}= W) aggregation (a) and fusion (b) kinetics. The time evolution of the total 90° scattered intensity is shown on the bottom of the figure.

$$i = \frac{R(i)^2 + [R(i) - 2\ell']^2}{R_o^2 + [R_o - 2\ell']^2} \tag{8}$$

for the fusion model, where $2\ell'$ represents the bilayer thickness below the phase transition, and

$$i = R(i)^3/R_o^3 \tag{9}$$

for the aggregation model.

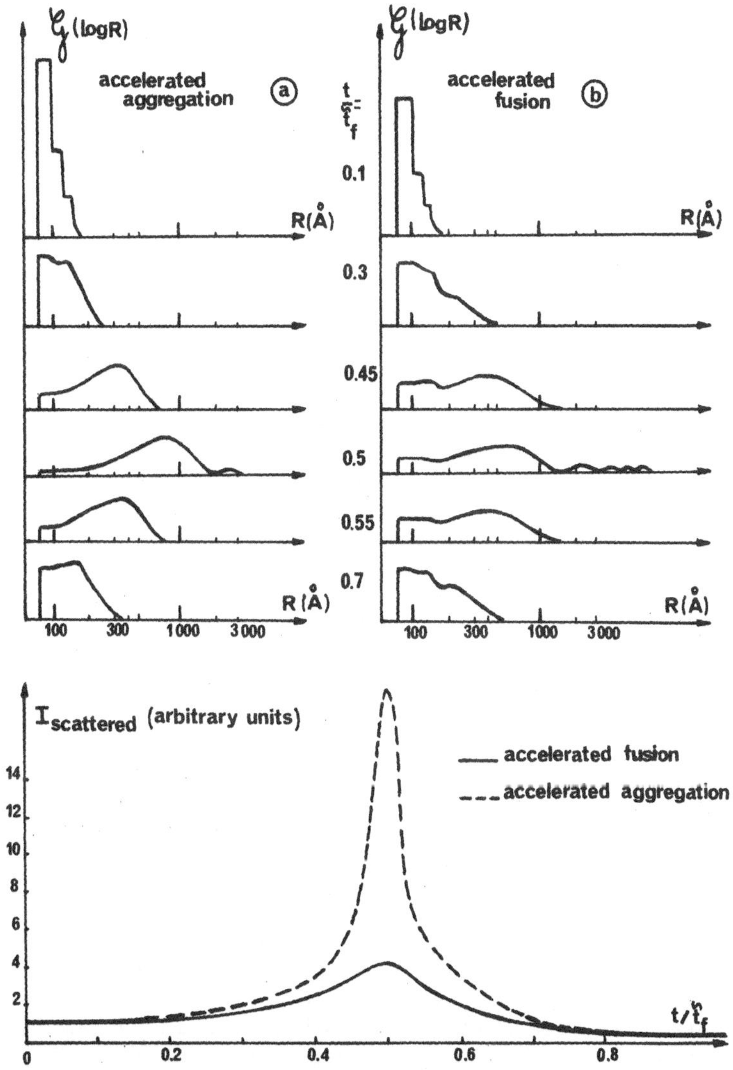

Figure 5. Simulated normalized size distributions for accelerated ($W_{ij} = ij\,\underset{\sim}{W}$) aggregation (a) and fusion (b) kinetics. The plot on the bottom represents the total 90° scattered intensity versus time.

It then follows that, at a given instant t,

$$\mathcal{G}(\log R)\, d(\log R) = n_i(t)\, di\, I(R), \qquad (10)$$

where the differential element di is to be computed from Equation (8) or (9). The R-dependence of I(R), for a given scattering vector q, is given by (see Figure 3)

$$I(R) \propto (\sin qR - qR \cos qR - \sin qR' + qR' \cos qR')^2 \qquad (11)$$

for the fusion model where the i-mer is a vesicle (hollow sphere of internal radius $R' = R - 2\ell'$), and

$$I(R) \propto \zeta(\sin qR - qR \cos qR)^2 \qquad (12)$$

for the aggregation model. ζ is a parameter less than 1 and independent of R, which accounts for the fact that an aggregate composed of vesicles is not totally filled with the amphiphilic molecules but contains a large part of water.

A comparison of Figure 2a and both Figures 4 and 5 shows clearly that the aggregation phenomenon (accelerated or not) cannot account for the observed spectrum which exhibits sizes up to several thousands angtröms with one or two minima in the intensity distribution function. On the other hand, the normal fusion process (Figure 4b) could account qualitatively for the large radius part of the observed spectra (obviously, the numerous oscillations shown on the last curves of Figure 4b cannot be resolved experimentally). Comparing the experimental time scale with the time between fusion events, $t_f = t_c/W$ appearing in Equation (6) yields for t_f a value of approximately 500 s, which means that 1 out of every 10^5 collisions would result in a fusion event.

However a striking feature shown by Figure 2a is that only a very small percentage of the initial vesicles will undergo this unlimited fusion process, whereas most of the vesicles will remain in the form of the first i-mers ($i \lesssim 4$). One can then question the first hypothesis that W_{ij} be a constant W.

To try and find the i-j dependence of W_{ij}, we computed the free energy μ_i of an amphiphilic molecule in an i-mer state, retaining only in μ_i the surface term $(\gamma/a_i)(a_i - a_o')^2$ of Equation (1), where a_i denotes the free surface per polar head in an i-mer state. This term is easily computed if one assumes that no flip-flop of amphiphilic molecules from one side of the bilayer to the other occurs during the fusion process. Introducing the dimensionless parameter

$$\lambda = (\ell'/\ell)\, \sqrt{a_o/a_o'} \qquad (13)$$

which characterizes the geometrical changes of the bilayer upon going through the phase transition, one gets:[8]

$$\mu_i = \frac{32\pi\gamma}{N_o}\frac{\ell^2}{\lambda^2}(\lambda/\sqrt{i} - 1)^2. \tag{14}$$

From the numerical values given in Equation (3) we see that this energy presents a minimum for i = 2. This simply states that all the constraints induced in the monomer when lowering the temperature below T_ϕ are annealed in the dimer state. This led us to the rather crude assumption that only fusion events involving the monomer could occur, that is,

$$W_{ij} = W\,\delta_{1j}. \tag{15}$$

Even with this oversimplified hypothesis, Equation (5) could not be solved analytically. We therefore used a numerical method and found, as expected, distributions which could account for the small radius part of the experimental data: a single distribution peaked on the quadrimer size ($R \simeq 140$ Å for $R_o = 90$ Å).

In order to interpret the whole radius range of the experimental data, we solved a model of accelerated fusion ($W_{ij} = ij\,\tilde{W}$) which we hoped would account for a decrease in stability of the i-mer as i increases [see Equation (7) and Figure (5)]. This model does show, at the shortest times, a tendency for the small vesicles to survive, but it does not account for the long time evolution of the experimental data. More work is being done to refine the model for μ_i and deduce a more realistic form for W_{ij}.

The temperature dependence of the kinetics was also studied in the range 8 to 20°C. It exhibits a sharp maximum around 13°C, which seems correlated with the important rise in the flip-flop rate as measured from NMR experiments.[9] This correlation is not surprising, as the assumption that no flip-flop occurs in a fusion event was essential to the derivation of Equation (14), which predicts a very limited size distribution for all times. We may therefore conjecture that the fusion is indeed triggered as soon as the temperature falls below T_ϕ, but the kinetics of the long range fusion may be limited by the flip-flop rate, which should therefore be taken into account in the calculation of μ_i.

In conclusion, our experiments have shown the importance of the phase transition in triggering the fusion phenomenon. The details of the kinetics are not yet completely worked out but one can infer that the great stability of the dimer must account for the fact that most of the vesicles will undergo a limited fusion. A small portion of the vesicles, however, seem to obey the normal fusion pattern, but with flip-flop rate limited kinetics. Different models for μ_i, taking into account in particular the flip-flop phenomenon, are currently being worked out to explain the entire radius range of the experimental data over a large time scale.

REFERENCES

1. J. N. Israelachvili, D. J. Mitchell, and B. W. Ninham, J. Chem. Soc. Far. Trans. II, 72: 1525 (1976); BBA 470: 185 (1977).
2. D. Chapman, R. M. Williams, and B. D. Ladbrooke, Chem. Phys. Lipids 1: 445 (1967); 3: 304 (1969).
3. A Tardieu, V. Luzzati, and F. C. Reman, J. Mol. Biol. 75: 711 (1973).
4. C. Taupin and H. M. McConnell, Mitochondria/Biomembrane (North-Holland, 1972), p. 219.
   J. H. Prestegard and B. Fellmeth, Biochem. 13: 1122 (1974).
   N. Ostrowsky and C. Hesse-Bezot, Chem. Phys. Lett. 52: 141 (1977).
5. N. Ostrowsky and D. Sornette, Light Scattering in Liquids and Macromolecular Solutions, V. Degiorgio, M. Corti, and M. Giglio, eds. (Plenum, 1980).
6. D. Sornette and N. Ostrowsky, "Numerical inversion of the Laplace transform", in this volume; N. Ostrowsky, P. Parker, E. R. Pike, and Didier Sornette, to be published in Opt. Acta (1981).
7. M. Z. Smoluchowski, Phys. Chem. 92: 129 (1917).
8. D. Sornette, C. Hesse-Bezot, and N. Ostrowsky, to be published.
9. B. De Kruijff and E. J. J. Van Zoelen, BBA 511: 105 (1978).

MICELLAR SIZE AS A FUNCTION OF PRESSURE, TEMPERATURE, AND SALT CONCENTRATION FOR A SERIES OF CATIONIC SURFACTANTS

D. F. Nicoli,* R. Ciccolello and J. Briggs

Physics Department
University of California at Santa Barbara
Santa Barbara, CA 93106

D. R. Dawson, H. W. Offen, L. Romsted and C. A. Bunton

Chemistry Department
University of California at Santa Barbara
Santa Barbara, CA 93106

CONTENTS

1. INTRODUCTION

The size, shape and structure of the micellar aggregates formed from ionic surfactants are believed to be influenced by a variety of variables[1,2]. These include the surfactant composition (hydrocarbon "chain length" and headgroup species) and molarity; the counterion species and concentration; and finally temperature and pressure. Pioneering studies of micelle size as a function of temperature and salt concentration have been made using the powerful technique of quasi-elastic light scattering spectroscopy by Benedek's group at MIT[3-5], Corti and Degiorgio[6,7], and others[8,9].

---

*A lecture presented at the NATO Advanced Study Institute on "Scattering Techniques Applied to Supramolecular and Non-Equilibrium Systems, Aug. 4-14, 1980, Wellesley College, Wellesley, Massachusetts, USA.)

One of the principal objectives of these measurements of micellar hydrodynamic radius by dynamic light scattering is the elucidation of the self-assembly mechanism of surfactant molecules.

This report focuses on two broad categories of photon correlation experiments. The first is a study of the growth of micelles as a function of surfactant composition, in which the hydrocarbon chain length is systematically perturbed, leaving all other chemical parameters constant. We employed the family of cationic surfactants $H(CH_2)_nN(CH_3)_3^+ - Br^-$ with n = 12,14,16. These are named, respectively, dodecyl-, myristyl- and cetyl- trimethylammonium bromide, or "DoTABr", "MyTABr" and "CTABr". Mazer, Carey and Benedek[4] ("MCB") have recently described a thermodynamic model for rod-like micellar growth of SDS (sodium dodecyl sulfate) which requires only two chemical potential parameters. Whenever possible, we have applied the MCB "ladder" model to our data to test the sensitivity of the sphere/cylinder thermodynamic equilibrium to surfactant chain length.

The second group of experiments consists of measurements of the micellar hydrodynamic radius R_h as a function of pressure (0-3000 bar) for each of these surfactants which differ only in chain length. Of all the variables known to influence micelle size, the one which has been studied the least is pressure. The pressure effects observed thus far are striking[10]; the P-dependence is not only large (up to 40% variations in R_h) but also often highly non-monotonic.

2. EXPERIMENTAL METHODS

Eastman surfactants were employed, recrystallized three times from absolute ethanol (lab of C.A. Bunton). Purity was established by the absence of a detectable minimum in a plot of surface tension vs (log) surfactant concentration. Elimination using silver oxide, followed by gas chromatography of the alkene products, yielded negligible concentrations of C_n alkenes for n different from the desired chain length.

In the experiments performed at atmospheric pressure, the samples were housed in 6 x 50 mm cylindrical glass culture tubes, centrifuged at 5000g for 10 minutes to settle out dirt. Temperature was regulated in the range 5-80°C ($\pm$0.1°C) using a Peltier controller. The high pressure experiments utilized a miniature cylindrical quartz scattering cell of volume 0.1 cc, sealed with a stainless steel pressure-transmitting piston. This cell is centered in a beryllium-copper pressure vessel containing three sapphire windows[11], with hexane used as the pressure-transmitting fluid. An oil hand pump combined with a 25:1 hydraulic intensifier was used to achieve hexane pressures in the range 1-4000 bar, monitored continuously by a manganin-wire strain gauge and digital pressure meter (± 5 bar). The pressure was always increased slowly in increments of 200-300 bar and the system allowed to equilibrate before data were accumulated; the pressure drop rate due to leaks was typically less than 1 bar min^{-1}.

The laser light source was a Spectra Physics Model 164 argon-ion laser (514.5 nm) operating at 80-150 mW output power. Scattered light was collected from roughly one coherence area at an angle of 90° and imaged onto the slit of an EMI 9789 PMT. For the pressure runs the laser beam was focused along the diameter of the cylindrical quartz cell and the 90° scattered light collected from the flat end, resulting in relatively clean optical imaging. The photocurrent fluctuations were analyzed by a 64-channel, 4-bit, 10 MHz computing autocorrelator (Nicomp Model 6864), which provides automatic quadratic and cubic least-squares fits (method of cumulants[12]) to the intensity autocorrelation data.

3. RESULTS AND DISCUSSION

A. Micellar Size vs Surfactant Chain Length

We first examine the 12-carbon chain surfactant, DoTABr. Figure 1 shows the micellar Stokes radius R_h plotted as a function of temperature T for 0.15M DoTABr at two different NaBr concentrations; 0.2M and 0.6M. (The surfactant molarity was arbitrarily fixed at 10 times the c.m.c. value in the absence of added salt.) At either salt concentration there is only a slight increase in R_h with decreasing T (e.g. for 0.2M NaBr, from 19-20 Å at 40°C to 22 Å at 5°C). This behavior is in sharp contrast to the large growth in micelle radius observed by MCB[4] for the anionic 12-C chain surfac-

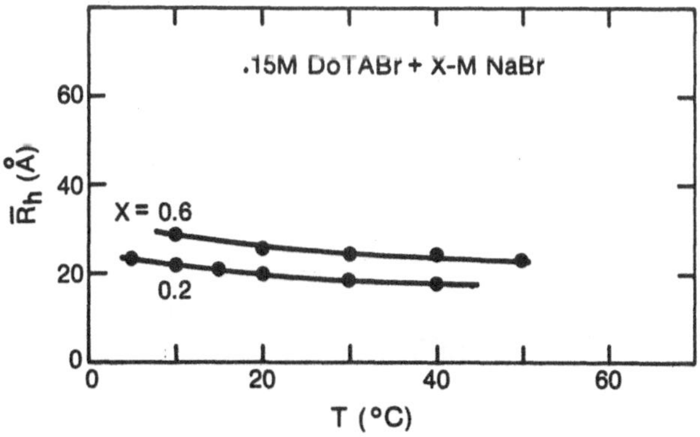

Figure 1. Hydrodynamic radius R_h (Å) versus temperature T(°C) for 0.15M DoTABr + X-M NaBr, where X = 0.2, 0,6.

tant, SDS. For example, for 0.069M SDS + 0.6M NaCl, MCB observed an increase in R_h from roughly 25 Å (the "minimum sphere" radius) at 70°C to over 150 Å at 20°C. Hence, we conclude that the ability of ionic micelles to grow with decreasing T or increasing salt concentration depends critically on the surfactant headgroup species and/or counterion type. Part of the apparent difference in the Stokes radii R_h at the two different salt concentrations may be the result of intermicellar interactions, as opposed to a true difference in the micelle sizes. That is, the decrease in R_h at 0.2M NaBr may be partly due to an enhancement in the diffusion coefficient because of less effective screening of the electrostatic repulsion between micelles at the lower salt molarity. More will be said about intermicellar interactions shortly.

We next examine R_h vs T for the 14-carbon chain surfactant, MyTABr, shown in Fig. 2 for NaBr concentrations ranging from 0.2M to 0.6M. (The MyTABr molarity, 0.035M, was again chosen to be 10 times the c.m.c. value in the absence of added salt.) One observes drastically larger micelle growth for the 14-C chain surfactant than was seen for the 12-C chain system, all other parameters being held constant. This is surprising, given a mere 17% increase in the hydrocarbon chain length.

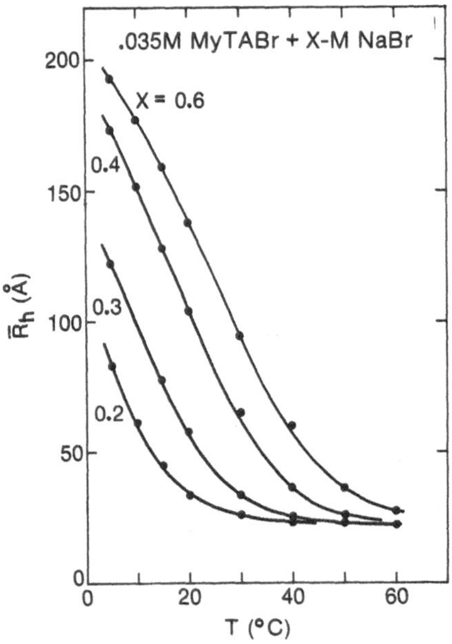

Figure 2. R_h(Å) versus T(°C) for 0.035M MyTABr + X-M NaBr, where X = 0.2, 0.3, 0.4 and 0.6.

We notice in Fig. 2 a feature of the R_h vs T data which was evidently not observed in SDS: a point of inflection at the highest salt concentrations (most apparent at 0.6M NaBr). That is, there is a slowing down of the rate of increase of R_h with decreasing T below 20-25°C. This behavior will be seen to significantly affect the thermodynamic analysis (i.e. van't Hoff plots).

In investigating the thermodynamics of micelle growth as revealed by our R_h vs T data, we follow closely the analysis of MCB[4]. In their "ladder" model, the chemical potential μ_n of a micelle of aggregation number n can be written simply as

$$\mu_n = \mu_{n_o} + (n-n_o)\mu \qquad (n \geq n_o) \qquad (1)$$

The micelle is presumed to grow in rod-like fashion[13], where μ_{n_o} is the free energy of the two hemispherical end caps (equivalent to a "minimum sphere") and μ is the free energy of a single surfactant molecule located in the cylindrical region. MCB find that the set of micelle mole fractions X_n is a function of the product KX, where X is the total surfactant mole fraction and K is the Boltzmann factor which either promotes or discourages micelle growth,

$$K = \exp\{(\mu_{n_o} - n_o\mu)/KT\} \qquad (2)$$

From the set of X_n and the appropriate intensity weighting and scattering form factors one can relate the measured mean Stokes radius R_h to quantity KX. MCB found the resulting lnK vs 1/T plots for their SDS data to be linear, thus supporting their free energy model of Eg'n 1.

In the case of MyTABr, our "van't Hoff" plots (lnK vs 1/T) are shown in Fig. 3. The K values were obtained from a theoretical calculation of R_h vs KX, assuming prolate ellipsoidal micelles of fixed semi-minor axis b = 21 Å (consistent with the high-T asymptotic value in Fig. 2) and micellar volume proportional to aggregation number. Interestingly, the lnK vs 1/T plots are by no means linear for every value of salt concentration. Only at the lowest molarity, 0.2M, where the micelle growth is smallest (barely a 60 Å at 10°C) is the dependence approximately linear. The curvature is more pronounced the higher the salt concentration. This is the regime in which the apparent micellar radius R_h is largest and also where the inflection in dR_h/dT is most evident. Such significant departures from linearity (inconsistent with the ladder model of Eq'n 1) were not observed by MCB for SDS micelles, whose measured minimum sphere radii were roughly comparable to the values measured for MyTABr.

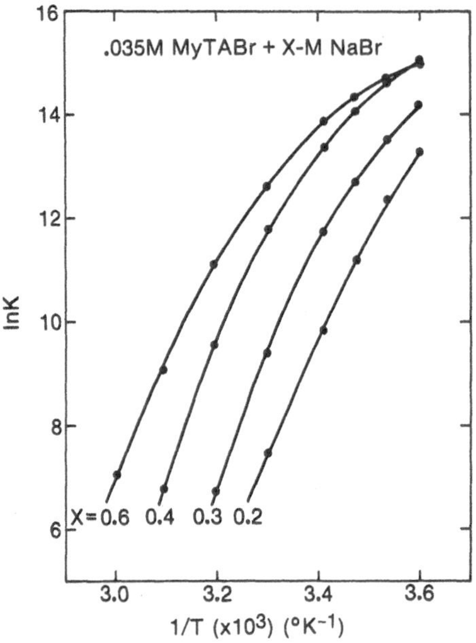

Figure 3. Van't Hoff plot, lnK versus 1/T, for data of Fig. 2 Assume: prolate ellipsoids (b = 21 Å), "ladder" model of MCB[4].

If we express Eq'n 2 in the form, $\ln K = \alpha(1/T) + \beta$, we find from Fig. 3 for the case of 0.2M NaBr the parameters $\alpha = 2.2 \times 10^4$ °K and $\beta = -65.4$ (same units as MCB). We now compare these parameters with those obtained by MCB for SDS (at 0.069M surfactant concentration): $\alpha = 1.71 \times 10^4$ °K and $\beta = -50$. At higher NaBr molarities it becomes meaningless to assign values to α and β due to nonideal behavior (i.e. highly non-linear van't Hoff plots). However, we can make the qualitative observation that the principal effect of adding salt is to increase (i.e. make less negative) the value of β, the enthalpy-like term.

The values of hydrodynamic radius claimed for micelles of MyTABr (Fig. 2), exceeding 150 Å for $T \leq 15$°C in 0.6M NaBr, are in qualitative agreement with earlier values obtained by MCB and others for SDS using the same technique. (The reported R_h values[3,4] for 0.035M SDS + 0.6M NaCl are somewhat smaller than those found for 0.35M MyTABr + 0.6M NaBr). All of these results suggest very large growth of the surfactant aggregates in the region of low temperature and high salt concentration. Nevertheless, it is clear that the fundamental quantity which is measured in dynamic light scattering is not R_h, but rather D_t, the translational diffusion coefficient. Fundamental to the claim that the associated

Stokes radius R_h ($= kT/6\pi\eta D_t$) does indeed represent the physical mean radius of a scattering particle is the assumption that the diffusion aggregates are sufficiently dilute that D_t is unaffected by interparticle interactions. We shall see that this assumption appears not to be valid for surfactant systems of relatively high surfactant concentration, such as are often investigated.

Micelle systems are highly concentration dependent as regards their physical characteristics. It is the very nature of these self-assembling systems, which contain monomer molecules in equilibrium with a distribution of micellar aggregates of various sizes, which makes the interpretation of light scattering experiments non-trivial. If significant intermicellar interactions (including van der Waals attractions and electrostatic repulsion) exist, their presence will certainly affect the apparent R_h values which are reported for these systems. If the net interaction is attractive, the apparent R_h will be too large; conversely, if electrostatic repulsion dominates, the R_h will be too small. We expect such effects to be most significant when micelle growth greatly exceeds the "minimum sphere" limiting size.

It is therefore instructive to estimate the maximum dimension of a micelle aggregate as well as the mean separation distance between micelles corresponding to values of R_h which are typically observed (i.e. 25-150 Å). For simplicity, we assume that each value of radius R_h results from a uniform distribution of micelles of prolate ellipsoidal (rod-like) shape. The polydispersity which is known experimentally to exist[3] (and which is central to the MCB ladder model[5]) will alter somewhat the numbers to be presented , but not the spirit of the results. Under these approximations, the mean separation distance S is given by

$$S = \{(n_o/M\cdot N_o)(a/b)\}^{1/3} \qquad (a>b) \qquad (3)$$

where n_o is the minimum sphere aggregation number, M is the surfactant molarity, N_o is Avagadro's number and a/b is the axial ratio of the prolate ellipsoids. The radius R_h is related to the semi-minor and -major axes (a,b) by the familiar Perrin result[15],

$$R_h = \frac{a(1-b^2/a^2)^{1/2}}{\log_e\{\frac{1 + (1-b^2/a^2)^{1/2}}{b/a}\}} \qquad (a>b) \qquad (4)$$

In Table I we list the micelle prolate length (2a) which corresponds (Eq'n 4) to values of the mean radius R_h in the range 25-150 Å. We have assumed a fixed semi-minor axis (b) of 21 Å, consistent with the high-T asymptotic value for MyTABr (Fig. 2). For prolates of large axial ratio a/b, R_h is relatively insensitive

TABLE I

R_h	Prolate Length (2a)	Mean Micelle Separation, S (x) = Surfactant Molarity		
		(0.035M)	(0.008M)	(0.004M)
25 Å	67 Å	185 Å	300 Å	380 Å
50	250	285	465	585
75	480	350	575	725
100	700	400	655	820
125	950	440	725	910
150	1200	480	780	985

Assume: Prolate ellipsoids (monodisperse), b = 21 Å; minimum sphere aggregation number = 80.

to the choice for b; hence, the values in Table I would not change significantly with the choice b = 25 Å (consistent with MCB's results for SDS). We also list the associated mean micelle separation S for three values of surfactant concentration: 0.035M, 0.008M and 0.004M.

The qualitative picture which emerges from Table I is indeed striking. First, one is reminded (cols. 1 and 2) that for prolate ellipsoids of large axial ratio there can be an enormous discrepancy between the length of the prolate (2a) and the apparent mean Stokes radius R_h corresponding to the translation diffusion coefficient, averaged over all orientations. Only when a is close to b does the resulting mean R_h accurately reflect the "size" of the particle. For our case of b = 21 Å, an apparent R_h of 50 Å (just 2.4 times b) corresponds to an actual prolate length which is <u>5 times</u> larger than R_h. At R_h = 150 Å the discrepancy becomes a factor of 8. Hence, from simple considerations of excluded volume, micelles which grow as prolate ellipsoids would appear to be very large indeed.

The second major feature which emerges from Table I involves the mean micelle separation, S. Clearly, as the nominal micelle size, as measured by R_h, grows, the value of S must increase as well. However, from Eq'n 3 we see that it grows very slowly, only as the 1/3-power of length a. Hence, at some point the prolate length exceeds the mean separation S. In the case of 0.035M MyTABr, this "crossover" is predicted to occur at the relatively small radius R_h of about 50 Å. The situation improves with decreasing total surfactant molarity (M), but only relatively slowly ($\sim M^{-1/3}$).

Hence, at 0.008M surfactant, the crossover 2a ≈ S ≈ 700 Å occurs at an R_h of about 100 Å.

Considering this question of micelle dimensions vs mean separation, it is not clear at what point intermicellar interactions become important in distorting the apparent measured value of R_h. We feel, however, that the arbitrary criterion of 2a = S as the crossover point is, if anything, probably conservative. That is, measured mean micellar radii may be significantly in error well below this point.

Corti and Degiorgio[16] have recently made very significant progress in elucidating the role of intermicellar interactions as they affect light scattering experiments. They measured both the scattered intensity and mean diffusion coefficient for SDS micelles as a function of surfactant concentration and salt molarity. They found that the interactions shift from being repulsive (higher D_t) to attractive (lower D_t) as the NaCl concentration is increased from 0.1M to 0.5M. Using the so-called DLVO theory of colloid stability, they obtain excellent theoretical fits to their data in a regime where they assume that the basic micelle size is small (e.g. ≈ 28 Å at 25°C) and essentially unchanging with either SDS concentration or salt concentration (within limits). Their claim (supported by the DLVO theory) is that the large changes observed in D_t, and hence in the apparent R_h values, with changing salt concentration are solely a consequence of intermicellar interactions. These effects actually exceed a factor of two at 0.069M SDS over the range 0.1-0.5M NaCl. In the future one should apply the Corti/Degiorgio calculations to MyTABr and CTABr to determine whether the apparent large micellar growth observed here is also in error, as is evidently the case for SDS.

We now proceed to examine R_h vs T for the 16-C chain surfactant, CTABr. We find that the "apparent" Stokes radii R_h are larger at low temperatures for CTABr than was observed for MyTABr, with values exceeding 200 Å at 20°C for 0.035M CTABr at the relatively low NaBr concentration of 0.2M. (CTABr precipitates out of aqueous NaBr solutions which are significantly more concentrated than 0.2M.) Given one's concerns about the effects of intermicellar interactions, it is useful to plot R_h vs T for a variety of surfactant molarities, from relatively concentrated (0.05M) to relatively dilute (0.004M), all at a fixed NaBr concentration of 0.2M.

Figure 4 shows the resulting Stokes radii associated with the measured values of D_t as a function of temperature for CTABr. First, we observe the familiar inflection in the slope dR_h/dT at values of R_h in the vicinity of 150 Å. The curvature is more pronounced than that previously seen for MyTABr (Fig. 2). This will be seen shortly to have obvious consequences for the lnK vs 1/T plots (i.e. causing them to be highly non-linear). However, the

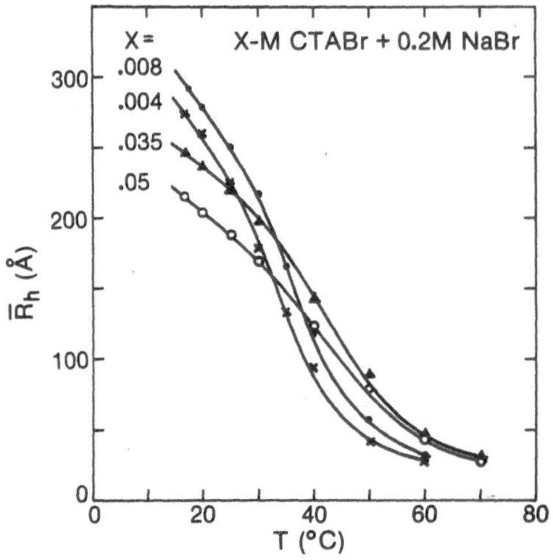

Figure 4: R_h(Å) versus T(°C) for X-M CTABr + 0.2M NaBr, where X = 0.004, 0.008, 0.035 and 0.05.

most unexpected feature of these CTABr results is a nonmonotonic dependence of the apparent R_h on surfactant concentration.

In the large-growth region at lowest temperatures (15-25°C), the value of R_h increases slightly as the CTABr concentration increases from 0.004M to 0.008M. This behavior is consistent with the simple notion that, above the c.m.c., an increase in the amount of surfactant produces not only more micelles, but also a larger average micelle aggregation number (i.e. size). However, much to our surprise, we find in Fig. 4 that a further increase in CTABr concentration from 0.008M to 0.035M results in a substantial decrease in the apparent mean radius R_h (but only for $T \leq 35$°C). The "radius" again drops by an equivalent percentage when the CTABr concentration is increased to 0.05M. At this relatively high surfactant concentration the apparent R_h is lower than that for 0.035M CTABr for all values of the temperature. For $T \geq 40$°C the "crossover" point in surfactant concentration has increased, so that the apparent R_h grows with CTABr molarity from 0.004M to 0.035M, before decreasing slightly at 0.05M.

Some qualitative conclusions can be tentatively drawn from the CTABr results of Fig. 4. It would appear that in 0.2M NaBr the dominant intermicellar interaction at low temperature is repulsive: i.e. D_t increases (R_h decreases) with increasing CTABr concentration. This agrees with the findings of Corti and Degiorgio[16] for SDS in the 0.2M NaCl. We attribute the initial decrease of D_t (increase of R_h) in going from 0.004M to 0.008M CTABr to an actual (albeit small) increase in the true micelle size. We further conclude that at higher temperatures (viz. $T \geq 40°C$) the crossover with surfactant concentration is pushed out to larger values because the mean micellar size is small enough relative to the mean separation distance that intermicellar repulsion becomes important only in the relatively concentrated case of 0.05M CTABr. Clearly, much more experimental data and theoretical modelling are needed to obtain a satisfactory quantitative explanation for this unusual behavior of the diffusion coefficient as a function of CTABr concentration and salt molarity.

Finally, in Fig. 5 we plot lnK vs 1/T, obtained from the MCB ladder model for the "R_h" vs T data of Fig. 4. Again, the choice of 21 Å for the semi-minor (b) axis of the prolates was dictated by the results of Fig. 4. As was the case for MyTABr (Fig. 3), we find highly non-linear van't Hoff plots for the larger surfactant molarities. Clearly, the most linear plot obtains at the lowest CTABr molarity, 0.004M, where the concentration of micelles is lowest. The resulting thermodynamic parameters for 0.004M CTABr + 0.2M NaBr are: $\alpha = 2.67 \times 10^4$ °K and $\beta = -67.7$. We recall

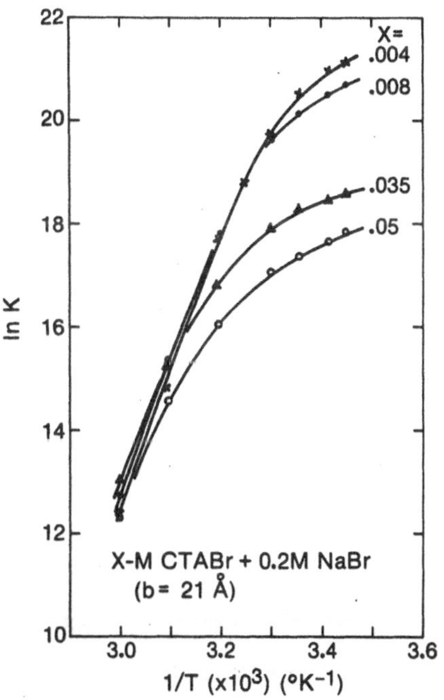

Figure 5. LnK versus 1/T for data of Fig. 4. Assume: prolate ellipsoids (b = 21 Å), "ladder" model of MCB[4].

that the corresponding parameters for 0.035M MyTABr + 0.2M NaBr are $\alpha = 2.2 \times 10$ °K and $\beta = -65$. We thus tentatively conclude that an increase in the surfactant hydrocarbon chain length from 14-C to 16-C has the principal effect of increasing (by about 20%) the magnitude of α, the entropy-like term.

B. Micellar Size vs Pressure

We have measured the intensity-weighted mean diffusion coefficient and associated hydrodynamic radius R_h as a function of increasing pressure at several discrete temperatures for the following systems: 0.15M DoTABr + 0.6M NaBr, 0.035M MyTABr + 0.2M NaBr, 0.035M MyTABr + 0.6M NaBr, 0.008 M CTABr + 0.2M NaBr, and 0.02M CTACl + 0.2M NaCl. Figures 6-10 show the pressure dependence on radius R_h for the above respective preparations. (In each case the surfactant concentration was arbitrarily chosen to be roughly 10 times the c.m.c. in the absence of added salt.)

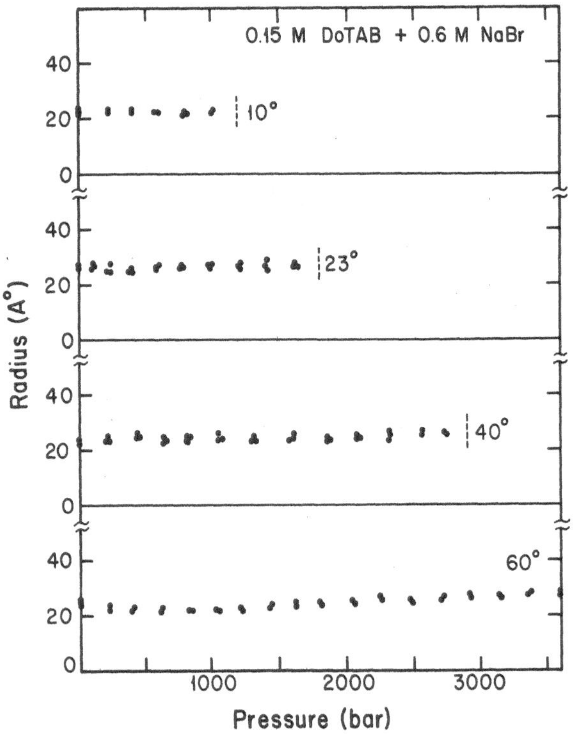

Figure 6. Hydrodynamic radius R_h(Å) versus pressure P(bar) for 0.15M DoTABr + 0.6M NaBr at 10°, 23°, 40° and 60°C.

Our previous R_h vs T studies established a clear trend: the apparent radius (i.e. $\overline{D_t^{-1}}$) of micelles was seen to increase greatly with surfactant chain length, for a given T, NaBr molarity. Similarly, we now find a steady progression in the magnitude of the P dependence of R_h with increasing chain length. The dependence ranges from essentially none for DoTABr to very large, nonmonotonic variations in R_h for MyTABr, and even larger changes for CTABr. In Fig. 6 we observe that the mean R_h for DoTABr (12-C) remains nearly constant at the minimum sphere value of approximately 21-23 Å.

The DoTABR results, however uninteresting with respect to pressure dependence, nevertheless reveal a universal feature of all our data: the existence of a critical pressure, P_c. At such a pressure (indicated by dotted lines in the figures) macroscopic aggregates of surfactant spontaneously precipitate out of solution. The transitions are highly hysteretic; to resume data taking the pressure must be lowered by 500-1500 bar to spontaneously redissolve the precipitate. We see from Fig. 6-10 that the critical pressures P_c are approximately the same at a given T for DoTABr (which exhibits no radius change) and MyTABr, but are substantially lower for CTABr. Hence, we conclude that critical behavior is not a consequence *per se* of micellar growth. In all cases P_c is observed to increase monotonically with increasing temperature.

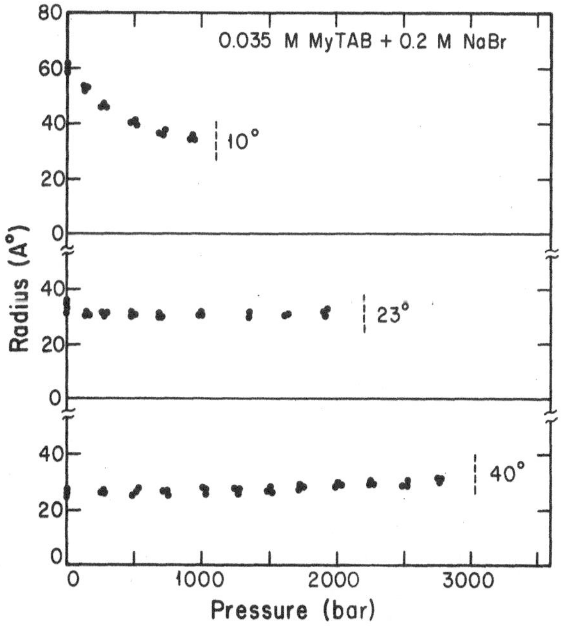

Figure 7. R_h(Å) versus P(bar) for 0.035M MyTABr + 0.2M NaBr at 10°, 23° and 40°C.

Radius R_h as a function of P for 0.035M MyTABr + 0.2M NaBr, seen in Fig. 7, shows a 40% decrease at 10°C before $P_c \approx 1000$ bar is reached; essentially negligible change at 23°C up to $P_c \approx 2000$ bar; and a 15-20% increase at 40°C before $P_c \approx 3000$ bar is reached. The finding that micelles can either shrink or expand, depending on the temperature employed, seems to be a most unusual result. (One naively might have expected either shrinkage or no size change at all with increasing P.) The pressure effects seen in MyTABr in Fig. 7 are much less pronounced than those reported earlier[10] in the higher salt environment of 0.6M NaBr, reviewed in Fig. 8. There, we find a complete variety of R_h vs P behavior, ranging from a strict decrease (25%) at 12°C to a strict increase (65%) at 41°C. At intermediate temperatures, 23°C for example, R_h is seen to decrease by 18% and then to change direction and increase by 16% to nearly the original value before P_c is reached. The reason for the much greater P dependence exhibited in Fig. 8 than in Fig. 7 is that at atmospheric pressure there is apparently more micelle growth and associated T dependence in 0.6M NaBr than in 0.2M NaBr solutions (Fig. 2). In Fig. 9 we observe a relatively wide range of R_h vs P behavior for 0.008M CTABr + 0.2M NaBr, reminiscent of the results of MyTABr at high salt concentration (Fig. 8). The occasional high data points at 40°, 50°C were due to the onset of the critical transition.

Finally, we measured R_h vs P for a 16-C chain surfactant identical to CTABr except for the replacement of the Br^- counterion with Cl^-: "CTACl". The Cl^- ion is known not to promote micellar growth from previous measurements of R_h vs T. Figure 10 contains a summary of our R_h vs P data for 0.02M CTACl + 0.2M NaCl. We see that at atmospheric pressure R_h barely increases as the temperature is lowered from 30° to 10°C. This behavior is in strong contrast to the enormous growth exhibited by CTABr + NaBr at comparable concentrations in the same temperature range (Fig. 4). However, the application of high pressure does cause substantial changes in apparent micellar size. At 10°C (below the Krafft point) there is the familiar drop in R_h with increasing P; however, since the starting value is already so close to the minimum sphere radius, only a small decrease in R_h is evidently possible. At 23°C the mean radius is virtually unchanged up to $P_c \approx 1500$ bar. At 30°C, however, the effects of pressure are large; growth in R_h is modest up to about 1250 bar, at which point it increases until $P_c \approx 1700$ bar is reached. Comparing Figs. 9 and 10, we observe two qualitative differences associated with the replacement of Br^- with Cl^-. First, the critical pressures P_c are higher for CTACl than for CTABr at the same temperatures. Second, the temperature at which R_h vs P changes from being strictly decreasing to strictly increasing is much lower for CTACl than for CTABr.

To summarize these pressure results, we have observed several universal features in the data. First, we find that whenever the

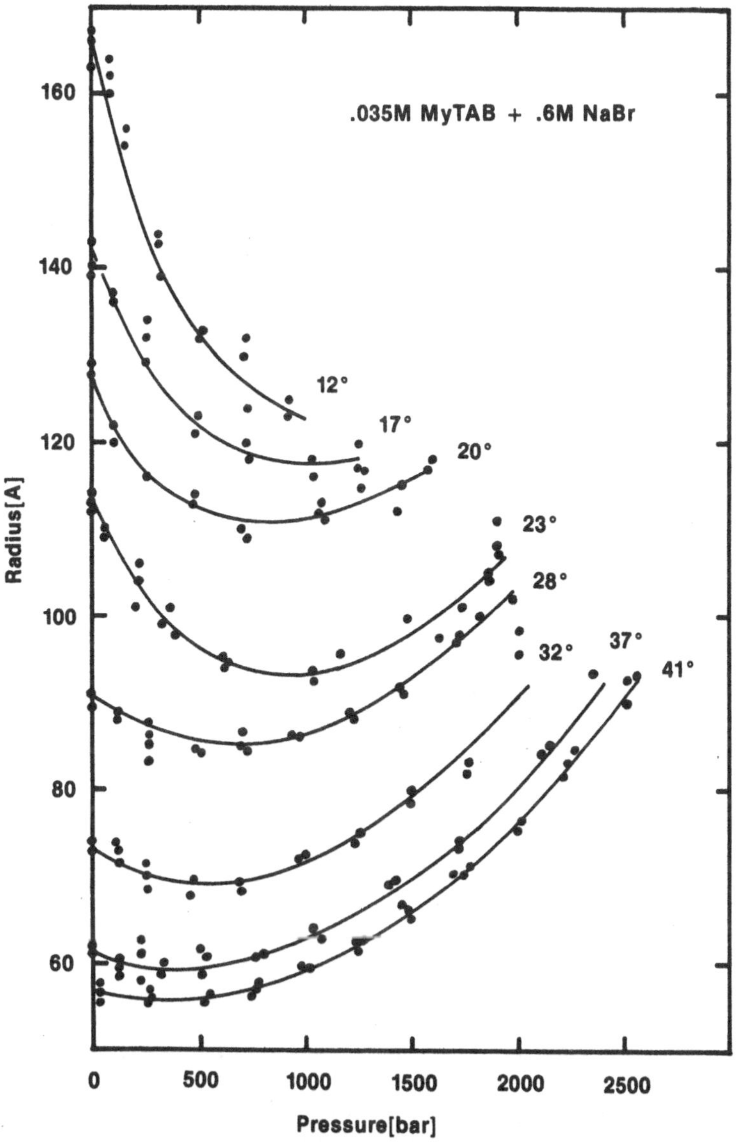

Figure 8. $R_h(\mathring{A})$ versus P(bar) for 0.035M MyTABr + 0.6M NaBr at eight temperatures between 12° and 41°C.

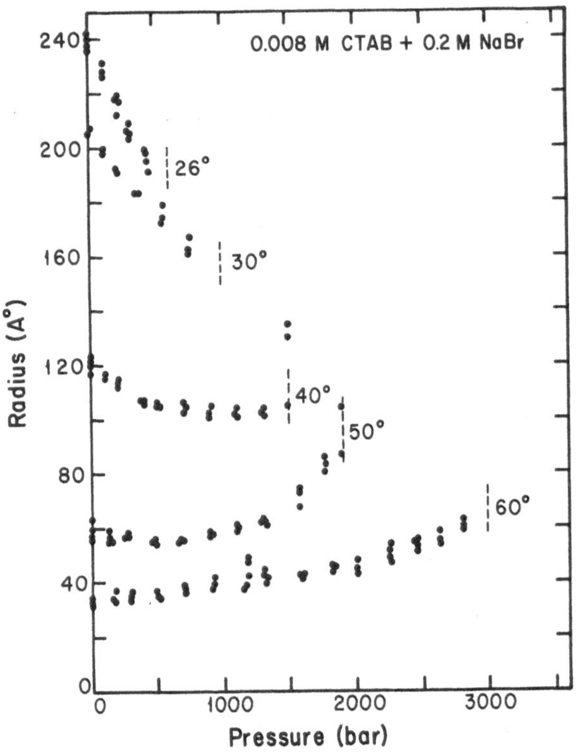

Figure 9. R_h(Å) versus P(bar) for 0.008M CTABr + 0.2M NaBr at 26°, 30°, 40°, 50° and 60°C.

micellar mean radius is close to that of the minimum sphere, application of pressure may or may not induce substantial growth, but it never leads to shrinkage of the micellar R_h. In a sense, this result tends to support the widely-held notion of the existence of a smallest, "minimum sphere" aggregate. At one extreme, 2500 bar pressure is required for DoTABr to yield barely a 15% increase in R_h at 60°C. At the other extreme, CTABr exhibits fully 45% growth in R_h at the same P, T.

Second, at low temperatures, below the Krafft point, application of high pressures always reduces the mean radius toward some minimum value; growth is never observed. For DoTABr, R_h starts at the minimum radius -- no further reduction can occur. In the case of MyTABr, the minimum value R_h = 35 Å is almost reached (10°C) when P_c is reached. For CTABr, R_h is still decreasing rapidly when $P_c \approx 700$ bar is reached.

Finally, at intermediate temperatures one observes a competing

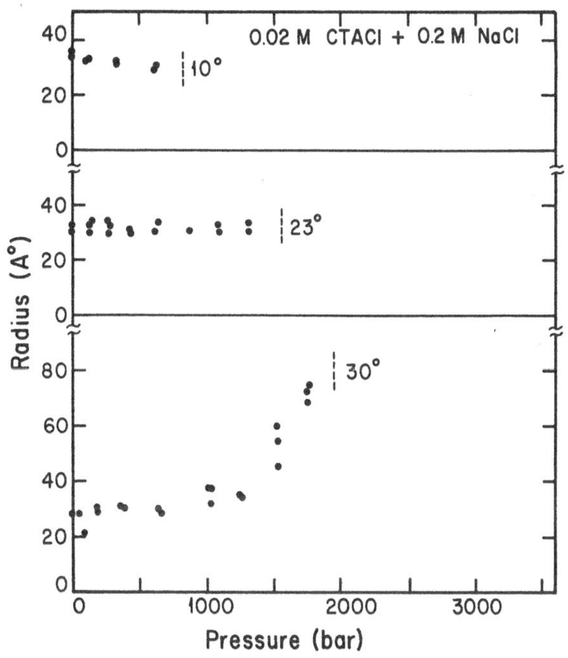

Figure 10. R_h(Å) versus P(bar) for 0.02M CTACl + 0.2M NaCl at 10°, 23° and 30°C.

behavior in which the initial response of the system is a decrease in the apparent R_h, followed by an increase with increasing P. Clearly, the immediate challenge is to obtain a free energy model for micellar growth (which obviously must include intermicellar interactions) which can successfully account for the fascinating pressure dependence observed for this family of cationic surfactants.

REFERENCES

1. H. Wennerstrom and B. Lindman, Phys. Reports 52, 1 (1979); B. Lindman and H. Wennerstrom, Topics Current Chem. 87, 3 (1980).
2. F. M. Menger, Accts. Chem. Res. 12, 111 (1979).
3. N. A. Mazer, G. B. Benedek and M. C. Carey, J. Phys. Chem. 80, 1075 (1976).
4. N. A. Mazer, M. C. Carey and G. B. Benedek in: Micellization, Solubilization and Microemulsions, Vol. 1, ed., K. L. Mittal (Plenum Press, New York, 1977).
5. P. J. Missel, N. A. Mazer, G. B. Benedek, C. Y. Young and M. C. Carey, J. Phys. Chem. 84, 1044 (1980).
6. M. Corti and V. Degiorgio, Opt. Commun. 14, 358 (1975).

7. M. Corti and V. Degiorgio, Ann. Phys. 3: 303 (1978).
8. D. McQueen and J. Hermans, J. Colloid Interface Sci. 39: 389 (1972); V. Cooper et al., Biochem. Biophys. Acta 363: 86 (1974).
9. A. Rohde and E. Sackmann, J. Colloid and Interface Sci. 70: 494 (1979).
10. D. F. Nicoli, D. R. Dawson, and J. W. Offen, Chem. Phys. Lett. 66: 291 (1979).
11. D. R. Dawson and H. W. Offen, Rev. Sci. Instrum, in press.
12. D. E. Koppel, J. Chem. Phys. 57: 4814 (1972).
13. C. Y. Young, P. J. Missel, N. A. Mazer, G. B. Benedek, and M. C. Carey, J. Phys. Chem. 82: 1375 (1978).
14. J. Briggs, D. F. Nicoli, and R. Ciccolello, Chem. Phys. Lett. 73: 149 (1980).
15. F. Perrin, J. Phys. Radium 7: 1 (1936); B. Chu, Laser Light Scattering (Academic Press, New York, 1974).
16. M. Corti and V. Degiorgio in: Light Scattering in Liquids and Macromolecular Solutions, ed. by V. Degiorgio, M. Corti and M. Giglio (Plenum Press, New York, 1980). Also - J. Phys. Chem., in press.

PART III

Nonequilibrium Systems

THE DYNAMICS OF PHASE SEPARATION NEAR THE CRITICAL POINT: SPINODAL DECOMPOSITION AND NUCLEATION

Walter I. Goldburg

Physics Department
University of Pittsburgh
Pittsburgh, PA 15260

CONTENTS

1. INTRODUCTION

It is useful to view phase separation as occurring either by nucleation or by spinodal decomposition. Nucleation is associated with metastability, i.e. the existence of an energy barrier and the occurrence of a rare but large energy fluctuation.[1-4] Spinodal decomposition, on the other hand, refers to phase separation in an initially unstable system; one in which even the smallest fluctuation will grow because an energy barrier is negligible or zero.[5-7]

The difference between nucleation and spinodal decomposition (SD) is illustrated in the temperature - composition phase diagram of a binary mixture in Fig. 1. This figure becomes applicable to a one-component system if the composition variable c is replaced by the density or molar volume of the fluid. The two phases, which are in thermal equilibrium at temperature $T<T_c$, contain c_A and c_B molecules/cm^3 of one of the components. The coexistence curve C-C, also labeled as $T_{cx}(c)$, has a maximum at the critical point (c_c,T_c).

As $T \to T_c$, the two phases become equivalent in every respect.

The spinodal line S-S, or $T_{sp}(c)$, encloses a region within which the system is thermodynamically unstable to the growth of composition fluctuations. In a simple fluid, the isothermal compressibility χ is negative within the spinodal region and vanishes along the spinodal line; in a binary mixture this line is defined by the condition $\chi_c^{-1} = (\partial c/\partial\mu)^{-1}_{P,T} = 0$, where μ is the difference in chemical potential of the two components. Sometimes χ and χ_c are called susceptibilities.

The region between C-C and S-S in Fig. 1 defines the metastable states of the system; when a fluid finds itself in this region, nucleation occurs. The nucleation line N-N, or $T_N(c)$, has a dynamic rather than a thermodynamic definition; a system which is supercooled to a temperature T lying just above this line may take hours, weeks, or years to form droplets of the stable phase, whereas a slight further decrease in temperature results in instantaneous and copious production of droplets.

In spite of the imprecision in this definition of the nucleation line, it had been thought until recently that its breadth would be almost impossible to measure. We will see, however, that near the critical point, nucleation does not occur abruptly as $T(c) \to T_N(c)$; in fact the line becomes so fuzzy as to lose its meaning.

While nucleation and spinodal decomposition theories have very different mathematical forms, there can, in reality, be no sharp distinction between them. Physically speaking, phase separation must proceed in the same way just above the spinodal line, where the energy barrier is small compared to k_BT, as just below it, where the barrier is zero.[8]

As it turns out, the most easily interpreted experiments are those in which the quench is either very deep or very shallow, i.e. $T_F \ll T_{SP}$ or $T_F \simeq T_N$, where T_F is the temperature to which the fluid is quenched. An extreme example of a deep quench is a so called "critical quench," in which the sample, of critical concentration or density, is cooled through the critical point, starting from an initial temperature T_i(see Fig. 1). The critical quench is especially simple, because the system does not have to pass through an intermediate range of metastable states. Most of the spinodal decomposition experiments to be discussed and analyzed in this review were in critically quenched fluid mixtures.

Section II will begin with a review of Cahn's linear theory of spinodal decomposition.[5] There then follows a summary of the nonlinear theory of Langer et al.,[23] the cluster dynamics analysis

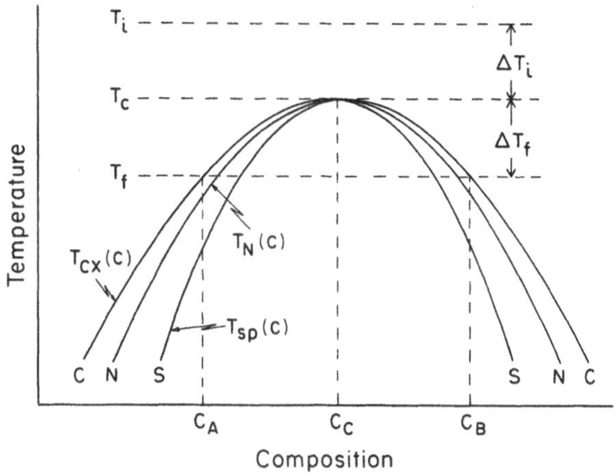

Figure 1. Phase diagram of a binary mixture showing the coexistence curve, $T_{cx}(c)$, the nucleation line, $T_N(c)$, and the spinodal line $T_{sp}(c)$. The initial and final temperatures are T_i and T_f, and the quench depth is ΔT_f.

of Binder[21] and of Binder and Stauffer,[3] and the hydrodynamic mode - coupling calculation of Kawasaki and Ohta[24].

Spinodal decomposition theories and experiments have been stimulated by the computer simulations of Lebowitz and his coworkers.[18,20,38] We will summarize some of their findings and also comment on the interesting mulecular dynamics studies of Abraham.[29]

The light scattering experiments will be discussed next. These measurements of the time dependent structure factor reveal two distinct domains of phase separation in critically quenched fluids. In the early stage, evolution of the system is governed by diffusion, though the coupling with hydrodynamics modes seems important. In the late stage, which is only qualitatively understood, coarsening seems to be controlled by surface-tension driven flow. This fact was first realized by Cahn and Moldover[35] and independently by Siggia.[34] We will present their ideas.

Evidence will be presented that the structure factor S(K,t) vs K has a simple, time-independent scaling form. This was proposed by Binder and Stauffer,[25] seen in the computer simulations of Marro et al.[38] and subsequently observed in light scattering experiments.[39]

Section III takes up recent experiments on anomalous supercooling near the critical point and the idea of Binder and Stauffer,[3] which appears to explain this phenomenon. Their explanation, which hinges on the concept of critical slowing down, has motivated Langer and Schwartz[48] to develop a more complete theory which combines nucleation (i.e. droplet birth) theory with the droplet growth theory of Lifshitz and Slyozov.[26] Their calculations will be compared with recent cloud-point experiments of Howland et al.[17] and with the microscope measurements of droplet growth by Krishnamurthy and Goldburg.[49]

2. SPINODAL DECOMPOSITION

A. Cahn's Linear Theory

It has long been known to metallurgists and ceramicists that deeply quenched metallic alloys and glasses may solidify into a structure whose composition varies quasiperiodically. This periodicity can be seen with the microscope or inferred from scattering experiments.[10,11] To explain these observations, J. W. Cahn assumed that solidification was taking place within the spinodal region, where the composition diffusivity D(in cm^2/sec) is negative. This follows from the fluctuation dissipation theorem;[12] which relates D and the susceptibility χ:

$$D = M/\chi = M\left(\frac{\partial\mu}{\partial c}\right)_{P,T} , \tag{1.1}$$

where the mobility M is always positive.

To understand why a negative value of D can produce periodicity, consider a mixture of average composition c and whose local composition $c(\underset{\sim}{r},t)$ obeys Fick's law of diffusion.

$$\underset{\sim}{J} = -D \underset{\sim}{\nabla} c (\underset{\sim}{r},t). \tag{1.2}$$

Combining this equation with the conservation law

$$\frac{\partial c(\underset{\sim}{r},t)}{\partial t} + \underset{\sim}{\nabla} \cdot \underset{\sim}{J} = 0, \tag{1.3}$$

one obtains the well known result,

$$\frac{\partial c(\underset{\sim}{r},t)}{\partial t} = D(T_f)\, \nabla^2 c(\underset{\sim}{r},t). \tag{1.4}$$

In these equations, $\underset{\sim}{J}(r,t)$ is the interparticle flux and $D(T_f)$ is the diffusivity at the quench temperature T_f. One can construct a spinodal decomposition theory for systems whose order parameter is not conserved. Such theories apply, for example, to order-disorder transitions[13] and to ^3He - ^4He.[14]

Let $c(\underset{\sim}{K},t)$ be a Fourier component of $c(\underset{\sim}{r},t)$. Then Eq. (1.4) has the solution

$$c(K,t) = c(K,0)e^{-\Gamma(K)t} \tag{1.5a}$$

where

$$\Gamma(K) = D(T_f)K^2 \tag{1.5b}$$

Since $D < 0$ in the spinodal region, all Fourier components, $c(K,t)$, will grow at an exponential rate. Assuming the Born approximation to be valid, the intensity $I(K)$ of light scattered by the mixture will be proportional to $|c(K,t)|^2$, where K is now the photon momentum transfer; it is related to the scattering angle Θ through the equation,

$$K = \frac{4\pi n}{\lambda_o} \sin(\theta/2), \tag{1.6}$$

where λ_o is the vacuum wavelength and n is the refractive index.

Equation (1.1) is an oversimplification in that it fails to take into account the K-dependence of D; when K is large, one might expect $\Gamma(K)$, and hence $D(K)$, to become positive, because rapid spatial variations in $c(K,t)$ are energetically costly to produce. As a result, Fourier components of composition with large wavenumber (say K greater than some $K > K_c$), would be expected to die out exponentially rather than grow with time. In Cahn's treatment this energy cost appears in the gradient term of the Ginsburg-Landau (GL) free energy density,

$$f_{GL}(r,t) = f(c) + 2\,B(\underset{\sim}{\nabla}\,c\,(r,t))^2, \tag{1.7a}$$

where

$$f = \text{const} + \chi^{-1}(c-c_c)^2 + A(c-c_c)^4, \tag{1.7b}$$

and A, B are constants.

From Eqs. (1.5) and (1.6), one can readily understand (see Fig. 2) why a spinodally quenched fluid will scatter light in the form of a ring or halo. When $K \gg K_c$, D is greater than zero and fluctuations decay. Also, when $K \ll K_c$, $c(K,t)$ will grow very slowly because $\Gamma \propto -K^2$. Consequently, the scattering will be dominated by the intermediate K-values. This implies that the scattering will be most intense in a small angular range, $\theta \simeq \theta_c$, corresponding to $K \simeq K_c$ in Eq. (1.6). In systems where the order

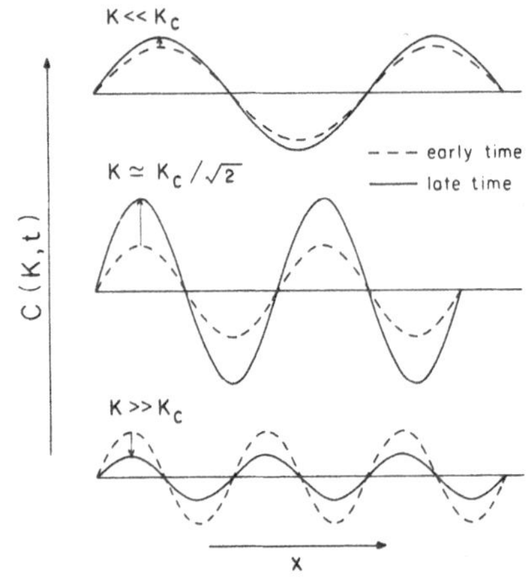

Figure 2. Schematic illustration of the slow growth of a composition fluctuation of very long wavelength (top) and rapid growth of a component of wavenumber $K \lesssim K_c$ (center). Fluctuations of short wavelength ($K > K_c$) decay (bottom). The mixture is assumed to have been quenched into the spinodal region.

parameter is not conserved, the intensity maximum develops at $K = 0$[8].

Putting Cahn's fundamental idea in mathematical terms, Eq. (1.2) is replaced by[6]

$$\underset{\sim}{J}(r,t) = -M\vec{\nabla}\,\frac{\delta F}{\delta C(\underset{\sim}{r})} = -M\vec{\nabla}[\tilde{\mu} - 2B\nabla^2 c] \qquad (1.8)$$

where $\tilde{F}$ is the volume integral of f_{GL}, and the local chemical potential $\tilde{\mu} = \partial f/\partial c$. (We continue to oversimplify by treating M as a constant.)[15] Combining Eqs. (1.8), and (1.3), taking $\chi^{-1} = (\partial\mu/\partial c)_{P,T}$, and linearizing about the average concentration c_1,

$$\frac{\partial u}{\partial t} = M\,\nabla^2[(\frac{\partial\tilde{\mu}}{\partial c})_{P,T} - 2\,B\,\nabla^2 u], \qquad (1.9)$$

where $u = c - c_1$. Fourier analyzing and solving for c(K,t) one again obtains Eq. (1.5a) with

$$\Gamma(K) = D(K)\,K^2 = M[\chi^{-1}(T_f) + 2B\,K^2]K^2. \qquad (1.10)$$

Within the spinodal region, $\chi(T_f) < 0$. Hence Γ changes sign at $K = K_c = (2B\chi)^{-1/2}$. By differentiating, one finds the most rapidly growing Fourier component $c(K_m,t)$ to have the wavenumber $K_m = K_c/\sqrt{2}$.

Taking the origin of time as the instant of quench, i.e., when $T = T_f$, assuming $<|c(K,0)^2|>$ to be of Ornstein-Zernike form, and taking the scattering intensity $I(K,t) \propto <|c(K,t)|^2>$, the above equations give

$$I(K,t) = I(K,0)e^{-2\Gamma t} \propto \frac{\chi}{1 + K^2\xi^2} e^{-2\Gamma(K)t}, \tag{1.11}$$

where $\Gamma(K)$ is given by Eq. (1.10). Since the system is plunged to a point within the coexistence curve at $t = 0$, the quantities χ and ξ in this equation should presumably be evaluated at a temperature $(T_c - T_f)$°K above T_c, i.e. in the one-phase region; it certainly seems incorrect to insert the two-phase thermal equilibrium values of $\chi(T_f)$ and $\xi(T_f)$ here, since this would give $I(K,\infty)$ rather than $I(K,0)$. Even in the more sophisticated theories of spinodal decomposition to be outlined below, there remain delicate theoretical questions associated with the evaluation of $I(K,0)$.†

Near the critical point, the susceptibility and the correlation length diverge strongly and have the form

$$\xi = \xi_o \, \xi^{-\nu} \tag{1.12}$$

$$\chi = g_o^{-1} \, \varepsilon^{-\gamma}, \tag{1.13}$$

where

$$\varepsilon \equiv |1 - T_f/T_c|. \tag{1.14}$$

In fluids, which belong to the same symmetry class of the Ising model, the critical exponents $\nu \simeq 0.63$ and $\gamma \simeq 1.24$.[16] The amplitudes g_o and ξ_o are system-dependent, with $\xi_o \simeq 1$Å and $g_o \simeq 1/10\; k_BT_c/c_c$.[17]

According to Eqs. (1.10) and (1.11), the intensity of the ring will increase with time, but its diameter, i.e. K_m will not change. However, inclusion of the nonlinear terms dropped from Eq. (1.9), will cause the ring to collapse; the coupling of the $c(K)$ generates components at sum and difference wavenumbers, say K_i and K_j. Since the terms $c(K_i + K_j,\, t)$ are more likely to be attenuated by the Ginsburg-Landau gradient energy in (1.7), the longer wavelength fluctuations are given increasing weight, which implies a decrease in $K_m(t)$ with increasing time. Both the formation of a scattering ring and the decrease of its diameter with time are experimentally observed. In the experiments and theories to be reviewed in subsequent sections, attention is usually focused on the time evolution of the "ring diameter" $K_m(t)$ and the "ring intensity",

$$I_m \equiv I(K_m(t),\, t) \propto S(K_m(t),\, t), \tag{1.15}$$

---

†The problem is complicated by the fact that for a short time after the quench, the system remembers its initial state, of temperature T_i.

where the structure factor,

$$S(K,t) \equiv \int < u\,(r + r_o,\ t)\ u\,(r_o,\ t) > e^{i\underset{\sim}{K}\cdot\underset{\sim}{r}}d\underset{\sim}{r}. \qquad (1.16)$$

Cahn's basic result, Eq. (1.11) can be expressed in scaled form by introducing the dimensionless variables

$$q \equiv K\ \xi \qquad (1.17)$$

$$\tau \equiv D\ \xi^{-2}t, \qquad (1.18)$$

where ξ^2/D is the composition diffusion time through a correlation length. Combining these definitions of q and τ with Eqs. (1.1), (1.4), (1.10) and (1.11) gives

$$\frac{I(q,\tau)}{I(0,0)} = \frac{1}{1 + q^2}\ e^{-2q^2(q^2-1)\tau}\ , \qquad (1.19a)$$

since

$$I(0,0) \propto \chi(T_f) \propto \varepsilon^{-\gamma}\ . \qquad (1.19b)$$

In Eq. (1.19), a quench into the spinodal region has been assumed; otherwise (q^2-1) is replaced by (q^2+1).

Equation (1.19) expresses Cahn's linear theory in system-independent form. The nonlinear theories described in the next section have the same scaling property as Eq. (1.19). Of course if additional parameters were to enter the problem, which seems likely in the late stage of spinodal decomposition, the theory must also involve other scaling parameters than g_o and ξ_o.

B. Recent Theoretical Developments

New light has been thrown on the spinodal decomposition problem by computer simulations of J. Lebowitz, M. Kalos, and their associates.[18-20] They have studied the time evolution of $K_m(t)$ and $S(K_m,t)$, for a cubic lattice of A(↑) and B(↓) spins, coupled by the Ising model interaction. The system is constrained to conserve the number of up-spins minus the number of down-spins, so that it simulates an alloy in which a low temperature favors the clustering of atoms (A and B) of similar type. The dynamics of the system is that of the Kawasaki spin-exchange model (see ref. 18 for details).

In one set of computer experiments, a critical mixture consisting of equal numbers of A and B spins is quenched from the disordered state, $T_i = \infty$ to various final temperatures $0.6 \leq T_f/T_c \leq 1.1$. The time evolution of S(k,t) in these critical quenches has the form of an initially diffuse ring which brightens, sharpens, and decreases in diameter with increasing time. Relatively far from the critical point ($\varepsilon = 0.4, 0.2, 0.1$), the data can be fitted to the algebraic form,

$$K_m(t) \simeq \alpha^1 (t + 10)^{-\phi} \qquad (2.1)$$

$$S(K_m, t) \simeq \alpha (t + 10)^{\theta}, \qquad (2.2)$$

where t has units of the spin exchange time.

The exponents ϕ and θ vary slightly with quench depth, but over the entire range, and in the time interval $0 < t < 10^3$,

$$\phi = 0.23 \pm 0.02 \qquad (2.3)$$

$$\theta = 0.69 \pm 0.04. \qquad (2.4)$$

Contrary to the Cahn theory, there was no significant time interval in which $S(K,t)$ increased at an exponential rate. Even when the final temperature is in the one phase region ($T_f/T_c = 1.1$), $S(K,t)$ has the form of a collapsing ring, but in this case, its intensity grows relatively slowly, i.e. θ is only 0.38. Subsequently this phenomenon was treated theoretically[21] and observed in the laboratory.[22]

The computer simulations stimulated a theoretical attack on the problem by Langer, Bar-on, and Miller.[23,6] They developed a decoupling scheme for evaluating the structure factor without resorting to perturbation theory. The starting point is a master equation for the probability distribution functional $\rho\{c(\underset{\sim}{r})\}$, in which the probability current density is driven by a chemical potential gradient as in the Cahn theory. A Langevin white noise term is included to drive the fluctuations. (In the simplified theory of the previous section, a fluctuation $c(K,t)$ was simply assumed to be present initially.) Langer et al.[23] then evaluate the correlation function $\langle u(r + r_o, t)\, u(r_o, t)\rangle$, using the weighting function ρ. A Fourier inversion then yields a differential equation for $S(K,t)$, involving all higher order correlation functions. To decouple these terms, Langer, Bar-on and Miller (LBM) make an approximation which, for technical reasons, is expected to lose its validity when the grain size exceeds several correlation lengths.[13] Since the correlation length ξ is such length, their computational scheme loses its validity when the size, $L(t) \sim K_m^{-1}(t)$, of a grain or cluster increases beyond several correlation lengths, i.e. for large τ.

In the range $0 < \tau < 100$, the results of LBM can be fitted to the power law, $q_m \propto \tau^{\phi}$, with $\phi \simeq 0.21$, in excellent agreement with the computer simulation value, Eq. (2.3). Since there are no flow terms in the starting equations of LBM, their results should not directly apply to fluids, except at the very outset of phase separation (the influence of flow is delayed by inertia).

To include the effect of flow on spinodal decomposition, Kawasaki and Ohta[24] (KO) replaced the partial derivative, $\partial c/\partial t$, with the convective derivative, $\partial c/\partial t + \underset{\sim}{v} \cdot \underset{\sim}{\nabla} c$. The second term, containing the velocity $\underset{\sim}{v}$, couples the Nāvier-Stokes equation with the diffusion terms treated in the Cahn and LBM theories. To solve these coupled equations, Kawasaki and Ohta resorted to the LBM decoupling scheme. Both theories will be compared with the experiment in the next section.

An entirely different approach to the phase separation problem has been undertaken by Binder[21] (see also Binder and Stauffer, Ref. 25). Here the Kawasaki spin-exchange model was invoked to calculate the growth rate of clusters. Obtaining an approximate solution to a master equation for the probability distribution of the spin state of the entire N-spin system, Binder recovers, in various approximations, the results of Cahn,[5] LBM, and Lifshitz and Slyozov.[26] The latter theory, to be discussed at some length below, describes the evolution of droplets which grow by diffusion.

In the cluster model a finite energy is always associated with an increase in cluster size, so the approach does not necessitate defining a spinodal line or negative diffusion constant. The spinodal region in Cahn's theory here becomes that portion of the phase diagram where the clustering energy is small compared with the thermal energy.

The approximations which Binder makes are better satisfied far from the critical point, where the clusters are of finite size and therefore have a clear geometrical meaning; the LBM theory works best in the opposite limit, where a wavenumber description is more appropriate. Langer's coarse-graining procedure for evaluation of the free energy[23,27] also restricts the domain of the LBM theory to the vicinity of the critical point, since the coarse-graining length must be much larger than an atomic spacing, yet no larger than ξ.[28] The Binder theory avoids this coarse graining assumption. On the other hand, the cluster dynamics description allows for only a crude determination of the scaled structure factor, $S(q,\tau)$.[8]

The above-mentioned computer studies and calculations discussed so far indicate that the linear theory of Cahn is quantitatively incorrect. However, somewhat contrary evidence comes from F. Abraham's molecular dynamics simulations of Lennard Jones fluid.[28,29] In his computer experiments, a system of 1372 atoms with Argon-like interaction parameters was critically quenched at a rate of $\sim 10^{14}$ K/sec to a reduced temperature, $\varepsilon = 0.8$. The results show S(K,t) vs t to grow initially at an approximately exponential rate. Abraham expressed time t in psec; when this dimensional time unit is translated into τ (using Eq. (1.18) and parameters appropriate to Argon), exponential growth in S continues out to $\tau \simeq 1$. The theories of LBM and KO, as well as the Monte Carlo calculations of Lebowitz et al., have been concerned with intervals much longer than this, as are the laboratory experiments to be described in the next section.

C. Spinodal Decomposition Experiments in Liquid Mixtures

So far, spinodal decomposition in fluids has been studied by light scattering only. The experiments are carried out on sealed samples, usually of critical composition. By scattering light from the system after it has been quenched, and measuring the angular distribution, one determines S(K,t) within a multiplicative constant; no one has measured the absolute value of S.

In some experiments the pre-quench temperature T_i is in the one phase region; in others, it is one of two-phase equilibrium. The quench may be produced by a sudden change in bath temperature or by a step-wise pressure change.[30] With the latter scheme, the temperature is ideally held constant. This pressure jump shifts the critical temperature and moves the entire coexistence curve with it. If the coexistence curve is normal, as in Fig. 1, the pressure must obviously be changed in a direction which raises T_c. So far, this technique has been successfully used with critical mixtures of isobutyric acid and water, (IW), in which $(\partial T_c/\partial P)_S \simeq -60$mK/atm. The speed of the quench is limited only by the velocity of sound and the size of the sample, so that an effective temperature shift from T_i to T_f need only take a few msec. In contrast, a conventional bath-temperature quench can take many seconds.[31,32]

Since liquids can be heated faster than they can be cooled, it becomes advantageous to study spinodal decomposition in systems with an inverted coexistence curve. The mixture 2,6-lutidine and water (LW), is an excellent candidate because a number of its critical parameters have already been measured.[31,32] Using a pulse of microwave radiation to produce dielectric heating, a LW sample is easily quenched at $\sim$5 mK/sec,[33] a rate which is competitive with that obtained by the pressure jump technique.

Each of these schemes has its own limitations. With microwave heating, the temperature of the surrounding bath must also be raised, or alternatively, the sample must be thermally isolated. On the other hand, the pressure jump method requires that the sample be judiciously chosen so that the pressure jump produces a relatively small change in the actual temperature of the

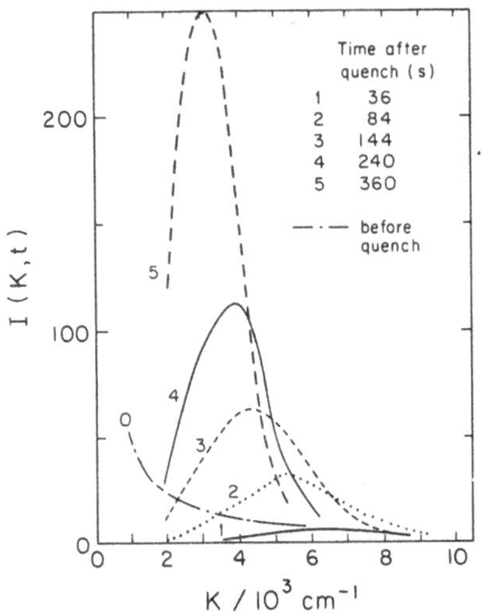

Figure 3. Scattered intensity versus wavenumber in a critically quenched mixture of isobutyric acid and water (IW), as measured by Wong and Knobler.[30]

sample. To minimize adiabatic heating or cooling, one should choose mixtures for which $|(\partial T_c/\partial P)_S| >> (\partial T/\partial P)_S$.

Figure 3 shows Wong and Knobler's measurements of the angular distribution I(K,t) vs K of light scattered from a critically quenched mixture of isobutyric acid and water.[30] The quench depth,

$$\Delta T_f \equiv |T_f - T_c|,$$

is 0.9 mK here, and the measurement times are as indicated. While I(K,t) is a strong function of ΔT_f, there is no evident dependence on $\Delta T_i \equiv |T_i - T_c|$.[31] Note that scattering was initially weak and diffuse, so that no data were reported before $t \equiv 36$ sec, even though the quench time was thousands of times smaller than this.

Wong and Knobler analyzed their data to extract the exponents θ and ϕ in Eqs. (2.3) and (2.4). Over the range 40 sec $\lesssim t \lesssim 10^3$ sec and 0.9 mK $\leq \Delta T_f \leq$ 8.0 mK they found ϕ to increase from 0.33 to 0.92 and θ to increase from 1.2 to 2.3. Their ring diameter measurements are replotted in reduced units in Fig. 4. Chou and Goldburg (CG)[32] have also studied spinodal composition in IW, and their data points appear in the figure as well. The quench depths shown in the inset are those of CG.

There is obviously excellent agreement between the results of WK and CG in Fig. 4, even though the quenching rates and sample geometries were very different; in the experiments of Chou and

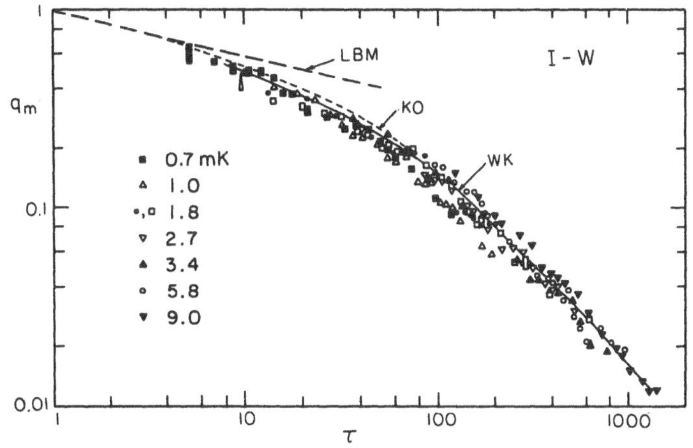

Figure 4. Ring diameter vs. time in a critically quenched IW mixture. The data are those of Wong and Knobler[30] (solid line) and Chou and Goldburg[32] (individual data points). The two theoretical curves are the results of Langer, Bar-on, and Miller[23] (LBM) and Kawasaki and Ohta[24] (KO). The reduced units q and τ are defined in Eqs. (1.17) and (1.18).

Goldburg, a quench took almost 10 seconds, and the samples were ten times thinner than those of Wong and Knobler. Multiple scattering motivated the choice of such a thin sample. Even at that, multiple scattering was large and therefore was only partially corrected for by dividing I(K,t) by the intensity of the unscattered laser beam, I_F.

Chou and Goldburg also studied spinodal decomposition in 2,6-lutidine and water, and their results are almost identical to those of Fig. 4. The observed equality of $q_m(\tau)$ in these two mixtures demonstrates the sample independence one hopes to find near the critical point.

The broken and dashed lines in Fig. 4 are the theoretical results of Langer, Bar-on and Miller[23] and Kawasaki and Ohta.[24] The two curves coincide in the limit $\tau \to 0$, and in this limit are consistent with the experiment. The data, however, are also consistent with $\phi = 1/3$, which is predicted by the droplet coalescence model of Binder and Stauffer[25] and the more detailed calculations of Siggia.[34] To obtain this value of ϕ, one assumes that the critically quenched mixture consists of interconnected domains of size L equal to their average separation. A typical domain should then double its size by diffusing through a distance L, according to Stokes' Law. Taking $D = k_BT/5\pi\eta L$ and $L \sim K_m^{-1}$ one finds[34,32]

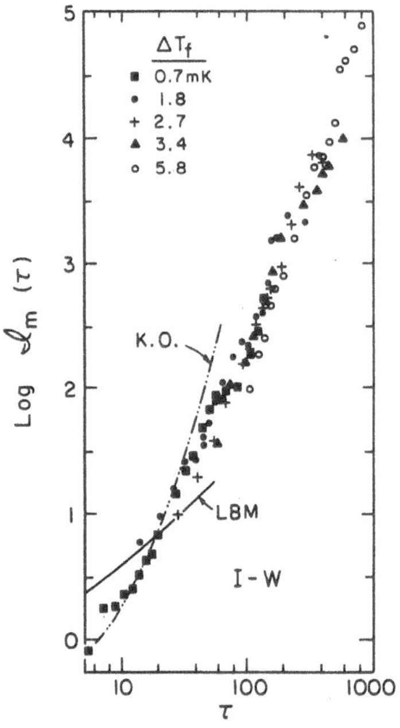

Figure 5. Ring intensity vs. time in IW (critical quench). The theoretical results of LBM and KO have been normalized to coincide with the measurements at τ = 20.

The data have also been corrected for multiple scattering. With the exception of one missing quench at T_f = 9.0 mK, the runs are the same as in Fig. 4.

The theoretical results of LBM and of KO are also plotted in Fig. 5. Since the intensity was measured in relative units only, it was necessary to normalize theory and experiment at an arbitrary point, which was chosen as τ = 20. While data in Fig. 5 agree better with the results of KO than with LBM, the reverse was found in LW. Nevertheless, the general trend in the two mixtures was the same; both systems exhibited a monotonic increase in θ with time, with the slopes approaching what appeared to be a limiting value, $\theta = 3$ in the range $300 \lesssim \tau \lesssim 1000$.

The following simple argument relates $K_m(t)$ to $I_m(t)$, and hence θ to ϕ. In the Born approximation (or more properly the Rayleigh-Gans approximation[32]), the intensity is proportional to $L^6 \times L^{-3}$. The first factor corresponds to the scattering from a single domain (regarded as a dipole), and the second, to the number of domains/cm^3 in the fluid. Thus $I_m \propto L^3 \sim K_m^{-3} \sim t^{3\phi} \propto t^{\theta}$,

$$q_m \simeq (\frac{5}{36})^{1/3}\ \tau^{-1/3} \text{ with } 1 \geq q_m \geq \frac{1}{10}\ . \tag{3.1}$$

This equation should hold only in the early stages of phase separation, which accounts for the lower limit on q_m. At $\tau = 10$ the disagreement between Eq. (3.1) and the measured value of q_m in Fig. 4 is only a factor of two!

It is apparent from Fig. 4 that (a) the hydrodynamic corrections of Kawasaki and Ohta are important, and (b) none of the theories discussed so far can account for the increase of ϕ to approximately unity at large τ. Cahn and Moldover,[35] and later, Siggia,[34] offered an explanation based on dimensional analysis. One must imagine the mixture to be deeply quenched, so that it develops interconnected domains of each phase. In different language, the system is said to be in the percolation regime.[36,37] Consider a particular region in the shape of a sausage. This structure is not hydrodynamically stable and will tend to pinch off, driving fluid from necks into bulges. The flow, whose effect is to cause the bulges to grow, is driven by a local pressure $\sim \sigma/r$, where σ is the surface tension and $1/r$ is the local curvature of the surface. The Navier-Stokes equation in the steady-state gives an estimate of $r(T) \sim L(t)$:

$$0 = \underset{\sim}{\nabla} P/\rho + \eta\, \nabla^2\, \underset{\sim}{v},$$

where ρ is the mass density and η the viscosity. Pursuing this dimensional argument, one sets $v = r/t$, $\nabla = 1/r$, $P = \sigma/r = k_BT/\xi^2 r$ and $r \sim K_m^{-1}$, giving $L \sim K_m^{-1} \sim t/\nu$. In dimensionless units $q_m \sim \tau^{-1}$. Estimating the proportionality constant, Siggia gets

$$q_m \simeq (6\pi\tau)^{-1} \qquad \frac{1}{10} \geq q_m \geq q_g . \tag{3.2}$$

The lower limit, $q_g = \xi g \Delta\rho/\sigma$, corresponds to a domain size at which gravitational effects become dominant. Here g is the gravitational constant and $\Delta\rho$ is the mass density difference of the two phases at $T = T_f$.

While Eq. (3.2) correctly gives $\phi - 1$ at large τ, it is quantitatively incorrect. At $\tau = 400$, for example, it predicts a dimensionless grain size, q_m^{-1}, which is a hundred times larger than that measured in IW and LW. At present there does not exist a quantitative treatment of domain growth in the late stage of spinodal decomposition.

The Kawasaki-Ohta-LBM theory is much more successful in explaining the coarsening rate, $q_m(\tau)$, than in predicting the time dependence of the scattering cross section. This is apparent from Fig. 5, which shows $I(K_m(\tau))$ vs τ in *IW*, where I is the measured intensity $I_m(\tau)$ divided by $\varepsilon^{-\gamma}$, as in (1.11) and (1.13).[39]

or $\theta = 3\phi$. This relationship is consistent with the experimental observations of Chou and Goldburg, who have found the product, $q_m{}^3 I_m(\tau)$, to be approximately time independent in these mixtures.[32]

Extrapolating a step further from the ring-intensity measurements, one might expect a similar result to hold at all scattering angles rather than at K_m alone. Indeed, Marro, Lebowitz and Kalos[38] have found in their computer simulations that the normalized structure factor

$$\overline{S}(K,t) \equiv \frac{S(K,t)}{\int S(K,t)^2 K^2 dK} \tag{3.3a}$$

develops the simple scaling form

$$\overline{S}(K,t) = K_1^{-3}\ F(K/K_1(t)), \tag{3.3b}$$

with the function $F(x)$ containing no explicit dependence on t or T_f. Here K_1 is the first moment of $S(K,t)$, and t is in units of the spin-exchange time; it should be of the order of $K_m(t)$, though it need not have exactly the same time dependence. The limits on the normalization factor in Eq. (3.3a) were, of necessity, finite in these computer experiments, but the studies spanned a wide enough interval in K to include a relatively sharp peak in $F(x)$, centered at $x = K/K_m = 1$. Only after an initial transient interval or "settling time", did the structure factor develop the simple scaling form of Eq. (3.3).

The scaling form of $S(K,t)$ embodied in Eq. (3.3) was first suggested by Binder and Stauffer[25] and appears in the subsequent theoretical studies of spinodal decomposition by Binder and his associates[21,8].

Stimulated by the computer experiments of Marro et. al.[38], Chou and Goldburg[39] reexamined their angular distribution measurements in IW and LW in order to determine if the function,

$$F(K/K_m) \equiv \frac{K_m{}^3\ I(K,\tau)}{\int_{K_a \ll K_m}^{K_b \gg K_m} I(K,\tau) K^2\ dK}\ , \tag{3.4}$$

was the same in both mixtures and was independent of τ and ε. Figure 6 is a semi-log plot of $F(x)$ vs $x = K/K_m$ in IW at four quench depths, $\Delta T_f = 1.0$, 1.8, 3.4 and 5.8 mK, and at various times uniformly spanning the interval 10 to 10^3 sec. A second set of

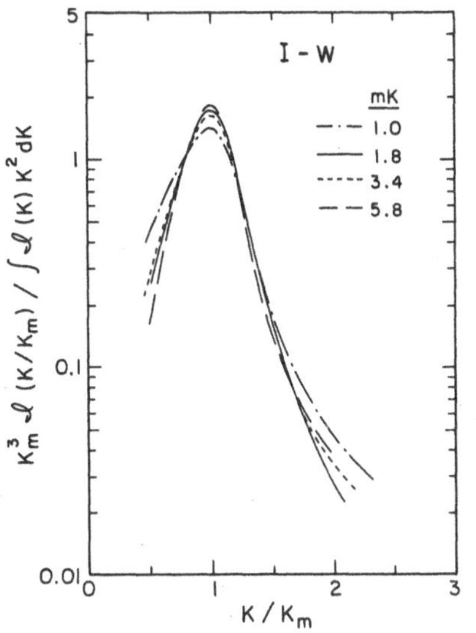

Figure 6. Angular distribution of light scattered from a critical IW mixture.[39] The ordinate is $F(K/K_m(t))$, defined in Eq. (3.4). This function appears to be independent of time and quench depth, as suggested by the computer simulations of Marro, Lebowitz, and Kalos.[38]

such curves for LW are almost indistinguishable from those shown here. Note the absence of adjustable parameters in the function F(x). [Eq. (3.4) is written correctly here, but there is a misprint in its counterpart, Eq. (4b), in ref. 39; it is corrected by replacing $\overline{K}^2 d\overline{K}$ with $K_m^3 \overline{K}^2 d\overline{K}$.]

When Wong and Knobler's measurements in Fig. 3 are analyzed in the above fashion, the resulting function F(x) is similar in height and half-width to the curves in Fig. 6.

An equation having the scaling form of Eq. (3.3) has been obtained by Schwartz[40] and is the starting point of an attack on the spinodal decomposition problem by Furukawa.[41] Schwartz has avoided some of the analytical approximations made by Kawasaki and Ohta, but his underlying approach is the same. Furukawa's results, which are apparently limited to off-critical quenches, agree very well with corresponding measurements of Wong and Knobler.[30] In this theory the dynamic behavior of the mobility, M, becomes crucial.

A notable feature of the critical-quench data in Fig. 6 is the insensitivity of F(x) to the mechanism of domain growth. This function retains its form from the early stage, where $\phi \simeq 1/3$, to the late stage where $\phi \simeq 1$. Neither the calculations of Schwartz nor those of Furukawa treat phase separation in this late stage.

3. NUCLEATION

A. Nucleation Theory

So far, discussion has focused on fluids, so deeply supercooled that the system consists of interconnected domains. Next we take up experiments and theories of nucleation in which the formation of a domain of the new phase is a statistically rare, highly localized event.

Classical nucleation theory, sometimes called Becker-Döring theory, is concerned with the rate of formation, J, of droplets of the stable phase when a fluid is supercooled by an amount δT or supersaturated by an initial amount δc_1. These initial parameters, and others to be needed later, appear in the inverted phase diagram of Fig. 7. Phase separation is imagined to proceed from an initial state at temperature T and of initial uniform composition c_1.

According to Becker-Döring (B-D) theory, J (in droplets per cm^2 per sec) is given by[1-3]

$$J = J_o e^{-\Delta G^*/k_B T} \ . \tag{4.1}$$

The prefactor J_o is proportional to the rate at which monomers impinge on the incipient nucleus, and $\Delta G^*(R^*)$ is the energy barrier to its formation. The embryo is presumed to be spherical and of radius R. If R exceeds the critical radius R*, growth of the droplet is energetically favored. A thermodynamic arguments gives

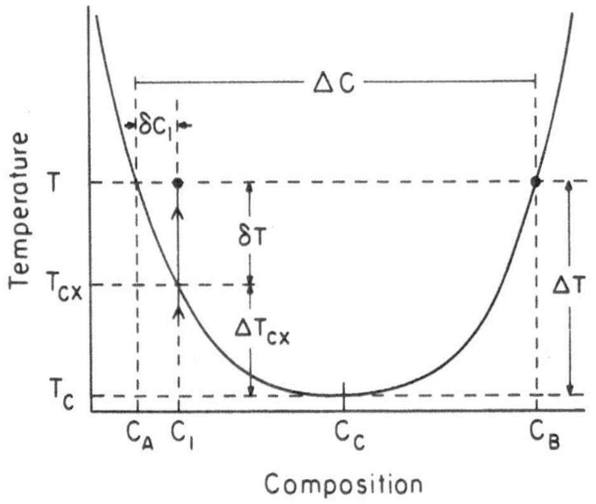

Figure 7. Phase diagram of a binary mixture with an inverted coexistence curve. The diagram introduces the notation appearing in Sec. III.

$$\Delta G^* = \frac{4}{3}\pi R^{*2}\sigma \quad . \tag{4.2}$$

This barrier is lowered as the supercooling, δT, increases, so that R^* varies inversely as δT. Near the critical point only one characteristic length should enter the problem, which suggests $R^* \simeq \Delta T_{cx}/\delta T\ \xi$. Taking $\sigma \simeq k_B T/\xi^2$ and inserting these values of R^* and σ into (4.2) gives $\Delta G^*/k_B T \simeq (\Delta T_{cx}/\delta T)^2$. A detailed calculation by Langer and Turski[42] gives

$$J = J_o^{B-D}\, e^{-b(\Delta T_{cx}/\delta T)^2} , \tag{4.3}$$

where b is expressible in terms of various critical parameters viz., the amplitudes of σ, ξ, χ, and the coexistence curve. All four constants are known for only a few simple fluids and binary mixtures, including CO_2[15,42] and IW.[17] They combine to give a value of b which is only weakly system-dependent.

A variant of the above equation was also derived by Langer and Turski.[42] Their approach takes statistical fluctuations into account and avoids the assumption that the droplet interface is sharp. Though two over-simplifications of the classical theory are thereby removed, their prediction of J vs $(\delta T/\Delta T_{cx})$ is almost the same as Eq. (4.3), the main difference being the acquisition of a strong temperature dependence of J_o in the Langer-Turski theory. Nevertheless, at all experimentally accessible values of reduced temperature, J_o remains very large ($\sim 10^{34}/cm^3 sec$). As a result, the nucleation rate, J, continues to be a very strong function of $\delta T/\Delta T_{cx}$. Conversely stated, this quantity is a weak function of J. For example, in CO_2 a large increase in J from $1\ cm^{-3}\ sec^{-1}$ to $10^5\ cm^{-3}\ sec^{-1}$ is associated with only a slight increase in $\delta T/\Delta T_{cx}$ from 0.15 to 0.16, according to the B-D theory. These values change by less than 25% in the Langer-Turski theory when

$$\varepsilon_{cx} \equiv \frac{\Delta T_{cx}}{T_c}$$

is greater than 10^{-5}

Because $\delta T/\Delta T_{cx}$ is a weak function of J_o, and because the constant b in Eq. (4.3) should vary little from one system to another, one expects that a sealed container of any simple fluid or binary mixture may be cooled approximately 15% of ΔT_{cx} before it suddenly becomes filled with droplets of the stable phase, which presents a cloudy appearance. The temperature T_N at which this phase transformation takes place is therefore referred to as the cloud point. The equation

$$\frac{|T_N(c) - T_{cx}(c)|}{\Delta T_{cx}(c)} = \frac{\delta T_N(c)}{\Delta T_{cx}(c)} = \text{const.} \simeq 0.15 \qquad (4.4)$$

implies that the nucleation line, $T_N(c)$, has the same shape as the coexistence curve, $|c-c_c| = \text{const.}\ |T_{cx} - T_c|^{\beta}$. According to Eq. (4.4), if a sample initially in two-phase equilibrium at T_{cx} is quickly supercooled by $\delta T = \delta T_N$, the reduced supercooling $\delta T_N/\Delta T_{cx}$ should be independent of its initial temperature T_{cx} or ε_{cx}. This prediction is at odds with the data in Fig. 8, which will be discussed next.

B. Anomalous Supercooling

Figure 8 is a plot of $\delta T_N/\Delta T_{cx}$ as a function of $\Delta T_{cx}/T_c$ as measured in CO_2(▲)[43,15], He^3 (X),[44] and the binary mixtures C_7H_{14} - C_7F_{14}(◇),[45] IW(■),[17] and LW(O,●).[46] The cloud point temperature was identified visually or by laser light scattering in the mixtures, while in the pure fluids nucleation was detected by measuring the latent heat released. In every experiment the supercooling was completed in a minute or less. This so-called "completion time", t_c, turns out to be a crucial parameter in the theories and experiments described below.

All of the data in Fig. 8 show an anomalous increase in $\delta T_N/\Delta T_{cx}$ as $T \to T_c$; in fact the LW measurements, replotted on a

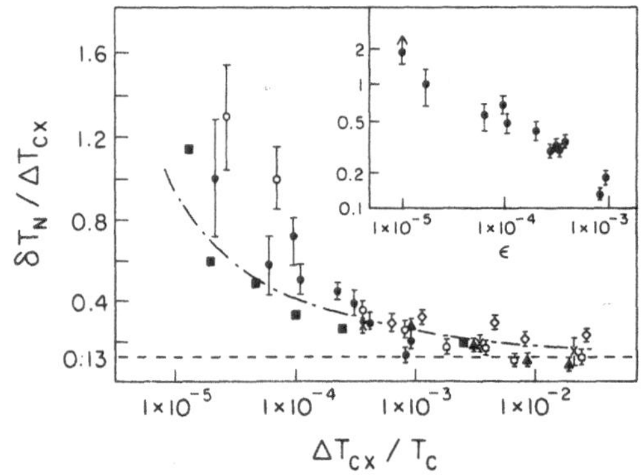

Figure 8. Reduced supercooling in the binary mixtures, isobutyric acid-water (■)[17], C_7F_{14} - C_7H_{14}(◇)[45], and 2.6-lutidine-water (O,●)[46], and in the simple fluids, He^3(X) and $^{44}CO_2$(▲)[43]. The horizontal dashed line is the Becker-Döring theory, with parameters relevant to CO_2. The broken line is the theoretical result of Binder and Stauffer;[3] it was calculated for a completion time (t_c) of 1 sec.

log-log scale in the inset, suggest that this parameter diverges at the critical point. The horizontal dashed line in the figure is the Becker-Döring value, $\delta T_N/\Delta T_{cx} \simeq 0.15$. The sloping, broken line in the figure is a theoretical curve to be discussed next.

4. THE EFFECT OF CRITICAL SLOWING DOWN

The probable origin of anomalous supercooling near the critical point was identified by K. Binder and D. Stauffer.[3] Crucial to their explanation is the vanishing of D near the critical point [when $K\,\xi \ll 1$, $D \propto 1/\xi$[47,9]]. The effect is to slow down the growth of droplets of radius $R > R^*$. As a result, the experimenter may fail to observe nucleation after a certain completion time t_c - not because nucleation hasn't occurred - but because the droplets are growing so slowly that they have not yet reached detectable size; he may therefore identify a larger value of $\delta T/\Delta T_{cx}$ as the reduced nucleation temperature, $\delta T_N/\Delta T_{cx}$. Since this effect will be accentuated near the critical point, the apparent reduced supercooling will also increase there.

Binder and Stauffer therefore proposed a time-dependent definition of δT_N, namely, that value of δT such that in a time t_c the stable droplets ($R > R^*$) will have grown large enough to occupy an appreciable fraction, f, of the sample volume. If δT_N is to have an operational meaning, it is essential, of course, that it be insensitive to the precise value of f. In Binder and Stauffer's model the droplets are assumed to grow independently, by diffusion. Their results appear as the broken line (- • - - • - •) in Fig. 8. The calculations were made using parameters appropriate to LW, with $f = 1$, and $t_c = 1$ sec. The result agrees well with the LW measurements,[46] but less so with those in IW.[17]

Stimulated by Binder and Stauffer's idea, Langer and Schwartz[48] then constructed a much more detailed theory, which included standard nucleation theory for the birth of the droplets and the Lifshitz-Slyozov (LS) theory for droplet growth.[26] The latter theory takes into account that the supersaturation $\delta c(t)$ is a decreasing function of time [in Fig. 7, $\delta c(0) \equiv \delta c_1$]. As phase separation proceeds (at constant temperature), the resulting decrease in $\delta c(t)$ increases the critical radius R^*. This inhibits further droplet formation and causes preformed droplets of radius $R < R^*$ to dissolve. The LS theory is obviously a nonlinear one.

The starting point of the Langer-Schwartz analysis is a differential equation for the droplet size distribution, $\nu(R,t)$, in which the above mentioned birth and growth terms appear. Making certain simplifying assumptions, they calculate the scaled supersaturation $y(\tau)$, the mean droplet radius $\rho(\tau)$, and droplet density $n(\tau)$. The parameters which naturally enter the theory are the reduced temperature,

$$\varepsilon \equiv \frac{\Delta T_{cx} + \delta T}{T_c},$$

and the initial scaled supersaturation $y(0) = y_1(\delta c_1$ in unscaled units). Since metastability is usually achieved by supercooling rather than by supersaturation, the parameter y_1 must be expressed in terms of $\delta T/\Delta T$ (see Fig. 7). It turns out that these two quantities are numerically almost equal for small supersaturations.[17,48]

With a second temperature renormalization, defined by

$$\varepsilon_c = G\,\varepsilon,$$

the behavior of all systems can be plotted on a single graph. The constant G, as always, is a combination of critical amplitudes.

To test the theory of Langer and Schwartz, imagine a series of cloud-point measurements, all of which utilize a certain experimental technique that fixes t_c. In the experiments of Howland et al,[17] $t_c \simeq 300$ sec, and in the LW measurements of Schwartz et al,[46] $t_c \simeq 1$ sec. The experiment begins with the fluid at its equilibrium temperature T_{cx}. The sample is then supercooled by δT, thus determining ε and y_1. After t_c sec, cloudiness may or may not be observed; if not, δT is increased, etc. By measuring δT as a function of T_{cx}, one obtains a plot of y as a function of ε or ε_c. Figure 9 displays a set of such curves calculated from the Langer-Schwartz theory with $f \simeq 0.5$. There is a curve for each of four completion times from $t_c = 1$ sec to 10^3 sec. The broken line is a variant of the Binder-Stauffer theory with $t_c = 1$ sec.[48]

The curves in Fig. 9 have the expected form; a large supersaturation (y_1) is measured near the critical point, especially if the completion time is small. Far from T_c, i.e., $\varepsilon_c > 10^{-2}$, the effect of critical slowing down is minimal and y_1 falls to its Becker-Döring value, becoming independent of t_c when $t_c > 1$ sec. According to the theory, anomalous supercooling, i.e. $y_1 > 0.2$, will be observed near the critical point, even for temperature steps lasting hundreds of seconds!

The circles and squares in Fig. 9 refer respectively to cloud measurements in IW (here labelled IBA - H_2O) and C_7F_{14} - C_7H_{14} (labelled PMCH + MCH).[17] As already noted, the completion time in these experiments was ~ 300 sec. As predicted, the data fall between the theoretical curves labelled $t_c = 10^2$ and 10^3 sec and lie close to an appropriately interpolated curve between these two.

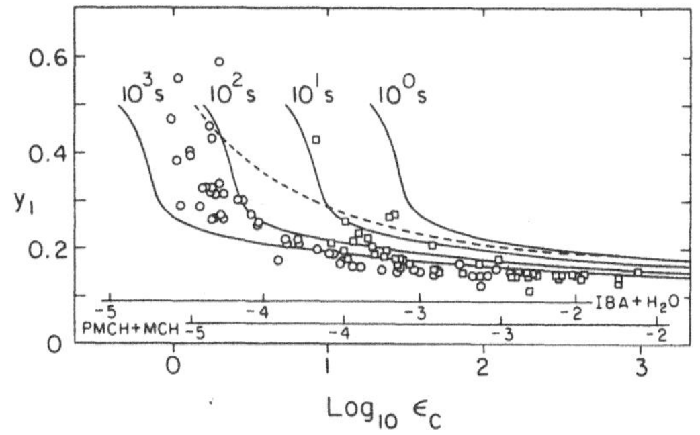

Figure 9. Cloud point measurements of Howland, Wong, and Knobler[17] in perfluoromethylcyclohexane-methylcyclohexane (☐) and isobutyric acid-water (O). The ordinate and abcissas are, respectively, the reduced initial supersaturation ($y(0) = y_1$) and the reduced temperatures ε and scaled reduced temperature ε_c. The solid curves show the theory of Langer-Schwartz[48] at the four indicated completion times. The dashed curve was calculated from the Binder-Stauffer theory, with $t_c = 1$ sec.

To put the Langer-Schwartz theory to a more severe test, Krishnamurthy and Goldburg[49] have used a microscope to measure the size distribution of nucleated droplets in various supercooled samples of LW. From their data, $y(\tau)$, $n(\tau)$, and $\rho(\tau)$ can be determined. Figure 10 shows a time sequence of droplet sizes as viewed at a magnification of 50. The scale in the photographs is 20 μm/small division. The initial conditions for this run were $y_1 = 0.21$ and $\delta T/\Delta T_{cx} = 0.18$; the sample cell was 1 mm thick. The photographs labelled (a) through (f) refer respectively to observations at t = 58 sec, 127 sec, 5.5 min, 13.4 min, 33 min, and 78 min after the quench. In (a) and (b) the droplets were too small to reproduce clearly and therefore were inked in to increase their visibility.

The qualitative features of the Binder-Stauffer and Langer-Schwartz calculations are clearly in evidence. Droplets are indeed formed at a relatively small value of $\delta T/\Delta T_{cx}$, but they initially occupy such a small fraction of the total sample volume that they surely went undetected in the earlier cloud-point experiments in Fig. 8. For example, as late as t = 13.4 min, the volume fraction of the sample occupied by droplets was only 1.6%.

Krishnamurthy and Goldburg (KG) compared their droplet density measurements with calculated values of $n(\tau)$ and observed wild

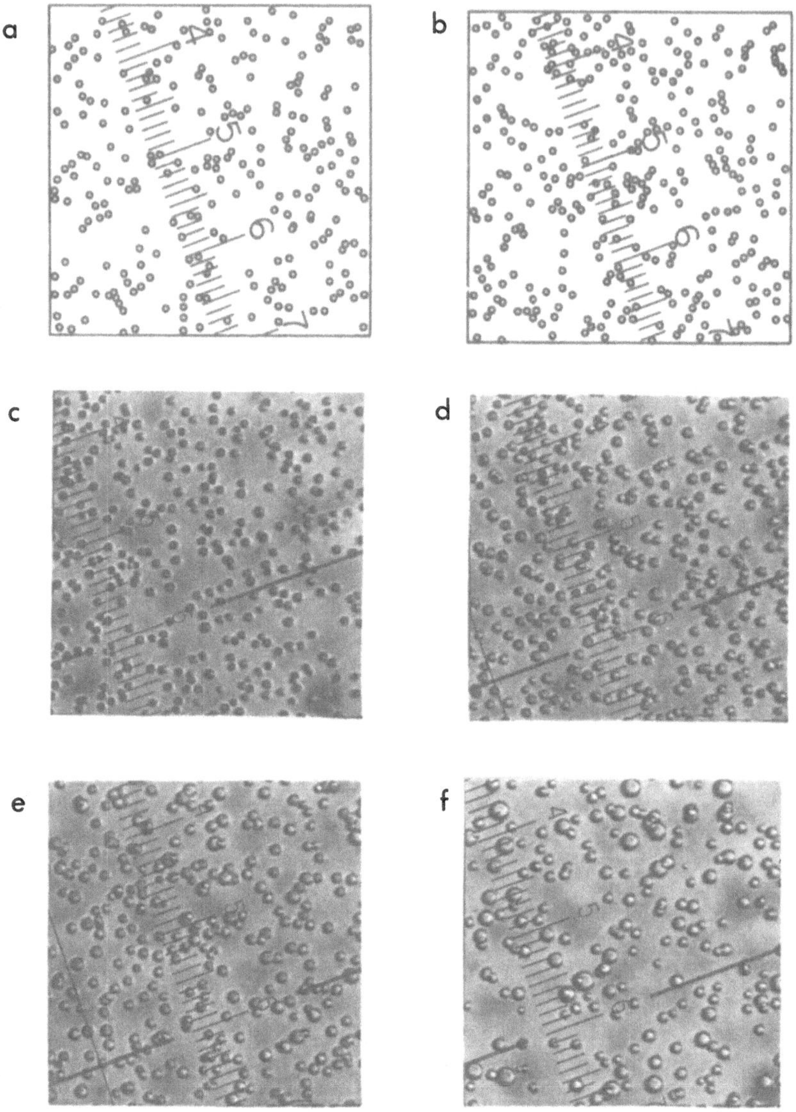

Figure 10. Microphotographs showing droplet growth in a quenched LW mixture. In this run, $\Delta T = 44$ mK and $\delta T = 8$ mK. The measuring times range from 58 sec (a) to 78 min (f) after the quench. The small scale divisions are 20 μm.

fluctuations about the Langer-Schwartz values. Presumably this lack of reproducibility was the result of the extreme sensitivity of $n(\tau)$ to δT; a theoretical estimate suggests that reproducible droplet densities would require control of the quench depth at the microkelvin level, whereas this temperature difference could be adjusted to no better than 0.1 mK.

To circumvent this reproducibility problem, KG therefore treated one of the early-time measurements of $n(\tau)$ as a parameter in the theory and calculated the subsequent evolution of $y(\tau)$ and $\rho(\tau)$. Since the measured droplet radii were almost always greater than R^*, only the Lifshitz-Slyozov growth term was retained in the Langer-Schwartz equations. The required numerical integrations were performed on a computer, and some of the results are shown in Fig. 11. In all the runs shown here, ΔT_{cx} = 35 mK. The three sets of data refer to the initial supersaturations, y_1 = 0.1, 0.21, and 0.27. The corresponding values of $\delta T/\Delta T_{cx}$ were 0.11, 0.25, and 0.33 respectively. The numerical integration of the Lifshitz-Slyozov equations (solid lines) were carried out for the two shallower quenches only, because droplet coalescence at the deepest quench complicated the interpretation of the data. Coalescence broadens the size distribution as a result of collision between droplets falling at different rates. This broadening may be seen in (e) and (f) of Fig. 10. In the deepest quenches this complication sets in very quickly, leaving little easily interpretable data.

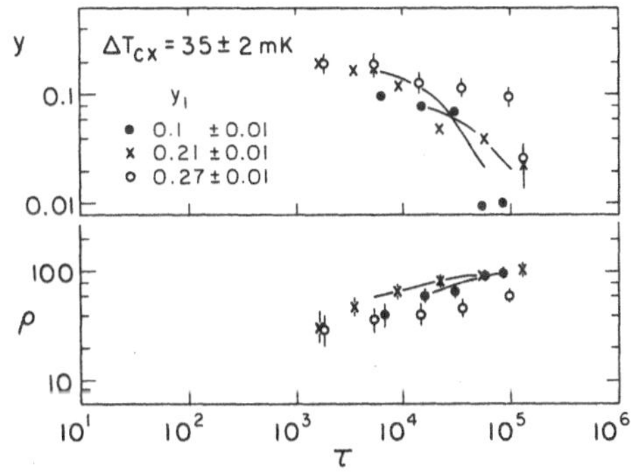

Figure 11. Reduced supersaturation (y) and mean droplet radius (ρ) vs time in a supercooled phase of LW for three initial supersaturations (y_1). The composition of this particular sample corresponds to ΔT_{cx} = 35 mK. The solid theoretical curves are obtained by integrating the Lifshitz-Slyozov equation of diffusive droplet growth[26] for y_1 = 0.1 and 0.21.

While there is satisfactory agreement between theory and experiment in Fig. 11, the results throw no light whatsoever on nucleation theory as opposed to the theory of diffusive droplet growth. In retrospect, the same thing can almost surely be said of all the anomalous supercooling experiments previously discussed.

To summarize, near the critical point nucleation can proceed for minutes to hours before becoming detectable by visual observation or crude light scattering experiments. While the droplets become observable through a microscope at earlier times, they have so far escaped detection until the growth stage, when their radii greatly exceed R*. A test of nucleation theory near the critical point must therefore be left to the future.

ACKNOWLEDGEMENTS

I have learned much from my many collaborators and have benefitted from conversations and correspondence with K. Binder, J. S. Cahn, D. Jasnow, K. Kawasaki, C. M. Knobler, J. S. Langer, J. L. Lebowitz, T. Ohta, J. V. Sengers, and E. Siggia. This work was supported by the National Science Foundation.

REFERENCES

1. J. Frenkel, Kinetic Theory of Nucleation (Dover, New York, 1955) Ch VII.
2. A. C. Zettlemoyer (Ed.) Nucleation (Marcel Dekker, New York, 1969).
3. K. Binder and D. Stauffer, Adv. in Physics, 25, 343 (1976).
4. F. F. Abraham, Homogeneous Nucleation Theory (Academic, New York, 1975).
5. J. W. Cahn, Trans. Metall. Soc. AIME 242, 166 (1968); J. W. Cahn and J. E. Hilliard, J. Chem. Phys. 28, 258 (1958).
6. J. S. Langer, in Fluctuations, Instabilities and Phase Transitions, edited by T. Riste (Plenum, New York, 1975) p. 19.
7. K. Binder, Lecture Notes for the VIth International School on Statistical Mechanics, Sitges, June 1980 (unpublished).
8. K. Binder, C. Billotet, and P. Mirold, Z. Physik B30, 183 (1978).
9. H. E. Stanley, Introduction to Phase Transitions and Critical Phenomena (Oxford, New York, 1971).
10. See for example, Mechanism of Phase Transformation in Crystalline Solids, No. 33 (Institute of Metals Monograph Series, London, 1969).
11. V. Gerold and J. Kostortz, J. Appl. Cryst. 11, 376 (1978).
12. P. C. Martin, in Many Body Physics, edited by C. DeWitt and R. Balian (Gordon and Breach, New York, 1968).
13. K. Kawasaki, M.C. Yalabik and J.D. Gunton, Phys. Rev. A17, 455 (1978), C. Billotet and K. Binder, Z. Phys. B32, 195 (1979).

14. P. C. Hohenberg and D. R. Nelson, Phys. Rev. B20, 2665 (1980).
15. W. I. Goldburg and J. S. Huang, in Fluctuations, Instabilities, and Phase Transitions, edited by T. Riste (Plenum, New York, 1975) p. 87.
16. J. C. LeGuillou and J. Zinn-Justin, Phys. Rev. B21, 3976 (1980).
17. R. G. Howland, N. C. Wong, and C. M. Knobler, J. Chem. Phys. 73, 522 (1980).
18. A. B. Bortz, M. Kalos, J. Lebowitz, and M. Zendejas, Phys. Rev. B10, 535 (1974).
19. J. Marro, A. B. Bortz, M. H. Kalos, and J. L. Lebowitz, Phys. Rev. B12, 2000 (1975).
20. K. Binder, M. H, Kalos, J. L. Lebowitz, and J. Marro, Adv. Coll. Interface Sci. 10, 173 (1979).
21. K. Binder, Phys. Rev. B15, 4425 (1977).
22. N. C. Wong and C. M. Knobler, Phys. Rev. Lett. 43, 1733 (1979).
23. J. S. Langer, M. Bar-on, and H. D. Miller, Phys. Rev. All, 1417 (1975).
24. K. Kawasaki and T. Ohta, Prog.Theor. Phys. 59, 362 (1978).
25. K. Binder and D. Stauffer, Phys. Rev. Lett. 33, 1006 (1974).
26. I. M. Lifshitz and V. V. Slyozov, J. Phys. Chem. Solids 19, 35 (1961).
27. J. S. Langer, Ann. Phys. (N.Y.) 65, 53 (1971).
28. M. R. Murzik, F. F. Abraham and G. M. Pound, J. Chem. Phys. 69, 3462 (1978).
29. F. F. Abraham, Phys. Repts. 53, 93 (1979).
30. N. C. Wong and C. M. Knobler, J. Chem. Phys, 69, 725 (1978).
31. W. I. Goldburg, C. H. Shaw, J. S. Huang and M. S. Pilant, J. Chem. Phys. 68, 484 (1978).
32. Y. C. Chou and Walter I. Goldburg, Phys. Rev. A20, 2105 (1979).
33. W. I. Goldburg, A. J. Schwartz, and M. W. Kim, Prog. Theor. Phys. Suppl. No. 64, 477 (1978).
34. E. D. Siggia, Phys. Rev. A20, 595 (1979).
35. J. W. Cahn and M. R. Moldover (unpublished).
36. S. Kirkpatrick, Rev. Mod. Phys. 45, 574 (1973).
37. K. Binder, Solid State Comm. 34, 191 (1980).
38. J. Marro, J. L. Lebowitz and M. H. Kalos, Phys. Rev. Lett. 43, 282 (1979).
39. Y. C. Chou and W. I. Goldburg, Phys. Rev. A23, 858 (1981).
40. A. J. Schwartz (unpublished).
41. H. Furukawa, Phys. Rev. Lett. 43, 136 (1979).
42. J. S. Langer and L. A. Turski, Phys. Rev. A8, 3230 (1973).
43. J. S. Huang, W. I. Goldburg and M. R. Moldover, Phys. Rev. Lett. 34, 639 (1975).
44. D. Dahl and M. R. Moldover, Phys. Rev. Lett. 27, 1421 (1973).
45. R. B. Heady and J. W. Cahn, J. Chem. Phys. 58, 896 (1973).
46. A. J. Schwartz, S. Krishnamurthy and W. I. Goldburg, Phys. Rev. A21, 1331 (1980).
47. K. Kawasaki, Ann. Phys. (New York), 61, 1 (1970).
48. J. S. Langer and A. J. Schwartz, Phys. Rev.A21, 948 (1980).
49. S. Krishnamurthy and W. I. Goldburg, Phys. Rev. A22, 2147 (1980).

NEW CRITICAL BEHAVIOR INDUCED BY SHEAR FLOW IN BINARY FLUIDS

D. Beysens

S.P.S.R.M.
C.E.N. Saclay
B. P. n°2 - 91190 Gif S/Yvette
France

CONTENTS

I. INTRODUCTION

Light scattering studies of critical fluctuations brought out of equilibrium by a shear flow started only about 2 years ago. It is therefore not surprising that many aspects are still unknown or, at least, not very well assessed. Experiments[1] have been performed and theories[2] developed in the same time - although quite independently - and a qualitative agreement has been found for a sample of Cyclohexane + Aniline submitted to a shear whose rate was continuously varying during the experiment. Only static properties have been investigated until now.

Together with the data on the Cyclohexane + Aniline system (C-A), we will present new measurements performed at nearly constant shear rate on the system Isobutyric acid + Water (I-W). Some aspects of the dynamics have been investigated. The following points will be discussed:

(i) The expected behavior, from phenomelogical assumptions, and a rigorous theory by Onuki and Kawasaki.

(ii) Experiments - Emphasis will be put on the difficulties.

(iii) Results and discussion.

In an appendix we will show as an application how one can instantaneously visualize the map of shears in any configuration of flow.

2. THEORETICAL CONSIDERATIONS

The fact that at the critical point (temperature T_c) the compressibility diverges prevents any experimental investigations of the behavior of density fluctuations in critical pure fluids. So only binary fluids near the liquid-liquid critical point (temperature T_c) were investigated. The order parameter in this case is the difference $C-C_c$ between the concentration C of one component and its critical values C_c (Fig. 1). The concentration fluctuations increase near T_c, and the correlation length ξ, and the osmotic susceptibility $\chi = (\partial\mu/\partial c)^{-1}_{p,T}$ diverges (Figs. 2 and 3). $\mu = \mu_1-\mu_2$ is the difference between the chemical potentials μ_1, μ_2 of the components; p is the pressure, T is the absolute temperature. The lifetime τ of the fluctuations correspondingly increases and diverges at T_c. τ can be simply written as the diffusion time of a droplet of size ξ,[4,5] in a medium of shear viscosity η : $\tau \simeq 6\pi\eta\xi^3/RkT$. With $R \simeq 1.20$ to be in agreement with critical dynamics theories[6], one finally obtains $\tau \simeq 5\pi\eta\xi^3/kT$. k is the Boltzmann constant.

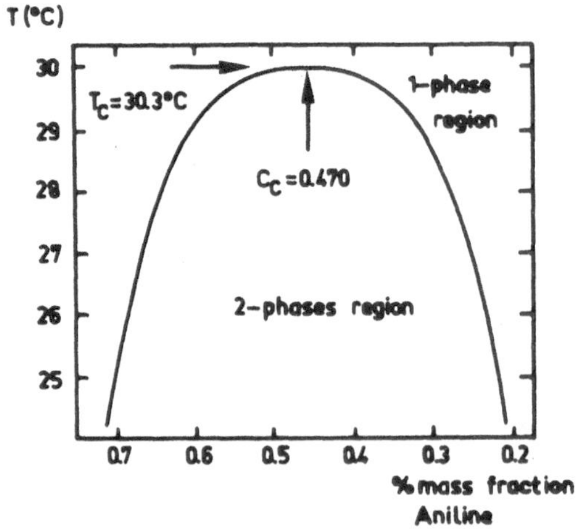

Figure 1. Coexistence curve

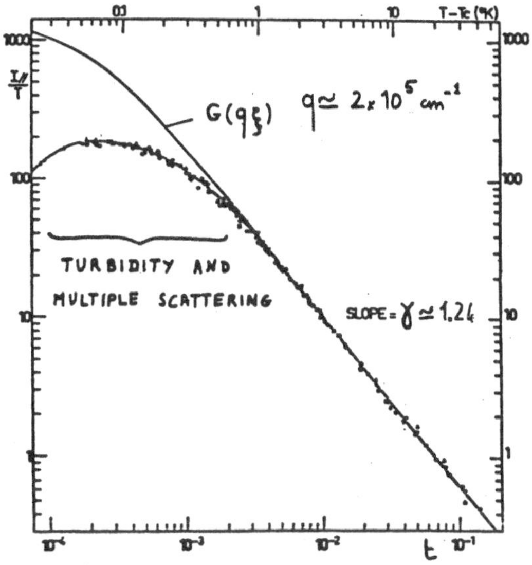

Figure 2. Susceptibility χq measured in the Nitrobenzene + n - Hexane system (from Reference 3) by light scattering. This system is very close to the Aniline + Cyclohexane mixture. Experiments were done at $q \simeq 2.10^5$ cm^{-1}. The full line represents the theoretical variation; the curvature occurs when $q\xi \simeq 1$, and depends generally on the experimental situation.

We can intuitively predict the main characteristics of the effect of shear on these concentration fluctuations. With S the shear rate (sec^{-1}):

(i) The effects will only be present in the region $S\tau > 1$. Indeed only when the lifetime τ is longer than the typical time (S^{-1}) associated with shear, will the fluctuation be able to "feel" the shear. A cross-over is therefore expected to occur in the region $S\tau \simeq 1$, from a standard behavior ($S\tau < 1$), to a new behavior ($S\tau > 1$).

(ii) In the direction perpendicular to the flow, the correlations will tend to be suppressed by the shear (Fig. 4) This means that the correlation volume is no longer isotropic, leading to an anisotropy in the susceptibility.

(iii) This reduction of the correlation extent should be connected to a change in the critical temperature. If $T_c(0)$ is the point where the fluctuations size diverges without shear, the effect of shear when $T = T_c(0)$ would prevent the correlation length from being infinite, at least in the shear direction. The system

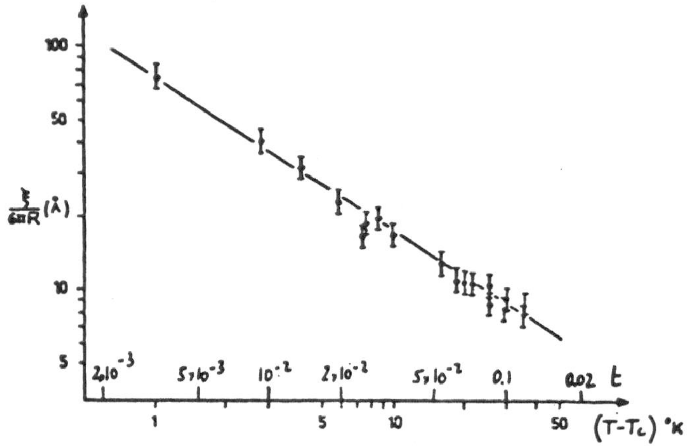

Figure 3. Correlation length divided by $6\pi R$, where $R \simeq 1.2$ in the Cyclohexane + Aniline system. ξ can be well described by: $\xi = \xi_0 t^{-\nu}$ with $\nu = 0.63$ and $\xi_0 = 2.45$ Å. From Reference 4.

would thus remain at a finite distance $T-T_c(S)$ from the new critical temperature $T_c(S)$.

These phenomenological predictions have been already verified by light scattering techniques[1] (Fig. 5) together with a less straightforward deduction concerning the effect of anisotropy, which leads, as in dipolar ferromagnets[7], to a cross-over in the region $S\,\tau \simeq 1$ from a standard susceptibility behavior $[\chi_q \propto (T-T_c)^{-1.24},\ S\tau < 1]$ to a mean-field behavior $[\chi_q \sim (T-T_c)^{-1},\ S\tau > 1]$. Also, the fact that the susceptibility, and thus the scattered light intensity, varies with shear was used to obtain an instantaneous mapping of the shear distribution for any geometrical configuration of the flow[(1-d)] (See also the Appendix).

From a theoretical point of view, Onuki and Kawasaki (O.K.) have derived nearly identical general conclusions, using a renormalization group approach. More precisely, they predicted:

(i) An effect visible in the region $S\,\tau > 1$. This leads to the temperature condition $T < T_S$. Using the above definition of τ, T_S is given by:

$$T_S = T_c \left[1 + \left(\frac{5\pi\eta\xi_0^3}{kT_c}\right)^{1/3\nu} S^{1/3\nu}\right] \qquad (1)$$

Here ξ_0 is the amplitude of the correlation length $\xi = \xi_0 \left[\frac{T-T_c}{T_c}\right]^{-\nu}$ with $\nu \simeq 0.630$.[8]

(ii) A lowering of the critical temperature

$$T_c(S) - T_c(0) = F\,[T_c(0) - T_S(S)]$$

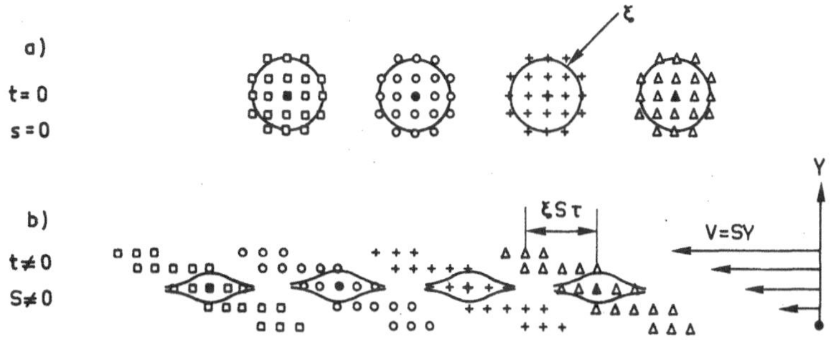

Figure 4. Naive picture of the effect of shear on critical fluctuations. At equilibrium: a) $t=0$, $S=0$, the correlations are isotropic, with a correlation length ξ. When the shear is applied: b) $t \neq 0$, $S \neq 0$, the volume elements of the liquid are shifted with time t in the shear direction as Vt, with V the local flow velocity. After a certain time, all correlations are therefore suppressed in the shear direction, and the correlations volumes become anisotropic.

Of course this time cannot be longer than the typical lifetime τ of the fluctuations. Moreover, in order to obtain an effect, the displacement ($=\xi S\tau$) of a point located within ξ of the correlation volume must be greater than ξ. These conditions give rise to $S\tau > 1$, i.e. to the unequality needed to obtain an effect from the shear on critical fluctuations.

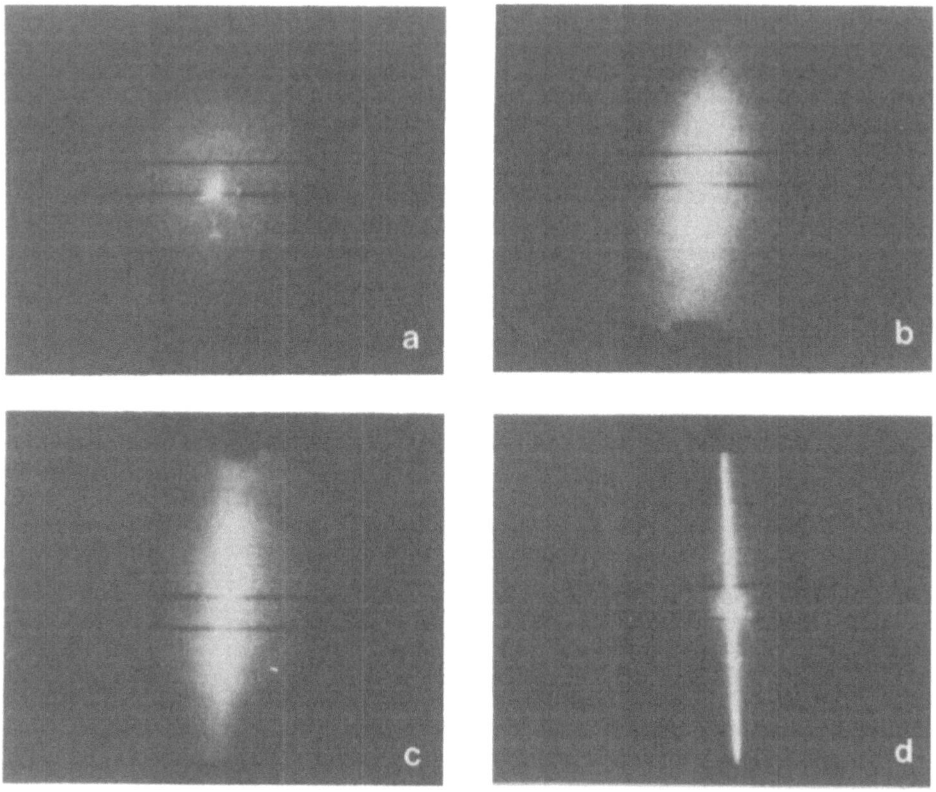

Figure 5. Scattered intensity at low angle from the mixture of cyclohexane-Aniline under shear flow. The laser beam is perpendicular to the plane of the figure, close to a side wall of the pipe where the flow is produced (the 2 horizontal black lines on the pictures). In a), the system is at equilibrium, and the scattered intensity is isotropic. In b), c), d), the anisotropy of the scattered light induced by the shear is quite visible. Also the light is more intense, due chiefly to the increase of transmission with shear.

With a renormalization group calculation, it was found F = 0.0832 to order ε[2-a], and 0.117 to order ε^2[2-f].

(iii) An anisotropy with respect to the flow direction (X axis) of the susceptibility $\chi_{\vec{q}}$ in the q-space, which would follow a mean-field behavior versus $T - T_c(S)$:

$$\chi_{\vec{q}}^{-1} \simeq A.S^{\frac{2\nu-1}{3\nu}} \left[\frac{T-T_c(S)}{T_c}\right]^{\gamma'} + B\, S^{8/15}\, |q_x|^{2/5} + q^2 \tag{3}$$

where $\gamma' = 1$, $A = \left(\frac{5\pi\eta}{kT_c}\right)^{\frac{2\nu-1}{3\nu}} \xi_o^{-1/\nu}$ and $B = \left(\frac{5\pi\eta}{kT_c}\right)^{8/15}$.

This is an approximate expression which, as stressed in Reference 2-f, can underestimate the results by 30 to 20%. Nevertheless this formula is very simple and we will use it to analyze our data semi-quantatively.

In the first sample, which is the cyclohexane + aniline mixture, shear rates were ranging from 0 to 1000 sec $^{-1}$. In the second sample, the binary mixture Isobutyric Acid + Water, the shear was kept constant within 20% at 1000 sec $^{-1}$.

We investigated both the transmission coefficient τ of the light through the sample, connected to the integrated susceptibility $\ell_N \tau \simeq \int_{\vec{q}} \chi_{\vec{q}} \cdot d\vec{q}$, and the scattered light intensity $I_{\vec{q}} = \chi_{\vec{q}}$, versus the 4 independent parameters of the phenomenon: temperature T, shear S, modulus of the transfer wavevector $|\vec{q}|$, and the component q_x along the flow direction.

3. EXPERIMENTAL

3.1 The Cyclohexane & Aniline System

The binary mixture used was the well-known cyclohexane + aniline system. Aniline was purified by fractioned distillation, cyclohexane was of spectroscopic grade. The experimental mass fraction of aniline $C = 0.4700 \pm 4 \times 10^{-4}$ was close to the critical composition (47% in weight).[9] The mixture was frozen in a special glass cell (Fig. 6) which was sealed under vacuum.

The shear flow is produced in a rectangular quartz pipe (C), as shown in Fig. 7, whose dimensions in cartesian axes OXYZ are $L_X = L = 15$ cm, $a_Y = a = 0.3$ cm, $b_Z = b = 0.5$ cm. O is the centre of symmetry and OY the vertical direction. This pipe (C) is set horizontal and, during the run, the liquid flowed along the X-axis through C from a graduated cylindrical reservoir A (with a maximum

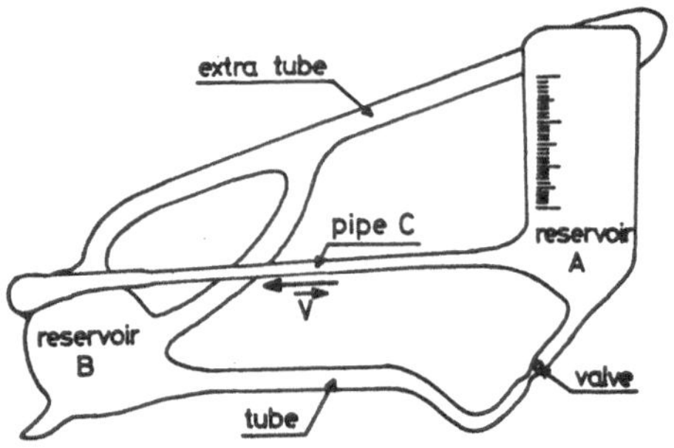

Figure 6. Cell used to produce the flow. The cell can rotate around the centre of C in the plane of A and B (see text).

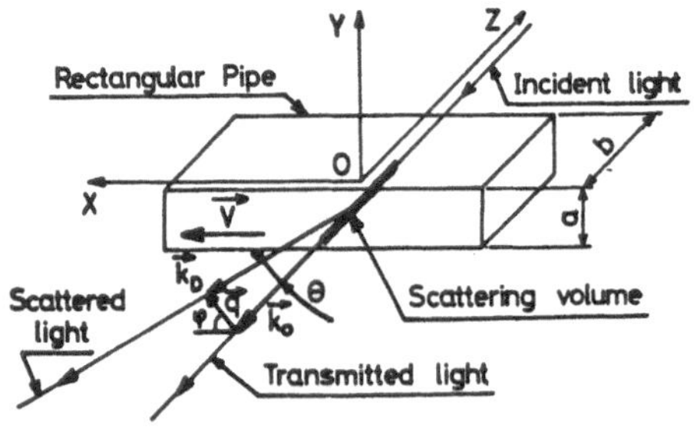

Figure 7. Scattering geometry. O is the centre of symmetry of the pipe.

fluid height difference H_o = 7.0 cm) into another reservoir B.

At the end of the run some liquid remained in C so that all experiments performed with or without shear could be directly compared under exactly the same geometrical and temperature conditions. The cell was mounted on a wheel which could rotate enabling the vessel A to be refilled at the end of the run. A mechanical locking system ensured that the pipe (C) returned to the same position for each run. A tube with a valve served to refill A while an extra tube served to equalize the pressures when flow took place from A to B or from B to A. This precaution was found to be necessary, since preliminary experiments carried out without this extra tube showed anomalous phenomena due to temperature variations associated with adiabatic pressure variations.

The whole apparatus (Fig. 8) was immersed in a large water bath where it could be seen through a double window. The thermal stabilization was $\pm$ 2 x 10^{-4}°C over more than one hour, as verified by a quartz thermometer. A laser beam (λ_0 = 6328 Å, diameter $\emptyset_0$ = 0.35 mm) directed parallel to the horizontal Z axis enters the pipe (C) at X = 0, Y = Y_0 = 0.1 cm, Z = -0.25 cm, as verified by a cathetometer. A lens centered on the beam images the scattering volume on the photomultiplier pinhole. The lens supports an off centre pin-hole, providing a choice of the scattering angle $\theta = (\vec{K}_D, \vec{K}_0)$, where $\vec{K}_0$ is the incident light wave vector (OZ) and $\vec{K}_D$ is the scattering wave vector. When rotating the pin-hole around the laser beam, θ or $|\vec{q} = \vec{K}_D - \vec{K}_0|$ remains constant while the azimuthal angle $\alpha = (\vec{q}, \vec{V})$ varies (see Fig. 7). Here $\vec{V}$ is the flow velocity vector, parallel to OX. We checked experimentally that for small scattering angles the (vertical) polarization of the laser beam does not modify the scattered intensity distribution which remains isotropic in the X-Y plane.

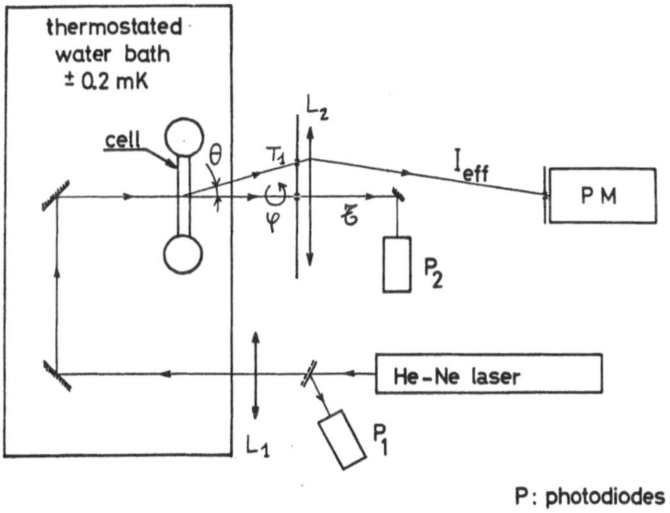

Figure 8. Schematic arrangement of the experiment -P_1P_2; Photodiodes. P_2 measures the transmitted beam intensity, P_1 measures the incident light intensity, L_1: focusing lens. L_2: imaging lens. C: Sample. M: Rotating mask with an off-centre pin-hole. T_1-T_2: pin-holes. P-M: photomultiplier tube.

In order to determine the velocity V, we used three methods:

(1) Measurement of the time variation of the liquid volume in (A)

For this, we recorded the liquid height H in A versus time t after the start of the run (Fig. 9). The following variation was found:

$$H = H_o \exp\left(-\frac{t}{t_o^{(1)}}\right) \text{ with } H_o = 7.0 \text{ cm and } t_o^{(1)} = 13.1 \text{ sec.}$$

We thus deduced the mean velocity $\overline{V}$ in (C), the ratio S_A/S_C of the crosssection areas of A and C being known ($S_A/S_C = 128$), so it follows:

$$\overline{V}_{(1)} = \overline{V}^o_{(1)} \exp\left(-\frac{t}{t_o^{(1)}}\right), \text{ with } \overline{V}^o_{(1)} = \frac{H_o}{t_o^{(1)}}\frac{S_A}{S_C} \simeq 68 \text{ cm/sec.}$$

(2) A calculation using a Poiseuille velocity distribution.

For the sake of simplicity we assumed that the actual mean velocity is not too far from the mean velocity $\overline{V}_{(2)}$ in a cylindrical pipe of diameter $\emptyset \sim 0.4$ cm. For a liquid with density $\rho = 0.87$ g cm^{-3}, (Ref. 15) with viscosity $\eta \simeq 1.8$ cP$_o$ (Ref. 10) flowing in a capillary of length $L_x = 15$ cm,

$$\overline{V}_{(2)} = \overline{V}^o_{(2)} \exp\left(-\frac{t}{t_o^{(2)}}\right), \text{ with } \overline{V}^o_{(2)} = \frac{\rho g\, \Phi^2\, H_o}{32\, \eta\, L_x} \simeq 110 \text{ cm/sec.}$$

The exponential variation is due to the proportionality which exists between the velocity and the volume variation of liquid:

$$S_C\, \overline{V}_{(2)} = S_A \frac{dH}{dt} \rightarrow H(t) = H_o \exp\left(-\frac{t}{t_o^{(2)}}\right) \text{ with}$$

$$t_o^{(2)} = \frac{S_A}{S_c} \cdot \frac{32\, L_x \eta}{\rho g\, \Phi^2} \simeq 8.0 \text{ sec.}$$

This value is of the same order of magnitude as the experimental value $t_o^{(1)} = 13.1$ sec.

(3) A laser doppler velocimetry determination, at the precise point where the measurements are to be performed. The homodyne spectrum of the light scattered in the X direction from the volume in $Y = Y_o = 0.1$ cm shows a very well defined cut-off frequency f_o which corresponds to the velocity at the measurement point. Fig. 9 shows that

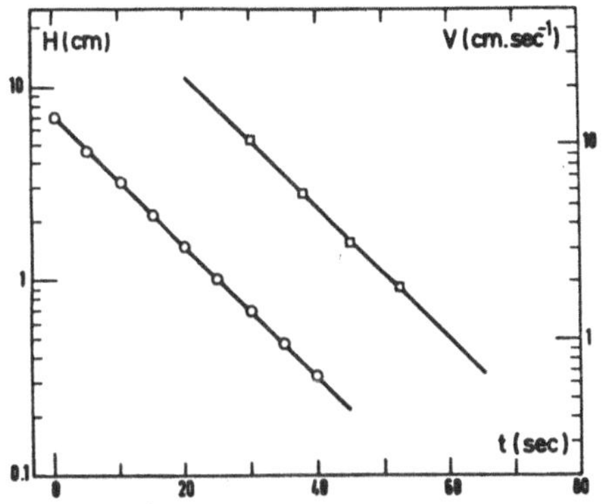

Figure 9. Time variation of the liquid height H in A, and of the mean velocity V at the measurement point (semi-log plot).

$$V_{(3)}\ (t,\ Y_o) = V^o_{(3)}\ (Y_o)\ \exp\ (-t/t_o^{(3)})$$

with $V^o_{(3)}\ (Y_o) = 93$ cm/sec and $t_o^{(3)} = t_o^{(1)} = 13.1$ sec. Assuming a parabolalike Poiseuille velocity distribution, the mean velocity in the pipe can be estimated to

$$\overline{V}^o_{(3)} = \frac{V^o_{(3)}\ (Y_o)}{2\left[1-2(\frac{Y_o}{a})^2\right]} = 84\ \text{cm/sec.}$$

The estimate of $\overline{V}_o^{(1)}$ (68 cm/sec) and $\overline{V}_o^{(2)}$ (110 cm/sec) are not very far from this experimental determination. We then determine the Reynolds number

$$R(t) = \frac{a.V_{(3)}\ (t,Y_o)}{\nu_o}$$

where $\nu_o = \eta/\rho \simeq 2 \times 10^{-2}$ Stokes[10] is the kinematic value. $R(o) \simeq 1400$ shows that turbulence is never reached. Nevertheless, we verified that all experiments which were started of various values of R(o) (i.e. various heights in A) gave equivalent results.

The length ℓ at which the parabolic velocity distribution is obtained within 1% is $\ell(t) = 0.03\ a\ R(t) \simeq 13\ \exp\ (-t/t_o)$ cm. At any time $t > 5$ sec the Poiseuille distribution is therefore established in the measurement region.

Therefore the velocity at the measurement point can be approximated by:

$$V(X,Y,Z,t) \simeq V(t,Y_o) = 2\overline{V}_o \left[1 - \left(\frac{2Y_o}{a}\right)^2\right] e^{-t/t_o} .$$

Here the influence of the velocity gradient parallel to OZ has been neglected. In fact the corresponding gradients are smaller [ratio $(b/a)^2 \simeq 3$], and are not well defined since they are integrated over all the scattering volume. Moreoever, the influence of this particular velocity gradient can be suppressed when imaging only the centre of the scattering volume.

The shear at $Y = Y_o$ can therefore be deduced,

$$S(Y_o \pm \Phi_0/2,t) = \frac{dV}{dY}(Y_o \pm \Phi_0/2,t) = (1600 \pm 280)e^{-t/t_o} .$$

The calculated uncertainty is due to the finite diameter Φ_0 of the laser beam in the direction Y, as verified in the appendix or in Ref. 1-d where the mapping of the shear was performed.

The lifetime of the concentration fluctuations without shear is $\tau = E(T-T_c)^{-3\nu}$ where $\nu = 0.63$ and $E = 5\pi\eta\, \xi_0^3 T_c^{3\nu}/kT$ (Ref 5-6) = 4.8×10^{-6} c.g.s. using the value $\xi_0 = 2.45$ Å (Ref. 11,18), T_c = 303 K and $\eta = 1.78$ cP_o (Ref. 10). For experimental values $T-T_c >$ 1×10^{-3}°C, then $\tau < 2.2$ sec. In these time intervals the velocity distribution varies by less than a few percent in (C).

The extra-heating due to the shear flow can be estimated as follows: the power dissipated in a volume element is proportional to S^2 through the viscosity. With C_P the specific heat,

$$C_P \frac{\Delta T}{\Delta t} = \eta\, S^2 .$$

During the run, the spurious heating occurs over the length $L_x/2$, i.e. during the time $\Delta t = L_x/2V(t,Y_o) \simeq 8L_x/S$. For time intervals lower than the thermal diffusivity time $a^2/D \sim 100$ sec ($D \sim 10^{-3}$ cm^2 sec^{-1} in liquids is the thermal diffusivity coefficient), the heat has not enough time to diffuse, and the adiabatic temperature increase can be evaluated as

$$\Delta T \simeq \frac{8\,\eta\, L_x}{C_P} S .$$

Thus ΔT is proportional to S. Measurements were performed for $t > 5$ sec, or $S < 900\ sec^{-1}$. In liquids, $C_p \sim 4 \times 10^7$ erg, and therefore $\Delta T < 5.10^{-5}$°K, and is completely negligible.

The contribution of multiple scattering[12] is also negligible, considering the small dimensions of the capillary tube. We verified this point visually. Finally, the gravitationally induced gradients do not have enough time to form,[13] and are thus also negligible.

3.2 The Isobutyric Acid + Water System

Another cell filled with the Isobutyric Acid-Water system has been used. The cell is similar to that used with the Cyclohexane-Aniline mixture, but during the run, the variation of the liquid height is minimized (Fig. 10), so that the shear remains roughly constant. The characteristics of the capillary are the same as described above.

The Isobutyric Acid was of a quality better than 99% pure. The quality of the water used corresponds to an ohmic resistivity of $18 \times 10^6\ \Omega$cm. The experimental mass fraction of the acid was $0.38900 \pm 9 \times 10^{-5}$, close to the critical mass fraction 0.3885.[14] The cell was sealed under the vapour pressure of the components, i.e. we pumped the air until the mixture began to vaporize (some bubbles appeared), then we sealed the cell. The critical temperature was found to be $T_c = 26.3$°C.

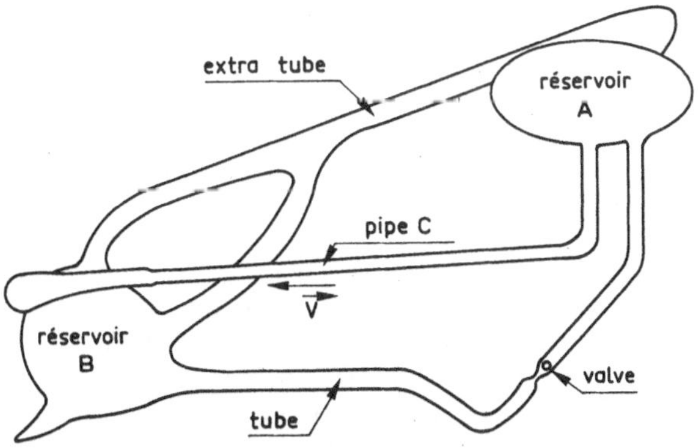

Figure 10. Cell used for the system Isobutyric Acid-Water. The run lasts 37 sec.

The shear was determined by extrapolating the data of the Cyclohexane + Aniline mixture at the density and the viscosity of the Acid-Water mixture. More precisely, we used the Poiseuille velocity distribution

$$V(Y_o) = 2\overline{V}_o \left[1 - \left(\frac{2Y_o}{a}\right)^2\right].$$

Since the matching of refractive indices between this mixture ($n \simeq 1.357$) and the capillary ($n \simeq 1.50$) is not so good as with the Cyclohexane-Aniline system ($n \simeq 1.52$), the position of the beam with respect to the walls of the capillary had to be slightly changed. The diameter is again $\Phi_o = 0.035$ cm, and the beam is at $Y_o = 0.065$ cm. Thus the shear is

$$S(Y_o \pm \Phi_{o/2}) = 16 \frac{\overline{V}_o Y_o}{a^2} \pm \frac{8\overline{V}_o \Phi_o}{a^2}.$$ $\overline{V}_o$ follows directly from:

$$\overline{V}_o = 93 \times \frac{\rho_2 H_o^{(2)} \eta^{(1)}}{\rho_1 H_o^{(1)} \eta(2)} \text{ cm.sec}^{-1}$$ where the subscripts 1 or 2

correspond to Cyclohexane-Aniline or Isobutyric Acid-Water. Using the data: $\rho_1 = 0.87$ g.cm^{-3}(Ref. 15), $\rho_2 = 0.993$ (Ref. 15), $\eta_1 = 1.8$ cP$_o$ (Ref. 10), $\eta_2 \simeq 2.9$ cP$_o$ (Ref. 16), $H_o^{(1)} = 7.00$ cm, and $H_o^{(2)} = 7.80 \pm 1.55$ cm, where the last uncertainty is only the height variation of the liquid during the run, we obtain: $\overline{V}_o = 90.8 \pm 18$ cm.sec^{-1}, and thus the mean shear $S(Y_o) = 1050 \pm 208$ sec^{-1}. The uncertainty due to the finite size of the beam is

$$S(\pm \frac{\Phi_o}{2}) = \pm 282 \text{ sec}^{-1}.$$

This geometrical uncertainty is certainly overestimated, the (Gaussian) intensity profile of the beam being not taken into account. So it seems reasonable to divide this uncertainty by a factor 3; the remaining total uncertainty can therefore be estimated to 300 sec^{-1}, and the shear is $S \simeq 1000 \pm 300$ sec^{-1}.

The Reynolds number can be estimated to be $\mathcal{R} = 940 \pm 190$. Turbulence is therefore not reached, as verified by the experiments. The parabolic velocity distribution is established at the distance $\ell = 0.03\, a\mathcal{R} = (8.5 \pm 1.7)$ cm, which corresponds about to the point of measurement ($\gtrsim 7.5$ cm). Extra-heating is of the same order of magnitude as in the Cyclohexane-Aniline system, i.e. remains negligible. Gravity induced gradients are not a problem in this mixture since the density of all components are matched. Finally, multiple scattering is also negligible for the same reasons as stated above, and also because the refractive indices of each components are close together (water: n = 1.33 (Ref. 15) - Isobutyric acid: n = 1.39 (Ref. 15).

On the other hand, this refractive index matching is rather troublesome since both turbidity and scattered intensity will remain low, precisely when large stray light phenomena are expected to occur due to the capillary refractive index which is no more matched with the refractive index of the mixture. Therefore, great care has to be taken when measuring the turbidity and the scattered intensity. Another aspect concerns the lifetime of concentration fluctuations $\tau = E'(T-T_c)^{-3\nu}$ with $E' = 5\pi\eta\xi_0^3 T_c^{3\nu}/kT$. Using ξ_0 = 3.62 Å (Ref. 17, 18), T_c = 299.5K, η = 2.9 cP_o (Ref. 16), and $T \simeq T_c$, we obtain E' = 2.50 x 10^{-5} c.g.s., that is a value more than 5 times great than for the Cyclohexane-Aniline mixture (E = 4.8 x 10^{-6} c.g.s.). For $T-T_c = 10^{-3}$K, τ = 12 sec. It is clear that a nearly constant shear must be used to avoid convolution effects due to instationary conditions.

4. RESULTS AND DISCUSSION

The experimental results are concerned with the following points:

(1) The critical temperature change $T_c = T_c(S)$.
(2) The cross-over temperature $T_S = T_S(S)$.
(3) The variation of the susceptibility $\chi_{\vec{q}}$ versus q_x at 2 different constant values of the modulus $|\vec{q}|$, and for various shears.
(4) The variations of $\chi_{\vec{q}}$ versus $|q|$ for various shears.
(5) The variation of χ_{q_x} versus T and S for 2 different constant values of the modulus $|\vec{q}|$.
(6) Reestablishment of equilibrium.

The variation of the susceptibility in the shear direction (Y) is not reported here, mainly due to the lack of accuracy. Experiments on the system Isobutyric acic + Water are in progress concerning this point. Of course, most of the results concern the system Cyclohexane-Aniline where the shear was varying.

The transmitted light intensity was chiefly used for studying point (1) and the scattered light intensity $I_{\vec{q}}$ for points (2) to (5).

4.1 T_c Change

a. The Cyclohexane & Aniline mixture

The critical temperature can be defined as the temperature at which $\chi_{\vec{q}}$ diverges, i.e. τ vanishes. In Fig. 11 the experimental variations of τ are shown versus shear rate S at various temperatures. Different behaviors to be seen in this figure are as follows:

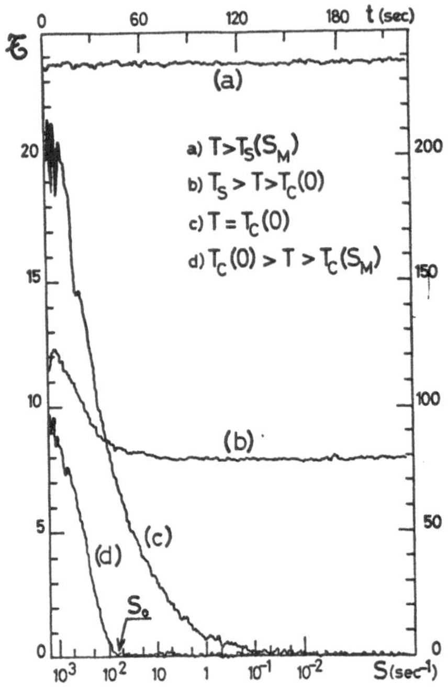

Figure 11. Behavior of the transmitted intensity versus shear S or time $t \propto - t_o \, \ell_N(S/S_M)$ at different temperatures (see text). The right hand scale corresponds to (a, b) and the left hand scale to (c, d).

(i) The introduction of shear has no influence for $T > T_S(S_M)$, which is the cross-over temperature T_S corresponding to the maximum shear S_M.

(ii) When $T < T_S(S_M)$, the shear is seen to increase the transmission and therefore to lower the susceptibility. This suggests that the critical temperature could be a function of shear, $T_c = T_c(S)$.

(iii) When $T = T_c(0)$ (the critical temperature at equilibrium), τ goes asymptotically to zero.

(iv) When $T_c(0) > T > T_c\ (S_M)$, i.e. when the run is started at a temperature slightly lower than $T_c(0)$, the transmission behaves as in (ii), i.e., as in the one-phase region. A new critical temperature $T_c(S)$ seems thus demonstrated. The transmission goes to zero for the shear S_o, revealing the critical temperature under shear S_o: $T = T_c(S_o) = T_c(0) - \Delta T_c(S_o)$. We noticed that the scattered light intensity diverged for the same temperature T at the same shear S_o, with $\chi_{q_y} >> \chi_{q_x}$, indicating that the first stage of the phase separating process could occur in stripes parallel to the

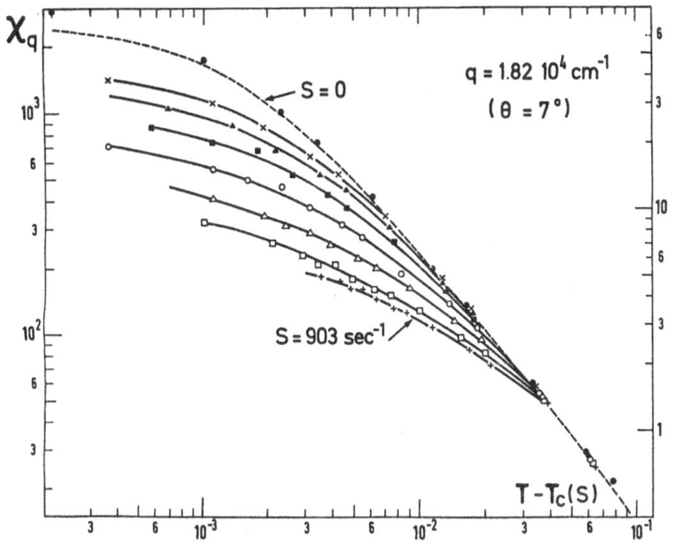

Figure 12. Transmitted intensity versus $T-T_c(0)$ at decreasing shear rates S=903, 510, 288, 163, 92, 52, 29, 16, 9.5, 5, 3, 0 sec^{-1}.

flow velocity. This is an interesting phenomenon, in accordance with the O-K predictions,[2] and which we intend to study more carefully.

From Fig. 11 we can infer the variations of τ versus $T-T_c(0)$ at various shears (Fig. 12). A striking feature is that a mere change in T_c is sufficient to reduce all the data on the single equilibrium curve, i.e. $\tau[T-T_c(0), S] \simeq \tau[T-T_c(S), 0]$. This allows the variation of $\Delta T_c(S) = T_c(0) - T_c(S)$ to be determined.

In Fig. 13 we show the variations of the change in T_c with S, evaluated in 2 different ways, first from Fig. 12 as we have just noted, and second from Fig. 11-d using $T_c(S_o)$. A close agreement exists between these two determinations. The simple power law

$$\Delta T_c(S) = T_o S^{\sigma_o}$$

is seen to hold, with

$$T_o = (1.8 \pm 0.2)\ 10^{-14} \text{ c.g.s.}$$

and,

$$\sigma_o = 0.53 \pm 0.03.$$

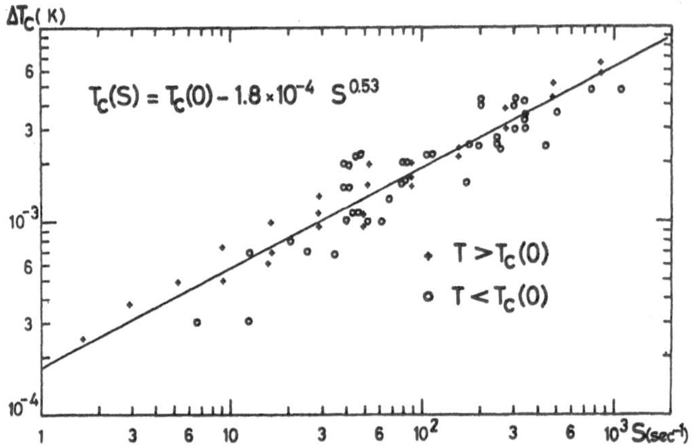

Figure 13. Critical temperature change versus shears.
+ From the variation of the transmission above $T_c(0)$ (Fig. 10).
0 From the determination of S_o below $T_c(0)$, (Fig. 9).

These experimental values have to be compared with the O-K calculation (formulas (1) and (2)): $\sigma_o = 1/3\nu = 0.529$, and $(T_o)_{OK} = 0.0832\ T_c\ (5\pi\eta\xi_0^3\ /\ k_BT_c)^{1/3\nu}$(Ref. 2-a) or $(T_o)_{OK} = 0.117\ T_c$ $5\pi\eta\xi_0^3\ /\ k_BT_c)^{1/3\nu}$(Ref. 2-f).

Using the numerical values of § III, the first evaluation gives $(T_o)_{OK} = 1.28\times10^{-4}$ c.g.s., which is low when compared to our data: the second is in full agreement, it gives $(T_o)_{O.K.} = 1.80\times10^{-4}$ c.g.s. This good agreement may be fortuitous, as will be discussed below.

b. The Isobutyric acid + Water system

Considering the results with the C-A system, we expect $T_o \simeq 3.06\times10^{-4}$ or 4.3×10^{-4} c.g.s. With a shear rate $S = 1000 \pm 300$ sec^{-1}, the effective change should be $T_c(0)-T_c(S) = (1.2 \pm 0.2)$ 10^{-2}K or (1.7 ± 0.7) 10^{-2}K. The last value is more probable.

Typical experimental recordings are shown in Fig. 14. Curve (a) corresponds to the temperature where the effect of shear becomes visible, i.e. T_S. Here $T_S-T_c(0) \simeq 0.1$K. Curve (b) is a recording within 0.006K of $T_c(0)$ and curve (c) corresponds to $T_c(0)$. Curve (d) is at -0.015K from $T_c(0)$, and curve (e) at -0.025K from $T_c(0)$.

The following Figure 15 is concerned with the transmission data versus $T-T_c(0)$ or $T-T_c(S)$. $T_c(S)$ is determined as shown above in Fig. 11 (C-A mixture), at the temperature at which the transmission

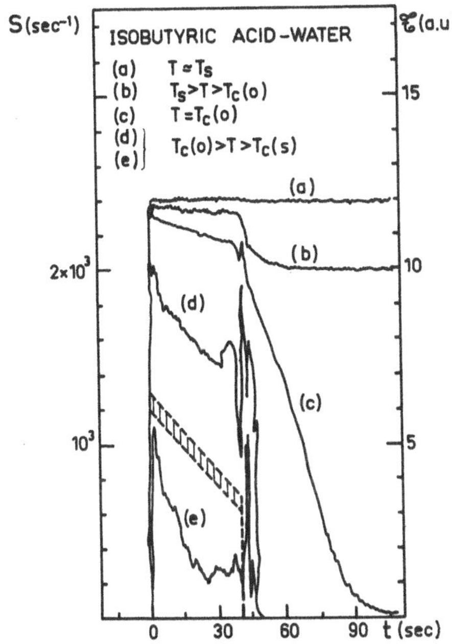

Figure 14. Behavior of the transmitted intensity (full line) versus time t or shear S (dashed line) at various temperature (see text).

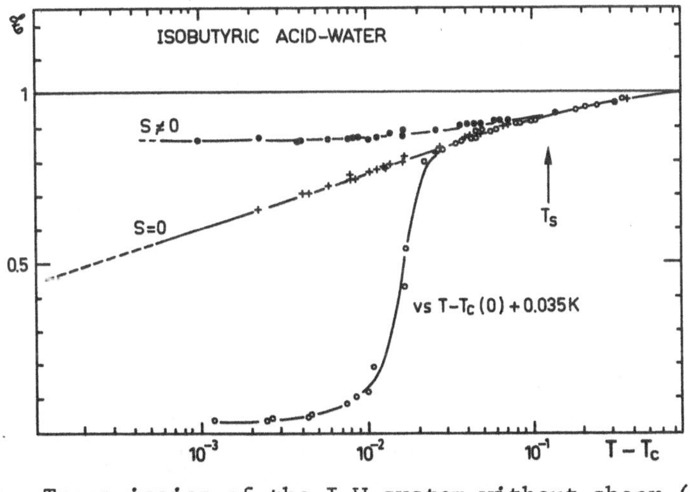

Figure 15. Transmission of the I-W system without shear (crosses), with shear (full points) when the critical temperature taken is $T_c(0)$, and with shear (open circles) when the critical temperature considered is lowered by 0.035K (see text). T_S is the temperature at which the effect of shear becomes visible.

under shear is no longer different from zero. This was found to be $T_c(S)-T_c(0) = -0.035K \pm 0.005$, which is about 2 times bigger than expected (at most $- 0.017K$ for the mean shear), but is in agreement with another determination of $T_c(S)$ from the difference between $\tau(S)$ and $\tau(0)$ at $T = T_c(0)$ (see 4.1 (a):

$$T_c(S) - T_c(0) = -(3.5 \pm 0.5)\ 10^{-2}K.$$

These results, which are only in semi-quantitative agreement with those of the C-A system, prompt the following remarks:

(i) The methods used, i.e. the variation of transmission with $T-T_c$, is much more sensitive for the C-A system than for the I-W mixture. This is largely due to the nearly matching of refractive indices in the last system, which prevents turbidity to become important.

(ii) The precise theoretical variation of τ with shear is not well known, chiefly when $T < T_c(0)$. This is particularly noticeable in Fig. 15 where the data $\tau(T,S)$ versus $T-T_c(S)$ cannot be surimposed on the corresponding data $\tau(T,0)$ versus $T-T_c(0)$, clearly showing a discrepancy in the behaviors. Both $\tau(T,S)$ and $\tau(T,0)$ lie however on the same curve when only the measurements at $T>T_c(0)$ are considered. This observation is also valid for the C-A system (Fig. 12).

(iii) The effect of shear is always instantaneously "felt" by the system. No transit time can be seen when one applies the shear. This is however not the case when the shear is stopped, some time is needed for the system to recover its equilibrium (see Fig. 14-b). We can intuitively predict that this time is related to the typical lifetime of fluctuations τ(see below, point 4.6.).

(iv) Finally, the definition of T_c as the temperature at which τ goes to zero is only obvious for a system at equilibrium: at this temperature the susceptibility (then the scattered intensity) goes to infinite. In a system which is expected to have different correlations lengths with regards to the directions of shear and velocity, leading to different transition temperatures, the problem is difficult. The scattered intensity with respect to the shear direction or to the flow direction would give useful information. Experiments are in progress concerning these points.

On the other hand, one can see on Fig. 15 that very close to T_c (within 10^{-4}K), the transmission is still 0.45. One could define T_c for this system by the temperature at which $\simeq \tau 0.45-0.50$. In this case, one finds $T_c(S)-T_c(0) = 1.75 \times 10^{-2}K$, which is then in full agreement with the C-A results and the O-K predictions.

4.2 The Cross-over Temperature

a. The Cyclohexane + Aniline system

In order to determine the temperature T_S below which the shear affects the fluctuations, it was found to be more reliable to use the scattered intensity data (Figs. 20-21) rather than the transmission data of Fig. 12. In these figures, the change of behavior is clearly shown and 2 scattering angles (wavevectors $q = 18200\ cm^{-1}$ and $5200\ cm^{-1}$) were investigated.

The corresponding variation $\Delta T_S(S) = T_S(S) - T_C(S)$ is reported versus S in Fig. 16. Large uncertainties related to the cross-over phenomenon itself prevent any convincing determination of amplitudes or exponents. Nevertheless, assuming the O.K. variation $T_S(S) = 13\ T_o S^{1/3\nu}$ and fitting the data with the imposed value $1/3\nu = 0.529$, we found $T_o \simeq 1.8 \times 10^{-4}$, to be compared with 1.8×10^{-4} found experimentally with the T_c change, and 1.28×10^{-4} or 1.80×10^{-4} from O.K. The agreement is remarkably good with the determination of T_o from $T_c(S)$.

b. The Isobutyric acid + water system

The transmission measurements of Fig. 15 show that the difference due to the shear becomes noticeable at $T - T_c(0) \simeq 0.12°C$, to be compared with $T_s - T_c(0) = 0.15 \pm 0.06$ K. Here the agreement seems also to be very good.

We will discuss now the scattered intensity data. They are only concerned with the C-A system.

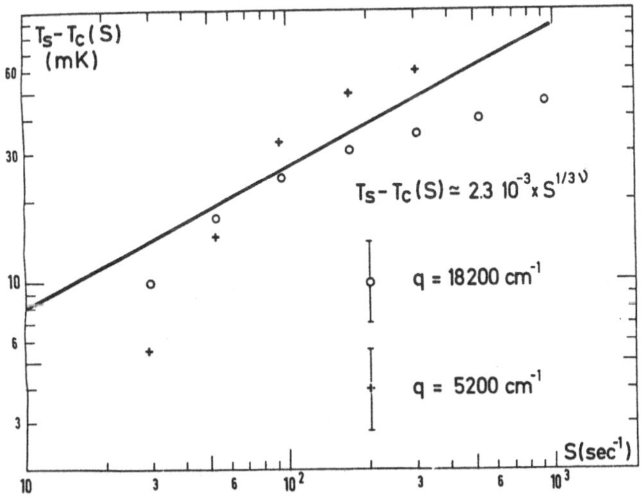

Figure 16. Cross-over temperature variation with shear S, from the scattered intensity data of Figs. 20 and 21. The straight line corresponds to a power law with an imposed exponent $1/3\nu = 0.53$.

4.3 Susceptibility Anisotropy With Respect to the Flow Direction:

$$\chi_{\vec{q}}\left(\frac{\vec{q}_x}{|q|}\right).$$

Besides the T_c change, the most striking feature of the shear-induced effect is the anisotropy of the scattered light, and therefore of the susceptibility with respect to the flow direction. Fig. 17 shows that the susceptibility is increased in the shear direction Y, revealing a considerable reduction of the correlation length in this direction. We notice that the effect increases with increasing shear and decreasing scattering wavevector.

Two scattering angles were carefully investigated with good angular resolution: $\theta_1 = 10^o \pm 2'30''$ and $\theta_2 = 2^o \pm 12'$ corresponding to $q_1 = (26000 \pm 100)\ cm^{-1}$ and $q_2 = (5200 \pm 500)\ cm^{-1}$. $\Delta T(0) = T-T_c(0) = 1.5\ 10^{-3}K$.

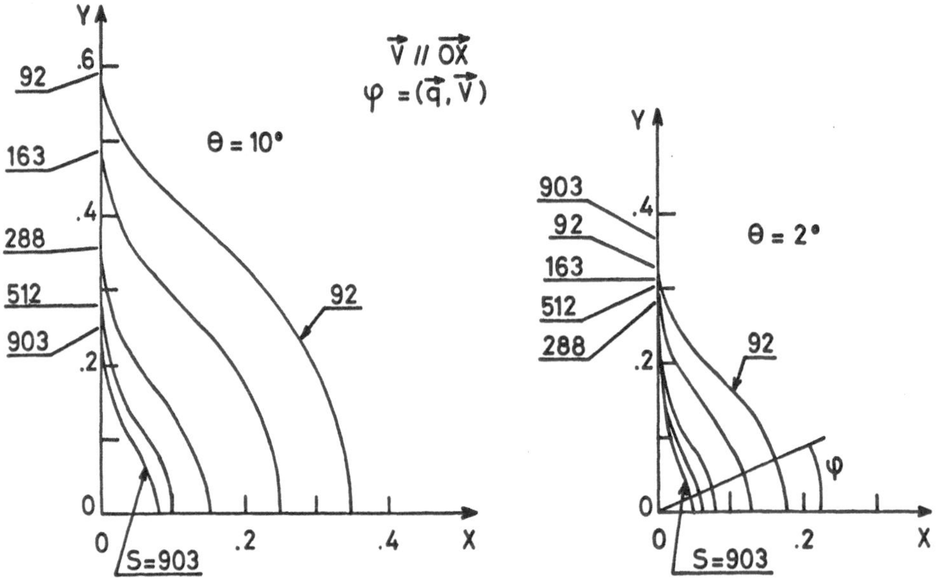

Figure 17. Reduced scattered intensity $\chi_q(T,S)\ /\ \chi_q(T,0)$ in the X-Y plane at $T-T_c(0) = 1.5$ mK (compensated for the transmission), and at 2 scattering angles $\theta=2^o$ and 10° ($q = 5200\ cm^{-1}$ and $2600\ cm^{-1}$). Polar coordinates have been used. Shear rates were S=903, 512, 288, 163 and 92 sec^{-1}. The best fits to experimental data were used (see Tables I-II) with the exponent X_3 imposed at the O.K. value 2/5.

The anisotropy in formula (3) (see II), is determined by the q variation of the 2nd term. Two numerical calculations can be made, first by simply substituting the numerical values in the formula (3) (O.K.), or second by taking into account the results (T_o) of the transmission data (O.K. exp). With $\varphi = (\vec{q}, \vec{V})$, this gives:

$$\chi_{\vec{q}}^{-1} = X_1 + X_2 \ |\cos \varphi|^{X_3}$$

$q_1 = 26000\ \mathrm{cm}^{-1}$: $X_1(\mathrm{O.K.}) = 2.28\mathrm{x}10^{11}\mathrm{x}S^{0.14}[\Delta T(0)+1.3\mathrm{x}10^{-4}S^{0.53}] +6.74\mathrm{x}10^{8}$

$X_1(\mathrm{O.K.\ exp}) = 2.53\mathrm{x}10^{11}S^{0.14}[\Delta T(0)+1.8\mathrm{x}10^{-4} S^{0.53}]+6.74\mathrm{x}10^{8}$

$X_2(\mathrm{O.K.}) = 4.02\mathrm{x}10^{8}\mathrm{x}S^{0.53}$

$X_2(\mathrm{O.K.\ exp}) = 5.71\mathrm{x}10^{8}\mathrm{x}S^{0.53}$

$X_3(\mathrm{O.K.}) = X_3(\mathrm{O.K.\ exp}) = 2/5$

$q_2 = 5200\ \mathrm{cm}^{-1}$: $X_1(\mathrm{O.K.}) = 2.28\mathrm{x}10^{11}S^{0.14}[\Delta T(0)+1.3\mathrm{x}10^{-4}S^{0.53}] +2.70\mathrm{x}10^{7}$

$X_1(\mathrm{O.K.\ exp}) = 2.53\mathrm{x}10^{11}\mathrm{x}S^{0.14}[\Delta T(0)+1.8\mathrm{x}10^{-4} S^{0.53}]+2.70\mathrm{x}10^{7}$

$X_2(\mathrm{O.K.}) = 2.11\mathrm{x}10^{8}\mathrm{x}S^{0.53}$

$X_2(\mathrm{O.K.\ exp}) = 3.00\mathrm{x}10^{8}\mathrm{x}S^{0.53}$

$X_3(\mathrm{O.K.}) = X_3(\mathrm{O.K.\ exp}) = 2/5$

The experimental temperature difference is $\Delta T(0) = 1.5\mathrm{x}10^{-3}$K for all data. The value X_1(O.K.) and X_1(O.K. exp) are close together so that in tables I and II, where the values of the ration X_2/X_1 are listed for various shears, we report only the mean value $X_2 / X_1 = \frac{1}{2}[(X_2/X_1)_{\mathrm{O.K.\ exp}} + (X_2/X_1)_{\mathrm{O.K.}}] \pm \frac{1}{2}[(X_2/X_1)_{\mathrm{O.K.\ exp}} - (X_2/X_1)_{\mathrm{O.K.}}]$

Also reported are the results concerning the experimental data. A program of statistical refining (M. Tournarie)[19] has been used, giving the standard error $\sigma(\chi_q)$ and a "quality coefficient" Q which measures the contribution of the statistical error to the total error. A=1 when no systematical distortions to the fit exist and

Q drastically decreases if there are systematic distortions.

The data (non-corrected for the transmission) were fitted to the following variation:

$$\chi_q^{-1} = X_1 + X_2 \, |\cos \varphi|^{X_3}$$

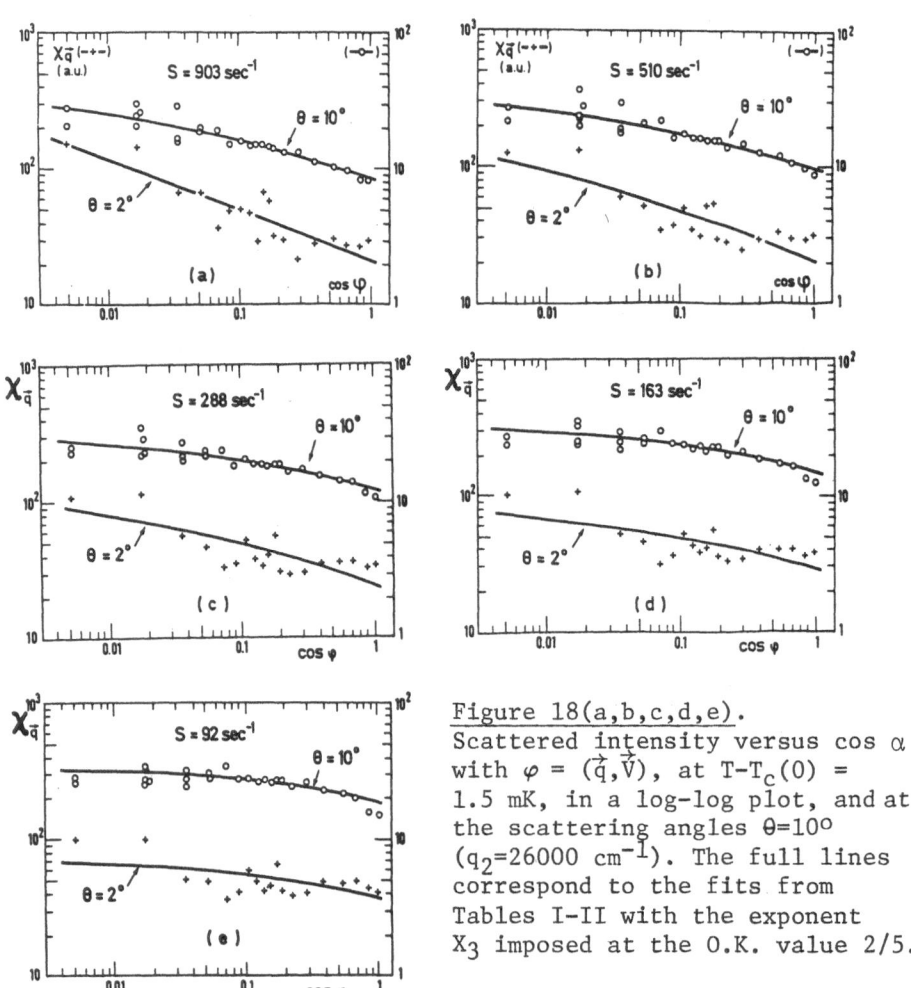

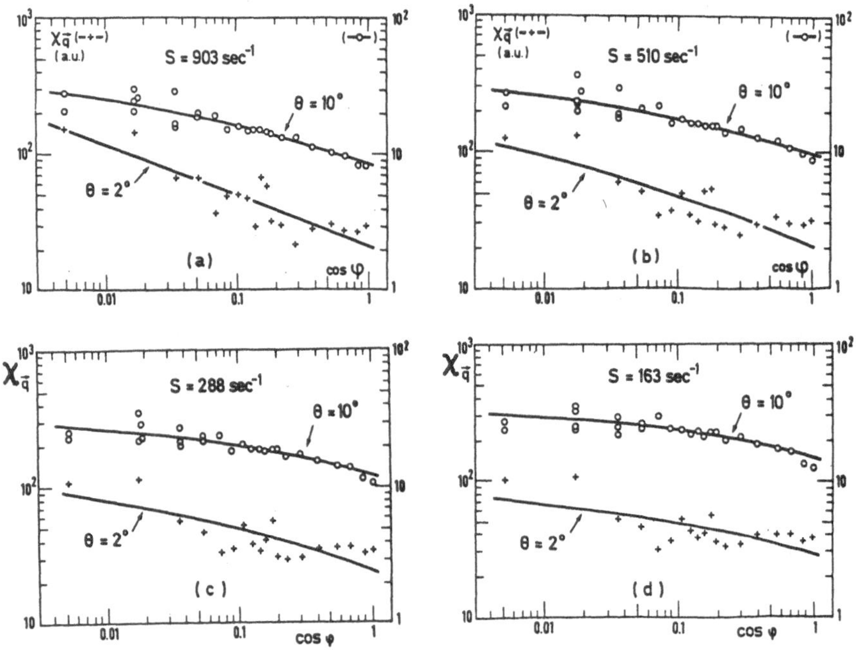

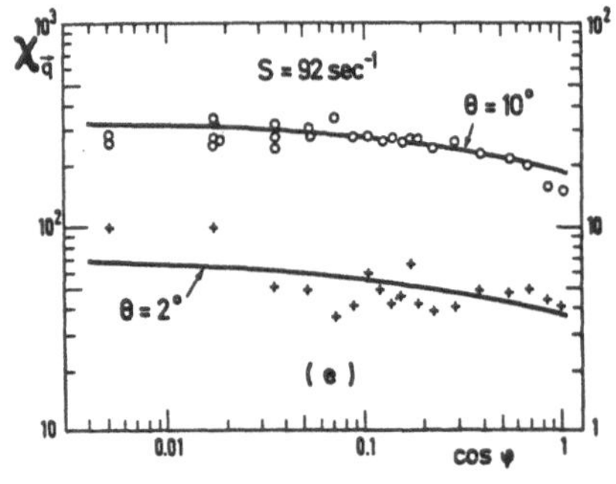

Figure 18(a,b,c,d,e).
Scattered intensity versus cos α with $\varphi = (\vec{q},\vec{V})$, at $T-T_c(0) = 1.5$ mK, in a log-log plot, and at the scattering angles $\theta=10^o$ ($q_2=26000$ cm^{-1}). The full lines correspond to the fits from Tables I-II with the exponent X_3 imposed at the O.K. value 2/5.

TABLE I. Fit of χ_q^{-1} to $X_1+X_2|\cos\varphi|^{X_3}$ for q=5200 cm^{-1} and $T-T_c(0)=$ 1.5 MK($\tau \simeq$1sec)·$(X_2/X_1)_{O.K.}$ is the amplitude expected from the O.K. theory. Brackets indicate that in the fit the parameter was held constant at the quoted value.

S(sec^{-1}) or Sτ	X_3	X_2/X_1	$(X_2/X_1)_{O.K.}$	Q	$\sigma(\chi_q)$%
903	0.36±0.05	-	2.07±0.01	0.92	28
	(0.40)	-		0.903	28
510	0.30±0.05	-	2.08±0.01	0.71	23
	(0.40)	12±10		0.65	23
288	0.25±0.04	-	2.05±0.03	0.67	21
	(0.40)	4.8±2.7		0.56	21
163	0.20±0.04	-	1.98±0.05	0.64	19
	(0.40)	2.2±1.1		0.53	20
92	0.14±0.04	-	1.88±0.08	0.66	18
	(0.40)	1.1±0.6		0.55	18
52	0.10±0.04	-	1.74±0.09	0.71	17
	(0.40)	0.6±0.4		0.63	18
29	0.07±0.03	-	1.56±0.10	0.74	15
	(0.40)	0.3±0.2		0.70	16
16	0.05±0.03	-	1.38±0.10	0.87	14
	(0.40)	-		0.84	14.5

with X_i the adjustable parameters. In a first step, all parameters were set free, so that both X_3 and X_2/X_1 could be compared to the O.K. values. Results are summarized in tables I and II. In Fig. 18 are reported the log-log variations of χ_q^{-1} versus $\cos\varphi$ for the various shears and the 2 values of q.

For X_3, we see that the experimental values are close to 2/5 for the highest shears (X_3 = 0.52±0.12 for S=903 sec^{-1}, q=2600 cm^{-1}, X_3=0.36±0.05 for S=903 sec^{-1}, q=5200 cm^{-1}). When the shear decreases, X_3 increases for the higher q value (X_3=3.8±1.8, S=16 sec^{-1}, q=26000 cm^{-1}) but decreases for the lower q value (X_3=0.05±0.03, S=16 sec^{-1}, q=5200 cm^{-1}). The amplitude ratio X_2/X_1 is well defined only for the highest q=2600 cm^{-1}. Its order of magnitude is correct for all shears, and agrees well with the O.K. value for the highest shears (S=903 sec^{-1}, X_2/X_1 = 2.8±0.8, $(X_2/X_1)_{O.K.}$=3.45±0.05).

In order to have a better determination of the amplitudes, we next impose X_3 to its O.K. value. The highest q, with the highest shears, gives the better agreement (a=2600 cm^{-1}, S=903 sec^{-1}, X_2/X_1=3.4±0.7, $(X_2/X_1)_{O.K.}$=3.45±0.05). In all cases the order of

TABLE II. Fit of χ_q^{-1} to $X_1+X_2|\cos\varphi|^{X_3}$ for q=26000 cm^{-1} and $T-T_c(0)$ =1.5mK(τ≃1sec). $(X_2/X_1)_{O.K.}$ is the amplitude expected from the O.K. theory. Brackets indicate that in the fit the parameter was held constant at the quoted value.

S(sec^{-1}) or Sr	X_3	X_2/X_1	$(X_2/X_1)_{O.K.}$	Q	$\sigma(\chi_q)$%
903	0.52±0.12	2.8±0.8	3.45±0.05	0.72	12.2
	(0.40)	3.4±0.7		0.69	12.2
510	0.52±0.14	2.24±0.6	3.3±0.1	0.80	12.5
	(0.40)	2.74±0.6		0.75	12.7
288	0.64±0.18	1.5±0.4	3.1±0.1	0.996	11
	(0.40)	1.90±0.35		0.87	11.5
163	0.9±0.2	1.1±0.2	2.8±0.2	0.99	9.8
	(0.40)	1.3±0.3		0.75	10.5
92	1.5±0.5	1.0±0.2	2.5±0.2	0.96	8.7
	(0.40)	0.9±0.2		0.61	9.9
52	2.0±1.0	0.73±0.2	2.1±0.2	0.98	7.3
	(0.40)	0.54±0.15		0.59	8.8
29	3.2±1.6	0.65±0.2	1.8±0.2	0.97	6.3
	(0.40)				
16	3.8±1.8	0.53±0.16	1.45±0.15	0.88	5.5
	(0.40)				

magnitude is correct. For the lowest q, a rather good agreement is obtained for the mean shears the highest shears giving rise to undetermined value (q=5200 cm^{-1}, S=163 sec^{-1}, X_2/X_1=2.2±1.1, $X_2/X_1)_{O.K.}$=1.98±0.05).

4.4 Variation of χ_{q_X} with $|\vec{q}|$.

Following formula (3), the same kind of variation as in §3 should be found when |q| is varied from 3900 cm^{-1} to 23400 cm^{-1}. We consider the experimental data concerning the scattered intensity in the flow direction, where the influence of q is preponderant. Moreover, In this direction, the Doppler effect very efficiently filters the intensity fluctuations. Here also, the intensity data are not corrected for transmission, so that the shear variation of χ_q cannot be determined.

The (O.K.) and (O.K. exp.) formulations can be written as:

$$\chi_q^{-1} = X_1 + X_2\, q^{X_3} + X_4\, q^2 .$$

TABLE III

$S(sec^{-1})$	X_1	X_2	X_3	X_4
903	$(1.65\pm1.34)x10^{-2}$	$(3.1\pm18)x10^{-5}$	0.74 ± 0.6	$(1.1\pm42)x10^{-12}$
	$(2.85\pm2.7x10^{-3}$	$(9.9\pm0.9)x10^{-4}$	(0.40)	$(2.0\pm0.6)x10^{-12}$
510	$(1.3\pm3.6)x10^{-4}$	$(1.16\pm0.24)x10^{-3}$	0.38 ± 0.02	$(2.8\pm8)x10^{-14}$
	$(1.8\pm2)x10^{-3}$	$(9.3\pm0.6)x10^{-4}$	(0.40)	$(2.7\pm7)x10^{-14}$
288	$(1.3\pm4.3)x10^{-4}$	$(1.0\pm0.17x10^{-3}$	0.38 ± 0.018	$(2.8\pm10)x10^{-14}$
	$(1.6\pm1.4)x10^{-3}$	$(8.0\pm3.6)x10^{-4}$	(0.40)	$(2.8\pm8)x10^{-14}$
163	$(7.9\pm6.5)x10^{-3}$	$(1.6\pm3.2)x10^{-4}$	0.5 ± 0.2	$(3\pm11)x10^{-14}$
	$(3.0\pm2.5)x10^{-3}$	$(6.2\pm0.8)x10^{-4}$	(0.40)	$(1.8\pm5)x10^{-12}$
92	$(11\pm3.5)x10^{-3}$	$(3.2\pm6.6)x10^{-5}$	0.6 ± 0.2	$(3\pm17x10^{-14}$
	$(6\pm2)x10^{-3}$	$(4.2\pm0.7)x10^{-4}$	(0.40)	$(4.3\pm4)x10^{-12}$
52	$(11\pm2)x10^{-3}$	$(8\pm16)x10^{-6}$	0.75 ± 0.2	$(5\pm20)x10^{-14}$
	$(7.7\pm1.7x10^{-3}$	$(2.9\pm0.6)x10^{-4}$	(0.40)	$(5\pm3)10^{-12}$
29	$(8.6\pm26)x10^{-3}$	$(3.2\pm53)x10^{-4}$	0.35 ± 1.5	$(9.6\pm40)x10^{-12}$
	$(9.5\pm1.3)x10^{-3}$	$(1.8\pm0.4)x10^{-4}$	(0.40)	$(9\pm2)x10^{-12}$

$$(X_1/X_4)_{O.K.} = 2.28x10^{11}S^{0.14}[\Delta T(0)+1.3x10^{-4}S^{0.53}]$$

$$(X_2/X_4)_{O.K.\ exp} = 2.53x10^{11}S^{0.14}[\Delta T(0)+1.8x10^{-4}S^{0.53}]$$

$$(X_2/X_4)_{O.K.} = 6.90x10^{6}S^{0.53}$$

$$(X_2/X_4)_{O.K.\ exp} = 9.79x10^{6}S^{0.53}$$

$$X_{3(O.K)} = X_3(O.K.\ exp) = 2/5$$

The experiments were carred out at $\Delta T(0) = 3.7x10^{-3}$°K and in the flow direction where $q_X = q$. The data (Fig. 19) were fitted to the same expression, where X_i are the adjustable parameters.

Table III summarizes the results. The exponent X_3 is rather well determined for shears 510 and 288 sec^{-1}: $X_3 = 0.38\pm0.02$ to be compared to the O.K. value: 2/5.

The other shears also gave values which are compatible with this determination. But no amplitudes could be extracted, so we fixed X_3 at the value 2/5 and again fitted the results. Table IV shows that the accuracy is sufficient only for 4 values of the shear. All the results agree with (O.K.) within the experimental errors. Here also, the mean value between (O.K.) and (O.K. exp) has been reported, the higher value corresponding to O.K. exp.

Other data which are not reported here, and which concern the temperature difference $\Delta T(0) = 9.2x10^{-3}$°K, support the same general conclusions, although with a lower accuracy.

Q	$\sigma(\chi_q)$%	S(sec^{-1})
0.534	2.0	903
0.543	2.1	
0.587	4.0	510
0.568	4.0	
0.631	3.2	288
0.609	3.2	
0.758	3.6	163
0.841	3.9	
0.641	3.1	92
0.704	3.4	
0.439	2.4	52
0.520	2.7	
0.341	1.4	29
0.331	1.7	

TABLE III (continued)
Fit of χ_{qx}^{-1} to $X_1+X_2\, q_x^{X_3}+X_4\, q^2$ for $T-T_c(0)=3.7$ mK and $q=q_x$ ranging from 3900 cm^{-1} to 23400 cm^{-1}. Brackets indicate that in the fit the parameter X_3 was held constant at the quoted value.

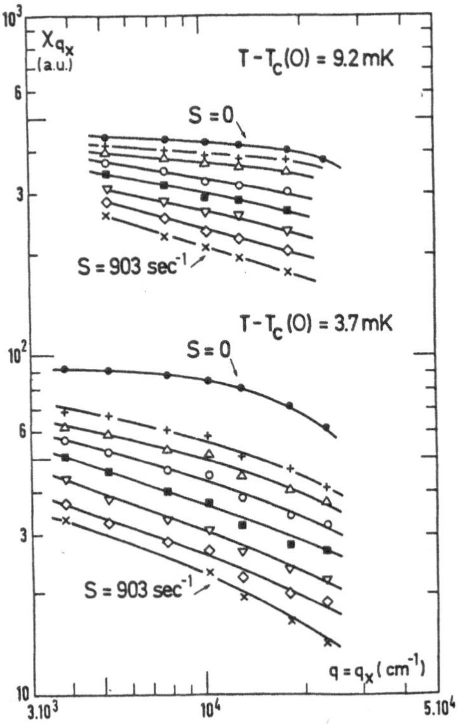

Figure 19. Scattered intensity in the flow direction, $\chi_{\vec{q}}$, versus $\vec{q}$, with $\vec{q}$ parallel to OX. Data are not corrected for the transmission. Two temperatures are reported. The analysis is made in tables III-IV. The shear rates are: S=0, 29, 52, 92, 163, 288, 510 and 903 sec^{-1}.

Table IV. Amplitude ratios from table III with X_3 imposed at the value 0.40. The values O.K. quoted are deduced from the Onuki-Kawasaki theory.

S	X_2/X_1 ($\times 10^{-2}$)	$(X_2/X_1)_{OK}$ ($\times 10^{-2}$)	X_2/X_4 ($\times 10^{-8}$)	$(X_2/X_4)_{OK}$ ($\times 10^{8}$)
903	-	-	5 ± 2	3.07 ± 0.53
92	7 ± 3.5	3.71 ± 0.70	1 ± 1	0.919 ± 0.016
52	4 ± 2	2.65 ± 0.25	0.58 ± 0.46	0.677 ± 0.0117
29	2.0 ± 0.7	2.76 ± 0.25	0.20 ± 0.09	0.347 ± 0.086

4.5 Temperature and Shear Variation χ_{q_x}

The temperature dependence of the susceptibility in the flow direction has been investigated at various shears and for two values of q (scattering angles 2° and 7°). The data are corrected for the shear dependence of the transmission (Fig. 12). Fig. 20 (small $q_1 = 5200\ cm^{-1}$) clearly shows a change of behavior in the region where $T \sim T_S$, that is χ_{q_x} varies as $[T-T_c(S)]^{-\gamma}$ with $\gamma = 1.24$ the standard value in fluids (Reference 8) when $T > T_S$, whereas it varies as $[T-T_c(S)]^{-\gamma'}$, with $\gamma' \simeq 1$ the mean-field exponent, when $T < T_S$. This is not so clear on Fig. 21 where we report χ_{q_x} for a bigger $q_2 = 18200\ cm^{-1}$, due to the standard rounding-off when $q\xi > 1$.

We fitted the data in the region $T < T_S$ to the following O.K. variation [formula (3)], where X_i are the adjustable parameters:

$$\chi_{q_x}^{-1} = X_1 [\Delta T_c(S)]^{X_2} S^{X_3} + X_4 S^{X_5} + X_6 .$$

The results of the fit are reported in tables V (q_1) and VI (q_2).

For both wavevectors it was impossible to determine X_i values when all parameters were set free. So we decided to impose the exponent X_3 which is expected to be very small from O.K. (0.14). For the lowest wavevector q_1 used, where X_6 is expected to be very small, we also tried a fit with $X_6 = 0$.

The temperature dependence exponent is seen to have the mean-field value 1.12 ± 0.05 or 1.11 ± 0.08 (q_1) and 0.984 ± 0.068 (q_2). The weak exponent is determined when $X_6 = 0$ is imposed, $X_3 = 0.14 \pm 0.05$ (q_1), in close agreement with the O.K. value. But the shear dependence exponent X_5 is found to be about twice the expected

Table V. Fit of χ_{qX}^{-1} to $X_1[T-T_c(S)]^{X_2}S^{X_3}+X_4S^{X_5}+X_6$ for $q_1 = 5200\ cm^{-1}$. Brackets indicate that in the fit the parameter was held constant at the quoted value.

Determination of:	X_1 ($\times10^{-6}$)	X_2	X_3	X_4 ($\times10^{-7}$)	X_5	X_6 ($\times10^{-9}$)	Q	$\sigma(\chi_q)$%
= all free	8 ± 2 8 ± 2	1.12 ± 0.05 1.11 ± 0.08	(0.14) 0.14 ± 0.05	7.0 ± 5.5 6.8 ± 6	1.22 ± 0.12 1.22 ± 0.13	1.6 ± 100 (0)	0.873 0.874	8.7 9
exponent X2	3.3 ± 1.7	1.3 ± 0.1	(0.14)	272 ± 56	(0.533)	1 ± 30	0.386	12
exponents X3, X5	9.7 ± 1.2	(1.0)	0.216 ± 0.003	2.6 ± 0.5	1.33 ± 0.04	1.3 ± 60	0.880	9.2
exponent X5	13.1 ± 0.3	(1.0)	(0.14)	2.6 ± 0.2	1.38 ± 0.02	1 ± 40	0.845	9.8
amplitudes	15 ± 1	(1.0)	(0.14)	107 ± 45	(0.533)	0.9 ± 20	0.243	11

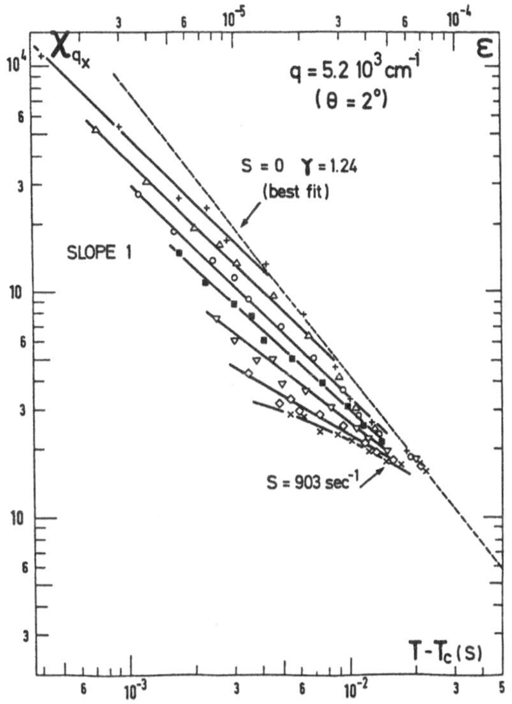

Figure 20. Scattered intensity in the flow direction, χ_{q_x}, versus $T-T_c(S)$ at various shears (see the caption of Fig. 19), for q = 5200 cm^{-1}. The dotted line is the best fit to all experimental data obtained with S=0, and corresponds to the normal behavior $\simeq (T-T_c)^{-1.24}$. These data are analyzed In table V.

value 0.533, X_5 = 1.22 ± 0.12 (q_1) and 0.84 ± 0.14 (q_2). We will discuss the amplitudes X_1, X_4, X_6 below.

We next tried to increase the accuracy of X_2 by fixing both exponents X_3 and X_5 to their O.K values. But the statistical quality of these fits was not very good, showing systematic discrepancies (q_1:Q=0.386 with σ = 12%, q_2: Q=0.569 with σ = 6%), especially for q_1. So this improvement is rather delusive and the values found: 1.3 ± 0.1 (q_1), 1.04 ± 0.04 (q_2) are not very reliable.

The next step was to fix X_2 to the mean-field value and determine both X_3 and X_5, or X_5 alone with X_3 fixed. X_3 was found to be not very far from the value predicted by O.K (0.14) : X_3=0.216 ± 0.003 (q_1) and 0.23 ± 0.05 (q_2). The X_5 exponent is again found to be larger than the O.K value (0.533) : X_5 = 1.38 ± 0.02 (q_1) and 0.84 ± 0.08 (q_2). For the amplitude ratios X_1/X_6, X_4/X_6 or X_4/X_1, we found only one value which is determined: X_4/X_1 = 3.9 ± 0.3 (q_2, exponents X_2, X_3,X_5 imposed), which is to be compared with 1.54 (O.K value) and 1.94 (O.K exp. value).

Table VI. Fit of $\chi_{q_x}^{-1}$ to $X_1[T-T(S)]^{X_2} S^{X_3} + X_4 S^{X_5} + X_6$ for $q = 18200\ cm^{-1}$. Brackets indicate that in the fit the parameter was held constant at the quoted value.

Determination of:	X_1 ($\times 10^{-5}$)	X_2	X_3	X_4 ($\times 10^{-5}$)	X_5	X_6 ($\times 10^{-4}$)	q	$\sigma(\chi_q)$%
= all free	2.3 ± 0.7	0.984 ± 0.068	(0.14)	1.2 ± 1.1	0.84 ± 0.14	2.9 ± 1.3	0.872	6
exponent X2	1.8 ± 0.3	1.04 ± 0.04	(0.14)	8.7 ± 0.4	(0.533)	$(1.5 \pm 90) 10^{-5}$	0.569	6
exponents X3, X5	1.4 ± 0.4	(1.0)	0.23 ± 0.05	1.5 ± 1.5	0.80 ± 0.15	3.5 ± 1.4	0.803	5
exponent X5	2.15 ± 0.05	(1.0)	(0.14)	1.3 ± 0.6	0.84 ± 0.08	2.9 ± 0.7	0.871	6
amplitude	2.17 ± 0.07	(1.0)	(0.14)	8.4 ± 0.25	(0.533)	$(1.4 \pm 75) 10^{-5}$	0.533	6

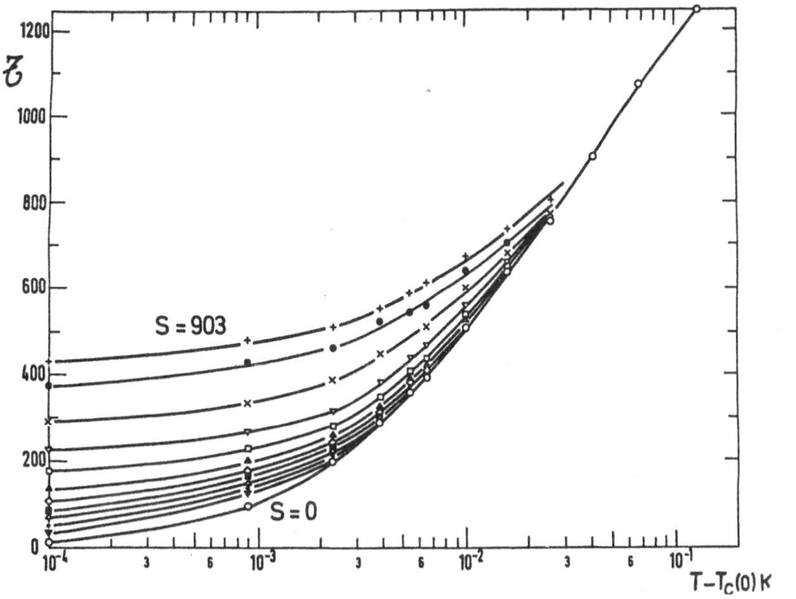

Figure 21. Scattered intensity in the flow direction χ_{q_x} versus $T-T_c(S)$ at various shears (see the caption of Fig.19) for $q = 18200\ cm^{-1}$. The dotted line corresponds to S=0, and behaves as $\simeq [\xi_0^2\ (T-T_c\ /\ T_c)^{1.24} + q^2]^{-1}$. These data are analyzed in table VI.

4.6 Reestablishment of Equilibrium

The typical timescale of the critical fluctuations in the I-W system is about 12 sec. at $T-T_c \simeq 10^{-3}$K. When the 2nd cell was used with the I-W system, it has been possible to show typical times related to the time necessary to the system for recovering its equilibrium after having been submitted to the shear, i.e., after the reduction of fluctuations by this shear. The qualitative results are the following (see also Figs. 22-24):

(i) The relaxation is exponential for both the transmitted intensity and the scattered intensity in the X direction (I_{q_x}).

(ii) The typical time associated to the transmission increases when $T-T_c$ is lowered. The order of magnitude of this time is $\tau \simeq 15$ sec. for $T-T_c \simeq 10^{-3}$K.

(iii) The typical time for the real scattered intensity I_{q_x}(i.e. recorded intensity divided by transmission) at a given $\vec{q}$ wavevector increases when $T-T_c$, or q, is lowered. For $q \simeq 15000\ cm^{-1}$, $T-T_c \simeq 10^{-3}$K, $\tau \simeq 60$ sec.

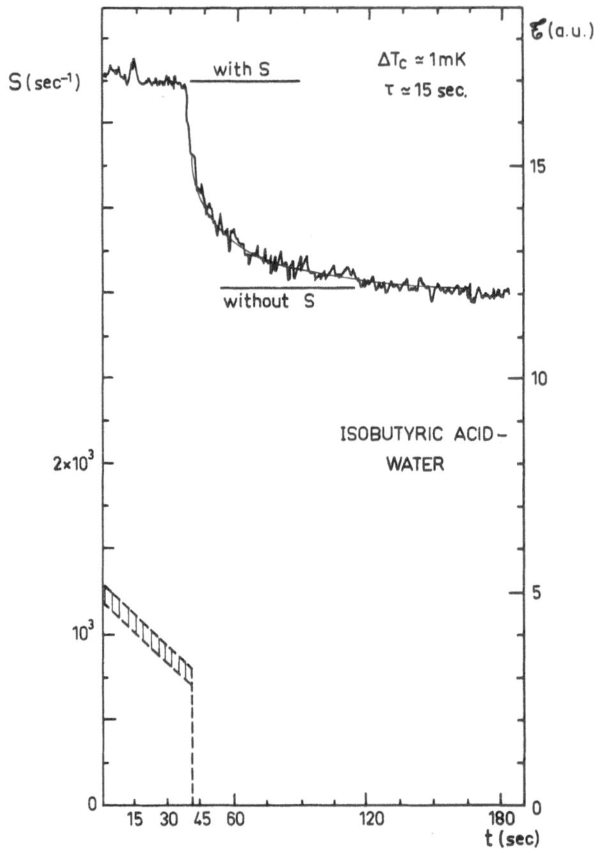

Figure 22. Transmission measurements versus time. Dashed line: Shear variation with time.

Experiments concerning these points are now in progress.

5. CONCLUSION

In table VII are summarized all our results concerning the transmission data and their relation to the T_c change, together with the scattered intensity data and their connexion to the susceptibility behavior of the system. The influence of the wavevector of the fluctuations, the direction of flow, the amplitude of shear and the distance to the critical temperature were investigated.

The qualitative agreement with the O.K theory is very good in both systems Cyclohexane-Aniline and Isobutyric acid-Water. A quan-

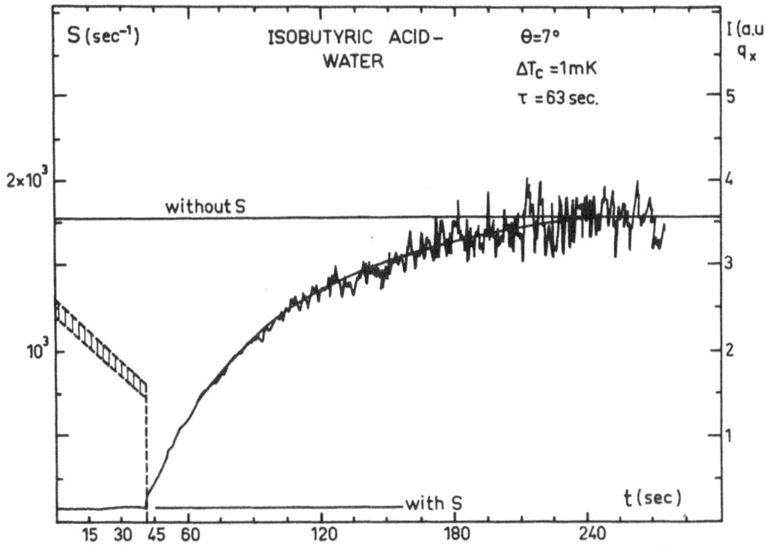

Figure 23. Scattered intensity measurements versus time. Dashed line: Shear variation with time.

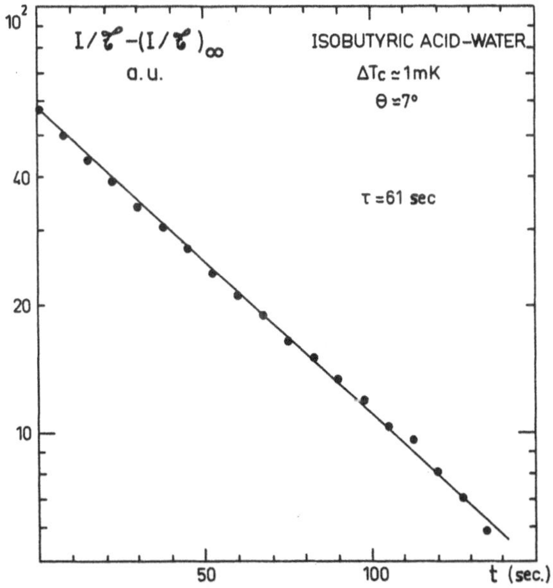

Figure 24. Ratio of the scattered intensity from Fig. 22 and of the transmission from Fig. 23 showing the exponential variation (semi-log plot).

Table VII. Principal results concerning the temperature change $\Delta T_c(S) = T_c(S) - T_c$ and the susceptibility $\chi_{\vec{q}}^{-1} = AS^{\sigma_1}[\Delta T_c(S)] + BS^{\sigma_2} q_x + q^2$ (see text). Brackets indicate the 0.K values. (See following page for letter reference explanations.)

Critical temp. change

$\Delta T_c(S) = -T_o \cdot S^{\sigma_o}$

exponent $\sigma_o(0.529)$	amplitude T_o (1.28×10^{-4}) (1.80×10^{-4})
0.53 ± 0.03 [a]	$(1.8 \pm 0.2)10^{-4}$ [a]

CYCLOHEXANE-ANILINE SYSTEM

Anisotropy term $B\,S^{\sigma_2} \cdot q_x^{\omega}$			Temperature term $A\,S^{\sigma_1}[\Delta T_c(S)]^{\gamma}$		
exponents		amplitude	exponents		amplitude
$\sigma_2(0.533)$	ω (0.400)	$B(6.85 \times 10^6)$	$\sigma_1(0.137)$	$\gamma(1.00)$	A/B (1.10×10^7)
1.22 ± 0.12 [j]	0.36 ± 0.05 [b]	$(13 \pm 5) \times 10^6$ [i]	0.14 ± 0.05 [l]	1.12 ± 0.05 [n]	$(0.40 \pm 0.03) \times 10^7$ [p]
0.84 ± 0.14 [k]	0.30 ± 0.05 [c]		0.23 ± 0.05 [m]	0.984 ± 0.068 [o]	
	0.52 ± 0.12 [d]				
	0.52 ± 0.14 [e]				
	0.64 ± 0.18 [f]				
	0.38 ± 0.02 [g]				
	0.38 ± 0.018 [h]				

Table VII. Letter Reference Explanations

a) from transmission measurements of § III-1.
b) from table I, $S=903$ sec^{-1}, $q=5200$ cm^{-1}.
c) from table I, $S=510$ sec^{-1}, $q=5200$ cm^{-1}.
d) from table II, $S=903$ sec^{-1}, $q=26000$ cm^{-1}.
e) from table II, $S=510$ sec^{-1}, $q=26000$ cm^{-1}.
f) from table II, $S=288$ sec^{-1}, $q=26000$ cm^{-1}.
g) from table III, $S=510$ sec^{-1}
h) from table III, $S=288$ sec^{-1}
i) from table IV, $S=903$; $\omega = 0.40$ imposed.
j) from table V, all free.
k) from table VI, all free.
l) from table V, all free but $X_6=0$
m) from table VI, X_3,X_5 imposed.
n) from table V, all free.
o) from table VI, all free.
p) from table VI, X_2,X_3,X_5 imposed.

tiative comparison indicates that the O.K theory describes the data well within 20-30%, the discrepancy being due to experiments, or theory, or both.

We did not investigate in detail the susceptibility in the shear direction. Experiments are in progress concerning the Isobutyric acid + water mixture under a nearly constant shear flow. The scattered intensity in the shear direction, the reestablishment of equilibrium and the dynamics of the shear-modified fluctuations are under study.

6. APPENDIX: INSTANTANEOUS MAPPING OF THE VELOCIGY GRADIENTS

We have just seen that when a binary fluid at critical composition is submitted to a shear S, it was established both experimentally and theoretically that (i) the critical temperature T_c is lowered $T_c=T_c(S)$ and (ii) the susceptibility is greatly modified in the region where the composition fluctuations have enough time to "feel" the shear, i.e. in the region $S\tau > 1$.

In a light scattering experiment the scattered light intensity I_q at constant transfer wavevector $\vec{q}$ is proportional to the susceptibility $\chi_{\vec{q}}$. Due to the strong scattering which usually occurs near T_c, the turbidity must be taken into account. The effectively measured scattered intensity is thus $I_{eff} = \tau . I_{\vec{q}}$, where τ is the transmission of light. τ is related to the turbidity $\delta \simeq \int_{\vec{q}} I_{\vec{q}} . dq$ through $\tau = \exp(-\delta L)$ where L is the path of light in the medium. In this particular experiment, we will use the variation of I_{eff} with shear to instantaneously obtain the map of the velocity gradients in the cross-section of any pipe.

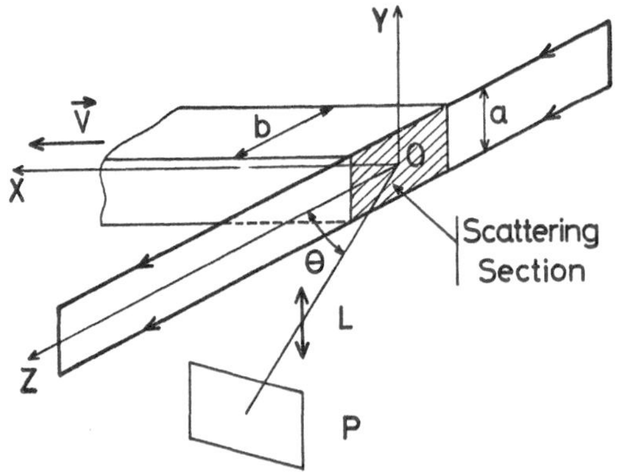

Figure 25. Experimental arrangement

The experimental set-up (Fig. 25) has been already described in § II-1, and concerns the Cyclohexane-Aniline system. A section of the pipe C(see Fig. 25) was illuminated by a thin plane He-Ne laser beam (vertical plan YZ), produced by a telescope with cylindrical lenses. A lens (L), whose principal axis in the horizontal plane (XZ) makes an angle θ (scattering angle) with the X-axis, images the scattering section onto a plane (P) where a photographic recording can be performed. We used a 400 ASA film, and with a 1/50 sec. time exposure, a good linearity between the darkness of the film and the light intensity is achieved.

The scattering angle (in the liquid) $\theta = 10^{o}$ was investigated ($q = 26000\ cm^{-1}$), with a resolution $\delta\theta/\theta \sim 5\%$. Fig. 26 shows pictures of the pipe section taken during the run, at various t. In d), where the fluid is at equilibrium, the intensity distribution is almost uniform. In c), the light is more intense in the centre of the pipe, revealing a cross-like structure. Other interesting features concern a) and b) where the steady state of the Poiseuille flow is not yet reached ($t < 5$ sec.). S-like and U-like spatial distributions can be seen, respectively in Figs. 26a) and 26b).

The microdensitometry of Fig. 26c reveals the line where the optical density is constant, i.e. where the scattered intensity and thus the shear is constant (Fig. 3). We verified that this map of shears is compatible with the classical calculation of a Poiseuille flow.[20]

We notice that the shear is nearly constant in most of the hatched region of Fig. 27 where our measurements of §III, IV, were done. Let us recall that we used a laser beam ($\lambda = 6328$ Å) of cross-section diameter $\Phi_o = 0.35$ mm, directed parallel to the Z-

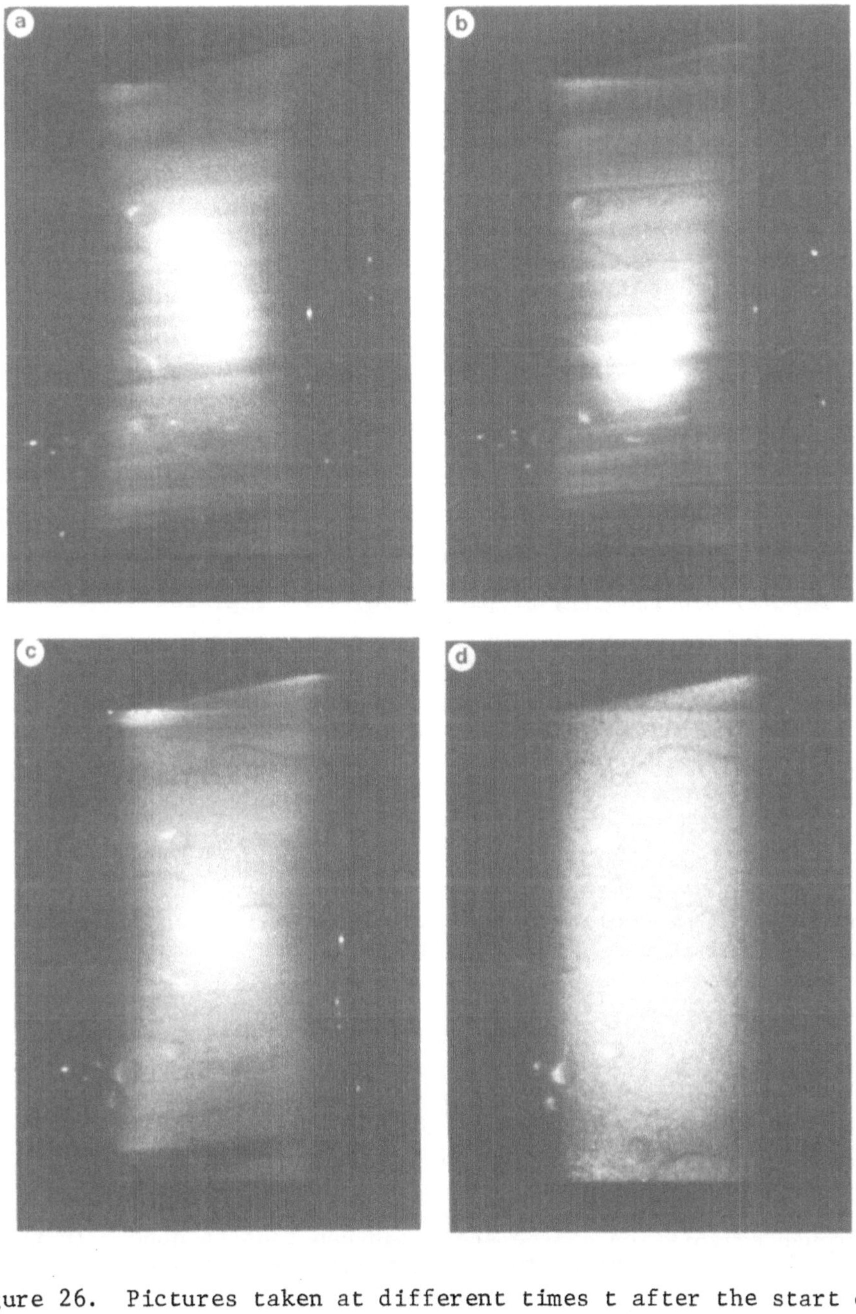

Figure 26. Pictures taken at different times t after the start of the run, at $T-T_c(0) \simeq 1.5$ mK.
a), b), t<5 sec. c) t=10 sec. d) t ≃300 sec.

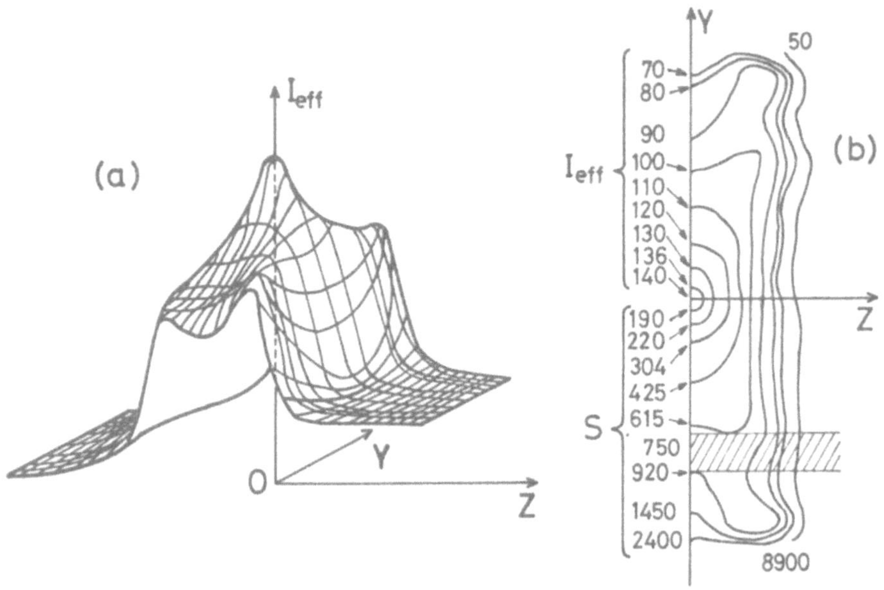

Figure 27. a) Densitometry of picture c) from Fig. 26.

b) Lines of constant intensity in picture c), from Fig. 26, i.e. map of the velocity isogradients. The units of I_{eff} are arbitrary, the units of S are sec^{-1}. The hatched region corresponds to the position of the beam in the above study (sections 3 and 4.)

axis in the XZ plane, and which entered the pipe at X=0 and Y=0.1 cm. With a lens we imaged the centre of this scattering volume on a photomultiplyer pin-hole. By a laser doppler velocimetry technique we measured at this point S_M=1600 $\pm$ 280 sec^{-1}. Therefore it is straightforward to connect the shear rate S to the brightness of the image. Fig. 28 shows the log-log plot of the measured scattered intensity I_{eff} versus S at two scattering angles θ = 9° and 11°, and two temperatures T such that $\Delta T = T-T_c(o)$ = 3.7 mK and 9.2 mK. As far as the condition $S\tau > 1$ holds (i.e. close to T_c), the experimental simple power law $I_{eff} = I_o S^{-\alpha}$ is seen to be valid. Here $\alpha = 0.26 \pm 0.02$ is a constant and I_o a coefficient which is not a function of S.

A rough calibration of the shear map can thus be done. In the hatched region of Fig. 27-b, 10 sec. after the run, the shear is S = (750 $\pm$ 130) sec^{-1}, giving $S = (I_{eff}/530)^{-3.85}$ with the units of Fig. 27-b. Let us note that the uncertainty on S is due chiefly to the extent of the beam, i.e., the Y-dimension of the hatched region. The uncertainty $\pm$ 130/750 = $\pm$ 17.3% estimated in Section 1 corresponds to that directly measured in Fig. 27-b ($\pm$ 18%).

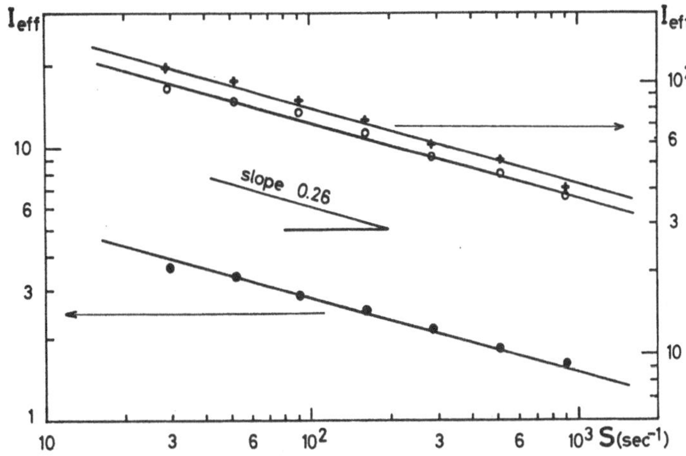

Figure 28. Variation of the experimental scattered intensity I_{eff} versus the applied shear S at 2 scattering angles $\theta_1 = 11^o$ (o: $\Delta T = 9.2$ mK) and $\theta_2 = 9^o$ (o: $\Delta T = 9.2$ mK; + :$\Delta T = 3.7$ mK).

The corrections to be considered are the following: i) the transmission τ through the liquid depends on the light path through θ. With the small angle $\theta = 10^o \pm 0.5^o$, the path vary from 0.5 cm to about $0.5/\cos\theta = 0.51$ cm. This leads to the variation $\Delta\tau / \tau < \pm 5\%$, and is negligible.

(ii) The illuminating beam is Gaussian, so corrections are necessary if precise data are required. The total corrections nevertheless do not exceed $\pm$ 5%, as we verified on pictures taken with the fluid at rest. However, the purpose of this work is not to obtain very precise values of the shears, but to show that an application of the "moustache effect" can give an instantaneous visualization and determination of the shear distribution in a flow of any geometrical shape. Of course the most interesting applications would be to the study of non stationary states, such as turbulence, etc... The limitations lie in the use of a mixture close to a critical point, i.e. in the region $S\tau > 1$. Other mixtures (including polymers), where τ is much bigger, would increase the region of interest, while requiring a much cruder temperature stabilization. Nevertheless, to our knowledge, this is the first time that such a map of shears has been recorded in real time.

ACKNOWLEDGEMENTS

I am grateful to M. Gbadamassi who contributed to all experiments presented here. It is a pleasure to thank P. Berge for fruitful discussions and A. Onuki and K. Kawasaki for having communicated their preprints and for very interesting discussions. I thank also L. Koschmieder for a critical reading of this manuscript.

REFERENCES

1. a) D. Beysens, M. Gbadamassi and L. Boyer, Phys. Rev. Lett. 43, 1253 (1979).

   b) D. Beysens and M. Gbadamassi, J. de Physique Lett. 40, L 565 (1979).

   c) D. Beysens, in "Ordering in strongly fluctuating condensed matter systems", N.A.T.O. Report (1979), Plenum Press, (N-Y), ed. by T. Riste.

   d) D. Beysens and M. Gbadamassi, Phys. Lett. 77A, 171 (1980).

   e) D. Beysens and M. Gbadamassi, Phys. Rev. A 22: 2250 (1980).

2. A. Onuki and K. Kawasaki,

   a) Annals of Physics 121, 456 (1979).

   b) Progr. Theor. Phys. Supplement 64, 436 (1978).

   c) Phys. Lett. A72, 233 (1979).

   d) K. Kawasaki and A. Onuki, in Lecture Notes in Physics 104, Dynamical Critical Phenomena and Related Topics, ed. by C. Enz (Springer, Heidelberg, 1979).

   e) A. Onuki and K. Kawasaki, Progr. Theor. Phys. 63, 122 (1980).

   f) A. Onuki, K. Miyazoe and K. Kawasaki, preprint.

3. D. Beysens, A. Bourgou and G. Zalczer, J. de Phys. Colloque C1, 37, 225 (1976).

4. P. Berge and M. Dubois, Phys. Rev. Lett. 27, 1125 (1971).

5. K. Kawasaki, Ann. Phys. (New-York) 61, 1 (1970).

6. E.D. Siggia, B.I. Halperin, and P.C. Hohenberg, Phys. Rev. B13, 2110 (1976).

7. J. Als Nielsen and R.J. Birgeneau, Am. J. Phys. 45, 554 (1977).

8. J.C. LeGuillou and J. Zinn-Justin, Phys. Lett. 39, 95 (1977).

9. J. Timmermans, "The physico-chemical constants of binary systems in concentrated solutions", vol. I, Intersec. Pub. Inc. N.Y. (1953) and ref. therein.

10. G. d'Arrigo, L. Mistura and P. Tartaglia, J. Chem. Phys. 66, 80 (1977).

11. See P. Calmettes, thesis (Paris 1978 Universite P.M. Curie), and P. Calmettes, I. Lagues and C. Laj, Phys. Rev. Lett. 28, 478 (1972), and P. Calmettes and C. Laj, Phys. Rev. Lett. 36, 1372 (1976)

12. See for instanceF D. Beysens and G. Zalczer, Opt. Comm. 26, 172 (1978).

13. See for instance, a) S.C. Greer, T.E. Block and C.M. Knobler, Phys. Rev. Lett. 34, 250 (1975) and b) M. Giglio and A. Vendramini, Phys. Rev. Lett. 35, 168 (1975)

14. S.C. Greer, Phys. Rev. A14, 1770 (1976).

15. American Institute of Physics Handbook, 2nd ed. edited by D.E. Gray, (McGraw-Hill, New York, 1957).

16. J.C. Allegra, A. Stein, and G.F. Allen, J. Chem. Phys. 55, 1716 (1971).

17. B. Chu, S.P. Lee, and W. Tscharnuter, Phys. Rev. A7, 353 (1977).

18. D. Beysens, A. Bourgou, and P. Calmettes, to be published.

19. M. Tournarie, J. Physique 30, 47 (1969).

20. See for instance: C.S. Yih, Fluid Mechanics (McGraw-Hill, New York, 1969).

FLUCTUATIONS IN NONEQUILIBRIUM CRITICAL PHENOMENA

Kyozi Kawasaki and Akira Onuki

Department of Physics
Kyushu University 33
Fukuoka, 812 Japan

CONTENTS

1. INTRODUCTION

Non-equilibrium steady states are expected to differ from equilibrium states in many important respects, which can show up in fluctuation spectra that occur in these states as are revealed for instance in light scattering experiments.[1] For instance, the absence of time-reversal symmetry in a steady state of fluid with heat flow leads to a difference in the peak heights of Stokes and anti-Stokes components of the Brillouin doublet.[1,2] However, effects of this type of dissipative perturbations on fluctuation spectra are rather small under ordinary conditions, and only very recently has experimental evidence come out.[2]

On the other hand, if the system is near the state of marginal stability such as the critical point, the effects of perturbations are dramatically enhanced. Here we describe some recent developments in this area with particular emphasis on critical fluids under incompressible flow fields.

Two features are common to this class of problems. First, we note that the dissipative perturbations under consideration such as the temperature gradient and non-uniform flow fields are macroscopic and hence have characteristic wave numbers denoted by k_c such that only those fluctuations with wave numbers k satisfying $k < k_c$ are drastically affected by dissipative perturbations. k_c can be regarded as a measure of the strength of perturbations. As long as the condition $k_c\xi \ll 1$ is satisfied where ξ is the correlation range of critical fluctuations, most critical fluctuations are very little affected by perturbations. If we approach the criticality $\xi = \infty$ with the fixed strength of a perturbation, we will eventually enter the regime $k_c\xi \gg 1$ where most critical fluctuations are strongly affected by the perturbation. It is this new regime that we will be concerned with in the following. We here list a few examples of k_c.

(1) Fluid near the liquid-gas critical point subjected to:

(a) shear flow with a constant rate of shear D

Here k_c is determined by

$$\Gamma_{k_c} = D \tag{1}$$

where

$$\Gamma_k = (k_B T_c/16\bar{\eta})k^3 \tag{2}$$

is the decay rate of the order parameter fluctuation with the wave number k for $k\xi \gg 1$, and $\bar{\eta}$ and T_c are the shear viscosity and the critical temperature, respectively.[3] Hence if we ignore the weak critical anomaly of $\bar{\eta}$, we find

$$k_c = (16\bar{\eta}D/k_B T_c)^{1/3} \tag{3}$$

(b) heat flow Q

Using

$$|Q| = \lambda|\nabla T| \sim \lambda\delta T k_c \sim \lambda\delta S k_c/C_p$$

where λ, C_p, δT and δS are the thermal conductivity, the specific heat at constant pressure, the sizes of the temperature and entropy fluctuations, respectively, and $\lambda \sim k_c^{-x_\lambda}$, $C_p \sim k_c^{-\gamma/\nu}$, $\delta S \sim k_c^{\beta/\nu}$, x_λ being the critical exponent for the thermal conductivity, we find

$$k_c \propto |Q|^{Y_Q} \tag{4}$$

with

$$Y_Q = \left(\frac{\beta + \gamma}{\nu} + 1 - x_\lambda \right)^{-1} \cong \frac{2}{5} \tag{5}$$

(2) Fluid mixture near the critical consolute point subjected to a temperature gradient*

$$k_c^{-1} \sim (k_c^{-1}|\nabla T|)^{-\nu} \quad \text{or} \quad k_c \sim |\nabla T|^{\nu/(1+\nu)} \tag{6}$$

The second feature that characterizes the present problem is the highly singular manner in which the strengths of dissipative perturbations enter the problem. This arises from the following circumstances: the dissipative perturbations generally suppress critical fluctuations with $k < k_c$. This affects fluctuation contributions to physical quantities such that in some cases k_c enters as a small wave number cut-off instead of ξ^{-1}. Thus the highly singular manner in which ξ^{-1} enters a physical quantity now shows up as a singularity in k_c, and hence, in the strengths of dissipative perturbations.

2. CRITICAL FLUIDS UNDER INCOMPRESSIBLE TWO_DIMENSIONAL FLOW FIELDS[4]

General incompressible two-dimensional flow field u(r) takes the form

$$\underset{\sim}{u}(\underset{\sim}{r}) = (S + A)y\underset{\sim}{e}_x + (S - A)x\underset{\sim}{e}_y \tag{7}$$

where $\underset{\sim}{e}_x$ and $\underset{\sim}{e}_y$ are unit vectors along the x and y axes, respectively, and S should not be confused with the entropy. The three qualitatively different cases can be distinguished: rotational flow ($|A| > |S|$), shear flow ($|A| = |S|$) and elongational flow ($|A| < |S|$), and will be discusses separately. Here, however, we are limited to the mean field theory.

(i) Rotational flow[4]

Here fluid elements move along the stream lines which form elliptic orbits with the angular frequency $\Omega_r \equiv |A|\ (1 - \sigma^2)^{1/2}$ with $\sigma \equiv |S/A|$. Critical fluctuations experience corresponding periodic distortions during their life times. This means that there are no secular effects of flow on critical fluctuations. Indeed, for $\Gamma_k \ll \Omega_r$, we obtain the following result for the variance $\underset{\sim}{\chi}_k$ (or the susceptibility) of the order parameter fluctuations:

$$\underset{\sim}{\chi}_k = (\tau + \ell^2 + \frac{1}{2}\sigma^2\ell^4/\ell^2)^{-1} \tag{8}$$

where $\tau = \xi^{-2}$ is a quantity proportional to $T - T_c$ and the vector ℓ is given by $\ell_x = (1 - \sigma)^{-1/2}k_x$, $\ell_y = (1 + \sigma)^{-1/2}k_y$; $\ell_z = k_z$ and $\ell_\perp^2 = \ell_x^2 + \ell_y^2$. Equation (8) is basically the Ornstein-Zernike type except for an anisotropy. Thus we expect that the critical behavior is essentially the same as for fluids at rest.

*Here, however, the Soret effect was ignored.

(ii) Shear flow[5]

A shear flow continuously distorts critical fluctuations from a spherical shpae into a cigar-like shape. Here we have

$$k_c = (D/\lambda)^{1/4} \tag{9}$$

where λ is the thermal conductivity for the case of liquid-gas critical point and $D = 2S$. For $\tau \lesssim k_c^2$ and $k \lesssim k_c$ we have a simple interpolation formule for the mean field variance:

$$\chi_k = (\tau + k_c^{8/5}|k_x|^{2/5} + k^2)^{-1} \tag{10}$$

The second term in the denominator acts to suppress fluctuations with $k_x \neq 0$ such that the critical dimensionality is lowered from 4 to 2.4. In particular, for three dimensions the critical behavior is described by the mean field exponents as far as the temperature dependence is concerned.

Suppression of fluctuations by a shear flow can be understood by considering a fluctuation of plane wave type with the wave vector k at time zero. As the time goes on the plane wave gets distorted by the shear flow such that the wave vector changes in time as[5]

$$\underset{\sim}{k}(t) = \underset{\sim}{k} - Dtk_x \underset{\sim}{e}_y \tag{11}$$

For $k_x \neq 0$ the second term of Equation 11 dominates at long times and $|\underset{\sim}{k}(t)|$ increases indefinitely. This, in turn, results in a rapid decay of the fluctuation, and hence suppression of fluctuations with $k_x \neq 0$ as in Equation 10. Another intuitive (and less precise) interpretation is that for an observer at rest situated at a position off the x-axis, fluctuations tend to be smoothed out by a flow along the x-direction.

(iii) Elongational flow[4]

Here the suppression of fluctuations is more drastic than the case with pure shear flow. For instance, for

$$\tau \ll k_e^2 \equiv (S^2 - A^2)^{1/4}\lambda^{-1/2},$$

we find

$$\chi_k \simeq \begin{cases} \frac{\pi}{2} k_e^{-2} & \text{for } (I/k_e^4)\ \ln|k_e/m_x| \ll 1 \\ (\tau + k_z^2)^{-1} & \text{for } (I/k_e^4)\ \ln|k_e/m_x| \gg 1 \end{cases} \tag{12}$$

where

$$I \equiv \tau(-2\delta m_x m_y + m_z^2) + 2m_x^2 m_y^2 + (-2\delta m_x m_y + m_z^2)^2 \tag{13}$$

with $\delta \equiv |A/S|$ and

$$m_x = \frac{1}{\sqrt{2}}\left(\frac{k_x}{\sqrt{1-\delta}} + \frac{k_y}{\sqrt{1+\delta}}\right), \quad m_y = \frac{1}{\sqrt{2}}\left(-\frac{k_x}{\sqrt{1-\delta}} + \frac{k_y}{\sqrt{1+\delta}}\right), \quad m_z = k_z \tag{14}$$

Thus, for most directions of $\hat{\underset{\sim}{k}}$, $\underset{\sim}{\chi}_k$ remains to be finite as $k \to 0$ even at criticality. We conjecture that her the critical dimensionality will be 2 as with fluids not at the criticality.

3. CRITICAL FLUIDS UNDER SHEAR FLOW

This section is devoted to a special case of incompressible flows treated in Section 2, namely, the case of shear flow where detailed theoretical[3-6] and experimental[7] investigations are available. Since these works have been reported elsewhere already, we will be rather brief.

We start by describing what was found in the light scattering experiments carried out by Beysens and his coworkers.[7] They found the following remarkable features that characterize the critical fluid mixture (aniline-cyclohexance mixture) when subjected to a shear flow with the rate of shear D

1. Lowering of T_c with increasing D as given by

$$T_c(D) - T_c(o) = -1.8 \times 10^{-4} D^{0.53} \qquad (D \text{ in cm}^{-1}) \tag{15}$$

2. Anisotropy in the variance $\underset{\sim}{\chi}_k$ which is more pronounced as D increases.
3. Reduction of turbidity with increasing D.

While this experimental work was in progress, this problem was being studied theoretically.[4-6] This theory which goes beyond the mean field theory will now be briefly described. Since we are dealing with critical fluids where only long wavelength fluctuations are relevant, we can start with the continuum model. The model can be expressed in the form of a Fokker-Planck equation for the distribution functional of the gross variables - here the local order parameter fluctuation $s(\vec{r})$ and the local velocity fluctuation $v(\vec{r})$ - as

$$\frac{\partial}{\partial t} P(\{s\},\{\vec{v}\};t) = L(\{s\},\{\vec{v}\})P(\{s\},\{\vec{v}\};t) \tag{16}$$

L is the Fokker-Planck operator given by

$$L = L^{(o)} + L^{(1)} \tag{17}$$

where

$$L^{(o)} = \int d\vec{r}\{\frac{\delta}{\delta s}(-\lambda_o \nabla^2)(\frac{\delta}{\delta s} + \frac{\delta\Phi o}{\delta s}) + \frac{\delta}{\delta\vec{v}}(-\bar{\eta}_o\nabla^2)(\frac{\delta}{\delta\vec{v}} + \vec{v})$$

$$+ \rho_o\left[\frac{\delta}{\delta s}\nabla\cdot(s\vec{v})\cdot + \frac{\delta}{\delta\vec{v}}\cdot(s\vec{v}\frac{\delta\Phi o}{\delta s})_\perp\right]\} \tag{18}$$

$$L^{(1)} = D\int d\vec{r}\,\frac{\delta}{\delta s}\,y\,\frac{\partial s}{\partial x} \tag{19}$$

and

$$\Phi_o = \int d\vec{r}\left[\frac{1}{2}(\tau + \tau_{oc})s^2 + \frac{1}{2}(\nabla s)^2 + \frac{1}{4!}u_o\Lambda^\varepsilon s^4\right] \tag{20}$$

Here τ is a quantity proportional to $T - T_c$, τ_{oc} is a constant, Λ is the upper wave number cut-off, λ_o and $\bar{\eta}_o$ are the bare transport coefficients, ρ_o and u_o are the mode coupling coefficients. In the absence of $L^{(1)}$, the model Equation 16 becomes equivalent to the one used in the ordinary critical dynamics of fluid.[8] The quantities imply that the fluctuation contributions to these quantities come only from fluctuations with wave numbers greater than Λ. Thus, in fact, $\lambda_o = \lambda(\Lambda)$, $\bar{\eta}_o = \bar{\eta}(\Lambda)$, etc.

Before embarking on full treatment of this nonlinear model, it is instructive to consider a simpler model that ignores all the nonlinearities in Equation 18 which decouples order parameter fluctuations from velocity fluctuations. Thus, in this case, we only have to consider the following simplified stochastic equation:

$$\frac{\partial}{\partial t}P(\{s\};t) = L_o(\{s\})P(\{s\};t) \tag{21}$$

where

$$L_o(\{s\}) = \int d\vec{r}\,\frac{\delta}{\delta s}(-\lambda\nabla^2)\left[\frac{\delta}{\delta s} + (\tau - \nabla^2)s\right] + Dy\frac{\partial s}{\partial x}\} \tag{22}$$

The steady state variance χ_k^o can readily be found as

$$\chi_k = \left[\lambda k^2(\tau + k^2) - \frac{D}{2}k_x\frac{\partial}{\partial k_y}\right]^{-1}k^2\lambda \tag{23}$$

which has Equation 10 as a simple interpolation formula. Equation 23 or Equation 10 already show the spatial anisotropy of χ_k which characterizes the critical fluids under shear flow (the moustache effect).

The nonlinearities that have been ignored above, however, play a fundamental role if we attempt a proper quantitative treatment of critical phenomena in three dimensions. Since a naive finite order perturbation theory is wholly inadequate here, a more sophisticated theoretical approach such as the renormalization group method is required. Since this is not a place to describe such a method in detail, we only explain it as applied to our problem in a few words. The renormalization group theory amounts to studying $\lambda(\Lambda)$, $\bar{\eta}(\Lambda)$ etc. as functions of Λ by taking into account more fluctuations, that is, lowering Λ to, say, Λ/b with $b > 1$. As a result of such RG treatment, we found the following. Apart from the rescaling factors associated with the RG treatment, both $u(\Lambda)$ and $f(\Lambda) \equiv \rho(\Lambda)^2/\lambda(\Lambda)$ $\bar{\eta}(\Lambda)\Lambda^{\varepsilon}$ approach the fixed point values u^* and f^* as Λ approaches k_c where $\varepsilon = 4 - d$, d being the dimensionality of space. For $0 < \varepsilon << 1$, these fixed point values are given by

$$u^* = \frac{16\pi^2}{3}\varepsilon, \quad f^* = \frac{24}{19} \cdot 8\pi^2\varepsilon \tag{24}$$

Then, the effects of nonlinear mode coupling in Equations 16-20 can be treated by a simple perturbation theory by replacing the mode coupling coefficients by their fixed point values in the spirit of the Feynman graph theory of Wilson.[9] In this way we obtain up to $O(\varepsilon)$ the following result for the variance χ_k for sufficiently small k and τ and $k_c\xi >> 1$;

$$\chi_{\vec{k}}(\tau) = \{k^2\lambda_\infty(\hat{k})[\bar{r}_\infty(\hat{k}) + (1+\varepsilon A_w)k^2] - \frac{D}{2}k_x \frac{\delta}{\delta k_y}\}^{-1}\vec{k}^2\lambda_\infty(\hat{k}) \tag{25}$$

with

$$\lambda_\infty(\hat{k}) = \lambda_o(k_c/\Lambda)^{-\frac{18}{19}\varepsilon}(1 + \varepsilon A_\lambda) \tag{26}$$

and

$$F_\infty(\hat{k}) = \tau(k_c/\Lambda)^{\frac{\varepsilon}{3}}(1 + \varepsilon A_T) + \varepsilon k_c^2 A_s \tag{27}$$

where the A's are the specifically nonequilibrium "amplitude corrections" which still depend on $\hat{k}$ and their explicit expressions are given elsewhere.

Let us now fix the length scale so that the equilibrium variance for k = 0 is given

$$\chi_k(\tau,\ D = 0) \quad = \quad \xi^2 \quad = \quad \xi_0{}^2 \left(\frac{T}{T_c} - 1\right)^{-2\nu} \tag{28}$$

where ξ_0 is the critical amplitude of ξ and is already known experimentally, and Fisher's small critical exponent η was ignored. Then, noting that $\varepsilon/3$ in Equation 27 is just the first term in the ε-expansion of $(\gamma - 1)/\nu$, we obtain in Equation 27

$$\tau(k_c/\Lambda)^{\varepsilon/3} \quad = \quad \tau(k_c/\Lambda)^{(\gamma - 1)/\nu}$$

$$= \quad \xi_0{}^{-2} \cdot (k_c\xi_0)^{(\gamma - 1)/\nu} \cdot \left(\frac{T}{T_c} - 1\right) \tag{29}$$

In equilibrium k_c is replaced by ξ^{-1} and Equation 29 reduces to $\xi_0{}^{-2} \cdot \left(\frac{T}{T_c} - 1\right)^{\gamma}$ which is just the inverse of the equilibrium susceptibility Equation 28 since $\gamma = 2\nu$ here. In equilibrium $\lambda_o(k_c/\Lambda)^{-\frac{18}{19}}$ reduces to $\lambda_o(\xi\Lambda)^{\frac{18}{19}\varepsilon}$ which is just the concentration conductivity (or the thermal conductivity for the liquid-gas critical fluid), and is already known experimentally. Thus, in comparing Equation 25 and its consequences with Beysens' experiments, no adjustable parameter enters. Before discussing this comparison we note that Equation 25 has the following scaling form with the two length scales ξ_{11} and $\xi_\perp$:

$$\chi_k(\tau) \quad = \quad \xi_\perp{}^2 F(k_x\xi_{11},\ k_y\xi_\perp,\ k_z\xi_\perp) \tag{30}$$

where

$$\xi_\perp \quad \equiv \quad \xi_0(k_c\xi_0)^{(1 - \gamma)/2\nu}\{[T - T_c(D)]/T_c(0)\}^{-1/2} \tag{31}$$

$$\xi_{11} \quad \equiv \quad \xi_\perp \cdot (k_c\xi_\perp)^4 \tag{32}$$

The ratio

$$\xi_{11}/\xi_\perp \quad = \quad (k_c\xi_\perp)^4 \quad = \quad (k_c\xi_0)^{2/\nu}\{(T - T_c(D))/T_c(0)\}^{-2} >> 1$$

gives the ratio of the two lengths characterizing a typical "cigar".

We now compare out theory with experiments for the critical fluid mixture of aniline and cyclohexane. The theoretical parameters used are those for this system.

The first prediction we can make is the lowering of T_c. Since inspection of Equations 25-27 reveals that the most dangerous fluctuations for small k are those with $\hat{k} = e_z$, T_c is determined by requiring $\bar{\gamma}_\infty(\vec{e}_z) = 0$. This yields*

$$T_c(D) - T_c(0) = -1.3 \times 10^{-4} D^{1/3\nu} \tag{33}$$

which is favorably compared with the experimental value Equation 15 where $1/3\nu = 0.53$ if we use the accepted value 0.63 for ν.

Next, we consider the anisotropy of χ_k as a function of ϕ at fixed values of θ as defined in Figure 1. The theoretical and experimental results for $\chi_k(D)/\chi_k(D = 0)$ are displayed in Figure 2 for $\theta = 2^\circ$. Agreements are satisfactory especially for large D.

Figure 3 displays the experimental and theoretical results of $\chi_k = \chi_{k_x}$ when $k_y = k_z = 0$ as a function of k_x at different values of D for a fixed temperature. They clearly demonstrate the power law $\chi_{k_x} \propto |k_x|^{-2/5}$ in a certain domain of k_x and D.

Finally, Figure 4 shows the theoretical and experimental results of χ_{k_x} as a function of temperature for a fixed value of k_x. Whereas the experimental χ_{k_x} shows wide regions in which χ_{k_x} behaves as $[T - T_c(D)]^{-1}$ in accordance with the mean field critical exponent $\gamma = 1$, such a behavior is not visible in the theoretical χ_{k_x}. This is the only qualitative discrepancy between the theory and experiments.

Very recently preliminary results of similar experiments for isobutyric acid and water solution have been reported[7], which show

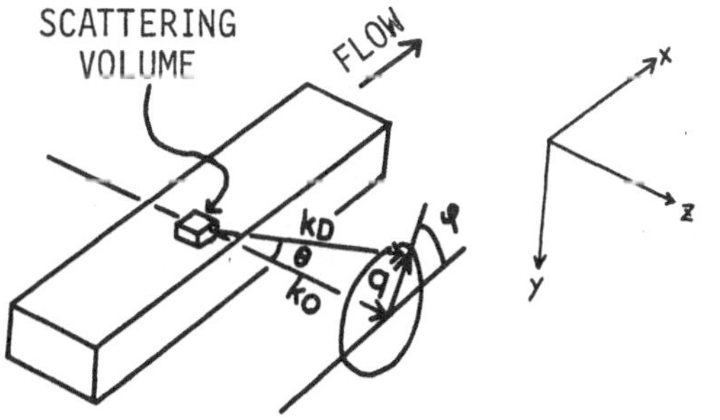

Figure 1. Experimental set up for measuring χ_k.

*In order to properly satisfy scaling, k_c^2 in the second term of Equation 27 should be replaced by $k_c^2 \cdot (k_c \xi_0)^{\gamma/\nu - 2}$.

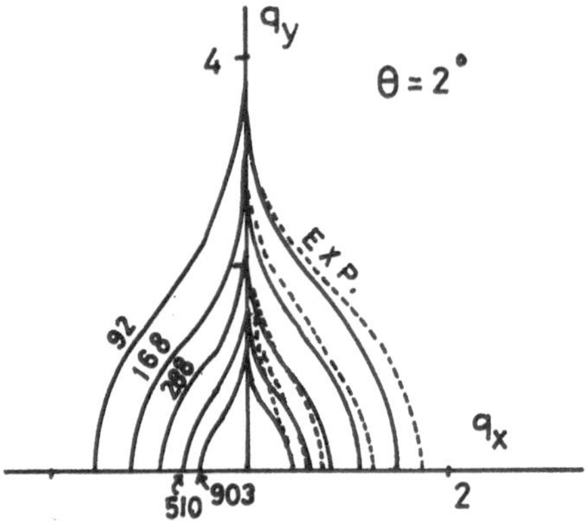

Figure 2. The normalized susceptibilities as a function of ϕ at different values of D in sec^{-1}. Solid and dashed lines represent theoretical and experimental results, respectively.

qualitatively the same behavior as aniline-cyclohexane mixture. However, there are more uncertain features such as the determination of T_c under shear flow, which require further experimental and theoretical investigations.

In summary, the theory presented here successfully accounts for main features of experimental findings except for the temperature dependence. The theory needs to be extended to cover the region of intermediate values of D and to higher orders in ε.

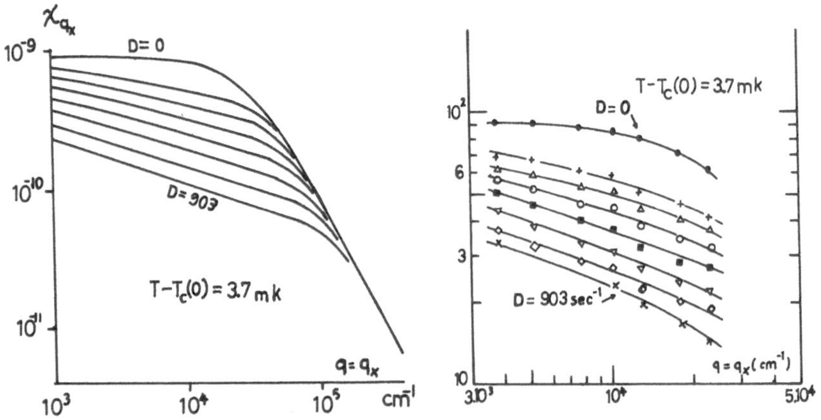

Figure 3. Theoretical (left) and experimental (right) χ_{q_x} vs q_x at various rates of shear.

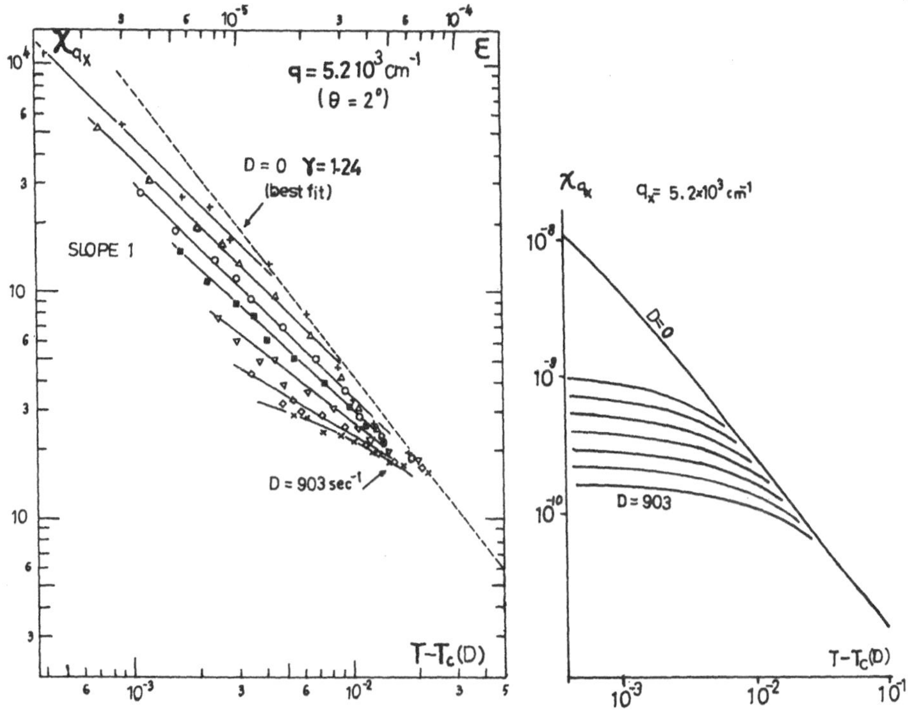

Figure 4. Experimental and theoretical χ_{q_x} vs $T - T_c(D)$ at different values of D.

4. DISCUSSION

In this article we have shown how a dissipative perturbation alters the character of ordinary critical point fluctuations for the case of two-dimensional incompressible flows. Although the mean field like behavior was recovered for the shear flow case, the resulting fixed point is different from the usual Gaussian fixed point in that there appear peculiar spatial anisotropies and high nonlinearities in the rate of shear. Such anisotropies and nonlinearities can be detected not only by light scattering but should show up in a flow birefringence[10], non-Newtonian and normal stress effects[11], and sound attenuation.

Other types of dissipative perturbations such as a heat flow are expected to reveal no less fascinating classes of phenomena, although so far very little has been reported.[12,13]

REFERENCES

1. I. Oppenheim, Lecture notes of this NATO Institute and the references quoted therein.
2. D. Beysens, Y. Garrabos, and G. Zalczer, Phys. Rev. Lett. 45: 403 (1980).
3. K. Kawasaki, Ann. Phys. (N.Y.) 61:1 (1980).
4. A. Onuki and K. Kawasaki, Progr. Theor. Phys. (Suppl.), to be published.
5. A. Onuki and K. Kawasaki, Ann. Phys. (N.Y.) 121:456 (1979).
6. A. Onuki, K. Yamazaki, and K. Kawasaki, to be published.
7. D. Beysens, Lecture notes of this NATO Institute and the references cited therein.
8. See for instance P. Hohenberg and B. I. Halperin, Rev. Mod. Phys. 49:435 (1977).
9. K. G. Wilson, Phys. Rev. Lett. 28:540 (1972).
10. A. Onuki and K. Kawasaki, Phys. Letters 78A: 354 (1980).
11. A. Onuki and K. Kawasaki, Phys. Letters, 75A:485 (1980).
12. A. Onuki, Prog. Theor. Phys. 63:1854 (1980).
13. A. Onuki, Prog. Theor. Phys. 64: 1902 (1980).

HYDRODYNAMIC INSTABILITY: STRUCTURE AND CHAOS

Jean-Pierre Boon

Faculté des Sciences
Université Libre de Bruxelles
B-1050 Bruxelles, Belgium+

and

Physique de la Matière Condensée
Université de Nice
F-06034 Nice, France

CONTENTS

1. MOTIVATION

It is about one century ago that Reynolds investigated hydrodynamic transitions in pipe and channel flows; active work in this area of fluid dynamics has been pursued continuously since. Yet turbulence which appears as the ultimate stage of hydrodynamic flow remains one of the puzzling problems of classical physics. Turbulence manifests itself by chaotic behavior; developped turbulence is characterized by complete loss of spatial and temporal correlations and by its extreme sensitivity to initial conditions. Fully developed turbulence as it appears in flows past an obstacle is

---

+Permanent address

probably the most familiar example of turbulent flow;[1] it is also the most complex and least well understood case of turbulence. It involves the excitation of a very large number of degrees of freedom according to Landau's 1944 original picture of turbulence.[2]

In the last decade an alternative view of the problem has led to a new concept: turbulence with few degrees of freedom. Initiated by Lorenz[3] in a model developped in - as early as - 1963 and substantiated on a more formal basis by Ruelle and Takens[4] in 1971, it was not until the mid-seventies that these new ideas were actually cast in a realistic framework: laboratory experiments[5-9] and numerical computations[10,11] provided confirmation of the validity of the concept of turbulence with few degrees of freedom.

One of the virtues of the Lorenz model is that it provides a simple guideline for the description of regime transitions in hydrodynamic systems like circular Couette flow and Rayleigh-Bénard convection. Although the model has a limited range of application, it is quite remarkable that it can explain, at least semi-quantitatively[12], a number of features characteristic of hydrodynamic transitions and of the onset of turbulence. In particular the Lorenz model is appropriate to describe the dynamics of systems with few degrees of freedom. Systems that belong to this class have been studied experimentally by light scattering techniques[13] and optical methods.[14] Exphasis will be put here on the study of the Rayleigh-Bénard problem and the analysis of the experimental results via the Lorenz model.

Among hydrodynamic instabilities that arise in fluid systems far from equilibrium, one of the simplest examples is the Rayleigh-Bénard problem where a fluid layer undergoes successive regime transitions under the influence of a vertical temperature gradient. For a system heated from below, convection sets in when the temperature gradient reaches a critical value; a macroscopic structure appearing as a convective pattern (rolls or cells) is formed. Close to this regime transition, subcritical hydrodynamic fluctuations are amplified and the transition point bears striking analogy with an equilibrium phase transition. For instance critical exponents of classical mean field theory have their analog in the linear theory of the first hydrodynamic regime transition. Further transitions occur at higher values of the temperature gradient where the system may exhibit periodic oscillations before turbulent behavior is observed. Similar effects are found in a rotating fluid, that is in circular Couette flow which is another thoroughly studied case of hydrodynamic instability. Here the fluid is contained between two co-axial cylinders and a sequence of flow regimes occurs with the increase of the angular velocity of the inner rotating cylinder, the outer one being at rest. The first transition is characterized by the appearance of horizontal toroidal vortices. At larger values of the angular velocity this vortex flow becomes time dependent: transverse waves superimpose on the horizontal vortices which eventually exhibit chaotic behavior when the rotating velocity is further increased.

The opposite situation can be realized by reversing the direction of the external force which has then a stabilizing effect on the system. For instance when, in the Rayleigh-Bénard problem, the fluid layer is heated from above no instability is expected. Nevertheless a transition occurs for a characteristic value of the temperature gradient beyond which thermoconvective waves propagage through the layer.

The hydrodynamic regime depends primarily on the value of a characteristic parameter, the Reynolds number. The term Reynolds number is often used in a generic sense to define the dimensionless measure of the strength of the external constraint, i.e., the temperature gradient in the Rayleigh-Bénard problem of thermal convection or the angular velocity in the Rayleigh-Taylor instability in circular Couette flow. In both cases stationary flow structures persist for low Reynolds numbers beyond a critical value; such structures belong to the class of "dissipative structures" which are found in a broad range of non-equilibrium systems.[15] As the Reynolds number is increased further, a second critical value will be reached beyond which the flow becomes time-dependent. Fluids will then often exhibit a sequence of distinct flow regimes leading ultimately to chaotic or turbulent behavior. A typical sequence is illustrated in Table 1.

The non-equilibrium transitions outlined above can be studied by hydrodynamic fluctuation theory provided the departure of the system from a known reference state does not involve high nonlinearity, which is the case when turbulent behavior becomes prominent. The latter situation can be described, at least qualitatively, on the basis of the Lorenz model. This is the guideline adopted here to present a simple analysis of typical sequences of hydrodynamics transitions as they have been observed in the laboratory by various optical methods: light scattering spectroscopy and laser Doppler velocimetry for the measure of the velocity field; differential interferometry for the mapping of local temperature gradients; forced Rayleigh scattering for the study of subcritical and propagating thermal modes. Some of these methods are described elsewhere in the present volume.[14]

We shall first deal with the analysis of the hydrodynamic system before a dissipative structure sets in, i.e., below the critical value R_c of the Reynolds number in which domain linearized hydrodynamic fluctuation theory is applicable. The second part is devoted to the study of the region above R_c where the system evolves from a time-independent dissipative structure to chaos via various routes to turbulence.

Table 1. Typical sequence of successive hydrodynamic regime transitions in convective and Couette flow

REYNOLDS NUMBER	VELOCITY FIELD	POWER SPECTRUM	HYDRODYNAMIC REGIME	PHASE DIAGRAM
$\|R\| > \|R^*\|$	---	---	thermoconvective	
$R = 0$	---	---	equilibrium	fixed point a the origin
$0 < R < R_c$	---	---	subcritical	
$R_c < R < R_1$	constant in time		stationary	fixed point
$R_1 < R < R_2$	periodic in time	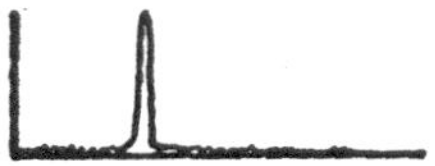	periodic	limit cycle
$R_2 < R < R_T$	biperiodic in time	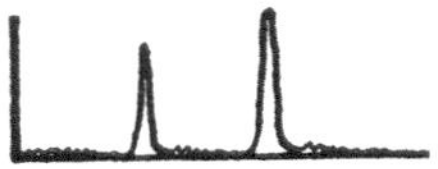	quasiperiodic	3 d-torus
$R_T < R$	chaotic	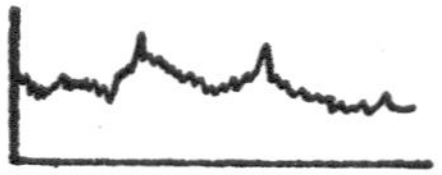	turbulent	strange attractor

2. THE RAYLEIGH-BENARD PROBLEM BELOW R_c

A. The Basic Equations

When a horizontal fluid layer, with thickness d, experiences a vertical linear temperature gradient

$$\nabla T^s = -\beta\hat{n}, \qquad \hat{n} = (0,0,1) \tag{1}$$

the steady state is characterized by the equations

$$T^s = T_o - \beta z, \qquad 0 \leq z \leq d$$

$$\nabla p^s = -\rho^s g\,\hat{n} \tag{2}$$

where g is the gravitational constant; the superscript s refers to the steady state and the subscript o denotes the reference state (z = 0). Thus the system is subject to the adverse effects of two constraints: the gravitational force and the temperature gradient whose combination constitutes the external force (Figure 1a) that has a destabilizing effect. Indeed when a fluid layer is heated from below, the system becomes topheavy and potentially unstable (provided that the fluid has a positive thermal expansion coefficient α). Therefore the fluid has a tendency to redistribute itself. This natural tendency is inhibited, however, by the viscosity of the fluid and its thermal conductivity. As a consequence the temperature gradient β must exceed a certain value β_c before the instability can manifest itself. When the temperature gradient reaches this critical value (which is a function of the properties of the fluid and of the geometry of the system) convective motion begins which permits the fluid to adjust. Most remarkable is the fact that convection arises in a peculiar way which results in an arrangement of spatially ordered convection cells or rolls, first demonstrated by Bénard in 1900[16] (Figure 1b). The temperature

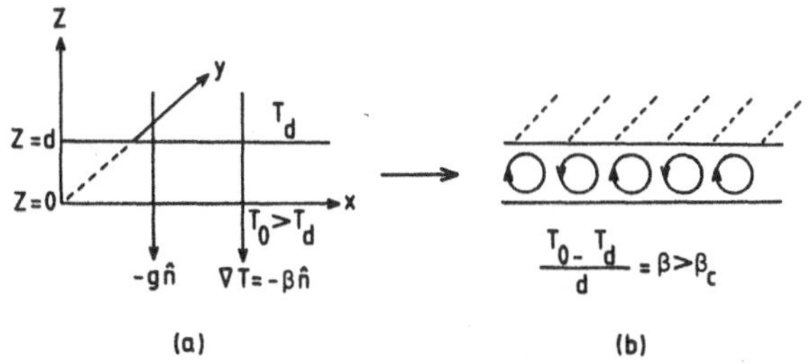

Figure 1. Schematic representation of the Rayleigh-Bénard system.

gradient is most conveniently measured by the dimensionless quantity

$$R = \frac{\alpha\beta g}{\nu\chi} d^4 \tag{3}$$

called the Rayleigh number which clearly expresses the ratio of the quantities responsible for destabilization (α, β, and g) to those seeking to relieve the effect of the imposed external force (ν the kinematic viscosity and χ the thermal diffusivity). The transition from the state of rest to the convective state is defined by the critical value R_c, which is a constant for all systems with the same boundary conditions.[17] In this section we consider those phenomena that occur below threshold ($\beta < \beta_c$ or $R < R_c$) whereas section 3 is devoted to a description of supercritical effects ($R > R_c$).

From hydrodynamic fluctuation theory, the Rayleigh-Bénard system can be described by the set of equations[18]

$$\partial_t \delta\rho + \nabla \cdot \underset{\sim}{j}_M = 0$$

$$\partial_t \underset{\sim}{v} + \nabla \cdot \underset{\approx}{P} - g\hat{n} = 0$$

$$\partial_t \delta T + \nabla \cdot \underset{\sim}{j}_T - \beta \underset{\sim}{v} \cdot \hat{n} = 0 \tag{4}$$

where $\underset{\sim}{j}_M$ is the mass current density, $\underset{\approx}{P}$ the pressure tensor, and $\underset{\sim}{j}_T$ the heat current. For a fluid with zero compressibility, i.e., $\delta\rho = -\alpha\rho^s \delta T$, and with properties that do not vary appreciably in the layer (Boussinesq conditions), the set of Equations (4) can be reduced to

$$\partial_t \nabla^2 v_z - \nu\nabla^4 v_z - \alpha g(\partial_x^2 + \partial_y^2)\delta T = 0$$

$$\partial_t \delta T - \chi\nabla^2 \delta T - \beta v_z = 0 \tag{5}$$

where v_z is the vertical component of the velocity field. When the last terms in Equations (5) are omitted ($g = \beta = 0$), the equations are decoupled and are representative of the shear mode and of the thermal diffusivity mode in a fluid at equilibrium. An important quantity here is the Prandtl number, $P = \nu/\chi$, whose value is indicative of which one of the modes is fast or slow. Thus in a fluid with high Prandtl number, say $P \sim 10$ or 100, the thermal diffusivity mode has the longest relaxation time.

The external force ($\beta \neq 0$, $g \neq 0$ in Equations (5)) induces mode coupling; therefore when increasing the strength of the external constraint, we expect a modification of the hydrodynamic modes as a manifestation of the dynamical behavior of the system, e.g. when it evolves from its equilibrium state towards the instability point ($\beta = 0 \rightarrow \beta_c$) or when the fluid layer is heated from above ($\beta < 0$).

The analysis will be simplified in the notation if we introduce reduced variables. Then, when space-Fourier transformed Equations (5) can be rewritten as

$$\dot{X} = -PX + PY$$

$$\dot{Y} = -Y + rX \qquad (6)$$

with

$$r = \frac{R}{R_c} = \frac{\alpha\beta g}{\nu\chi}\,\frac{K^2}{k^6} \qquad (7)$$

where $k^2 = K^2 + k_z^2$, $K^2 = k_x^2 + k_y^2$, and $k_z^2 = \pi^2 d^{-2}$ define the wavewavelength of the characteristic mode of the system; $X = \frac{Kv_z}{\chi k^2}$ and $Y = (r/\beta)KT$ are the dimensionless measures of the vertical component of the velocity field and of the temperature variation respectively; the reduced time variable is $\chi k^2 t$. Equations (6) have the steady state solution $X^s = Y^s = 0$ which characterizes the state of rest. The stability of the system can be investigated by perturbation analysis, that is by considering the dynamical behavior of small superposed perturbations $x(t)$ and $y(t)$ which are then governed by the Laplace transformed linearized equations obtained from Equations (6)

$$(z + P)\,\tilde{x}(z) - P\tilde{y}(z) = x(0)$$

$$(z + 1)\,\tilde{y}(z) - r\tilde{x}(z) = y(0) \qquad (8)$$

The dispersion equation constructed from Equations (8) has the solutions

$$z_{\pm} = -\tfrac{1}{2}(P + 1) \pm \tfrac{1}{2}\left[(P + 1)^2 - 4P\varepsilon\right]^{\frac{1}{2}}\ ; \quad \varepsilon = 1 - r \qquad (9)$$

which shows that at equilibrium ($\varepsilon = 1$) the two modes are decoupled, i.e.

$$z_-^o = -P, \qquad z_+^o = -1 \qquad (10)$$

or in dimensional variables ($s = z\chi k^2$)

$$s_-^o = -\nu k^2 \quad \text{(shear mode or vorticity mode)}$$

$$s_{+}^{o} = -\chi k^2 \quad \text{(thermal diffusivity mode)} \tag{11}$$

B. The Pretransitional Regime $(0 < R < R_c)$

When the system approaches the transition point $R = R_c$, Equation (9) yields

$$s_{-} \simeq -(\nu + \chi)k^2 + \frac{\chi k^2 \varepsilon}{1 + P^{-1}} \xrightarrow[\varepsilon \to 0]{} -(\nu + \chi)k^2$$

$$s_{+} \simeq -\frac{\chi k^2 \varepsilon}{1 + P^{-1}} \xrightarrow[\varepsilon \to 0]{} 0 \quad \text{(soft mode)} \tag{12}$$

as illustrated in Figure 2. The soft mode behavior is characteristic of the critical slowing down of thermal fluctuations on the approach to the transition point.[18] It expresses that the slow mode (the thermal diffusivity mode for a fluid with high Prandtl number, $P > 1$) has a relaxation time that increases like ε^{-1} when the system approaches the convection threshold.

So far we have treated ε as an independent parameter. In fact, it is easily realized from Equation (7) that ε is a quantity which depends explicitly on the wavenumber, $\varepsilon = \varepsilon(k)$. In practice one can write that close to the transition point[19]

$$\varepsilon(k) = f(\varepsilon, K - K_c) = \varepsilon[1 + \xi^2 (K - K_c)^2] \tag{13}$$

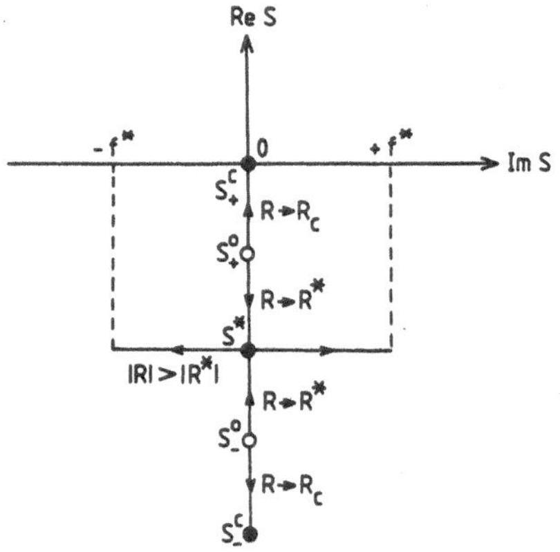

Figure 2. Schematic behavior of the thermal diffusivity mode (s_+) and of the vorticity mode (s_-) in the Rayleigh-Bénard system.

where K_c^{-1} corresponds to the wavelength ($\sim$ d) of the first mode to become unstable; then ε is the parameter that measures the distance from the critical point, i.e. $\varepsilon = (R_c - R)/R_c$ where R_c denotes the minimum value of R where the mode with wavenumber k_c becomes unstable.[17] The quantity ξ is the correlation length which growths like $\sim d\xi^{-\frac{1}{2}}$ in the vicinity of R_c. So the above results, Equations (12), are valid with $\varepsilon(k) = \varepsilon$ in so far as the mode considered has exactly the critical wavelength $2\pi K_c^{-1}$. In general, however, Equations (12) must be combined with (13).

The two major results from the above analysis are that the relaxation time of the critical mode increases like $s_+^{-1} \propto \varepsilon^{-1}$ close to the transition point and correspondingly that the correlation length growths like $\xi \propto \varepsilon^{-\frac{1}{2}}$. Such a behavior is typical of mean field theory predictions; a summary is given in Table 2 which gives clear indication of the analogy between a hydrodynamic regime transition and an equilibrium second-order phase transition.

Thus, the theory predicts that pretransitional effects should manifest themselves as precursors of the new hydrodynamic state by the fact that thermal fluctuations become long lived and long ranged close to the transition point. Because thermal fluctuations in fluid systems can be probed by light scattering spectroscopy[20] it was logical to infer that this experimental technique could be used to investigate pretransitional phenomena in the Rayleigh-Bénard instability.[21] Now because the wavelength of the critical mode is of the order of the characteristic dimension of the system (i.e. the thickness of the layer, $d \sim 1.0 - 0.1$ cm), the experiments would have to be performed at scattering angles as small as 1 mrad; reaching the corresponding required instrumental resolution cons-

Table 2. Summary of critical behaviour in the convective instability. The last column gives the critical exponents. The subcritical state ($R < R_c$; $\varepsilon = (R_c - R)\, R_c^{-1}$) is analyzed in Section 3. A discussion of the behavior above threshold ($R > R_c$; $\varepsilon = (R - R_c)\, R_c^{-1}$) is given in Section 3.

$R < R_c$	Susceptibility	$s_+^{-1} \propto \varepsilon^{-\gamma}$	$\gamma = 1$
	Correlation length	$\xi \propto \varepsilon^{-\nu}$	$\nu = \frac{1}{2}$
$R > R_c$	Order parameter Velocity field	$v \propto \varepsilon^{\beta}$	$\beta = \frac{1}{2}$

titutes a real tour de force.[22] The difficulty can be circumvented however by using the method of forced Rayleigh scattering,[23] where a superposed heat pulse with a temperature pattern that has the appropriate wavelength enhances the amplitude of the spontaneous thermal fluctuations. Their spectrum can then be more easily detected and yields a measure of their characteristic relaxation time. Experimental evidence was so obtained of the critical behavior of the thermal mode as predicted by the theory.[24] This is shown in Figure 3 where the "effective" thermal diffusivity $\chi_{eff} = s_+ k^{-2}$ is plotted as a function of the temperature difference; the expected power law $s_+ \propto \varepsilon$ is well observed.

Simultaneously with the growth of the characteristic relaxation time of critical thermal fluctuations ($\tau_+ = s_+^{-1} \propto \varepsilon^{-1}$) their correlations length increases as $\xi = \Lambda_c \varepsilon^{-\frac{1}{2}}$ where Λ_c is the wavelength of the critical mode ($\Lambda_c \sim d$). Experiments were performed to measure the correlation length by induced pretransitional Rayleigh-Bénard convection,[25] that is by measuring by laser Doppler velocimetry[14] the velocity field induced in that part of the fluid layer that was maintained in the subcritical region ($R < R_c$) while the other part of the layer is above threshold ($R > R_c$). The convective motion in the supercritical part of the cell then induces subcritical rolls which are exponentially damped in space as shown schematically in Figure 4a. Measurements of the spatial decay rate as a function of the temperature gradient in the subcritical region yields data that are plotted as $\xi = f(\varepsilon)$ in Figure 4b. One observes that the predicted power law $\xi \propto \varepsilon^{-\frac{1}{2}}$ is well verified.[26]

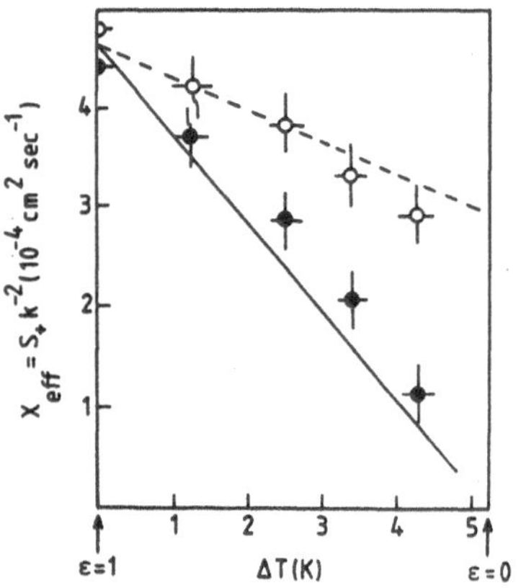

Figure 3. Critical slowing down in ethanol. The full line represents the behavior of the critical thermal mode with wavelength $2\pi k_c^{-1}$. The black dots are the data obtained with a grid with wavelength $\Lambda = 0.30$ cm $\simeq 2\pi k_c^{-1}$ whereas the circles are the data corresponding to $\Lambda = 0.15$ cm. (After Allain, Cummins, and Lallemand, 1978).

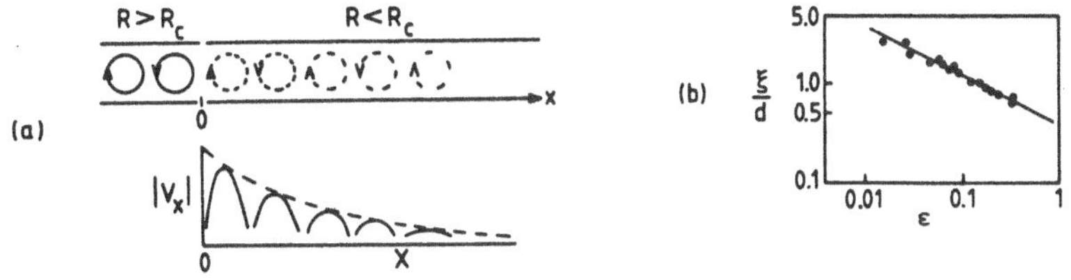

Figure 4. (a) Illustration of the experimental measurement of the correlation length in silicone oil (ν = 0.56 stokes, d = 0.6 cm); the envelope of the amplitude of the component $|V_x|$ of the velocity is $\exp(-x/\xi)$; (b) The reduced correlation length as a function of ε; the solid line represents the theoretical prediction $\xi/d = 0.38\varepsilon^{-\frac{1}{2}}$. (After Wesfreid, Bergé and Dubois, 1979).

C. The Propagative Thermoconvective Regime ($R < R^* < 0$)

The analysis presented so far describes the evolution of the Bénard system from its equilibrium state to the convection transition point. Referring to Table 1, we have thus covered the domain where $0 \leq R \leq R_c$. Consider now the reverse situation where the fluid layer is heated from above, in which case the external force has no destabilizing effect and $R < 0$. Then the discriminant of Equation (9) vanishes when r has the value

$$r^* \equiv \frac{R^*}{R_c} = -\frac{(P-1)^2}{4P} \tag{14}$$

At $R = R^*$ the two modes are degenerate: $s_+ = s_- = s^* = -\frac{1}{2}(\nu + \chi)k^2$, and when $|R| > |R^*|$ they become complex conjugates[27] (see Figure 2). The complete solution of the dispersion equation is shown in Figure 5. More explicitely, from Equation (9), one finds for $|R| > |R^*|$

$$s_{\pm} = s^* \pm if^* \quad \text{with} \quad f^{*2} = \nu\chi k^4\varepsilon \tag{15}$$

where ε is defined as $\varepsilon = (|R| - |R^*|)|R^*|^{-1}$. The physical meaning of the result, Equation (15), is that the thermal mode becomes propagating.[28] Thus there is a threshold $R = R^*$ beyond which thermoconvective waves develop with damping constant s^* and propagating velocity $\propto \varepsilon^{\frac{1}{2}}$. (Note that the power law is the same as for the convective velocity above the transition $R = R_c$; see Section 3.) The propagating behavior of such thermal waves was observed experimentally by forced Rayleigh scattering as shown in Figure 6. Suc-

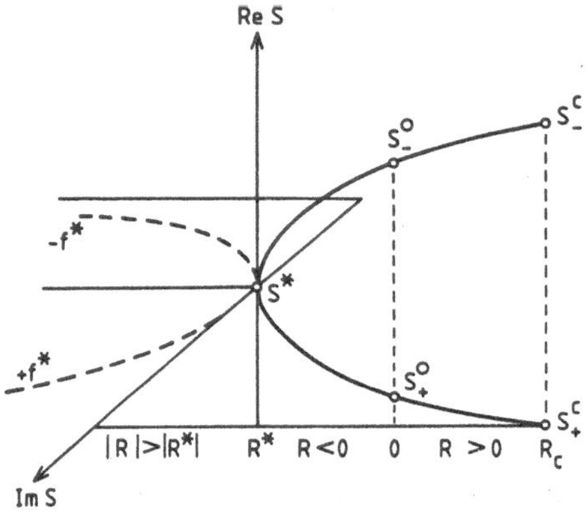

Figure 5. Representation of the behavior of the hydrodynamic modes in the region below threshold ($R \leq R_c$).

cessive experiments performed at various values of the temperature gradient yield the corresponding values of f* as a function of $|R|$. The predicted power law for the propagating velocity is compared to the experimental results in Figure 7. The quantitative discrepancy between the experimental data and the theoretical predictions could be caused by nonuniform heating by the heat pulse inherent to the forced Rayleigh scattering method.[28] The power law

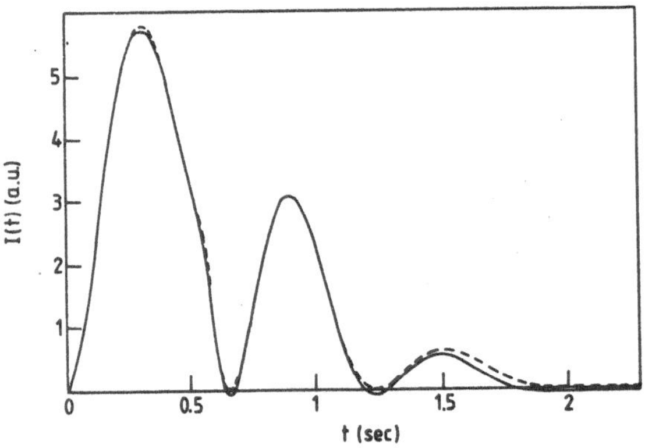

Figure 6. Typical spectrum showing the propagative behavior of the thermal mode in a fluid layer heated from above at $|R| > |R^*|$. The experiment was performed by the forced Raleigh scattering method in acetone with d = 0.29 cm, Λ = 0.3 cm and $|R| = 69 \times 10^3$. $I(t) = \text{const} + E(t)^2$ where E(t) is the scattered field. The solid curve is experimental and the dashed line is obtained by a least square fit calculation (After Boon, Allain & Lallemand, 1979).

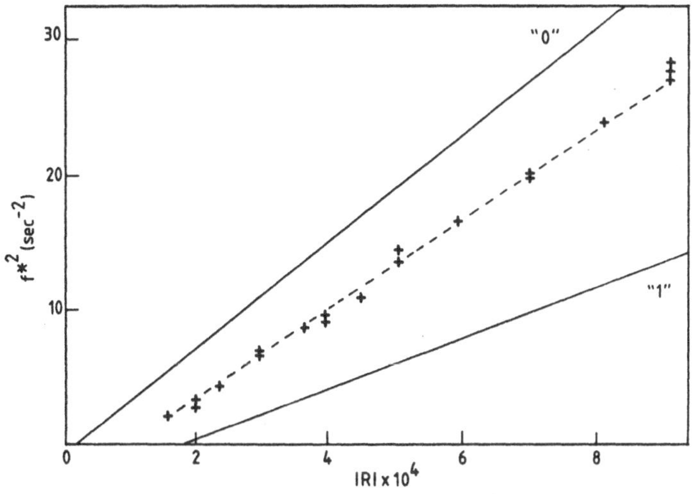

Figure 7. The squared frequency of the propagating thermal mode as a function of the Rayleigh number. The experimental data (crosses) are obtained from spectral measurements as shown in Figure 6. The solid lines represent the theoretical predictions for the first even ("0") and first odd ("I") modes (After Boon, Allain and Lallemand, 1979).

$f^{*2} \propto |R|$ predicted from Equation (15) is nevertheless qualitatively verified.

3. THE RAYLEIGH-BÉNARD PROBLEM ABOVE R_c

A. The Convection Equations

When the Rayleigh number reaches and exceeds its critical value ($r \geq 1$), convection sets in and, as a result, fluid motion distords the temperature profile, which effect must now be accounted for in the basic equations of section 2.A. By introducing a new variable that expresses the perturbation $\delta\beta$ to the linear temperature gradient β_o and considering the symetry of the system and its boundary conditions, one obtains the basic set of equations that was initially developped by Lorentz in 1963. With all quantities expressed in reduced units (see Section 2.A.) and $Z = (r/\beta_o)\delta\beta$ denoting the dimensionless measure of the distorsion of the temperature profile from linearity, the Lorentz model reads[9]

$$\begin{aligned} \dot{X} &= -PX + PY \\ \dot{Y} &= -\ Y + rX - XZ \\ \dot{Z} &= -bZ \qquad + XY \end{aligned} \tag{16}$$

where $b = 4k_z^2/k^2$. On the rhs of Equations (16), the terms in the first column describe linear mode damping (regular dissipation; see the solutions of the dispersion equation in the equilibrium state, Equation (11)), the terms in the second column express mode coupling (see the behavior of s_+ and s_- for $0 < r < 1$ in Section 2) and the terms in the third column are responsible for the nonlinear transfer of energy to the mode Z. This is most easily seen by considering the dispersion equation as obtained from the set of equations (16), when the equations are linearized and Laplace transformed, i.e.

$$\begin{vmatrix} z + P & -P & 0 \\ Z^s - r & z + 1 & X^s \\ -Y^s & -X^s & z + b \end{vmatrix} = 0 \qquad (17)$$

In the non-convective steady state ($r < 1$), $X^s = Y^s = Z^s = 0$ and Equation (17) reduces to the analysis of the second degree dispersion equation discussed in Section 2. For steady convection ($r > 1$), the steady state solutions of Equation (16) are

$$X^s = Y^s = \pm\sqrt{b(r - 1)} \quad , \quad Z^s = r - 1 \qquad (18)$$

From the former relation, one has the explicit expression for the convective velocity field which yields the power law $v_z \propto \varepsilon^{1/2}$ where here $\varepsilon = (R - R_c)R_c^{-1}$. The fluid velocity can be measured by seeding the fluid in the Bénard cell with small test particles (latex spheres with diameter $\sim 1\mu$m) and using laser Doppler anemometry[14]) or the laser speckle method[29]. Heterodyne light scattering spectroscopy has also been used to perform velocity measurements in circular Couette flow to study the Taylor instability[30] and in the convective motion of the Rayleigh-Bénard instability.[31] An illustration of the results obtained by the crossed beam technique[32] is given in Figure 8; one observes that the convective velocity grows

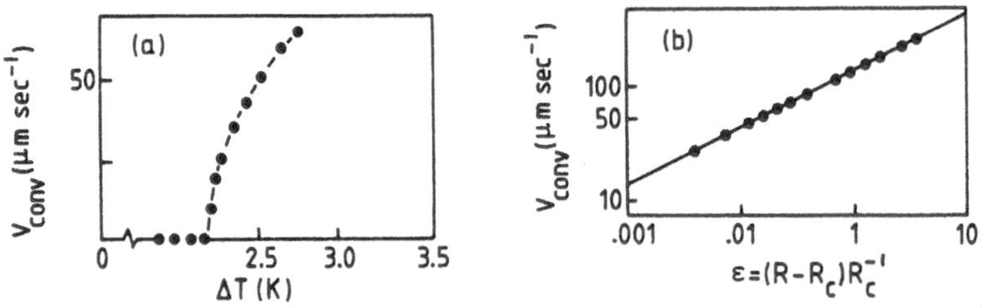

Figure 8. Laser Doppler anemometry measurements of the convective velocity in silicone oil. The solid line in (b) represents the power law $V_c \propto \varepsilon^\beta$ with $\beta = 0.5$ (after Dubois and Bergé, 1978).

when the temperature gradient is increased (Figure 8a) according to the power law predicted from mean field theory (Figure 8b). For a given value of the Rayleigh number the velocity stays constant; such a time independent regime persists in the domain $r > 1$ where the steady state solutions (Equation (18)) are stable.

When these steady state solutions are substituted in the dispersion equation (17), the latter becomes

$$z^3 + (P + b + 1)z^2 + b(P + r)z + 2bP(r - 1) = 0 \tag{19}$$

which possesses one real root and two pure imaginary roots when

$$r = \tilde{r} \equiv P\,\frac{P + b + 3}{P - b - 1} \tag{20}$$

This expression gives the critical value of R beyond which steady convection becomes unstable. The corresponding solutions to the cubic equation (19) are

$$\tilde{z}_o = -(P + b + 1)$$

$$\tilde{z}_\pm = \pm\, i\tilde{w}, \qquad \tilde{w}^2 = b(P + \tilde{r}) \tag{21}$$

Note that this result requires that P be larger than $(b + 1)$. Indeed it is clear that if $P < b + 1$, there is no positive value of r that satisfies Equation (20), in which case steady convection would always remain stable. So it is for fluids with sufficiently high Prandtl number $(P > b + 1)$ that the Lorenz model predicts the existence of complex roots and shows that if unstable steady convection is disturbed, oscillatory motion should set in at $r = \tilde{r}$. However, such a periodic regime corresponds to a limit cycle which can be shown to be unstable in the present model,[33] and as a result the Lorenz model has a limited range of application. However the major, and most interesting, feature of the Lorenz model appears when r is increased further. What happens when the disturbances to the steady convection regime become large can no longer be treated by linear theory; then appeal must be made to numerical computation to solve the set of equations (16). What the calculation reveals is that, when $r > \tilde{r}$, the system evolves in time by passing through a transient oscillatory regime which leads to chaotic behavior. This is illustrated in Figure 9 for the variable Z which measures the distorsion of the linear temperature profile. A detailed study of Degiorgio who generalized the model[12] shows the following interesting feature: beyond the threshold $r = 1$, the temperature profile deviates from linearity but gradient locking occurs at midheight of the cell where ∇T remains at its critical value ∇T_c. Note also that a study of the Lorenz model with rigid boundary conditions including side walls[34] leads essentially to the same qualitative results as to the sequence to chaotic behavior.

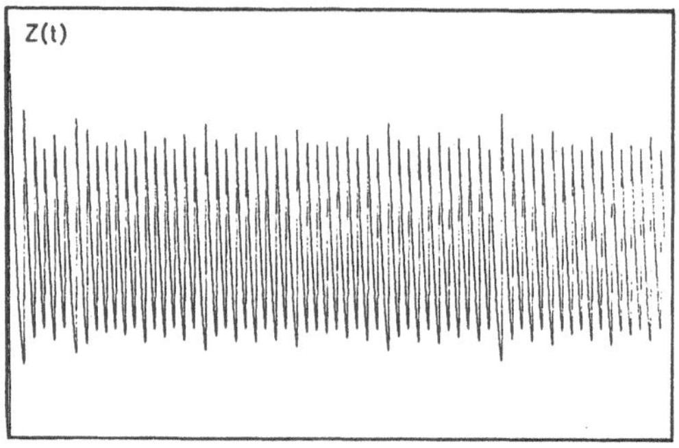

Figure 9. The time behavior of the distorsion of the temperature profile in the Bénard system, $Z = Z(t)$, as obtained from the Lorenz model, Equations (16), when r exceeds the threshold $\tilde{r}$, Equation (20), with $b = 8/3$ and $P = 130$.

The most important feature of the Lorenz model is that simple deterministic equations with only three degrees of freedom can produce chaos, which for the convection problem means turbulent motion. The chaotic behavior is conveniently shown by using a phase space representation: projections on the (X, Y) and (Z, Y) planes of the trajectory in phase space are given in Figure 10. The origin $0 = (X = 0, Y = 0, Z = 0)$ is the fixed point representative of the state of rest $(r < 1)$; the fixed points $C_+, C_- = (X = Y = \pm\sqrt{b(r-1)}, Z = r - 1)$ represent the steady convection state $(1 < r < \tilde{r})$; when r becomes larger than $\tilde{r}$, steady convection is unstable and the system travels along a complicated irregular path confined in a limited region of phase space without ever going back to a position encountered earlier: such a trajectory along which a system with chaotic behavior travels in phase space was given the name "strange attractor".[4] How representative such a model would be of the behavior of actual laboratory systems must now be examined.

B. Routes to Turbulence

The behavior of the simple system discussed above leads to a chaotic regime when steady convection becomes unstable: this shows one route to turbulence. The example illustrated schematically in Table 1 reveals that a periodic regime and even a biperiodic regime may occur before chaotic behavior, i.e. turbulent convection, ultimately sets in. Actually it is observed that the way a real system evolves towards chaos is strongly dependent on the experimental conditions. In fact mathematical models have been constructed that show the existence of a number of routes to turbulence;[11, 33]

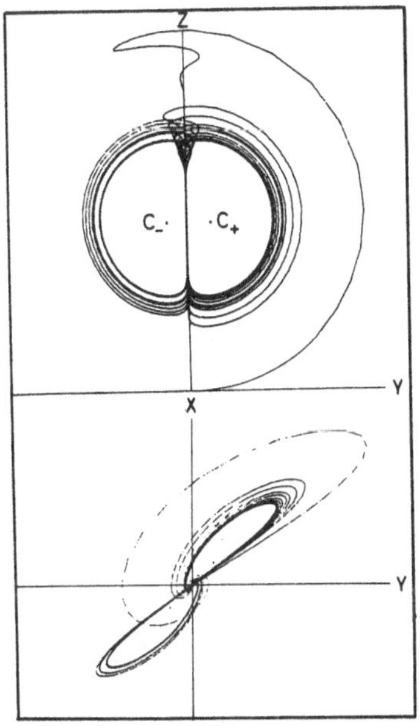

Figure 10. (X, Y) and (Z, Y) projections of the phase space trajectory of the Bénard system represented by the Lorenz model with b = 8/3 and r = 150. Here $\tilde{r}$ = 139.6 as obtained from Equation (20) with P = 130, the value of the Prandtl number for the experimental system discussed in the text.

several have been discovered experimentally.[35] An illustration is given in Table 3. The remainder of our discussion will be devoted to a description of the physical features of some of these routes.

Despite the fascinating aspect of model calculations that lead to the characterization of the turbulent regime as a strange attractor, such computations do not provide information on the physical nature of the mechanisms that occur on the way to turbulence. Questions like e.g. what are the oscillators that are responsible for the periodic regimes in a real system, can only be answered at present from experimental observations. In particular those experiments based on optical and vizualization methods are extremely useful in this respect. Here we shall restrict our discussion to the supercritical regimes of the Rayleigh-Bénard system for fluids with high Prandtl number and which have been investigated by the optical techniques described in this volume.[14]

Referring to our discussion of Section 1, we must note that the routes to the turbulent regime are strongly dependent on the experimental conditions and in particular on the geometry of the

Table 3. Routes to turbulence. Symbol M refers to those routes for which a model has been developped and symbol E indicates the routes that have been observed experimentally (After Coullet, 1980).

Stationary state (dissipative structure) fixed point	→ Instability with inverted bifurcation	→ Turbulence with hysteris	
	→ Stationnary solution disappears	→ Turbulence with hysteris or intermittency	
Periodic Regime (single frequency oscillations) limit cyle	→ Instability with inverted bifurcation	→ Turbulence with hysteris M	
	→ Subharmonic instabilities	→ Noise → Turbulence M E	
	→ Periodic solution disappears	→ Turbulence with hysteresis or intermittency M E	
Quasi-periodic regime (two-frequency oscillations) 3-D torus	→ Alternation of frequency locking and biperiodic oscillations	→ Frequency locking	→ Turbulence with hysterssis or intermittency M
		→ Subharmonic instabilities	→ Turbulence M
	→ Biperiodic solution disappear	→ Turbulence with hysteresis or intermittency E	

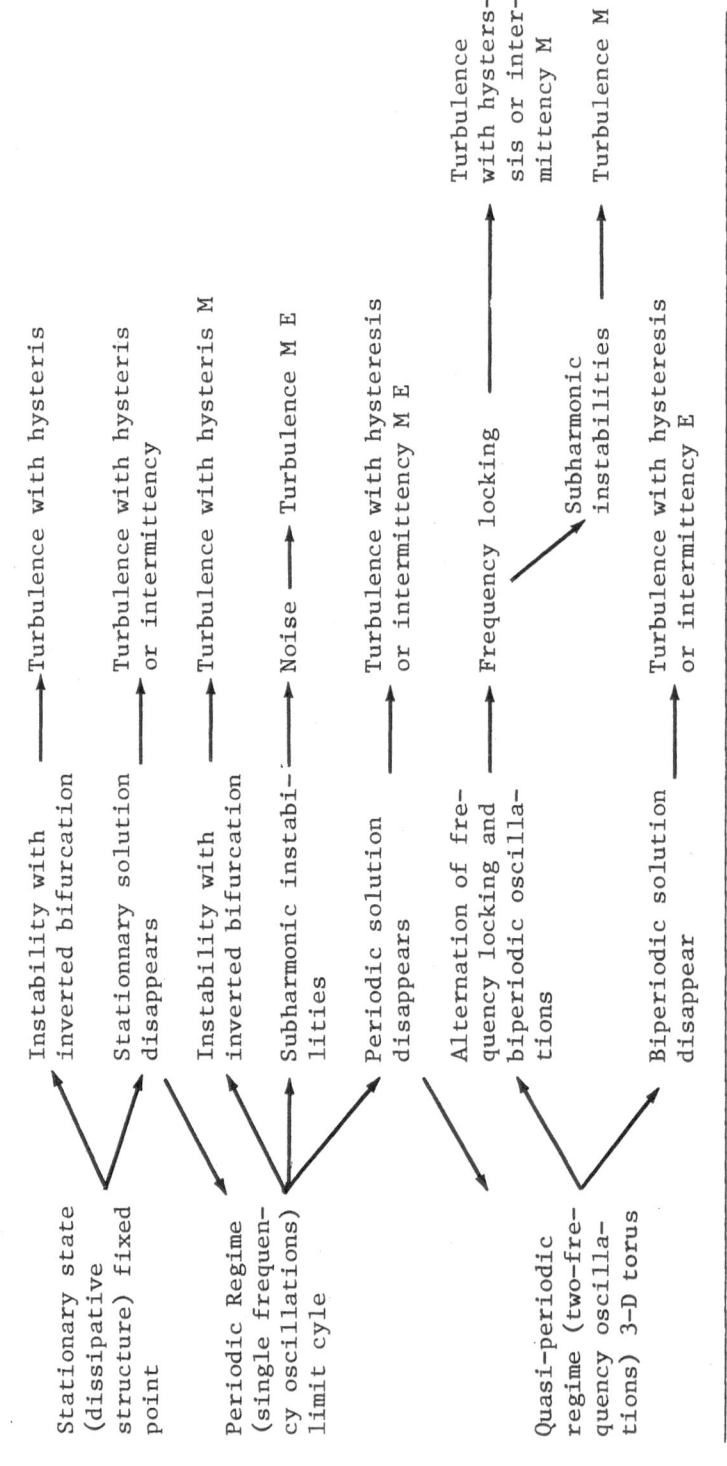

system. This brings in an important parameter, the aspect ratio $\Gamma_i = L_i/d$ (i = x, y), which measures the horizontal dimension of the box with respect to its thickness. In large boxes ($\Gamma \sim$ 10-100), the chaotic regime tends to set in near (or even at) convection threshold, a feature characteristic of systems with large number of degrees of freedom: the breaking of spatial structure occurs at relatively low Rayleigh number and leads to "phase turbulence" (see Figure 13 of the paper by Bergé and Dubois in this volume.) In contradistinction, we refer to turbulence with few degrees of freedom as the phenomenon that occurs in systems with confined geometry $\Gamma \sim 1$, i.e. small boxes where the convective structure contains only a couple of rolls such that boundary effects inhibit phase turbulence. Yet in such small boxes various structures can occur and several routes to turbulence are possible. This is illustrated in Figure 11 which summarizes the routes that have been observed experimentally in a Rayleigh-Bénard system with silicone oil (P = 130) and dimensions $\Gamma_x = 2$, $\Gamma_y = 1.2$. The various sequences that occur beyond steady convection can be represented schematically as follows:

A ⟶ Monoperiodic regime ⟶ Biperiodic regime ⟶ Turbulence

B ⟶ Monoperiodic regime with subharmics ⟶ Turbulence with intermittency

C ⟶ Monoperiodic regime ⟶ Turbulence

D ⟶ Turbulence

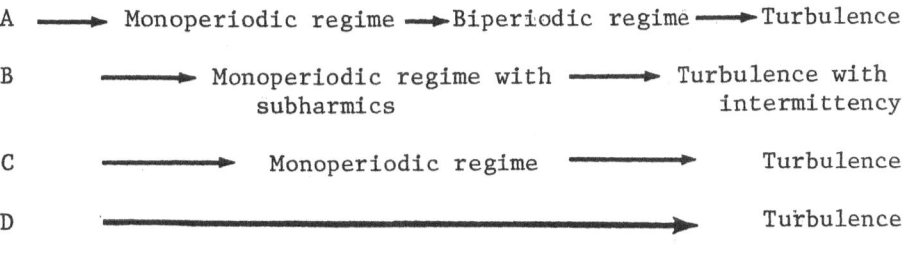

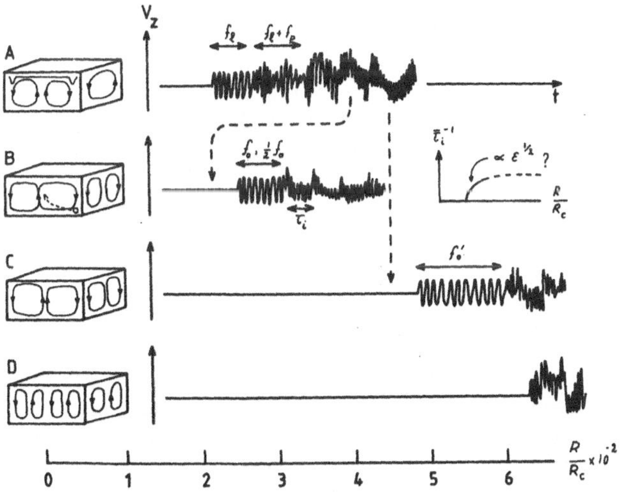

Figure 11. The various routes to turbulence as observed experimentally in the Bénard system with P = 130 (silicone oil), $\Gamma_x = 2$ and $\Gamma_y = 1.2$ (After Bergé and Dubois, 1980).

In each case the velocity was measured as a function of time by laser anemometry and the data were Fourier analyzed to obtain the power spectrum.[14] Then the presence of a well defined peak in the spectrum indicates without ambiguity the existence of an oscillatory regime whereas the appearance of a noisy broad band is characteristic of chaotic behavior. In the periodic regimes the motor of the oscillation could be physically identified by interferometry measurements.[14] The value of the external constraint at which a regime transition occurs is given in terms of $r = R/R_c$ which measures the distance from the convection critical threshold ($r = 1$).

In a small box system (here $\Gamma_x = 2$, $\Gamma_y = 1.2$), the convective structure that sets in at the threshold $r = 1$ is composed of two horizontal rolls with their axes oriented paralell to the shorter side of the box (X-rolls). When r reaches and exceeds the value 20, the convective structure becomes tridimensional in that a second horizontal roll pattern is formed that is oriented in the direction perpendicular to the former (Y-rolls). Such a tridimensional convective structure may remain steady up to high values of r. However, there is a tendency for the wavelength of the rolls to grow when the Rayleigh number increases. As a result, distorsion in the symetry of the structure can occur which leads to various possible sequences.

Route A. Below $r = 215$ steady convection remains stable with a convective structure composed of two symetrical X-rolls (with wavelength $\Lambda_x = 2d$) and one Y-roll (with wavelength $\Lambda_y = 1.2\ \Lambda_x$) (see Figure 11A). At $r = 215$, the velocity field becomes time dependent and oscillates with frequency f_ℓ (see Figure 12). It was possible to identify the physical nature of the oscillator by Mach-Zehnder interferometry: Dubois and Bergé[36] found that the oscillations of the velocity amplitude are driven by the pulsation of the upper boundary layer with the cold front progressing from the upper plate and vanishing in the convective fluid. A photograph is shown in Figure 15B in the paper by Bergé and Dubois in this volume.

The oscillating mechanism as suggested by Howard[37] should be governed by the thermal diffusivity mode where from it follows that the frequency would be given by $f_\ell = \frac{1}{2}\pi\chi\ell^{-2}$, where ℓ is the depth of the boundary layer. When r reaches and exceeds the value $r = 250$, the hydrodynamic regime becomes biperiodic: the power spectrum of the velocity field has two well defined sharp peaks at f_ℓ and f_p respectively as shown in Figure 12. As identified with the same interferometric method, the second oscillator appears as a couple of pulsating bumps in the cold boundary layer (see the locus A and B in Figure 15B of Reference 14); each of them originates from a local thermal perturbation that grows as a "thermal plume" along the downward streamline of the convective roll where it vanishes. The "plume" frequency seems to increase with r according to a power law: $f_p \propto r^{2/3}$, whereas the variation $f_\ell = f(r)$ is not well esta-

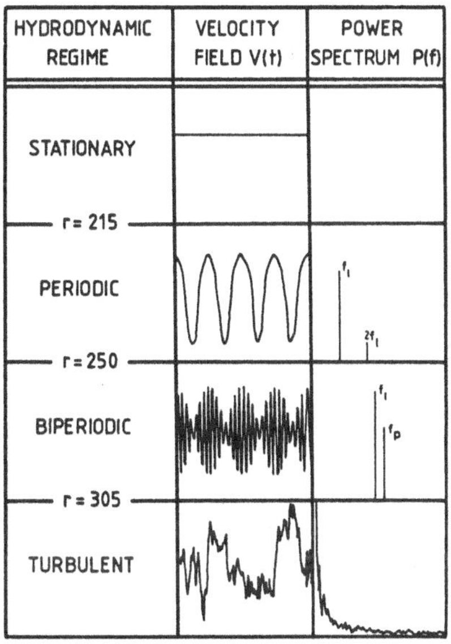

Figure 12. Local velocity as a function of time and its power spectrum in the successive regimes of the route to turbulence noted A in Figure 11. (After Bergé and Pomeau, 1980.)[38]

blished.[37] The biperiodic regime persists in the domain $250 < r < 305$ with the following sequence: frequency locking ($f_\ell/f_p \simeq 6/7$) for $250 \lesssim r \lesssim 270$, region of incommensurate frequencies ($270 \lesssim r \lesssim 290$), frequency locking ($f_\ell/f_p \simeq 7/9$) for $r > 270$. A major transition occurs at $r = 305$. No longer does the velocity behave regularly in time and a broad band centered about $f = 0$ appears in the power spectrum (see Figure 12): convective motion has become turbulent. Whether turbulence would be generated by non-linear coupling of the two oscillators or by the onset of a third oscillator could not be elucidated experimentally.

Route A can be characterized as a cascade of four successive bifurcations from the state of rest ($r < 1$) to the turbulent regime ($r > 305$). In phase space representation the behavior of the system can be described by a fixed point at the origin in the motionless state ($r < 1$), two fixed points for steady convection ($1 < r < 215$), one limit cycle in the monoperiodic regime ($215 < r < 250$), a 3-dimensional torus in the biperiodic regime ($250 < r < 305$) and ultimately a strange attractor in the state of turbulence ($r > 305$) (see Table 1). In order to obtain the strange attractor experimentally, one need measuring simultaneously at least two of the three governing variables of the system, e.g. the local velocity field

(~X) and the local temperature (~Y) or the local variations in the temperature gradient (~Z). In principle such measurements are possible by means of the experimental techniques described elsewhere in this volume.[14] Alternately, the strange attractor could also be obtained from the time behavior of one of the variables and its time derivates (here e.g. X, $\dot{X}$, and $\ddot{X}$) as it has been used to represent the chaotic regime in chemical instabilities.[39]

While route A is characteristic of one of the routes to turbulence for the Rayleigh-Bénard system described above, this type of sequence is more general as it also occurs in other systems. The route to turbulence via periodic regimes has indeed been observed in Rayleigh-Bénard experiments conducted in low Prandtl number fluids like water ($P \sim 2.5$)[40] and helium ($P \sim 1$),[35,41] and in circular Couette flow experiments.[42,43] As far as the mathematical analysis is concerned, route A cannot be modeled into the Lorenz scheme as presented in Section 3A because it exhibits an inverted bifurcation and immediate transition to turbulence. Models with direct bifurcation have been developped where time periodic regimes precede turbulence in qualitative[44] or semiquantitative[10] agreement with experimental observations.

Route B differs from the route A in two ways (see Figure 11): no biperiodic regime occurs on the way to the turbulent state and the latter manifests itself as intermittent chaotic bursts in the velocity amplitude. This type of sequence takes place in the Rayleigh-Bénard system when the convective structure has two asymetrical X-rolls and two Y-rolls (see Figure 11B). Steady convection maintains up to $r \simeq 250$, beyond which value the motion becomes periodic with frequency f_o. When the Rayleigh number is further increased, the subharmonic $f_o/2$ appears in the spectrum. Photographs of interferometric images allow the identification of the oscillator (see Figure 20 in the paper by Bergé and Dubois in this volume). A "hot drop" forms from a local temperature perturbation in one of the lower corners of the cell; the hot drop then moves along the convective stream line and vanishes when it is advected in the ascending current in the center of the cell (see Figure 11B). The process occurs with a period $\tau_o = f_o^{-1}$. The appearance of the subharmonic is interpreted as the manifestation of the longer period corresponding to the advection of a larger thermal perturbation.[45] Which one of the corners of the box is preferential to the drop formation is determined by two factors[46]: the drop forms in the larger X-roll and is advected in the central streamline i.e. a hot (cold) drop forms in the lower (upper) corner of the larger X-roll when the flow is oriented downwards (upwards) along the vertical edges of the cell.

The regular drop oscillations illustrated in the upper part of Figure 13 persists up to $r = 290$ where a regime transition occurs. The regular behavior of the oscillator then becomes disrupted: ir-

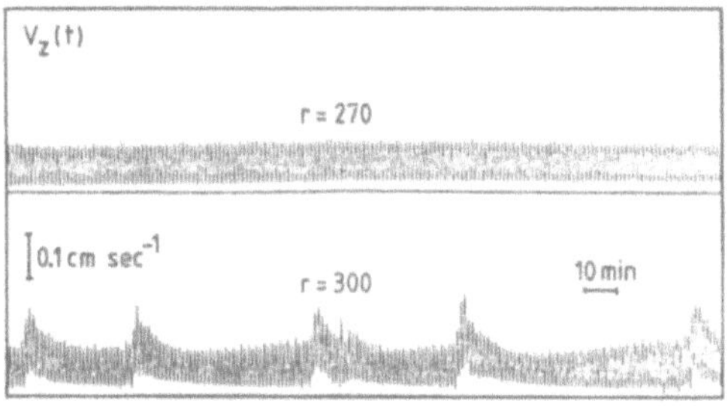

Figure 13. Vertical velocity as a function of time measured at the center of the cell in the flow configuration noted B in Figure 11, below and above intermittency threshold. The vertical and horizontal bars indicate the velocity and time scales. (After Bergé, Dubois, Manneville and Pomeau, 1980).

regular amplitude variations appear in the velocity field as shown in the lower part of Figure 13, and the time intervals between two bursts are randomly distributed. Simultaneously the power spectrum now exhibits broadening of the peak centered about f_0 and a noisy broadband centered at $f = 0$. The system has reached the state of "turbulence with intermittency". The further r is increased, the more often will bursts come to disrupt the regularity of the oscillatory behaviour. The physical origin of such bursts could be related by interferometry measurements to the formation of abnormal larger drops inducing strong perturbations to the temperature and velocity fields.[46] When the average time interval between bursts, τ_i, is measured as function of the Rayleigh number, there is indication that τ_i would vary according to the power law: $\tau_i^{-1} \propto (R-R_i)$, where R_i is the intermittency threshold. Such a power law was predicted by Manneville and Pomeau on the basis of a transformation of the Lorenz model.[47] Experimentally it is found that τ_i^{-1} varies indeed as the predicted power law near R_i, but seems to level off to a plateau at larger values of R (see Figure 11B). Finally we note that the sequence that characterizes route B can be interpreted qualitatively on the basis of a model developped on recent results obtained on the Lorenz system which then exhibits an intermittency phase as transition state between a limit cycle and a strange attractor.[45]

When the three-dimensional structure has two Y-rolls and two symetrical X-rolls (Figure 11C), steady convection remains stable up to higher values of r and there is only one intermediate periodic regime before chaotic behavior sets in. Turbulent convection can even be delayed further when four symetrical X-rolls form which tend to stabilize the structure (Figure 11D); then turbulence follows immediately steady convection, which case would correspond to the original Lorenz model.

The turbulent regime can even be reached as soon as motion sets in. This occurs in large boxes ($\Gamma = 57$), where it was found from experiments on liquid helium that turbulent flow starts very close to convective threshold, suggesting that the transition to turbulence occurs at $r = 1$ in the limit of infinite Γ.

Other sequences have been observed in small boxes and low Prandtl number fluids (water[48] and helium[49]) with intermediate regimes involving subharmonic bifurcations for which a model was developped by Feigenbaum[50] who obtains quantitative agreement between his theoretical predictions and the experimental results of Libchaber and Maurer.[49]

Whether in Rayleigh-Bénard systems, with various geometrical and fluid properties, in circular Couette flow, or in even more complex systems, many routes to turbulence have been, and are being, discovered, which calls for a rational classification. Attempts via mathematical modelization are being made in this direction,[11] but are beyond the scope of this section. Here we have restricted our discussion to some typical examples which illustrate the variety of possible routes in a simple, well defined system. Such sequences have two main common features: (1) the spatial modes are fixed, and chaotic behavior arises as a result of the time dependence of a limited number of modes; (2) there are at most four regime transitions between the state of rest and the turbulent state. These experimental observations provide convincing support to the concept of turbulence with few degrees of freedom.

ACKNOWLEDGEMENT

P. Bergé, M. Dubois and P. Lallemand, and A. Arneodo, P. Coullet, and C. Tresser are acknowledged for stimulating discussions. Thanks are due to B. Herpigny for handling numerical computations. Part of this work was supported by the Fonds National de la Recherche Scientifique (F.N.R.S., Belgium).

REFERENCES

1. R. P. Feynman, R. B. Leighton, and M. Sands, The Feynman Lectures in Physics, (Addison-Wesley, Reading, MA, 1964). Vol. II, Chapter 40.
2. L. D. Landau, and E. M. Lifschitz, Fluid Mechanics, (Pergamon Press, 1959). Chapter 3.
3. E. N. Lorenz, J. Atmosp. Sci. 20: 448 (1963).
4. D. Ruelle, and F. Takens, Comm. Math. Phys. 20: 167 (1971).
5. G. Ahlers, Phys. Rev. Lett. 33: 1185 (1974).
6. J. P. Gollub, and H. L. Swinney, Phys. Rev. Lett. 35: 927 (1975).
7. P. Bergé, and M. Dubois, Optics Comm. 19: 129 (1976).
8. G. Ahlers, and P.P. Behringer, Phys. Rev. Lett.040: 712 (1978).

9. A. Libchaber, and J. Maurer, J. Phys. Lett. 39: L-369 (1978).
10. J. B. McLaughlin, and P. C. Martin, Phys. Rev. A12: 186 (1975).
11. P. Coullet, Ph.D. Thesis, University of Nice (1980) and reference therein.
12. V. Degiorgio, Phys. Rev. A20: 2193 (1979).
13. J. P. Gollub, and M. H. Frielich, Phys. Rev. Lett. 33: 1465 (1974).
14. P. Bergé, and M. Dubois, in the present volume.
15. G. Nicolis, and I. Prigogine, Self-Organization in Nonequilibrium Systems, (Springer-Verlag, Berlin, 1977).
16. For an introduction to the problem of convection, see M. G. Velarde, and C. Normand, Sci. Amer. 243: 79 (1980).
17. S. Chandrasekhar, Hydrodynamic and Hydromagnetic Stability, (Oxford University Press, 1961). Chapter 2.
18. H. N. W. Lekkerkerker, and J. P. Boon, Phys. Rev. A10: 1355 (1974).
19. V. M. Zaitsev, and M. I. Shliomis, Sov. Phys. JETP 32: 866 (1971).
20. B. J. Berne, and R. Pecora, Dynamic Light Scattering, (Wiley-Interscience, New York, 1976). Chapter 10.
21. J. P. Boon, and P. Deguent, Phys. Lett. A39: 315 (1972).
22. J. B. Lastovka, Bell. Syst. Tech. J. 55: 1225 (1976).
23. D. W. Pohl, IBM J. 23: 605 (1979).
24. C. Allain, H. Z. Cummins, and P. Lallemand, J. Phys. Lett. 39: L-473 (1978).
25. J. Weisfreid, P. Bergé, and M. Dubois, Phys. Rev. A19: 1231 (1979).
26. A more complete discussion of pretransitional phenomena is given by the author in "Les Instabilités Hydrodynamiques", (Springer Verlag, Heidelberg, 1978).
27. G. Z. Gershuni, and E. M. Zhukovitskii, Convection Stability of Incompressible Fluids, (Keter Publishing House, Jerusalem, 1976). Chapter 1.
28. J. P. Boon, C. Allain, and P. Lallemand, Phys. Rev. Lett. 43: 199 (1979).
29. P. G. Simpkins, and T. D. Dudderar, J. Fluid Mech. 89: 665 (1978); R. Meynart, Applied Optics Lett. 19: 1385 (1980).
30. J. P. Gollub, and M. H. Freilich, Phys. Rev. Lett. 33: 1465 (1974).
31. H. L. Swinney, and J. P. Gollub, Phys. Today 31: 41 (1978), and references therein.
32. M. Dubois, and P. Bergé, J. Fluid Mech. 85: 641 (1978).
33. P. C. Martin, in "Proceedings of the International Conference on Statistical Physics" (North Holland, Amsterdam, 1975).
34. C. Baesens, M. D. Thesis (University of Brussels, 1980).
35. H. L. Swinney, Prog. Theor. Phys. Suppl. 64: 164 (1978); J. Maurer and A. Libchaber, J. Physique Lett. 40: L-419 (1979); G. Ahlers, in "Systems Far From Equilibrium" (Springer-Verlag, Heidelberg, 1980); M. Dubois, Colloque P. Curie,

to be published in J. Physique.
36. M. Dubois, and P. Bergé, to appear in J. Physique.
37. L. N. Howard, in "Proceedings of the 11th International Conference of Applied Mechanics", (Springer-Verlag, Heidelberg, 1966). p. 1109.
38. P. Bergé, and Y. Pomeau, La Recherche, 11: 422 (1980).
39. J. C. Roux, A. Rossi, S. Bachelart, and C. Vidal, Phys. Lett. 77A (1980).
40. J. P. Gollub, and S. Benson, Phys. Rev. Lett. 41: 948 (1978).
41. G. Ahlers, and R. P. Behringer, Prog. Theor. Phys. Suppl. 64: 186 (1978).
42. P. R. Fenstermacher, H. L. Swinney, and J. P. Gollub, J. Fluid Mech. 94: 103 (1979).
43. M. Gorman, and H. L. Swinney, Phys. Rev. Lett. 43: 1871 (1979).
44. P. Coullet, C. Tresser, and A. Arneodo, Phys. Lett. 77A: 327 (1980).
45. P. Bergé, M. Dubois, P. Manneville, and Y. Pomeau, J. Physique Lett. 41: L-341 (1980).
46. M. Dubois, Communication at the "Colloque Pierre Curie" (Paris, 1980), to be published in J. Physique.
47. P. Manneville, and Y. Pomeau, Phys. Lett. 75A: 1 (1979).
48. J. P. Gollub, and S. Benson, to be published (1980).
49. A. Libchaber, and J. Maurer, J. Physique C-3: 41 (1980).
50. M. J. Feigenbaum, Phys. Lett. 74A: 375 (1979).

STUDY OF RAYLEIGH-BÉNARD CONVECTION PROPERTIES THROUGH OPTICAL MEASUREMENTS

P. Bergé and M. Dubois

Division de la Physique
Service de Physique du Solide et de Resonance Magnetique
Commissariat a l'Energie Atomique
C. E. N. Saclay, France

CONTENTS

1. INTRODUCTION

Rayleigh-Bernard instability takes place in a fluid layer heated from below, when the temperature difference ΔT applied to the layer exceeds a critical value ΔT_c. The parameter of the instability is the Rayleigh number Ra.[1]

When $\Delta T > \Delta T_c$ (or $Ra > Ra_c$), motion sets in the form of rolls, the periodicity of which is Λ, the wavelength of the organized motion. The two variables which describe directly the convective state are the local velocity V and the local temperature perturbation θ. In the steady state, θ and V_z (velocity component along the z-axis) are simply related, but their relation becomes more complicated and dependent upon Prandtl number in the unsteady regimes.

The main problem, under study nowadays, concerns the evolution of convective motion from steady to turbulent. The properties of the time-dependent convection are generally studied through the Fourier spectra of some variables of the system, e.g., the Nusselt number,[2,3] a local velocity amplitude,[4,5] or a local temperature perturbation.[6] But, in a certain sense, these measurements are blind for they do not give, at any instant, any spatial correlation or global "visualization" of the convective structure; these two last characteristics need to be known in order to understand the time-dependent phenomena. The experiments show that there are two kinds of turbulence, one in great boxes which is directly reached from the steady state (large aspect ratios $\Gamma x = Lx/d$ and $\Gamma y = Ly/d$), the other one in small containers which is obtained as the ultimate step of different periodic regimes when the Rayleigh number is increased.[3,5] The aim of this paper is to show how one may measure both velocity and temperature perturbations through optical means; furthermore, we wish to show how simultaneous measurements of both quantities, together with the knowledge of the overall structure through visualization, provides a new understanding of the time dependent effects in the particular case of Rayleigh-Benard convection.

2. EXPERIMENTAL SITUATION

The experiments we report in this paper have been done with silicone oil, i.e., on high Prandtl number fluid: $Pr = \nu/D_T > 100$, where ν is the kinematic viscosity and D_T the thermal diffusivity of the fluid. The top and the bottom of the cell which contains the fluid are made of massive copper plates - except in the experiment reported at the end of the Chapter III - and thermally regulated within 10^{-2}°C by circulating water. The lateral boundaries consist of a rectangular plexiglass frame, dimensions of which are adjusted depending on the properties to be measured. In most cases (except for the experiment of Chapter III) all the optical beams necessary for velocity measurements as well as for temperature observations enter and exit through the lateral plexiglass frame.

One can see on Figure 1 the simplest arrangement of convective rolls in a rectangular frame.

3. MEASUREMENT OF THE VELOCITY

We do not want to enter into the details of the Laser Doppler Velocimetry (LDV) technique. This is now conventional and we wish to refer the reader to classical reference.[7]

We use the crossed beam technique (see Figure 2) in which two focused beams B_1B_2, issued from the same laser, cross at the point P of the fluid at which the velocity has to be determined. A small amount of latex spheres is added into the fluid. The scattered light from the crossed beams is collected by focusing it by a stan-

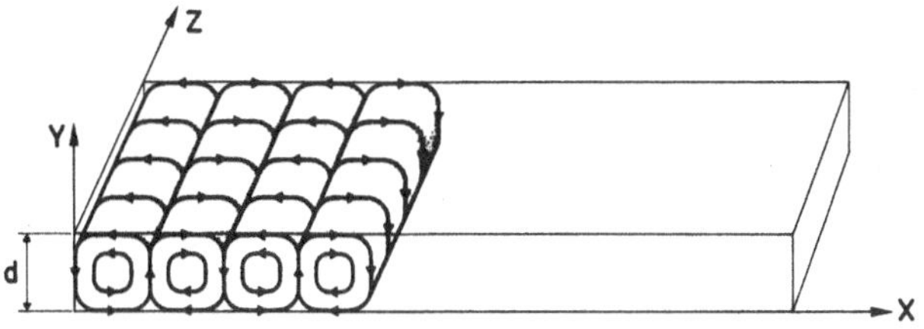

Figure 1. Schematic arrangement of two dimensional rolls in a rectangular frame.

dard photographic lense L; one then gets a cross as image. A translatáble pinhole H placed in front of a photomultiplier PM is carefully adjusted to be exactly at the crossing point of the image. The fluctuating photocurrent contains the Doppler frequency f_d. From this frequency we can obtain the velocity component V cosα, where α is the angle between the velocity and the exterior bisector of the angle formed by the two crossed beams.

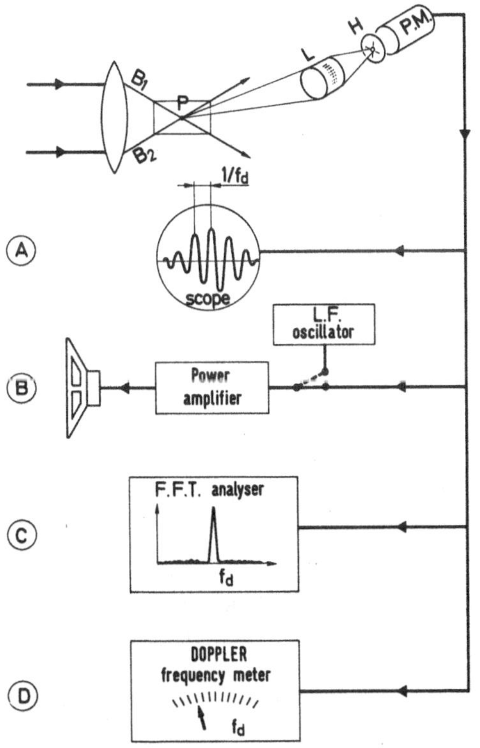

Figure 2. Principles of LDV measurements.

It is not straightforward to detect the Doppler frequency without ambiguity. Although the photo-current fluctuates with the Doppler frequency, both the amplitude and the phase of this frequency are at random and, furthermore coexist with fluctuations of other origins (mechanical vibrations, mains or laser ripple, photon noise etc.) Because the Doppler signal in the photo-current is rarely pure and regular in time, a conventional frequencimeter does not work.

The use of a storage oscilloscope is the simplest method to determine roughly the Doppler frequency f_d, by choosing a sample of photo-current fluctuations containing an intense burst of Doppler signal [Figure 2(A)].

Another method of very low cost which is very useful for many purposes is the following. If the Doppler frequency ranges between 100 H_z and few kilocycles, i.e., in the audiofrequency, which is the case of the convective problem, it is possible to hear directly the Doppler frequency just by using a power amplifier and a loudspeaker [Figure 2(B)]. This is very useful for optimizing any settings and, with some practice, for quickly checking the overall LDV installation. Furthermore, in the particular case of steady convection (or slowly time dependent), one can determine with great accuracy the Doppler frequency just by comparing the height of the tone with that emitted from of a tunable oscillator. As surprising as it will appear, this kind of determination has the same accuracy as that obtained by sophisticated electronic devices.

However, the best method for determining the Doppler frequency, even if this signal is hidden among shot noise and spurious signals, is fast Fourier analysis FFT. [Figure 2(C)]. The corresponding peak appears well resolved and one immediately can know the actual frequency, with the spread in frequency either due to finite volume effect (presence of spatial gradients in the velocity) or due to turbulent effects. With respect to the FFT method, another method which has been alternatively proposed is the correlation technique. We believe, however, that this technique, very useful for detecting correlation functions which are generally exponential, is not interesting* in the field of LDV for it does not directly give the required information, namely, the characteristic frequencies. Indeed, one can Fourier transform the autocorrelation function but, generally, this is expensive and produces a lack of resolution.

Note that FFT determination of the Doppler frequency is only interesting in the steady state (or in slowly time dependent state); nevertheless, in that case it is very useful in order to determine with great accuracy the convective amplitude in any point and whatever the velocity may be. For example, we can measure steady velocities lower than 1 $\mu m\ s^{-1}$ with reasonable accuracy, the standard accuracy for velocities $\geq 10\ \mu m\ s^{-1}$ being better than 1%. See

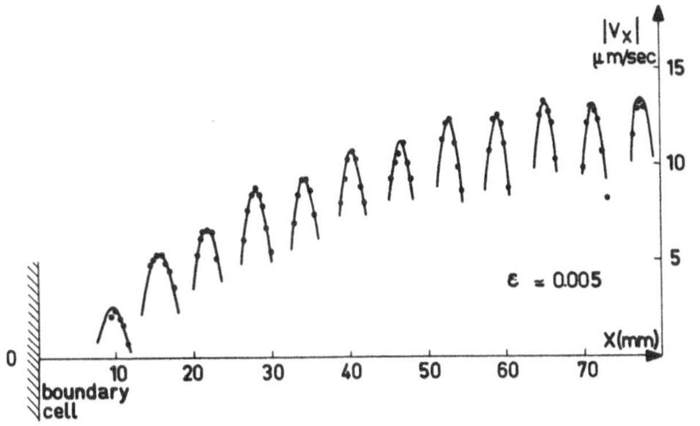

Figure 3. Velocity profile very near the onset of convection showing the long range influence of a lateral boundary.

examples of such FFT determinations of velocity fields on Figures 3 and 4.

For the measurement of time dependent velocity, many kinds of automatic Doppler frequencimeter have been designed to give quasi-instantaneously the frequency or period of the Doppler signal [Figure 2(D)]. In such devices many precautions are taken in order to reject spurious signals always present in the photo-current fluctuations. Such an automatic frequencimeter is used in the work described here (ALEP Grenoble, France.) The analog output is proportional to the Doppler frequency, computed by comparison between a preselected number of consecutive periods; spurious signals and parasites are automatically not taken into account. This kind of LDV frequencimeter is of great interest in two cases: study of unsteady convection and automatic recording of a velocity field. The only disadvantage is the lack of accuracy (5%) against less than 1% for a FFT determination.

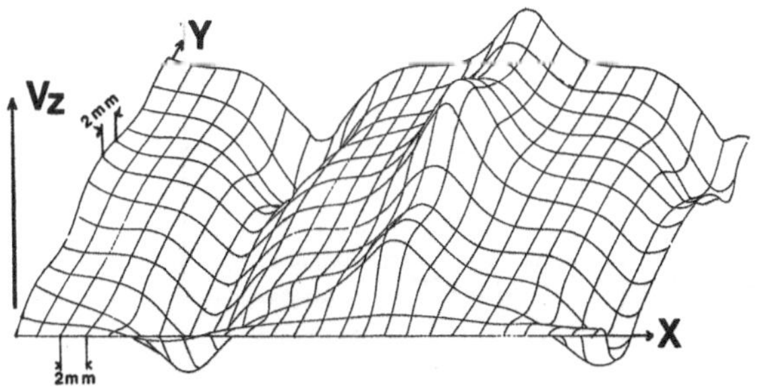

Figure 4. Velocity field illustrating three dimensional (or bimodal) convection at $Ra/Ra_c = 15$ (Reference 18).

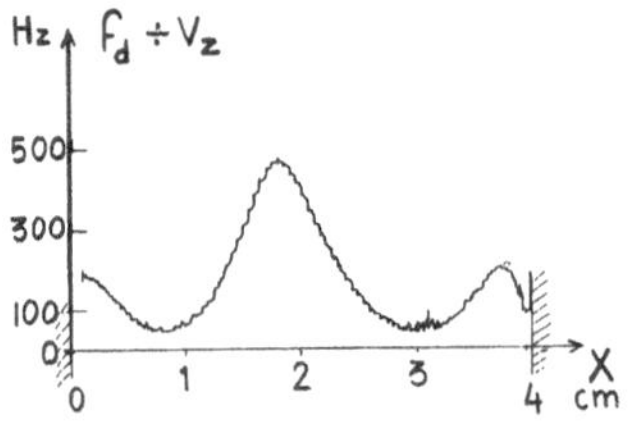

Figure 5. Photograph of a direct recording of a velocity profile versus X in a box Lx = 2d, d = 2 cm, at Y = $L_Y/2$, Z = d/2, Ra/Ra_c = 143.

As illustration of the use of an automatic Doppler frequencimeter one can see on Figure 5 the convective velocity profile determined in a cell of small aspect ratio. The principle of the method consists of using an X, Y plotter (see Figure 6). The X-axis represents any horizontal coordinate of the cell. This is achieved by using a linear potentiometer LP attached to the cell. The variation of the position of the cell with respect to the measurement point P (which is fixed in space) produces a proportional variation of the voltage sent on the X input. The output of the frequencimeter (here proportional to the vertical component V_Z) is directly sent to the Y-axis of the plotter. By gently moving the cell, the velocity profile is directly obtained on the X, Y plotter. The other classical application of the automatic Doppler frequencimeter is the study of the time dependent velocity. In that case the output of the frequencimeter is sent on a y = f(t) recorder. One can see on Figure 7 some typical examples of time-dependent behaviour in the Rayleigh-Benard convection. The study of time records is very useful but not sufficient to characterize these behaviours. One has also to study the Fourier analysis of the velocity fluctuations. This is done by using a FFT analyser with an external clock to adapt the sampling time to the very low frequencies to be analyzed. One can see in Figure 8 the scheme of the experimental set up and, on

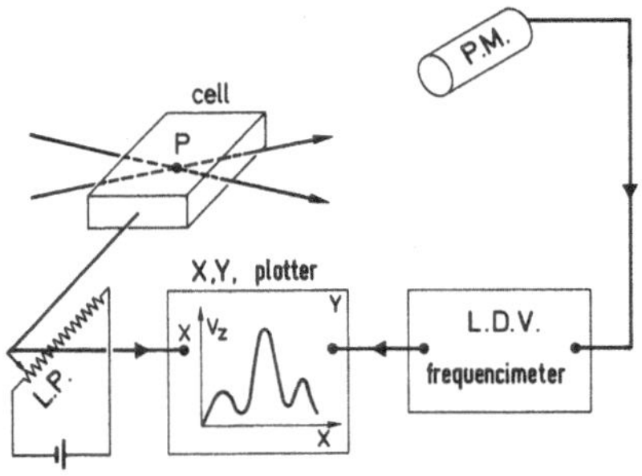

Figure 6. Principle of the method of direct recording of velocity profiles.

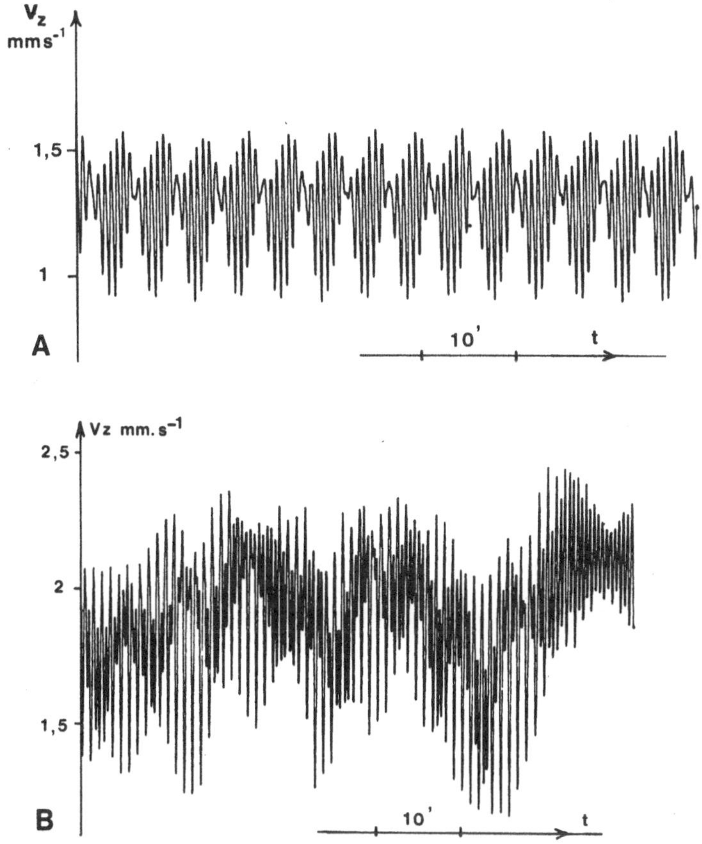

Figure 7. Time dependent velocity in a small box. (A) Biperiodic regime Ra/Ra_c = 253; (B) turbulent regime Ra/Ra_c = 310.

Figure 9, spectra obtained in some typical situations of the R.B. convection.

At this point we have to mention a very useful trick. We have already mentioned that Doppler frequencies for velocities ranging below a few mm/s correspond to audiofrequencies. A natural and reliable way to memorize any velocity behaviour is to use a conventional magnetic tape recorder, (see Figure 8). The analysis of the behaviour can be made later by reading the output of the recorder eventually at a higher speed in order to decrease the characteristic time of the time-dependent velocities, and to gain time for the FFT analysis of the velocity fluctuations. The great advantage is that all information of any experiment is recorded reliably on the tape. Subsequent checks or different treatments are always possible.

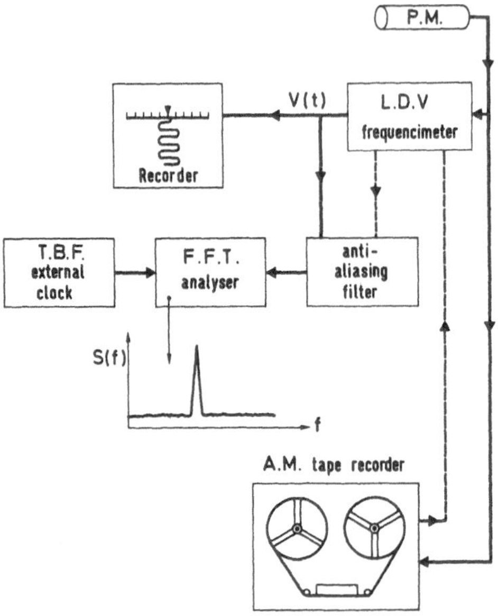

Figure 8. Principle of determination of time dependent behaviour of the velocity.

We note that in many cases of unsteady behaviour of the velocity, the corresponding change of V implies a change of the sign (upward velocity changes into downward one for example.) The experimental set up has to be modified in order to produce, for example, a continuous phase shift of the fringes pattern system existing at the crossing point P of the two beams; this is achieved by shifting the optical frequency of these beams. This shift is usually achieved by two "Bragg modulators" commercially available (see Figure 10). For the low velocities considered here the limitation is given by the frequency drift of the two oscillators driving the Bragg tanks. Then, one has to improve the frequency stability of commercial equipment down to 10^{-8}.

An alternative solution to shift the frequency of the two beams is to use a rotating transparent disc on which a radial grating is scatched (see Figure 10). Here the stability of the shift $2\delta f$ is determined by the speed of rotation, then $2\delta f$ may be of order of kilocycles ± 1 Hertz.

4. STRUCTURE VISUALIZATION

One of the most relevant properties of the convective motion is its spatial organization. When the fluid is in a rectangular cell, it is "well known" that the rolls are in principle organized

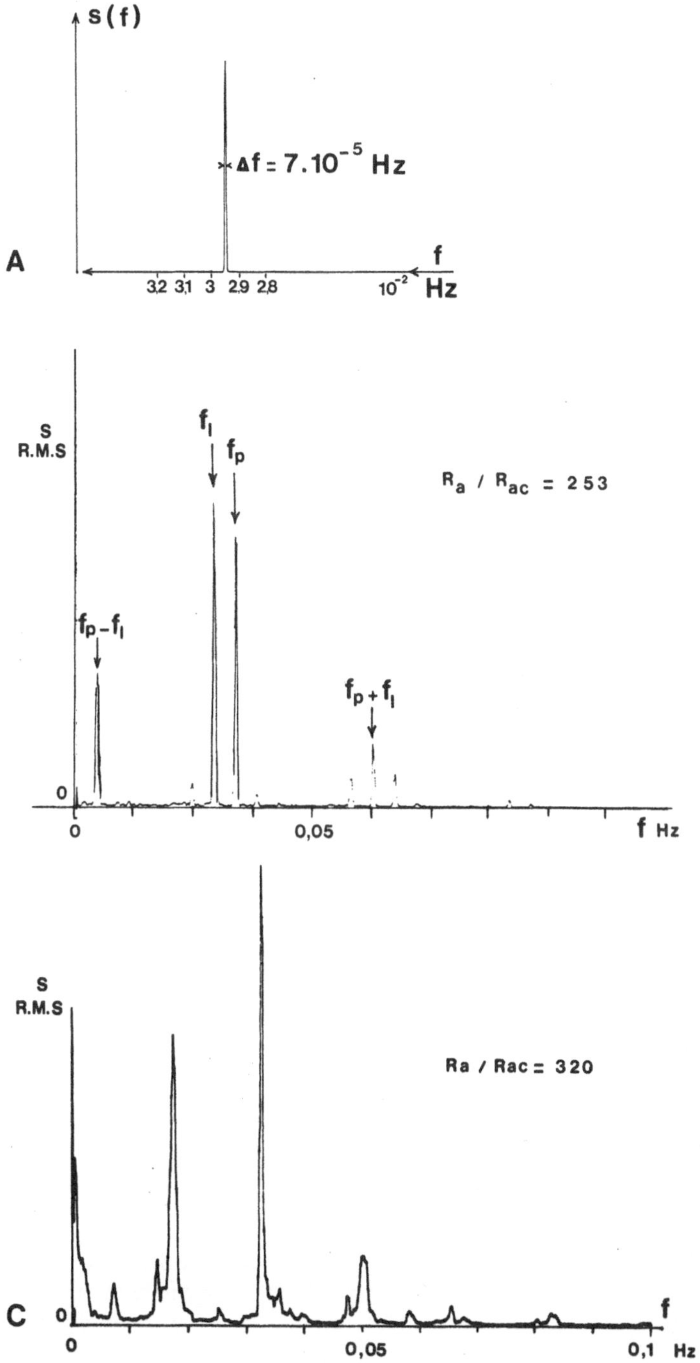

Figure 9. Some typical spectra of time dependent velocity. (A) Very high resolution spectrum showing the monochromatism of a convective oscillator; (B) biperiodic regime; (C) turbulent regime.

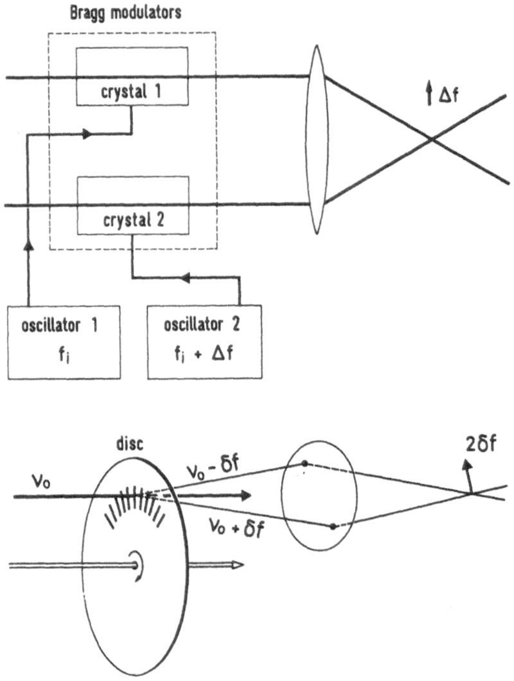

Figure 10. Principle of experimental set up to remove the Doppler ambiguity.

parallel to the short side of the cell. We observed this remarkable order in cells with $L_x \geq 10$ d and $L_y \simeq 3$ to 5d. A schematic illustration of that is shown in Figure 1 and verified in Figure 11 where the structure is revealed (in projection) through the "focusing effect". When one sends a large horizontal beam through the convecting layer parallel to the axis of the rolls (Y-axis), the colder

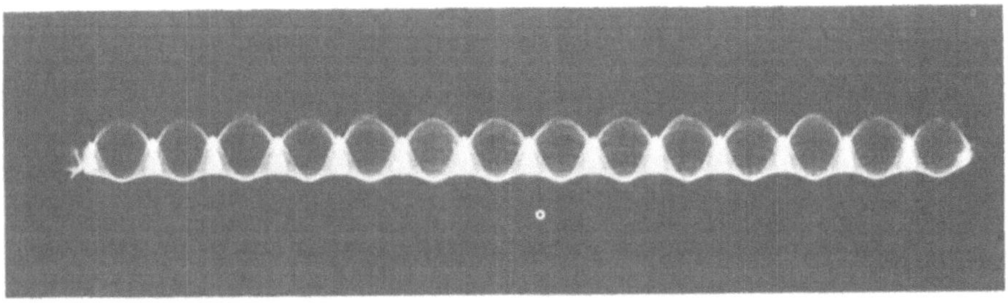

Figure 11. Visualisation of a bidimensional pattern of 29 rolls through the focusing effect.

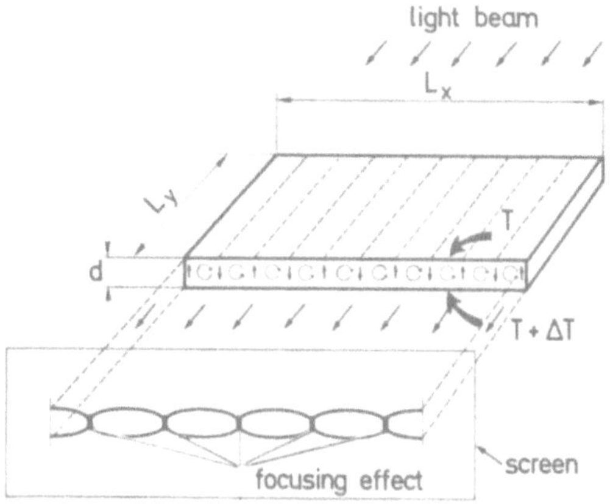

Figure 12. Principle of the focusing effect method.

descending fluid acts as a converging lens while the warmer ascending fluid makes the light diverge. Thus, we observe on a screen placed at a suitable distance from the cell (dependent on the Rayleigh number) bright lines corresponding to descending motions and dark lines corresponding to ascending motions (see Figure 12).

Such a picture gives instantaneously the wavelength of the convective pattern (the number of rolls actually present in a given cell). The picture shown in Figure 11 has been taken through the frame and gives the periodicity of the structure in a ZX plane, i.e., with integration along the Y-axis.

When the structure is not perfect (for example rolls not parallel to Y-axis), or when it is three-dimensional, the knowledge of the pattern cannot be obtained only by illumination through the frame; on the contrary, knowledge of the actual structure can be obtained by illuminating parallel to the vertical Z-axis as done in Reference 8. The problem here is the fact that in this Reference the convecting layer was confined between two glass plates, the finite heat conductivity of which changes the convecting behaviour.[9] In order to avoid this kind of perturbation we arranged a set up as described in Figure 13. The bottom plate consists of a massive copper plate, the upper side of which was very carefully machined and then polished to a mirror finish. The top plate consists of a sapphire plate, the thermal conductivity of which is approximately 300 times that of the oil used in the experiments. One can see on Figure 14 a photograph of the experimental set up, and on Figure 15 two photographs of the convecting pattern seen from above through the sapphire and using the focusing effect and the reflection on the copper mirror. One of these corresponds to a regular three-dimensional pattern because we forced the structure by initiating the rolls through a spatial periodic heating of the layer done as in Reference 8. The other

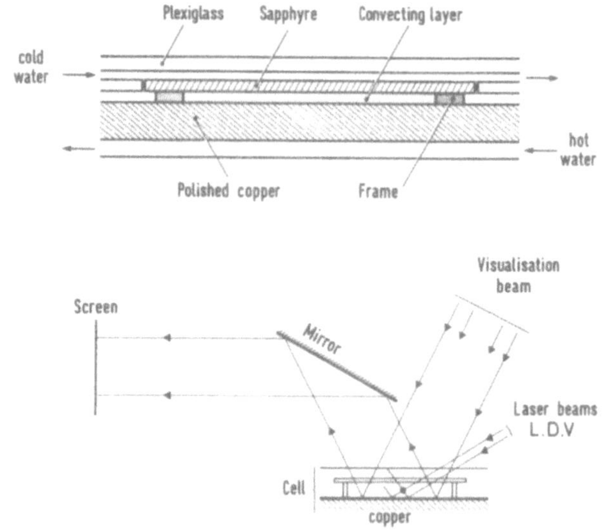

Figure 13. Schematic representation of the copper-sapphire cell.

photograph corresponds to a spontaneous pattern which can develop naturally by a very slow increase of ΔT through the critical onset.[10]

Note that in the same sapphire cell, velocity measurements are performed simultaneously with visualization (see the LDV laser beam arrangement on Figure 13); in that case the vertical V_Z component is measured.

Figure 14. Photograph of the copper-sapphire cell ($Lx = 20d$, $Ly = 12d$, $d = 0.5$ cm).

5. OPTICAL METHODS TO MEASURE TEMPERATURE PERTURBATIONS

As we said before, it is essential to know the temperature together with the velocity field of the convective structure to point out the physical mechanisms of the time-dependent phenomena, such as, for example, that of the oscillatory regimes. Local measurements of the temperature perturbations in three-dimensional structures are available,[11] but through sophisticated methods and delay times due to computer treatment. Here we present a very simple (and not expensive) set up which gives directly, in real time, quantitative information on the whole temperature field. Moreover, the velocity can be measured simultaneously.

A. Differential Interferometry

The principle of the method is differential interferometry through the cell, in the Z-X plane for example. The corresponding scheme of the set up is shown in Figure 16. An expanded parallel laser beam crosses the cell along the Y'Y axis; at the end of the cell it is reflected on the two sides of a glass plate, flatness and parallelism of which is approximately equal to λ (wavelength of the light). So, after reflection, interferences are constructed between two rays, such as 1 and 2 on the figure, which have crossed the cell at the distance

$$\delta e = 2e.\mathrm{tgr}.\cos\varphi$$

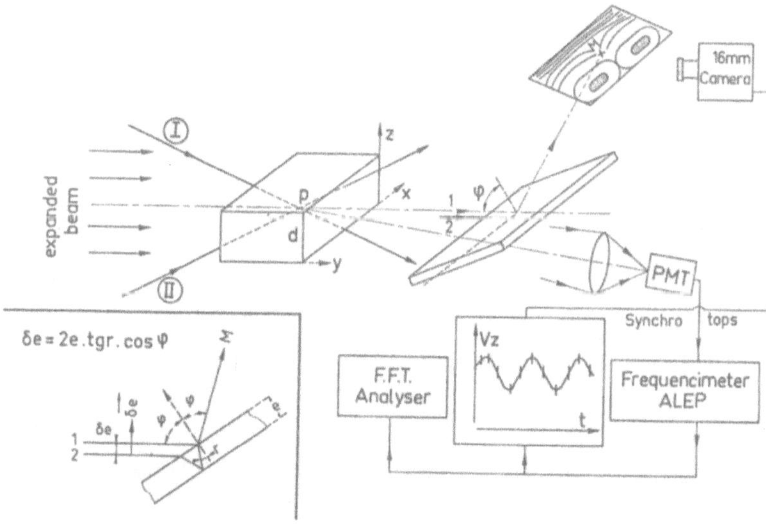

Figure 16. Principle of the differential interferometry working simultaneously with LDV.

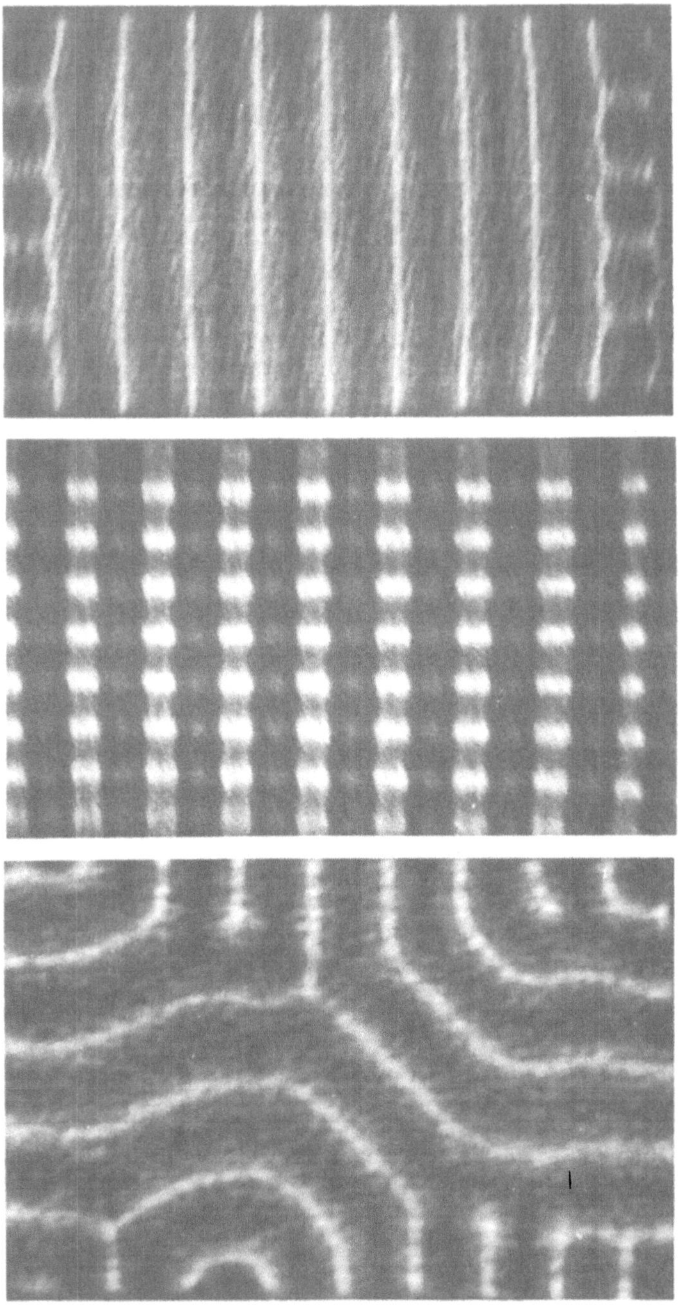

(a)

(b)

(c)

Figure 15. Photographs of rolls pattern seen from above (focusing effect). (A) Straight induced rolls $Ra/Ra_c \simeq 2$; (B) bimodal convection from straight induced rolls $Ra/Ra_c = 12$; (C) spontaneous irregular structure $Ra/Ra_c \simeq 2$.

e is the width of the glass plate (e = 2mm in our experiment), φ is the incident angle of the beam on the glass plate and $\sin \varphi = n \sin r$, where n is the refractive index of the glass plate.

The interference order δk of the two rays 1 and 2 is directly related to the difference between the mean refractive indexes along the path of the two rays, and therefore to their mean temperature difference δT. So, we get, by imaging the interferences on a screen, the map of the absolute value of the temperature derivatives in the δe direction. In practice, we choose δe to be parallel to the X'X or to the Z'Z direction and we observe consequently the map of the temperature gradients $\nabla_X T$ or $\nabla_Z T$. An example is shown on Figure 17B where we can see simultaneously the isogradient lines $\nabla_Z T = f(X,Z)$ and $\nabla_Z T = f(Y,Z)$. The spatial variation of the temperature gradients is obtained through the variation of δk:

$$\Delta(\delta k) = \frac{\ell(\delta e)_Z}{\lambda}\left(\frac{dn}{dT}\right)\cdot \overline{\Delta\left(\frac{dT}{dZ}\right)}$$

ℓ is the length of the path in the cell, λ is the wavelength of the laser beam and $\overline{\Delta(dT/dZ)}$ is the mean value of the variations of the local temperature gradient. In the case of our observations,

$$(dn/dT) = 4\,10^{-4}\deg^{-1}, \qquad \lambda = 0.6328\ \mu m$$

and with $\ell = 2$ cm and $(\delta e)_Z \simeq 1$ mm

$$\overline{\Delta\left(\frac{dT}{dZ}\right)} = 0.08\ \Delta(\delta k)\ \text{deg. mm}^{-1}.$$

The importance of this method is first that it gives at any instant an overall quantitative picture of the structure, and then that it is relatively insensitive to the optical imperfections of the frame and to mechanical vibrations.

A restriction which has to be mentioned is that the observed image corresponds to the integration along the path of the laser beam, and therefore does not give local information. Nevertheless, as shown in Figure 17B by simultaneously mapping the mean temperature gradients in the two planes ZY and ZX we get a good idea of some local temperature heterogeneity, then semi-quantitative measurements are allowed in small aspect ratio cells. Another point is that integration of the observed gradients to obtain the temperature perturbation is no longer possible when the local temperature gradients are sufficiently strong to produce a deformation of the initially parallel beam.

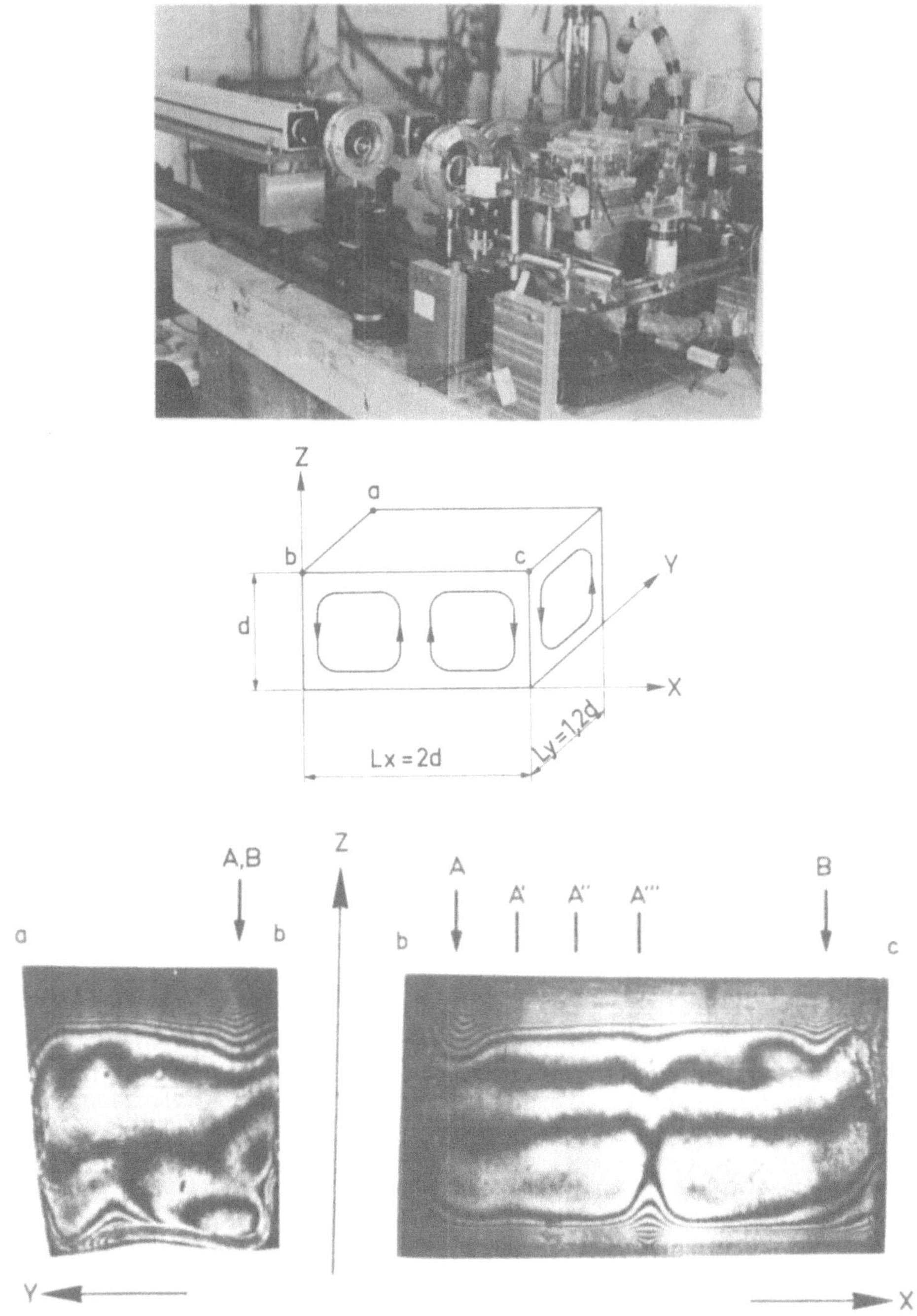

Figure 17. (A) Photograph of the interferometric + LDV set up; (B) Differential interferometric pattern ($\nabla_Z T$) projected in the (Z,X) and (Z,Y) planes. One can see the two boundary layers and the bumps A and B.

B. Mach-Zehnder Interferometry

To obtain more quantitative temperature measurements, we have in certain cases used Mach-Zehnder interferometry. The principle is described in Reference 12. Before crossing the cell, the expanded laser beam is separated in two parts by a semi-transparent plate. One part crosses the cell, the other goes outside. They are recombined at the end of the cell through an other semi-transparent plate. The interferences, so obtained by the recombination, are directly related to the mean temperature field T = f(X,Z). An example is shown Figure 18, for a single roll, where we can see very well in particular the temperature variation in an ascending and a descending motion. The interferometric pattern is very sensitive to mechanical vibrations of the set-up as well to imperfections of the optical path. Nevertheless we made, with this technique, significant measurements of the temperature variations in time-dependent convection (see Figure 19). We took a film of the dynamical interferometric pattern, image after image (1 sec between two images), and then we exploited this film by following the displacement of the interference fringes in some characteristic points of the layer.

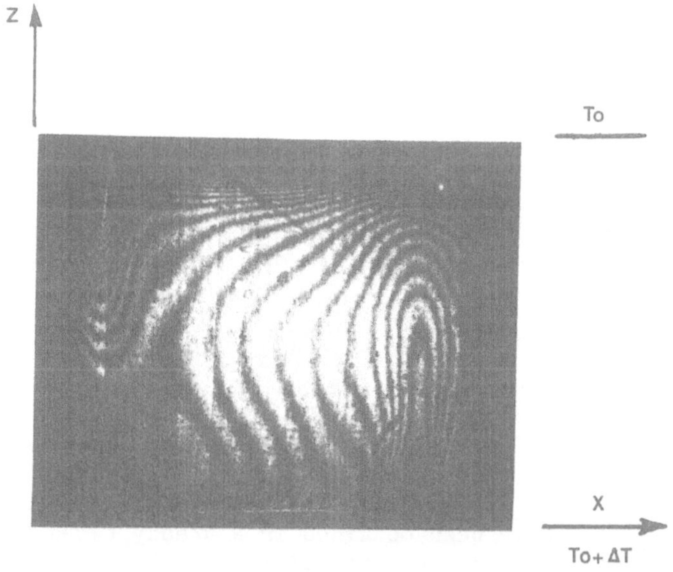

Figure 18. Photograph of Mach-Zehnder fringes pattern corresponding to a single roll - on the left part of the picture one can see the bump - $Ra/Ra_c = 295$.

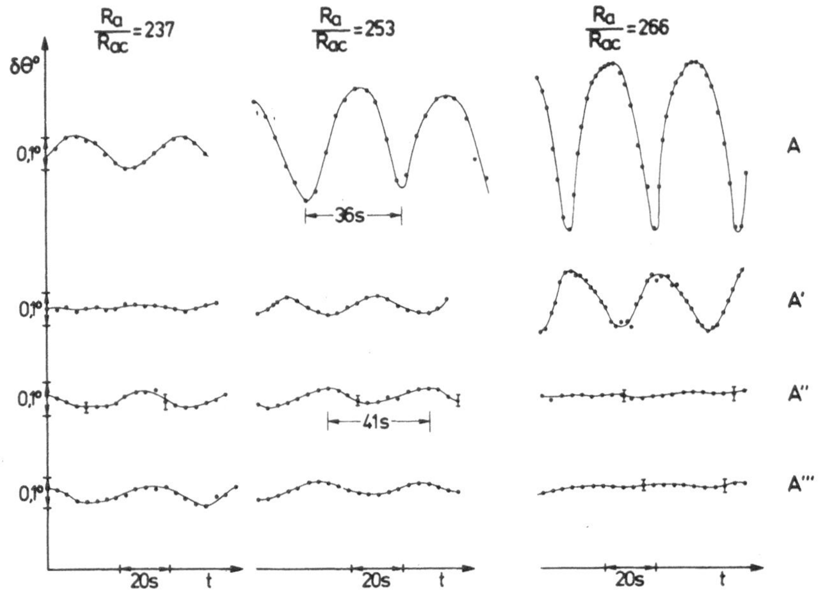

Figure 19. Temperature modulations deduced from Mach-Zehnder interferometry. A, A', A", A"' refer to different points in the cold boundary layer represented in Figure 17B.

C. Deflection of a Narrow Beam

This technique is not related to interferometric measurements, but it is very simple and gives direct information on local temperature gradients (averaged along the beam path). The principle is the same as that used for structure visualization by the "focusing effect" and is due to the fact that a ray, crossing a material of non-constant refractive index, bends towards the region of higher refractive index. If R is the radius of curvature of this ray,

$$1/R = \vec{\nu} \cdot \overrightarrow{\text{grad}}(\log n)$$

where $\vec{\nu}$ is the unit vector normal to the ray.

Thus, by measuring the deflection of a narrow laser beam sent through the cell, we get the mean value of the temperature gradient along the path of the ray. An example of such measurements is shown in Figure 20. These have been obtained in a steady two-dimensional structure and we can observe the sensitivity to the presence of spatial harmonics (here the third harmonic) of the fundamental mode. In time-dependent phenomena, the deflection will follow the time variations of the temperature gradients. The results presented in Figure 21 show such variations in a biperiodic regime. In this

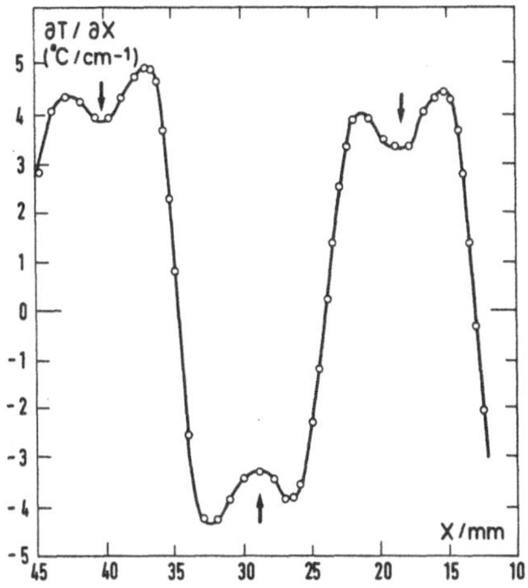

Figure 20. Temperature gradient $\nabla_X T$ deduced from the beam deflection technique in steady bidimensional convection. The beam crosses parallelly to the Y-axis at Z = d/2. The arrows show the manifestation of the third harmonic, Ra/Ra_c = 3.2.

experiment, at a given instant, a small diode is placed near the center of the deflected beam. Further displacement of the beam causes a change in the intensity detected by the diode and the periodicity of the diode signal gives the periodicity of the variation of the local temperature gradient undergone by the ray.

By using several beams and diodes it is possible to perform spatial correlation and other interesting measurements.

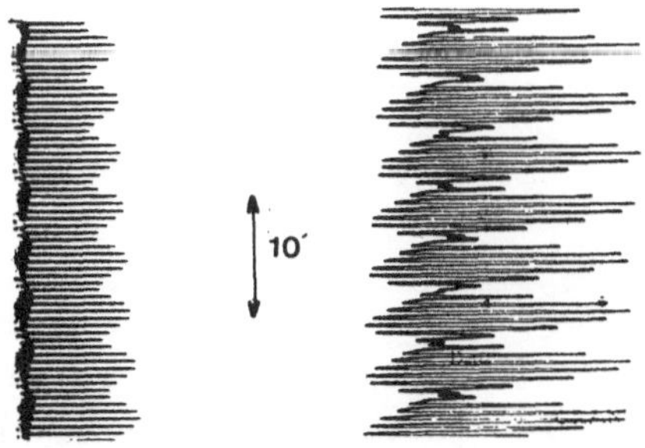

Figure 21. Time dependent deflections of a beam in the biperiodic regime Ra/Ra_c = 256.

6. EXPERIMENTAL RESULTS/EVIDENCE OF THERMOCONVECTIVE OSCILLATORS

All the optical measurements discussed in the previous part are better suited for the study of convective properties in small cells, at least when the structure is three-dimensional. For illustration we give here some results which have been obtained in a cell with Lx = 2d, Ly = 1.2d, and d = 2cm, and in which we have followed the evolution from the steady state to the turbulent state in different convective structures when the Rayleigh number is increased. Indeed, the experimental observation shows that the evolution strongly depends on the structure. Nevertheless, a current picture of the development gives the following sequence: steady state, oscillatory regimes, turbulence. Through all the optical methods described above, we will show how we were able to discern the physical oscillators responsible for different oscillatory regimes.

A. Formation of a Droplet

In a structure, consisting of two X rolls and two Y rolls (see Figure 22), at Ra/Ra_c = 250, the previously steady velocity becomes

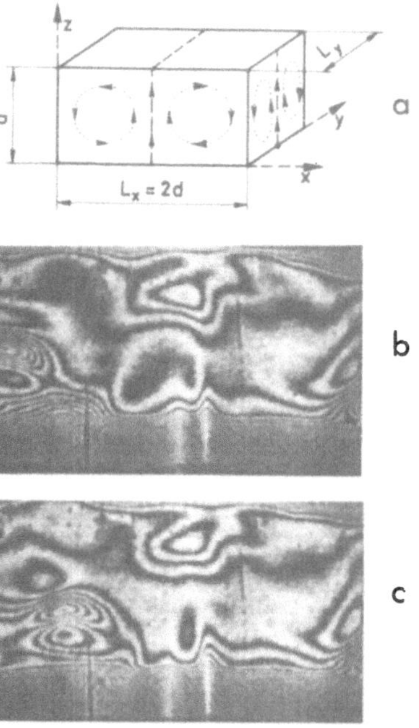

Figure 22. Differential interferometric pattern in the case of the presence of a thermal droplet; (a) schematic of the corresponding structure; (b) droplet at its birth, at the left lower corner of the image; (c) droplet being advected. Ra/Ra_c = 280.

periodic. At the same time, if we look at the differential interferometer images, we notice the periodic formation of a "droplet" (a local temperature heterogeneity of the fluid) in a corner of the cell, and its advection and thermal relaxation.[14] We observed always a single droplet. The degeneracy of the eight corners of the cell was raised, by the following considerations:

- The droplet is born in the corner corresponding to the greatest roll; in general, the two X rolls are asymmetric one being greater than the other; if they are strictly equal, no droplet is formed and convection remains steady.

- If the motion is ascending in the centre of the cell, the droplet is formed near the hot boundary layer (at the bottom); it is a hot droplet.

- If the motion is descending in the centre of the cell, the droplet is cold, and formed near the cold boundary layer.

The knowledge in real time of the behaviour of the droplet allowed us to construct the corresponding velocity field versus time; we recorded the dependence of the velocity with time, $V_Z = f(t)$, for different X and Y values, with the same phase reference given by the evolution of the droplet; from that we constructed the time dependent velocity field $V_Z = f(X)$ and $V_Z = f(Y)$ around the center of the cell. A typical example is shown on Figure 23 where the time $t = 0$ corresponds to the beginning of the droplet advection. We notice how the velocity field is perturbed and strongly enhanced by the droplet advection, and its crossing near the centre of the cell. The action of the droplet is so strong that it induces a displacement of the stream lines like a spatial oscillation of the frontiers between the rolls (where V_Z is maximum). This behaviour is to be compared with that observed in Reference 15 by the focalization effect.

B. Cold Boundary Layer Instabilities

In a structure where two X rolls and one Y roll are developed (see Figure 17B) the time-dependent phenomenon sets in at $Ra/Ra_c \simeq 215$, in the form of a periodic motion, the period T of which is approximately two times smaller than in the previous case (droplet). Here, we were able to point out the physical oscillator responsible of this regime, by looking at the time-variation of the isotherms observed by Mach-Zehnder interferometry. We noticed that the cold front of the cold boundary layer advances in the convective part of the fluid, then regresses and so on, giving a periodic modulation of the velocity amplitude. This instability in the cold boundary layer was first described theoretically by Howard[16] but, until now, not observed. The corresponding temperature variations were measured through Mach-Zehnder interferometry in different points

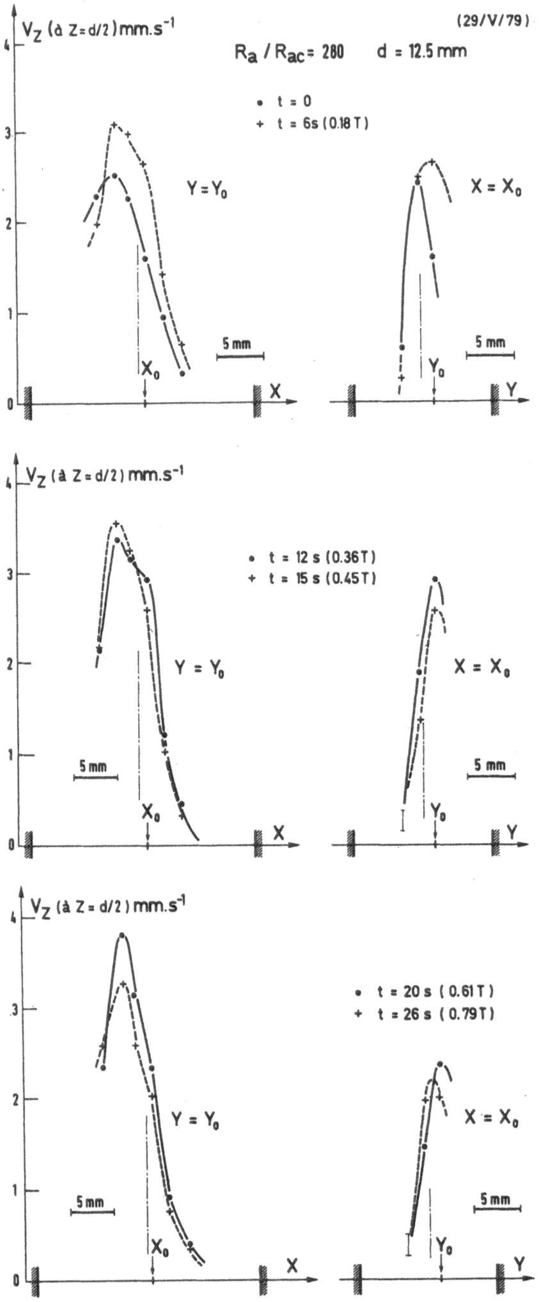

Figure 23. Time dependent velocity field (near the center of the cell) in the case of the droplet oscillator. The different velocity profiles are taken at some characteristic times of the period T of the droplet cycle.

in the front of the cold boundary layer; the results (see Figure 19) show how the local temperature modulations are small.

When the Rayleigh number is increased, a biperiodic regime appears for $Ra/Ra_c \simeq 250$. Looking at the Mach-Zehnder images, we notice that a new oscillator appears in the form of the periodic growing of the thermal bumps, located in the descending motions (see Figure 17). When the Rayleigh number is sufficiently high, this periodic perturbation gives rise to the formation of a cold "plume" which crosses the fluid from the cold boundary layer to the "hot" one, the corresponding variations of the temperature near the cold boundary layer are shown in Figure 19. We notice the existence of two periods, one related to the global "oscillation" of the cold boundary layer (τ = 41 S), the other given by the periodic growing of the localized thermal bump (τ = 36 S).

So in this manner, the two physical oscillators responsible for the periodic and biperiodic behaviours have been identified separately. This information is supplemented with the beam deflection technique described in Chapter III-C. One of the beams crosses the layer around the central part of the cold boundary layer, the other one crosses the "bump". Each diode signal reveals a different period, and the corresponding spectra exhibits clearly the two different periods of the two oscillators, the bulk boundary layer and the bump (see Figure 24).[17]

As the two oscillators can be measured separately, it is possible to draw their relative phase diagram. This one shown on Figure 25 was related to a locked regime of the two oscillators (f_1/f_2 = 6/7).

7. CONCLUSION

The Rayleigh-Benard convection problem, when studied near room temperature, lends itself naturally to be studied by optical means, which have the advantage to be non-perturbing and to give quantitative local informations of the convective properties.

The best, the "ideal" would be to have, in particular when the properties are time-dependent, instantaneous and simultaneous measurements of the amplitudes of the velocity field and of the temperature perturbations in a great part of the structure. We have described some approaches to this aim and the results obtained by different methods.

ACKNOWLEDGEMENT

We wish to thank V. Croquette and E. L. Koschmieder for helpful discussions and B. Ozenda and M. Labouise for technical assistance.

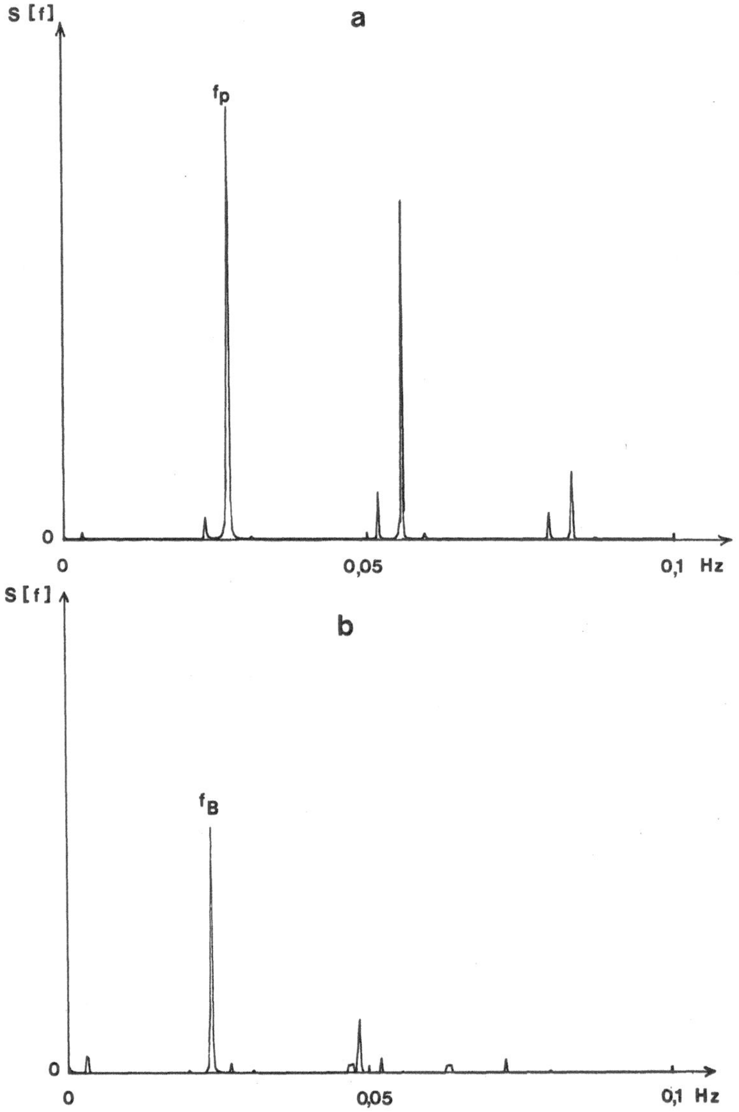

Figure 24. Spectra of the deflection of the beam corresponding to the bump (a) and to the boundary layer (b).

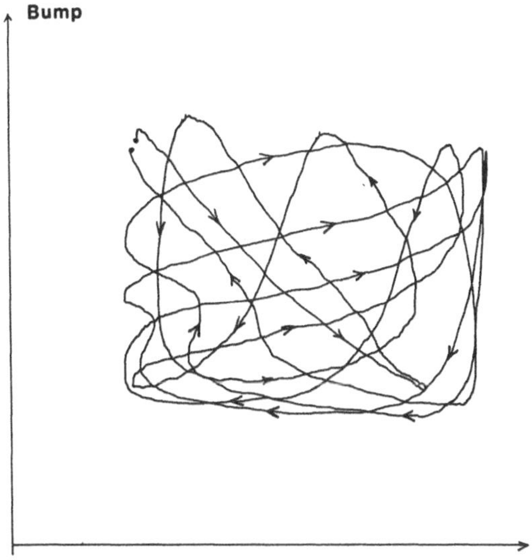

Figure 25. Phase space diagram between the two oscillators "bump" and "boundary layer" as detected by the beam deflection technique.

REFERENCES

1. S. Chandrasekhar, Hydrodynamic and Hydromagnetic Stability, (Oxford: Clarendon, 1961).
   C. Normand, Y. Pomeau, M. G. Velarde, Rev. Mod. Phys 49: 581 (1977).
2. E. L. Koschmieder, and S. G. Pallas, Int. J. Heat Mass Transfer 17: 991 (1974).
3. G. Ahlers, and R. P. Behringer, Prog. Theoretical Phys. Supp. No. 64 (1978).
4. J. P. Gollub, and S. V. Benson, Phys. Rev. Lett. 41: 948 (1978).
5. P. Bergé, Dynamical Critical Phenomena and Related Topics, ed. by H. P. Enz, (Berlin, Springer Verlag, 1979).
6. A. Libchaber, and J. Maurer, J. Phys. Lett. 39: L369 (1978).
7. F. Durst, A. Melling, and J. H. Whitelaw, Principles and Practice of Laser - Doppler Anemometry, (New York: Academic Press, 1976).
8. F. H. Busse, and J. A. Whitehead, J. Fluid Mech. 47: 305 (1971).
9. P. Bergé, and M. Dubois, Phys. Rev. Lett. 32: 1041 (1974).
   I. Procaccia, "System far from Equilibrium" Sitges (Barcelona, Juin 1980).
10. P. Berge, V. Croquette, and M. Dubois (to be published).
11. H. Oertel, Physics - Chemical Hydrodynamics, P. CH 80 (Madrid, 1980).

12. R. Farhadieh, and R. S. Tankin, J. Fluid Mech. 66: 739 (1974).
13. H. Nataf, private communication.
14. P. Bergé, and M. Dubois, J. Phys. Lett. 40: L505 (1979).
15. F. H. Busse, and J. A. Whitehead, J. Fluid Mech. 66: 67 (1974).
16. L. N. Howard, Proceedings of the Eleventh International Congress on Applied Mechanics, (Springer Verlag, 1966). p. 1109.
17. M. Dubois, and P. Bergé, Phys. Lett. A76: 53 (1980); and J. de Phys. (submitted).
18. V. Croquette, private communication.

*Photon correlation technique becomes interesting in LDV when the studied fluid is a gas, poorly seeded and studied via backward scattering.

CONVECTIVE INSTABILITIES IN A HORIZONTAL BINARY LIQUID LAYER

Marzio Giglio - CISE, P.O. Box 12081
Milano 20100, Italy

Vittorio Degiorgio - Instituto di Fisica Applicata
Università di Pavia
Pavia, Italy
and
CISE, P.O. Box 12081
Milano, 20100, Italy

CONTENTS

ABSTRACT

The Soret-driven instability in a thin horizontal layer of a two-component liquid is studied both experimentally and theoretically. The experiment is performed on a dilute polymer solution by using the laser-beam deflection technique. At steady state above threshold, the horizontally-averaged concentration gradient observed at mid-height of the layer shows a saturation effect, and the mass convective flux is found to depend linearly on the temperature difference between the plates. The approach to steady-state shows slowing down effects near threshold, whereas it is characterized by relaxation oscillations somewhat above threshold. The experimental results are in good agreement with a self-consistent single-mode dynamic model of the Soret-driven instability.

1. INTRODUCTION

In this lecture we will describe some experimental results on a convective instability in a two component system and at the same time we will give an account of the relevant theoretical work on the subject. Most of the material to be presented here is related to work done at CISE over the last few years.[1-5]

The physical system we consider is very similar to that encountered in the classical one-component convective instability, the Rayleigh-Bénard instability (RBI). It consists of a thin horizontal layer of fluid, confined between two highly conducting plates. The fluid slab eventually goes into a convective mode, once the temperature difference between the plates exceeds a well defined threshold.[6-9] Two component systems are in general more complex to study than pure fluids, because the density changes and the associated buoyancy forces can be generated both by temperature and concentration spatial variations. In the discussion of the stability and dynamic behavior of these systems a relevant role is played by the Soret effect[10] which couples the mass flow to the temperature gradient (convective instabilities in two component systems go under the name of Soret Driven Instabilities (SDI)). Let us recall here that quite generally the mass flow J is given by

$$J = -\rho D \text{ grad } c - \rho D(k_T/T) \text{ grad } T \tag{1}$$

where D is the mass diffusion coefficient, c the mass fraction of the denser component and k_T is the Soret coefficient. At steady state, when J = 0, one then has a stable concentration gradient given by

$$(\text{grad } c)_{t\to\infty} = -(k_T/T) \text{ grad } T \tag{2}$$

The actual value of k_T depends on the solute and on the solvent and in general it depends on the value of the concentration. Furthermore, it can be either positive or negative. According to the definition for c, a positive value implies that the denser component migrates to the cold plate, and the reverse is true for a negative Soret coefficient.

Therefore, there are four possible cases to be considered according to whether the temperature gradient is stabilizing or destabilizing and whether the Soret induced concentration gradient is stabilizing or destabilizing. We will discuss here the case in which both effects are destabilizing (heating from below, and a positive Soret coefficient). In such a system, the theory[7-9] predicts that convective motion will set in much earlier than the RBI threshold, because the unstable concentration profile can induce convection very efficiently, and the temperature profile plays a very inessential role. This is due to the fact that concentration spatial inhomogeneities are more long lived than temperature inhomogeneities, and therefore persistent buoyancy forces are mainly due to the concentration distribution. The depression of the threshold value is more pronounced for systems characterized by a small value of (D/χ) and large values of k_T (χ is the thermal diffusivity of the sample). Furthermore, large values for k_T guarantee that the concentration modulations introduced by the convection state will be more readily observable. As a sample, we have therefore chosen a dilute polymer solution (c = 0.01 weight fraction) of polystyrene in methylethylketone). The polystyrene sample is monodisperse and the molecular weight is MW = 4000.

The study reported here is restricted to a region not too far from the threshold. As the threshold is exceeded, the two relevant variables, the fluid velocity and solute concentration assume a well defined space dependence. While the velocity does not produce per se any change in the index of refraction, concentration modulations do produce such changes and are therefore observable with classical optical methods. The technique we have used is a laser beam deflection technique, which gives the value of the vertical concentration gradient averaged along the length of the path traversed by the beam. Beam deflection data provide us with quantitative information on the convective flow, which in turn depends on the product of the amplitudes of the concentration mode and velocity mode.

The measurements have not been confined to the steady state behavior for different settings of the temperature gradient above threshold. We have also investigated the time evolution of the concentration gradient when the temperature difference between the plates is applied suddenly starting from an isothermal condition. Very close to the threshold these data provide us with a fairly good insight on the slowing down process for the buildup of the two lowest order velocity and concentration modes. Further away from threshold a more complex dynamical behavior is observed, and the system exhibits damped relaxation oscillations. All the results are interpreted in terms of the available theoretical predictions.

2. EXPERIMENTAL SETUP

The experimental setup consists of the following parts: a) the optical system necessary for the evaluation of the concentration gradient at the midplane of the sample, b) the cell, c) the temperature controller. We will briefly describe these parts. More details can be found in Reference 3.

A. The Optical Setup

A nice way to perform measurements of the concentration gradients is via the determination of the index of refraction gradient. Unfortunately temperature gradients will also produce refractive index gradients

$$\frac{dn}{dz} = \frac{\partial n}{\partial T}\frac{dT}{dz} + \frac{\partial n}{\partial c}\frac{dc}{dz} \tag{3}$$

If the temperature contribution can be accounted for and $(\partial n/\partial c)$ is known, then one can determine quantitatively (dc/dz). Although many optical techniques could be devised for the determination of (dn/dz), we found it very convenient to exploit the laser beam deflection technique, already mentioned in other publications.[11] With this technique one determines the refraction index gradient (dn/dz) by measuring the angular deflection θ suffered by a laser beam traversing a cell of length ℓ :

$$\theta = \ell \frac{dn}{dz} \tag{4}$$

By so doing one obtains a value for (dn/dz) which is the average value along the path traversed by the beam. The shaping optics consists of a spatial filter and a focusing lens. In this way we can produce a beam with a waist diameter $2w = 0.1$ mm, located at mid-height of the sample. The associated Rayleigh range $2Z_R = 2\pi w_o^2/\lambda \simeq 20$ mm is roughly equal to the length of the sample.

After passing through the sample, the beam is intercepted by a beam sensor located one meter away from the sample. The sensor is a solid state position sensing device which gives an output proportional to the displacement of the beam centroid from a fixed position. Unfortunately the output is also proportional to the actual beam intensity. By proper wiring of the sensor however, it is possible to extract an output proportional to the beam intensity. These outputs are fed to a ratiometer with a 0.1 % accuracy and therefore the data are truly proportional to beam centroid displacement and insensitive to laser intensity level drifts. The deflection θ can be measured with an accuracy of a few microradians.

B. The Cell

The cell has a cylindrical geometry, so that the design reduces to a three piece assembly, consisting of two aluminum alloy plates and a cylindrical glass window. The two plates are of square cross section, are identical in shape, and have two cylindrical expansions which slide into the glass window which was ground and polished from a solid piece of optical glass. Sealing between the plates and the flat portions of the cylindrical window was achieved by means of preformed, annular flat indium gaskets, which were compressed by means of six bolts clamping together the two plates and the glass window.

The actual dimensions of the cell are: cylindrical window OD equal to 40 mm, ID 20 mm, 10 mm high. The cylindrical expansions on the plates are slightly less than 20 mm in diameter and 4.60 mm high. Once the indium gaskets are compressed the fluid height is a = 1.03 mm.

Filling of the cell is performed through two skew holes positioned on the top plate, terminating on a side of the top plate and closed by an indium gasketed flange. Both top and bottom plates are flat, with no protruding parts. Two small grooves are machined on these surfaces to allocate two fast response thermistors which are the sensing elements of the electronic servo to be discussed later.

C. The Temperature Controller

The temperature controller is a two stage system. The first stage consists of a cubic box, 27 cm by the side, made of 2 cm thick aluminum alloy plate. All the plates of the box have channels through which we circulate temperature controlled fluid provided by a commercially available machine. The temperature stability of the box is better than ± 30 mdeg over days. The second servo is a homemade one. It operates as follows. The two cell plates are in contact with two Peltier pumps. Above and below the Peltier elements we have two aluminum blocks through which we circulate the temperature-controlled fluid which also circulates in the large aluminum box. These two blocks act as heat sinks for the Peltier elements, which are connected in series in such a way that when current flows across them, if one element removes heat from one of the cell plates the second one actually injects heat into the other. The magnitude and sign of the current is controlled by the electronic servo. The error signal to the servo is provided by a d. c. bridge made of the two thermistors in contact with the cell plates, a low temperature coefficient resistor and a six digit decade box (also of low temperature coefficient, less than ± 10 ppm/^{o}C). Typical performances of the servo can be specified as follows: the time constant for the temperature buildup is 30 sec, long term stability: ± 0.2 mdeg over at least 12 hours. The time constant was determined by watching the evolution of the temperature difference as measured by the two quartz probes inserted in the cell plates and by deconvolving with the time constant of the probes themselves.

3. EXPERIMENTAL RESULTS

The data have been taken according to the following procedure. Before each run the system was prepared in an isothermal state by

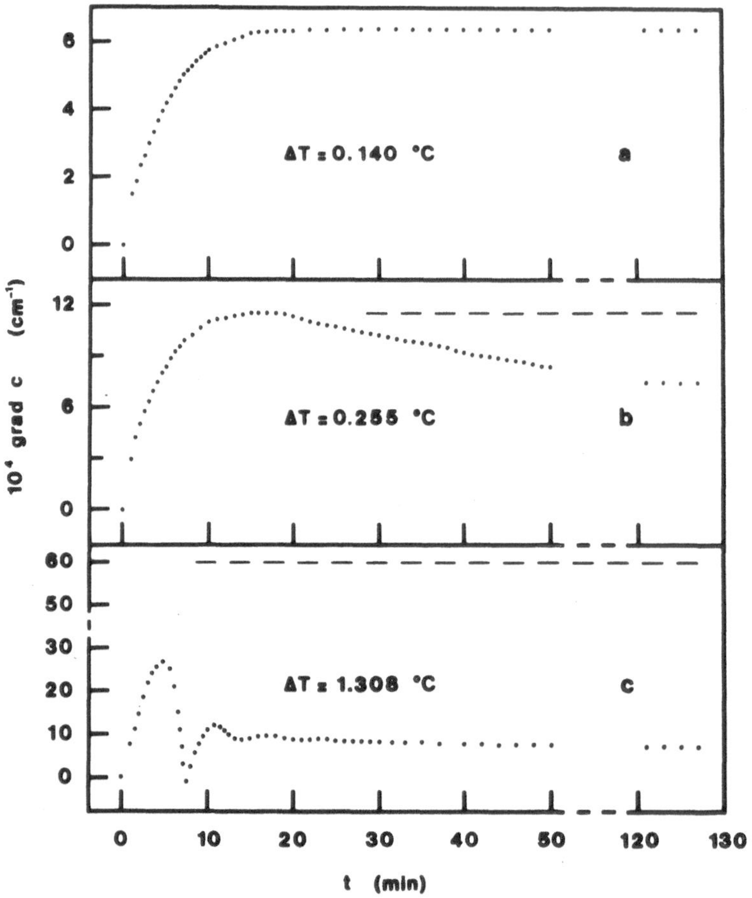

Figure 1. Time evolution of the concentration gradient evaluated at midheight of the sample for different values of the temperature difference between the plates.

setting the temperature difference between the plates equal to zero with the electronic servo. The beam center position was then determined and its stability controlled in order to assess the absence of any drift in the unstressed sample. A chosen temperature difference was suddenly applied by selecting the appropriate value on the resistance decade box and measurements of the time evolution of the beam deflection were then performed. From these data, values for (dc/dz) were obtained. Typical results are reported in Figure 1. Time evolutions like the one reported in Figure 1a are typical of the Soret-generated concentration gradients in the absence of instability. For time delays of the order of $(a^2/\pi^2 D)$ the time evolution of the concentration gradient at midheight is well represented by a single exponential decay

$$\frac{dc(t)}{dz} = \frac{dc(\infty)}{dz}\left(1 - \frac{4}{\pi} e^{-t/\tau_o}\right) \tag{5}$$

where $\tau_o = a^2/\pi^2 D$. Measurements under stable conditions, although not very interesting to the discussion of the onset of the convective instability are however useful since they provide us with an independent estimate of both the diffusion coefficient D and the Soret coefficient k_T. We find $D = 4.63 \times 10^{-6}$ cm^2/sec and $k_T = 4.87$ in good agreement with the data available in the literature.[12] As the temperature difference is made larger, one encounters time evolutions of the type shown in Figure 1b. Notice that now the concentration gradient at first approaches the value one expects in the absence of the instability, but such value is abandoned, and the gradient then approaches a new, true, steady-state value. By making the temperature gradient somewhat larger, the growth of the concentration gradient is no longer fast enough to allow even a temporary attainment of the value expected with no instability (see the dashed line in Figure 1c) and the gradient falls rather rapidly to the steady state value via heavily damped oscillations.

We shall see now how these results can be interpreted and what kind of information can be extracted. The basic question is why one observes at steady state a reduction of the concentration gradient, provided that the temperature difference between the plates becomes large enough. The explanation lies in the fact that above a well-defined threshold a convective instability sets in, leading to the formation of a spatial structure for both the concentration and the velocity field. We will postpone for the moment the theoretical description, but we will give an intuitive description of the mechanism leading to the reduction of the concentration gradient. It is particularly instructive to describe what happens in the horizontal plane at midheight for the following two variables, the concentration c and the vertical component of the fluid velocity v_z. The velocity v_z attains a negative value in correspondence with the position where the concentration is higher, while it is positive where the concentration is lower than average. As a consequence the onset of the convective mode leads to the formation of an average convected mass flow J_c, directed downward and given by

$$J_c = \int v_z\,(x,y)\; c\,(x,y)\; dxdy \tag{6}$$

The total mass flow across the horizontal plane, given by

$$J = -\rho D \text{ grad } c - \rho D\,(k_T/T) \text{ grad } T + J_c \tag{7}$$

will of course be zero once the steady state is attained. Consequently the steady state concentration gradient will be

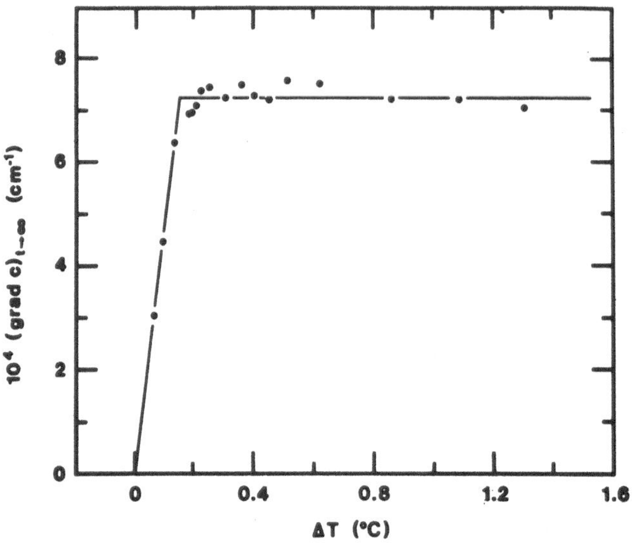

Figure 2. Steady state value of the concentration gradient at mid-height of the sample as a function of the temperature difference between the plates.

$$(\text{grad } c)_{t\to\infty} = -(k_T/T) \text{ grad } T + \frac{J_c}{\rho D} \tag{8}$$

Notice that a quantitative measure of the difference between the value expected with no instability (Equation 2) and the value actually attained at a steady state allows a precise determination of the vertically convected mass flow above threshold. The actual position of the threshold can be found by plotting the value of the steady state concentration gradient as a function of ΔT since a break is expected at $\Delta T = \Delta T_c$. Such a break is indeed observed in Figure 2 which shows very clearly an abrupt change in the slope at $\Delta T = (155 \pm 3)$ mdeg. We take this as the critical temperature difference for our system. Furthermore, one can notice that, rather surprisingly, beyond the break the gradient remains constant, locked to the threshold value and insensitive to any further increase in the temperature gradient. From the data shown in Figure 2, using Equation 8 in conjunction with our values for k_T and D, we can derive the values of the mass flow as a function of $\Delta T - \Delta T_c$. These data are presented in Figure 3 on a log-log plot. The best fit line gives an exponent $\gamma = 1.01 \pm 0.02$.

Besides steady state properties, dynamic quantities can also be extracted from the plots presented in Figure 1. First, one can study if there is any change in the time constant τ_o for the growth of the concentration gradient immediately after the temperature

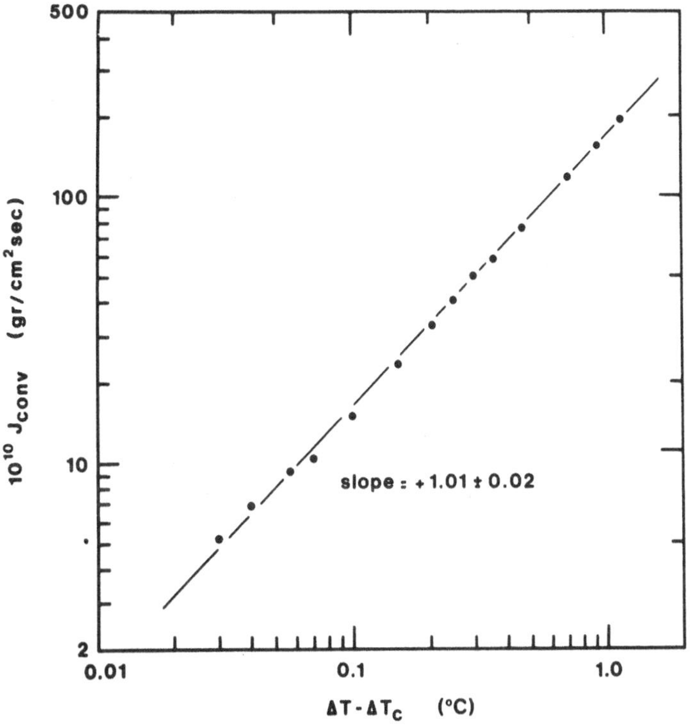

Figure 3. Vertically convected mass flow across the midplane of the cell as a function of ΔT-ΔT_c

gradient is applied to the system. Second, and more interestingly, one can study the time constant τ_{sd} which characterizes the approach to the convective steady state.

In Figure 4 we report the values of τ_o and τ_{sd} as a function of $\Delta T - \Delta T_c$ in a log-log plot. As one can see, τ_o does not change appreciably. On the contrary τ_{sd} exhibits a fairly pronounced dependence on $\Delta T - \Delta T_c$. If a power law fit is attempted for τ_{sd} we find that τ_{sd}=A $[(\Delta T - \Delta T_c)/\Delta T_c]^{-\nu}$ with A ≑ (571.8 ± 17.1)s and $\nu = 1.02 \pm 0.03$.

Finally, let us describe the results related to the plots exhibiting oscillatory behaviour. As the temperature difference is made progressively larger, we noticed that the period of the oscillations became shorter and shorter. The number of oscillations which are clearly detectable is rather limited, and the period can be evaluated as an average over a few cycles (typically two or three cycles only). As we pointed out, oscillatory behaviour is observed for values of ΔT somewhat larger than ΔT_c. We assumed,

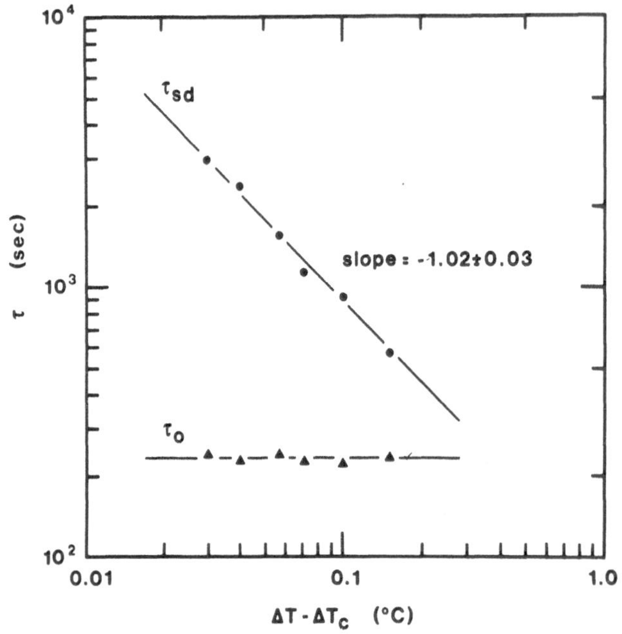

Figure 4. Behaviour of τ_o and τ_{sd} as a function of $\Delta T - \Delta T_c$.

as it is plausible, that there exists a well defined second threshold ΔT_{osc} at which the system is critically damped and we have attempted a rather crude fit of the type $\tau_{osc} = B\,(\Delta T - \Delta T_{osc})/\Delta T_{osc}^{-\theta}$. We obtain $B = (221.7 \pm 6.6)$s, $\Delta T_{osc} = (337 \pm 5)$mdeg, $\theta = 0.50 \pm 0.02$, keeping both ΔT_{osc} and θ floating. These data are reported in Figure 5.

We shall now proceed to the interpretation of all these experimental results on the basis of the available theories and to the comparison of some of the observed features with those reported in the literature for other types of instabilities.

4. THEORETICAL CONSIDERATIONS AND DISCUSSION

The well known conservation equations for a two-component liquid mixture are:[8]

$$\frac{\partial c}{\partial t} + \sum_{j=1}^{3} v_j \frac{\partial c}{\partial x_j} = D\nabla^2 c + \frac{Dk_T}{\bar{T}} \nabla^2 T \tag{9}$$

$$\frac{\partial v_i}{\partial t} + \sum_{j=1}^{3} v_j \frac{\partial v_i}{\partial x_j} = g\,[1-\alpha(T-\bar{T})+\gamma(c-\bar{c})]\lambda_i - \frac{1}{\bar{\rho}}\frac{\partial p}{\partial x_i} + \nu\nabla^2 v_i \tag{10}$$

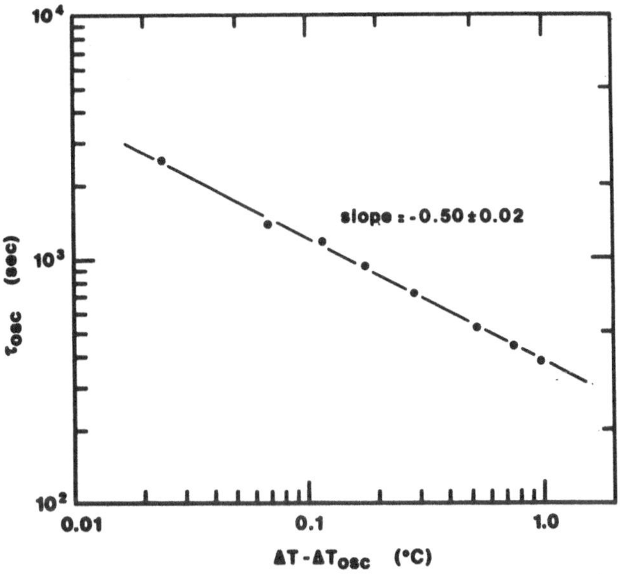

Figure 5. Damped oscillation period t_{osc} as a function of $\Delta T - \Delta T_{osc}$.

$$\frac{\partial T}{\partial t} + \sum_{j=1}^{3} v_j \frac{\partial T}{\partial x_j} = \chi \nabla^2 T \tag{11}$$

$$\sum_{j=1}^{3} \frac{\partial v_j}{\partial x_j} = 0 \tag{12}$$

where c is the mass fraction of the heavier component, v_i is the i-th component of the velocity, T the temperature, and P the pressure. The constant ν represents the kinematic viscosity, g the gravity acceleration, $\alpha = -(1/\rho)\,(\partial\rho/\partial T)$, and $\gamma = (1/\rho)\,(\partial\rho/\partial c)$, where ρ is the density. The symbols $\bar{T}$, $\bar{c}$, $\bar{\rho}$ represent average quantities over the layer. Taking the z-axis (i = 3) coincident with the vertical axis, $\lambda_i = 0$ for i = 1, 2, and $\lambda_3 = 1$. A discussion of the assumptions involved in the derivation of Equations (9)-(12) can be found, for instance in Reference 8. The equations for a single-component system are immediately obtained from Equations (9)-(12) by putting $c = \bar{c} = 0$.

The situation considered here is that of a liquid mixture, having $k_T > 0$, which is heated from below. At small temperature differences ΔT, a constant vertical concentration gradient will arise, according to the Soret effect, so that $c(z) = \bar{c} + (k_T/T)\,(\Delta T/a)z$. Above the critical temperature difference ΔT_c, the system goes unstable. Since no convective heat flux is originated by the Soret-driven instability if $\chi >> D$, it is reasonable to assume that the temperature distribution inside the cell is not modified by the convective mass motion, and therefore, for any value of ΔT,

$$T = \bar{T} - (\Delta T/a)\ z \tag{13}$$

It is easy to verify with Equations (9)-(12) that this assumption makes the SDI problem formally identical to the RBI problem, with the concentration c playing the role of the temperature. The vertical component of the velocity v_z and the concentration c are expressed in the single-mode approach[4,5] (cylindrical rolls parallel to the y-axis) as

$$v_z = B(t)\ v(z)\ \cos kx \tag{14}$$

$$c = \bar{c} + A(z,\ t) + C(t)\ w(z)\ \cos kx \tag{15}$$

The dimensionless functions v(z) and w(z) are taken here as known functions. They are indeed the solutions of the eigenvalue problem which defines the neutral stability curve. The two functions have their maximum value at the midplane $z = 0$. We assume that $v(0) = w(0) = 1$, without loss of generality because the dependence on ΔT can be included into the time-dependent amplitudes B(t) and C(t). We write A(z, t) as

$$A(z,\ t) = \frac{k_T}{\bar{T}} \frac{\Delta T}{a} z - A(t)\ f(z) \tag{16}$$

where the dimensionless function f(z) must satisfy the boundary conditions $f(0) = 0$ and $df/dz = 0$ for $z = \pm\ a/2$, and is given by

$$f(z) = \int_0^z \frac{v-w}{a}\, dz \tag{17}$$

By substituting Equations (13)-(17) into Equations (9)-(12), we obtain after a calculation described in detail in Reference 5, the set of equations

$$\dot{\Delta} = [h_1/(2a^2)]\ BC - (h_2 D/a^2)\ (\Delta - \Delta_o) \tag{18}$$

$$\dot{C} = -\Delta B - h_3 D k^2 C \tag{19}$$

$$\dot{B} = -h_4 \nu k^2 [B + (h_3 D k^2/\Delta_c)\ C] \tag{20}$$

where $\Delta(t) = \frac{k_T}{T}\frac{\Delta T}{a} - A(t)\,(df/dz)_{z=0}$, $\Delta_o = \frac{k_T}{T}\frac{\Delta T}{a}$, $\Delta_c = \frac{k_T}{T}\frac{\Delta T_c}{a}$, and the h_j's are dimensionless constants, whose expressions are given in Reference 5.

An approximate evaluation of the h_j's for the SDI can be performed by choosing for v(z) and w(z) the lowest order polynomial expression which satisfies the boundary conditions, that is, $v(z) = 1 - 8(z/a)^2 + 16(z/a)^4$, $w(z) = 1$. One finds $h_1 = h_2 = 10.91$, $h_3 = 1$. Furthermore, for small ka, $h_4 = 24/(ka)$.

As mentioned in the Introduction, the Soret-driven instability may also occur in a binary mixture having $k_T < 0$, but heated from above, that is, with a negative temperature difference ΔT. It can be immediately verified that Equations (18)-(20) describe equally well this latter case.

The steady-state solutions Δ_s, C_s, and B_s of Equations (21)-(23) are

$$\Delta_s = \Delta_c = \frac{k_T}{\overline{T}}\frac{\Delta T_c}{a} \tag{21}$$

$$C_s = \left(\frac{2h_2}{h_1 h_3}\right)^{\frac{1}{2}} \frac{\Delta_c}{k}\,\varepsilon^{\frac{1}{2}} = 1.41\,\frac{k_T}{ka}\frac{\Delta T_c}{\overline{T}}\,\varepsilon \tag{22}$$

$$B_s = -\left(\frac{2h_2 h_3}{h_1}\right)^{\frac{1}{2}} kD\varepsilon^{\frac{1}{2}} = -1.41\,ka\,\frac{D}{a}\,\varepsilon^{\frac{1}{2}} \tag{23}$$

where $\varepsilon = (\Delta T - \Delta T_c)/\Delta T_c$.

Equation (21) predicts that the horizontally-averaged concentration gradient at the midplane of the cell takes, above threshold, a value independent of ΔT and coincident with the critical concentration gradient; this prediction is in very good agreement with the measurements performed for the SDI in the mixture ethanol-toluene[1] and with the results reported in Figure 2.

The steady-state convected mass flow is

$$J_c = \frac{B_s C_s}{2} = \frac{1.99}{2}\,k_T\,\frac{D}{a}\,\Delta T_c\,\varepsilon \tag{24}$$

The ε dependence of the convected mass flow determined experimentally is in very good agreement with the prediction of Equation (24) since the exponent for the ε dependence determined experimentally is very close to unity (see Figure 3). Furthermore, the numerical coefficient in Equation (24), as calculated from our determination of D and k_T, is equal to 32.1 x 10^{-10} (gr/cm^2 sec), while the experimental value derived from the data of Figure 3 (when replotted as a function of ε instead of $\Delta T - \Delta T_c$) is 25.3 x 10^{-10} (gr/cm^2 sec). As one can notice the agreement is better than just an order of magnitude agreement. This is rather remarkable and perhaps fortuitous since the calculations are made for a rectangular geometry, while in our case the geometry is cylindrical.

Equations (18)-(20) fully describe the transient evolution of the SDI toward the steady-state starting from an arbitrary initial condition at t = 0. In many relevant cases the description of the transient can be considerably simplified by using the so-called adiabatic approximation which is based upon the comparison of the different time scales involved in the problem and the consequent elimination of the faster variables. We discuss here for simplicity only the case $\nu/D \gg 1$ since this is the case investigated in the experiments about SDI transients. Under this assumption the fast variable is B and the variable showing critical slowing down is c. Also, in the region very close to threshold ($\varepsilon \ll 1$), the dynamics of Δ is fast in comparison with that of c, and therefore both B and Δ can be adiabatically eliminated by putting $\dot{B} = \dot{\Delta} = 0$. The resulting dynamic equation for c is

$$\dot{C} = h_3 D k^2 \varepsilon C - \frac{h_1 h_3^2}{2h_2} \frac{Dk^4}{\Delta_c^2} C^3 \tag{25}$$

Equation (25) is a Van der Pol equation, such as the equation describing the dynamics near threshold of electronic and optical oscillators. The solution of Equation (25) is well-known and it will not be discussed here. We simply note that the time constant τ_{sd} for the decay of small deviations from the steady-state, as derived from Equation (25), is

$$\tau_{sd} = \frac{\varepsilon^{-1}}{h_3 k_c^2 D} \tag{26}$$

Considering the data reported in Figure 4, we see that the experimentally determined power law dependence on ε is in very good agreement with the theoretical prediction. The numerical coefficient in Equation (26) depends on the actual value of the wavevector of the mode. Since k_c is not known in this case, no meaningful comparison can be made between the theoretical and the experimental values of the coefficient. We simply note that if we take for h_3 the value appropriate for rectangular geometry, $h_3 = 1$, by comparing the experimental τ_{sd} with Equation (26), we derive $k_c = 20\ \text{cm}^{-1}$.

Slightly above threshold, when τ cannot be considered much larger than the characteristic evolution time of the concentration gradient which is $(Dh_2/a^2)^{-1}$, it is not possible to eliminate adiabatically Δ. The single conditions $\dot{B} = 0$ leads to the following pair of equations:

$$\dot{\Delta} = -(h_1/a^2)J_c - (h_2D/a^2)(\Delta - \Delta_o) \qquad (27)$$

$$\dot{J}_c = (2h_3Dk^2/\Delta_c)(\Delta - \Delta_c)J_c \qquad (28)$$

where $J_c = BC/2$ represents the convective mass flux at the midplane of the liquid layer. Note that, under the assumption $\dot{B} = 0$, J_c is proportional to c^2 and to B^2.

The evolutions of Δ and J_c have been obtained by numerical computation of Equations (27) and (28) with the initial conditions $\Delta(0) = \Delta_o$ and $J_c(0) = J_{ci}$, where J_{ci} is a small value simulating the noise which triggers the onset of the instability. Some evolutions, obtained for three distinct values of ε, and for h = 1, are shown in Figures 6 and 7. The most interesting and new feature of the transients is the damped oscillatory approach to steady state shown for not too small values of ε. The curves relative to large values of ε are particularly striking because the gradient Δ goes through a negative peak and the convective flux has an overshoot several times larger than the steady-state value.

Some information about the damped oscillating behavior can be be gained by a linearization of Equations (27) and (28) around the steady-state. The associated characteristic equation possesses two complex conjugate roots when $\varepsilon > \varepsilon_t$. The threshold for the appearance of oscillations is $\varepsilon_t = h_2/(8h_3k^2a^2)$, the expression of the period of oscillations is $\varepsilon_{osc} = (2a^2k^2h_2h_3/\pi^2)^{-\frac{1}{2}}(a^2/D)(\varepsilon-\varepsilon_t)^{-\frac{1}{2}}$, and the damping time of the oscillations is $\tau_D = 2(Dh_2/a^2)^{-1}$.

By comparing the experimental results with Equations (27) and (28) the following conclusions can be drawn. The computer generated solutions reproduce fairly well the qualitative features of the experimental curves shown in Figure 4. The data confirm the existence of a threshold $\varepsilon_{osc} > 0$ for the appearence of transient oscillations. The order of magnitude of ε_{osc}, taking for h_2 the value $h_2 = 10.9$, is correctly predicted by theory. The data confirm also that τ_{osc} follows a power law dependence on $(\varepsilon - \varepsilon_{osc})$ with exponent $(-\frac{1}{2})$.

This work was partially supported by CNR.CISE contract n.80. 00016.02.

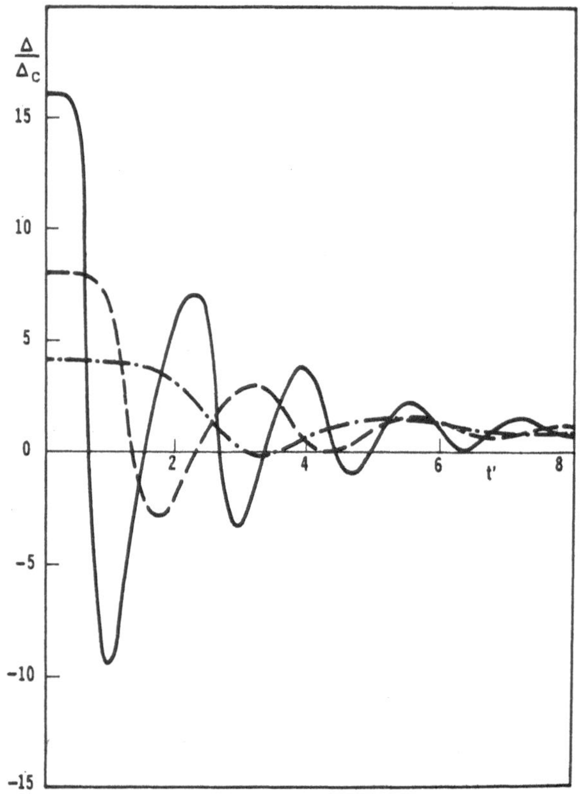

Figure 6. Theoretical time evolution of the normalized temperature gradient Δ/Δ_c starting from the initial condition $\Delta(0) = \Delta_o$ for three distinct values of $\varepsilon = (\Delta_o - \Delta_c)/\Delta_c$; $\varepsilon = 3$ (dot-dashed line), $\varepsilon = 7$ (dashed line), and $\varepsilon = 15$ (full line). The normalized time t' is defined as $t' = (h_2 D/a^2)t$.

REFERENCES

1. M. Giglio, and A. Vendramini, Opt. Comm. 20: 438 (1977).
2. M. Giglio, and A. Vendramini, Phys. Rev. Lett. 39: 1014 (1977).
3. M. Giglio, and A. Vendramini, to be published.
4. V. Degiorgio, Phys. Rev. Lett. 41: 1293 (1978).
5. V. Degiorgio, Phys. Rev. A. 20: 2193 (1979).
6. G. Z. Gershuni, and E. M. Zhuckhoviskii, Prikl. Mat. Mech. (USSR) 27: 1197 (1963).
7. D. T. J. Hurle, and E. Jakeman, J. Fluid Mech. 47: 667 (1971).
8. R. S. Schechter, I. Prigogine, and R. Hamm, Phys. Fluids 15: 379 (1972).
9. M. G. Velarde, and R. S. Schechter, Phys. Fluids 15: 1707 (1972).

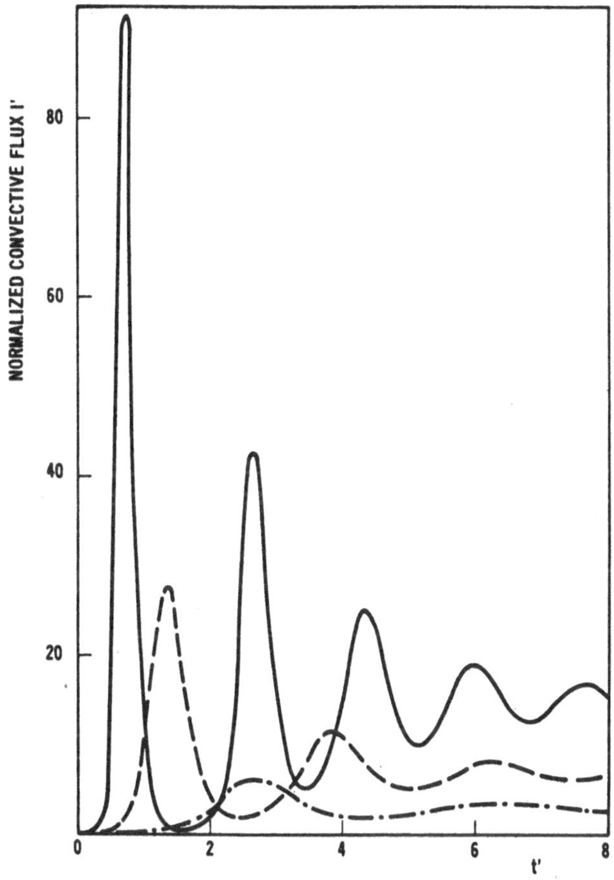

Figure 7. Theoretical time evolution of the normalized convective flux $I' = (h_1/2h_2\chi\Delta_c)J_c$, starting from the initial condition $I'(0) = 10^{-2}$ for the same three values of ε used in Figure 6.

10. S. R. DeGroot, and P. Mazur, Non-Equilibrium Thermodynamics. (North-Holland, Amsterdam, 1962).
11. M. Giglio, and A. Vendramini, in Photon Correlation Spectroscopy and Velocimetry, edited by H. Z. Cummins, and E. R. Pike (Plenum, New York, 1977). p. 471.
12. J. C. Giddings, F. J. F. Yang, and M. N. Myers, Macromolecules 8: 106 (1976).

LOW ANGLE BRILLOUIN SCATTERING IN SYSTEMS UNDER TEMPERATURE GRADIENT

Daniel Beysens

C.E.N. Saclay
SPSRM, BP. No. 2
F91190 Gif-sur-Yvette, France

CONTENTS

1. INTRODUCTION

The modification of the scattering of light by the application of a temperature gradient has been first proposed by Mandel'shtam (1934)[1] and Leontovitch (1939)[2] as a new method for measuring the damping of sound. The idea was the following: the less the damping of the thermal waves, the more the light scattered at a given point of a non-uniformly heated crystal differs from light scattered by the system at equilibrium at the same temperature. An attempt to verify experimentally this effect gave no result within a 1% precision.[3]

If we consider the spectral intensity distribution in the Brillouin spectrum of a crystal submitted to a thermal gradient, the main features were first obtained by Vladmirskii (1942).[4] Griffin[5], in 1968, came about to the same result with a more microscopic calculation. More precisely (Figure 1), in a crystal submitted to a

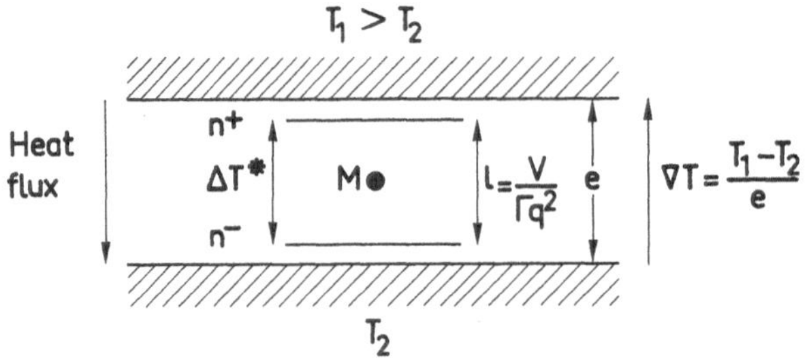

Figure 1. A naive expanation of the Brillouin intensity asymmetry ε (see text). $\ell = v/\Gamma q^2$ is the mean free path of the phonons which can act at the measurement point M. ℓ determines an effective temperature difference $\Delta T^* = \ell \cdot \nabla T$, giving rise to a phonon population difference $n^+ - n^- \propto \Delta T^*$, itself proportional to ε, which thus should vary as $v \cdot \nabla T/\Gamma q^2$.

temperature gradient, there are more phonons propagating in the direction of the heat flux (number n^+) than in the opposite direction (number n^-), giving rise to an asymmetry in the Brillouin lines. At temperature T, the difference $n^+ - n^-$ can be expressed simply as $n^+ - n^- = (dn/dT) \cdot \Delta T^*$, The phonons which contribute to the effect are created within one mean free path ℓ. With v the velocity of the sound wave with momentum q and lifetime $(\Gamma q^2)^{-1}$, ℓ is expressed as $\ell = v/\Gamma q^2$. The relevant ΔT^* is related to the thermal gradient ∇T in the sample by $\Delta T^* = \nabla T \cdot \ell$, showing that the asymmetry in the Brillouin lines is expected to be proportional to $\nabla T/q^2$. Let us note that this effect is calculated in crystals where the phonons which are involved in light scattering contribute to the heat current. This asymmetry should not occur in ordinary fluids where sound waves are hydrodynamic modes, which should make only a negligible contribution to the heat flux.

However very recently a number of theoretical predictions have been worked out about fluids in non-equilibrium steady state. They are sufficiently general to apply to other problems than the Brillouin asymmetry in liquids (such as the transport coefficients in fluids and the logarithmic term in its Virial development), which makes still more important the tests of their validity. They can be divided, as shown in Table I, in molecular theory,[6] kinetic theory[7] for low density fluids, and hydrodynamic theories,[8-10] some using mode-coupling.[11-12] In the near-equilibrium approximation, and letting aside the temperature dependence of the Brillouin parameters (see nevertheless Table 1), they give the same result, which is also the same as in crystals, When a fluid is brought out of equilibrium by a steady temperature gradient ∇T, long range correlations should appear, leading to a modification of the structure factor $S(\vec{q}, \omega)$ which would exhibit an extra $1/q^2$ term. This can be tested by a Brillouin light scattering experiment.

Table 1. Different formulations of the ratio $\varepsilon = [I(\omega<0)-I(\omega>0)]/[I(\omega<0)+I(\omega>0)]$ with I the integrated intensity. Only the difference f has been expressed in: $\varepsilon = (v/2\Gamma T)(\hat{q}\cdot\nabla T/q^2)(1+f)$. Numerical results concern liquid water. Parameters are defined in Table 2 and in the text.

THEORY				f	$\lvert\varepsilon\rvert$ Numerical application : Water at 40°C, q = 2000 cm^{-1} (from table II)	Authors
CRYSTALS /				0 a)b)	0.28	Vladimirskii (1942) Griffin (1968)
FLUIDS	MOLECULAR			0 c)	0.28	Procaccia, Ronis, Oppenheim (1979)
	KINETIC (low density)			$-\frac{2T}{\Gamma}\left(\frac{\partial\Gamma}{\partial T}\right)_p + \frac{1}{2}\cdot\cdot\frac{3}{8}\frac{\mathcal{D}T}{\Gamma} - 2\left(\frac{1}{\rho}\frac{\delta\rho}{\partial T}\right)_p T$ d)	3.22	Kirkpatrick, Cohen, Dorfman (1979)
	HYDRODYNAMIC			$-2\left(\frac{1}{\rho}\frac{\partial\rho}{\partial T}\right)_p T + 4\left(\frac{1}{v}\frac{\partial v}{\partial T}\right)_p T$ e)	0.75	Zwan, Mazur (1980)
				0 f)	0.28	Tremblay, Siggia, Arai (1980)
				0 g)	0.28	Ronnis, Putterman (1980)
		+ MODE COUPLING		0 h)	0.28	Kirpatrick, Cohen, Dorfman (1980)
			far from equilibrium	Solved numerically i)	0.078	Kirpatrick, Cohen, (1980)

a) Reference 4; b) Reference 5; c) Reference 6; d) Reference 7; e) Reference 8; f) Reference 9; g) Reference 10; h) Reference 11; and i) Reference 12.

Table 2. Some useful parameters of liquid water at 40°C: thermal diffusivity ratio D_T, density ρ, velocity of sound V, damping of sound Γ, and their thermal derivative.

D_T $(cm^2.sec^{-1})$	1.510×10^{-3}	*
ρ $(g.cm^{-3})$	0.99224	*
$(\frac{\partial\rho}{\partial T})_p$ $(g.cm^{-3}.K^{-1})$	-3.825×10^{-4}	*
v $(cm.sec^{-1})$	1.530×10^{5}	**
$(\frac{\partial V}{\partial T})_p$ $(cm.sec^{-1}.K^{-1})$	175	**
Γ $(cm^2.sec^{-1})$	1.300×10^{-2}	
$(\frac{\partial\Gamma}{\partial T})_p$ $(cm^2.sec^{-1}.K^{-1})$	-2.040×10^{-4}	**

*Reference 14.
**Reference 13.

The Brillouin spectrum can be expressed as (see Figure 2)

$$S(\vec{q}, \omega) = \frac{1 + \varepsilon(\vec{q},\omega)}{(\omega+vq)^2 + (\Gamma q^2)^2} + \frac{1 - \varepsilon(\vec{q},\omega)}{(\omega-vq)^2 + (\Gamma q^2)^2} .$$

Here $\vec{q}$ is the transfer wavevector, v is the sound velocity, and ω is the angular frequency. Γq^2 corresponds to the sound damping and is the half-width at mid-height of the Brillouin lines at equilibrium. $\varepsilon(\vec{q}, \omega)$ is in general a complex - and debated - function of $\vec{q}$ and ω. For reasons explained below, we will have access only to the static (integrated) intensities of the Brillouin lines, and in this case ε reduces to

$$\varepsilon_q = \frac{v}{2\Gamma T} \cdot \frac{\hat{q}\cdot\vec{\nabla}T}{q^2} [1 + f(\hat{q}\cdot\frac{\vec{\nabla}T}{q^2})],$$

with $\hat{q} = \vec{q}/q$. In most theories, $f[\hat{q}(\vec{\nabla}T/q^2)] = 0$ (see Table 1). Another formulation[12] concerns liquid water far from equilibium (high values of $\nabla T/q^2$). It had to be solved numerically, and is reported in Figure 3.

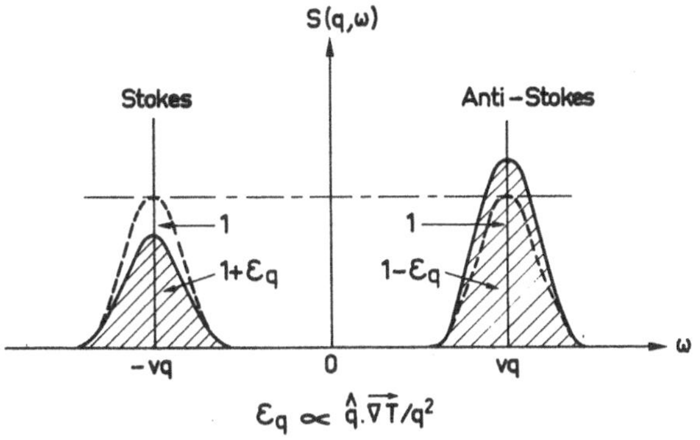

Figure 2. Brillouin spectrum of a fluid. When submitted to a temperature gradient, the lines become asymmetric, the areas are changed from 1 to $1 \pm \varepsilon_q$. Let us note that $\varepsilon_q < 0$ here.

2. EXPERIMENTAL (Figure 4)

The long range contribution $1/q^2$ implies that experiments had to be performed at very low scattering angles in order to obtain a perceptible effect. Water near room temperature and at atmospheric pressure was used, despite the fact it is a poor light scatterer (the Brillouin intensity is about five times lower than in ordinary fluids such as Benzene) and that it is difficult to remove small impurities, bacterias, dust, etc. which scatter very much at small angles. On the other hand the velocity of sound is higher than in

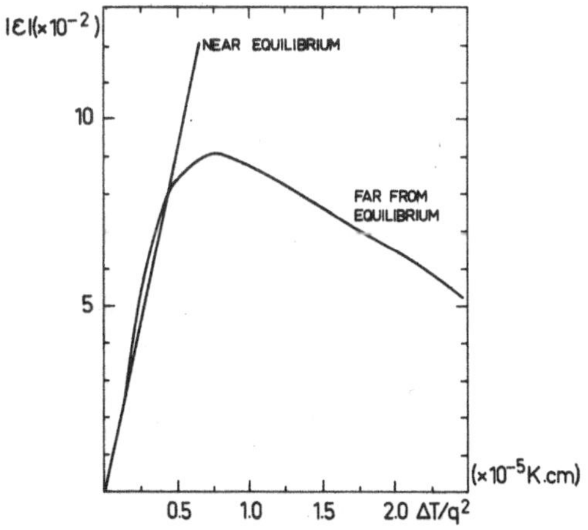

Figure 3. Asymmetry ratio ε versus the reduced variable $\nabla T/q^2$ for liquid water at 40°C (from References 12 and 15.)

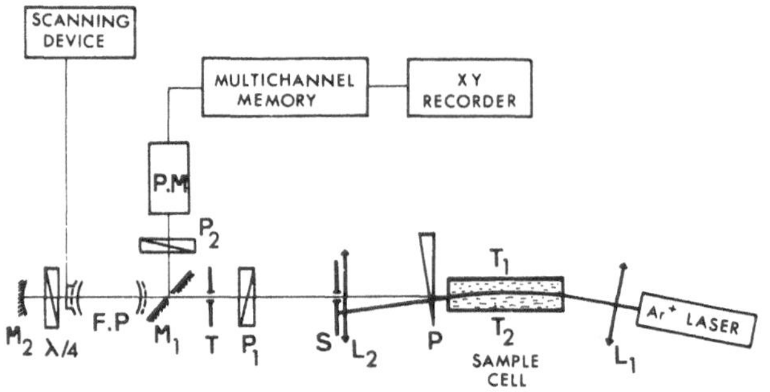

Figure 4. Experimental set up. L_1, L_2 = lenses. P = prism. P_1, P_2 = polarizers. M_1, M_2 = mirrors. S = slit. T = pin-hole. (From Reference 18.)

other liquids (see Table 2 for the numerical values of parameters involved here), and the sound damping is lower. More practically, the small variation of the refractive index with temperature, though it still bends the beam and gives rise to defocusing (thermal lens effect[16]) allows high temperature gradients to be applied.

The cell was composed of two copper plates (12 cm x 5 cm) maintained at constant temperatures within 0.1 K by circulating water from thermostatic baths. The upper plate was at a temperature higher than that of the lower plate in order to prevent convection phenomena. The water was confined between the plates by means of a transparent plexiglass frame. The spacing between the copper plates was $e = 0.515$ cm, allowing temperature gradients of about 100 $K.cm^{-1}$ to be obtained. The temperature difference was measured and controled by thermocouple junctions placed inside the two copper plates. During an experiment (recording time: 1 or 2 days) the gradient was maintained constant within 1%. Deionized water was used. Between runs the water was filtered through 0.2 μm teflon filters.

The experiments were carried out with the mean temperature kept near 40°C. At this temperature $v = 1.53 \times 10^5$ $cm.sec^{-1}$ and $\Gamma = 1.30 \times 10^{-2}$ $cm^2.sec^{-1}$ (see Table 2). The smallest angle attainable was 0.7° ($q \simeq 2000$ cm^{-1}). The Brillouin shift is thus 3×10^8 $rd.sec^{-1}$ (50 MHz), the half-width is 5.2×10^4 $rd.sec^{-1}$ (0.008 MHz), corresponding to a mean free path of pressure fluctuations $\ell = v/\Gamma q^2 \simeq 3$ cm. With a thermal gradient $\nabla T \simeq 100$ $K.cm^{-1}$, the expected asymmetry in the Brillouin intensity is about 50%.

The scattering wavevector was selected by means of a slit whose larger dimension was always maintained perpendicular to $\vec{q}$. The uncertainty on $\vec{q}$ is typically 10%, due to the finite dimensions of both the slit and the beam. We used a 213.545 MHz free spectral range confocal Fabry-Pérot spectrometer which was piezo-scanned and could be used either in a single-passed or a double-passed arrangement, the latter providing a much better contrast. The light source was a monomode Argon ion laser of 200 mW power whose beam (diameter 0.02 cm) was slightly focused in the sample. The frequency drifts of both the laser and the Fabry-Pérot are compensated by a triggering technique associated to a multichannel memory.[17] Due to the vibrations of the laser tube ("jitter"), the resolution corresponding to the half-width at midheight of the instrumental profile (5.7 MHz), was about four times the resolution expected from the reflection factor (1.3 MHz), but this spurious effect had no influence on the contrast which was found to be larger than 5×10^5 in the double-pass arrangement. Since the natural Brillouin linewidth is 0.008 MHz at $q = 2000\ cm^{-1}$, it is impossible to resolve the spectrum. Even with a perfect instrument there are some other intrinsic limitations (see below), which prevent any such analysis. Finally, the sign of the frequency shift with respect to the incident laser frequency was determined by using, in place of the sample, an electrooptic modulator (see Figure 5). A piezo transducer sent 40 MHz sound waves from one end of a crystal which was specially cut to avoid reflexion. Thus they were only progressive unidirectional sound waves in the crystal, which gave rise to diffraction of light with the frequency change ω_B, allowing the Stokes or anti-Stokes side of the spectrum to be unambiguously determined.

3. PROBLEMS RELATED TO THE THERMAL GRADIENT

In such a low angle and strongly temperature dependent experiment, we have to consider carefully the following points.

A. The sound Velocity Variation in the Observed Volume

The variation is $\Delta v = (\frac{\partial v}{\partial T})_p \cdot \nabla T \cdot \Delta Z$, where ΔZ is the full height

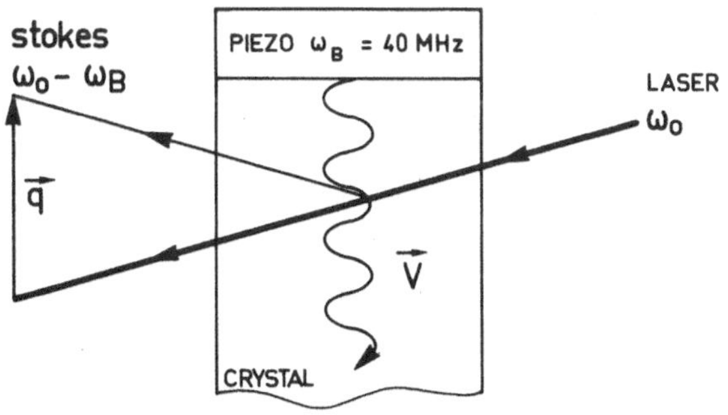

Figure 5. Determination of the Stokes side of the experimental spectrum. The crystal is specially cut to have no reflected sound waves.

difference of the illuminating beam in the same direction as $\vec{\nabla}T$; ΔZ is partly due to the beam diameter and partly due to the bending of the beam in the temperature gradient. Typically $\Delta Z \lesssim 0.1$ cm, and with $(\partial v/\partial T)_p \simeq 200$ cm.sec^{-1}K^{-1} (Table 2), $\nabla T \lesssim 100$ K.cm^{-1}, the uncertainty is $\Delta v \lesssim 2 \times 10^3$ cm.sec^{-1}, leading to an apparent linewidth broadening. $\Delta\omega = q.\Delta v \simeq 4 \times 10^6$ rd.sec^{-1} (0.6 MHz) at $q = 2000$ cm^{-1}. This is negligible with respect to the Brillouin shift and the Fabry-Perot resolution, but it is 100 times the true linewidth.

B. The Linewidth Variation in the Observed Volume

The variation of linewidth is $\Delta\Gamma = (\partial\Gamma/\partial T)_p.\nabla T.\Delta Z$. With the data of Table 2, $\Delta\Gamma/\Gamma \simeq 15\%$, which is negligible when considering the linewidth broadening from Section 3A; its consequences remain weak when the value of ε has to be estimated.

C. Influence of Aperture Angle, Beam Diameter, and Gradient-Induced Defocusing

The finite aperture angle, the finite beam diameter and the defocusing due to the temperature gradient lead to some broadening. Indeed, in water the second order derivative of the refractive index $(\partial^2 n/\partial T^2)_p$ is not negligible at high temperature gradient, leading to a cylindrical lens effect. Typically, the uncertainty Δq varies from 5% ($\nabla T = 0$), to 10% ($\nabla T \simeq 100$ K.cm^{-1}). At $q = 2000$ cm^{-1}, the broadening varies from 1.5×10^7 rd.sec^{-1} (2.5 MHz, $\nabla T = 0$), to 3.10^7 rd.sec^{-1} (5 MHz, $\nabla T \simeq 100$ K.cm^{-1}). This effect is weak enough to alter the detection of the Brillouin lines only little, but it is sufficiently high to be measured (see Table 3).

Finally, we can conclude that the limitations from point 1 to point 3 prevent any spectral analysis to be made, even with a perfect spectrometer.

D. Influence of the Finite Dimension of the Sample

The mean free path ℓ of the sound waves compared to the dimension e of the cell in the gradient direction is $\ell/e \simeq 6$ at $q \sim 2000$ cm^{-1}. It is not easy to estimate the influence of this on ε, whose value should be reduced. Moreover, ℓ varies strongly with T. At the top of the cell, $\ell \sim 8.5$ cm, and at the bottom, $\ell \sim 1.5$ cm (from $\ell = v/\Gamma q^2$, and Table 2).

Nevertheless, we can state that a wave which "feels" the gradient in a direction cannot be efficiently reflected so as to feel the gradient in the opposite direction, which, in principle, could cancel the effect of the gradient. The sound wavelength is small ($\simeq 30$ μm) compared to the surface roughness of the copper plates (~ 100 μm), and the efficiency of the reflexion can be expected to be weak.

Table 3*

q cm^{-1} a)	∇T K.cm	$\vec{q}.\vec{\nabla}T$ sign	ε ($\times 10^{-2}$) b)	$10^{-6} \times \Gamma_B$ $rd.s^{-1}$ c)
3360	59	+	+ 4.3 d)	4
3420	59	+	+ 7.5 d)	4
3130	59	+	+ 3 d)	8.2
2760	0	0	0 d)	3.5
3220	59	-	- 3.5 d)	7.8
2110	59	+	+ 8.84	9.7
2380	59	-	-11.74	11.9
2080	0	0	0	3.5
2120	59	0 e)	0	6.6
3060	83	+	+ 6.25	17.3
2720	92	+	+ 7.38	24.5
2590	93	+	+ 3.93	24.8
2100	84	-	- 7.92	16.6

*from Reference 18.

a) q is deduced from the measured Brillouin shift.

b) $\varepsilon = [I(\omega<0) - I(\omega>0)]/[I(\omega<0) + I(\omega>0)]$ where I is the integrated Brillouin intensity. The accuracy is about 1 x 10^{-2}.

c) Half-width a midheight of the Brillouin lines, obtained by subtracting the apparatus frunction half-width (35.8 x 10^6 $rd.sec^{-1}$) from the experimental half-width.

d) The Febry-Pérot was used in single-passed arrangement.

e) $\vec{q}$ is perpendicular to $\vec{\nabla}T$.

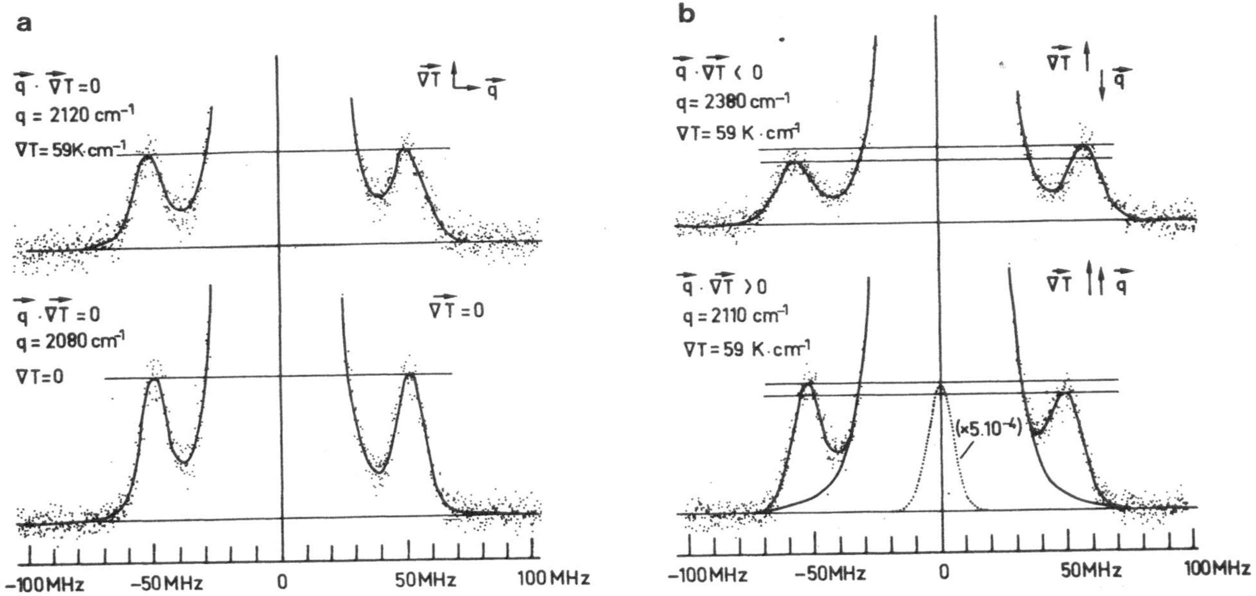

Figure 6. Experimental spectra. The solid line is the best fit. The strong central peak corresponds to an elastic scattering by dust and windows. It is clear that the intensities asymmetry varies as $\vec{q}.\vec{\nabla}T$.

4. RESULTS AND DISCUSSION

The results of our measurements are given in Figures 6-9 and Table 3. The check of the asymmetry was performed by changing the sign of $\vec{q}.\vec{\nabla}T$ under nearly the same experimental conditions, and verifying that the asymmetry in the Brillouin intensities is reversed. The lines remained symmetrical in intensity whenever $\vec{q}.\vec{\nabla}T = 0$, that is for $\vec{\nabla}T = 0$ or $\vec{q} \perp \vec{\nabla}T$. The measurement of this asymmetry was performed by integrating the lines after having substracted the base line. The accuracy is usually 10^{-2}.

In order to check the variation of the experimental value of ε, i.e., $\varepsilon_{exp} = [I(\omega<0) - I(\omega>0)]/[I(\omega<0) + I(\omega>0)]$, with respect to the reduced variable $\hat{q}.\vec{\nabla}T/q^2$, we determined the actual $|\vec{q}|$ vector from the experimental Brillouin shift $\omega_B = vq$. The accuracy was better than 0.5%, and the position of the lines was found symmetrical with respect to the incident frequency. As shown in Figure 8, the linear dependence of ε_{exp} with $\hat{q}.\vec{\nabla}T/q^2$ is verified, but the experimental slope $\varepsilon_{exp}/[\hat{q}.\vec{\nabla}T/q^2] \simeq 6700\ K^{-1}cm^{-1}$ is about three times smaller than the expected value $\simeq 19000\ K^{-1}.cm^{-1}$. This may be due at least to two reasons. The first is the finite size of the sample ($\ell/e \sim 6$), as noted above, and the second is that the near equilibrium theories can no longer be applied at such high values of $\hat{q}.\vec{\nabla}T/q^2$. The first effect can be studied separately, and experiments are in progress concerning this point. The second has been just considered by Kirpatrick and Cohen (Figure 3).[12,15] The comparison made in Figure 9 between their numerical solution and our experimental data shows a better agreement than with near equilibrium theories.

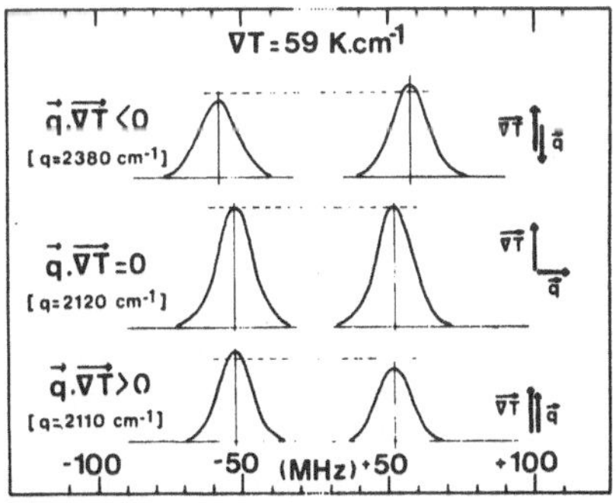

Figure 7. Best fit of the Brillouin lines from Figure 6. Units are arbitrary for each spectrum (from Reference 18).

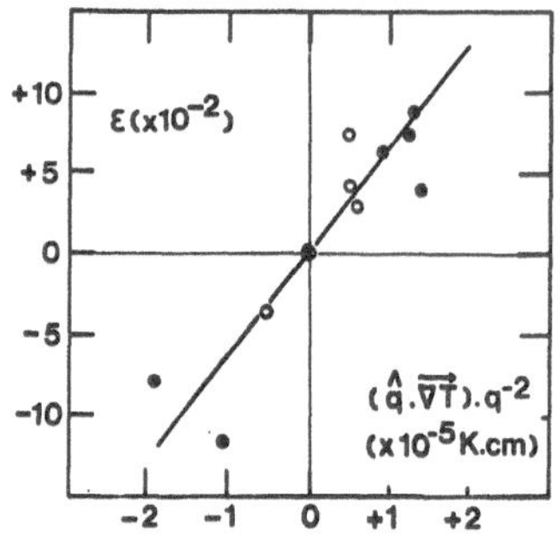

Figure 8. Intensity asymmetry $\varepsilon = [I(\omega<0) - I(\omega>0)]/[I(\omega<0)+I(\omega>0)]$ versus $\hat{q}.\vec{\nabla}/q^2$. The straight line corresponds to a slope of $\simeq$ 6700 $K^{-1}.cm^{-1}$. Full points: double-passed Fabry-Pérot. Open circle: single-passed Fabry-Pérot (from Reference 18).

5. CONCLUSION

We have shown experimentally that when a fluid is brought out of equilibrium by a temperature gradient, the structure factor determined by light scattering is changed by a long range contribution, following $\hat{q}.\vec{\nabla}T/q^2$, which makes the Brillouin lines intensities asymmetric. Both the sign and the order of magnitude of the effect have

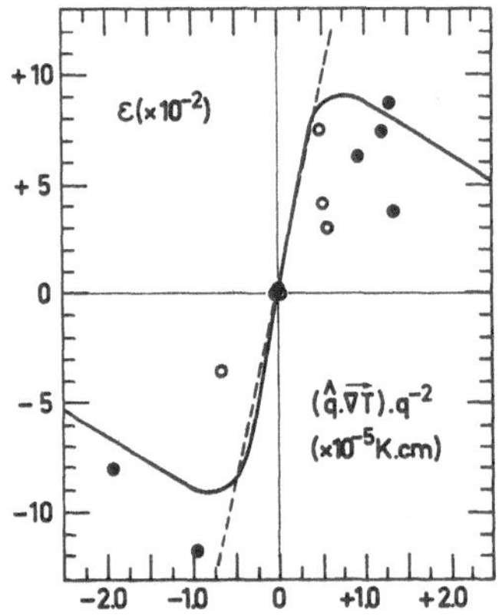

Figure 9. Intensity asymmetry $\varepsilon = [I(\omega<0) - I(\omega>0)]/[I(\omega<0)+I(\omega>0)]$ versus $\nabla T/q^2$. The dashed line is the near equilibrium result, with slope $\simeq$ 19000 $cm^{-1}.K^{-1}$. The full line is the far from equilibrium prediction. Full points refer to the double-passed Fabry-Pérot, open points to the single-passed Fabry-Pérot.

been found to be in agreement with the theoretical expectations, and thus provide a direct check of the statistical methods used in non-equilibrium physics. The effect has been found to be significantly lower than expected from near equilibrium theories. We cannot state if this disagreement found with near equilibrium calculation is due to the finite size of the sample of actually stresses the relevancy of the far from equilibrium theory.

Finally spurious effects, apparently unavoidable and due to the temperature gradient itself or to the geometry of the experiment, prevent any further investigation of the spectral shape.

ACKNOWLEDGEMENTS

I am grateful to Y. Garrabos and G. Zalczer with whom I performed this experiment. I thank P. Bergé, I. Oppenheim and P. Mazur for fruitful discussions, I. Procaccia for having indicated to us the opportunity of this experiment, T. Kirkpatrick, E. G. D. Cohen and J. R. Dorfman for very interesting discussions and for having communicated their preprints.

REFERENCES

1. L. I. Mandel'shtam, Dokl. Akad. Nauk. SSSR 2: 219 (1934), cited by I. L. Fabelinskii, Molecular Scattering of Light, Plenum Press, New York, 1968.
2. M. A. Leontovich, Dokl. Akad. Nauk. SSSR 1: 93, 1335; and Zhur. Eks. i Teor. Fiz. 9: 1314 (1939).
3. G. S. Landsberg, and A. A. Shubin, Zhur. Eks. i Teor. Fiz. 9: 1309 (1939).
4. V. V. Vladimirskii, Dokl. Akad. Nauk. SSSR 38: 229 (1942).
5. A. Griffin, Can. J. Phys. 46: 2843 (1968).
6. I. Procaccia, D. Ronis, and I. Oppenheim, Phys. Rev. Lett. 42: 287 (1979).
7. T. Kirkpatrick, E. G. D. Cohen, and J. R. Dorfman, Phys. Rev. Lett. 42: 862 (1979).
8. G. V. D. Zwan, and P. Mazur, Phys. Lett. 75A: 370 (1980).
9. A. M. S. Tremblay, E. D. Siggia, and M. R. Arai, Phys. Lett. 76A: 57 (1980).
10. D. Ronis, and S. Putterman, Phys. Rev. A22: 773 (1980).
11. T. Kirkpatrick, E. G. D. Cohen, and J. R. Dorfman, Phys. Rev. Lett. 44: 471 (1980).
12. T. Kirkpatrick, and E. G. D. Cohen, Phys. Lett. 78A: 350 (1980).
13. J. Rouch, C. C. Lai, and S. H. Chen, J. Chem. Phys. 65: 4016 (1976).
14. Handbook of Chemistry and Physics, 44th ed., Chemical Rubber Co., Columbus, Ohio, 1962-3.

15. T. Kirkpatrick, and E. G. D. Cohen, private communication (1980).
16. P. Calmettes, and C. Laj, Phys. Rev. Lett. 27: 239 (1971).
17. D. Beysens, Rev. Phys. Appl. 8: 175 (1973); J. Chem. Phys. 64: 2579 (1976).
18. D. Beysens, Y. Garrabos, and G. Zalczer, Phys. Rev. Lett. 45: 403 (1980).

FLUCTUATIONS IN STEADY STATE SYSTEMS[+]

Irwin Oppenheim and Jonathan Machta[*]

Department of Chemistry
Massachusetts Institute of Technology
Cambridge, Massachusetts

In this paper, a summary of a multimode theory of fluctuations in steady state systems is presented. A detailed development of this theory is presented in a manuscript which will appear in Phys. Rev. A.

We consider a classical system of N identical point particles in a volume V. The dynamical variables whose averages are the densities of the hydrodynamic variables are the number density, $N(\mathbf{r},X(t)) \equiv N(\mathbf{r},t)$, the energy density, $E(\mathbf{r},t)$, and the momentum density, $\mathbf{P}(\mathbf{r},t)$. Here, $X(t)$ represents the phase point of the system at time t. These densities are the components of the vector $\mathbf{A}(\mathbf{r},t)$.

We introduce the vector of the multimode quantities, $\mathbf{Q}(\{\mathbf{r}\},t)$, whose components are:

$$Q^{(0)} = 1$$

$$\mathbf{Q}^{(1)} \equiv \hat{\mathbf{A}}(\mathbf{r},t) \equiv \mathbf{A}(\mathbf{r},t) - \langle\mathbf{A}(\mathbf{r},t)\rangle \tag{1}$$

$$\mathbf{Q}^{(2)}(\mathbf{r},\mathbf{r}',t) \equiv \hat{\mathbf{A}}(\mathbf{r},t)\hat{\mathbf{A}}(\mathbf{r}',t) - \langle\hat{\mathbf{A}}(\mathbf{r})\hat{\mathbf{A}}(\mathbf{r}')\rangle$$

[+]Supported in part by the National Science Foundation under Grant CHE 79-23235

[*]Present Address: University of Maryland
Institute for Physical Science and Technology
College Park, Maryland

$$- <\hat{A}(\underset{\sim}{r})\hat{A}(\underset{\sim}{r}')\hat{A}(\underset{\sim}{r}_1)> * <\hat{A}\hat{A}>^{-1}(\underset{\sim}{r}_1,\underset{\sim}{r}_2) * \hat{A}(\underset{\sim}{r}_2) \quad \text{etc.}$$

Here, the notation $\{\underset{\sim}{r}\}$ stands for a set of position coordinates, the brackets <> denote an equilibrium average, and the * denotes a sum over the components of the vectors as well as an integration over the repeated coordinates $\{r\}$ on each side of the star. Finally, $\underset{\sim}{A}(\underset{\sim}{r}) \equiv \underset{\sim}{A}(\underset{\sim}{r},t=0)$. The multimode quantities Q, are assumed to span the space of all the slow variables in the system of interest. They are constructed to be orthogonal in the sense that:

$$<\underset{\sim}{Q}^{(\ell)}\underset{\sim}{Q}^{(m)}> = \delta_{\ell m}<\underset{\sim}{Q}^{(\ell)}\underset{\sim}{Q}^{(\ell)}> \tag{2}$$

Note the important property that any product of elements of Q can be written as a linear sum of elements of $\underset{\sim}{Q}$.

Any dynamical variable, B(t), can be written in the form:

$$B(t) \equiv \underset{\sim B}{\dot{M}}(t) * \underset{\sim}{Q}(t) + I_B(t) \tag{3}$$

where $B(t) = e^{iLt}B$, iL is the Liouville operator of the system of interest,

$$\underset{\sim B}{M}(t) \equiv <B(t)\underset{\sim}{Q}> * <\underset{\sim}{Q}(t)\underset{\sim}{Q}>^{-1} \tag{4}$$

Equation (4) defines $\underset{\sim B}{M}(t)$, and Equation (3) defines $I_B(t)$. The quantity $I_B(t)$ has the important properties,

$$<I_B(t)\underset{\sim}{Q}> = <I_B(t)F(\underset{\sim}{Q})> = 0 \tag{5}$$

where $\underset{\sim}{F}(Q)$ is any function of Q which can be expanded in a power series. Since the variables $\underset{\sim}{Q}$ span the slow set of variables, we can assume that $I_B(t)$ is quickly varying in some sense.

It will be convenient to rewrite Equation (4) in the form:

$$\underset{\sim B}{M}(t) = \underset{\sim B}{M}(0) + \int_0^t \frac{d}{d\tau} \underset{\sim B}{M}(\tau)d\tau$$

$$= <B\underset{\sim}{Q}> * <\underset{\sim}{Q}\underset{\sim}{Q}>^{-1} - \int_0^t <I_B(\tau)\underset{\sim}{\dot{Q}}> * <\underset{\sim}{Q}(\tau)\underset{\sim}{Q}>^{-1}d\tau \tag{6}$$

For times $t > \tau_m$, where τ_m is a molecular relaxation time, we can rewrite Equation (6) in the approximate form:

$$\underset{\sim}{M}_B(t > \tau_m) \simeq \underset{\sim}{M}_B(\infty) \equiv \underset{\sim}{M}_B$$

$$\simeq [<B\underset{\sim}{Q}> - \int_0^\infty <I_B(\tau)\dot{\underset{\sim}{Q}}> d\tau] * <\underset{\sim\sim}{QQ}>^{-1} \qquad (7)$$

where we have assumed that

$$<I_B(\tau)\dot{\underset{\sim}{Q}}> \xrightarrow[\tau>\tau_m]{} <I_B(\tau)><\dot{\underset{\sim}{Q}}> = 0 \qquad (8)$$

Equation (7) is valid to some order in the slowness parameter λ which characterizes the time dependence of Q. The order depends on the particular variable B.

A particular case of Equation (3) is the one in which $B(t) = \dot{\underset{\sim}{Q}}(t)$. Then Equation (3) becomes

$$\dot{\underset{\sim}{Q}}(t) \equiv \underset{\approx}{M}(t) * \underset{\sim}{Q}(t) + \underset{\sim}{I}(t) \qquad (9)$$

where

$$\underset{\approx}{M}(t) \equiv <\dot{\underset{\sim}{Q}}(t)\underset{\sim}{Q}> * <\underset{\sim}{Q}(t)\underset{\sim}{Q}>^{-1}$$

$$= <\dot{\underset{\sim\sim}{Q}Q}> * <\underset{\sim\sim}{QQ}>^{-1} - \int_0^t <\underset{\sim}{I}(\tau)\underset{\sim}{I}> * <\underset{\sim}{Q}(\tau)\underset{\sim}{Q}>^{-1} d\tau \qquad (10)$$

Here, $\underset{\sim}{I}(t)$ is defined by Equation (9) and has the properties of Equation (5). To Navier-Stokes order, i.e. through order λ^2, Equation (10) for $t > \tau_m$ can be written

$$\underset{\approx}{M}(t>\tau_m) \simeq \underset{\approx}{M} = [<\dot{\underset{\sim\sim}{Q}Q}> - \int_0^\infty <\underset{\sim}{I}(\tau)\underset{\sim}{I}> d\tau] * <\underset{\sim\sim}{QQ}>^{-1} + O(\lambda^3) \qquad (11)$$

The first term on the RHS of Equation (11) is the convective or Euler part of the hydrodynamic matrix, M; the second term on the RHS is the Navier-Stokes or dissipative part. Equation (9) can be formally solved to yield:

$$\underset{\sim}{Q}(t) \simeq \underset{\approx}{e}^{Mt} * \underset{\sim}{Q} + \int_0^t \underset{\approx}{e}^{M(t-\sigma)} * \underset{\sim}{I}(\sigma)d\sigma \qquad (12)$$

In a steady-state system, the average of $\underset{\sim}{Q}(t)$ should be independent of time. We wish to find a prescription for determining an initial distribution function, $\rho(0)$, of the form

$$\rho(0) = \rho_0 H(\underset{\sim}{Q}) \tag{13}$$

where ρ_0 is the equilibrium distribution function and $H(\underset{\sim}{Q})$ can be expanded as a power series in $\underset{\sim}{Q}$, such that

$$\mathrm{Tr}\{\underset{\sim}{Q}(t)\rho(0)\} \equiv \overline{\underset{\sim}{Q}(t)} = \underline{\underline{e}}^{Mt} * \overline{\underset{\sim}{Q}} = \overline{\underset{\sim}{Q}} \tag{14}$$

Thus, $H(\underset{\sim}{Q})$ is determined by the condition:

$$\overline{\underset{\sim}{Q}(t)} = \overline{\underset{\sim}{Q}} \tag{15}$$

and therefore

$$\underline{\underline{M}} * \overline{\underset{\sim}{Q}} = 0 \tag{16}$$

The steady-state average of Equation (3) yields:

$$\overline{B(t)} = \underset{\sim}{M}_B(t) * \overline{\underset{\sim}{Q}} \tag{17}$$

which for $t > \tau_m$, becomes

$$\overline{B(t)} = \underset{\sim}{M}_B * \overline{\underset{\sim}{Q}} \tag{18}$$

Our tasks now are to find $H(\underset{\sim}{Q})$ (Equation (13)) and the form of $\underset{\sim}{M}_B$ for particular choices of B.

We are primarily interested in the behavior of the time dependent fluctuations of the dynamical variables $\underset{\sim}{A}$ in the steady-state. We choose B to be:

$$B = \tilde{A}_\alpha(\underset{\sim}{r},\sigma)\tilde{A}_\beta(\underset{\sim}{r}') \tag{19}$$

where

$$\tilde{A} \equiv A - \overline{A} \tag{20}$$

Equation (17) becomes:

$$\begin{aligned}\overline{\tilde{A}_\alpha(\underset{\sim}{r},\sigma+t)\tilde{A}_\beta(\underset{\sim}{r}',t)} &= \langle\tilde{A}_\alpha(\underset{\sim}{r},\sigma+t)\tilde{A}_\beta(\underset{\sim}{r}',t)\underset{\sim}{Q}\rangle*\langle\underset{\sim}{Q}(t)\underset{\sim}{Q}\rangle^{-1}*\overline{\underset{\sim}{Q}} \\ &= \langle Q_\alpha^{(1)}(\underset{\sim}{r},\sigma+t)Q_\beta^{(1)}(\underset{\sim}{r}',t)\underset{\sim}{Q}\rangle*\langle\underset{\sim}{Q}(t)\underset{\sim}{Q}\rangle^{-1}*\overline{\underset{\sim}{Q}} \\ &\quad - \overline{Q_\alpha^{(1)}(\underset{\sim}{r})}\;\overline{Q_\beta^{(1)}(\underset{\sim}{r}')}\end{aligned} \tag{21}$$

It is straightforward to show using Equations (5), (6), and (12) that, to Navier-Stokes order,

$$<\underset{\sim}{Q}_{\alpha}^{(1)}(\underset{\sim}{r},\sigma + t)\underset{\sim}{Q}_{\beta}^{(1)}(\underset{\sim}{r}',t)\underset{\sim}{Q}> * <\underset{\sim}{Q}(t)\underset{\sim}{Q}>^{-1}$$

$$= \quad <\underset{\sim}{Q}_{\alpha}^{(1)}(\underset{\sim}{r},\sigma)\underset{\sim}{Q}_{\beta}^{(1)}(\underset{\sim}{r}')\underset{\sim}{Q}> * <\underset{\sim}{Q}\underset{\sim}{Q}>^{-1} \tag{22}$$

Again, using Equations (12) and (5), we find that Equation (21) reduces to:

$$\overline{\tilde{A}_{\alpha}(\underset{\sim}{r},\sigma+t)\tilde{A}_{\beta}(\underset{\sim}{r}',t)} = \sum_{n} \underline{\underline{e}}_{\alpha n}^{M\sigma} * <\underset{\sim}{Q}^{(n)}\underset{\sim}{Q}_{\beta}^{(1)}(\underset{\sim}{r}')\underset{\sim}{Q}>*<\underset{\sim}{Q}\underset{\sim}{Q}>^{-1}*\overline{\underset{\sim}{Q}}$$

$$- \overline{Q_{\alpha}^{(1)}(\underset{\sim}{r})}\ \overline{Q_{\beta}^{(1)}(\underset{\sim}{r}')} \tag{23}$$

Equation (23) is a special case of the more general result

$$\overline{\underset{\sim}{Q}(\sigma + t)\underset{\sim}{Q}(t)} = \overline{\underset{\sim}{Q}(\sigma)\underset{\sim}{Q}} = \underline{\underline{e}}^{M\sigma} * \overline{\underset{\sim}{Q}\underset{\sim}{Q}} \tag{24}$$

Equation (24) represents a generalization of the Onsager hypothesis to steady-state systems.

We shall now specialize our results to steady-state systems which are linearly displaced from equilibrium. In this case H(Q) from Equation (13) can be written:

$$H(\underset{\sim}{Q}) = 1 + \underset{\sim}{Q}*\underset{\sim}{\Phi} \tag{25}$$

where $\underset{\sim}{\Phi}^{(0)} = 0$ and the other conjugate forces $\underset{\sim}{\Phi}$ are to be determined using Equation (15). We find

$$\underline{M}*\overline{\underset{\sim}{Q}} = \underline{M}*<\underset{\sim}{Q}> + \underline{M}*<\underset{\sim}{Q}\underset{\sim}{Q}>*\underset{\sim}{\Phi} = 0 \tag{26}$$

where the first term on the RHS of Equation (26) is zero. Substitution of Equation (11) into Equation (26) yields:

$$\underline{\underline{N}}*\underset{\sim}{\Phi} = 0 \tag{27}$$

where

$$\underline{\underline{N}} = <\dot{\underset{\sim}{Q}}\underset{\sim}{Q}> - \int_0^{\infty} <\underset{\sim}{I}(\tau)\underset{\sim}{I}>d\tau \tag{28}$$

We now make use of smallness arguments to simplify the set of Equations (27). In Fourier space,

$$N_{\ell n} = <\dot{Q}^{(\ell)}_{\{k\}}Q^{(n)}_{\{k\}}> - \int_0^\infty <I^{(\ell)}_{\{k\}}(\tau)I^{(n)}_{\{k\}}>d\tau, \tag{29}$$

where

$$Q^{(\ell)}_{\{k\}} = \int e^{i\sum_{j=1}^{\ell} k_j \cdot r_j} Q^{(\ell)}(r^\ell)dr^\ell, \tag{30}$$

is of order $N^{\min(\ell,n)}$. Thus, the set of Equations (27) can be written:

$$N_{11} * \Phi^{(1)} = 0 \tag{31a}$$

$$N_{21} * \Phi^{(1)} + N_{22} * \Phi^{(2)} = 0, \text{ etc.} \tag{31b}$$

The set of variables, $\Phi^{(1)}$, are the usual conjugate variables to the set A, i.e. $\beta\mu$, βv, $-\beta$, where $\beta = (k_BT)^{-1}$, μ is the chemical potential, and v is the velocity. These variables are determined by Equation (31a) and by the boundary conditions imposed on the steady state system.

The set of variables, $\Phi^{(2)}$, are determined from Equation (31b) since the $\Phi^{(1)}$ are now known. Note that the $\Phi^{(2)}$ are of order N^{-1} compared to the $\Phi^{(1)}$. Because of the operations denoted by the * in Eq. (27), the true ordering parameter is $k_c a$ where k_c is a hydrodynamic cut off wave vector and a is the correlation length in the equilibrium ensemble. Thus, by writing Equations (31) we are neglecting long-time tail phenomena.

We now return to Equation (23) which becomes

$$\overline{A_\alpha(r,\sigma+t)A_\beta(r',t)} = \sum_n e^{M\sigma}_{\alpha n} * <Q^{(n)}Q^{(1)}_\beta(r')Q> * <QQ>^{-1} * [<Q> + <QQ> * \Phi] \tag{32}$$

to 1st order in deviations from equilibrium. In Fourier space, Equation (32) becomes

$$\overline{A_{\alpha k}(\sigma + t)A_{\beta k'}(t)} = \sum_\gamma e^{M\sigma}_{\alpha\gamma}<Q^{(1)}_{\gamma k}Q^{(1)}_{\beta k'}>$$

$$+ \sum_n e^{M\sigma}_{\alpha n} [<Q^{(n)}_k Q^{(1)}_{\beta k'} Q^{(n-1)}> * \Phi^{(n-1)}$$

$$+ <Q^{(n)}_k Q^{(1)}_{\beta k'} Q^{(n)}> * \Phi^{(n)}$$

$$+ \ <Q_k^{(n)} Q_{\beta k'}^{(1)} \ Q^{(n+1)}> *\underset{\sim}{\Phi}^{(n+1)}]$$ (33)

where we have used the orthogonality property of the Q's, Equation (2). Equation (33) can be simplified using the N ordering scheme. We find $e_{\alpha n}^{M\sigma} \sim O(N^{1-n})$, $\Phi^{(n)} \sim O(N^{1-n})$, $<Q^{(n)}Q^{(1)}Q^{(n-1)}> \sim O(N^n)$, $<Q^{(n)}Q^{(1)}Q^{(n)}> \sim O(N^n)$, $<Q^{(n)}Q^{(1)}Q^{(n+1)}> \sim O(N^{n+1})$. Thus the leading terms, $O(N)$, in Equation (33) are

$$\overline{\check{A}_{\alpha k}(\sigma + t)\check{A}_{\beta k'}(t)} \ = \ e_{\alpha 1}^{M\sigma} * <Q_{\sim k}^{(1)} Q_{\beta k'}^{(1)}>$$

$$+ \ e_{\alpha 1}^{M\sigma} * <Q_k^{(1)} Q_{\beta k'}^{(1)} Q_{-k-k'}^{(1)}> \cdot \underset{\sim}{\Phi}_{k+k'}^{(1)}$$

$$+ \ e_{\alpha 2}^{M\sigma} * <Q_{\sim k}^{(2)} Q_{\beta k'}^{(1)} Q_{\sim -k-k'}^{(1)}> \cdot \underset{\sim}{\Phi}_{k+k'}^{(1)}$$

$$+ \ \dot{e}_{\alpha 1}^{M\sigma} \cdot <Q_k^{(1)} Q_{\beta k'}^{(1)} Q_{\sim -k-k'}^{(2)}> \cdot \underset{\sim}{\Phi}_{k+k'}^{(2)}$$ (34)

The physical interpretation of the terms on the RHS of Equation (34) is the following: the first term is the **equilibrium**, time dependent correlation function; the second term is the local equilibrium correction to the equal time correlation function; the third term is the local equilibrium correction to the time dependence of the correlation function; the fourth term is due to the difference between the true steady state distribution function and the local equilibrium form. It is this term which yields new and interesting results for fluctuations in steady-state systems. These new results depend on the fact that the equal time correlation function has the form:

$$\overline{\check{A}_{\alpha k}(t)\check{A}_{\beta k'}(t)} \ = \ <\hat{A}_{\alpha k}\hat{A}_{\beta k'}> + <\hat{A}_{\alpha k}\hat{A}_{\beta k'}\hat{A}_{-k-k'}> \cdot \underset{\sim}{\Phi}_{k+k'}^{(1)}$$

$$+ \ <\hat{A}_{\alpha k}\hat{A}_{\beta k'}Q_{\sim -k-k'}^{(2)}> \cdot \underset{\sim}{\Phi}_{k+k'}^{(2)}$$ (35)

where again the last term on the RHS contains the interesting new results.

As a special case of Equation (34), we present the results for density-density fluctuations in a heat conducting steady state. The result is:

$$S_{NN}(\underset{\sim}{k},\omega) \ = \ 2 \ \mathrm{Re} \int_0^\infty e^{-i\omega t} \ \frac{\overline{N_k(t)N_{-k}}}{V} \ dt$$

$$= \frac{2k_BTn}{mc_0^2}\{(\frac{c_p}{c_v} - 1)\frac{\Gamma_T k^2}{\omega^2+(\Gamma_T k^2)^2} + \frac{\Gamma_s k^2}{(\omega+kc_o)^2+[\Gamma_s k^2(1-\varepsilon^+(\underset{\sim}{k})]^2}$$

$$+ \frac{\Gamma_s k^2}{(\omega-kc_o)^2 + [\Gamma_s k^2(1-\varepsilon^-(\underset{\sim}{k}))]^2}\} \tag{36}$$

where

$$\varepsilon^{\pm} = \pm \frac{c_0}{2k^2\Gamma_s}\hat{k}\cdot\frac{\nabla T}{T} \tag{37}$$

For $\varepsilon^{\pm} = 0$, Equation (36) becomes the equilibrium structure factor. Here, c_0 is the velocity of sound; c_p and c_v the heat capacities at constant pressure and volume, respectively; Γ_T is the heat mode attenuation coefficient and Γ_s is the sound attenuation coefficient. Equation (36) is valid to first order in ε. The most striking feature of Equation (36) is that the two Brillouin peaks are modified differently and the resulting spectrum is asymmetric in ω.

The results presented here are in agreement with those in references 1-4 although the approach is quite different. The details of this approach are based on the thesis of J. Machta, Fluctuation in Nonequilibrium Fluids, M.I.T., September 1980 and will appear in Phys. Rev. A.

REFERENCES

1. T. Kirkpatrick, E. G. D. Cohen, and J. R. Dorfman, Phys. Rev. Lett. 44:472 (1980).
2. A. -M. S. Tremblay, E. D. Siggia, and M. R. Arai, Phys. Lett. 76A:57 (1980).
3. I. Oppenheim, in Molecular Structure and Dynamics, ed. by M. Balaban (International Science Services, Philadelphia, 1980), p. 3.
4. D. Ronis and S. Putterman, Phys. Rev. A 22: 773 (1980).

MODELS OF STOCHASTIC BEHAVIOR IN NON-EQUILIBRIUM STEADY STATES

Robert Graham

University Essen, Fachbereich Physik
West Germany

CONTENTS

I. INTRODUCTION

Macroscopic systems are frequently described in terms of a set of macroscopic variables staisfying a given system of deterministic differential equations. The fluctuations of macroscopic variables being small, such a deterministic description is, in many cases, sufficiently accurate and already contains a wealth of information: e.g., for systems which are driven by stationary external forces and interacting with time independent reservoirs, the deterministic description allows to discuss the steady state as a fixed point or a limit cycle or a more complicated 'strange attractor' in the space of the macroscopic variables. Within the same description it is also possible to discuss the stability of the various steady states, in which the system may be found, and the time dependent relaxation towards such states from a given initial state.

However, there are a number of cases, when a deterministic description is not adequate and a statistical description is called for even though the systems considered are macroscopic. We mention four typical and physically very different situations where this is the case:

(i) In thermodynamic equilibrium, the spectrum of the thermal fluctuations of the macroscopic variables is related to the susceptibilities and transport parameters of the system by the fluctuation dissipation theorem. A measurement of that spectrum, e.g. by light scattering, is often possible and is then very useful. In non-equilibrium steady states with thermal fluctuations a measurement of the fluctuation spectrum may still be possible. However, the fluctuation-dissipation theorem in its usual form does not apply to non-equilibrium steady states; its application has to be replaced by an explicit calculation using a stochastic description, cf. e.g. Ref. (1), including thermal fluctuations explicitly, even though they are small. For recent theoretical work on light scattering in non-equilibrium steady states cf.[2]

(ii) If the steady state of the system is close to a limit of instability, fluctuations may be enhanced and may then no longer be negligible. For systems in thermodynamic equilibrium, well known examples are systems near critical points of continuous phase transitions. Lasers near threshold are well known examples of noisy non-equilibrium steady states, which require a statistical description.[3]

(iii) If the deterministic steady state is a limit cycle, even weak fluctuations have a profound effect if the system is observed over sufficiently long times. Indeed, in this case all deterministic variables are periodic functions of time, while all variables averaged over the fluctuations become time independent in the steady state. Of course, the time average by which expectation values etc. are defined has to be extended over longer and longer time intervals if the fluctuations become weaker and weaker. The mechanism which wipes out the periodicity of the deterministic process over sufficiently long times is the temporal diffusion of the phase of the limit cycle under the influence of the fluctuations. The phase fluctuations are diffusive, since no restoring force exists holding the phase in place due to the stationarity of all external forces and reservoirs acting on the system. Examples of this effect occur in lasers and other self-sustained oscillators above threshold like the chemical limit cycles used to model the Belousov-Zhabotinsky reaction. For reviews which emphasize general properties of this kind cf.[4,5,6]

(iv) If the steady state in the deterministic description is neither a fixed point nor a limit cycle but a strange attractor[6,7,8] a description in terms of a deterministic trajectory on the strange attractor usually becomes impractical. It may then be more convenient to consider ensembles of trajectories and to use a statistical description from the start. Furthermore, the deterministic strange attractors, as a rule, are complicated point sets with a Cantor set substructure.

If fluctuations are taken into account, even if they are weak, the sharp Cantor set substructure is presumably broadened a bit, and may therefore be expected to disappear. At the end of these lectures we will discuss the Lorenz model from this point of view, as an example.

In the four cases mentioned, the motivation for including fluctuations in the description of the steady state is rather different. However, in all these cases one is rather naturally led to a description of the process in terms of a Fokker Planck equation. Its validity requires that a clear separation of time scales exists between the macroscopic variables and all the microscopic variables, and that the macroscopic variables are continuous functions of time. These assumptions are usually well satisfied in the kind of applications referred to above.

In section 2a we discuss the Fokker Planck description of steady state ensembles by considering examples in hydrodynamics, chemically reacting systems and quantum optics.

In section 2b we discuss the mathematical aspects of the Fokker Planck description (cf. e.g. Ref. 9). Usually one has to deal with a non-hermitian boundary value problem. In some cases the boundary value problem may be reduced to a hermitian one. As we show in section 2b a key role for the boundary value problem is played by the time independent stationary ensemble.[10,11] This ensemble is easily obtained in case the system satisfies detailed balance[12] as discussed in section 2c. In the same section we discuss physical examples of systems with detailed balance in equilibrium and non-equilibrium steady states.

However, in non-equilibrium steady states detailed balance is usually absent. In section 2d we present a method by which it is sometimes possible to obtain the steady state ensemble in the limit of weak noise.[13,14]

In section 3 the general methods are applied to a discussion of the modification of a bifurcation point by noise. In section 3a we consider the influence of additive noise on the point of a forward Hopf bifurcation, a problem which satisfies detailed balance.[4] The analysis of noise in single mode lasers near threshold[15,16] solves this problem generally. In section 3b we consider the influence of multiplicative noise on forward Hopf bifurcation,[17,18] again a problem with a detailed balance symmetry.

In section 4 two problems without detailed balance are considered. The first example is dispersive optical bistability.[19] We review recent work in collaboration with A. Schenzle[20] in which the steady state probability density of this system has been worked out. The second example is the Lorenz model [7] under the influence of weak external noise. Again very recent work in collaboration with H.J. Scholz[21] is reviewed. An analytical approximation is obtained of the surface of maximum probability in the steady state. Then, a local approximation of the probability distribution over that surface in the vicinity of its lower boundary is worked out. The probability density is shown to vanish by a power law on that boundary.

2. GENERAL STATISTICAL DESCRIPTION

(a) Fokker Planck Equation

The macroscopic variables of the system are $q^\nu (\nu = 1 \ldots n)$. We use a description by discrete variables rather than fields, assuming that fields are discretized either by subdividing the system into small volume cells or by expanding the fields in terms of a complete set of modes satisfying the boundary conditions. The Fokker Planck equation for the conditional probability density $P(q|q_o,\tau)$ reads

$$\begin{aligned} \frac{\partial P}{\partial t} &= L_F P \\ L_F &= -\frac{\partial}{\partial q^\nu} K^\nu(q) + \frac{1}{2} \frac{\partial^2}{\partial q^\nu \partial q^\mu} Q^{\nu\mu}(q) \end{aligned} \tag{1}$$

We assume 'natural boundary conditions', i.e. $P(q|q_o\tau)$ vanishing at the boundaries. For $\tau=0$, P reduces to the n-dimensional δ-function $\delta(q-q_o)$ as an initial condition. We consider some physical examples of Equation (1).

(i) Hydrodynamics of an incompressible fluid

In order to arrive at a description in terms of fields, we let the volume cells approach zero. The derivatives $\partial/\partial q^\nu$ then become functional derivatives and the sum over the index ν contains a volume integral. For an incompressible isothermal fluid we then have[22]

$$\begin{aligned} &q^\nu,\ q^\mu \rightarrow v_i(x),\ v_j(x) \\ &K^\nu(q) \rightarrow -v_j\ \partial_j\ v_i - \frac{1}{\rho} \nabla_i p + \nu \nabla^2 v_i - f_i \\ &Q^{\nu\mu}(q) \rightarrow 2\ \frac{k_B T}{\rho}\ \nu\ \nabla\cdot\nabla'\ \delta^{tr}_{ij}\ (x-x') \end{aligned} \tag{2}$$

where $v(x)$ is the Eulerian streaming velocity, ρ is the uniform density, f_i are external mass forces,

$$p(x) = \frac{\rho}{4\pi}\int d^3x' \frac{(\nabla_i' v_j(x'))(\nabla_j' v_i(x')) + \nabla_i' f_i(x')}{|x-x'|} \tag{3}$$

is the pressure (in an infinite system), ν is the kinematic viscosity, k_BT is the temperature in energy units and $\delta_{ij}^{tr}(x-x')$ is the transverse δ-function

$$\delta_{ij}^{tr}(x-x') = \int \frac{d^3k}{(2\pi)^3} (\delta_{ij} - \frac{k_i k_j}{k^2}) \exp(ik\cdot(x-x')) \tag{4}$$

The drift $K^\nu(q)$ is thus given by the Navier-Stokes equation. The diffusion matrix $Q^{\nu\mu}(q)$ is completely fixed by the assumption of local thermal equilibrium.[22,23] This assumption is of course already contained in the Navier-Stokes equation. The Fokker Planck equation then reads explicitly

$$\frac{\partial P}{\partial t} = +\int d^3x \frac{\delta}{\delta v_i(x)} \left\{ (v_j(x)\nabla_j v_i(x) + \frac{1}{\rho}\nabla_i p(x) + f_i(x) - \nu\nabla^2 v_i(x))P \right\} + \frac{k_BT\nu}{\rho}\int d^3x (\nabla_j \frac{\delta}{\delta v_i(x)}) (\nabla_j \frac{\delta}{\delta v_i(x)}) P \tag{5}$$

where the functional derivatives are taken with respect to transverse velocity fields

$$v_i(x) = \int \frac{d^3k}{(2\pi)^3} (\delta_{ij} - \frac{k_i k_j}{k^2}) v_j(k) \exp ikx \tag{6}$$

only.

(ii) Chemically reacting continuous systems

The stochastic description of such systems linearized around a steady state has recently been considered in detail by Grossmann.[24] As an example, we consider a single-phase two component system with one chemical reaction $Y \rightleftarrows Z$ at constant density ρ, pressure p and vanishing streaming velocity v.

The macroscopic variables are the mass concentration c of component Y (component Z has concentration 1-c) and the internal energy per unit mass U.

$$q^\nu \rightarrow c(x), U(x)$$

The drift $K^\nu(q)$ is represented by the deterministic equations

$$\dot{q}^\nu = K^\nu(q) \tag{7}$$

as

$$\dot{C} = \operatorname{div} \ell + \Gamma \qquad \dot{U} = \operatorname{div} q \tag{8}$$

with the diffuse particle current

$$\ell = \rho D \nabla c \tag{9}$$

and the energy current

$$q = -\lambda \nabla T + \left(\mu - T \left.\frac{\partial \mu}{\partial T}\right|_{c,p} + k_T \left.\frac{\partial \mu}{\partial c}\right|_{T,p}\right) \ell \tag{10}$$

The mass production rate per unit volume is given in terms of the rate coefficient ε of the reaction

$$\Gamma = - \varepsilon \nu_c m_c \; A(c) \tag{11}$$

where the affinity A is

$$A = \nu_c \; m_c \cdot \mu \tag{12}$$

m_c is the molecular mass of the component Y, ν_c its stoichiometric coefficient in the chemical reaction, $\mu = \mu_c/m_c - \mu_{1-c}/m_{1-c}$ is the difference of the chemical potential of the two components. D is the diffusion constant, $\gamma$the thermal conductivity and k_T the thermodiffusion coefficient.

The diffusion matrix $Q^{\nu\mu}$ of the Fokker Planck equation for this system is again determined by the local equilibrium assumption[22,23] which is already implicit in the form of the hydrodynamic equations. In an obvious notation we may write Grossmann's result[24] in the form

$$Q^{cc} = (\nabla \cdot \nabla' \quad k_B T \rho D \left.\frac{\partial c}{\partial \mu}\right|_{T,p} + k_B T \; \nu_c^2 \; m_c^2 \; \varepsilon) \; \delta(x-x')$$

$$Q^{uu} = \nabla \cdot \nabla' \quad k_B T \left[T\lambda + \rho D \left.\frac{\partial c}{\partial \mu}\right|_{T,p} \left(\mu - T \left.\frac{\partial \mu}{\partial T}\right|_{p,c} + k_T \frac{\partial \mu}{\partial c}_{T,p}\right)^2\right] \delta(x-x')$$

$$Q^{uc} = Q^{cu} = \nabla \cdot \nabla' \; k_B T \rho D \left[k_T + \left.\frac{\partial c}{\partial \mu}\right|_{T,p} \left(\mu - T \left.\frac{\partial \mu}{\partial T}\right|_{c,p}\right)\right] \; \delta(x-x') \tag{13}$$

Similar results may be derived from phenomenological master equations for the chemical reactions, by considering the limit of large particle numbers.[25] However, the master equation often contains other mathematical and physical assumptions which are often more serious than the Fokker Planck approximation (applicability of birth and death formalism, ideal gas approximation, neglect of reaction heat etc.), which may be avoided in a Fokker Planck desscription, however, at the cost of restricting oneself to large particle numbers in each volume cell.

(iii) Maxwell Bloch equations in Quantum Optics

We consider the resonant interaction of the electromagnetic field with the electronic transitions of two-level atoms in dipole-coupling and rotating wave approximation.[3] The macroscopic variables are then the complex slowly varying amplitude of the electric field $E^{\pm}$, the complex amplitude of the polarization of the atoms $P^{\pm}$, and the density of occupation number difference of the two atomic variables.

$$q^{\nu} \rightarrow E^{\pm},\ P^{\pm},\ W$$

The drift K^{ν} is given by the deterministic coupled Maxwell-Bloch equations in rotating wave approximation in the form $\dot{q}^{\nu} = K^{\nu}(q)$

$$\begin{aligned} \dot{E}^{-} &= [+i\,(\omega_o-\omega)-\kappa]\ E^{-} + igP^{-} + \mu E_0 \\ \dot{P} &= [i(\nu-\omega) - \gamma_{\perp}]\,P^{-} - igWE^{-} \\ \dot{W} &= -\gamma_{\parallel}(W-W_o) - 2ig\,(P^{-}E^{+} - P^{+}E^{-}) \end{aligned} \qquad (14)$$

Here ω is the frequency of a driving external electric field with amplitude $\sqrt{2\pi h\omega_o}\, E_0$, ω_o is the main frequency of the internal electric field with amplitude $\sqrt{2\pi h\omega_o}\,(E^{+}\, e^{-i\omega t} + c.c.)$, $h\nu$ is the energy difference of the atomic transition with dipole moment $e_o d$, polarization $e_o d\,(P^{+}(t)\, e^{-i\omega t} + c.c)$ and density of occupation number difference W_o in equilibrium, $\gamma_{\perp} = 1/T_2$, $\gamma_{\parallel} = 1/T_1$ are the transverse and longitudinal relaxation rates, respectively, $\kappa = C/L(1-R)$ is the inverse resonator lifetime, determined by the length L and reflectivity R of the resonator, $g = e_o d\,(2\pi\nu^2/h\omega_o)^{1/2}$ is the dipole coupling constant of the two-level transition and the resonator mode.

The diffusion matrix of the Fokker Planck equation is determined by an analogue of the local equilibrium assumption for the present case.

The electric field and the atoms are coupled to each other and to different heat reservoirs which may be at different temperatures. The latter coupling gives rise to dissipation and fluctuations. It is assumed, that there is no influence of the coupling between $E^{\pm}$ and $P^{\pm}$ effected by g on the dissipation and fluctuation of $E^{\pm}$, $P^{\pm}$ and W due to their respective reservoir couplings. This assumption is analogous to the local equilibrium assumption, where the influence of the coupling between volume cells on local properties is neglected. This assumption cannot be rigorously satisfied, since dissipation and fluctuation is influenced by the total energy stored in the system, and energy is partially stored in the form of interaction energy of atoms and field. However, the fraction of the total energy residing in the interaction energy is small (of the order q/ν) and may therefore be neglected.

The diffusion matrix is presented in terms of the real variables

$$E^+ = x + iy, \quad P^+ = u + iv, \quad W \tag{15}$$

and given by[3] (V = beam volume, n = number density of atoms)

$$\begin{aligned} Q^{xx} &= Q^{YY} = \frac{\kappa}{V}\left(\exp\frac{\hbar\omega_o}{k_B T} - 1\right)^{-1} \\ Q^{uu} &= Q^{vv} = \gamma_\perp \frac{(n+W_o)}{2V} \\ Q^{WW} &= 2\gamma_\parallel \frac{n}{V} \end{aligned} \tag{16}$$

In all the concrete examples considered in this section, the diffusion matrix was determined under the assumption that the fluctuations are generated by thermal noise; in the preceding example quantum noise is also included, and is, in fact, more important than thermal noise at optical frequencies. Thermal fluctuations and quantum fluctuations are universal noise sources. However, in specific situations they are often eclipsed by stronger but non-universal noise sources. In such non-universal cases the validity of the Fokker Planck equation (1) and its specific form depend on the special properties of the noise sources present.

(b) A General Transformation of the Fokker Planck Equation

We assume that the solution of Equation (1) for $\tau \to \infty$ approaches a unique time independent probability density W which we write in the form

$$W = \exp(-\Phi) \tag{17}$$

In case the steady state distribution, and hence Φ, are known, the Fokker Planck operator L_F (1) and its adjoint may be transformed into L and L^+ given by

$$\begin{aligned} L &= e^{\frac{1}{2}\Phi} L_F \, e^{-\frac{1}{2}\Phi} \\ L^+ &= e^{-\frac{1}{2}\Phi} L_F^+ \, e^{\frac{1}{2}\Phi} \end{aligned} \tag{18}$$

where

$$\begin{aligned} L &= L_H + L_A \\ L^+ &= L_H - L_A \end{aligned} \tag{19}$$

and

$$L_H = \frac{1}{2} \frac{\partial}{\partial q^\nu} Q^{\nu\mu}(q) \frac{\partial}{\partial q^\mu} - V(q) \tag{20}$$

is hermitian, while

$$L_A = -r^\nu(q) \frac{\partial}{\partial q^\nu} - \frac{1}{2} r^\nu(q) \frac{\partial \Phi}{\partial q^\nu} \tag{21}$$

is antihermitian.

The hermitian part L_H of L is reminiscent of the quantum mechanical Hamiltonian in q^ν -representation of a system described in the generalized variables q^ν. The 'potential energy', V(q) is given in terms of Φ by

$$V = -\frac{1}{4} \frac{\partial}{\partial q^\nu} (Q^{\nu\mu}(q) \frac{\partial \Phi}{\partial q^\mu}) + \frac{1}{8} Q^{\nu\mu}(q) \frac{\partial \Phi}{\partial q^\nu} \frac{\partial \Phi}{\partial q^\mu} \tag{22}$$

The antihermitian part L_A contains the quantity $r^\nu(q)$ defined in terms of Φ by

$$r^\nu(q) = K^\nu(q) + \frac{1}{2} Q^{\nu\mu} \frac{\partial \Phi}{\partial q^\mu} - \frac{1}{2} \frac{\partial Q^{\nu\mu}}{\partial q^\mu} \tag{23}$$

By their definitions, $r^\nu(q)$ and Φ satisfy the relation

$$\frac{\partial r^\nu(q)}{\partial q^\nu} - r^\nu \cdot \frac{\partial \Phi}{\partial q^\nu} = 0 \tag{24}$$

which is equivalent to the Fokker Planck equation for W. Using Equation (24), L_A may be expressed entirely in terms of $r^\nu(q)$

$$\begin{aligned} L_A &= -r^\nu \frac{\partial}{\partial q^\nu} - \frac{1}{2} (\frac{\partial r^\nu}{\partial q^\nu}) \\ &= -\frac{1}{2} (\frac{\partial}{\partial q^\nu} r^\nu(q) + r^\nu(q) \frac{\partial}{\partial q^\nu}) \end{aligned} \tag{25}$$

which makes its antihermiticity explicit.

The form of L_A is again reminiscent of quantum mechanics in that it resembles a vector potential term in the Hamiltonian, apart from a missing factor of i which gives rise to its antihermiticity. The transformation we made here was discussed for the special case of systems in detailed balance by Risken[10] and for the general case in 11. Indeed, it is important to realize that the restriction to systems with detailed balance is not necessary in principle. Surely, Φ is not easily obtained if detailed balance is absent, which in practice often leads to a restriction of the present transformation to systems in detailed balance. In the course of these lectures we will describe a method by which Φ can sometimes be found for systems lacking detailed balance, in the limit of weak noise. If Φ has been determined, $r^\nu(q()$, $V(q)$ and, therefore, L_A and L_H are known explicitly.

The Fokker Planck equation and its adjoint (the backward equation) now read in an obvious notation

$$\frac{\partial \tilde{P}(q|q_o,\tau)}{\partial \tau} = L[q]\; \tilde{P}(q|q_o,\tau)$$
$$\frac{\partial \tilde{P}(q|q_o,\tau)}{\partial \tau} = L^+[q_o]\; \tilde{P}(q|q_o,\tau) \tag{26}$$

with

$$\tilde{P}(q|q_o,\tau) = \exp\left(\frac{\Phi(q) - \Phi(q_o)}{2}\right) P(q|q_o,\tau) \tag{27}$$

The conditional probability $P(q|q_o,\tau)$ may be expanded in terms of the biorthogonal set $\psi_\mu(q)$, $\tilde{\psi}_\mu(q)$ satisfying

$$L[q]\; \psi_\mu(q) = -\lambda_\mu\; \psi_\mu(q)$$
$$L^+[q]\; \tilde{\psi}_\mu(q) = -\lambda_\mu^*\; \psi_\mu(q) \tag{28}$$
$$\int [\tilde{\psi}_\mu(q)]^*\; \psi_\nu(q)\; dq = \delta_{\mu\nu}$$

assuming that these functions form a complete set

$$\sum_\mu \psi_\mu(q)\; \tilde{\psi}_\mu^*(q_o') = \delta(q-q_o) \tag{29}$$

The expansion of P is then given by

$$P(q|q_o,\tau) = \exp\left(-\frac{1}{2}(\Phi(q) - \Phi(q_o))\right)\sum_{\mu}\tilde{\psi}^*_{\mu}(q_o)\ \psi_{\mu}(q)\ e^{-\lambda_{\mu}\tau} \quad (30)$$

An argument given by Risken[10] for systems in detailed balance may now be generalized to show that, for systems with a non-degenerate steady state distribution W, the real parts of all eigen-values (except $\lambda_o = 0$) are positive

$$\mathrm{Re}\ \lambda_{\mu} > 0 \text{ for } \mu \neq 0,$$

and that the real part of the eigenvalue λ_1 with the smallest non-vanishing real part can only increase if the drift $r^{\nu}(q)$ is switched on from zero,

$$\mathrm{Re}\,\lambda_1\ (\{r\}) \geq \mathrm{Re}\ \lambda_1\ \ (\{0\}).$$

Correlation functions in the steady state are very simply expressed in terms of the eigenfunctions ψ_{μ}, $\tilde{\psi}_{\mu}$. From Equations (17), (30) we obtain

$$<f(q(\tau))\ g(q(0))> \ = \sum_{\mu} f_{\mu}\ \tilde{g}^*_{\mu}\ e^{-\lambda_{\mu}\tau} \quad (31)$$

with

$$\begin{aligned} f_{\mu} &= \int dq\ f(q)\ \psi_{\mu}(q)\ e^{-\frac{1}{2}\Phi(q)} \\ \tilde{f}_{\mu} &= \int dq\ f(q)\ \tilde{\psi}_{\mu}(q)\ e^{-\frac{1}{2}\Phi(q)} \end{aligned} \quad (32)$$

A Fokker Planck operator is particularly simple if all $r^{\nu}(q)$ turn out to be zero. Then the transformed operator L is hermitian. By Equation (23) this can only happen if $K^{\nu}(q)$ satisfies

$$K^{\nu}(q) - \frac{1}{2}\frac{\partial Q^{\nu\mu}}{\partial q^{\mu}} = -\frac{1}{2}Q^{\nu\mu}(q)\frac{\partial\Phi}{\partial q^{\mu}} \quad (23)$$

with a potential Φ approaching infinity at the boundaries in q-space. Another class of Fokker Planck operators, which is nearly as simple, is defined by the class.

$$[L_A, L_H] = 0. \quad (34)$$

If Equation (34) is satisfied, one may diagnalize L_H and L_A simultaneously by the same eigenfunctions and thereby obtain the real and imaginary part of λ_{μ} separately.

(c) Detailed Balance

(α) Definition

Detailed balance is always defined relative to a given time reversal transformation, which we write in the form $q \to \tilde{q}$

$$\text{with} \quad \tilde{q}^{\nu} = \varepsilon^{\nu} q^{\nu} \tag{35}$$

where $\varepsilon^{\nu} = 1, -1$ if q^{ν} is even, odd under time reversal. The special form of (35) assumes that the variables have been split into even and odd under time reversal. We emphasize that a system may well be in detailed balance with respect to one time reversal transformation, but out of detailed balance with respect to another one. The time reversal transformation required to achieve detailed balance may be more complicated than Equation (35) and may even require that parameters λ of the system (e.g. external magnetic fields) are also changed

$$\begin{aligned} \lambda &\to \tilde{\lambda} \\ \tilde{\lambda}^{\alpha} &= \varepsilon^{\alpha} \lambda^{\alpha} \end{aligned} \tag{36}$$

where $\varepsilon^{\alpha} = 1$, if λ^{α} is even, odd under time reversal. In the following the dependence of all quantities on λ is made explicitly, if needed.

A system described by a Fokker Planck equation is in detailed balance, with respect to a given time reversal transformation, if and only if the following conditions are satisfied:[12,10]

(i) $\Phi(q,\lambda)$ defined by Equation (17) is invariant under time reversal

$$\Phi(\tilde{q},\tilde{\lambda}) = \Phi(q,\lambda) \tag{37}$$

(ii) $r^{\nu}(q)$ defined by Equation (23) consists of the reversible part of $K^{\nu}(q)$

$$r^{\nu}(q,\lambda) = \frac{1}{2}\left(K^{\nu}(q,\lambda) - \varepsilon^{\nu} K^{\nu}(\tilde{q},\tilde{\lambda})\right) \tag{38}$$

$$\text{(iii)} \quad Q^{\nu\mu}(q,\lambda) = \varepsilon^{\nu}\varepsilon^{\mu} Q^{\nu\mu}(\tilde{q},\tilde{\lambda}). \tag{39}$$

The listed conditions are equivalent to the symmetry relations

$$\begin{aligned} W(q,\lambda) &= W(\tilde{q},\tilde{\lambda}) \\ P(q|q_o,\tau,\lambda)\, W(q_o,\lambda) &= P(\tilde{q}_o|\tilde{q},\tau,\tilde{\lambda})\, W(\tilde{q},\tilde{\lambda}) \end{aligned} \tag{40}$$

which are usually taken as a definition of detailed balance[28]. By making the time reversal transformation sufficiently complicated, one can always arrange for the fulfillment of the above listed conditions and, in this sense, the fulfillment of detailed balance may always be arranged by inventing a suitable time reversal transformation[11]. However, the required time reversal transformations will become extremely complicated as a rule, and we will therefore define detailed balance only for simple time reversal transformations, which one can write down before having any explicit knowledge of $\Phi(q)$ and $r^{\nu}(q)$ ("manifest detailed balance").[11]

β Examples

(1) Systems in thermodynamic equilibrium have the property of detailed balance with respect to the time reversal transformation, which leaves the microscopic equations of motion underlying the macroscopic dynamics invariant.[28] We consider the examples discussed in section 2a.

Incompressible fluid:

For $f_i(x) = 0$ the system can reach equilibrium. Time reversal is defined by $v \to -v$. The potential Φ is given by

$$\Phi = \frac{1}{2k_B T} \int d^3x \, \rho \, v^2(x) \tag{41}$$

The drift $r^{\nu}(q)$ is given by the reversible part of the Navier-Stokes drift,

$$r^{\nu}(q) \to -v_j \, \partial_j \, v_i - \frac{1}{\rho} \nabla_i p \tag{42}$$

Chemically reacting system:

Time reversal is defined by $c \to c$; $u \to u$. Since there are only even variables, the drift $r^{\nu}(q)$ in a state with detailed balance must vanish (cf. Equation (38)). The potential Φ is given by the thermodynamic potential with the natural variables u, c, which is the entropy of the system

$$\Phi = -\frac{1}{k_B} S(u,c) + \text{const.} \tag{43}$$

Maxwell Bloch equations

Time reversal is defined by $E^{\pm} \to E^{\mp}$, $p^{\pm} \to p^{\mp}$, $W \to W$. Thermal equilibrium may be reached if all reservoirs are at the same temperature and no external field is applied, $E_0 = 0, \omega = 0$. The drift r^{ν} in detailed balance is given by the reversible part of the total drift K^ν. Writing $\dot{q}^\nu = r^\nu$ we may represent r^ν by

$$\begin{aligned}\dot{E}^- &= i\omega_o E^- + igP^- \\ \dot{P}^- &= i\nu P^- - igWE^- \\ \dot{W} &= -2ig(P^-E^+ - P^+E^-)\end{aligned} \tag{44}$$

The potential Φ is given by

$$\Phi = V\left(\frac{E^+E^-}{n_{th}} + \frac{2P^+P^-}{n+W_o} + \frac{1}{2}\frac{(W-W_o)^2}{n}\right). \tag{45}$$

(2) Non-equilibrium steady states are generally not in detailed balance. However, there are important special cases of detailed balance in non-equilibrium steady states.

Single variable systems with natural boundary conditions always satisfy detailed balance with respect to the time reversal transformation, which treats the single variables as even. One has, by Equations (38), (33),

$$\begin{aligned} r(q) &= 0 \\ \Phi(q) &= -2\int^q dq'\left[\frac{K(q')}{Q(q')}\right] + \ln Q(q) + \text{const.}\end{aligned} \tag{46}$$

Some two-variable systems in resonance with an external driving field also show detailed balance in the steady state on a long time scale. A simple example is a harmonic oscillator driven on resonance. For this discussion we consider the Langevin equation, which is stochastically equivalent to the Fokker Planck equation.[9] In an obvious notation

$$\begin{aligned}\dot{x} &= v \\ \dot{v} + 2\gamma v + \omega_o^2 x &= f_o \cos\omega_o t + \xi(t)\end{aligned} \tag{47}$$

with a Gaussian white noise Langevin force whose first two moments are

$$\langle\xi(t)\rangle = 0; \qquad \langle\xi(\tau)\,\xi(0)\rangle = 2\gamma Q\delta(\tau). \tag{48}$$

For $f_o = 0$ there is detailed balance under $x\to x$, $v\to -v$ with

$$\Phi = \frac{1}{2Q}v^2 + \frac{\omega_o^2}{2Q}x^2, \quad r_x = v, \quad r_v = -\omega_o^2 x.$$

Due to the presence of the time dependent force $f_o \cos \omega_o t$, detailed balance is destroyed. However, if we consider the system on time scales much longer than $1/\omega_o$ we may introduce the complex amplitudes $b(t)$, $b^*(t)$ by

$$x(t) = \frac{1}{2}\left(b(t)\, e^{-i\omega_o t} + b^*(t)\, e^{i\omega_o t}\right) \tag{49}$$

and assuming $|\dot{b}| << \omega.|b|$

$$v(t) = \frac{\omega_o}{2i}\left(b(t)\, e^{-i\omega_o t} - b^*(t)\, e^{i\omega_o t}\right) \tag{50}$$

Restricting ourselves to long time scales, we may apply the rotating wave approximation to obtain

$$\dot{b}(t) + \gamma b(t) = \frac{if_o}{2\omega_o} + \tilde{\xi}(t) \tag{51}$$

and the complex conjugate equation with

$$< \tilde{\xi}(t) > = 0, < \tilde{\xi}(\tau)\, \tilde{\xi}^*(0) > = 2\frac{\gamma Q}{\omega_o^2}\, \delta(\tau). \tag{52}$$

Equation (51) satisfies detailed balance under the transformation $b \to b^*$, $b^* \to b$, $f_o \to -f_o$ with the potential

$$\Phi = \omega_o^2 \frac{\left|b - \frac{if_o}{2\gamma\omega_o}\right|^2}{Q} \tag{53}$$

and the drift $r^{\nu} = 0$. Thus, detailed balance in this example is destroyed on the short time scale $1/\omega_o$, but restored in the average over many cycles on a time scale $>> 1/\omega_o$. The assumed exact resonance between the external force and the oscillator was essential for obtaining this result.

A physically more interesting example where the same phenomenon happens is provided by the Maxwell Bloch equations (14) driven on resonance, i.e. where $\Delta \equiv \omega - \nu/\gamma_\perp = 0$, $\delta \equiv \omega_o - \omega/\kappa = 0$ is satisfied. For $\gamma_\perp, \gamma_{||} >> \kappa$ one may eliminate $P^{\pm}$ and W from Equation (14) by adiabatic approximation.[3,29] The resulting equation for $E^{\pm}$ in dimensionless units

$$\dot{E}^+ = -E^+ + E_o - \Gamma^2 \frac{E^+}{1+|E|^2} \tag{54}$$

$$Q_{xx} = Q_{yy} = 0$$

holds only on time scales >> $\gamma_\perp^{-1}$, $\gamma_{||}^{-1}$.
In Equation (54) we introduced rescaled electric field strengths by $\sqrt{4g^2/\gamma_{||}\gamma_\perp(1+\Delta^2)}\, E^+ \to E^+$ and similarly for E_0, scaled time by $\kappa t \to t$, and defined the parameters

$$\Gamma^2 = -\frac{gW_0}{2\kappa(1+\Delta^2)}\sqrt{\frac{\gamma_{||}}{\gamma_\perp}},$$

$$Q = \frac{4g^2}{\gamma_{||}\gamma_\perp\kappa(1+\Delta^2)V}\left(\frac{\kappa}{e^{(h\omega_0/k_BT)}-1} + \frac{1}{2}\gamma_\perp (n+W_0)\right).$$

In the present case we have $\Delta = 0$. The definitions for $\Delta \neq 0$ will be needed later. Due to the resonance condition, Equation (54) satisfies detailed balance[29] with

$$r^\nu(q) = 0$$

$$\Phi(q) = \frac{1}{Q}\left[|E^+ - E_0|^2 + \Gamma^2 \ln(1+|E^+|^2)\right] \tag{55}$$

even though the original equations (14) which were not restricted to time scales >> $\gamma_\perp$, $\gamma_{||}$, are not in detailed balance.

For sufficiently strong external driving fields E_0 and $\Gamma^2 > 8$ the potential Φ has two local minima and thus exhibits bistability. The present model has been used for a statistical description of absorptive optical bistability[29,30].

Systems near bifurcation points of a continous forward Hopf bifurcation are a particularly important class of two-variable systems exhibiting detailed balance on a long time scale[4]. The laser near threshold[15,16] and Benard convection near the critical Rayleigh number[31,32] have been treated explicitly. The general mechanism which is responsible for detailed balance in this case is very simple. First, the vicinity of the bifurcation point implies the slowing down of that mode of the system which becomes unstable at the bifurcation point, compared with all other modes. Therefore, only the slow mode has to be considered on sufficiently long time scales.

Second, the amplitude of that mode is still small in the close vicinity of the bifurcation point. Hence, a Landau expansion is possible and the equation of motion of the slow mode assumes the standard form

$$\dot{b} = \alpha b - \beta |b|^2 b + \xi(t) \tag{56}$$

where b is the complex amplitude, and α, β are complex parameters with

$$\mathrm{Re}\,\alpha \sim (\lambda-\lambda_c) \tag{57}$$

where λ is the bifurcation parameter. We assume $\mathrm{Re}\beta > 0$, otherwise higher orders in b have to be included. The Langevin force in the Fokker Planck approximation is Gaussian white noise with

$$< \xi(\tau) > = 0; \quad < \xi^*(\tau)\ \xi\ (0) > = 2Q\,\delta(t) \tag{58}$$

and Q independent of b near λ_c.

Equation (56) satisfies detailed balance under the transformation $b \to b^*$ with

$$\Phi = \frac{1}{Q}\,(-\mathrm{Re}\ \alpha|b|^2 + \mathrm{Re}\beta\ |b|^4) \tag{59}$$

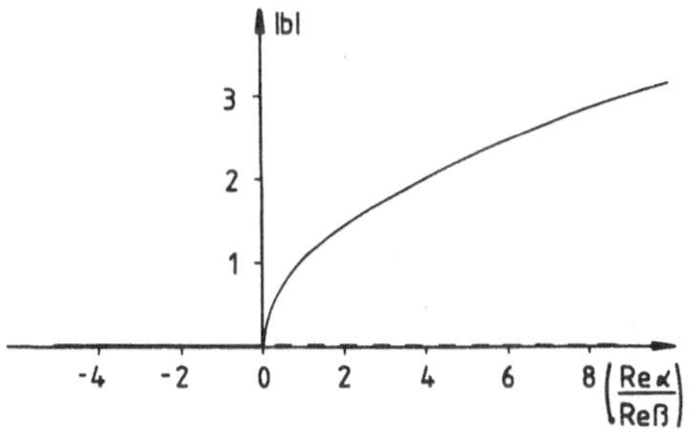

Figure 1. Forward Hopf bifurcation of steady state amplitude $|b|$ near the bifurcation point $\lambda = \lambda_c$, without noise, $\xi(t) = 0$. Full line stable steady state, dashed line unstable steady state.

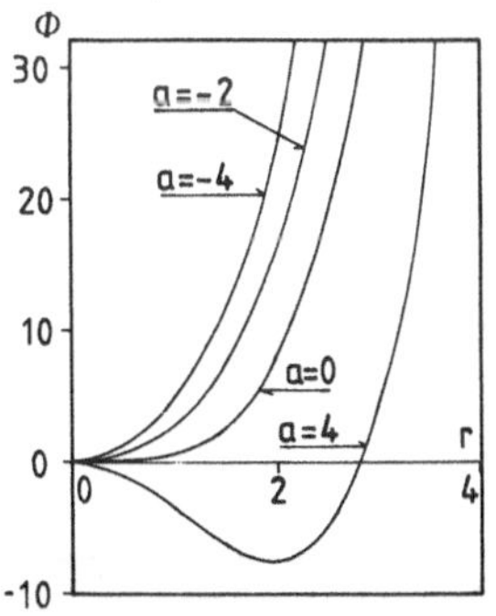

Figure 2. The potential (59), $\Phi = -ar^2 + \frac{1}{2}r^4$ for λ below at, and above the bifurcation point. $r = (2\mathrm{Re}\beta/Q)^{\frac{1}{4}}|b|$. $a = (2Q\mathrm{Re}\beta)^{-\frac{1}{2}}\,\mathrm{Re}\alpha$.

and

$$r_b = i\left((\mathrm{Im}\,\alpha) - (\mathrm{Im}\,\beta)\ |b|^2\right) b \qquad (60)$$
$$r_{b*} = (r_b)^*$$

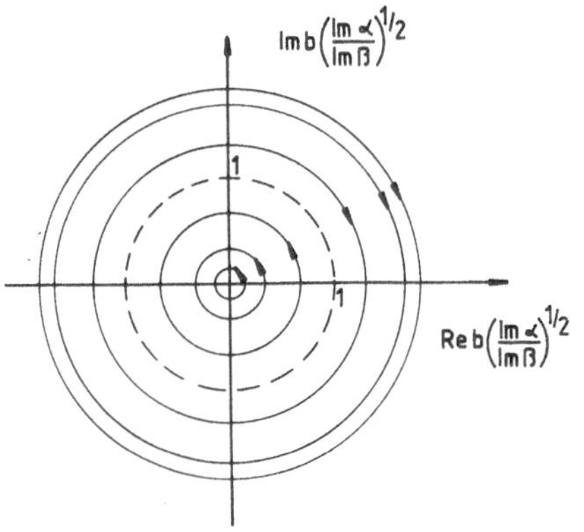

Figure 3. Flow lines of r_b, Equation (60), schematically for $\mathrm{Im}\,\alpha>0$, $\mathrm{Im}\,\beta>0$. Along dashed line $r_b=0$.

Detailed balance is preserved if the noise source in Equation (56) is replaced by

$$\tilde{\xi}(t) = b\ \xi(t) \qquad (61)$$

in the Stratonovich calculus,[9] i.e.

$$< \tilde{\xi}(t) > = 0, \quad < \tilde{\xi}^*(\tau)\ \tilde{\xi}(0) > = 2Q\ |b|^2\ \delta(\tau). \qquad (62)$$

Such fluctuations arise if λ in Equation (57) has a fluctuating component. The potential Φ in that case is given by

$$\Phi = \frac{1}{Q}\left(\left(-\mathrm{Re}\,\alpha + \frac{Q}{2}\right) \ln\ |b|^2 + \mathrm{Re}\,\beta\ \ |b|^2\right) + \mathrm{const.} \qquad (63)$$

while r_β, $r_{\beta*}$ remain unchanged compared to (60).

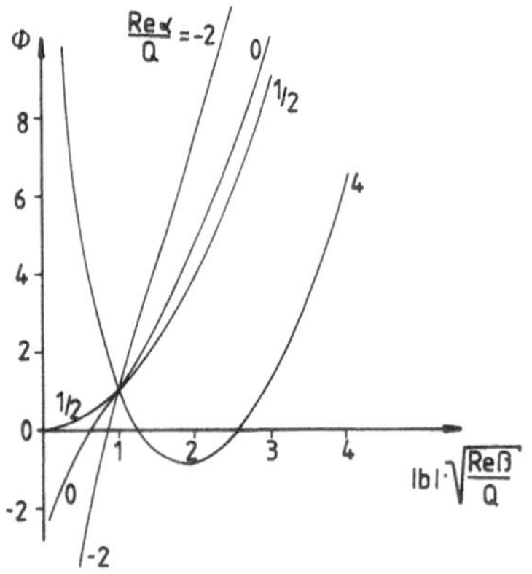

Figure 4. The potential (63) for Re α smaller, equal and larger $Q/2$. ($Q=1$, $\mathrm{Re}\beta=1$).

At $b = 0$ the probability density $W = \exp(-\Phi)$ defined by Equation (63) is only normalizable if $\alpha + \alpha^* > 0$. For $\alpha + \alpha^* < 0$, the steady state distribution of this model reduces to a δ-function at the origin

$$W = \delta(\mathrm{Re}\ b)\ \delta\ (\mathrm{Im}\ b) \tag{64}$$

because deterministically the origin is attracting and the noise source vanishes there.

If the system is removed from the vicinity of the bifurcation point the normal form (56) is destroyed and detailed balance will no longer be satisfied.

(d) Absence of detailed balance

If the system is lacking detailed balance, Φ and r^ν must be determined from the equations

$$r^\nu \frac{\partial\Phi}{\partial q^\nu} - \frac{\partial r^\nu}{\partial q^\nu} = 0 \tag{65}$$

$$K^\nu - \frac{1}{2}\frac{\partial Q^{\nu\mu}(q)}{\partial q^\mu} = -\frac{1}{2} Q^{\nu\mu}(q) \frac{\partial\Phi}{\partial q^\mu} + r^\nu. \tag{66}$$

Eliminating r^ν we obtain

$$-\frac{1}{2} Q^{\nu\mu}(q) \frac{\partial^2\Phi}{\partial q^\nu q^\mu} + (K^\nu(q) - \frac{\partial Q^{\nu\mu}(q)}{\partial q^\mu}) \frac{\partial\Phi}{\partial q^\nu}$$

$$+\frac{1}{2} Q^{\nu\mu}(q) \frac{\partial\Phi}{\partial q^\nu}\frac{\partial\Phi}{\partial q^\mu} = \frac{\partial K^\nu(q)}{\partial q^\nu} - \frac{1}{2}\frac{\partial^2 Q^{\nu\mu}(q)}{\partial q^\nu \partial q^\mu} \tag{67}$$

This equation is equivalent to the time independent Fokker Planck equation for W and is not accessible to exact solution. However, a considerable simplication is achieved if the limit of weak noise is considered. In this limit $Q^{\nu\mu}(q)$ is replaced by $\varepsilon Q^{\nu\mu}(q)$, $\Phi(q)$ is replaced by $\Phi(q) = \Phi_0(q)/\varepsilon + \Phi_1(q) + \ldots$, $K^\nu(q)$ is replaced by $K^\nu{}_o(q) + \varepsilon K^\nu{}_1(q) + \ldots$, $r^\nu(q)$ is replaced by $r^\nu(q) = r_o{}^\nu(q) + \varepsilon r_1{}^\nu(q) + \ldots$, and the limit $\varepsilon \to 0^+$ is taken. There are cases where, even though $\Phi_0(q)$ exists, the corrections are not analytical in ε for $\varepsilon \to 0$. In all instances where we consider Φ_1, in the following, we have to exclude such cases explicitly.

In the lowest order in ε we obtain

$$\frac{1}{2} Q_o^{\nu\mu}(q) \frac{\partial\Phi_o}{\partial q^\nu} \frac{\partial\Phi_o}{\partial q^\nu} + K_o^\nu(q) \frac{\partial\Phi_o}{\partial q^\nu} = 0. \tag{68}$$

to first order in ε we obtain

$$\begin{aligned} &(Q_o^{\nu\mu}(q) \frac{\partial\Phi_o(q)}{\partial q^\mu} + K_o^\nu(q)) \frac{\partial\Phi_1(q)}{\partial q^\nu} = \\ &-K_1^\nu \frac{\partial\Phi_o(q)}{\partial q^\nu} - \frac{1}{2} Q_1^{\nu\mu} \frac{\partial\Phi_o(q)}{\partial q^\nu} \frac{\partial\Phi_o(q)}{\partial q^\mu} + \\ &\frac{\partial K_o^\nu(q)}{\partial q^\nu} + \frac{\partial Q_o^{\nu\mu}(q)}{\partial q^\mu} \frac{\partial\Phi_o(q)}{\partial q^\nu} + \frac{1}{2} Q_o^{\nu\mu}(q) \frac{\partial^2\Phi_o(q)}{\partial q^\nu \partial q^\mu} \end{aligned} \tag{69}$$

In the limit $\varepsilon \to 0^+$ the problem of finding Φ has therefore been reduced to solving a nonlinear first order partial differential equation for Φ_o and a sequence of linear inhomogeneous partial differential equations for Φ_1, Φ_2 etc. The equation for Φ_o has the form of a time independent Hamilton-Jacobi equation in classical mechanics. Unfortunately, this equation is not easy to solve apart from special cases. However, the present approach becomes a powerful one if a small parameter λ in $Q_o^{\nu\mu}(q)$ or $K_o^\nu(q)$ exists, which, when taken to zero, allows us to find the solution for Φ_o. We may then expand

$$\begin{aligned} K_o^\nu(q) &= K_{oo}^\nu + \lambda K_{o1}^\nu \\ Q_o^{\nu\mu}(q) &= Q_{oo}^{\nu\mu} + \lambda Q_{o1}^{\nu\mu} \\ \Phi_o &= \sum_{n=0}^{\infty} \lambda^n \Phi_{on} \end{aligned} \tag{70}$$

and obtain a hierarchy of linear first order partial differential equations of Φ_{o1}, Φ_{o2}, etc., and a nonlinear equation for Φ_{oo} which is assumed to be solvable. So far this approach has not yet been applied widely. There is also no mathematical theory which specifies the conditions under which Φ_o is a reasonable approximation of Φ. Therefore one must test this approach in specific examples.

In the last section we will discuss some results obtained recently by an application of this approach to the dispersive optical bistability, which does not satisfy detailed balance.[20] We will also discuss the applicability of this approach and its limitations to the Lorenz model, which describes a system without detailed balance with a strange attractor.[21] If Φ_o can be obtained from Equation (68) it is a very valuable function for the following reasons:

- It determines the steady probability density in the limit of weak noise.
- It is a Lyapunoff function of the deterministic equations $\dot{q}^\nu = K_o^\nu(q)$. This can be seen from the fact that $\Phi_o(q)$ must be bounded from below (otherwise the probability density would diverge) and can only decrease

$$\frac{d\Phi_o}{dt} = \frac{\partial\Phi_o}{\partial q^\nu}\dot{q}^\nu = \frac{\partial\Phi_o}{\partial q^\nu} K_o^\nu(q) = - \frac{1}{2} Q_o^{\nu\mu} \frac{\partial\Phi_o}{\partial q^\nu} \frac{\partial\Phi_o}{\partial q^\mu} \leq 0 \tag{71}$$

In this derivation we used the facts that, for $\varepsilon \to 0^+$, Equation (65) reduces to

$$K_o^\nu = - \frac{1}{2} Q_o^{\nu\mu} \frac{\partial\Phi_o}{\partial q^\nu} + r_o^\nu. \tag{72}$$

that Q_o (q) is nonnegative, and that for $\varepsilon \to 0^+$ Equation (65) reduces to

$$r_o^\nu \frac{\partial\Phi_o}{\partial q^\nu} = 0. \tag{73}$$

- Due to the preceding two relations Φ_o plays a role for the non-equilibrium steady state which is similar to the role played by a thermodynamic potential for thermodynamic equilibrium. Φ_o has therefore been called a non-equilibrium thermodynamic potential.[13]

3. EXAMPLES WITH DETAILED BALANCE

In this section we want to apply the methods outlined in Section 2 to systems in the vicinity of a normal Hopf bifurcation point, restricting ourselves to long time scales. As we discussed in Section 2, such systems have the property of detailed balance. We want to study how fluctuations influence the properties of the system near its bifurcation point. To this end we consider separately the influence of external fluctuations or 'additive' noise and the influence of internal or 'multiplicative' noise.

(a) Forward Hopf bifurcation under the influence of additive Gaussian white noise

The Langevin equation of this class of models is given by Equation (56) with the noise source (58). It is useful to introduce normalized variables by

$$z = r\,e^{i\psi} = x + iy = \sqrt[4]{\frac{(\beta+\beta^*)}{Q}}\; e^{+\,i\Omega t}\, b$$

$$\tau = \sqrt{\frac{(\beta+\beta^*)}{4}}\; Q\, t \tag{1}$$

with

$$\Omega = \frac{i\,(-\,\alpha^*\beta+\alpha\beta^*)}{\beta+\beta\,^*} \tag{2}$$

The potential Φ in polar coordinates r, ψ may then be written as+)

$$\Phi = -\frac{a}{2}r^2 + \frac{r^4}{4} - \ln r + \text{const.} \tag{3}$$

with

$$a = \frac{\alpha+\alpha^*}{\sqrt{Q(\beta+\beta^*)}} \tag{4}$$

The operators L_H and L_A of Equations (20), (21) are obtained as

$$L_H = \frac{\partial^2}{\partial r^2} + \frac{1}{r}\frac{\partial}{\partial r} + \frac{1}{r^2}\frac{\partial^2}{\partial \psi^2} - V \tag{5}$$

with

$$V = a + \left(\frac{a^2}{4} - 2\right) r^2 - \frac{ar^4}{2} + \frac{r^6}{4} \tag{6}$$

and

$$L_A = -\,\delta(a - r^2)\,\frac{\partial}{\partial\psi} \tag{7}$$

with

$$\delta = \frac{\beta-\beta^*}{i\,(\beta+\beta^*)} \tag{8}$$

The only surviving parameters are a and δ. The potential V for all a admits only 'bound state' solutions with a discrete spectrum. Upon integration over ψ, the antihermitian operator L_A drops out. These equations have first been derived for a single model laser near threshold by Risken and Vollmer.[15] These authors worked out the numerical solution of the eigenvalue problem for L_H, while the

---

+) Note that the probability $(r, \psi, dr, d\psi)$ is $P\,dr\,d\psi$

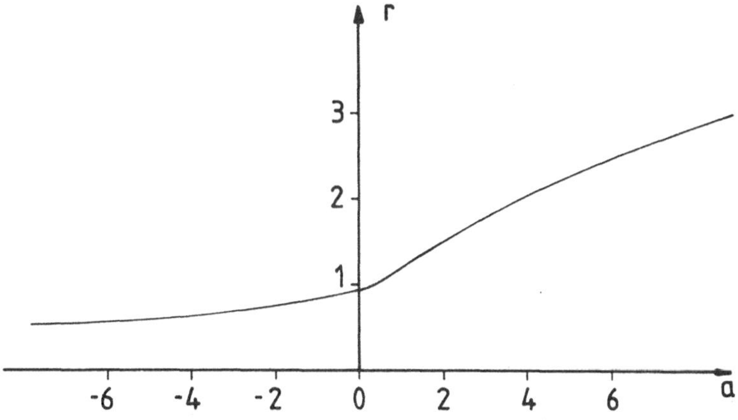

Figure 5. Modification of forward Hopf bifurcation (Figure 1) of steady state amplitude $<|b|>$ near the bifurcation point under the influence of additive Gaussian white noise. $r = (2\mathrm{Re}\beta/Q)^{1/4} <|b|>$; $a = (2Q\mathrm{Re}\beta)^{-1/2}\,\mathrm{Re}\alpha$.

eigenvalue problem of $L_H + L_A$ was solved numerically by Seybold and Risken.[16] We refer the reader to these papers and do not go into further detail here, except to note that the sharp bifurcation point of the deterministic description at $\alpha+\alpha^* = 0$ disappears in the presence of arbitrarily small but finite fluctuations and is replaced by a finite threshold region near $\alpha+\alpha^* = 0$ of size $(Q(\beta+\beta^*))^{1/2}$ where the auto-correlation time of $|b|^2$ for small Q becomes large $\sim((\beta+\beta^*)Q)^{-1/2}$ but remains finite. The fluctuation intensity $<|b|^2>$ in the threshold region is of the order $\sqrt{Q/(\beta+\beta^*)}$. These results are independent of δ since they do not involve the phase ψ of b. The influence of δ on the correlation time of b was calculated by Seybold and Risken.[16] These results originally obtained for the single mode laser near threshold, are applicable to other bifurcation problems with additive noise, as well.

(b) Forward Hopf Bifurcation Under the Influence of Multiplicative Gaussian White Noise

The Langevin equation of this class of models is given by Equation (56) with the Langevin force (61). We introduce normalized variables by

$$z = x + iy = re^{i\psi} = \sqrt{\frac{(\beta+\beta^*)}{Q}}\, e^{+i\Omega t}\, b \qquad (9)$$

$$\tau = \frac{Q}{2} t$$

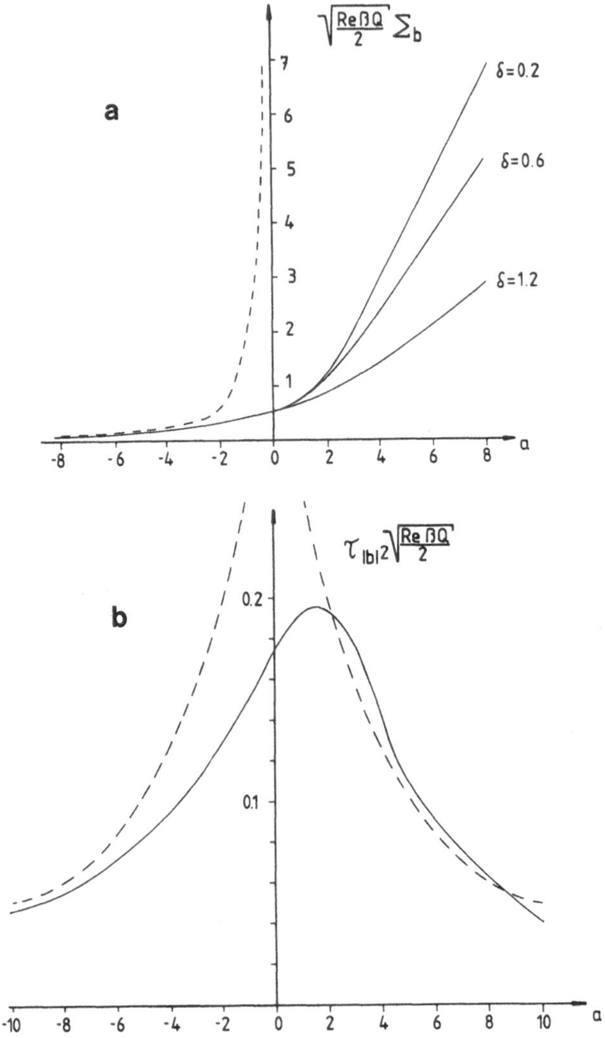

Figure 6a. Average relaxation time of b, Equation
6b. (56), in the steady state near the bifurcation point. Full line after[16] with additive Gaussian white noise, dashed lint without noise. (6a) Average relaxation time of $|b|^2$ in the steady state near the bifurcation point. Full line after[15] with additive Gaussian white noise, dashed line without noise. (6b)

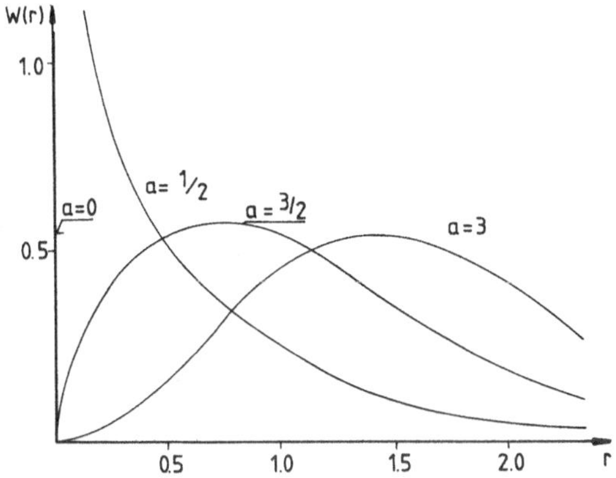

Figure 7. The steady state distribution W=exp (- Φ) obtained from (11) (cf. Vol. II of Ref. 9).

with

$$\Omega = \frac{i\,(-\alpha^*\beta+\alpha\beta^*)}{\beta+\beta^*} \tag{10}$$

The potential Φ is now given by

$$\Phi = (1-a)\ \ln r + \frac{1}{2}\, r^2 + \text{const.} \tag{11}$$

$$a = \frac{\alpha+\alpha^*}{Q}$$

For $a > 0$, the steady state distribution is given by $W = \exp(-\Phi)$. For $a < 0$, the potential Φ can still be defined by Equation (11). However, the steady state distribution is now given by $W = 1/\pi\ \delta(r)$. With this proviso, the following analysis also applies to the case $a < 0$.

The operators L_H, L_A defined by Equations (20), (21) are easily obtained in the form

$$L_H = \frac{\partial}{\partial r}\ r^2\ \frac{\partial}{\partial r} + \frac{\partial^2}{\partial\psi^2} - V(r) \tag{12}$$

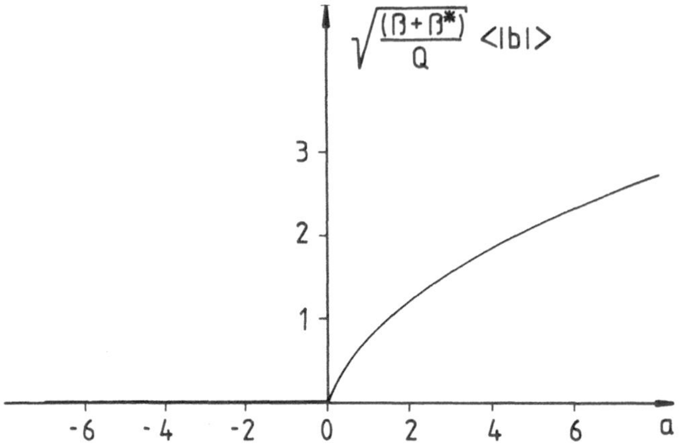

Figure 8. Forward Hopf bifurcation of steady state amplitude $\sqrt{<|b|>}$ near the bifurcation point under the influence of multiplicative Gaussian white noise, $<|b|> = (Q/2\ (\beta+\beta^*))^{1/2}\ \Gamma(a+1/2)\ /\ \Gamma(a/2)$.

with

$$V(r) = \frac{1}{4} r^2 \left(\frac{\partial\Phi}{\partial r}\right)^2 - r\frac{\partial\Phi}{\partial r} - \frac{r^2}{2}\frac{\partial^2\Phi}{\partial r^2} \tag{13}$$

$$= \frac{1}{4}(a^2-1) - \frac{a+2}{2} r^2 + \frac{r^4}{4}$$

and

$$L_A = -\delta(a-r^2)\frac{\partial}{\partial\psi} \tag{14}$$

The eigenfunctions $\psi_{nm}(r, \psi)$ corresponding to the eigenvalue λ_{nm} may be separated in r, ψ. Writing

$$\psi_{nm}(r, \psi) = \frac{1}{r^{3/2}} \psi_{nm}(r^2)\, e^{im\psi} \tag{15}$$

and introducing $x = r^2$, we obtain

$$0 = \left[\frac{\partial^2}{\partial x^2} + \frac{1}{4}\left(\frac{1-\frac{a^2}{4}+\lambda_{nm}-m^2-ia\delta m}{x^2} + \frac{a+2+2im\delta}{2x} - \frac{1}{4}\right)\right]\psi_{nm}(x). \tag{16}$$

Equation (16) is the Whittaker differential equation with complex coefficients, if $\delta = 0$. In the special case $\delta = 0$ Equation (16) has already been obtained and solved by Schenzle and Brand.[17] The general non-hermitian problem for $\delta \neq 0$ has recently been solved[18]. The solutions of Equation (16) can be expressed in terms of confluent hypergeometric functions with complex parameters.[33,34]

$$\psi_{nm}(x) = x^{\nu} e^{-\frac{x}{4}} \; {}_1F_1(\nu - \tfrac{1}{2} - \tfrac{a}{4} - \tfrac{im\delta}{2},\; 2\nu, \tfrac{x}{2}) \tag{17}$$

where

$$\nu = \tfrac{1}{2} (\pm) \tfrac{1}{2} (\tfrac{a^2}{4} + m^2 + iam\delta - \lambda_{nm})^{1/2} \tag{18}$$

The spectrum λ_{nm} has a discrete and a continuous part. In the discrete part ${}_1F_1$ can only grow like a polynominal for large x, which requires that we put

$$\nu - \tfrac{1}{2} - \tfrac{a}{4} - \tfrac{im\delta}{2} = -n \tag{19}$$

for integer positive n. This condition fixes the discrete eigenvalues

$$\lambda_{nm} = 2an - 4n^2 + m^2 (1+\delta^2) + 4imn\delta . \tag{20}$$

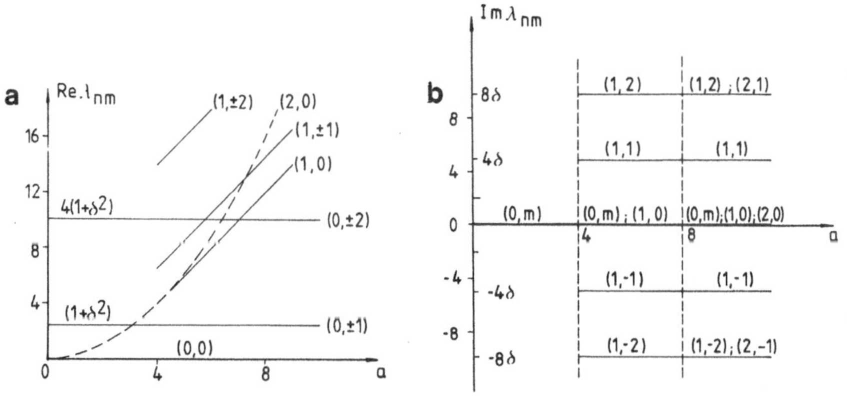

Figure 9a. Real part of the discrete spectrum (20)
9b. for n = 0,1,2; m = 0, ± 1 near the bifurcation point, parametrized by (n,m). Lower boundary of continuous spectrum given by dashed line. (9a) Imaginary part of the discrete spectrum (20) for n = 0,1,2; m = 0, ± 1, ± 2 near the bifurcation point. (9b)

We note that $\lambda = 0$ for $n = m = 0$, as it should be, and that λ_{nm} is real for $\delta = 0$, and complex for $\delta \neq 0$ and m, n both different from zero. The antihermitian part of L increases the real part of the eigenvalues with $m \neq 0$. The square integrability of $\psi_{nm}(r,\psi)$ at the origin requires

$$\frac{1}{2}(\nu+\nu^*) > \frac{1}{2} \tag{21}$$

which can only be satisfied by the positive sign in Equation (18), and requires

$$n < \frac{a}{4}, \ \mathrm{Re}\ \lambda_{nm} < \frac{a^2}{4} + m^2 (1+\delta^2)$$

due to Equation (19).

The (unnormalized) eigenfunctions in the discrete spectrum are given by the complex polynomials ${}_1F_1(-n, \alpha, x/2)$

$$\psi_{nm}(x) = x^{\frac{1}{2}+\frac{a}{4}-n+\frac{im\delta}{2}} e^{-\frac{x}{4}} {}_1F_1\left(-n, -2n+1+\frac{a}{2}+im\delta, \frac{x}{2}\right) \tag{22}$$

The eigenfuctions of the adjoint problem are obtained from this result by replacing δ by $-\delta$. We now turn to the continuous part of the spectrum. The linear combination of eigenfunctions (17) which does not diverge exponentially at infinity is given by[34].

$$\psi_{nm}(x) = \left(\frac{x}{2}\right)^{\frac{1}{2}} e^{-\frac{x}{4}} \left\{ \left(\frac{x}{2}\right)^{\mu} \frac{{}_1F_1\left(\mu - \frac{a}{4} - \frac{im\delta}{2}, 2\mu+1, \frac{x}{2}\right)}{\Gamma\left(-\mu - \frac{a}{4} - \frac{im\delta}{2}\right) \Gamma(2\mu+1)} - (\mu \to -\mu) \right\} \tag{23}$$

where

$$\mu = \nu - \frac{1}{2} = \frac{1}{2}\left(\frac{a^2}{4} + m^2 + iam\delta - \lambda_{\varkappa m}\right)^{1/2} \tag{24}$$

Square integrability of $\psi_{nm}(r, \psi)$ at the origin requires that μ be purely imaginary

$$\mu + \mu^* = o, \quad \mu = i\varkappa \tag{25}$$

Therefore, we have the condition

$$\lambda_{\kappa m} = \lambda_{\kappa m}^{(r)} + i\ am\delta = \frac{a^2}{4} + m^2 + 4\kappa^2 + i\ am\delta \tag{26}$$

with
$$\lambda_{\kappa m}^{(r)} > \frac{a^2}{4} + m^2 \tag{27}$$

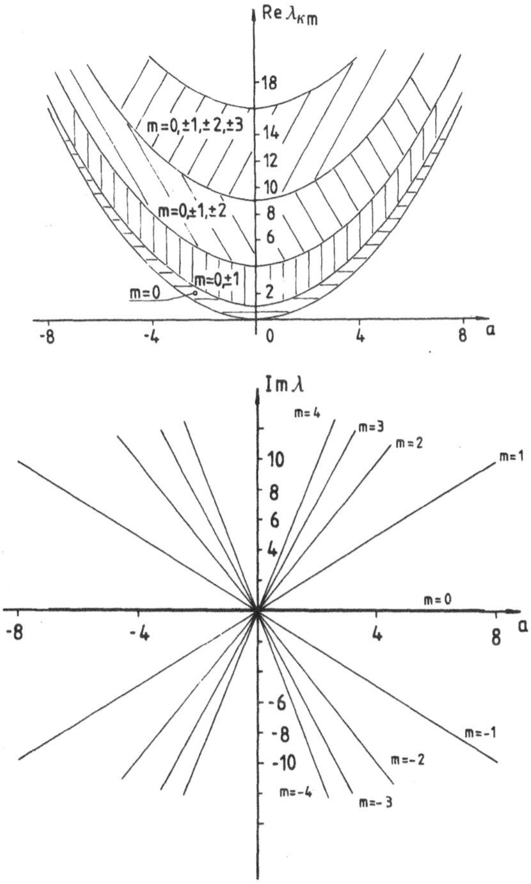

Figure 10a. Real part of the continuous spectrum Equation (26) for $m = 0, \pm 1, \pm 2, \pm 3, \pm 4$.

Figure 10b. Imaginary part of the continuous spectrum Equation (26) for $m = 0, \pm 1, \pm 2, \pm 3, \pm 4$.

and continuous. We see therefore, that the imaginary part of $\lambda_{\varkappa m}$ remains discrete, due to angular momentum quantization in the language of the Schrodinger equation. For $\delta = 0$ all these results reduce to those obtained by Schenzle and Brand.[17] It is interesting to note, that for fixed m the real parts of the discrete spectrum and the continuous spectrum overlap. This overlapping region is defined by

$$0 < \frac{a}{2} - 2n < |m\delta| \tag{28}$$

and only exists in the case $\delta \neq 0$, i.e. it is due to the non-hermiticity of the eigenvalue problem we have solved here.

Since the eigenfunctions are complicated we do not enter into a discussion of averages here but refer to the literature Ref. 17 for $\delta = 0$ and Ref. 18 for $\phi \neq 0$. However, the eigenvalues are very simple and allow to give a simple qualitative picture of the behavior as $\underline{a}$ approaches zero from above. For $m = 0$, λ_{no} gives the relaxation constants of fluctuations of $|z|^2$. This part of the spectrum is unchanged by $\delta \neq 0$. Approaching the bifurcation point $\underline{a} = 0$ from large positive $\underline{a}$, we see that there is at first a large number of discrete real eigenvalues λ_{no} which decrease linearly with $\underline{a}$ and disappear one by one when they hit their upper bound, which decreases more rapidly with a^2. Correlation functions decay exponentically for long times in this regime but show a slowing down of fluctuations as we decrease $\underline{a}$. The lowest discrete eigenvalue above $\lambda_{oo} = 0$ disappears for $\underline{a} = 4$. For smaller values of $\underline{a}$ there is only a continuous spectrum above $\lambda_{oo} = 0$, whose lower boundry is given by $a^2/4$. Therefore, there is still at least exponential decay of correlation functions in this regime, as long as $\underline{a} \neq 0$, however with slowing down proportional to $\underline{a}^2$. At the point $\underline{a} = 0$ the continuous spectrum reaches down to $\lambda_{oo} = 0$, and a non-exponential decay of correlation functions occurs at this point. At this point, the steady state distribution reduces to a δ - function around the origin and remains so for $\underline{a} < 0$. For negative $\underline{a}$ there is only a continuous spectrum with a finite gap given by $a^2/4$ (cf. Fig. 11).

The correlation function of z contains the eigenvalues $_{n1}$, $\lambda_{\varkappa 1}$ which are influenced by δ. The imaginary parts lead to a frequency shift. This shift has a fixed value $\underline{a\delta}$ for the continuous part of the spectrum. In the discrete spectrum for $\underline{a} > 0$ it increases by equidistant steps of size 4δ from 0.

The discrete eigenvalue

$$\lambda_{01} = 1 + \delta^2 \tag{29}$$

exists down to $\underline{a} = 0$. However, for $\underline{a} < 2|\delta|$ the eigenvalue with the smallest real part is

$$\lambda_{\varkappa 1} = \frac{a^2}{4} + 1 + \varkappa^2 + ia\delta \tag{30}$$

taken for $\varkappa = 0$, i.e., the discrete eigenvalue λ_{01} enters the continuous part of the spectrum $\lambda_{\varkappa 1}$. The frequency shift of the lowest eigenvalue with m = 1, n = 0, $\varkappa = 0$ is therefore 0 for $\underline{a} \geq 2|\delta|$ and $\underline{a}$ for $\underline{a} \leq 2|\delta|$. There is always a finite gap for finite m which separates the eigenvalue λ_{oo} from the real part of λ_{nm}, even at $\underline{a} = 0$. Therefore the correlation functions of any functions of z and z* which depend on the phase ψ always decay at least exponentially for long times. We see that the transition from below to above threshold in this example is much less smooth than in the case of additive noise and exhibits interesting structure.

The bifurcation point $\underline{a} = 0$ of the deterministic description is still distinguished in the statistical description, as the point where the continuous spectrum starts with the eigenvalue 0 and where the steady state probability density changes from a δ-function to a smooth function.

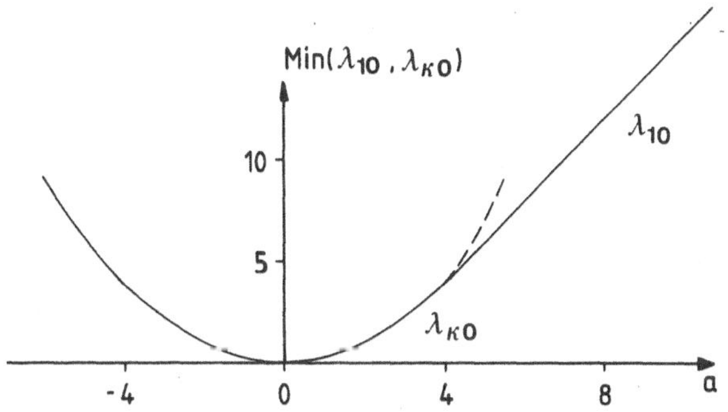

Figure 11. Eigenvalue λ_{no}, $\lambda_{\varkappa o}$ with smallest non-vanishing real part according to (20), (26). Dashed line gives lower boundary of continuous spectrum for a > 4.

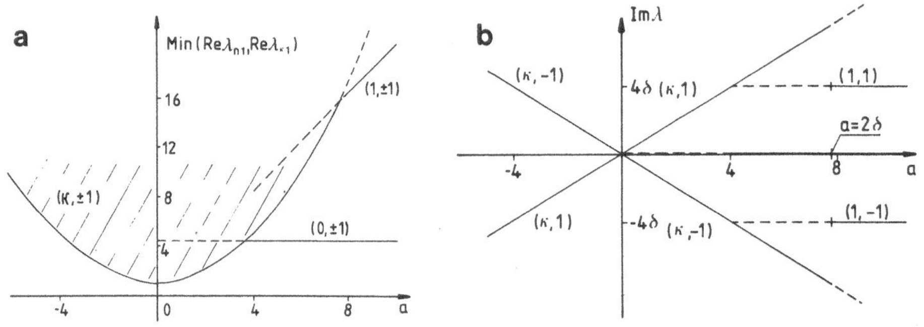

Figure 12a. Real part of eigenvalues λ_{n1}, $\lambda_{\kappa 1}$ with
12b. smallest nonvanishing real part according to (20, (21) (full time). (12a) Imaginary part of the eigenvalues with Lowest real part (full line). (12b)

4. EXAMPLES WITHOUT DETAILED BALANCE

The only class of models without detailed balance where the entire scheme of section 2 can be carried through are linear Gaussian processes (Ornstein-Uhlenbeck processes). Such processes occur naturally, if the system has a stable deterministic steady state given by a fixed point, and if the stochastic equations can be linearized around that fixed point. This is the usual procedure one applies if one wants to derive fluctuation-dissipation theorems and linear response for such a non-equilibrium steady state.[1,2] As this theory is well known, we will not elaborate it here, in spite of its great practical importance. However, many systems have deterministic steady states with a more interesting structure than a simple fixed point. For example, the system may exhibit regimes of multistability, where two or more locally stable deterministic fixed points coexist. The question then arises how the system is distributed over the various fixed points, a question which cannot be answered by investigating only the close vicinity of a single fixed point. The first model we want to treat here is bistability in optical systems discovered some years ago.[19] The discussion here is based on recent work performed in collaboration with A. Schenzle.[20] In a second example we will consider the Lorenz model[7] under the influence of external noise, which is a model deterministically described by a strange attractor in the steady state. The work we report here was performed in collaboration with H.J. Scholz.[21]

(a) A Model of Dispersive Optical Bistability

The starting point of the model is the Maxwell-Bloch equations (14), which we consider in the adiabatic approximation already applied in eq. (54). However, we do not assume resonance between the driving frequency ω, the atomic transition ν, and the resonator frequency ω_0. In the adiabatic approximation we obtain the Langevin equation

$$\dot{E}^+ = (i\delta-1)\,E^+ + E_o - \Gamma^2\,\frac{(1-i\Delta)}{1+|E|^2}\,E^+ + R^+ \tag{1}$$

with the Gaussian force $R^+(t)$

$$< R^+(t)\;\; R(0) > \; = \; 2\,Q\delta(t),$$

$$< R\,(t)\;\; R(0) > \; = \; 0 \; = \; < R^+(t)\;\; R^+(0) >. \tag{2}$$

We introduced dimensionless variables and parameters as in section 2.

Deterministically, Equation (1) has three stationary fixed points within a certain region of the parameter space (Γ^2, Δ, δ, E_o). For $\delta = \Delta = 0$ a necessary condition is $\Gamma^2 > 8$. The Fokker Planck equation of this model takes the form

$$\frac{\partial P}{\partial t} = \frac{1}{r}\frac{\partial}{\partial r}\, r\,\left(\frac{r}{1+r^2}\,(1+\Gamma^2+r^2) - r_o \cos\psi + \frac{Q}{2}\frac{\partial}{\partial r}\right)P$$

$$+ \frac{1}{r}\frac{\partial}{\partial\psi}\,\left(\delta r + \Delta\Gamma^2\,\frac{r}{1+r^2} + r_o \sin\psi + \frac{Q}{2}\,\frac{1}{r}\,\frac{\partial}{\partial\psi}\right)P \tag{3}$$

where we have used polar coordinates

$$E^{\pm} = r\,\exp\,(\pm i\psi), \quad E_o = r_o \tag{4}$$

and normalized the probability density such that $P\, r\, dr\, d\psi$ is the probability for (r, ψ). The latter convention differs from the one adopted in sections 2 and 3 (where $P\, dr\, d\psi$ was the probability for (r, ψ)) but is the convention adopted in Ref. 20. The steady state distribution of Equation (2) can be given in two cases. For $\delta = 0 = \Delta$ we have detailed balance under the transformation defined in section 2. The potential Φ for this case has already been given in section 2. The hermitian eigenvalue problem for this case can therefore easily be constructed along the lines exemplified in section 3. (Ritz's variational principle was applied to this problem in Ref. 30.)

For $\delta = \Delta$ there is no detailed balance under the time reversal transformation defined in section 2. Nevertheless an exact potential has been found also in this case.[20] It is given by

$$\Phi = \frac{1}{Q} \left(r^2 - \frac{2rr_o}{\sqrt{1+\Delta^2}} \cos (\psi-\psi_o) + \Gamma^2 \ln (1+r^2)\right) + \text{const.} \tag{5}$$

with $\tan \psi_o = -\Delta$. (6)

One may use this Φ to redefine time reversal in such a way that detailed balance is recovered. The transformation achieving this goal is

$$r \to r, \ \psi \to - \psi \tag{7}$$

as before, but in addition instead of $E_o \to E_o$, $\delta \to \delta$, $\Delta \to \Delta$ which we used before (if only implicitly) we now define

$$\overline{E}_o = E_o e^{i \arctan \delta} = E_o e^{-i\psi_o} \tag{8}$$

and require

$$\overline{E}_o \to \overline{E}_o^*, \ \delta \to \delta, \ \Delta \to \Delta. \tag{9}$$

Under this somewhat artificial transformation Φ is indeed reversible and detailed balance is reobtained. In any case, since Φ is known, it is straightforward to derive L_H and L_A of section 2.

A more difficult situation has to be faced if one is not so lucky to know an exact expression for Φ. Unfortunately, this is the typical case if detailed balance with respect to a simple time reversal transformation is not satisfied. At present the explicit study of such models is confined to the steady state distribution function. If one wants to find analytical results one may apply the method of section 2d, valid for weak fluctuations, in order to obtain an approximate expression for Φ. This has been done for the present example in Ref. 20. As described in section 2d, an additional small parameter (in addition to Q) is needed in order to solve the Hamilton-Jacobi equation approximately. Unfortunately, it is difficult, for the present model, to identify a small parameter which would generate a perturbation analysis for $\Phi(r, \psi)$ which approximates the exact potential uniformly in the entire (r, ψ) plane. In this respect, the best parameter for such an expansion seems to be r_o, also in view of the fact that the exact solution for $\delta = \Delta$ depends only linearly on r_o and can therefore be recovered in first order in r_o. An additional advantage of this choice is the fact that, for $r_o = 0$ the potential Φ_o obtained from the Hamilton-Jacobi equation exactly solves the time independent

Fokker Planck equation. Therefore, no problems arise with exchanging the order in which our two small parameters Q and r_o are taken to zero.

For $r_o = 0$ we obtain

$$\Phi_{oo} = \frac{1}{Q}(r^2 + \Gamma^2 \ln (1+r^2)). \tag{10}$$

To first order in r_o we obtain the equation for Φ_{01}

$$\frac{\partial \Phi_{01}}{\partial r} \left(- \frac{r}{1+r^2}(1+\Gamma^2+r^2)\right) + \frac{1}{r}\frac{\partial \Phi_{01}}{\partial \psi} r \left(\delta + \frac{\Delta r^2}{1+r^2}\right) =$$

$$- \frac{1}{Q}\frac{rr_o}{1+r^2}(1+\Gamma^2+r^2) \cos \psi \tag{11}$$

The ansatz

$$\Phi_{01} = \tilde{\Phi}_{01} - \frac{1}{Q}\frac{2rr_o}{\sqrt{1+\delta^2}} \cos (\psi-\psi_o), \tag{12}$$

with ψ_o defined by Equation (6), leads to an equation for $\tilde{\Phi}_{01}$ which one can solve by the method of characteristics. The characteristics satisfy

$$\frac{d\psi}{dr} = - \frac{\delta(1+r^2) + \Delta\Gamma^2}{r(1+\Gamma^2 + r^2)} \tag{13}$$

Upon integration we find

$$\psi = \psi(0) + \frac{(\delta-\Delta)}{2(1+\Gamma^2)} \Gamma^2 \ln \frac{r^2}{1+\Gamma^2+r^2} - \delta \ln r \tag{14}$$

$\tilde{\Phi}_{01}$ can be found by integration along the characteristics

$$\tilde{\Phi}_{o1} = - \frac{1}{Q}\int \frac{2(\delta-\Delta)\, r_o\, \Gamma^2}{\sqrt{1+\delta^2}\,(1+\Gamma^2+r^2)} \sin (\psi(r) - \psi_o)\, dr \tag{15}$$

which yields the total result

$$\Phi_0 = \Phi_{oo} + \Phi_{01}$$

$$= \frac{1}{Q}\left(r^2 - 2\frac{rr_o}{\sqrt{1+\delta^2}} \cos(\psi-\psi_o) + \Gamma^2 \ln(1+r^2)\right)$$

$$- 2\frac{\Gamma^2}{Q}(\delta-\Delta)\frac{rr_o}{\sqrt{1+\delta^2}} \text{Im} \frac{e^{i(\psi-\psi_o)}}{1+\Gamma^2-i(\Delta\Gamma^2+\delta)} F\left(-\frac{r^2}{1+\Gamma^2}\right) \tag{16}$$

where F(z) is a hypergeometric function in the notation of Ref. 33

$$F(z) = {}_2F_1\left(1, \frac{1}{2} - \frac{i\delta}{2}, \frac{3}{2} - \frac{i}{2}\frac{\delta+\Gamma\Delta^2}{1+\Gamma^2}; z\right) \tag{17}$$

The solution (16) is single valued and well behaved in the entire (r, ψ) plane. For a detailed discussion of this result in various limiting cases we refer to the original papers - Ref. 20. Exactly contained in this result are the potential Φ for the single mode laser (for $r_o = 0$, $\Gamma^2 < -1$, in order to have a laser above threshold), the absorptive bistability for $\delta = \Delta = 0$, and dispersive bistability for $\delta = \Delta \neq 0$. In order to check the quality of the approximation to Φ_0 in other cases we discuss the case $\delta = 0$, $\Delta \neq 0$ where Equation (16) reduces to

$$\Phi_0 = \frac{1}{a}\left(r^2 - 2rr_o \cos\psi + \Gamma^2 \ln(1+r^2)\right)$$

$$+ \frac{2}{Q}\frac{\Gamma^2}{1+\Gamma^2} rr_o \text{Im}\left(\frac{e^{i\psi}}{1-i\frac{\Delta r^2}{1+\Gamma^2}}\right.$$

$$\left. {}_2F_1\left(1, \frac{1}{2}; \frac{3}{2} - \frac{i\Gamma^2\Delta}{2(1+\Gamma^2)}, -\frac{r^2}{1+\Gamma^2}\right)\right) \tag{18}$$

For further simplification we consider the limit of small Δ where (18) reduces to

$$\Phi_0 = \frac{1}{Q}\left(r^2 - 2rr_o \cos\psi + \Gamma^2 \ln(1+r^2)\right)$$

$$+ 2\frac{\Delta\Gamma^2 r_o}{\sqrt{1+\Gamma^2}} \sin\psi \arctan\frac{r}{\sqrt{1+\Gamma^2}} \tag{19}$$

The corresponding probability density is shown in Figure 13 for $\Delta = 0.5$, $\Gamma^2 = 25$, $r_o = 10.1$.

As a check of the quality of the approximation (18) we compared the minima of Φ_0 (Fig. 14) with the deterministic steady states shown in Fig. 15. For the exact potential both sets of curves would agree exactly.

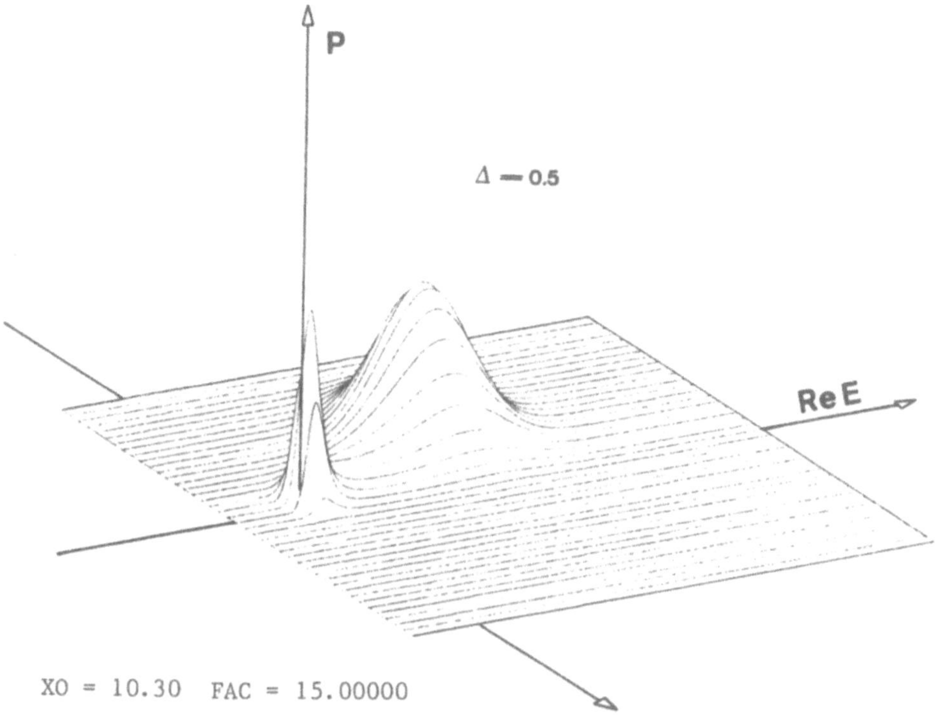

Figure 13. The probability density $W \sim \exp(-\Phi)$ obtained from (19) for $\Delta = 0.5$, $\Gamma^2 = 25$, $r_o = 10.1$, after[20].

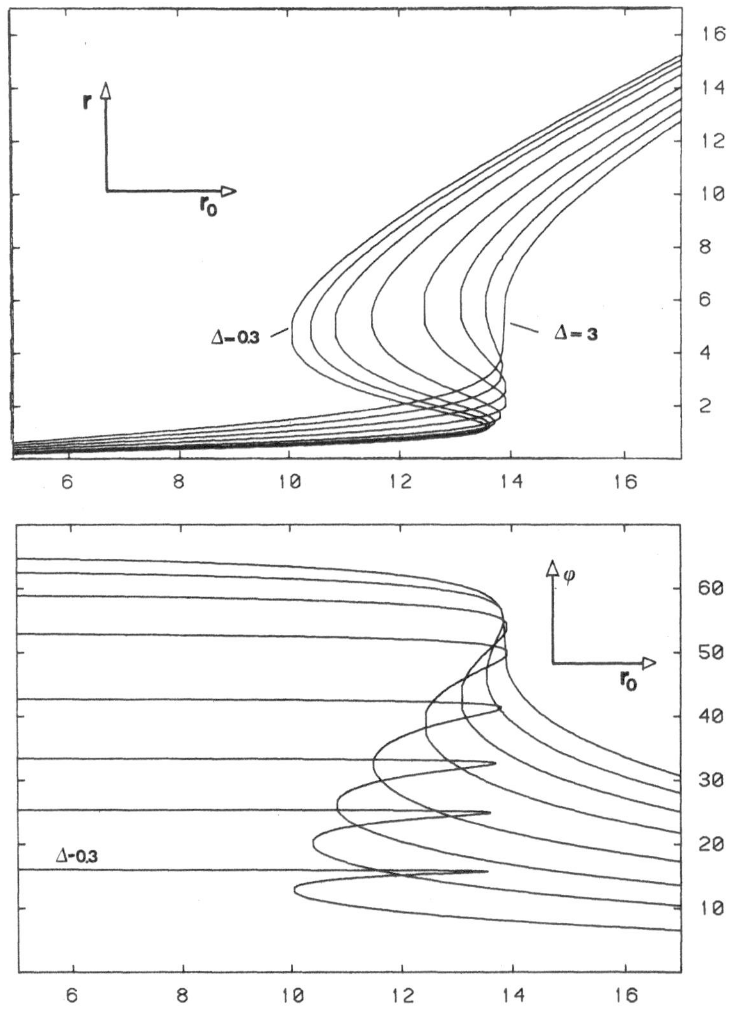

Figure 14. The minima of the potential (18) as a function of r_0 for $\Gamma^2 = 25$, $\Delta = 0.3$ 0.5, 0.7, 1.0, 1.5, 2.0, 2.5, 3.0, after[20].

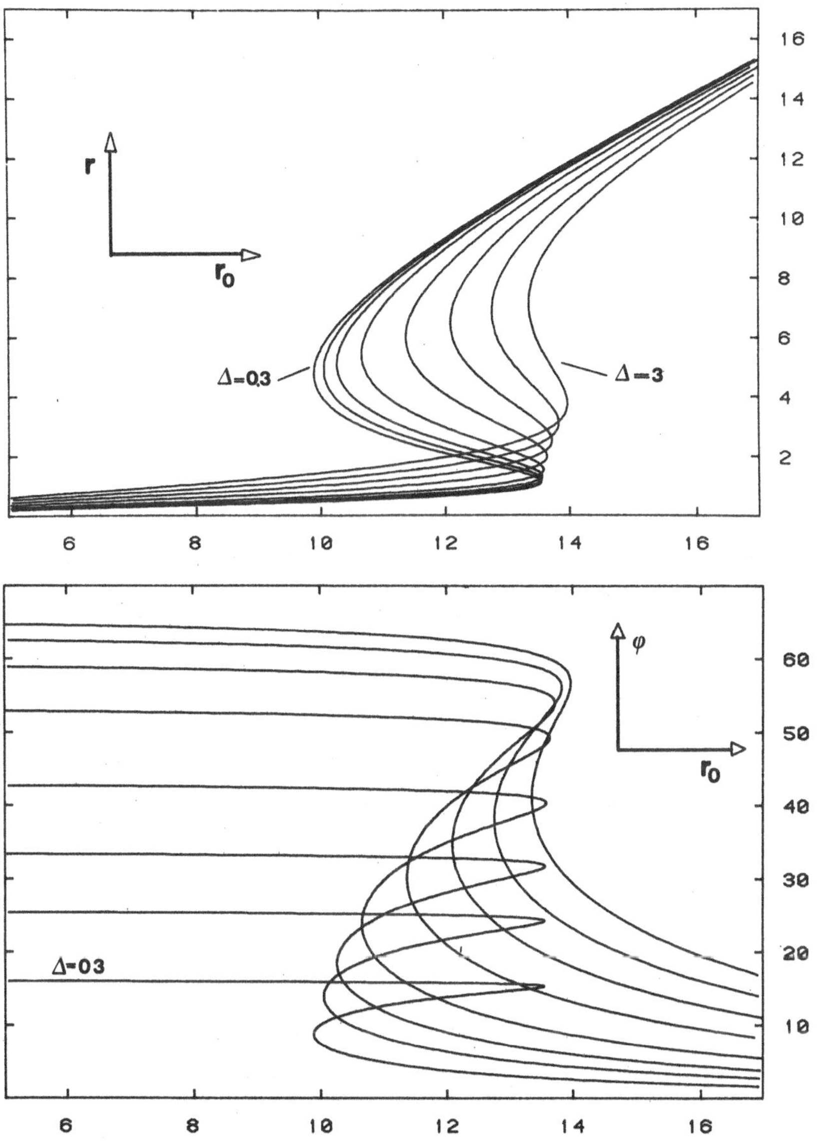

Figure 15. The deterministic fixed points of Equation (1) as a function of r_o for $\Gamma^2 = 25$, $\Delta = 0.3, 0.5, 0.7, 1.0, 1.5, 2.0, 2.5, 3.0$, after[20].

As we can see the agreement is qualitatively good with very good quantitative agreement on the unsaturated branch and qualitative agreement on the saturated branch. The size of the bistable domain is reproduced also in reasonable quantitative approximation. Unfortunately, the approximation deterioriates if one uses, instead of the hypergeometric function, the analytically much simpler solution (19) for small Δ.

By our solution we can now predict the relative probability of the occurrence of the two coexisting deterministic steady states (Figure 16). We can also determine the generalization of the Maxwell construction for first-order phase transitions in thermodynamic equilibrium for our non-equilibrium steady state. Since Φ_0 is a Lyapunoff function of the deterministic equations of motion, stability is exchanged between the two coexisting branches, where the corresponding two minima of Φ_0 have equal depth. Examples for the application of that rule are shown in Figure 17.

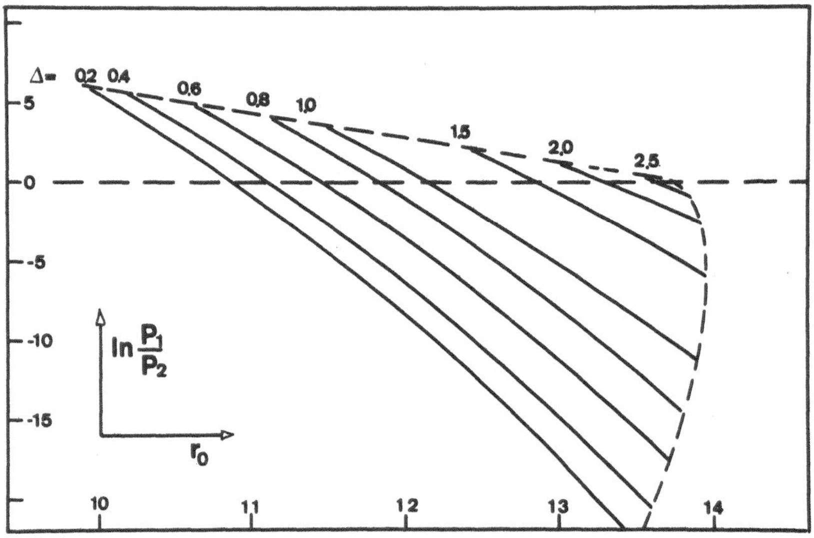

Figure 16. Ratio of the depths of the potential (18) on the two branches in the bistable domain, after[20].

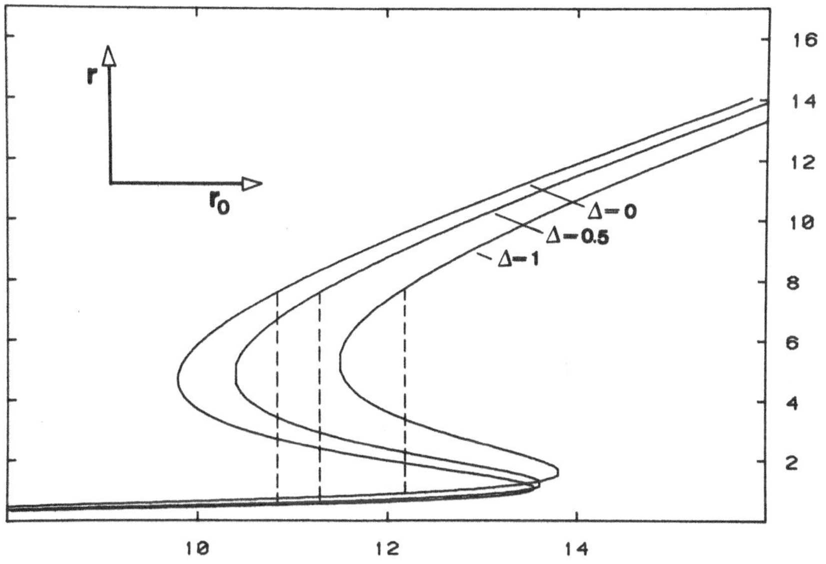

Figure 17. Exchange of stability between the extrema of the potential (18) after[20].

(b) Models With A Limit Cycle

We now want to consider the case, where the solutions of the deterministic equations $\dot{q}^\nu = K^\nu(q)$ are attracted by a limit cycle for $t \to \infty$. Since the potential Φ_0 obtained from Equation (68) is a Lyapunoff function, it must be a minimum and degenerate on the limit cycle. Because of the degeneracy of Φ_0 the zero order probability density $W(q) \sim \exp(-\Phi_0(q)/\varepsilon)$ is also degenerate along the limit cycle. The distribution along the limit cycle is given by the first order correction Φ_1, and does not become sharp as $\varepsilon \to 0^+$. However, since Φ_1 is only needed to lift the degeneracy of Φ_0 along the limit cycle, it is sufficient in lowest order to determine Φ_1 along the limit cycle. In this case, Equation (69) reduces to a one-dimensional problem, which can easily be solved by quadrature, once the zero order problem (68) has been solved and the limit cycle has been determined.

We proceed as follows. We consider the case $K_1^{\nu} = 0$, $Q_1^{\nu\mu} = 0$. Equation (72) is used in order to rewrite Equation (69) in the form

$$\left(\frac{1}{2} Q_o^{\nu\mu} \frac{\partial\Phi_0}{\partial q^\mu} + r_o^{\nu}\right) \frac{\partial\Phi_1}{\partial q^\nu} = \frac{\partial r_o^{\nu}}{\partial q^\nu} + \frac{1}{2} \frac{\partial Q_o^{\nu\mu}}{\partial q^\nu} \frac{\partial\Phi_0}{\partial q^\mu} \tag{20}$$

This equation is considered only on the limit cycle, where the first order derivatives of Φ_0 vanish. The drift r_o^{ν} carries the system through the limit cycle. It is convenient to introduce a phase - like variable q'', which varies along the limit cycle, and variables $q_{,}^{\nu}$ ($\nu = 1 \ldots n-1$) which are locally transverse to the limit cycle. In these variables Equation (20) on the limit cycle reduces to

$$r_o'' \frac{\partial\Phi_1}{\partial q''} = \frac{\partial r_o''}{\partial q''} \tag{21}$$

This one-dimensional problem is easily solved by quadrature

$$\Phi_1\Big|_{\substack{\text{limit}\\ \text{cycle}}} = \ln|r_o''|\Big|_{\substack{\text{limit}\\ \text{cycle}}} \tag{22}$$

The steady state probability density is then given by

$$W \sim \frac{1}{|r_o|} \exp\left(-\frac{\Phi_0}{\varepsilon}\right) \tag{23}$$

On the right hand side the restriction of $|r_o|$ to the limit cycle need not be indicated explicitly, since the factor $\exp(-\Phi_0/\varepsilon)$ is sharply concentrated there for $\varepsilon \to 0^+$.

As an illustrative example+) let us consider the deterministic model

$$\dot{r} = (\alpha - \beta r^2) r$$

$$\dot{\psi} = \omega(r, \psi); \quad (\omega > 0) \tag{24}$$

for α, β real and positive under the influence of weak additive noise (58) of strength Q. This model is a generalization of the one studied in section 3a. Physical examples where $\dot{\Phi}$ depends explicitly on ψ occur in lasers with phase locking (cf. (3)) and in the synchronization of self sustained oscillators (cf. (9)). The frequency $\omega(r, \psi)$ now depends explicitly on the phase ψ, and detailed balance under the transformation $r \to r$, $\psi \to -\psi$ is no longer present. Nevertheless, it is straightforward to determine Φ_o from Equation (68), which takes the form

$$\frac{1}{2}\left(\frac{\partial \Phi_o}{\partial r}\right)^2 + \frac{1}{2r^2}\left(\frac{\partial \Phi_o}{\partial \psi}\right)^2 + [(\alpha - \beta r^2) r]\frac{\partial \Phi_o}{\partial r} + \omega(r, \psi)\frac{\partial \Phi_o}{\partial \psi} = 0. \tag{25}$$

We find

$$\Phi_o = -\frac{1}{2}\alpha r^2 + \frac{1}{4}\beta r^4. \tag{26}$$

The drift r_o^{ν} is obtained as

$$r_o^{\nu} = \begin{pmatrix} 0 \\ \omega(r, \psi) \end{pmatrix} \tag{27}$$

The potential Φ_o takes on a minimum at $r = \sqrt{\alpha/\beta}$ and is degenerate with respect to ψ. Thus, there is a stable limit cycle of radius $\sqrt{\alpha/\beta}$, along which the system drifts with the rate $r_o'' = r_o^{\psi} = \omega(\sqrt{\alpha/\beta}, \psi)$. The steady state probability density in the limit of weak noise $Q \to 0^+$

$$W \sim \frac{1}{\omega(\sqrt{\alpha/\beta}, \psi)} \exp\left[+\frac{1}{2Q}\left(\alpha r^2 - \frac{\beta}{2} r^4\right)\right] \tag{28}$$

and depends on the phase . It is clear, in this example, that the limit of weak noise may only reasonably be taken, if the deterministic radius of the limit cycle is sufficiently large compared to its fluctuations due to the presence of noise, i.e. if

$$\sqrt{\frac{\alpha}{\beta}} \gg \sqrt[4]{\frac{2Q}{\beta}} \tag{29}$$

+) I am grateful to Dr. H.J. Scholz, who suggested this example

(c) The Lorenz Model With Weak External Noise

The Lorenz model[7] is described by the equations

$$\dot{x} = -\sigma x + \sigma y$$
$$\dot{y} = -y + rx - xz$$
$$\dot{z} = -bz + xy. \quad (30)$$

It was invented as a model of convection, but also models other systems like a single mode laser,[35] a mode locked pulse train in a laser,[36] or a dynamo system.[37] In the application to convection x is the convecting velocity, y and z are the fundamental and the second harmonics of the temperature (all in dimensionless units), b is a geometrical parameter, σ the Prandtl number and r is the Rayleigh number (proportional to the temperature difference forced on the convecting system). This system was studied numerically in great detail by many authors (cf. 38)). For $0 \leq r \leq 1$ the origin $P_o = (0,0,0)$ is a stable fixed point. For $1 \leq r \leq r_T$ with

$$r_T = \frac{\sigma(\sigma+b+3)}{\sigma-b-1} \quad (31)$$

assuming $\sigma > b+1$, P_o becomes unstable, and two new stable fixed points

$$P_{\pm} = (\pm\sqrt{b(r-1)}, \pm\sqrt{b(r-1)}, r-1) \quad (32)$$

exist.

For $r > r_T$ there is a region with a strange attractor,[7] where the deterministic trajectories in the steady state fill out a region in phase space with volume 0, which satisfies an equation of the form

$$x = f(y,z,c) \quad (33)$$

where c is a variable which runs through all members of an infinite set C of measure 0. An example of such a set is the Cantor set C_a, defined by all real numbers between 0 and 1 which may be written as

$$C_a = \left\{ \sum_{k=1}^{\infty} \frac{a_k}{3^k} \right\} \quad (34)$$

with $a_k = 0,2$.

Since the set C (like the Cantor set) has measure 0 in the unit interval, the volume of the attractor satisfying eq. (33) vanishes. The explicit analytical form of the function f in Equation (33) and of the set C has never been exhibited. Instead, the attractor has been studied numerically. These numerical studies show that x in

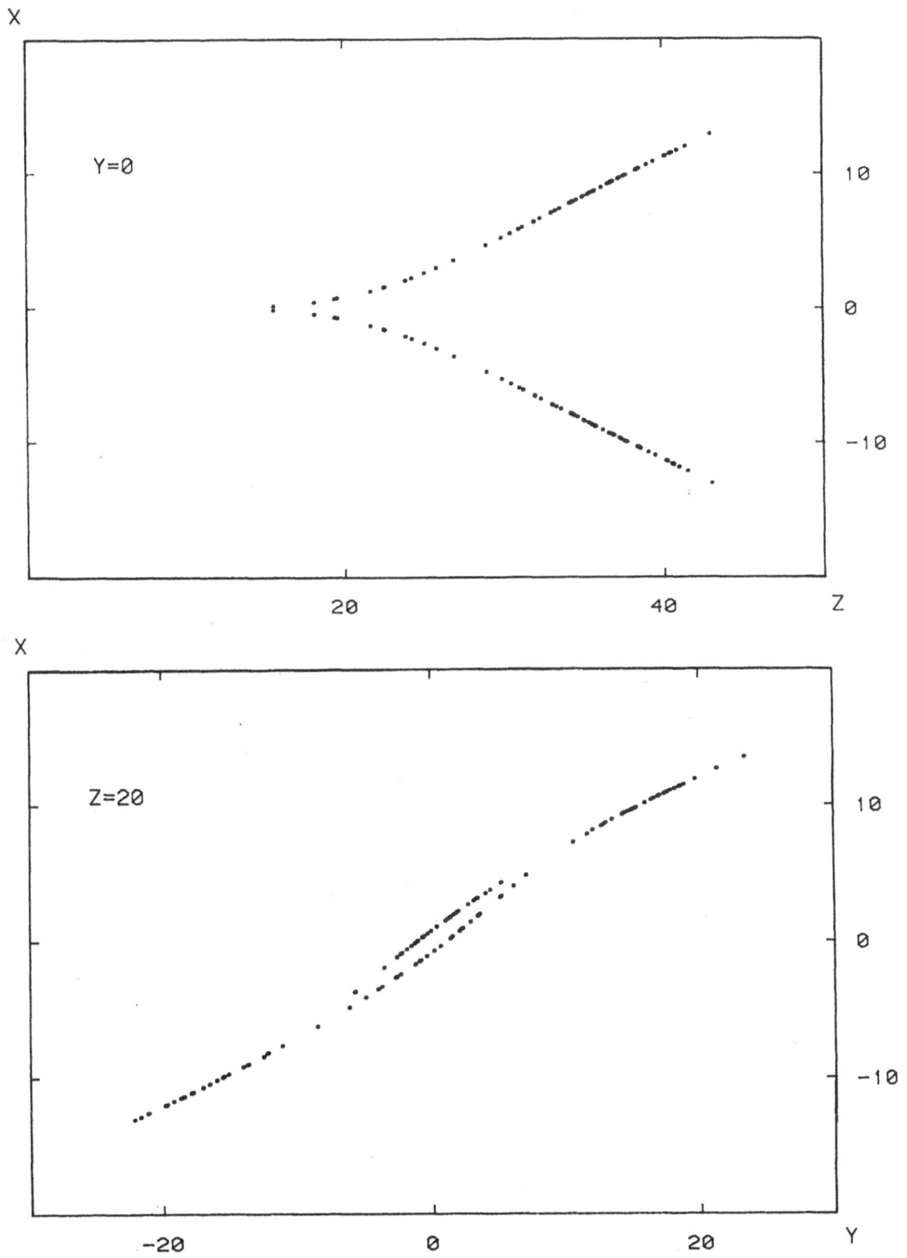

Figure 18. Cut of the strange attractor (23) of the Lorenz model with the planes y = 0 (Figure (18a) and z = 20 (Figure 18b) for $\sigma = 10$, b = 8/3, r = 28 (I am indebted to Dr. H.J. Scholz for these plots).

Equation (33) has essentially two branches but is nearly independent of the number c which gives the attractor its complicated topological structure. A cut of the attractor with the planes y=0, Z=20 is shown in Figures 18a, b for $\sigma = 10$, $b = 8/3$, $r = 28$. The same parameter values are taken in all subsequent plots.

We now add Langevin forces to the Lorenz equations and obtain the Fokker Planck equation

$$\begin{aligned}\frac{\partial P}{\partial t} &= \frac{\partial}{\partial x}\left(\sigma(x-y) + \frac{1}{2} Q\sigma \frac{\partial}{\partial x}\right)P \\ &+ \frac{\partial}{\partial y}\left(y - rx + zx + \frac{1}{2} Q \frac{\partial}{\partial y}\right)P \\ &+ \frac{\partial}{\partial z}\left(bz - xy + \frac{1}{2} Qb \frac{\partial}{\partial z}\right)P\end{aligned} \tag{35}$$

The relative strength of the Langevin forces has been fixed arbitrarily in such a way that, for $r = -\sigma$, detailed balance with respect to $(x,y,z) \rightarrow (x,y,z)$ is satisfied. Later, we will only be interested in quantities which are independent of that choice. For $r = -\sigma$, the equilibrium distribution is given by

$$W \sim \exp\left(-\frac{1}{Q}(x^2 + y^2 + z^2)\right). \tag{36}$$

For $r > -\sigma$, and in particular for $r > 0$, detailed balance is destroyed. As long as the deterministic steady state is a fixed point, the steady state distribution may be approximated by a Gaussian around that fixed point. The Gaussian may be easily worked out from Equation (35) by linearizing the drift around the fixed point. However, in the region of the strange attractor (or the regions for still larger r where limit cycles exist), linearization of the drift is not possible. We now consider the limit of small Q and apply the method of section 2d in the strange attractor regime of r.

Writing the steady state distribution in the form $W = \exp(-(\Phi_o/Q + \Phi_1 + \ldots))$ we obtain, according to section 2d,

$$\begin{aligned}&-\sigma(x-y)\frac{\partial\Phi_0}{\partial x} - (y-rx+zx)\frac{\partial\Phi_0}{\partial y} - (bz-xy)\frac{\partial\Phi_0}{\partial z} \\ &+ \frac{1}{2}\sigma\left(\frac{\partial\Phi_0}{\partial x}\right)^2 + \frac{1}{2}\left(\frac{\partial\Phi_0}{\partial y}\right)^2 + \frac{1}{2} b\left(\frac{\partial\Phi_0}{\partial z}\right)^2 = 0\end{aligned} \tag{37}$$

and

$$\left(-\sigma(x-y)+\sigma\frac{\partial\Phi_o}{\partial x}\right)\frac{\partial\Phi_1}{\partial x} + \left(-y+rx-zx+\frac{\partial\Phi_o}{\partial x}\right)\frac{\partial\Phi_1}{\partial y}$$

$$+\left(-bz+xy+b\frac{\partial\Phi_o}{\partial z}\right)\frac{\partial\Phi_1}{\partial z} =$$

$$-(\sigma+1+b)+\frac{1}{2}\left(\sigma\frac{\partial^2\Phi_o}{\partial x^2}+\frac{\partial^2\Phi_o}{\partial y^2}+b\frac{\partial^2\Phi_o}{\partial z^2}\right). \tag{38}$$

We first consider the equation for Φ_0. In the region of parameter space which is of interest here, there is no small parameter in the problem. The only obvious possibility of solving Equation (37) is, then, by constructing local power series expansions of Φ_0 in x,y,z around different points in (x,y,z) - space and fitting these different expansions together.

In the strange attractor region the minimum of the potential Φ_0 is degenerate on the deterministic attractor, Equation (33). Therefore, Φ_0 contains the dependence on c and cannot be a smooth function. However, upon adding noise to the system we expect that dependence of Φ on c is smoothed by diffusion. Therefore, we expect that the correction Φ_1 and the higher order corrections to Φ will wipe out the Cantor set structure in the potential Φ, which <u>will</u> be a smooth function. Instead of having a minimum on the strange attractor, Equation (33), Φ will have a minimum on a smooth manifold of volume 0, i.e. on a two-dimensional surface

$$x = f(y,z) \tag{39}$$

in the vicinity of (33). Unfortunately, these assumptions are without proof and can only be based on plausibility.

On the basis of these assumptions we neglect the Cantor set structure in Φ_0 from the very beginning, arguing that this structure will be wiped out by the higher order corrections anyway. We therefore look for a smooth local power series expansion of Φ_0 in the form

$$\Phi_0 = \alpha^2(y,z)\ (x-f\ (y,z))^2 + \ldots \tag{40}$$

We will neglect the higher order terms in (x-f). Inserting this expansion in Equation (37) and comparing coefficients of $(x-f)^n$ we obtain, in lowest order, an equation for f independent of α^2

$$-\sigma\,(f-y) + (y-rf+zf)\,\frac{\partial f}{\partial y} + (bz-fy)\,\frac{\partial f}{\partial z} = 0 \tag{41}$$

From the coefficient of $(x-f)^2$ we can obtain also an equation for $\alpha^2(y,z)$. However, the form of this equation and its solution depend on the details of our assumptions about the coefficients of the Langevin noise sources and are therefore without general interest. We note that Equation (41) could also be obtained from the Lorenz equations without noise sources, assuming that $x=f(y,z)$ is a surface which is left invariant by these equations.[21] The derivation of Equation (41) given here is intended to make clear that Equation (41) is not exact, but based on a local approximation to Φ and the assumption that the Cantor set structure is negligible in Φ. Indeed, an exact global constant of the motion $x-f(y,z)$ of the Lorenz equations certainly does not exist.

Such invariant surfaces only exist locally, i.e. if a trajectory comes back to a point after having made an excursion outside the local vicinity of that point, it will be on a different surface. We now look for local solutions of Equation (41), again in the form of power series expansions around points $P_i = (x_i, y_i, z_i)$ on the surface.

$$f(y_i + \eta, z_i + \xi) = x_i + \sum_{n=0}^{\infty} \sum_{m=0}^{\infty} a_{mn}^{(i)} \eta^m \xi^{n-m} \tag{42}$$

with $a_{00}^{(i)} = 0$. As points P_i we have chosen the unstable stationary points $P_o = (0,0,0)$ and $P_\pm$ given by Equation (32), assuming that the minimum surface of ϕ_o passes through these points. The details of the analysis are reported in Reference 21. We find that the attracting patches of 2-dimensional surfaces around P_o, $P_\pm$ constructed analytically in this way up to n=10 grow together (numerically) in a region between these points, and give a very good numerical approximation of the two main branches of the Lorenz attractor, Equation (33). In Fig. 19 we give a contour line plot of the surfaces obtained in this way, together with numerically determined points on the Lorenz attractor. The boundary of that surface is given by the Lorenz trajectory through P_o, a result which will emerge at the end of the following calculations.

We now consider the next approximation, Φ_1, in order to learn something about the distribution of the system over the surface $f(y,z)$. Since we are not interested in the details of the sharp Gaussian distribution orthogonal to the surface we may integrate the probability density

$$W(x,y,z) \sim \exp\left[-\frac{\alpha^2(y,z)}{Q}(x-f(y,z))^2 - \Phi_1(x,y,z) - \ldots\right] \tag{43}$$

over x, using a saddle point approximation, to obtain

$$W(y,z) \sim \exp\left[-\Phi_1(f,y,z) - \ln \alpha(y,z) - \ldots\right] = \exp(u(y,z)). \tag{44}$$

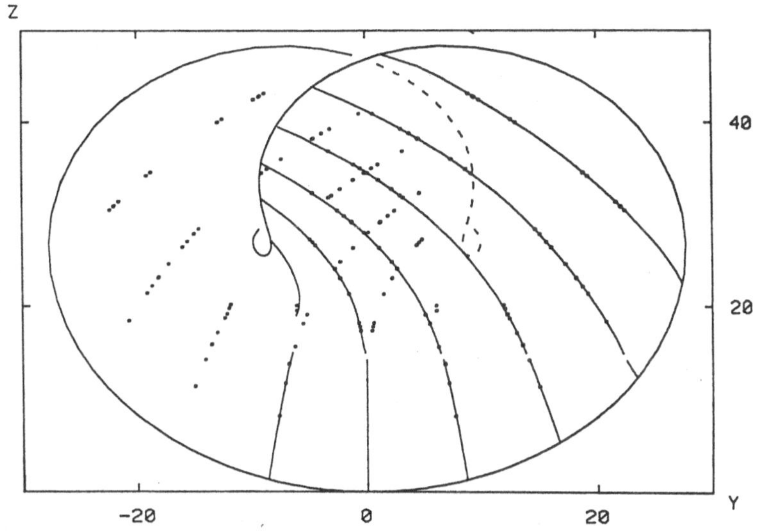

Figure 19. Contour line plot of the invariant surface, Equation 29. The points have been obtained numerically and lie on the Lorenz attractor. The boundary is determined from $\zeta = 0$ according to Equation (36), after Ref. 21.

Therefore, the distribution over the surface f(y,z) turns out to be independent of Q, up to higher order corrections. In particular, it does not become sharp, as Q approaches 0, in contrast to the distribution orthogonal to the surface x = f(y,z). Hence, we may as well evaluate W(y,z) for Q=0. where all the higher order terms vanish exactly. From Equation (44) we see that $\Phi_1 + \ln \alpha\,(y,z) = u(y,z)$ must be independent of the relative strength of the Langevin sources, even though Φ_1 and $\alpha(y,z)$ depend on these sources separately. In order to obtain a differential equation for u(y,z) it is therefore easiest to return to Equation (35) for W(x,y,z) with $\partial W/\partial t = 0$, integrate that equation over x with the assumption that W(x,y,z) is sharply concentrated at x = f(y,z), insert W(y,z) = exp (u (y,z)) and take Q=0. We obtain immediately by inspection

$$(-\,y + (r-z)\,f)\,\frac{\partial u}{\partial y} + (-bz+fy)\,\frac{\partial u}{\partial z} = 1-(r-z)\,\frac{\partial f}{\partial y} + b - y\,\frac{\partial f}{\partial z}. \quad (45)$$

With more labor, the same equation can be obtained using the separate equations of $\Phi_1(x,y,z)$ and $\alpha(y,z)$. Equation (45) is a linear first order partial differential equation.

Its characteristics locally are just the Lorenz trajectories on the local surface $x = f(y,z)$. In order to construct global solutions, the different local patches of $f(y,z)$ we obtained before and the characteristics on them have to be pieced together in the regions where they were found to overlap in good approximation. The various patches of $x = f(y,z)$ cannot meet exactly. Therefore, the joining of the characteristics on their borders involves an approximation in principle, which is only justified by the presence of external noise. By the approximate joining of trajectories in the overlap region it is possible that trajectories with a different past are joined and have a common future. Thus, a "forgetting" mechanism is introduced into the system.

In Reference 21 a local solution of Equation (45) on the surface patch around P_o has been constructed explicitly. It is useful for that purpose to introduce the new variable

$$\zeta = z - R(y) \tag{46}$$

where $z = R(y)$ is the equation of the characteristic of Equation (45) emanating from the origin P_o. This characteristic may also be constructed in a local approximation near P_o, Reference 21, and takes the form $z = c_o y^2 + \ldots$, where c_o may be expressed by the parameters of the system. The local approximation of u near the z-axis is assumed to be of the form

$$u(y,\zeta) = u_o(\zeta) + u_2(\zeta)\, Y^2 + O(y^4) \tag{47}$$

which makes use of the symmetry $x \rightarrow -x$, $y \rightarrow -y$ of the model.

Inserting this ansatz in Equation (45) and using the local approximation for $f(y,z)$ near P_o, we obtain two ordinary differential equations for $u_o(\zeta)$ and $u_2(\zeta)$ which can be solved by quadrature. The two constants of integration can only be fixed by global conditions, e.g. requiring single-valuedness of the global solution. Therefore, they have to be left undetermined here and are fixed by fitting to the numerically determined probability density. The fit is shown in Figure 20. Near $\zeta = 0$ the result simplifies and we obtain

$$\begin{aligned} u_o(\zeta) &= \ln \zeta^q + \text{const} + O(\zeta) \\ u_2(\zeta) &= -u_{2o} + O(\zeta, \zeta^{q+1}) \end{aligned} \tag{48}$$

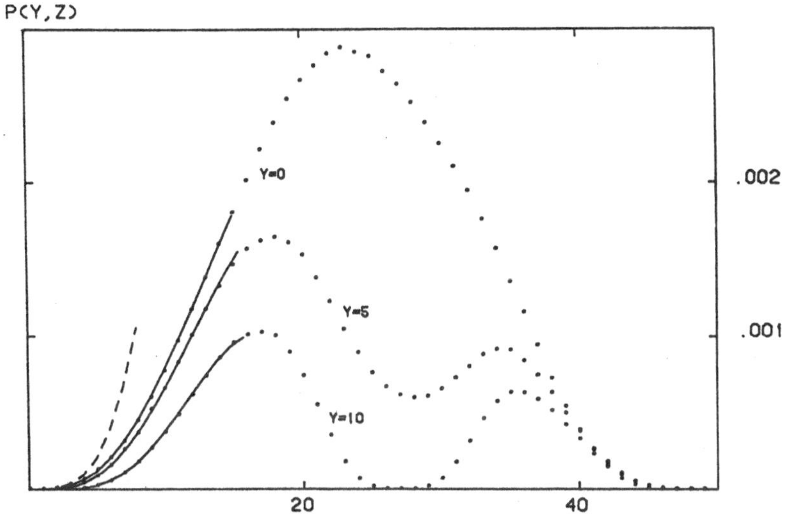

Figure 20. Probability density W = exp (u(x,y)) in the lower part of the invariant surface analytically, using two fit parameters, and numerically, summing over the Cantor set of the attractor. The dashed line represents the result (48) near the boundary $\zeta = 0$. After[21].

where the exponent q is obtained a

$$q = \frac{1}{2b} \left(\sqrt{(\sigma-1)^2 + 4r\sigma} - \sigma - 1\right) - 1 \tag{49}$$

and u_{2o} is a constant which can be expressed by σ, r, b, **Ref.** 21. For our numerical examples we have

$$q = 3.43, \quad u_{2o} = 0.002 \tag{50}$$

Our result (48) shows that, near the characteristic $\zeta = 0$, the probability density is given by

$$W(y,z) = \text{const.}\ (z-R(y))^q \exp(-u_{2o}y^2 + \ldots) \tag{51}$$

i.e., it vanishes by a power law on the trajectory $z = R(y)$ emerging from the origin.

This defines the boundary of the surface $f(y,z)$ near P_0. The probability density remains zero on that trajectory until it is joined with another trajectory with a different past, which carries finite probability density. This happens only after the trajectory has reached the vicinity of P_+ or P_-, where trajectories on the invariant surface, 'outside' the boundary $\zeta = 0$ may enter the region 'inside'. The local solution we have obtained compares well with the numerically generated probability density near P_0. A global probability density on the surface $x = f(y,z)$ remains to be worked out, which is a nontrivial task.

The surface $x = f(y,z)$ we have calculated also exists for $r < r_T$ and remains attracting. However, the points $P_\pm$ on that surface are attracting for $r < r_T$, i.e., the steady state probability density $W(y,z)$ collapses (for $Q = 0$) to a pair of δ-functions at these points.

The limitations of the approximation of weak noise become apparent in this example. As soon as the attractor of the system is higher than one-dimensional, it is necessary to determine the correction Φ_1 to Φ_0 in more than one dimension. This may be possible in special cases, but is difficult generally.

REFERENCES

1. M. Lax, Rev. Mod. Phys. 32: 25 (1960); 38:359 (1966); 38:541 (1966).

2. T. Kirkpatrick, E.G.D. Cohen, J.R. Dorfman, Phys. Rev. Lett. 44: 472 (1980). I. Procaccia, D. Ronis, I. Oppenheim, Phys. Rev. A20: 2533 (1979).

3. H. Haken, Encyclopedia of Physics 25/2c, Springer, New York 1970.

4. R. Graham, Springer Tracts in Mod. Phys. 66, Springer, New York 1973.

5. H. Haken, Rev. Mod. Phys. 47: 67 (1975).

6. R. Graham, in "Order and Fluctuations in Equilibrium and Nonequilibrium Statistical Mechanics", ed. Nicolis et al., John Wiley, New York (1981), p. 235.

7. E. N. Lorenz, J. Atmos. Sci. 20: 130 (1963).

8. D. Ruelle, F. Takens, Commun. Math. Phys. 20: 167 (1971).

9. R.L. Stratonovich, "Topics in the Theory of Random Noise," Vol. I, II, Gordon and Breach, New York 1963.

10. H. Risken, Z. Physik 251: 231 (1972).

11. R. Graham, Z. Physik B40: 149 (1980).

12. R. Graham, H. Haken Z. Physik 243: 289 (1971); 245: 141 (1971).

13. R. Graham, in Coherence and Quantum Optics, ed. L. Mandel, E. Wolf, Plenum, New York 1973.

14. R. Graham, in "Fluctuations, Instabilities and Phase Transitions," ed. T. Riste, Plenum, New York 1975.

15. H. Risken, H.D. Vollmer, Z. Physik 201: 323 (1967); 204:240 (1967)

16. K. Seybold, H. Risken, Z. Physik 267 : 323 (1974).

17. A. Schenzle, H. Brand, Phys. Rev. A4 : 1628 (1979).

18. R. Graham, Phys. Lett. 80A: 351 (1980).

19. S.L. McCall, Phys. Rev. A9 : 1515 (1974); H.M. Gibbs, S.L. McCall, T.N.C. Venkatesan, Phys. Rev. Lett. 36 : 1135 (1976).

20. R. Graham, A. Schenzle, in "Proceedings of the International Optical Bistability Conference," Asheville, N.C. USA, 1980; R. Graham, A. Schenzle, Phys. Rev. A23, to appear.

21. R. Graham, H.J. Scholz, Phys. Rev. A22: 1198 (1980).

22. L.D. Landau, E.M. Lifshitz, "Fluid Mechanics", Pergamon, London 1958.

23. L.D. Landau, E.M. Lifshitz, "Statistical Physics", Pergamon, London 1959.

24. S. Grossmann, J. Chem. Phys. 65 : 2007 (1976).

25. C. Gardiner, J. Stat. Phys. 14 : 309 (1976).

26. N. van Kampen, in "Topics in Statistical Mechanics and Biophysics," ed. R.A. Piccirelli, 1976.

27. W. Horsthemke, L. Brenig, Z. Physik B27 : 341 (1977).

28. S.R. DeGroot, P. Mazur, "Non equilibrium thermodynamics," North Holland, Amsterdam 1962.

29. A. Schenzle, H. Brand, Optics Commun. 23: 151 (1978).

30. A. Schenzle, H. Brand, Optics Commun. 31: 401 (1979).

31. R. Graham, Phys. Rev. Lett. 31: 1479 (1973); Phys. Rev. A10: 1762 (1974).

32. W.A. Smith, Phys. Rev. Lett. 32: 1164 (1974).

33. I.S. Gradsteyn, I.M. Ryshik, "Table of Integrals, Series and Products," Academic, New York 1965.

34. M. Abramowitz, I. Stegun, "Handbook of Mathematical Functions," Dover, New York 1965.

35. H. Haken, Phys. Lett. 53A: 77(1975).

36. R. Graham, Phys. Lett. 58A: 440 (1976).

37. K.A. Robbins, Proc. Natl. Acad. Sci. USA 73: 4297 (1976).

38. J.B. McLaughlin, P.C. Martin, Phys. Rev. A12: 186 (1975); O. Lanford in "Turbulence Seminar", Lecture Notes in Mathematics 615: 113, Springer, New York 1977.

PART IV

Biological Applications

APPLICATIONS OF SMALL ANGLE NEUTRON SCATTERING TO BIOLOGICAL SYSTEMS

G. Zaccai

Institut Laue-Langevin
156X centre de Tri
38042 Grenoble Cedex, France

CONTENTS

1. INTRODUCTION

The physics of neutron scattering experiments in biology has to be simple because the biochemistry and molecular biology involved are usually very difficult. A first aim of structural studies in biology could be called anatomy at the molecular level; it is to provide an image, always a necessary but seldom a sufficient step in understanding a biological system. Then, one should relate that image or structure to the physico-chemical properties of the system. The most ambitious aim is to relate the structure to biological function in a one-to-one correspondance. The chemical components of living material are relatively few and well-known. The central theme in molecular biology is the "structure-function" relationship, which states, essentially, that the diversity of biological function has its roots in the way the same few chemical components are put together in different molecular assemblies. When one talks of a protein, for example, one means a molecule of a certain chemical composition (a polymer of aminoacid residues), which performs a given biological function. There are two parts to that definition: the first tells us that all proteins are similar since they have similar compositions, and the second tells us that each protein is rather special, since it has a given specific biological function. This duality is important to bear in mind in any physical experiment in biology. Unlike an experiment in physics or chemistry, we are not investigating laws or effects which are general to a class of substances, but looking for hints as to why a given member of a class behaves in a specific way.

The physics of neutron scattering experiments on biological material is based on the fact that as far as their interactions with neutrons are concerned, we can consider proteins, lipids, nucleic acids etc. in a general way. In order to design meaningful experiments and interpret their results, however, one should know the biological system very well. Its biochemistry must be well-characterized to start with. The hope is that its molecular biology will be understood when its structure is defined in the context of its biochemistry. In these lectures, I shall describe aspects of neutron scattering which make it one of the useful physical techniques in molecular biology.

These lectures are not a critical review of neutron scattering in biology because these studies are best reviewed in the context of their molecular biology rather than the technique. I shall describe the different aspects of neutron technique in some detail, and each time refer to published applications, which the interested student could pursue further.

All the approaches developed in X-ray diffraction, from small angle scattering[1] to protein crystallography[2] could be applicable with neutrons, and have been used in biological work.[3] The main difference between X-ray and neutron studies arises from the nature of their interaction with matter. In particular, neutrons are scattered by atomic nuclei, with different isotopes having different scattering lengths.[4]

2. CONTRAST

To understand the special usefulness of neutrons it is essential to understand the notion of contrast and contrast variation. In any diffraction imaging experiment, contrast allows us to see the object in question. It is a physically well-defined quantity which expresses the fact that the object scatters differently from its surrounding medium.

A. Scattering by A Single Particle

Consider a particle made up of atoms i of neutron scattering amplitude b_i and coordinates $\underline{r}_i$. The amplitude of radiation scattered by that particle is given by:

$$F_{part.}(\underline{Q}) = \sum_i b_i \exp i\underline{Q}\cdot\underline{r}_i \qquad (1)$$

$\underline{Q} = \underline{k}_1 - \underline{k}_o$ is the scattering vector, where $\underline{k}_o$, $\underline{k}_1$ are the incident and reflected wave-vectors respectively (Figure 1); $|\underline{Q}| = 4\pi \sin\theta/\lambda$ where 2θ is the scattering angle and λ is the wavelength of the radiation. The intensity of the scattered radiation is proportional to $|F(\underline{Q})|^2$.

B. Resolution

If $|Q_{max}|$ corresponds to the maximum angle to which $I(\underline{Q})$ has been measured, the length $2\pi/Q_{max}$ is the shortest distance which can be "seen" in the particle by the experiment; waves coming from points which are closer than $2\pi/Q_{max}$ will not differ sufficiently in phase relationship to interfere with each other. The quantity $2\pi/Q_{max}$ is the resolution of the experiment. In practice, the right hand side of Equation (1) can be evaluated by taking atoms in groups with a mean coordinate $\underline{r}$ for each group provided the volumes of the groups are small compared to $(2\pi/Q_{max})^3$.

C. Contrast for One Particle in Solution

Consider the particle of Equation (1) in solution. At experimental resolutions, which are usually much worse than 10 Å, the solvent can be considered as a continuous medium of mean scattering density ρ_s (ρ_s is calculated by adding the scattering amplitudes of

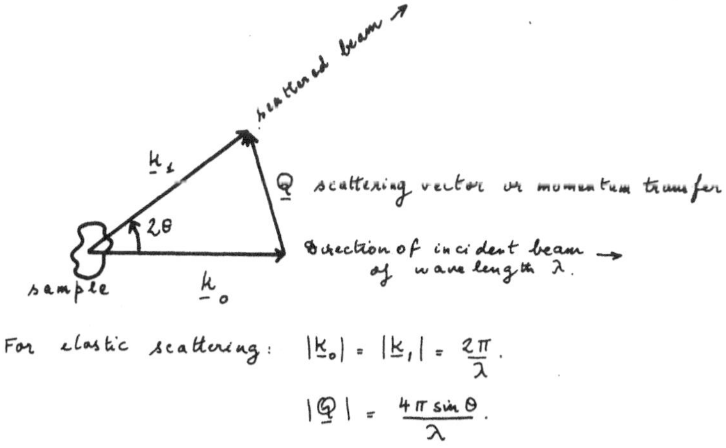

Figure 1. Definition of the scattering vector.

all the atoms in a unit volume of solvent). The amplitude of radiation scattered by the particle in solution is

$$F_{sol}(\underline{Q}) = \sum_i m_i \exp i\underline{Q}.\underline{r}_i \quad (2)$$

where $m_i = b_i - \rho_s v_i$, v_i is the volume associated with the atom i. The effective scattering amplitude from each atom is now given by the amplitude of that atom less the scattering amplitude of an equal volume of solvent. Note the analogy with Archimedes' principle.

The quantity m_i has been called, variously, excess scattering mass, excess scattering amplitude, excess scattering length or contrast length. Here also, within the limits imposed by resolution the summation of Equation (2) is more conveniently performed over groups of atoms in the particle; a protein could be divided up into aminoacid residues; a ribosome into protein subunits, etc., provided the volumes of the groups are much smaller than the resolution of the experiment.

The scattering density of the solvent is defined far away from the particle. The boundary between particle and solvent is not sharp and there is a region where solvent molecules are not packed with the same density as bulk solvent. This region should be considered to be part of the particle and included in the summation (Figure 2). To be on the safe side, the particle boundary should be taken far enough out so as to include some bulk solvent. The bulk solvent contribution to the summation will be zero because its excess scattering mass is zero, by definition.

Excess scattering mass or contrast length depends on three independent parameters: absolute scattering length (constants of the atoms in the groups), excluded volume (depends on the chemistry of the group) and solvent scattering density (depends on the chemical composition of the solvent). Excess scattering mass can be positive or negative with well-defined physical meaning.

Contrast is defined, in general, as a difference in scattering density. It is important to specify, then, between which components this difference is taken.

3. PRINCIPLES OF CONTRAST VARIATION

Table 1 shows the neutron scattering lengths of different nuclei. Note that except for hydrogen they are similar in magnitude. The values for X-rays are also tabulated for comparison. In neutron experiments, contrast is changed by replacing H by D. Figure 3 shows the scattering density of water as a function of $[D_2O]$:$[H_2O]$ ratio. Also plotted are mean density values for a few chemi-

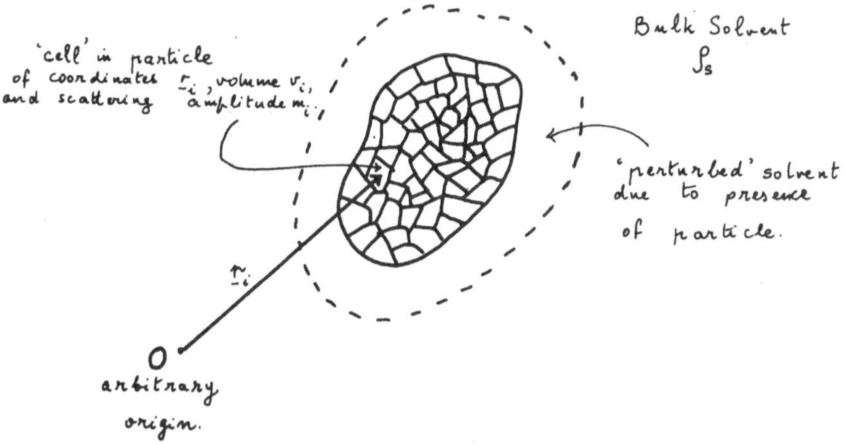

Figure 2. Scattering by a particle in solution. It can be divided up into "cells" of volume v_i and absolute scattering length b_i. In a solvent of scattering density ρ_s, the amplitude of radiation scattered by cell i is proportional to $m_i = b_i - \rho_s v_i$. All solvent volumes which do not have the same density as the bulk will also contribute to the scattering.

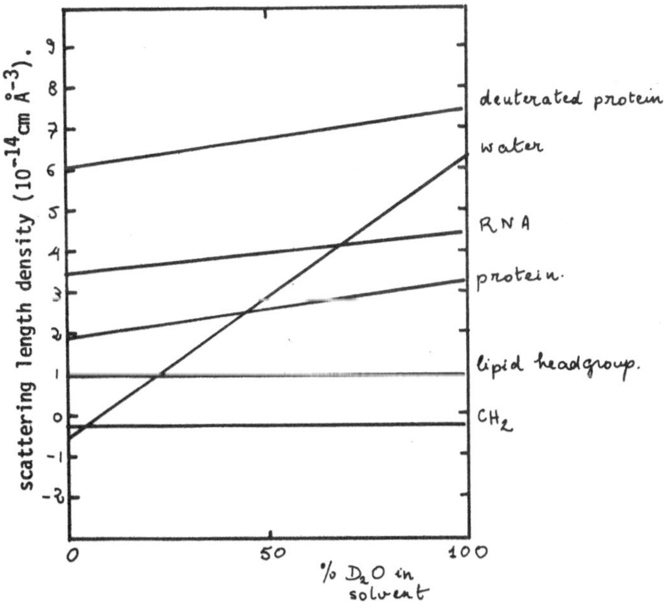

Figure 3. The neutron scattering density of different biological materials as a function of $H_2O:D_2O$ in the solvent. In part from Reference 1.

Table 1. Neutron scattering lengths in units of 10^{-12} cm

Atom	b (Neutron)	f(X-Ray)(θ = 0°)
Hydrogen	-0.3742	0.28
Deuterium	0.6671	0.28
Carbon	0.6651	1.69
Nitrogen	0.9400	1.97
Oxygen	0.5804	2.25
Sodium	0.3600	3.08
Magnesium	0.5200	3.36
Phosphorus	0.5200	4.20
Sulphur	0.2800	4.50
Chlorine	0.9600	4.76
Potassium	0.3700	5.32
Calcium	0.4700	5.60

cal types of biological molecule in solution. Their densities are not constant with $H_2O:D_2O$ ratio because of exchange between their labile H and the D in the solvent. The figure shows very clearly how contrast can be varied almost at will between any two components, by manipulating the H:D content of either the solvent or the molecules, or both. For example, the contrast between a particular protein and others in a complex can be increased from close to zero to $\sim 4 \times 10^{-14}$ cm Å^{-3} by deuteration. The classic example is that of a protein nucleic acid complex in aqueous solution (a ribosome or a virus), for which at ∿70 % D_2O the contrast between solvent and nucleic acid is ∿0 and at ∿42% D_2O, the contrast between protein and solvent is ∿0.

In this section, we are justified in considering proteins, nucleic acids etc. in a general way and can claim Figure 3 to be a guide-line for experiments on any biological system. Even though proteins are made up of 20 aminoacids with quite different scattering densities, their compositions are not random mixtures of these aminoacids, and the scattering densities of all soluble proteins are similar. The same is true for nucleic acids and nucleotides. In fact, we must not forget that there are two parameters to scattering density: scattering mass and excluded volume. The latter is a function of solvent conditions especially for polyelectrolytes such as nucleic acids.[5] Variations in scattering density between nucleic acids are as likely to arise from volume as from composition effects. Tables 2 and 3 give scattering amplitude data on amino acids, nucleotides, proteins.

Table 2. Scattering lengths of aminoacid residues and nucleotides calculated from their chemical compositions at pH = 7. The values in D_2O assume all labile hydrygen atoms (Hex.) are exchanged with solvent. Σb (deuterated) is for when all hydrogen atoms have been replaced by deuterium. M_w is the molecular weight of the residue, and V is its partial volume. The table is adapted from Reference 3a.

Amino Acid Residues

Amino Acid	Mw	Chemical composition	H(ex)	Σb(H_2O) (10^{-12}cm)	Σb(D_2O) (10^{-12}cm)	Σb(deuterated) (10^{-12}cm)	V($\mathring{A}^3$)
Glycine	57	C_2NOH_3	1	1.728	2.769	4.850	66.4
Alanine	71	C_3NOH_5	1	1.645	2.686	6.852	91.5
Valine	99	C_5NOH_9	1	1.479	2.520	10.854	141.7
Leucine	113	C_6NOH_{11}	1	1.396	2.437	12.850	167.9
Isoleucine	113	C_6NOH_{11}	1	1.396	2.437	12.850	168.8
Phenylalanine	147	C_9NOH_9	1	4.139	5.180	13.510	203.4
Tyrosine	163	$C_9NO_2H_9$	2	4.719	6.802	14.090	203.6
Tryptophan	186	$C_{11}N_2OH_{10}$	2	6.035	8.118	16.450	237.6
Aspartic acid	114	$C_4NO_3H_4$	1	3.845	4.886	8.010	113.6
Glutamic acid	128	$C_5NO_3H_6$	1	3.762	4.803	10.010	140.6
Serine	87	$C_3NO_2H_5$	2	2.225	4.308	7.432	99.1
Threonine	101	$C_4NO_2H_7$	2	2.141	4.224	9.432	122.1
Asparagine	114	$C_4N_2O_2H_6$	3	3.456	6.580	9.704	135.2
Glutamine	128	$C_5N_2O_2H_8$	3	3.373	6.497	11.700	161.1
Lysine	129	$C_6N_2OH_{13}$	4	1.586	5.752	15.120	176.2
Arginine	157	$C_6N_4OH_{13}$	6	3.466	9.714	17.000	180.8
Histidine	136.5	$C_6N_3OH_{6.5}$	1.5	4.959	6.521	11.730	167.3
Methionine	131	C_5NOSH_9	1	1.764	2.805	11.140	170.8

Table 2. (Continued)

Cysteine	103.1	C_3NOSH_5	2	1.930	4.013	7.137	105.6
Proline	97	C_5NOH_7	0	2.227	2.227	9.516	129.3

Nucleotides

Base		Mw	Chemical Composition	H(ex)	$\Sigma b(H_2O)$ (10^{-12}cm)	$\Sigma b(D_2O)$ (10^{-12}cm)	Σb(Deuterated) (10^{-12}cm)
Adenine	RNA	328	$PN_5C_{10}O_6H_{11}$	3	11.23	14.35	22.68
	DNA	312	$PN_5C_{10}O_5H_{11}$	2	10.65	12.73	22.10
Guanine	RNA	344	$PN_5C_{10}O_7H_{11}$	4	11.81	15.98	23.26
	DNA	328	$PN_5C_{10}O_6H_{11}$	3	11.23	14.35	22.68
Cytosine	RNA	304	$PN_3C_9O_7H_{11}$	3	9.26	12.39	20.72
	DNA	288	$PN_3C_9O_6H_{11}$	2	8.68	10.77	20.14
Uracil	RNA	305	$PN_2C_9O_8H_{10}$	2	9.28	11.36	19.69
Thymine	DNA	303	$PN_2C_{10}O_7H_{12}$	1	8.61	9.65	21.11

4. SMALL ANGLE SCATTERING FROM SOLUTIONS

Consider a solution of particles, which is dilute enough that there are no correlations between the particles in either position or orientation. Only waves scattered from atoms of the same particle interfere. There is no interference between waves scattered by different particles. To calculate the radiation scattered in a given angle, we add amplitudes for waves from within a particle and intensities for waves from different particles. We thus have

$$I(Q) = N \langle |\sum_i m_i \exp i\underline{Q}.\underline{r}_i|^2 \rangle \tag{3}$$

for N identical particles (monodisperse solution). The averaging is over all orientations. If there are N_1 particles of one type, N_2 of another etc. (polydisperse solution), one then finds

$$I(Q) = N_1 \langle |..|^2 \rangle + N_2 \langle |..|^2 \rangle + \ldots \tag{3a}$$

Information can be obtained from I(Q) in two regions of Q. The small angle region where the Guinier approximation is valid, and the more extended region in Q.[1] For a single particle, it can be

Table 3. The scattering length per dalton of several soluble proteins in H_2O and D_2O solvents, assuming all labile hydrogen atoms (Hex.) exchange with solvent (from Reference 6).

Protein	$\Sigma b(H_2O)/Mw$ ($\times 10^{-14}$ cm dalton^{-1})	Hex./Mw ($\times 10^{-2}$ dalton^{-1})	$\Sigma b(D_2O)/Mw$ ($\times 10^{-14}$ cm dalton^{-1})
Elastase (pig)	2.35	1.69	4.11
Subtilisin (*Bacillus amyloque faciens*)	2.32	1.55	3.93
Ribonuclease (bovine)	2.33	1.79	4.19
Lysozyme (chicken)	2.37	1.86	4.30
Lysozyme (I4 bacteriophage)	2.23	1.76	4.06
Tryptophan synthetase α chain (*E. Coli*)	2.22	1.47	3.75
Glyceraldehyde 3-phosphate dehydrogenase (pig)	2.27	1.53	3.86
Penicillinase (Staphylococcus aureus)	2.14	1.71	3.92
Tobacco mosaic virus protein	2.33	1.67	4.07
MS2 bacteriophage protein	2.25	1.64	3.96
Hemoglobin α-chain (human)	2.21	1.46	3.73
Cytochrome c_2 (human)	2.17	1.69	3.93
Cytochrome c_2 (*Rhodospirillum rubrum*)	2.25	1.62	3.93
Cytochrome c_3 (*Desulfovibrio vulgaris*) flour...	2.29	1.74	4.10
Azurin (*Pseudomonas fourescens*)	2.28	1.55	3.89
Ferredoxin (*Clostridium Pasteurianum*)	2.30	1.46	3.82
Myoglobin (horse)	2.27	1.51	3.84
Bence Jones λ chain (human)	2.34	1.63	4.04
Trypsinogen (bovine)	2.28	1.69	4.04
Mean	2.27	1.63	3.97

shown that

$$I(Q) = \sum_{ij} m_i m_j \cdot \frac{\sin Q\, r_{ij}}{Q\, r_{ij}}. \qquad (4)$$

This is the Debye formula.

A. Contrast Variation in the Guinier Approximation

Provided Q is sufficiently small, any scattering curve can be approximated by a Gaussian:

$$I(Q) = I(0) \exp\left(- \frac{1}{3} R_G^2 Q^2\right). \qquad (5)$$

I(0) is the forward scattered intensity and R_G is the radius of gyration of excess scattering mass (for a mono-disperse solution),

$$I(0) = N \left(\Sigma\, m_i\right)^2 \qquad (6)$$

$$R_G^2 = \frac{\Sigma m_i r_i^2}{\Sigma\, m_i}\,, \qquad (7)$$

where r_i is the distance to the centre of mass of m_i, R_G follows all the rules of its mechanical counterpart, e.g., the parallel axes theorem[7] (Figure 4). The approximation is a good one for

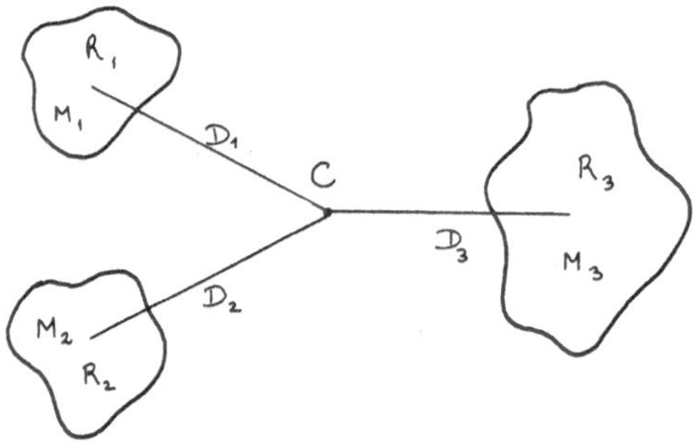

$$(M_1 + M_2 + M_3 + \ldots)\, R_G^2 = M_1\left(R_1^2 + D_1^2\right) + M_2\left(R_2^2 + D_2^2\right) + M_3\left(R_3^2 + M_3^2\right) + \ldots$$

Figure 4. The parallel axes theorem. C is the centre of mass of the bodies; R_1, R_2, R_3 are the respective radii of gyration for each about its own centre of mass; M_1, M_2, M_3 are their respective masses.

$R_GQ \lesssim 1$. An analysis of $I(Q)$ according to the Guinier approximation can lead to accurate experimental values of the two parameters R_G^2 and $I(0)$. In the polydisperse case, the measured values are related to the R_{Gj} and $I_j(0)$ of the individual components by

$$I(0) = \sum_j C_j (\Sigma m_i)_j^2 \tag{8}$$

$$R_G^2 = \frac{\sum_j C_j (\Sigma m_i)_j^2 R_{Gj}^2}{\sum_j C_j (\Sigma m_i)_j^2} \tag{9}$$

where C_j is the molar concentration, $(\Sigma m_i)_j$ is the excess scattering amplitude and R_{Gj} is the radius of gyration of component j.[1]

The variation of $I(0)$ and R_G as the contrast between the mean particle density and solvent is changed by varying the $[H_2O]:[D_2O]$ ratio, X, has been analysed in detail by Stuhrmann,[8] and for specific cases by Serdyuk[9] and Engelman and Moore.[10] Equations (6) and (7) are the basis of the analysis. Recall that $m_i = b_i - \rho_s v_i$ and that both b_i and ρ_s are linear with X. In the monodisperse case, therefore, the square root of $I(0)$ is linear with X:

$$\sqrt{I(0)} = \sum_K b_K + hb_H - V\rho_{H_2O} + X[h(b_D - b_H) - V(\rho_{D_2O} - \rho_{H_2O})] . \tag{10}$$

In Equation (10), the atoms b_i in the particle have been divided into the h labile hydrogen atoms which exchange with solvent and the others, Σb_K; b_H, b_D, are the scattering amplitudes of hydrogen and deuterium, respectively; ρ_{H_2O} and ρ_{D_2O} are the scattering densities of H_2O and D_2O, respectively, and $V = \Sigma v_i$ is the partial volume of the particle. The intercept of $\sqrt{I(0)}$ versus X is Σm_i in H_2O and the gradient is the term $h(b_D - b_H) - V(\rho_{D_2O} - \rho_{H_2O})$, which depends on both the exchange with solvent and the partial volume. The value of X where $I(0)$ vanishes is called the match point of the particle. There, its mean scattering density $[\Sigma b_i / \Sigma v_i]$ is the same as that of the solvent. Soluble proteins have a match point close to 40%, whereas nucleic acids are close to 70% D_2O.[11] This can be seen on Figure 3. Because in aqueous solution ρ_s ranges from a small negative value to a large positive value the sensitivity of $I(\theta)$ to the scattering mass on the one hand and to partial volume on the other depends on ρ_s. Thus for a protein a ~1% change in partial volume affects $I(0)$ by ~0.5% in H_2O solvent, and by ~4% in D_2O solvent. Contrast variation has been used to identify partial volume changes in interactions,[14] while the relative insensitivity of the data from H_2O solutions to partial volume is useful for molecular weight determinations.[15]

Equation (7) shows that the R_G is also a function of contrast between mean particle density and solvent. With different values of ρ_s, not only are the m_i different but their centre of mass also changes.

The simplest way of using R_G in contrast variation is for a two component particle, when the resolution is such that each component can be considered to be of homogeneous density. Measuring at the matchpoint of one component yields the radius of gyration of the other component in the particle. The relation of R_G^2 to the contrast can be derived by a direct application of the parallel axes theorem to give

$$R_G^2 = \frac{R_1^2 M_1}{M_1+M_2} + \frac{R_2^2 M_2}{M_1+M_2} + \frac{L^2 M_1 M_2}{(M_1+M_2)^2} \qquad (11)$$

where R_1, R_2, M_1, M_2 are the radii of gyration and Σm_i of components 1 and 2, respectively; L is the separation between the centres of mass of the two components.

This approach has been used on ribosomal subunits to determine the radii of gyration of the protein and nucleic acid distributions, separately. Serdyuk and Grenader[9] used the naturally different scattering densities of protein and nucleic acid in different solvents and for different types of radiation. They found that the centres of mass of protein and RNA coincided within the errors and that, on average, proteins were further from the centre of mass than RNA. A disadvantage involved in the use of Equation (11) is that, to derive reliable values of R_1, R_2 and L, the ratios (M_1/M_1+M_2) and (M_2/M_1+M_2) should be known with accuracy; i.e., the relative composition of the particle, and the partial volume of each component within the particle must be known.

The formula of Stuhrmann and his co-workers[8] relates fluctuations in scattering density in a particle to the contrast between the mean particle density and the solvent $\bar{\rho}$:

$$R_G^2 = R_C^2 + \frac{\alpha}{\bar{\rho}} - \frac{\beta}{\bar{\rho}^2}\,, \qquad (12)$$

where R_G is the observed radius of gyration at contrast ρ. R_C is the radius of gyration at infinite contrast, and α and β are coefficients related to the second and first moments of the fluctuations in density within the particle, respectively. It is convenient to plot R_G^2 versus $1/\rho$. One such plot from Reference (12) is shown in Figure 5. A positive gradient α implies the particle to be denser further away from the centre of mass and vice-versa. The value of β is related to the separation between the centre of mass of the

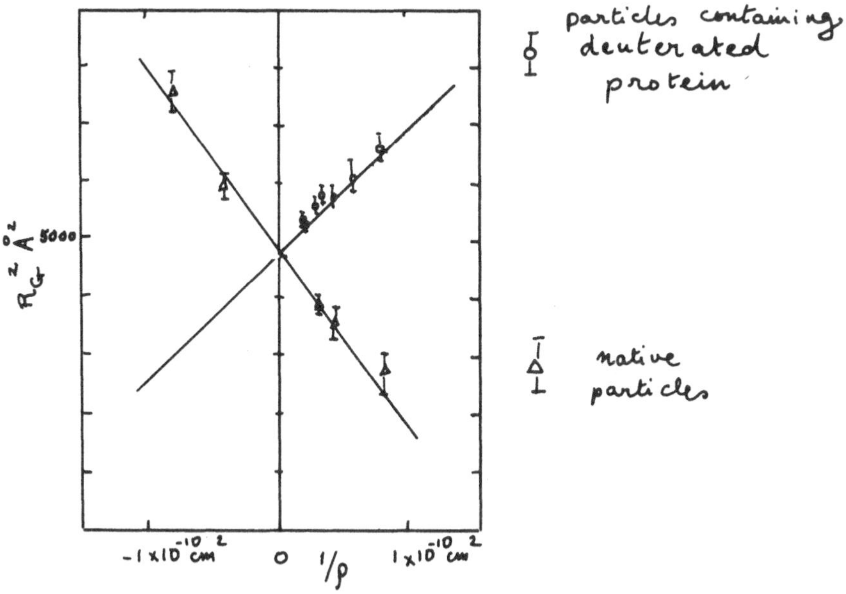

Figure 5. A plot of R_G^2 versus $1/\rho$ for native ribosomal subunits (the line with negative slope), and ribosomal subunits reconstituted with deuterated proteins (positive slope). Note that the intercept on the R_G^2 axis is the same for both; it is equal to R_C^2, the square of the radius of gyration at infinite contrast which depends only on the shape of the particle volume and not on the distribution of scattering density within it, from Reference 12.

particle considered as homogeneous (the particle shape), and that of the scattering density distribution. Figure 5 shows data for native ribosomal subunits and subunits in which the contrast between protein and RNA has been enhanced by reconstitution using deuterated components. Note the large negative α for the native particle and the position α for particles containing deuterated proteins, showing the protein to be further away from the centre of mass than the RNA. The value of β is zero within errors, showing the centres of mass of shape and scattering density to coincide, i.e., that the centres of mass of protein and RNA coincide, since their densities are different.

Equations (11) and (12) can be derived from each other. Equation (11) refers to the special case of the two component particle M_1, M_2 separated by L, whereas Equation (12) refers to the general case for a particle divided up into a large number of components of volume dv with coordinates $\underline{r}$ and of scattering mass $[\rho(\underline{r}) - \rho_s]$ dv.

A Stuhrmann plot such as Figure 5 allows the derivation of well-defined parameters R_C, α and β, which are related to the shape and density distribution of the particle. If, however, one needs to go further and interpret these parameters in terms of a biological model (e.g., protein and RNA in a ribosome), then the approach becomes equivalent to the direct use of Equation (11), and has the same limitations.

A number of experiments have been based on the variation of the two Guinier parameters, I(0) and R_G, with contrast. I have already mentioned experiments on ribosomes. Additionally, the arrangement of DNA and histone proteins in the nucleosome (a particle of chromatin) has been deduced.[16] Also, several soluble proteins have been studied by contrast variation;[17] it appears to be a general result that they have a positive value of α, which is attributed to the denser hydrophilic groups on the outside surfaces. A recent study on a glycoprotein (transcortin) showed that the sugars are isotropically distributed on its surface, and that the polysaccharide chains are extended outwards.[18] Protein interactions with detergent micelles also have been studied; a large β was observed for a complex of colipase with bile salt micelles showing the association to be a lateral one.[19] Rhodopsin, a membrance protein, was studied in detergent micelles because it is not soluble in aqueous solution.[20,21] Osborne et al.[21] used specifically deuterated detergent molecules and carefully examined a range of conditions in order to investigate the effects of inhomogeneous density distributions. This is because the resolution condition, that the detergent micelle and protein could be considered as homogeneous, is not satisfied by these complexes.

Analyses of I(0) and R_G have also been used to study the interactions between amminoacyl-tRNA synthetases and tRNA as a function of external conditions and concentrations of enzyme and substrate.[22,23] These form protein-nucleic acid complexes of varying stoichiometry and organization. Equation (8) was used to derive stoichiometry and concentrations of the different species in the solution to yield association constants. Full use was made of contrast variation, since the same interactions were observed in solvents where the relative excess masses of the components were very different. A set of data is analyzed in Figure 6, and Figure 7 is a Stuhrmann plot showing a major conformational change of the enzyme in a complex with respect to its free conformation. In Figure 6, the two Guinier parameters I(0) (top panels) and R_G (bottom panels) are plotted for each point in the titrations. The basic assumption of the method is that the same interactions are observed in the three solvents. The only difference is the fact that the contrasts with solvent of tRNA and protein are different. In H_2O the excess scattering mass ratio for tRNA and protein is largest; about twice the value in D_2O. In 77% D_2O, it is close to zero so that, in the middle panels, only the protein parts of the complexes are seen. The two

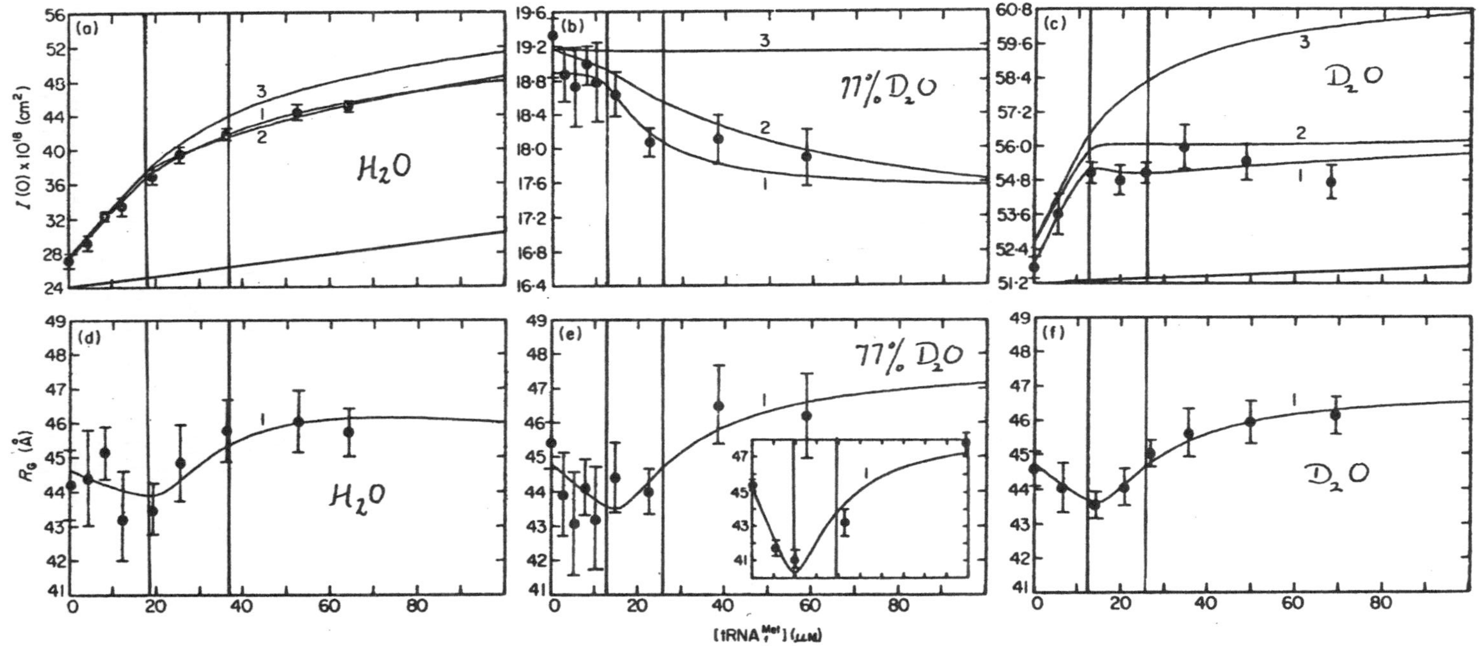

Figure 6. Neutron scattering titration of methionyl-tRNA synthetase from E. Coli by $tRNA_F^{Met}$ in buffer containing 10 mM-$MgCl_2$.[22] The enzyme solution was titrated by a solution containing the same enzyme concentration and excess $tRNA_F^{Met}$. Three sets of experiments were performed: in H_2O buffer (left panels); in 77% D_2O buffer (middle panels); and in D_2O buffer (right panels), from Reference 22.

vertical lines show the concentration values for 1:1 and 2:1 [tRNA]: [enzyme] ratio. The curves were calculated using Equations (8) and (9) for different equilibrium schemes between the complexes formed. An interesting feature which must be accounted for by the models is the drop in intensity in 77% D_2O beyond an enzyme: tRNA ratio of 1. This is due to partial dissociation of the protein (which is a dimer) upon complex formation with a second tRNA molecule, and has provided an insight into the mechanism of anticooperativity towards tRNA binding, which this enzyme was known to exhibit. The effects of dissociation are masked by the scattering of tRNA in H_2O and D_2O buffers; neutron scattering using contrast variation has provided a so far unique way of identifying it. A minimum in R_G is observed at a ratio of 1 in 77% D_2O (the inset shows data using a different wavelength with smaller error bars). This is not due to the dissociation, which I(0) shows to occur beyond this ratio, but to a decrease in the radius of gyration of the dimer itself when it binds one tRNA molecule. We see how R_G and I(0) measurements give complementary information and how contrast variation tells us which parts of the complex we are in fact looking at.

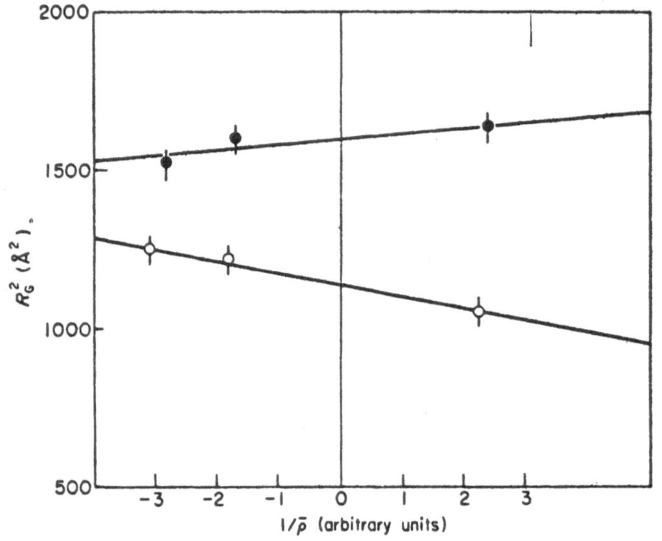

Figure 7. From a neutron study of the complex formation between valyl-tRNA synthetase of yeast and tRNA.[23] R_G^2 plotted versus $1/\bar{\rho}$ for the protein alone (●), and for its 1:1 complex with $tRNA^{val}$(O). The negative slope for the complex shows the tRNA to be bound close to its centre of mass. The positive slope for the protein is due to the denser hydrophilic groups on its surface. The match point of tRNA is at about $1/\rho$ = -3 in the arbitrary units of the figure. The plot shows a significant conformational change of the protein when in the complex, since, otherwise, the two lines should have crossed at the match point of tRNA.

B. Contrast Variation and the Extended I(Q)

Except for the special case of a monodisperse solution of particles with spherical symmetry, the I(Q) cannot be interpreted directly in terms of a scattering density distribution. This is because of the averaging over all orientations in Equation (3). The spherical case is considered in Section 4C.

A way to tackle the interpretation of I(Q) is to compare the observed curve to curves calculated for very simple shapes, ellipsoids or cylinders, for example. At very low resolution this procedure gives a qualitative idea of the shape of the particle. Figure 8 shows neutron data and an ellipsoid model for the lac-repressor protein. The model was obtained from the radii of gyration of the different subunits of the repressor and from the I(Q).[24] By plotting I(Q) as a function of R_GQ, the curves are scaled to compare the shapes of different particles rather than their sizes. In a study of BF_1 - ATPase of E. Coli, a comparison of the scaled curves in H_2O and D_2O suggested solvent regions inside the protein.[37]

A small number of well-defined parameters can be derived from I(Q), for example, the maximum distance in the particle and the surface area (this has been analyzed extensively for X-ray experiments.[1] Recently, Tardieu and Luzzati have discussed the information content of scattering curves.[25]

Contrast variation arguments apply to the extended I(Q) in the same way as to measurements of R_G. There have been 3 approaches:

The first applies to a two component particle of known composition for which there are good arguments to suppose that at the resolution of the experiment each of the two components is homogeneous in density. Such is the case of ribosomal subunits. RNA is quite homogeneous in density and extends the length of the particle. Single proteins are not homogeneous, but ribosomal proteins are very small compared to the RNA and there are several proteins in a resolution volume, so that the protein parts of the particle can be considered as homogeneous. Also, the aminoacid compositions of the different proteins are known and have similar scattering density. In such a particle, the I(Q) can be observed in solvents which match the protein (∿40% D_2O) where it will be dominated by the RNA contribution, and in solvents which match the RNA (∿70% D_2O) where it will be dominated by the protein. This was done for both ribosomal subunits of E. Coli.[35] The curves were analyzed by comparing with detailed models which had been proposed for the arrangements of RNA within the ribosome. For each contrast, parameters such as longest distance and occupied volume were derived, but they were assumed to apply to only one of the components within the particle. The validity of this assumption depends on the condition of homogeneity and resolution mentioned above, and also on accuracy of

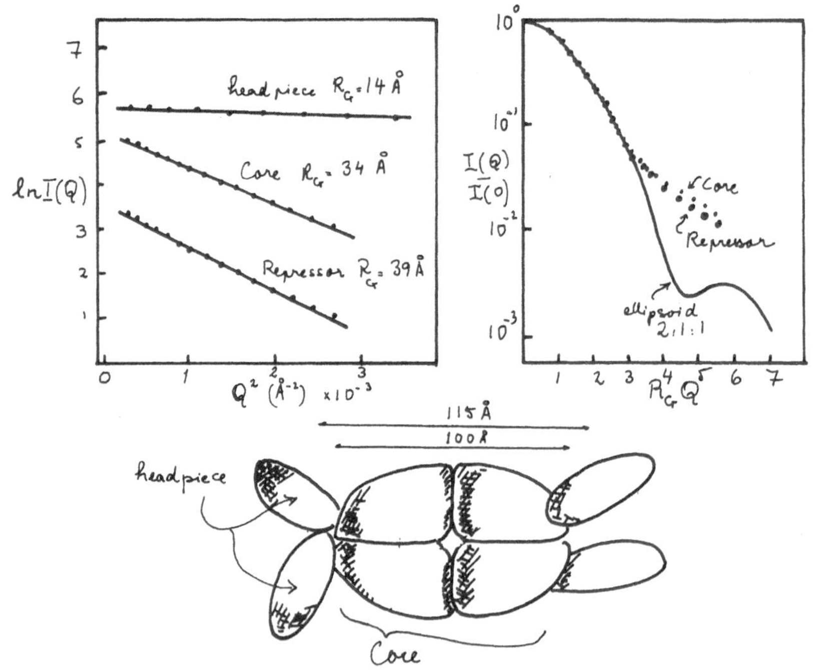

Figure 8. Guinier approximation and extended scattering curve data for the lac-repressor protein; and its different subunits. In part from Reference 24 and from Charlier, Maurizot and Zaccai, to be published. The information on the subunits taken together and using the parallel axes theorem led to the model shown. The model is qualitative in that it is not a precise image of the repressor. On the other hand, it has quantitative features. The radii of gyration are correct as well as the distances from the centre of mass of the different subunits. The maximum dimensions are also correct to ± 10 Å [obtained from the I(Q)]. It is also deduced from the I(Q) that the protein is slightly elongated and not spherical in shape.

the scattering density taken for each component since they determine the contrasts where the measurements should be made.

The second appraoch to analyzing I(Q) with contrast variation is due to Stuhrmann.[8] By writing the scattering density at each point in the particle as a sum of the mean particle density and a fluctuation about that mean, it is possible to separate the expression for I(Q) of Equation (4) into three terms

$$I(Q) = I_s(Q) + \bar{\rho}\, I_{cs}(Q) + \bar{\rho}^2\, I_c(Q) \qquad (13)$$

where $\bar{\rho}$ is the contrast between the mean particle density and the solvent density. The functions I_s, I_{cs}, I_c are called the characteristic functions (in Reference 3, the notation is I_F, I_{VF} and I_V, respectively; in Reference 25, the notation is i_ρ, $i_{\rho v}$, i_v).

The term I_c is given by

$$I_c(Q) = \Sigma\, \delta v_i \delta v_j \frac{\sin Qr_{ij}}{Qr_{ij}} \tag{14}$$

by comparing with Equation (4), we see that $I_c(Q)$ is the scattering curve of a particle in vacuum which has the same shape as the observed particle but which is homogeneous in density. The Stuhrmann analysis, therefore, allows us to extract from the set of I(Q) as a function of contrast the information pertaining to the shape of the particle. The term I_s depends on the internal density fluctuations, and the term I_{cs} is a cross-term. There is an important complication in the divison of I(Q) into its characteristic function for experiments with H_2O: D_2O exchange; the implicit assumption of this analysis, that the density distribution inside the particle volume is independent of the solvent density,[25] is not valid because of the exchange of labile hydrogen atoms between the particle and the solvent. This has been discussed by Ibel and Stuhrmann[26] and in the review by Jacrot.[3] Also the use of the characteristic functions requires very good data at several contrasts.

Measuring neutron scattering curves with precision is a long process (typically, ten times as long as for a radius of gyration measurement) and it is an important economic question how much time should be spent in collecting data on the extended I(Q). Because of this and the problem of H-D exchange there are few examples of the application of the characteristic function approach in neutron scattering. A good one is the determination of the shape of the 50S ribosomal subunit.[27] The $I_c(Q)$ was obtained from a set of I(Q) measured at different contrasts. The particle shape was then expressed as a multipole expansion of spherical harmonics, and a calculated $I_c(Q)$ was derived. The multipole expansion was terminated when a good agreement was obtained between the calculated and observed $I_c(Q)$. The final shape was in good agreement with models which had been proposed from electron microscopy.

The third approach to interpreting I(Q) with contrast variation is to simplify the right hand side of Equation (4) by labelling the particle in some way. This is the basis of triangulation experiments. For example, if there were only two points in the particle, I(Q) would be given by

$$I(Q) = \frac{2m_1m_2 \sin Qr_{12}}{Qr_{12}} + m_1^2 + m_2^2 \tag{15}$$

and r_{12} could be measured directly.

If a multicomponent biological particle can be reconstituted from its subunits, then it might be possible to use this approach. Deuterated and native (protonated) subunits are prepared. To measure the distance between two particular subunits (A and B, say), four preparations are necessary: the particle with A, B, both protonated (A_H, B_H); A deuterated (A_D, B_H); B deuterated (B_D, A_H); A and B, both deuterated (A_D, B_D). It can be shown, by expanding Equation (4) in terms of A, B and the rest of the particle, that the I(Q) due to A, B only is given by

$$I_{AB}(Q) = I_{A_D B_D}(Q) + I_{A_H B_H}(Q) - I_{A_D B_H}(Q) - I_{A_H B_D}(Q) \qquad (16)$$

If the resolution and the shapes of A and B are such that A and B can be assumed to be point-like, then $I_{AB}(Q)$ can be expressed by a $(\sin Qr_{12}/Qr_{12})$ function. Figure 9 shows such a curve from a study on the 30S subunit of the ribosome.[28] Equation (16) is independent of the contrast with the solvent, and depends only on the contrast between the different subunits. The quality of the data is improved, nevertheless, if a solvent density is chosen to match the rest of the particle. A triangulation experiment consists of several pair measurements to build up a geometrical picture of the subunits in the particle. These are the principles of this approach and they are quite simple. Apart from the difficulties associated with the biochemical preparation and characterization of samples, however,

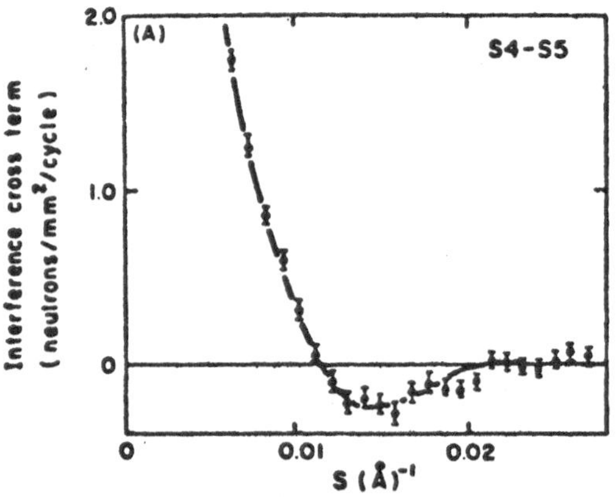

Figure 9. The interference curve $I_{AB}(s)$ for two proteins in the 30S subunit of the ribosome (s is related to Q by $2\pi s$ = A). In the point-like approximation the function is $(\sin 2\pi sr_{12}/2\pi r_{12})$ which crosses zero at $s = 1/2r_{12}$ (from Reference 29).

there have been problems to be solved in the interpretation of the scattering experiments. These have been due to relatively low signal to noise levels, the possibility that A and B could have shapes which would make the point approximation invalid, and the possibility that A and B could exist in more than one copy each in the particle. These problems have been tackled, often in ingeneous ways. They are idscussed in the series of articles of Reference 29 and by Stöckel et al.[30]

C. Spherical Particles

To a resolution of about 40 Å, certain viruses are perfect monodisperse spheres. The orientational average of Equation (3) can be dropped and I(Q) is the square of the Fourier transform of the radial density function. The data from such preparations is quite striking, showing the characteristic series of maxima of the diffraction pattern of a sphere. See, for example, the figures in the paper in this volume by Zulauf et al.[32] Figure 10 shows the relationship between the measured data on spherical viruses and the Fourier transform of a sphere. Neutron work on spherical viruses has been reviewed recently by Jacrot in the context of structural analysis of viruses by different methods.[31]

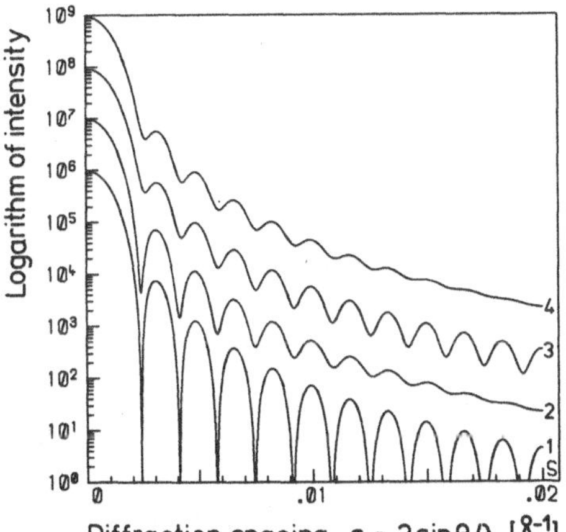

Figure 10. The effects of wavelength and aperture convolution on the I(s) of a sphere of 300 Å radius. Curve 1 is the square of the structure factor. Curve 2 shows the smearing due to 8% wavelength resolution and curve 3 the smearing due to finite aperture. Curve 4 is due to both effects combined and is very similar to observed data, from Schneider and Zulauf.[36]

The radial distribution of excess scattering density in a sphere in solution is given by

$$\rho(r) - \rho_s = \frac{1}{2\pi^2 r} \int_0^\infty QF(Q) \sin Qr dr \qquad (17)$$

where the structure factor F(Q) is obtained from

$$I(Q) = |F(Q)|^2.$$

In the spherical approximation F(Q) is a well behaved function with oscillating signs and can be obtained from the square root of I(Q) without ambiguity.

$$F(Q) = \frac{2}{r} \int_0^\infty r[\rho(r) - \rho_s] \sin Qr dr \qquad (18)$$

Using Equations (17) and (18), contrast variation provides a unique way of probing the internal structure of spherical viruses. There are two ways of using the experimental data: either by using Equation (17) and calculating $\rho(r) - \rho_s$ from F(Q) from each value of ρ, or by calculating F(Q) from spherical shell models, using Equation (18), and comparing it to the observed curve. The advantage of the first method is that it assumes no a-priori knowledge of the virus structure; the disadvantages arise from Fourier termination errors. As was mentioned above, the viruses are spherical only to 40 Å resolution and data beyond this cannot be used. Departure from spherical symmetry can be seen very clearly in the data presented in this volume by Zulauf et al.[30] The second approach, using spherical shells and calculating F(Q) has been developed by Schneider et al.[33] and Chauvin et al.[34] It has the advantage that only the data in the spherical approximation is used fully and that instrumental smearing effects can be taken into account easily (Figure 10). The number of parameters which it is reasonable to vary for a given range of Q and size of particle can be estimated from the formula in Reference 25. These parameters are the inner and outer radii of each shell and the mean value of $\rho(r)$ within it.

The main chemical components of viruses are water, protein, nucleic acids and sometimes lipids. Using either approach, together with a knowledge of the matching points of these components in H_2O:D_2O solvents, has provided a very reliable method of identifying the radial distributions of these components separately. Essentially, for either approach, the excess density at each radius is determined as a function of ρ_s. If plotted against ρ_s it would yield a straight line. From the match point $[\rho(r) - \rho_s = 0]$ the relative dry chemical composition at this radius is obtained, whereas at $\rho_s = 0$ the value of $\rho(r)$ depends on the total chemical composition including hydration at this radius. For example, a value of $\rho(r)$ with a match point of 40% D_2O shows the presence of protein. If at $\rho_s = 0$ the value for $\rho(r)$ is half what one would expect for pure protein, a ratio of protein to water of 1:1 can be proposed for that radius. A very good example of a spherical virus study by neutron scattering is given in this volume.[32]

5. CONCLUSION

These lectures were strictly limited to one aspect of neutron scattering studies in biology - small angle scattering from solution. The main message I have tried to convey is that the method has been unique in probing the internal structure of certain biological molecular organizations. The principles which have been discussed, and especially the notions of contrast and contrast variation by deuteration, are not only applicable to solution studies; much work has been done on membranes, single crystals, and fibres. What I have not mentioned here could fill a book; one, indeed, is in preparation and will be published by Elsevier-North Holland in the coming year, edited by D. L. Worcester.

REFERENCES

1. A. Guinier, and G. Fournet, Small-Angle Scattering of X-rays, John Wiley, New York (1955).
2. T. L. Blundell, and L. N. Johnson, Protein Crystallography, Academic Press, New York (1976).
3a. B. Jacrot, Rep. Prog. Phys. 39: 911-953 (1976).
3b. G. Zaccai, in Topics in Current Physics, vol. VI. Neutron Diffraction, ed. by H. Dachs, Springer-Verlag, Berline (1978).
4. G. E. Bacon, Neutron Diffraction, 3d ed., Oxford University Press, Oxford, England (1975).
5. H. Eisenberg, Biological Macromolecules and Polyelectrolytes in Solution, Clarendon Press, Oxford, England (1976).
6. B. Jacrot, and G. Zaccai, Biopolymers, in press.
7. H. Goldstein, Classical Mechanics, Addison-Wesley, Reading, Mass. (1962).
8. H. B. Stuhrmann, J. Appl. Cryst. 7: 173-178 (1974).
9. I. N. Serdyuk, and A. K. Grenader, Eur. J. Biochem. 79: 495-504 (1977).
10. P. B. Moore, D. M. Engelman, and B. P. Schoenborn, Proc. Nat. Acad. Sci. USA 71: 172-176 (1974).
11. Same as Reference 3a.
12. R. R. Crichton, D. M. Engelman, J. Haas, M. H. J. Koch, P. B. Moore, R. Parfait, and H. B. Stuhrmann, Proc. Nat. Acad. Sci. USA 74: 5547-5550 (1977).
13. G. Zaccai, P. Morin, B. Jacrot, D. Moras, J. C. Thierry, and R. Giegé, J. Mol. Biol. 129: 483-500 (1979).

15. Same as Reference 6.
16. J. F. Pardon, D. L. Worcester, J. C. Wooley, K. Tatchell, K. E. Van Holde, and B. M. Richards, Nucleic Acids Res. 2: 2163-2176 (1975).
17. Same as Reference 8.
18. Z. G. Li, B. Jacrot, F. Le Gaillard, and M. H. Loucheux-Lefebvre, FEBS Lett. 122: 203-206 (1980).
19. M. Charles, M. Semeriva, and M. Chabre, J. Mol. Biol. 139: 297-317 (1980).
20. M. J. Yeager, in Neutron Scattering for the analysis of Biological Structures, ed. by B. P. Schoenborn, Brookhaven Symposia in Biology No. 27, Brookhaven National Laboratory, New York (1976).
21. H. B. Osborne, C. Sardet, M. Michel-Villaz, and M. Chabre, J. Mol. Biol. 123: 177-206 (1978).
22. P. Dessen, S. Blanquet, G. Zaccai, and B. Jacrot, J. Mol. Biol. 126: 293-314 (1978).
23. Same as Reference 14.
24. M. Charlier, J. C. Marrizot, and G. Zaccai, Nature 286: 423-425 (1980).
25. V. Luzzati, and A. Tardieu, Ann. Rev. Biophys. Bioeng. 9: 1-29 (1980).
26. K. Ibel, and H. B. Stuhrmann, J. Mol. Biol. 93: 255-265 (1975).
27. H. B. Stuhrmann, M. H. J. Koch, R. Parfait, J. Haas, and R. R. Crichton, Proc. Nat. Acad. Sci. USA 74: 2316-2320 (1977).
28. J. A. Langer, D. M. Engelman, and P. B. Moore, J. Mol. Biol. 119: 463-485 (1978).
29. P. B. Moore, and D. M. Engelman, in Methods in Enzymology, vol. 59, ed. by K. Moldave and L. Grossman, Academic Press, New York (1979). pp. 629-638.
P. B. Moore, ibid, pp. 639-656; D. M. Engelman, ibid, pp. 656-669.
30. P. Stöckel, R. May, I. Strell, Z. Cejka, and W. Hoffe, J. Appl. Cryst. 12: 176 (1979).
31. B. Jacrot, in Comprehensive Virology, vol. 17, ed. by H. Fraenkel-Conrat, and R. R. Wagner, Plenum Press, New York (1980).
32. M. Zulauf, M. Cuillel, and B. Jacrot, (in this volume).
33. D. Schneider, M. Zulauf, R. Schafer, and R. M. Franklin, J. Mol. Biol., 124: 97-122 (1978).
34. C. Chauvin, J. Witz, and B. Jacrot, J. Mol. Biol. 124: 641-651 (1978).
35. I. N. Serdyuk, A. K. Grenader, and G. Zaccai, J. Mol. Biol. 135: 691-707 (1979).
36. D. Schneider, and M. Zulauf, J. Appl. Cryst., in press.
37. M. Satre, and G. Zaccai, FEBS Lett. 102: 244-248 (1979).

LIGHT SCATTERING FROM CELLS

Sow-Hsin Chen

Nuclear Engineering Department
Massachusetts Institute of Technology
Cambridge, Massachusetts, U.S.A.

Contents

I. PARTICLE VELOCIMETRY IN TIME-DOMAIN[1,2]

There is an aspect of application of quasielastic light scattering which deals with measurements of directed motions of particles. The particles can be motile bacteria, swimming spermatozoa or streaming water droplets. The simplest objective in these experiments is to measure the directed speed or speed distribution of the particles. For simplicity let us ignore initially the size effects in the light scattering spectra. This simplication is allowed for instance when the particle dimension is much smaller compared to the wave length of light used, or when the particle has a spheric symmetry so that its light scattering property is independent of its orientation. Using the photon correlation technique we are in effect measuring the Doppler shift in time domain.

Referring to Figure 1, suppose the incident laser beam has a frequency ω, wave vector $\underset{\sim}{k}_i$ ($|\underset{\sim}{k}_i| = 2\pi/\lambda$) and polarization vector $\hat{\varepsilon}$ perpendicular to the page, i.e., $\vec{E}_i(t) \sim \hat{\varepsilon}e^{-i\omega t}$:

The particle P is moving with an instantaneous velocity $\vec{V}$. Then:

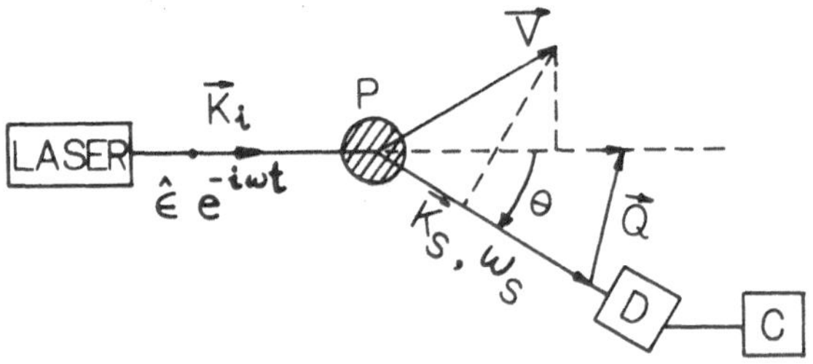

Figure 1. A schematic set up for the laser particle velocimetry using a photon correlation technique. Incident field polarization $\hat{\varepsilon}$ is perpendicular to the plane of the page.

$$-\Delta\omega_i = \omega\left(\frac{V_{||}}{c}\right) = \underset{\sim}{k}_i \cdot \underset{\sim}{V} = \text{frequency lowering as seen by the moving particle (P)}$$

$$+\Delta\omega_s = \omega\left(\frac{V_{||}'}{c}\right) = \underset{\sim}{k}_s \cdot \underset{\sim}{V} = \text{frequency increase as seen by the observer (D)}$$

Therefore the scattered frequency ω_s is

$$\omega_s = \omega + \Delta\omega_i + \Delta\omega_s = \omega - (\underset{\sim}{k}_i - \underset{\sim}{k}_s) \cdot \underset{\sim}{V} \equiv \omega - \underset{\sim}{Q} \cdot \underset{\sim}{V}$$

and

$$\vec{E}_s(t) \sim \hat{\varepsilon} e^{-i\omega_s t} = \hat{\varepsilon} e^{-i\omega t}\, e^{i\underline{Q} \cdot \underline{V} t} \tag{1}$$

The measured photon correlation function is proportional to

$$C_s(t) = \left| \frac{\langle \underset{\sim}{E}_s{}^*(0) \cdot \underset{\sim}{E}_s(t) \rangle}{\langle |E_s|^2 \rangle} \right|^2 \tag{2}$$

$$= |\langle e^{i\underset{\sim}{Q} \cdot \underset{\sim}{V} t} \rangle|^2 \equiv |F_s(\underset{\sim}{Q}, t)|^2 \tag{3}$$

where the bracketed average is over all particles in the scattering volume. We have from Equation (3), which defines the intermediate scattering function $F_s(Q, t)$, an explicit relation:

medium. We use notations $k_v = \omega/c$, wave number in vacuum; and $k = n_o k_v$, wave number in the medium; and $\hat{\varepsilon}$ the incident electric field polarization vector.

Figure 2 illustrates a typical scattering geometry. The incident wave vector $\underline{k}$ in the y direction with the polarization vector $\hat{\varepsilon}$ along the $\hat{z}$ direction. The scattered wave vector $\underline{k}'$ is taken to be in the x-y plane with a scattering angle θ. A unit vector $\hat{v}$ with direction cosines (ℓ_1, ℓ_2, ℓ_3) denotes the symmetry axis of the particle with its direction specified by polar angles (θ_s, ϕ_s). In a most general case involving an arbitrary incident polarization, one can decompose the incident field into a transverse component (perpendicular to the scattering plane) and a parallel component (parallel to the scattering plane, in this case the x-y plane) each specified by a unit vecotr $\hat{t}_o$ and $\hat{\ell}_o$, respectively. The scattered fields are then also decomposable into the transverse and parallel components each specified by unit vectors $\hat{t}$ and $\hat{\ell}$. Since the Maxwell's equations are linear, we have in general a linear relation between the incident and the scattered fields expressible as:[4]

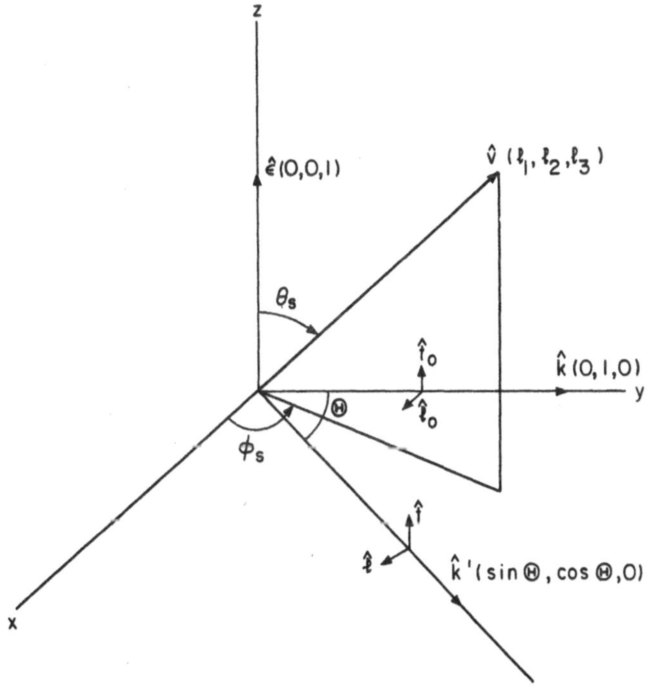

Figure 2. Scattering geometry used in a typical laser light scattering experiment. Scattering angle is denoted by Θ in this figure while in the text it is denoted as θ.[9]

2. LIGHT SCATTERING FROM A MACROSCOPIC PARTICLE - RIGOROUS THEORY

Living cells are macroscopic particles in the sense that their typical dimensions are of the same order of magnitude or larger than wave lengths of light. For example, prokaryotic cells such as Escherichia coli or Salmonella typhimurium have an outer dimension of about 1.5 μm x 0.7 μm and are shaped like a spheroid with an outer membrane surrounding cytoplasm inside. Another example is eukaryotic cells such as bull spermatozoa which have a head dimension of 10 μm x 4.6 μm x 1.0 μm, a flat disc like object, joining to a long tail of about 30 μm.

These cells are therefore extended particles. They have not only macroscopic dimensions and non-spherical shape but also possess internal structures which are optically non-homogeneous. In order to deal with light scattering properties of these objects one has to consider a multi-region problem which is obviously complicated. But one should realize that just as conventional light scattering experiments (measurement of angular distribution of scattered intensity) can give shape and structural information of cells, the dynamical light-scattering using the photon correlation technique can also give the shape and structural information. In fact, the latter is more sensitive to the shape because a deviation from sphericity immediately reflects itself in the non-scaling property of the correlation function.

A complete solution to the problem of scattering of light by a homogeneous nonmagnetic particle of index of refraction n (called region 1) immersed in a homogeneous dielectric medium of index of refraction n_o (called region II) can be obtained from solving Maxwell's equations

$$\nabla \times \overline{E} = ik_v\overline{B} \tag{6}$$

$$\nabla \times \overline{B} = ik_v n^2(\underline{r})\overline{E} \tag{7}$$

in regions I and II with following boundary conditions:

(a) in region I where $n(\underline{r}) = n$, there are in internal fields $(\overline{E}_i, \overline{B}_i)$;

(b) in region II where $n(\underline{r}) = n_o$, there are the scattered fields $(\overline{E}_s, \overline{B}_s)$ plus the incident fields

$$\overline{E}_0(\underline{r},t) = E_0\hat{\varepsilon} \exp [i(\underline{k}\cdot\underline{r} - \omega t] \tag{8}$$

and its magnetic counterpart;

(c) the tangetial components of the fields in I and II are continuous at the boundary between the particle and the

$$F_s(\underset{\sim}{Q},t) = \int d^3V P(\underset{\sim}{V}) e^{i\underset{\sim}{Q}\cdot\underset{\sim}{V}t} . \tag{4}$$

In the case where the velocity distribution function is isotropic, namely, $P(\underline{V}) = P(V)$, we can integrate out the angular part in Equation 4 to get

$$F_s(Q,t) = \int_0^\infty dV\ 4\pi V^2 P(V) \left(\frac{\sin QVt}{QVt}\right). \tag{5}$$

Equation 5 is a formula first given by Nossal, Chen and Lai[3] in 1971. One can easily see from Equation (5) and (2) that there is an intimate connection between the measured photon correlation function $C_s(t)$ and the particle speed distribution function P(V). One observes from Equation (5) that in the case of isotropic velocity distribution, $F_s(Q,t)$ is a function of a scale variable X = Qt only. This simple property, which we shall call "scaling", can be used to test whether the particles in the system have a spherical symmetry. This is particularly true in the case where their dimensions are comparable to the wave length of light - a case which is almost universally true for cells. On the other hand, if the particle motion is not isotropic, then the scaling property would not exist. $F_s(\underset{\sim}{Q},t)$ in this case depends on $\underset{\sim}{Q}$ and t separately. A good example is when there is a collection of particles each making an independent helical motion along some arbitrary direction. Since in this case the motion is not rectilinear we have to extend the definition of the intermediate scattering function as

$$F_s(\underset{\sim}{Q},t) = \langle e^{i\underset{\sim}{Q}\cdot\underset{\sim}{R}(t)}\rangle. \tag{6}$$

We have then for the center of mass coordinates of a typical particle

$$\underset{\sim}{R} = [R\cos(\omega t+\alpha),\ R\sin(\omega t+\alpha),\ Vt]$$

and the $\vec{Q}$ vector

$$\underset{\sim}{Q} = [Q\sin\theta\cos\phi,\ Q\sin\theta\sin\phi,\ Q\cos\theta].$$

As a result, the intermediate scattering function averaged over all particle motions becomes the following, which, except for a case QR<<1, does not have the scaling property.

$$F_s(Q,t) = \int_0^\infty dV P(V) V^2 \int_0^{2\pi} \frac{d\alpha}{2\pi} \int_0^{2\pi} d\phi \int_{-1}^{1} d\mu\ e^{iQVt\mu} \exp[iQR\sqrt{1-\mu^2}\cos(\omega t-\alpha+\phi)].$$

$$\begin{bmatrix} E_{st} \\ E_{s\ell} \end{bmatrix} = \begin{bmatrix} S_1 & S_4 \\ S_3 & S_2 \end{bmatrix} \begin{bmatrix} E_{ot} \\ E_{o\ell} \end{bmatrix} \frac{e^{ikr}}{-ikr} \exp(-i\underline{k}\cdot\underline{r}) \tag{9}$$

where the elements of the scattering matrix S_i are dimensionless scattering amplitudes which depend on relative orientations of the four vectors: $\hat{v}$, $\hat{\varepsilon}$, $\hat{k}$ and $\hat{k}'$. In the usual laser scattering geometry depicted in Figure 2, $\bar{E}_{o\ell} = 0$ and $\bar{E}_{ot} = \bar{E}_o(\underline{r},t)$ as given in Equation (8). The polarized scattering amplitude is given by S_1, and the depolarized scattering amplitude by S_3. This is the so called "VV" scattering geometry. In the case of "HH" geometry where $\bar{E}_{ot} = 0$ and $\bar{E}_{o\ell} \neq 0$, S_2 is the polarized scattering amplitude and S_4 the depolarized scattering amplitude. The scattering amplitude S_i can be determined for a given particle geometry by solving the Maxwell's equations as a boundary value problem as mentioned above.

A beautiful analytical solution of the above problem was achieved for a sphere by Mie[5] (1908) and Debye[6] (1909). Their solutions for $S_1(\theta)$ and $S_2(\theta)$, which are functions of scattering angle θ only due to the spherical symmetry, are given by[4]

$$S_1(\theta) = \sum_{n=1}^{\infty} \frac{2n+1}{n(n+1)} \left[a_n \Pi_n(\cos\theta) + b_n \tau_n(\cos\theta)\right] \tag{10}$$

$$S_2(\theta) = \sum_{n=1}^{\infty} \frac{2n+1}{n(n+1)} \left[a_n \tau_n(\cos\theta) + b_n \Pi_n(\cos\theta)\right] \tag{11}$$

where

$$\Pi_n(\cos\theta) = P_n^1(\cos\theta)/\sin\theta \tag{12}$$

$$\tau_n(\cos\theta) = d[P_n^1(\cos\theta)]/d\theta \tag{13}$$

$$a_n = [\psi_n(x)\psi_n'(y) - m\psi_n'(x)\psi_n(y)]\,[S_n(x)\psi_n'(y) - mS_n'(x)\psi_n(y)]^{-1} \tag{14}$$

$$b_n = [\psi_n'(x)\psi_n(y) - m\psi_n(x)\psi_n'(y)]\,[mS_n(x)\psi_n'(y) - S_n'(x)\,\psi_n(y)]^{-1} \tag{15}$$

and the two functions ψ_n and S_n are given by $\psi_n(z) \equiv z\, j_n(z)$, $S_n(z) = zh_n^{(1)}(z)$ in terms of Besel functions. Important parameters in the problem are: $m = n/n_o$, relative index of refraction; $x = ka(=2\pi n_o a/\lambda$, a radius of the sphere) and $y = mka$, relative size of the particle to wave length.

(A) Rayleigh Limit: ka<<1, no limitation on m

This limiting case applied to macromolecules in solutions which have typical dimension of a few hundred angstrom or less. In this case it can be shown[4]

$$|S_1(\theta)|^2 \doteq (\frac{m^2-1}{m^2+2})\ (ka)^6 \tag{16}$$

$$|S_2(\theta)|^2 \doteq (\frac{m^2-1}{m^2+2})^2\ (ka)^6\ \cos^2\theta. \tag{17}$$

Using these amplitudes the differential crossections can be obtained as

$$\frac{d\sigma vv}{d\Omega} = \frac{1}{k^2}\ |S_1(\theta)|^2 = k^4\alpha^2$$

$$\frac{d\sigma HH}{d\Omega} = \frac{1}{k^4}\ |S_2(\theta)|^2 = k^4\alpha^2\ \cos^2\theta \tag{19}$$

where $\alpha = a^3(m^2-1)\ /\ (m^2+2)$ is the polarizability of a sphere of radius a. In this case the small particle radiates like an electric dipole with a dipole moment $\vec{p} = \hat{\varepsilon}E_o\alpha$.

(B) Geometrical optics limit: ka>>1, m>1.

This case applies to most of large cells and water droplets. For example, for a 10 μm cell suspended in water ka = 66. In this limit, one can use geometrical optics ray tracing method to derive[7]

$$S_i(\theta) = S_{diff}(\theta) + S_i^{(1)}(\theta) + S_i^{(2)}(\theta)\ ,\quad i = 1,2 \tag{20}$$

where $S_{diff}(\theta)$ is the diffraction contribution at small angles.

$$S_{diff}(\theta) = x^2\left[\frac{J_1(x\sin\theta}{x\sin\theta}\right]. \tag{21}$$

$S_i^{(1)}(\theta)$ is contribution from rays once reflected from the spherical surface (we give i=1 case here, for i=2, see Reference 7):

$$S_1^{(1)}(\theta) = x\left[\frac{\sin\theta/2 - \sqrt{m^2-\cos^2\theta/2}}{\sin\theta/2 + \sqrt{m^2-\cos^2\theta/2}}\right]\frac{1}{2}\exp\ [\frac{\pi}{2}+2x\sin\frac{\theta}{2})] \tag{22}$$

and $S_i^{(2)}$ is the contribution from twice refracted rays from the surface

$$S_1^{(1)}(\theta) = x\left[1-\left(\frac{1+m^2-2m\cos\theta/2}{1-m^2}\right)^2\right]\sqrt{\frac{m^2\sin\frac{\theta}{2}\;(m\cos\frac{\theta}{2}-1)(m-\cos\frac{\theta}{2})}{2\sin\theta(1+m^2-2m\cos\frac{\theta}{2})^2}} *$$

$$* \exp i(\frac{3\pi}{2} - 2x\sqrt{1+m^2-2m\cos\frac{\theta}{2}}) \tag{23}$$

Glantsching and Chen[7] showed that the three term expression in Equation (20) is a very good approximation in the case of water droplets by comparing results of this expression with numerical calculation based on the Mie theory.[8] In Reference 7 it is shown that the geometrical optics limit is already reached for a water droplet of 10 μm size. By observing the scattering at sufficiently large angles, large enough to avoid the diffraction effect (for 10 μm water droplets this requirement amounts to $\theta > 10^o$), and by integrating over a suitable angular range (such as $\Delta\theta \approx 2^o$) it is shown[7] that the angular averaged differential scattering crossection can be made to be proportional to (size)^2, i.e.

$$\bar{I}(\theta) = K(\theta)\; x^2.$$

This conclusion has a practical implication in using the pulse height analysis technique for large particle sizing.

3. INTEGRAL FORMULATION OF THE SCATTERING THEORY AND THE RAYLEIGH-GANS-DEBYE APPROXIMATION

Maxwell's equations (6) and (7) can be converted to an integral equation. This can be accomplished by first eliminating $\bar{B}$ from Equations (6) and (7) to get

$$\nabla X(\nabla x\bar{E}) - k_v^2 n_o^2 \bar{E} = k_v^2[n^2(\underline{r})-n_o^2]\bar{E} \tag{25}$$

This equation shows that the scattering is due to inhomogeneity of the index of refraction as expressed by the right hand side of Equation (25). Equation (25) can be converted to an integral equation by introducing the Green's function $\bar{G}(\underline{r}-\underline{r}')$ which represents a solution due to a point source situated at $\underline{r}'$. Aside from a time factor exp $(-i\omega t)$ the total field at position $\underline{r}$ satisfies then an integral equation

$$\bar{E}(\underline{r}) = \bar{E}_o(r) + \int_V d^3r'\; \bar{\bar{G}}\;(|\underline{r}-\underline{r}'|)\cdot\bar{E}(\underline{r}')k_v^2(n^2-n_o^2) \tag{26}$$

where the integration is over the volume V of the particle. The Green's tensor in the far field position $\underline{r}$ is known to be[9]

$$\bar{\bar{G}}\ (|\underline{r}-\underline{r}'|) \underset{r>>\lambda}{=\!=\!=} (\bar{\bar{1}}-\hat{r}\hat{r})\ \frac{\exp(ikr-i\underline{k}'\cdot r')}{4\pi r} \tag{27}$$

where $\underline{k}' = k\hat{r}$ is the scattered wave vector.

Looking at the right hand side of Equation (26), we estimate that the ratio of the magnitude of the second term to the first term is of the order $d^2k^2(m^2-1)$ where d is the range of integration in the second term or the dimension of the particle. Therefore when this quantity is small we may approximate the second term by substituting $\bar{E}_o(\underline{r}')$ for $\bar{E}(\underline{r}')$ in it.

This is called the Rayleigh-Gans-Debye approximation[4] in the literature. It corresponds to the Born approximation in the case of Schrodinger equation. The result for the scattered field is[9]

$$E_s^{(1)}(\underline{r},t) = E_o k^2 (m^2-1) V \bar{a}\ [\hat{v}(t),\hat{Q}]\ \frac{e^{ikr}}{4\pi r} \exp\,[iQ\cdot R(t)-i\omega t] \tag{28}$$

where the dynamic form factor $\bar{a}$, has a simple form

$$\bar{a}\ [\hat{v}(t),\hat{Q}] = (\bar{\bar{1}}-\hat{r}\hat{r})\cdot\hat{\varepsilon}\frac{1}{V} \int_V d^3\rho\ \exp(i\underline{Q}\cdot\underline{\rho}). \tag{29}$$

We have introduced the scattering vector $\underline{Q} = \underline{k}-\underline{k}'$, the center of mass coordinate of the particle $\underline{R}(t)$ and the relative coordinate $\underline{\rho}(t)$ which are related by $\underline{r}' = \underline{R} + \underline{\rho}$.

In Equation (9) the scattering matrix $\bar{\bar{S}}$ is defined with respect to a particle situated at origin $\underline{R} = 0$. For a particle whose center of mass is situated at $\underline{R}$ we showed[9] that the scattering matrix $\bar{\bar{S}}(\underline{R})$ is given by a relation

$$\bar{\bar{S}}(\underline{R}) = \bar{\bar{S}}(0)\ \exp\ (i\underline{Q}\cdot\underline{R}). \tag{30}$$

This simple relation allows us to use the result of rigorous light scattering theory in calculation of the particle self correlation functions. By comparing results of the integral formulation of scattering theory of this section with rigorous formulation given in the last section we can identify the scattering matrix elements in the RGD approximation as[9]

$$S_1^{(RGD)}(\hat{v}\cdot\hat{Q}) = \frac{i}{4\pi} k^3(m^2-1) \int_V d^3\rho \exp(i\underline{Q}\cdot\underline{\rho}) \tag{31}$$

$$S_2^{(RGD)}(\hat{v}\cdot\hat{Q}) = -\cos\theta\ S_1^{(RGD)} \tag{32}$$

$$S_1^{(RGD)} = S_4^{(RGD)} = 0 \tag{33}$$

Above equations show that in the RGD approximation there is no depolarization in the scattering and furthermore, all the scattering matrix elements are functions of an angle between the scattering vector $\underline{Q}$ and the symmetry axis of the particle $\hat{v}$ only.

4. CALCULATION OF TIME CORRELATION FUNCTIONS

Once the scattering matrix is obtained by solving the light scattering problem of a particle situated at the origin according to the procedure of section 2, the scattered field at the detector position $\underline{r}$ due to N particles in the scattering volume uniformly illuminated by the incident light can be written down using Equation (9) and (30).

$$\overline{E}_s(\underline{r},t) = \sum_{i=1}^{N} \exp(i\underline{Q}\cdot\underline{R}_i(t))\ S_1 E_o \hat{\varepsilon}\ \exp(-i\omega t)\ \frac{e^{ikr}}{-ikr} \tag{34}$$

where we limit ourselves only to the VV scattering for simplicity. The normalized scattered field correlation function is defined as the intermediate scattering function:

$$F(\underline{Q},t) = \frac{\langle \overline{E}_s^*(o)\cdot \overline{E}_s(t)\rangle}{\langle |E_s|^2\rangle} . \tag{35}$$

We shall assume that cells move independently so that there are no interference terms. This gives (with the subscript s to denote the self term)

$$F_s(\underline{Q},t) = \frac{\langle \exp\ [i\underline{Q}\cdot(\underline{R}(t)-\underline{R}(0)]\ S_1^*(0)S_1(t)\rangle}{\langle S_1^*(0)S_1(0)\rangle} \tag{36}$$

where the time dependence of $S_1(t) = S_1(\hat{v}(t))$ comes about because of rotational motions of the particle. We notice from Equation (36) that if the particle has a spherical symmetry then the time dependence in S_1 goes out. Since the center of mass motion is independent of the orientation of the spherical particle, Equation (36) simplifies to

$$F_s(\underline{Q},t) = \langle \exp\ [i\underline{Q}\cdot(\underline{R}(t)-\underline{R}(0)]\rangle \tag{37}$$

This is identical to the point particle result where only center of mass motion enters into the self correlation function (or the self-intermediate scattering function). Even if the spherical particle has an internal structure, this conclusion would also be true. We shall give in the following some examples taken from calculations for the case of bacteria.

(A) Scattered Intensity

From the basic scattered field expression, Equation (34) we can immediately write down the scattered field correlation function as:

$$<\overline{E}_s^*(t)\cdot\overline{E}_s(t)> = \frac{N|E_o|^2}{k^2r^2} <\frac{1}{N}\sum_{i,j}^{N} \exp[i\underline{Q}\cdot(\underline{Ri}-\underline{Rj})]\ S_i^*\ S_j> \tag{38}$$

We shall call the bracketed quantity in Equation (38) the intensity factor I(Q). Then, for a mono-dispersed system of particles

$$I(Q) = P(Q)\ S(Q) \tag{39}$$

where S(Q) is the interparticle structure factor:

$$S(Q) = <\frac{1}{N}\sum_{i,j}^{N} \exp[i\underline{Q}\cdot(\underline{Ri}-\underline{Rj})]> \tag{40}$$

and P(Q) is the intra-particle structure factor:

$$P(Q) = \frac{1}{4\pi}\int_0^{2\pi} d\phi_s \int_0^{\pi} d\theta_s \sin\theta_s\ \ |S_1(\theta_s\phi_s\theta)|^2 \tag{41}$$

which is the angular-averaged, polarized scattering intensity component from a particle. The depolarized component can be similarly written down in terms of S_2 function. We note here that for a system of isotropically moving particles the scattered intensity is independent of the state of motion.

(B) Self-Correlation Function for a Diffusing Particle[9]

Starting from the basic expression Equation (36) we can carry out the motional average as follows. For a micron-size particle the ratio of rotational to translational diffusion time constants is: $(\tau_r/\tau_r) \sim (Qd)^2 \sim 100$. Thus, for the short time component of the correlation function, one can neglect the rotational diffusion and consider the particle undergoing only a translational diffusion.[10] Since the orientation of the particle does not change within this time scale we can neglect the time dependence in $S_1(t)$ factor. The motional average for the center of mass motion then gives[10]

$$<\exp\ [i\underline{Q}\cdot(R(t)-\underline{R}(0))]> = \exp\ [\frac{\chi}{\beta} - \frac{\chi}{3} - \chi\mu^2] \tag{42}$$

where

$$\chi = \bar{D} Q^2 \beta t \tag{43}$$

$$\bar{D} = (D_{\parallel}+2D_{\perp})/3 \tag{44}$$

$$\beta = (D_{\parallel}-D_{\perp})/\bar{D}$$

$$\mu \equiv \hat{Q}\cdot\hat{v} = \sin\theta_s \, [-\cos\phi_s \sin\theta + \sin\phi_s (1-\cos\theta)][2(1-\cos\theta)]^{-\frac{1}{2}} \tag{45}$$

The self-correlation function for the polarized component of the scattering is then obtained by the angular average

$$F_s^D(Q,t) = \exp\left[-\frac{\chi}{\beta}+\frac{\chi}{3}\right] \cdot \frac{A}{B} \tag{46}$$

$$A = \int_0^{2\pi} d\phi_s \int_0^{\pi} d\theta_s \sin\theta_s \, |S_1(\theta_s\phi_s\theta|^2 \exp[-\chi^2\mu^2(\theta_s\phi_s\theta)]$$

$$B = \int_0^{2\pi} d\phi_s \int_0^{\pi} d\theta_s \sin\theta_s \, |S_1(\theta_s\phi_s\theta)|^2.$$

An example of such a calculation for a spheroid modeling a diffusing bacterium was given by Kotlarchyk at al.[10] The size parameters were chosen as a = 0.7 μm, a/b = 1.94 and m = 1.007. Calculation was made of the self correlation function using for input $S_1(\theta_s\phi_s\theta)$ both numerical solutions of the rigorous theory and the RGD approximation. The result is summarized in Figure 3 in terms of an angular dependence of the scaled half-width: $\chi_{1/2} \equiv \bar{D} Q^2 \beta t_{1/2}$. We see that for this case the RGD approximation is indeed very good and the scaling relation is very well satisfied.

(C) <u>Self-Correlation Function for a Freely Translating Particle</u>[9]

This corresponds to a case when bacteria move along straight lines for distances much longer than Q^{-1}. We shall take a case of isotropic motions with a Maxwellian distirbution.[11] Since the particle moves in a straight line along its figure axis, $S_1(t)$ has no time dependence during the motion. The motional average can again be made of the translational factor.

$$\langle \exp[i\underline{Q}\cdot(\underline{R}(t)-\underline{R}(0))]\rangle = (1-2W^2\mu^2)\exp(-W^2\mu^2). \tag{47}$$

where W is the scaled time variable in terms of the mean square speed $\langle V^2\rangle$,

$$W = Q\,t\,(\langle V^2\rangle/6)^{1/2}, \tag{48}$$

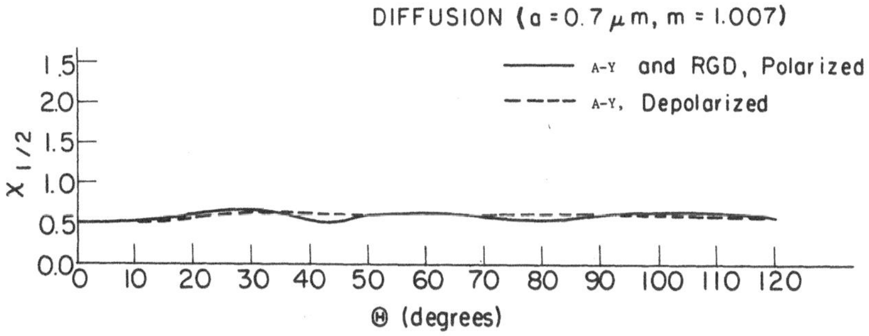

Figure 3. The sealed half-width of the self-correlation function, $X_{1/2} = \bar{D}Q^2\beta t_{1/2}$, plotted as a function of scattering angle, Θ. The calculation is for a diffusing spheroid of the semi-major axis a, and semi-minor axis b, a/b = 1.94. Solid line refers to the results of a rigorous calculation (denoted by A-Y) and the RGD approximation, both giving the same results for the polarized component. Dashed line refers to the depolarized component.[9]

and μ is $\hat{Q}\cdot\hat{v}$. The self-correlation function of running bacteria is therefore:

$$F_S^R(Q,t) = \frac{\int_0^{2\pi} d\phi_s \int_0^{\pi} d\theta_s \sin\theta_s \, |S_1(Q_s\phi_s\theta)|^2 \exp(-W^2\mu^2)(1-2W^2\mu^2)}{\int_0^{2\pi} d\phi_s \int_0^{\pi} d\theta_s \sin\theta_s \, |S_1(\theta_s\phi_s\theta)|^2} \tag{49}$$

An example[9] is given in Figure 4 again for the case of bacteria. The angular dependence of the scaled half-width, $W_{1/2} = Q\, t_{1/2}(\langle V^2\rangle/6)^{1/2}$ was calculated for size paramters a = 0.7 μm, a/b = 1.94, m = 1007 and m = 1.03. One sees the oscillatory dependence on θ showing the structural effect in motility measurement which has been noted experimentally also.[11] Again one observes that for these size parameters, the RGD approximation is good, especially when m is closed to unity.

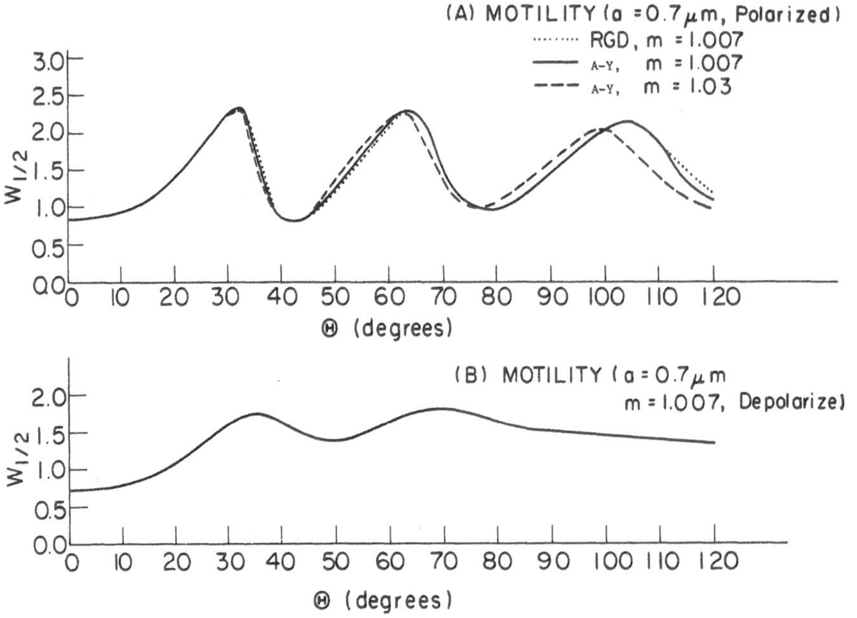

Figure 4. The scaled half-width of the self-correlation function: $W_{1/2} = Qt_{1/2}\ (<V^2>/6)^{1/2}$, plotted as a function of scattering angle H. Calculation was made for the case of spheroid with the same size parameter as in Figure 3, moving with a Maxwell speed distribution. A-Y refers to the rigorous calculation for the scattering matrix element S_1.[9] Notice that when m is increased to 1.03, deviation of the RGD approximation from the rigorous calculation becomes noticable.

(D) Self-Correlation Function for a Non-Spherical Particle Moving in a Complicated Trajectory[12]

For this more complicated case one can still compute the self correlation function if $S_1(\theta_s\phi_s\theta)$ is known. Since the motional average in this case is complicated, one would like to evaluate S_1 in the RGD approximation. This has been done in Reference 12 for a bacterium moving along a helical path. In fact, one can write down a general expression for $F_s(Q,t)$ in RGD approximation as:

$$F_s(Q,t) = \frac{1}{C_n} < \int_0^{2\pi} \frac{d\psi}{2\pi} \int_{-1}^{1} d\nu \int_0^{\infty} dV (2\pi V^2) P(V)\ a\ [\hat{Q}\cdot\hat{v}(t)]\ a\ [\hat{Q}\cdot\hat{v}(0)] \cdot \exp\,(iQV\ \nu t)> \qquad (50)$$

$$C_n = < \int_0^{2\pi} \frac{d\psi}{2\pi} \int_{-1}^{1} d\nu\ |a\ [\hat{Q}\cdot\hat{v}]|> \qquad (51)$$

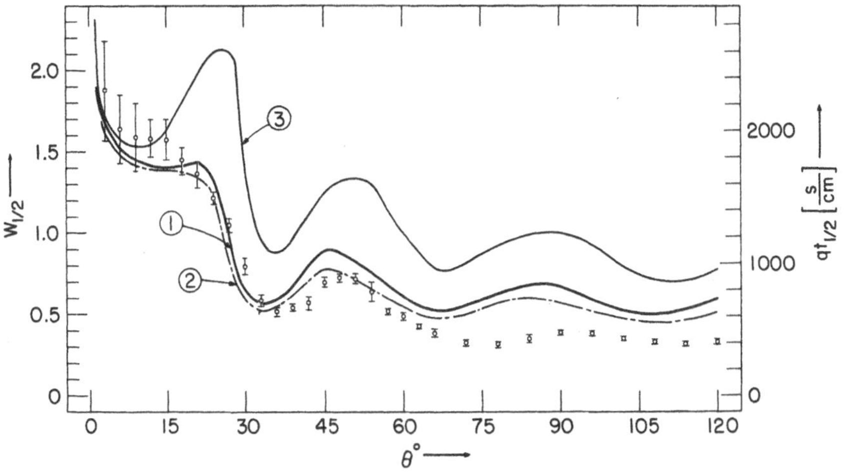

Figure 5. The scaled half-width vs. scattering angle plot for motile E. coli In motility buffer. Curve 1 and 2 are calculations based on RGD approximations for a control spheroid making slightly different helical motions. Curve 3 is for the same coated spheroid making straight line motions with Maxwell speed distribution. For details of the calculation, see Reference 12.

where the time-dependent form factor $a[\hat{Q}\cdot\hat{v}(t)]$ can be evaluated analytically for many geometries. Figure 5 shows an example of $W_{1/2}$ vs θ calculation.[12] It is interesting to see that by taking into account "wobble" or helical motion in their runs, it is possible to obtain much better agreement with experiment. Another similar calculation for the case of bull spermatozoa was done by Craig et al.[13] and a full description of it is given in the next article.

5. CONCLUSION[14]

The self correlation function for a diffusing or translating cell depends, besides on the center of mass motion, also on the shape and structure of the cell. When the shape of the cell deviates from sphericity, the structural effect is immediately reflected in the angular dependence of the half-width of the correlation function. It is shown that this shape and structure effects in the correlation function can be rigorously taken into account if one computes the scattering matrix element by solving the Maxwell's equations. In practice, depending on whether one is treating bacteria (∿1μm) or large cells like spermatozoa (∿10μm), one can make either the RGD approximation or the geometrical optics approximation. In most cases the numerical problem is quite tractable.

ACKNOWLEDGEMENT

This research is supported by NSF PCM 7815844

REFERENCES

1. Pike, E.R., "Photon Correlation Velocimetry", in Photon Correlation Spectroscopy and Velocimetry, Ed. H.Z. Cummins and E.R. Pike, NATO Advanced Study Institutes Series (1977), p.246.

2. Cummins, H.Z., "Intensity Fluctuation Spectroscopy of Motile Organisms", ibid, p. 200.

3. Nossal, R., S.H. Chen and C.C. Lai, Optics Commun. 4, 35 (1971); Nossal, R., Biophys. J. 11, 341 (1971).

4. van de Hulst, H.C., Light Scattering By Small Particles, Wiley, New York (1957).

5. Mie, G., Ann Physik. 25, 377 (1908).

6. Debye, P., Ann. Physik. 30, 57 (1909).

7. Glantsching, W.J. and S.H. Chen, "Light Scattering From Water Droplets in the Geometrical Optics Approximation", to appear in Appl. Optics (1981).

8. Dane, J.V., "Subroutines for Computing the Parameters of the Electromagnetic Radiation Scattered by a Sphere", IBM Data Processing Division, Palo Alto Scientific Center (1968).

9. Kotlarchyk, M., S.H. Chen and S. Asano, Appl. Opt. 18, 2470 (1979).

10. Chen, S.H., M. Holz and P. Tartaglia, Appl. Opt. 16 187 (1977).

11. Holz, M. and S.H. Chen, Appl. Opt. 17, 1930 (1978).

12. Holz, M. and S.H. Chen, Appl. Opt. 17, 3197 (1978).

13. Craig, T. and F.R. Hallett and B. Nickel, Biophy. J. 28 457 (1979).

14. Chen, S.H. and Hallett, F.R.. See a review article on "Determination of Motility Behavior of Prokaryotic and Eukaryotic Cells by Dynamic Light Scattering" to appear in Quarterly Reviews of Biophysics.

MOTILITY STUDIES OF LARGE CELLS

F. Ross Hallett

Department of Physics
University of Guelph
Guelph, Ontario, Canada

CONTENTS

1. INTRODUCTION

Most of us in the light scattering discipline are familiar with the general characteristics of autocorrelation functions obtained from motile cells. First of all the decay times of these functions are 10 to 100 times shorter than would be expected from diffusing particles of the same size. Secondly the functions, especially those determined at low scattering angles, exhibit a shoulder in the first few channels before they tail off to background. Beyond this, however, there is a tremendous variation in decay times and shapes of correlation functions from one motile system to the next. In the case of bull spermatozoa one often sees significant changes from one sample to the next. Figure 1 shows the electric field autocorrelation functions (dots) from four different bull semen samples. Over the years we have learned that the source of these variations is the presence of differing relative populations of three classes of spermatozoa within the samples. The first class is most common, namely the normal motile spermatozoa. These cells are beautiful to watch by slow motion cinematography. As they swim they follow an extremely regular helical path the axis of which can persist in a straight line for several millimeters, a distance which is roughly forty or fifty times the length of the cell. The whole spermatozoa including the highly scattering head region rotates 360° in every turn of the helix.[1] The second

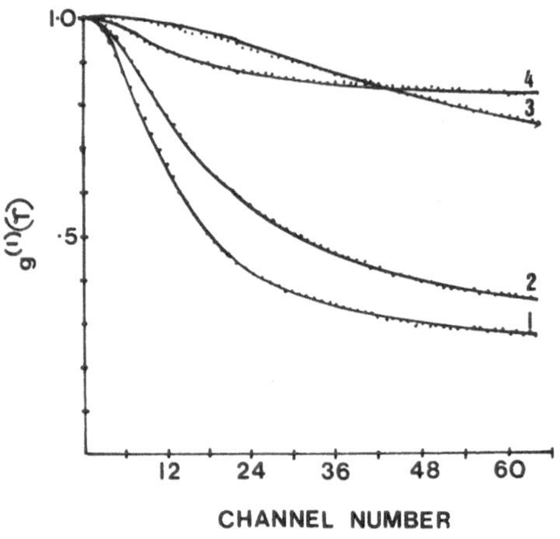

Figure 1. Experimental electric field autocorrelation functions from four different semen samples (dots) and fits by appropriate sums of the empirical functions (Equations 1 - 3). The channel width is 120 μs.

class of spermatozoa have been termed defective swimmers or circular swimmers. In this motion, the cell does not rotate. Instead it oscillates in an approximately sinusoidal way along a track which is nearly circular. It is known that pH changes can induce defective swimming, but other factors such as thermal shocks or cellular aging may also be involved. Significant populations of defective cells have been present in approximately 30% of the samples we have studied. The third population is non-motile, consisting mostly of dead spermatozoa, but some fat globules and other cells as well. All samples contain a significant fraction of this class. The dead spermatozoa settle vertically through the solution, rotating slowly as they move.

Each type of motion described above leads to a characteristic electric field autocorrelation function at low scattering angle ($\Theta = 15°$). Figure 2 illustrates two simple empirical functions for normal and defective cells plus a third more complex function, also empirical for the non-motile component. These empirical functions;

$$f_{Normal} = \frac{1}{1 + (x\tau)^2} \quad (1)$$

$$f_{Defective} = \frac{1}{x\tau} \cot^{-1}(x\tau) \quad (2)$$

$$f_{Dead} = \ell + m\tau + n\tau^2 \quad (3)$$

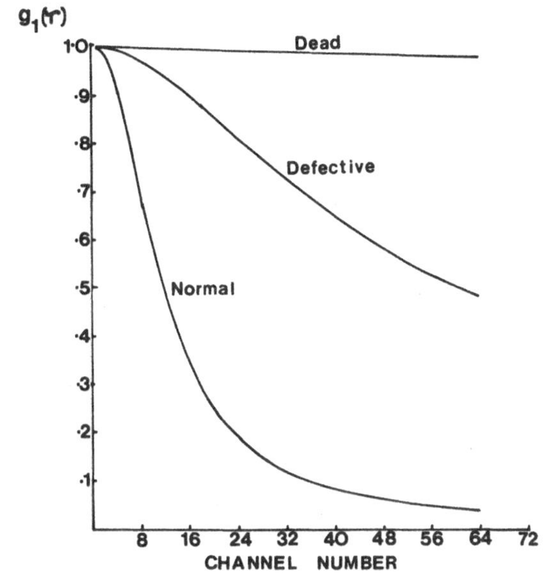

Figure 2. The empirical functions of equations 1 - 3 are shown with half width times typical to those observed experimentally from each cell population. The half width time of the dead function was fixed experimentally from a study of totally dead samples. It was not an adjustable parameter in least squares fits to data from samples containing motile cells.

can be fit extremely well to experimental functions from each population. As well, all of the approximately 150 semen samples which we have examined in the last few years yielded functions which could be fit by appropriate sums of these three empirical functions. Examples of such fits are shown in Figure 1 (solid lines).

Two important questions arose from these observations. First, is there any theoretical justification for the choice of these empirical functions and secondly, what information can be extracted from the data ? These questions have consumed most of our time in recent months. At the present time we feel that we have a fairly sound understanding of the processes involved in the scattering from normal cells and dead cells. Work on defective cells is not yet complete. The following sections will principally deal with our investigation of normal bull spermatozoa cells. Some of the conclusions, however, will be substantiated by data from dead cells and from another swimming system, *Chlamydomonas reinhardtii*.

2. MODEL CALCULATIONS

The electric field autocorrelation function $g^{(1)}(\tau)$ from a motile system is given by

$$g^{(1)}(\tau) = N < e^{i\vec{k}\cdot[\vec{r}(\tau)-\vec{r}(0)]} e^{i\vec{k}\cdot[\vec{R}(\tau)-\vec{R}(0)]} A(\vec{k},\tau)A^{*}(\vec{R},0)> \quad (4)$$

where $\vec{r}(\tau)$ and $\vec{R}(\tau)$ are the instantaneous position vectors of the centre of mass of the scatterer parallel to and perpendicular to the average direction of motion (see Figure 3). The form factor $A(\vec{k},\tau)$ is the instantaneous scattering amplitude of the cell, determined by its shape and orientation with respect to the scattering vector, $\vec{k}$. The geometry of this situation is demonstrated in Figure 4. Essentially all motile cells exhibit rhythmic movements such as the beat of a flagellum as they progress through the solution. The instantaneous position of a cell on its track can, therefore, be characterized in part by a phase angle, ψ. If the swimming motion is isotropic then one must average over all directions and starting angles ψ_o. Thus Equation (4) becomes [2,3]

$$g^{(1)}(\tau) = C_N^{-1} < \frac{1}{4\pi} \int_o^{2\pi} d\psi_o \int_{-1}^{1} d\nu \; e^{ik\nu v\tau} e^{i\vec{k}\cdot[\vec{R}(\tau)-\vec{R}(0)]} A(\vec{k},\tau)A^{*}(\vec{k},0)> \quad (5)$$

where $\nu = \cos\Theta$.

The displacement along the average direction of motion has been replaced by $v\tau$ where v is the speed of the scatterer. This assumption is valid if the scatterer travels in a straight line for times of the order of $(kv)^{-1}$, a situation which holds for most motile cells.

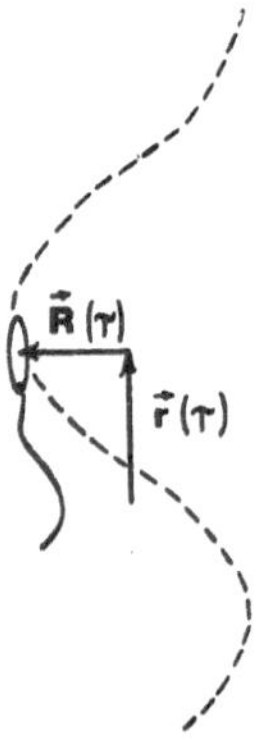

Figure 3. Orientations of $\vec{r}(t)$ and $\vec{R}(t)$ with respect to the average direction of motion.

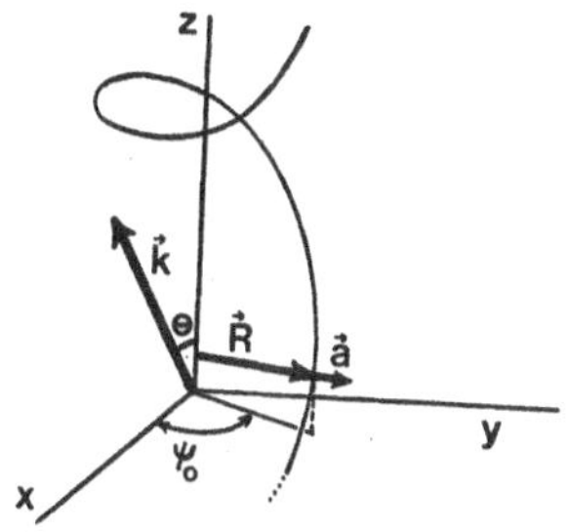

Figure 4. The scattering geometry, showing the average direction of motion, z, the scattering vector, $\vec{k}$, the position vector, $\vec{R}$, and the semi-axis, $\vec{a}$.

In order to proceed further one must obtain information on structural properties of the cell such as shape, size, and refractive index as well as dynamic information on the swimming trajectory. Therefore, fairly comprehensive photomicroscopic and cinematographic studies must be performed. In the case of bull spermatozoa most of the cellular mass is located in the head region which is disk-shaped with dimensions 1.0 μm x 4.6 μm x 9.0 μm.[4] Most of the remaining mass is located in the midpiece which is a cylinder 1.0 μm in diameter and 15 μm long. The long thin flagellum contributes negligibly to the total scattering power of the cell. We have modelled the spermatozoal cell as an ellipsoid of semi-axis a, b, c as shown in Figure 5. Using the Rayleigh-Gans-Debye formulation the corresponding form factor becomes[5]

$$A(\vec{k},\tau) = \frac{3j_1(\kappa)}{\kappa} \tag{6}$$

$$\text{where } \kappa^2 = (\vec{k}\cdot\vec{a})^2 + (\vec{k}\cdot\vec{b})^2 + (\vec{k}\cdot\vec{c})^2 \tag{7}$$

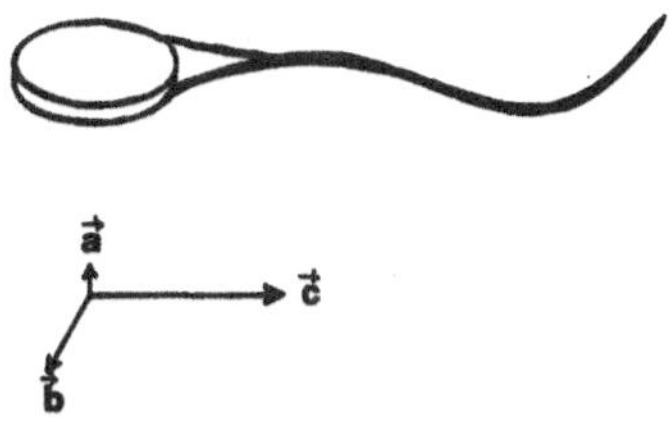

Figure 5. A schematic drawing of a bull spermatozoon showing the relative orientations of the $\vec{a}$, $\vec{b}$, $\vec{c}$ semi-axes.

The helical motion of normal cells can be represented if the centroid of this ellipse travels on a helical path of radius $\vec{R}$. Since $\psi(t)$ represents the instantaneous angle between $\vec{R}$ and the xz plane then

$$\psi(t) = \omega t + \psi_o \tag{8}$$

introduces the appropriate time dependence.

The long axis of the head is not parallel to the average direction of motion. Instead it is tipped so that the c axis projects along the helical track. As well the head is tipped "outwards" from the track leading to the so-called "duck footed" pattern seen by either cinematographic or time lapse photography. These tipping angles must be included in the dot products of Equations (5) and (7).

The final averaging of Equation (5) is over the distributions of v and ω. These distributions also are available from cinematographic analysis although several hours of tedious measurement on projected film are required. In the case of normal bull spermatozoa the distribution functions have the form

$$P_s(v) = \frac{4v}{\bar{v}^2} e^{-2v/\bar{v}} \tag{9}$$

$$P_s(\omega) = \frac{4\omega}{\bar{\omega}^2} e^{-2\omega/\bar{\omega}} \tag{10}$$

Their identical form results from the fact that the average progressive speed $\bar{v}$ is linearly proportional to $\bar{\omega}$. The final computational form of Equation (5) becomes

$$g^{(1)}(\tau) = C_N^{-1} R \int_o^1 d\nu \sum_n [1 + i(q_\nu + p_n)]^{-2} |B_n|^2 \tag{11}$$

where

$$q_\nu = \frac{1}{2} k\nu\bar{v}\tau \tag{12}$$

$$p_n = \frac{1}{2} n\bar{\omega}\tau \tag{13}$$

and the B_n are the Fourier coefficients of all of the periodic terms in the original integral. Usually 64 such coefficients were sufficient for the evaluation of Equation (11). The remaining integral in Equation (11) is not analytic. It was obtained by a Simpson's rule computation with 50 segments. Using these techniques a 64 point autocorrelation function could be calculated in about 5 minutes CPU time on an Amdahl V5 computer.

As we have seen the use of cinematographic techniques have been extremely important for providing many of the parameters required for the evaluation of Equation (11). One parameter which cinematography could not specify with any accuracy was the magnitude of the "c" semi-axis. This is because of the uncertainty of the extent to which the midpiece of the cell contributed to the total scattered intensity. Hence "c" was adjusted until the half-width time (HWT) of the calculated function matched that of the data (Figure 6). This matching occurred for a c-value of 9.2 μm, implying that the midpiece does play a significant role (a distance of 18 μm from the anterior end of the spermatozoon encompasses all of the head region and about two thirds of the midpiece). A comparison of this calculated function with the Lorentzian which fit the data (Figure 7) is shown in Figure 8. The close agreement indicates that Equation (11) behaves much like a Bessel transform of the distribution function at least at low values of k. This justifies then, our choice of a Lorentzian as a simple empirical function for fitting the scattering function from normal cells. At higher scattering angles both the calculated and the experimental scattering functions lose their Lorentzian shape.

It was interesting, once Equation (11) could be calculated routinely to systematically vary the magnitude of the structural and dynamic parameters and monitor the changes to the calculated functions. For this purpose we modelled hypothetical swimmers as disk-shaped ellipsoids of thickness 1.0 μm, but having diameters ranging from 1.0 to 15.0 μm. Figure 9 illustrates dependence of the half-width time (HWT) on the rotational frequency ($\bar{\nu}$) of these disks. In these calculations the helical track of the centre of

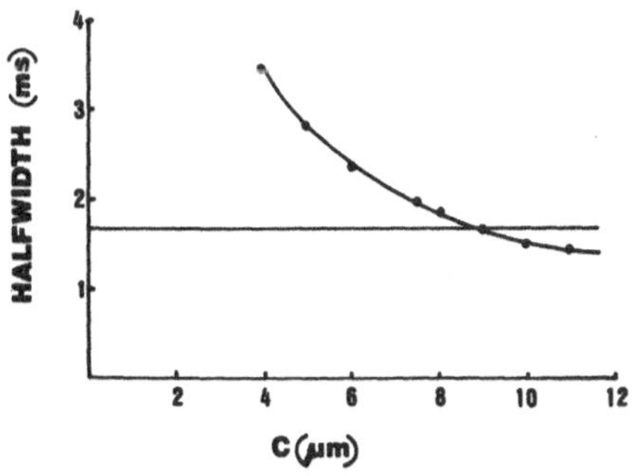

Figure 6. Dependence of the HWT on the magnitude of the "c" semi-axis, calculated using Equation (11).

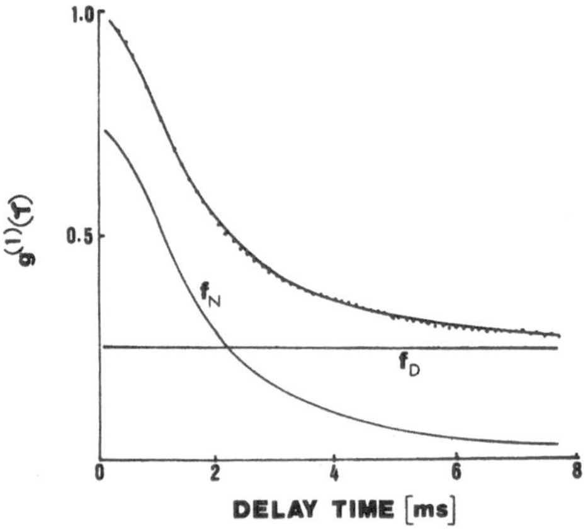

Figure 7. Least squares fit of f_{Normal} (a Lorentzian) and f_{Dead} to a normalized experimental electric field autocorrelation function from a sample containing normal swimmers and some dead cells, but no defectives.

mass was reduced to linear motion by setting R = 0. The centre of mass speed was held at 110 μm s^{-1} for all model disks. The HWT was clearly inversely proportional to the rotational frequency for the larger disk diameters. This implied that rotational effects could dominate translational effects in these systems. This implication was verified by the calculations illustrated in Figure 10. In this study the same range of disk sizes was used, but in each case $\bar{\nu}$ and $\bar{v}$ were altered so as to obtain a fixed HWT of 1.44 ms. When the particle is spherical rotational effects are unimportant, and the width is determined solely by the translational speed. At the high axial ratios the rotational frequency becomes the dominant parameter and translational speeds become unimportant. We have found that the a, b, c ellipsoids which we used to model spermatozoa

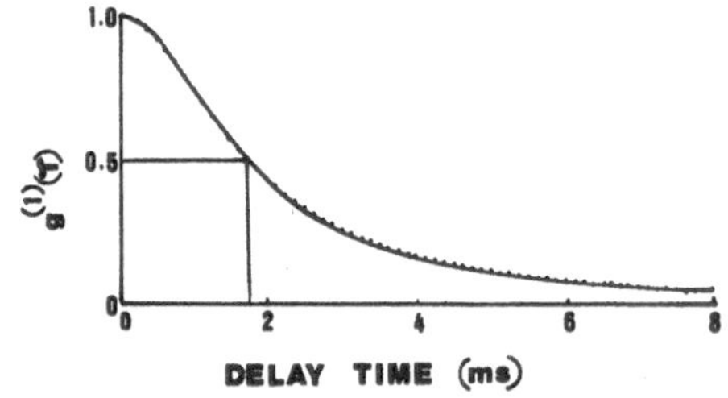

Figure 8. A comparison of the best fit Lorentzian in Figure 7 (line) to a function calculated using Equation (11) with all parameters except c determined from cinematographic or microscopic measurements. The value of the parameter c was 9.2 μm from Figure 6.

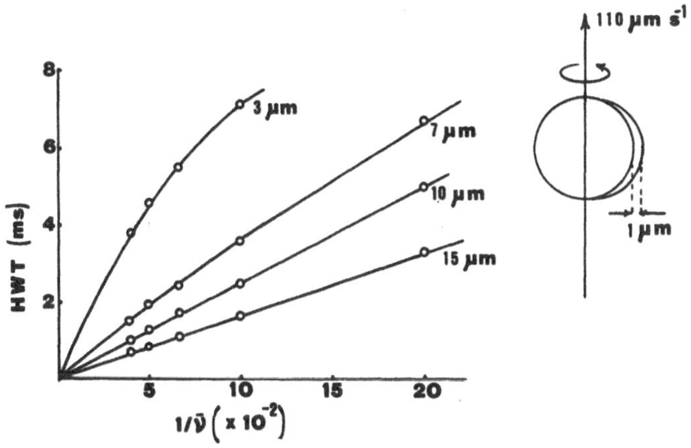

Figure 9. Model calculations using Equation (11), for translating and rotating disks having a thickness of 1.0 µm, R = 0, a speed of 110 µm s^{-1} and a range of diameters as shown.

behave in this same way. Thus both the off axis displacement R and the translational speed distribution in Equation (11) are unimportant. Indeed, setting R = 0 and eliminating the velocity averaging yields functions identical to that shown in Figure 8. On the basis of these calculations one can conclude that the importance of rotational motion is a strong function of the cellular shape. Rotational motion can be neglected only in cases where the cell shape is spherical. Good examples of this special case are the Chlamydomonas reinhardtii reported by Racey and Hallett in these proceedings and in a study of echinoderm spermatozoa.[6]

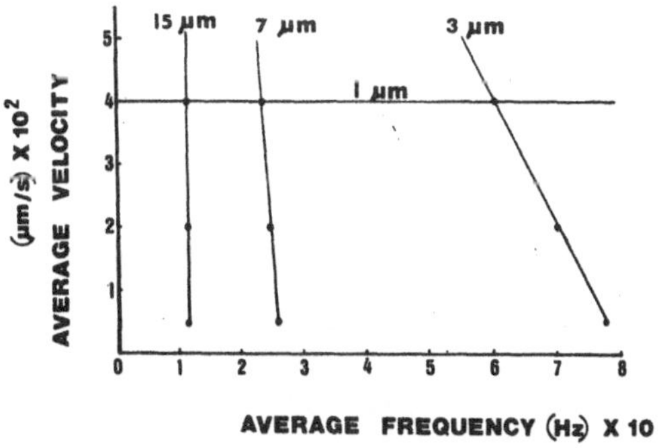

Figure 10. A figure showing the combinations of average speed and average rotational frequency required to produce functions of identical HWT of 1.44 ms for a range of disk sizes as in Figure 9. The spherical case is 1.0 µm.

Because samples of motile spermatozoa usually contain significant fractions of dead and defective cells in addition to the normal cells, experimental studies of k-scaling have been extremely difficult to perform. The only case in which the sample chamber contains a uniform population is when all the cells are dead. For this reason our first scaling experiments were performed on dead samples. As mentioned earlier dead spermatozoa settle vertically with a slow rotation much as a falling leaf. With some modification to the dynamic terms Equation (11) can be used to model calculated autocorrelation functions from this motion. The structural parameters in $A(\vec{k},\tau)$ were maintained at the identical values used in calculations for normal cells.

Figure 11 illustrates the scaling curves predicted by the modified Equation (11) and obtained experimentally. There is striking agreement in the general shape of these curves and in the positions of the peaks, suggesting that the Rayleigh-Gans-Debye formulation which we have used is a suitable one for this system. The origin of the peaks is primarily due to diffractive effects occurring in light scattered from the front and back surfaces of the cells. The positions of the peaks in model scaling curves could be changed by varying the magnitude of the half-thickness of the cell, the semi-axis "a". Changes in the magnitudes of dynamic

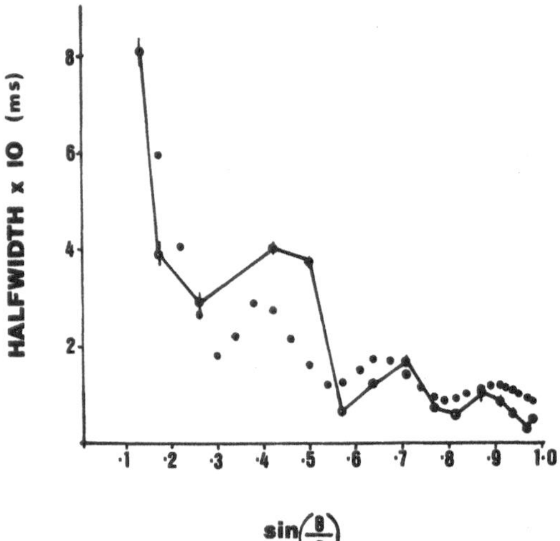

Figure 11. A figure showing the scattering angle dependence of the HWT of experimental functions and functions calculated by Equation (11) modified to describe the motion of dead cells.

parameters in "b" or "c" minimally affected the peak positions. Further evidence for the diffractive nature of the peaks is contained in the calculated zero delay function, $S(\vec{k})$, shown in Figure 12. Comparison of Figures 11 and 12 indicates that peaks occur at similar scattering angles in each curve. The knowledge of the scaling curves for dead cells greatly assisted our attempts to determine the scaling curves for normal and defective cells. Our progress in the normal case is discussed elsewhere in these proceedings.[7]

3. CONCLUSION

We have demonstrated that electric field autocorrelation functions from both dead and motile spermatozoa can be calculated from realistic models of the cell's structure and dynamics. The best source of structural and dynamic information for the construction of these models is cinematography. The various model calculations which were performed indicated that the rotation of the normally swimming spermatozoon was the dominant factor in determining the half-width time of the scattering functions. At low angle both the theory and the data from normal swimmers can be well represented by a Lorentzian whose half width is inversely proportional to the average rotational frequency of the normal cells such as

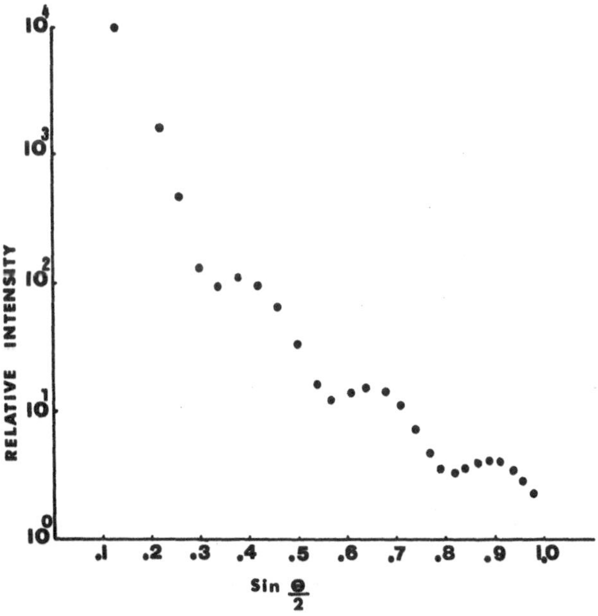

Figure 12. The "zero delay" function, $S(\vec{k})$, calculated for dead cells.

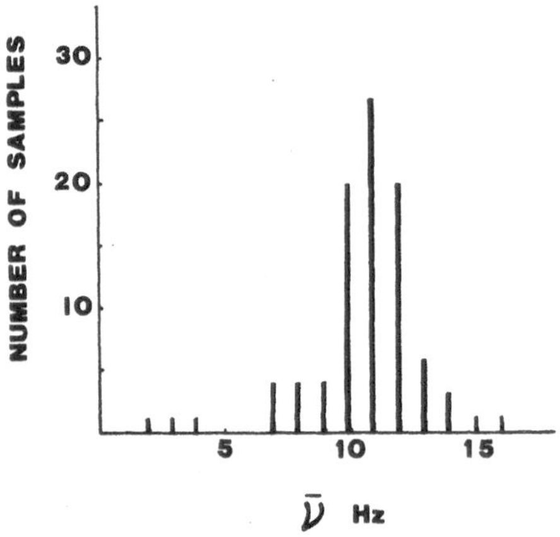

Figure 13. Distribution of $\bar{\nu}$, average frequency determined by least squares fits of sums of Equations (2), (3) and (14) to experimental functions from approximately 90 different samples.

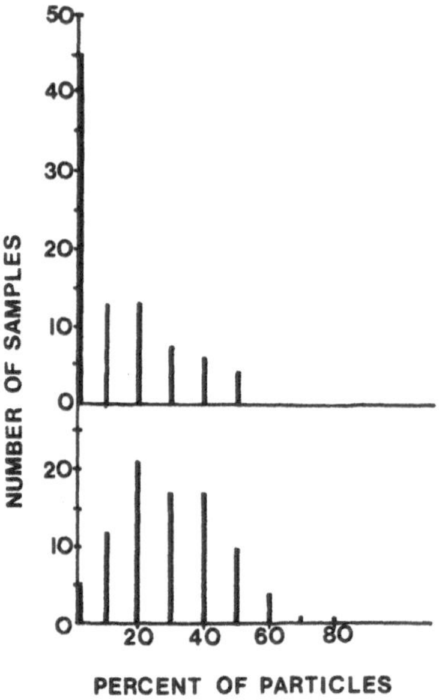

Figure 14. Distributions of fractions of normal cells (lower graph) and defective cells (upper graph) determined by least squares fits of sums of Equations (2), (3) and (14) to experimental functions from approximately 90 different samples.

$$f_N = \frac{1}{1 + (2\pi C k \bar{\nu} \tau)^2} \tag{14}$$

where C ≃ 2.75 μm is the effective "length" of the rotating cells. If this equation is used in place of Equation (1) in least squares fits to experimental data then one can extract the fraction and the mean rotational frequency of the normal cells in each sample. Figures 13 and 14 show the results of many such analyses.

ACKNOWLEDGEMENTS

I am indebted to my graduate students Mr. T. Craig and Mr. T. Racey whose work and enthusiasm were central to this research. I am also indebted to Dr. B. Nickel for his excellent theoretical guidance.

This research was supported by a grant from the Natural Science and Engineering Council of Canada.

REFERENCES

1. R. Rikmenspoel, G. van Herpen, and J. Eijkhout, Cinematographic observations of the movement of bull sperm cells, Phys. Med. Biol. 5:167-181 (1960).
2. M. Holz and S.-H. Chen, Rotational-translational models for interpretation of quasi-electric light scattering spectra of motile bacteria, Appl. Opt. 17:3197-3204 (1978).
3. T. Craig, F. R. Hallett, and B. Nickel, Quasi-elastic light-scattering spectra of swimming spermatozoa: rotational and translational effects, Biophys. J. 28:457-472 (1979).
4. C. van Duijn, Jr. and C. van Voorst, Precision measurements of dimensions, refractive index, and mass of bull spermatozoa in the living state, Mikroscopie 27:142-167 (1971).
5. Craig, op. cit.
6. B. Herpigny and J.-P. Boon, Photon correlation study of spermatozoa motility, J. de Physique 40:1085-1088 (1979).
7. Craig, op. cit.

DYNAMIC LASER SCATTERING STUDY OF SPERM MIGRATION THROUGH CERVICAL MUCUS+

Wylie I. Lee

Center for Bioengineering and
Department of Biological Structure
University of Washington
Seattle, Washington 98195

CONTENTS

ABSTRACT

Migration of spermatozoa through the cervix is of great interest in human reproductive biology, and is important in fertility studies. The penetration of spermatozoa into cervical mucus depends on sperm motility and also on the molecular structure of cervical mucus. We have used the technique of dynamic laser scattering to measure sperm motility and to study the structure of cervical mucus by analyzing the molecular dynamics of the mucus. The correlation of the two is evaluated by a modified Kremer test. The clinical significance of the test is quantitatively expressed by the penetration factor and the sperm-mucus interaction coefficient, which are derived from theory of transport phenomena.

+Supported by NIH Grant HD-12629.

1. INTRODUCTION

In a number of mammalian species, including man, sperm must migrate through the uterine cervix before reaching the site of fertilization within the ampullae of the oviducts. However, the penetration of spermatozoa through the cervix depends not only on the motility of the spermatozoa but also on the physical and chemical properties of mucus produced within the cervix. Measurements of human sperm penetration and survival in human cervical mucus are a valuable tool to evaluate infertility.[1] Sperm penetration into cervical mucus in vitro can be performed by using either a plane glass slide[2] or capillary tube system.[3] The capillary tube developed by Kremer provides an objective measurement and has gained broad clinical acceptance in infertility evaluation.[4] This "capillary tube sperm penetration meter" consists of a capillary tube filled with cervical mucus and a semen container. One end of the tube is sealed with wax and the open end is immersed in semen. The distance traveled by "vanguard" spermatozoa during the pre-set time interval can be observed and the numbers of spermatozoa at locations along the tube can be estimated. Although this method has been modified and used extensively in clinics,[5-7] there is a lack of quantitative analysis of the sperm penetration test that correlates the quality of the sperm specimen and the properties of the cervical mucus.

Recently, dynamic laser light-scattering has been used to evaluate the viscoelastic properties of cervical mucus[8] and to measure the motility of spermatozoa.[9,10] Because the method is non-invasive and the measurements are usually rapid, and also require only a small amount of sample, laser light-scattering provides a new approach for evaluating the in vitro sperm penetration test. In these lectures, we discuss a study of the effects of freezing on bovine cervical mucus in which laser light-scattering has been used. Because estrous bovine cervical mucus is similar in its supra-molecular arrangement to human midcycle mucus and allows rapid penetration by human spermatozoa,[11] it could be developed as a possible substitute for human cervical mucus in clinical treatment[12] or tests. Since after collection mucus specimens are frozen and kept for future use, it is critical to evaluate the effects of freezing on bovine mucus and the penetrability of human spermatozoa into it.

2. BACKGROUND

Although there is no complete description of normal gamete transport mechanisms for any animal, the basic mechanisms of gamete transport are remarkably similar in all mammals so far studied. In primates, ovulated eggs are transported from the fimbria through the ostium to the ampulla, and then to the ampullaristhmic junction (AIJ) (Figure 1). This phase of egg transport requires approximately

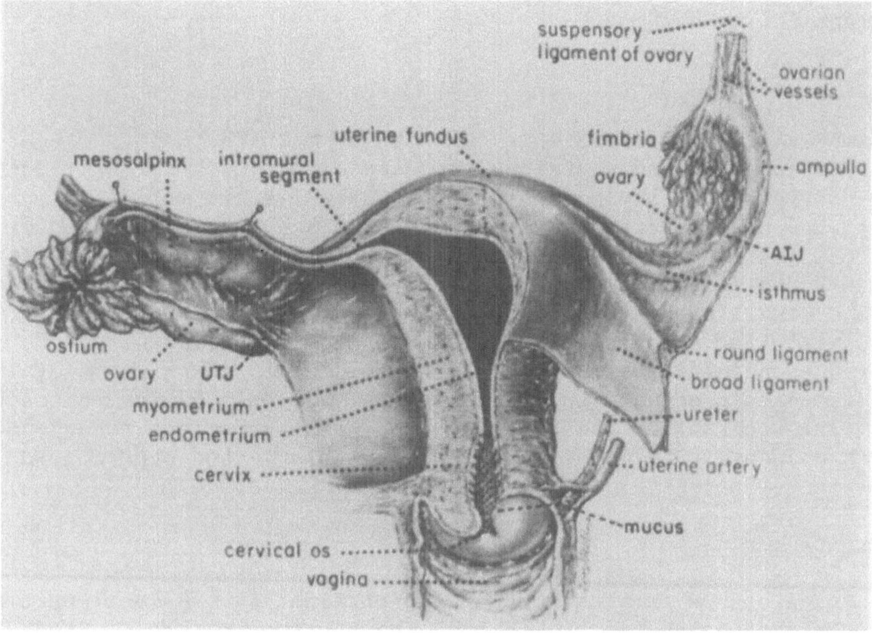

Figure 1. Drawing of the human female reproductive tract (the original figure was provided by Dr. Richard J. Blandau).

30 minutes and is accomplished by the action of both ampullary muscle and cilia covering the fimbrial surface and lining the ampullary lumen.[13,14] The egg is retained at the AIJ for a period of 15-25 hours and is then transported through the isthmus, the uterotubal junction (UTJ) and into the uterus. This last phase of egg transport requires approximately 40 hours.

Spermatozoa are deposited on the cervix and in posterior vaginal fornix. In approximately five minutes they start to migrate through cervical mucus with their own motility and even pass through the cervical canal and enter the uterus. By a complicated combination of sperm motility and myometrial contraction, spermatozoa are transported from the uterus through the UTJ to the isthmus, where they must ascend to the site of fertilization at the AIJ (see Reference 15 for an extensive review.)

Recently, laser light-scattering techniques have been employed to evaluate important factors involved in mechanisms of gamete transport such as the ciliary activity of the epithelium of fimbria,[16] the physical properties of cervical mucus,[8] and sperm motility.[9,10] It is our hope that this new technique will provide greater knowledge of the factors regulating gamete transport.

3. MATERIALS AND METHODS

A. Spermatozoa

Samples of human semen were obtained by masturbation from healthy male donors who had abstained from sexual activity for at least 3 days previously. Only samples showing normal sperm motility, morphology, and concentration were used. The ejaculate was diluted with one-half volume of Krebs-Ringer-phosphate solution to which fructose had been added at a final concentration of 200mg/100ml. The diluted semen was centrifuged at 40g for 10 min, and then at 140g for a further 10 min. The supernatant was removed and fresh diluent added. After resuspending the spermatozoa, the sample was centrifuged again by the two-step procedure outlinedabove but for only 5 mins. at each speed. The supernatant was discarded and the spermatozoa resuspended in fresh diluent. Sperm samples washed by this method showed good motility and viability.

B. Cervical Mucus

Samples of cow cervical mucus were obtained from a local cattle breeding and research center (through the courtesy of Dr. E. Studer of Carnation Research Farm, Carnation, Washington). The specimens were collected by a skilled veterinarian from cows in the estrous phase of the reproductive cycle. Care was taken to collect the mucus as cleanly as possible; for this purpose a glass tube was inserted directly into the cervical canal. The mucus was transported to the laboratory within 2-3 hours of collection. Each sample of cervical mucus obtained from different animals was divided into five approximately equal aliquots. Each aliquot was about 2ml in volume and was drawn up into a 5ml disposable syringe. One aliquot from each mucus sample was used to examine sperm penetration on the day of collection (prior to freezing) to serve as a control. The remaining aliquots were placed in the freezing compartment of a refrigerator. The temperature of this compartment was kept at -12°C. At specific intervals after the day of collection, an aliquot from each mucus specimen was removed from the freezer, allowed to return to room temperature and tested with washed human spermatozoa from a fresh semen sample.

C. Capillary Tube System

Capillary tubes for sperm penetration tests were set up in the same manner as reported earlier.[11] Briefly, estrous bovine cervical mucus was drawn up into flat capillary tubes approximately 4.5 x 1mm in cross-sectional area and 50mm in length (Vitro Dynamics, Inc., Rockaway, NJ, Model 2540). The path length of tube is 400μl. When a column of mucus approximately 40mm long had been drawn up into a tube, the "trailing mucus" emerging from the lower end of the tube was cut with scissors. The tube was then positioned vertically with its lower end immersed in a suspension of washed human spermatozoa, about 500μl in volume, in a small beaker. The upper end was sealed with Vaseline or Critoseal. All measurements were performed at room temperature. The observations were performed under a binocular dissecting microscope at timed intervals.

D. Dynamic Laser Scattering Measurements

In an earlier lecture in this NATO ASI, S. H. Chen presented a review of light scattering from macroscopic particles and cells. The chapter based on his presentation can be referred to for a general theoretical background. The spectrometer has been described elsewhere,[8] except that a real-time spectrum analyzer (Princeton Applied Research, Model 4512) was here used in addition to the previously existing signal correlator. The scattering angle employed in the sperm motility study was 7°. At this small scattering angle, the spectrum of light scattered from a suspension of motile spermatozoa can be interpreted in a form of exponential[9,10]

$$S(\omega) = A \exp(-\omega/c), \tag{1}$$

where c = aK, K is the magnitude of scattering vector and a is the most probable swimming speed in the Poisson-like swimming speed distribution function for spermatozoa

$$P_s(v) = (1/a^2)v \exp(-v/a). \tag{2}$$

The most probable swimming speed a was calculated from the slope of log $S(\omega)$ and is used as the index of motility for the sperm sample. Note that the average swimming speed of the sperm sample is $2a$.

Light scattering from estrous cow and human mid-cycle cervical mucus is similar to the scattering from a solution of entangled macromolecules.[8] The autocorrelation function of light scattered from these mucus specimens can be best fitted to the sum of two exponentials rather than to a single decaying exponential. Furthermore, the inverse decay times do not have a linear relation to the square of scattering vector as they do in the systems of dilute solutions of microspheres or the polymeric gels. The faster decay time, which is typically 10ms for estrous mucus when the measurement is made at 30° of scattering angle, is a function of the stage in the estrous or menstrual cycle. It is slowest during estrus, or at midcycle in the human, and becomes faster sharply when the luteal phase begins. These changes mimic variations in the concentration of glycoprotein in cervical mucus during the estrous cycle.[8]

After observations of sperm penetration have been completed, the capillary tube is mounted on a three-directional translation stage so that the entire tube can be scanned by the laser beam. Light scattered by spermatozoa is much stronger than that by glycoproteins in the mucus, so that the motility of spermatozoa in the mucus can be measured without the interference of light scattered by cervical mucus.

E. Sperm Transport in a Mucus-Filled Capillary Tube

Whether it is a sample of liquified semen or a washed specimen, the concentration of spermatozoa in a typical sperm sample is about 70 millions per ml. The motion of spermatozoa is analogous to the classical motion of gas molecules enclosed in a box, where each molecule moves in a straight line until a collision with another molecule occurs. Thus the transport phenomena of spermatozoa can be approximated to the transport phenomena of gases.[16] This formulation was used to measure the swimming speed of *chlamydomonas*[17] and also for the sperm penetration test.[7] We present here a modification of the analysis of Katz et al.[7] Our derivation enables us to correlate sperm motility and the penetrability of mucus in a closed form. When the unsealed end of the mucus-filled capillary tube was immersed in the suspension of washed spermatozoa, the net flux of spermatozoa along the tube (direction z, as shown in Figure 2) is

$$J_z(t) = N(t) \langle u_z \rangle / V_s, \tag{3}$$

where N(t) is the total number of spermatozoa in the suspension of volume V_s. $\langle u_z \rangle$ is the mean speed of spermatozoa that cross the cross-sectional area A of the tube, and it is related with the average swimming speed $\langle v \rangle$ such that

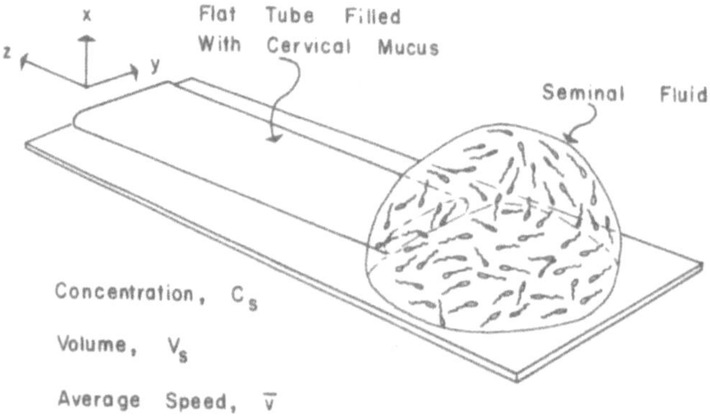

Figure 2. *In vitro* sperm penetration test with a flat capillary tube.

$$\langle u_z \rangle = \langle v \rangle/4 \equiv a/2 \quad (4)$$

since $\langle v \rangle = 2a$.

For a fixed cross-sectional area A and a fixed volume of sperm suspension, the number of spermatozoa that remain in the suspension can be derived from

$$\frac{dN(t)}{dt} = -\frac{N(t)a}{2V_s} A$$

and

$$N(t) = N(0) \exp\left[-(aA/2V_s)t\right]. \quad (5)$$

The number of spermatozoa that penetrate into the tube is

$$n(t) = \Pi[N(0) - N(t)] = \Pi N(0)[1 - \exp(-\beta t)], \quad (6)$$

where β is used to denote $aA/2V_s$ and Π is a factor signifying the per cent of sperm transport into cervical mucus. Π has value ranging from 0 and 1.

Sperm concentration inside the capillary tube can be approximated as[18]

$$C_m(z) = C_m(0,t) \exp[-z/L_m(t)]. \quad (7)$$

Thus, the total number of spermatozoa in the tube is

$$n(t) = A\int_0^\infty C_m(z)dz = A\, C_m(0,t)\, L_m(t). \quad (8)$$

$L_m(t)$ is the characteristic depth of penetration. L_m can be evaluated if we assume that the speed of spermatozoa inside the tube takes the following form:

$$dz/dt = \langle u_z \rangle \exp(-t/\tau_m). \quad (9)$$

τ_m is called the sperm penetration time and is a measurement of sperm-mucus interaction. For example, if mucus contains spermaglutinnis then sperm will soon cease their progressive movement and change to a stationary, shaking motion.[19] From the relation in Equation (9), we have

$$L_m(t) = \int_0^t dz/dt'(dt') = \langle u_z \rangle \int_0^t \exp(-t'/\tau_m)dt'$$

or

$$L_m(t) = a/2\ \tau_m\ [1- \exp(-t/\tau_m)]. \quad (10)$$

From Equations (6), (8), and (10), we can derive a transport equation for sperm penetration into the mucus-filled tube:

$$\Pi[1 - \exp(-\beta t)] = \beta\tau_m[1 - \exp(-t/\tau_m)]\ \exp(-\beta t), \quad (11)$$

which can be arranged as the following

$$\frac{\Pi\alpha}{1 - \exp(-\alpha t)} = \frac{\beta\ \exp(-\beta t)}{1 - \exp(-\beta t)}, \quad \text{for } t > 0 \quad (12)$$

Again, Π = sperm transport factor, α = inverse of sperm penetration time τ_m^{-1}, and $\beta = aA/2V_s$, sperm transport rate. Once the characteristic swimming speed, a, is measured by dynamic laser scattering, β can be calculated with the known values of cross-sectional area of the tube A and the total volume of sperm suspension V_s. α can be calculated from observations using a microscope, taking half of the distance traveled by vanguard spermatozoa at the end of 10 minutes as L_m.

4. RESULTS AND DISCUSSION

The results of our preliminary tests are shown in Table I. It is not clear that the sperm penetration test demonstrates any significant change in the rheological properties of frozen bovine estrous cervical mucus. The test is strongly dependent on the motility of spermatozoa used for the test. Therefore, unless the transport rates (β) are identical, the variation in the penetration distance (L_m) and the sperm penetration time (τ_m) may not indicate the differences in the rheological properties of the mucus. Samples from mucus 1 and mucus 3 show that sperm were penetrating more easily in 7-week frozen mucus than 1-week frozen mucus. However, the results obtained by laser light-scattering showed that there was no change in the rheological properties of the mucus when it was kept frozen at -12°C. But after six weeks of storage in the frozen state, the faster decay time, τ_f, of the measured autocorrelation function become four times slower than that of the control samples. This slow down in decay time may indicate a transition of gel state from the "many chain state" to the "single chain state,"[20] or to the sol state. Because the mucus lost its viscoelastic properties and became liquid, spermatozoa in this liquified mucus would behave as though they were penetrating into a solution of glycoprotein instead of penetrating into an entangled network of glycoprotein.

The sperm penetration factor (Π) shows a distinct correlation

Table 1. Freezing Effect on Bovine Estrous Cervical Mucus

Frozen Period (Weeks)	0 (control)			1			3			6			7		
Sperm Transport Rate, $\beta\ 10^6$, s^{-1}	10.90			12.88			2.02			2.76			12.02		
Mucus Sample	$\tau_m^{(a)}$	Π	$\tau_f^{(b)}$	τ_m	Π	τ_f	τ_m	Π	τ_f	τ_m	Π	τ_f	τ_m	Π	τ_f
1	781	0.69	14.2	621	0.64	-	236	0.88	14.0	135	0.81	40.0	709	0.67	43.5
2	735	0.68	12.0	431	0.52	-	236	0.88	12.0	116	0.78	27.6	221	0.34	44.8
3	833	0.71	10.0	909	0.73	-	185	0.85	15.2	212	0.87	35.0	970	0.74	42.1

(a) Sperm transport time, in seconds; (b) Faster decay time of the autocorrelation function obtained by laser scattering, at 30° in milliseconds.

with sperm motility and the sperm transport rate. Its value becomes closer to 1 for spermatozoa of lower motility. When penetration is difficult due to low sperm motility or immunological interaction between the mucus and spermatozoa, the tests result in a short penetration length, a short penetration time, and a high value for the penetration factor. It is important to point out that the penetration factor calculated from the controls has an average value of 0.69, while a similar assessment from sperm penetration tests using human midcycle cervical mucus yielded values that are less than 0.5.[7] It has been reported that human spermatozoa penetrate bovine estrous cervical mucus more readily than human midcycle cervical mucus,[11] and our present analysis seems to provide quantitative confirmation of that study.

The method of analysis presented here is readily applicable to laboratories which perform in vitro sperm penetration tests routinely. It is pertinent to the tests, and requires fewer restrictions than the previously published method,[7] such as the evaluation of sperm count in both semen and cervical mucus and an accurate measurement of the ratio of the velocity in semen and the velocity in mucus. Our formula can be applied to tests of various penetration lengths and times. Since the sperm transport process is not a linear process, there is no simple scaling factor for these tests. Consequently, a standardization of tube length and observation time, such as those we used in these studies, is recommended for routine tests.

Migration of spermatozoa through cervical mucus has previously been described in great detail by Moghissi[21] and the unidirectional progression of spermatozoa in stretched cervical mucus was reemphasized recently by Gaddum-Rosse et al.[11] Since cervical mucus was drawn up into capillary tubes in the present study, alignment of mucins may have occurred, thus providing linear paths for the spermatozoa. When mucus is drawn up into tubes rapidly, and the penetration test is performed soon after the preparation, the spectrum of scattered light from a troop of spermatozoa often shows a peak at 25Hz. From the relation of velocity and Doppler shift frequency, one can estimate that the forward speed of the troop is 20μ/s. It has been suggested that the light-scattering technique can be an objective velocimetry for studying sperm motion in cervical mucus;[22] however, precaution should be taken for the "shaking phenomena" due to anti-sperm antibodies.[23] Light scattering from these agglutinated spermatozoa also presents a peak in the spectrum at a frequency close to the beating frequency of flagella, which is about 10Hz. Most of the capillary tubes we prepared showed exponential spectra from scattering sperm in mucus, and the value of the characteristic swimming speed of spermatozoa remained the same along the tube except near entrance, where it was slightly higher (shown in Figure 3).

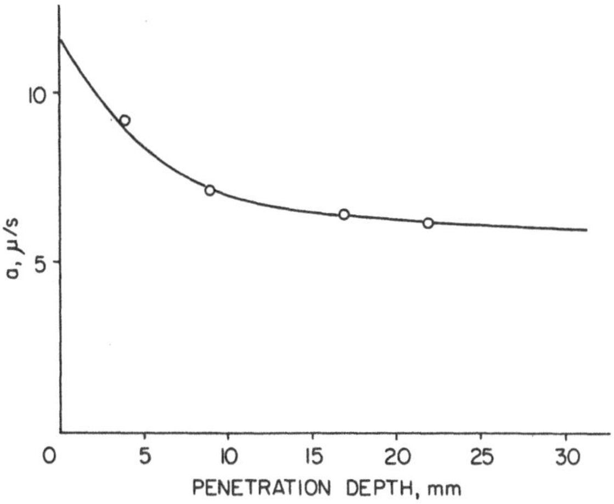

Figure 3. Characteristic swimming speed (a, in μ/s) of spermatozoa in bovine estrous cervical mucus, as a function of the distance of penetration.

In conclusion, the theoretical model presented in this paper provides a standard formula for quantitative analysis of in vitro sperm penetration into cervical mucus. The parameters are easily measurable by non-invasive and rapid laser light-scattering techniques. An accurate and comprehensive evaluation of sperm penetration is important in the diagnosis of infertility. It can also be used for routine semen analysis, using either normal donor cervical mucus or substitutes such as estrous cow cervical mucus or synthetic hydrogels. Other applications include evaluation of male infertility and response to therapy, evaluation of semen quality before artifical insemination, and monitoring of ovulation in fertility studies.

ACKNOWLEDGMENT

The author would like to acknowledge the assistance of Dr. P. Gaddum-Rosse and Mrs. Lynn Langley, Department of Biological Structure, in performing the in vitro penetration test. The figure of the female reproductive tract was kindly provided by Dr. Richard J. Blandau.

NOTE

In the Poster Session of this conference, the paper entitled "Human Cervical Mucus: Preliminary Data by Light Scattering Techques" presented by A. M. Cazabet, B. Volochine, J. M. Kuntsmann, and F. C. Chretien, concluded that the mucus is a hydrogel instead of an entangled network as we have proposed.[8] Their data in fact fit very well to our mathematical formula,[24] which predicts K^2-dependence of τ_{fast}^{-1} at large K region. Because both the weak bonds in a hydrogel and the entanglement in an entangled network of highly-branched macromolecules are not distinctly defined, it is difficult to make a conclusion based on the measurements presented by these investigators. In both models, there are two decay times and the correct relation of τ_{fast} with scattering vector K is $\tau_{fast}^{-1} \propto K^2$, at large K. Our previous claim[8] that mucus is not a cross-linked network of fibrillar micelles is still correct.

REFERENCES

1. J. M. Sims, "On the microscope, as an aid in the diagnosis and treatment of sterility", New York Med. J. 8: 393 (1869).
2. E. G. Mill, Jr., and R. Kurzok, "Biochemical studies of human semen. III. Factors affecting migration of sperm through the cervix," Amer. J. Obstet. Gynec. 24: 19 (1932).
3. J. Lamar, L. Shettles, and E. Delfs, "Cyclic penetrability of human cervical mucus to spermatozoa in vitro," Amer. J. Physiol. 129: 234 (1940).
4. J. Kremer, "A simple sperm penetration test," Int. J. Fertil. 10: 201 (1965).
5. M. Ulstein, "Evaluation of a capillary tube sperm penetration method for fertility investigations," Acta Obstet. Gynecol. Scand. 51: 287 (1972).
6. J. Reichman, V. Insler, and D. M. Serr, "A modified in vitro spermatozoal sperm penetration test. I. Method," Int. J. Fertil. 18: 232 (1973).
7. D. Katz, J. W. Overstreet, and F. W. Hanson, "A new quantitative test for sperm penetration into cervical mucus," Fert. and Steril. 33: 179 (1980).
8. W. I. Lee, P. Verdugo, R. J. Blandau and P. Gaddum-Rosse, "Molecular arrangement of cervical mucus: A re-evaluation based on laser light-scattering spectroscopy," Gynec. Invest. 8: 254 (1977).
9. P. Jouannet, B. Volochine, P. Degnent, C. Serres, and G. David, "Light-scattering determination of various characteristic parameters of spermatozoa motility in a series of human sperm," Andrologia 9: 36 (1977).
10. W. I. Lee, and R. J. Blandau, "Laser light-scattering study of the effect of progesterone on sperm motility," Fert. and Steril. 32: 320 (1979).
11. P. Gaddum-Rosse, R. J. Blandau, and W. I. Lee, "Sperm penetration into cervical mucus in vitro. II. Human spermatozoa in bovine mucus," Fert. and Steril. 33: 644 (1980).
12. R. J. Blandau, P. Gaddum-Rosse, and W. I. Lee, "Letter to the Editor on the possibility of using bovine cervical mucus in the human," Fert. and Steril. 29: 707 (1976).

13. R. J. Blandau, "Gamete transport-comparative aspects," in The Mammalian Oviduct, Hafez/Blandau, eds., The University of Chicago Press, Chicago, (1969). pp. 129-162.
14. P. Verdugo, W. I. Lee, S. A. Halbert, R. J. Blandau, and P. Y. Tam, "A stochastic model for oviductal egg transport," Biophys. J. 29: 257 (1980).
15. E. S. E. Hafez, "Transport and survival of spermatozoa in the female reproductive tract," in The Biology of Spermatozoa, Hafez/Thibault, eds., S. Karger AG, Basel, (1975). pp. 107-129.
16. F. Rief, "Statistical and Thermal Physics," McGraw-Hill, New York, (1965). Chapter 13, p. 494.
17. G. K. Ojakian, and D. F. Katz, "A simple technique for the measurement of swimming speed of chlamydomonas," Exptl. Cell Res. 81: 487 (1973).
18. E. Kesseru, "In vivo sperm penetration and in vitro sperm migration tests," Fert. and Steril. 24: 584 (1973).
19. J. Kremer, and S. Jager, "The sperm-cervical mucus contact test: A preliminary report," Fert. and Steril. 27: 335 (1976).
20. P. G. DeGennes, "Dynamics of entangled polymer solutions. I. The Rouse Model," Macromolecules 9: 587 (1976).
21. K. S. Moghissi, "Sperm migration through the human cervix," in The Uterine Cervix in Reproduction, Insler/Bettendorf, eds., Georg Thieme Publishers, Stuttgart, (1977). pp. 146-165.
22. M. Dubois, P. Jouannet, P. Berge, and G. Davis, "Spermatozoa motility in human cervical mucus," Nature 252: 711 (1974).
23. J. Kremer, S. Jager, and J. Kuiken, "The meaning of cervical mucus in couples with antisperm antibodies" in The Uterine Cervix in Reproduction, Insler/Bettendorf, eds., Georg Thieme Publishers, Stuttgart, (1977). pp. 181-186.
24. W. I. Lee, K. S. Schmitz, S. C. Lin, and J. M. Schurr, "Dynamic Light scattering studies of DNA. I. The coupling of of internal modes with anisotropic translational diffusion in congested solutions", Biopolymers 16: 583 (1977).

LASER-DOPPLER CONTINUOUS REAL-TIME MONITOR OF PULSATILE AND MEAN BLOOD FLOW IN TISSUE MICROCIRCULATION

R. F. Bonner and T. R. Clem

National Institutes of Health
Biomedical Engineering and Instrumentation Branch, DRS
Bethesda, Maryland 20205

and

P. D. Bowen and R. L. Bowman

National Institutes of Health
Laboratory of Technical Development, NHLBI
Bethesda, Maryland 20205

CONTENTS

ABSTRACT

A noninvasive instrument capable of instantaneous (∿ 30 msec resolution) and continous evaluation of local flow in tissue microcirculation has been developed based on laser Doppler velocimetry. Laser light is delivered to and detected from the region to be studied by flexible, graded-index fiber optic light guides. The Doppler-broadening of laser light scattered by moving red blood cells within the tissue is analyzed in real-time by an analogue processor which provides a continuous output of the instantaneous mean Doppler frequency in the photocurrent detected by a square-law detector. The mean Doppler frequency is predicted by theory to be linearly correlated with microcirculatory blood flow in a variety of tissues, and the Laser-Doppler signal has been correlated with measurements of average tissue blood flow by other techniques. The local spatial ensemble of moving red blood cells within $1mm^3$ of the microcirculation allows an instantaneous assessment of flow pulsations during the cardiac cycle as well as slower changes in average flow. The instrument has been designed to facilitate simple rapid clinical measurements of local spatial and temporal variations in microcirculatory flow. Measurement of local tissue blood flow in normal volunteers and patients has demonstrated the ease of accurately measuring basal and perturbed flow (reactive hyperemia, thermal hyperemia, position changes).

1. INTRODUCTION TO LDV BLOOD FLOW MEASUREMENTS

The ability to measure local blood flow noninvasively and continuously has many potential applications in physiology, pharmacology, and clinical medicine. Laser Doppler velocimetry was first applied to this problem by Riva, Ross, and Benedek,[1] who studied blood flow in 200 μm diameter tubes. Subsequently, measurements of flow in single human retinal vessels were made by Tanaka, Riva and Ben-Sira.[2] Although interpretation of data from concentrated blood cell suspensions in large vessels is complicated by absorption and multiple scattering, the technique can be calibrated with *in vitro* model systems and *in vivo* indicators. Stern,[3] first, observed changes in the power spectra of laser-light scattered from skin which were correlated with changes in microvascular blood flow. Later work demonstrated the feasibility of analogue processing of the detected photocurrent in order to provide a continuous laser Doppler flow value, verified by simultaneous measurement of ^{133}Xe washout in the forearm[4] and radioactive microsphere uptake in the kidney cortex.[5] Bonner et al[6] examined the scattering process in tissue in order to assess the information content of the power spectrum. In this paper we summarize investigations with a new laser Doppler tissue blood flow monitor. The optics and analogue processor of the instrument have been optimized for routine clinical measurement of tissue blood flow with high temporal resolution and linearity over the two orders of magnitude in flow observed from various tissues in normal and diseased states. Additionally, we summarize a theory developed to describe the complex Doppler process of photon diffusion in a microcirculatory bed (Bonner and Nossal).[7] This theory clarifies the information contained in the light scattering signal and assists in the interpretation of our analogue laser-Doppler blood flow signal.

2. THEORY OF DOPPLER SHIFTS OF LIGHT DIFFUSING THROUGH TISSUE

A theory of Doppler shifts associated with light diffusing through perfused tissue is necessary in order to evaluate the relaship between the detected Doppler shifts and the velocity and number

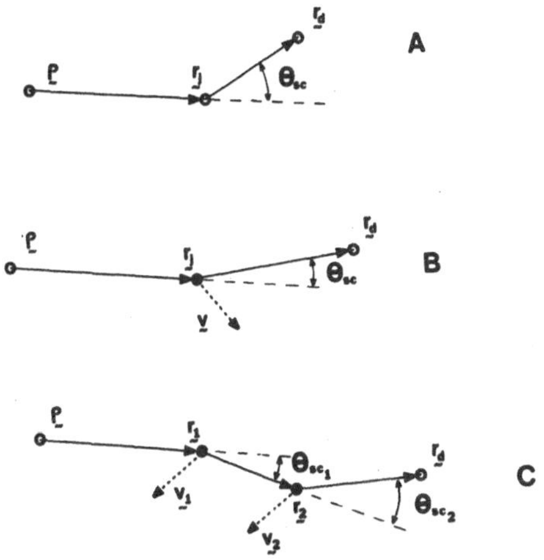

Figure 1. The diffusion of photons within the tissue can be represented as a series of scattering steps of types A, B, or C. Step A represents scattering from a static tissue element, which does not impart any Doppler shift. The distance between static scattering will depend on tissue structure, but typically is 100 μm. Step B represents small-angle Doppler scattering from a moving red blood cell. Step C represents sequential scattering from two moving red blood cells which occurs in larger vessels (diameters > 50 μm).

density of red blood cells within the tissue. Bonner et al. (1978) made a preliminary attempt to present a theoretical model and verify some of its predictions by experiments with a physical model approximating tissue with small vessel perfusion. A more general theoretical development is presented in the Bonner and Nossal paper. The theoretical treatment of laser Doppler scattering within perfused tissue is based on the following physical assumptions (Figure 1):

(1) Visible photons diffuse through the tissue being scattered many times per mm without Doppler shift by static index of refraction variations associated with tissue cells, organelles, etc.

(2) The detected photons which diffuse back to the surface of the tissue have been scattered by moving red blood cells (RBC) with a distribution of angles determined by the highly-anisotropic, angular scattering cross section of the RBC.

(3) Typically, a photon diffusing between two points on the surface of a well-perfused tissue separated by 1 mm will be scattered by one or more red blood cells, resulting in small Doppler shifts (0-20000 Hz).

(4) The mean number of red blood cells by which a single photon is scattered increases linearly with local blood volume within the tissue. Multiple scattering from moving red blood cells increases the Doppler shift.

Modifying the formulation of the swimming bacteria problem of Nossal et al. (1971) to the case of the microcirculatory bed in tissue, one can write the general intermediate scattering function $I_1(t)$ for a single scattering step (Figure 1b) in the tissue as

$$I_1(t) = \frac{\int_0^{\pi} S(Q(\theta)) \langle e^{i\underset{\sim}{Q}(\theta) \cdot \Delta\underset{\sim}{R}(t)} \rangle \sin\theta d\theta}{\int_0^{\pi} S(Q(\theta)) \sin\theta d\theta} \tag{1}$$

where S(Q) is the structure factor of an average scatterer (red blood cell) and $\Delta\underset{\sim}{R}(t)$ is the displacement of a moving cell during time t. As shown for the case of "straight swimming" bacteria with $\underset{\sim}{V}$ random relative to $\underset{\sim}{Q}$

$$\langle e^{i\underset{\sim}{Q} \cdot \Delta\underset{\sim}{R}(t)} \rangle = \int_0^{\infty} j_o(QVt)P(V)dV \tag{2}$$

where P(V) is the speed distribution of the moving cells. To apply these equations to the microcirculation one assumes that

(1) the red blood cells change shape slowly compared with the rate at which they traverse 1/Q along $\underset{\sim}{Q}$ (as demonstrated by microcinematography), and

(2) the orientation of $\underset{\sim}{Q}$ and $\underset{\sim}{V}$ are random, which is the result of both the random diffusion of the incident light by static tissue structures and the random network of microvessels.

For the case of the red blood cell of effective radius "a", we have approximated the structure factor by

$$S(Qa) = \left[\frac{3}{(Qa)^3} (\sin(Qa) - Qa \cdot \cos(Qa))\right]^2 \approx e^{-2\xi(Qa)^2} \tag{3}$$

where $\xi = 0.1$. Using Equations (2) and (3) in Equation (1) with the substitutions $Z = (1 - \cos\theta)/2$ and $\gamma = 2\xi(2ka)^2$, one obtains

$$I_1(t) = \frac{\int_0^1 e^{-\gamma Z^2} \int_0^{\infty} \frac{\sin(2kZVt)}{2kZVt} P(V)dV\, ZdZ}{\int_0^1 Ze^{-\gamma Z^2} dZ} \tag{4}$$

For large particles with $\gamma \gg 1$, Equation (4) can be expanded in a power series

$$I_1(t) = \sum (-1)^n \left(\frac{t}{(2\xi)^{\frac{1}{2}}a}\right)^{2n} \langle V^{2n} \rangle \frac{n!}{(2n+1)!} \,. \tag{5}$$

For any given speed distribution of red blood cells within the microvasculature, the intermediate scattering function of the single scattering events can be computed. The simplest form of $I_1(t)$ is obtained for the physiologically plausible Gaussian speed distribution

$$P(V) = \left(\frac{2}{\pi}\right)^{\frac{1}{2}} \left(\frac{3}{\langle V^2 \rangle}\right)^{3/2} V^2 \, e^{-3V^2/2\langle V^2 \rangle} \tag{6}$$

for which, given $T = t\langle V^2 \rangle^{\frac{1}{2}}/(12\xi)^{\frac{1}{2}}a$, one finds

$$I_1(t) = 1/(1 + T^2). \tag{7}$$

Uniform, single, and two speed distributions give similar monotonically decaying $I_1(t)$ which in practice would be difficult to distinguish from one another.

In tissues with very low blood volumes, this "single scattering" component occurs in the presence of a stronger unshifted "heterodyne" reference arising from photons which are only scattered by static tissue elements. On the other hand, for most tissues the typical detected photon will be scattered by one or more moving red blood cells. We treat the case of multiple scattering by assuming that the sequential Doppler shifts are uncorrelated and the resultant intermediate scattering function can be written

$$I(t) = \sum_{m=1}^{\infty} P_m (I_1(t))^m / (1 - P_o) \tag{8}$$

where P_m is the probability of a photon interacting with "m" red blood cells. We assume the many step diffusion of photons through 1 mm of tissue results in a Poisson distribution of P_m with the average degree of multiple scattering $\bar{m}$ proportional to the local tissue blood volume. In this case, the normalized autocorrelation function $g^{(2)}(t)$ can be written as[7]

$$g^{(2)}(t) = 1 + [e^{2\bar{m}(I_1(t)-1)} - e^{-2\bar{m}}]. \tag{9}$$

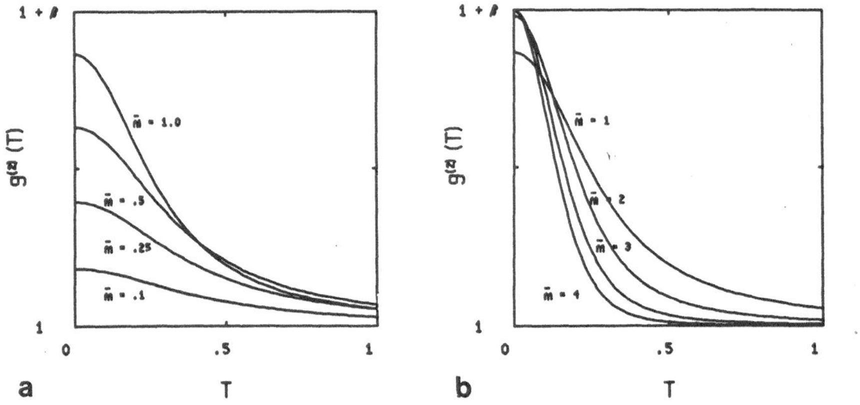

Figure 2. The normalized intensity autocorrelation function $g^{(2)}(T)$, plotted as a function of the reduced time variable, $T = \langle V^2\rangle^{\frac{1}{2}}\, t/(12\xi)^{\frac{1}{2}}\, a$, for different degrees of multiple scattering. (a) For small values of $\bar{m}$, the predominant effect on $g^{(2)}(T)$ is an increase in amplitude with increasing $\bar{m}$. (b) For larger $\bar{m}$, the amplitude of $g^{(2)}(0)$ approaches a maximum value, $1 + \beta$, determined by spatial coherence at the detector, while the decay rate increases and approaches $\bar{m}^{\frac{1}{2}}$ scaling. The shape of $g^{(2)}(T)$ changes from Lorentzian at small $\bar{m}$ to Gaussian at large $\bar{m}$.

The dependence of $g^{(2)}(t)$ on the average degree of multiple scattering (see Figure 2) shows an increasing amplitude as $\bar{m}$ increases up to $\lesssim 1$ ("heterodyne") and an increasing decay rate as $\bar{m}$ increases from 1 to 4 (multiple scattering). For a large degree of multiple scattering ($\bar{m} \geq 3$) the form of $g^{(2)}(t)$ and $P(\omega)$ will be nearly Gaussian regardless of the speed distribution of the red blood cells. From Equations (9) and (5) one observes that the detected photocurrent contains information on the even moments of the speed distribution and the local blood volume (Figures 2 and 3).

The power spectra $P(\omega)$ can be related to $g^{(2)}(t)$ by:

$$P(\omega) = i_o^{\,2}\delta(\omega) + ei_o/\pi + i_o^{\,2}S(\omega) + D(\omega) \tag{10}$$

where

$$S(\omega) = \frac{1}{\pi}\int_o^\infty \cos \omega t[g^{(2)}(t) - 1]dt,$$

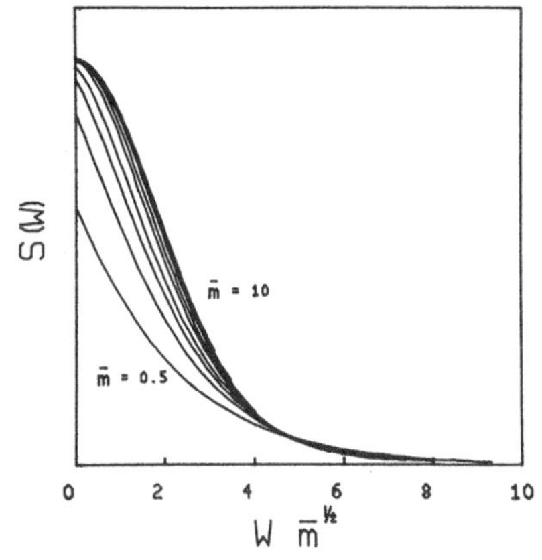

Figure 3. The power spectrum $S(\omega)$ of the detected photocurrent, computed as a function of $\omega\bar{m}^{\frac{1}{2}}$ for values of $\bar{m}$ = 0.5, 1, 1.5, 2, 3, 4, 6, 8, 10 (left to right respectively). For $\bar{m} < 1$, the "normalized" first moment of these spectra increases approximately linearly with $\bar{m}$, whereas for large $\bar{m}$ the spectra scale increases with $\bar{m}^{\frac{1}{2}}$.

(see Figure 3), i_o is the mean detected intensity and $D(\omega)$ is the amplifier noise (which in the case of our photodiode detector dominates the shot noise term). In our instrument we have chosen to process the photocurrent with a real-time analogue algorithm which gives $\langle\omega\rangle$, the mean frequency, times an amplitude term $f(\bar{m})$ dependent on the blood volume

$$\langle\omega\rangle \equiv \int_{-\infty}^{\infty} |\omega| S(\omega) d\omega = \frac{\langle V^2\rangle^{\frac{1}{2}}}{(12\xi)^{\frac{1}{2}} a} \beta f(\bar{m}). \tag{11}$$

The right-hand equality in Equation (11) is for the Gaussian speed distribution (Equation (6)) and is a good approximation for any other physiological speed distribution. The expression for $f(\bar{m})$,

$$f(\bar{m}) = \frac{2e^{-2\bar{m}}}{\pi^{\frac{1}{2}}} \sum_{j=1}^{\infty} \frac{(2\bar{m})^j \Gamma(j + \frac{1}{2})}{\Gamma(j + 1)\Gamma(j)}, \tag{12}$$

is approximated by $2\bar{m}$ for $\bar{m} < 1$ and $\bar{m}^{\frac{1}{2}}$ for $\bar{m} >> 1$. Thus our "mean frequency" signal $\langle\omega\rangle$ will be linear with red blood cell RMS velocity for all tissues and in the limit of low tissue blood volumes ($\bar{m} < 1$) will be linear with blood volume as well. In tissues with larger blood volumes, the measured flow value will vary as the product of the RMS velocity and the square root of the local red cell number density. For diffuse tissue scattering, the Doppler signal samples the flow in the different microvessels (arterioles, capillaries, and venules) in proportion to the number of red blood cells they contain (vessel volume = cross section times length). Thus the vessels with the large capacitance or length within sampled volume (capillaries and venules) and lower velocities will dominate the signal in those tissues without significant high velocity arterio-venous shunt flow.

The theory derived here for photon diffusion in tissue (i.e. diffuse illumination) must be modified for the case of directly illuminated large surface vessels (such as on the retina, Tanaka et al.[2] or cerebral cortex, Williams et al.[9]). In the case of normal incidence, $\underset{\sim}{Q}$ is nearly in the plane of $\underset{\sim}{V}$ for each scattering event as the photon transits the large vessel (Figure 1C). Assuming this forward diffusing light is backscattered by the vessel wall or surrounding tissue, the power spectrum of the detected photocurrent will be Gaussian distributed with $\langle\omega\rangle \approx 0.1\ \bar{m}^{\frac{1}{2}}\langle Q^2\rangle^{\frac{1}{2}}\langle V^2\rangle^{\frac{1}{2}}$. For human blood $\langle Q^2\rangle^{\frac{1}{2}} \sim 0.83\ \mu m^{-1}$ and $\bar{m}^{\frac{1}{2}} \sim (D/6\ \mu m)^{\frac{1}{2}}$. These large Doppler shifts ($\langle\omega\rangle \sim 3$kHz for a 130 μm vessel with $\langle V^2\rangle^{\frac{1}{2}} \sim 7.5$ mm/sec) will, if present, dominate the signal from the nearby microvasculature.

3. FIBER OPTICAL SYSTEM USED IN CLINICAL BLOOD FLOW MONITOR

The optical criteria for a tissue blood flow monitor were to provide a light delivery and detection system which would reproducibly sample a small tissue volume with good signal-to-noise, without perturbing the flow in the sampled volume, and to allow an accurate reading from a variety of tissues in different clinical circumstances. HeNe lasers (5-15 mW)were chosen on the basis of cost, safety, and minimum tissue absorption (HeNe, Ar and Neodynium lasers have all been tested with our system). Because the argon laser wavelengths are strongly absorbed by hemoglobin, the volume of the tissue sampled by the instrument may vary significantly with changes in blood volume. These absorption changes may invalidate theoretical assumption (4) (see Section 2) and lead to deviations from linearity if the distribution of microvessels within the sampled region is not uniform or if the blood volume changes are large. For HeNe and Neodynium lasers these sample volume changes are much smaller due to significantly smaller blood absorption at 633 and 1000 nm. The light delivery and detection system incorporates flexible 125 μm diameter graded-index fibers which provided a high efficiency, reproducable geometry with maximum ease of measurement from a variety of tissues (Figure 4).

The fiber optic probes for use with the laser Doppler blood flow monitor are individually constructed utilizing commercially available materials. Two 4 meter graded-index glass fibers (Corning 10200) are passed through 3.5 meters of black teflon tubing (ID.030" and O.D.055"). On one end the two fibers are passed separately through 0.3 m of teflon tubing which forms a Y with the longer common section. The two separate ends of the fibers are terminated with Amphenol fiber

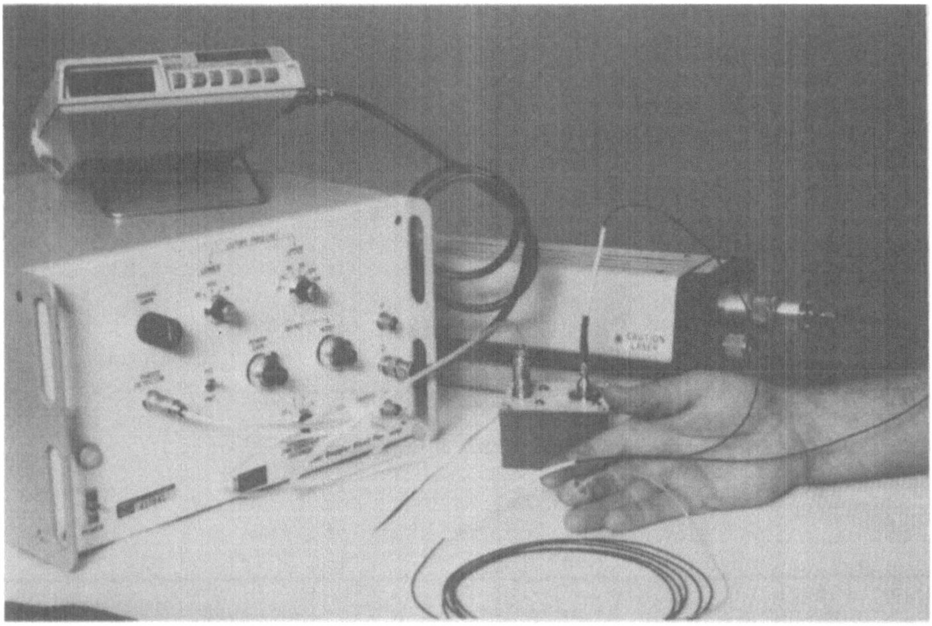

Figure 4. The laser Doppler blood flowmeter measuring fingertip flow.

optics terminations (906-110-5000 and 905-119-5000) which couple to the laser-objective mount and the photo diode (UDC 040A), respectively. The laser beam is coupled to the exciting fiber by a 5X, 0.10 N.A. microscope objective mounted in a cylindrical tube which screws directly into the laser and has a precision Amphenol female termination at the opposite end. The optical fiber can thus be reproducibly and quickly mounted to the laser system at the focal point (beam waist) of the focussed laser beam (waist radius - 38 μm). Typically, 85% of the laser output is emitted at the tissue end of the fiber optic. At the tissue, the probe consists of the two parallel exciting and receiving, graded index fibers separated by 0.5 mm. The objective of fiber separation is to introduce a significant minimum photon diffusion path within the tissue, thus avoiding detection of surface reflection from tissue interface, as well as providing a more homogenous distribution of diffusion paths of the detected photons through the tissue (which determines the degree of multiple scattering from moving red blood cells). For fiber separations less than 0.25 mm significant unshifted light is detected from surface reflections which reduces the signal to noise in the detected photocurrent. On the other hand, for fiber separations > 1 mm the efficiency of detection decreases rapidly, resulting in a significantly lower signal level. Within these limits, as the fiber separation increases, tissue absorption is not dominant and the mean depth at which blood flow is probed increases. The volume sampled by the typical probe is ~ 1 mm^3.

The characteristics of the graded index fibers are 125 μm diameter, numerical aperture of 0.204, 870 MHz bandwidth, and 4.2 dB/km attenuation at 820 nm. A test of other fibers (larger, step-index plastic fibers) demonstrated considerably worse signal to noise (3-5 fold less intensity per coherence at the detector). Additionally, other investigators report large 1 mm step-index fibers that have been plagued with a high frequency (~ 15 kHz) resonance in the detected photocurrent which is difficult to overcome (Holloway et al.[10], Nilsson et al.[11]). HeNe laser beam coupled into the core of the exciting graded-index fiber removes this large extraneous variable signal in the detected photocurrent and maximizes signal-to-noise ratio, as well as providing a highly flexible probe ideally suited for nonperturbing flow measurements at tissue sites difficult to access (e.g. muscle during biopsy, gums of the molars, kidney cortex).

4. ELECTRONIC PROCESSOR YIELDING MEAN DOPPLER FREQUENCY

As developed in the theoretical section and in Bonner and Nossal[7], the mean frequency of the Doppler shift of light scattered from moving red blood cells is proportional to the local tissue blood flow. Since the frequencies of the Doppler shifts are in the range of 30-30000 Hz and the probe geometry samples are all red blood cells (~ 10^5 to 10^6) within the microvasculature in a 1 mm^3 volume, $\bar{\omega}$ (the mean frequency in the detector noise spectrum sampled for 0.1 sec) may be expected to be a good statistical sample of the instantaneous local blood flow. The ideal electronic processor will economically provide an accurate, continuous (0.1 sec average) estimate of the mean frequency corrected for all extraneous noise sources (detection noise, amplifier noise, tissue motion noise) (Figure 5). Since the theoretical analysis suggests that the low frequency 30-1000 Hz Fourier components of the spectrum (Figure 6) are associated with the typical scattering through small angles by red blood cells in the capillaries (V ~ 0.5 mm/sec), the analyzer must be able to analyze the entire tissue light scattering spectrum in order to be linear over the large physiological and pathological range of tissue blood flows (1 to 100 ml/100 g/min).

The basic algorithm computes a normalized first moment of the power spectrum corrected for intrinsic detector and processor noise. The block diagram (Figure 5) shows the elements of the analogue circuitry. The signal is detected by a square-law detector of the self-beat frequency noise. On the left of the schematic the first moment of the signal noise spectrum is computed. First the detector voltage output is amplified (AC/DC gain = 10 to maximize range of processing stages utilized), then bandpass-filtered in order to maximize signal to noise by including only those frequencies associated with RBC Doppler shifts. This band-limited noise signal is fed through a passive lattice with a gain proportional to $\omega^{\frac{1}{2}}$ within 2% over the range of 30 to 30000 Hz. The time average of the squared output of this lattice is the first moment of the power spectrum of the photo detector

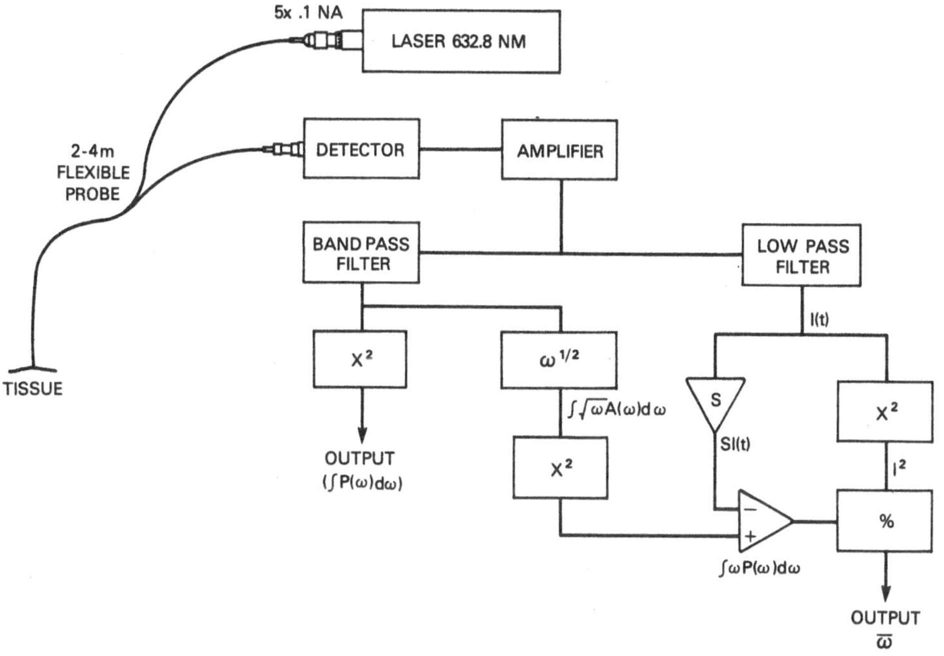

Figure 5. Block diagram $\bar{\omega}$ processor for the laser Doppler blood flow monitor.

voltage noise. This output contains the contributions of detector noise and amplifier noise as well as the noise associated with the self-beat Doppler signal. To correct for these extraneous noise sources, the output is nulled for a condition of no flow (such as reflection of a static diffuser or tissue with occlusion of supplying vessels). In the case of a PMT, the extraneous noise source is dominated by the shot noise of the detection process and its contribution to the noise in the first moment of the power spectrum will be proportional to the mean detected intensity. A variable (setable) fraction of the mean voltage, sufficient to null a signal from a static sample, is subtracted from the averaged first moment of the photovoltage power spectrum. This variable fraction will be dependent on the bandwidth selected. In the case of photodiode detection at low power levels the detection noise is dominated by dark-noise contributions independent of detected light intensity. In this case a variable (setable) fraction of a constant reference voltage is subtracted from the first moment of the power spectrum in order to give an output of zero for a no flow condition.

This output, the corrected first moment of the power spectrum of the Doppler shift noise signal, will be the product of the mean frequency of detector noise arising from RBC Doppler shifts and the instantaneous power in the detected signal. This output varies as

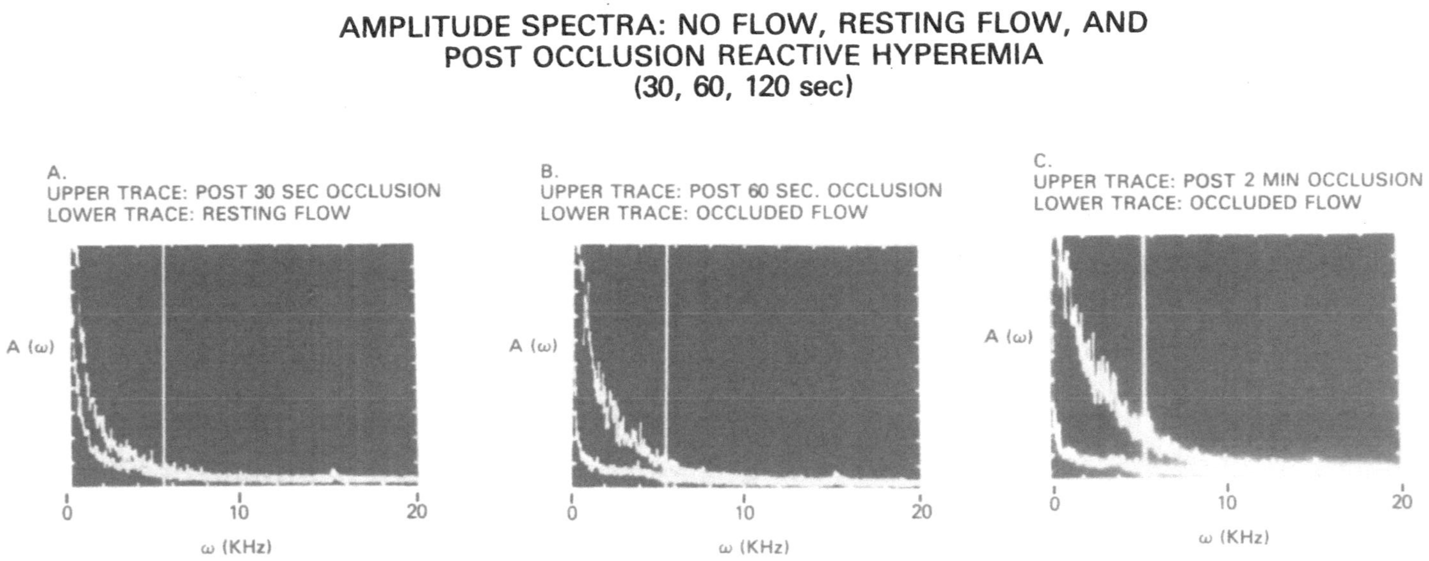

Figure 6. Amplitude spectra (no flow, resting flow, and post occlusion reactive hyperemia, at 30, 60, and 120 sec respectively) of laser Doppler signal from human muscle. The flowmeter output during these particular responses is recorded in Figure 10.

the mean-squared detected intensity (varying with tissue pigmentation or exciting light flux). In order to normalize the corrected first momentof the power spectrum the instantaneous photodetector power is monitored by squaring the mean photovoltage. A variable fraction (s_2I^2) of this power measure is divided into the first moment of the power spectrum to give the normalized first moment of the power spectrum or the mean frequency of the Doppler shift corresponding to the tissue blood flow (with an arbitrary instrument scale factor S_2/β).

5. USE OF LASER DOPPLER BLOOD FLOW MONITOR

In order to use the Laser Doppler blood flow monitor, the excitation fiber is positioned at the focal point of the laser focussing objective in order to couple the maximum laser intensity to the tissue (~ 85% incident laser power). The receiving fiber is attached to the photodetector. A processor bandwidth (usually 30-30000 Hz) is then selected which will detect all Doppler frequencies to be analyzed with minimum non-Doppler electronic noise. The signal gain is set to avoid saturation while maximizing the linear range utilized in the processor electronics. The non-Doppler noise offset is adjusted to give a .00V output of $\bar{\omega}$ for a condition of no-flow. The output signal voltage can be adjusted from 0 to +10V by varying the power gain (output divisor) from 0 to 10. The output flow signal is linear with flow on an arbitrary scale determined by the optics and processing electronics, with an adjustable factor given by the reciprocal of the power gain. The placement of the probe tip within 5 mm of the surface of a tissue to be measured will insure reproducible output signal. The output signal is a continuous real-time indicator of tissue blood flow.

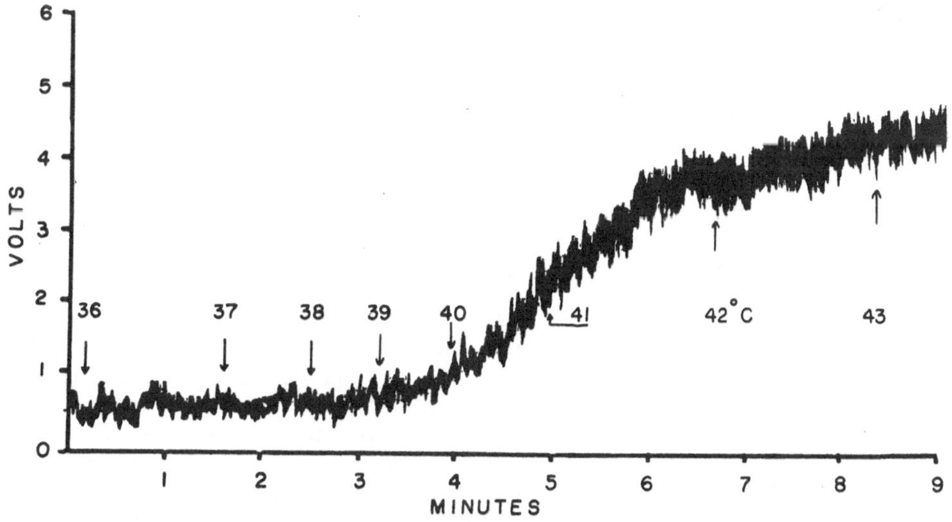

Figure 7. Forearm skin blood flow response to local heating

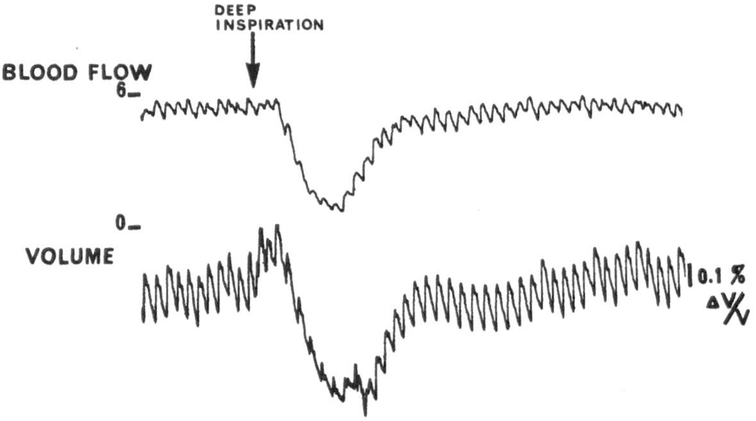

Figure 8. Simultaneous measurements of fingertip blood flow and volume during Valsalva maneuver which transiently occludes A-V shunts. The pulsations in blood flow (0.15 of mean) are much larger than the volume pulsations with heartbeat.

For human clinical use, the ideal blood flow monitor will be continuous and noninvasive, linear over 2 orders of magnitude (1-100 ml/100 g/min), and have sufficient temporal resolution and sensitivity to follow pulsatile changes in flow with cardiac output. Figures 7 and 8 show typical blood flow from human skin on the forearm and fingertip (with flow values on order of 1-10 ml/100 g/min and 30 ml/100 g/min respectively). The most difficult Doppler signal to process will be one associated with very low flow which is pulsatile with cardiac output requiring the greatest temporal resolution and sensitivity. A good example of such a signal is that obtained from human calf or forearm skin, which in the resting state has a small blood volume and pulsatile flow arising from the surface capillary and venule bed. Figure 9 shows a typical real-time laser Doppler flow algorithm output for human forearm skin at rest, during occlusion, and during subsequent reactive hyperthermia. For tissues with higher flow such as the human fingertip the real-time resolution of flow pulse shape is even greater (Figure 8).

The instrument output "flow values" have been compared with average values reported in the literature for a variety of human skin and muscle sites. Over a range of tissue blood flow between 2-60 ml/100 g/min the laser Doppler output correlates well with average literature values obtained by a variety of alternative methods. For our instrument (with power gain s_2 set equal to 1) the conversion factor for all these human tissues is roughly 1 volt = 10 ml/100 g/min. An even better correspondence is found at a particular site during physiolo-

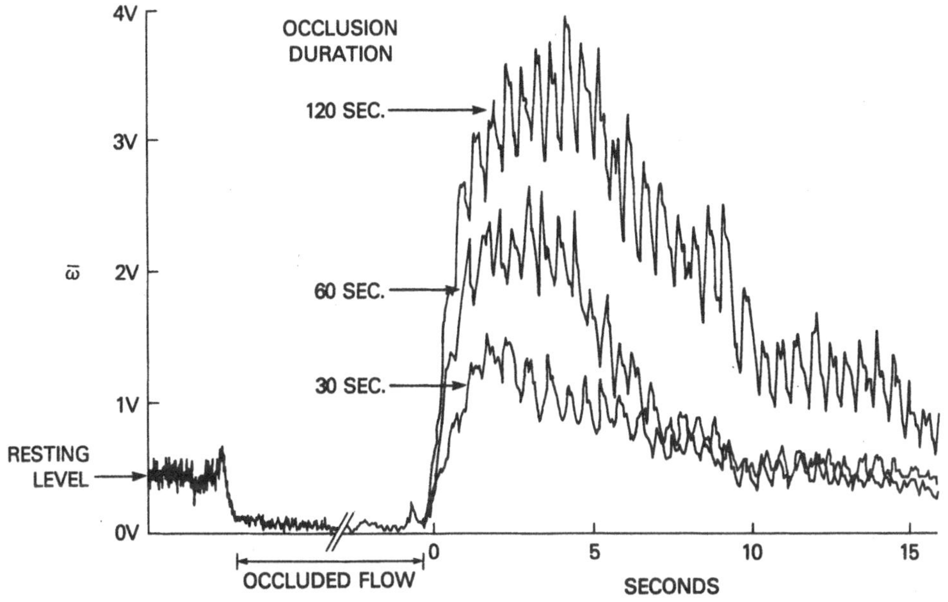

Figure 9. Forearm skin blood flow during reactive hyperemia. Time scale is compressed prior to occlusion. As the microvasculature dilates after release of the occlusion, the mean flow and pulsatile flow increase dramatically.

gical maneuvers which induce large changes in local blood flow. For example, laser Doppler measurements during the Valsalva maneuver (Figure 8) clearly demonstrate that 80-90% of the resting flow is through A-V shunts, which are occluded following deep inspiration. Since the changes in this case are largely in velocity rather than blood volume (for fingertip $\bar{m} > 1$), the Doppler flow value $\bar{\omega}$ is expected to be nearly linear with flow. During the hyperemias of forearm skin (Figures 7 and 9) and muscle (Figure 10), the blood flow increases 6 to 10 fold with a significant change in microvascular volume as well as red blood cell velocity. In these cases $\bar{\omega}$ is again expected to be nearly linear with flow because the tissue blood volumes are lower ($\bar{m} < 1$). The instrument may, however, underestimate the change in flow associated with a large change in blood volume in a tissue with a large blood volume ($\bar{m} > 1$), such as during the blanching of the cortex of the kidney.

In summary, this laser Doppler blood flow monitor is particularly suited for local blood flow measurements which must be

(1) noninvasive,

(2) of temporal resolution sufficient to follow dynamics associated with cardiac output or local regulation,

(3) continuous, or

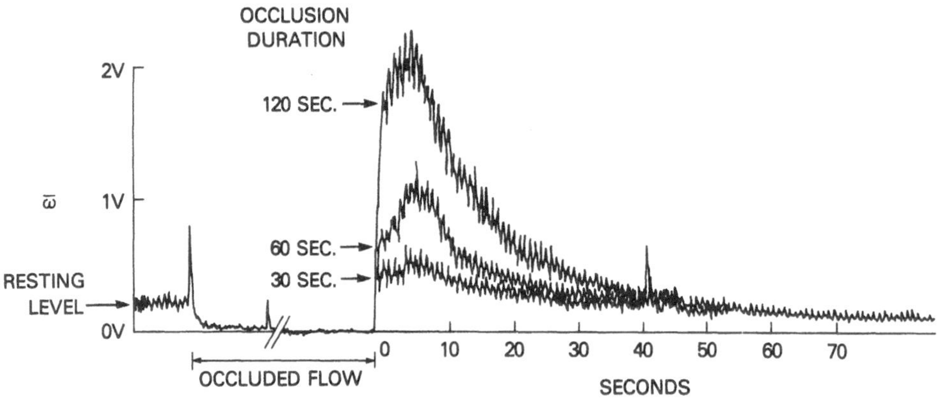

Figure 10. Reactive hyperemia in human muscle with neuromuscular disease. Spectra obtained during these records are presented in Figure 6.

(4) of high spatial resolution (~ 1 mm).
The output is approximately linear with volume flow for tissue blood flow over the entire physiologic range (0-100 ml/100 g/min) for a variety of tissues.

6. CLINICAL APPLICATIONS

Normal and pathological microvascular blood flow dynamics in humans are poorly understood since comparable techniques for continuous microvascular flow dynamics do not exist or are applicable only to individual capillaries of the human finger nailfold (i.e. which can be resolved in vivo by light microscopy). For this reason the clinical diagnostic potential of the technique is only now being examined. Potential areas of impact are the microvascular functional changes associated with diabetes, sickle-cell anemia, dermatological abnormalities; burns, some neuromuscular disorders, and other diseases directly altering microvasculature. The microvasculature may be directly monitored noninvasively at different sites in the peripheral circulation as a diagnostic of the progress of disease and the effect of therapy.

REFERENCES

1. C. Riva, B. Ross, and G. B. Benedek, Invest. Opthal. 11: 936 (1972).
2. T. Tanaka, C. Riva, and I. Ben-Sira, Science 186: 830 (1974).
3. M. Stern, Nature 254: 56 (1975).
4. M. D. Stern, D. L. Lappe, P. D. Bowen, J. E. Chimosky, G. A. Holloway, Jr., H. R. Keiser, and R. L. Bowman, Am. J. Physiol. 232: H441 (1977).

5. M. D. Stern, P. D. Bowen, R. Parma, R. W. Osgood, R. L. Bowman, and J. H. Stein, Am. J. Physiol. 236: F80 (1979).
6. R. F. Bonner, P. Bowen, R. L. Bowman, and R. Nossal, in "Proc. Electrooptics Laser '78 Conf." (Indust. and Sci. Conf. Mgm't, Inc., Chicago, 1978), p. 539.
7. R. F. Bonner and R. Nossal, Applied Optics (June 1981).
8. R. Nossal, S. H. Chen, and C. C. Lai, Optics Comm. 4: 35 (1971).
9. P. C. Williams, M. D. Stern, P. D. Bowen, R. Brooks, M. K. Hammock, R. L. Bowman, and G. DiChiro, Med. Res. Eng. 13: 3 (1980).
10. D. Watkins and G. A. Holloway, Jr., IEEE Trans. Biomed. Engr. BME-25: 28 (1978).
11. G. E. Nilsson, T. Tenland, and P. A. Oberg, IEEE Trans. Biomed. Engr. BME-27: 12 (1980).

PHASE SEPARATION OF CYTOPLASMA IN THE LENS

Toyoichi Tanaka, Izumi Nishio and Shao-Tang Sun

Department of Physics and
Center for Materials Science and Engineering
Massachusetts Institute of Technology
Cambridge, Massachusetts 02139, USA

CONTENTS

ABSTRACT

We would like to present our study on the phase separation of cytoplasma in the animal lens with application to cataract diseases. The following subjects will be discussed: 1) The protein-water mixture does undergo phase separation and shows critical phenomena; 2) the cold cataract, which is the opacification of an excised lens of a young vertebrate upon lowering the temperature, is a manifestation of the phase separation of the lens cytoplasma; 3) the nuclear opacity, which appears in the lens of a galactosemic rat is due to such phase separation; and 4) the diffusivity of protein molecules in the lens cytoplasma can be determined *in vivo* using the technique of laser light scattering spectroscopy. The protein diffusivity is a direct measure of the susceptibility of the lens to the phase separation.

1. INTRODUCTION

Only recently has the physical basis of the lens opacity of the cataract disease been given in terms of the spacial fluctuations

The lysozyme molecule (Mw ≅ 14388) is approximately ellipsoidal, with dimension of 40 x 30 x 30 Å.[9] The dimensions of the individual molecules of the two components in a binary mixture of lysozyme and salt water thus differ by as much as an order of magnitude. The lysozyme molecule carries a net positive charge of 6.5 at the experimental conditions, pH 5.4.[10] Such a mixture of compact globular macromolecules and solvent, or a mixture in which one of the components carries net charge, has not previously been investigated as a critical binary mixture. The salt water is treated as a single component although it contains Na^+ and Cl^- ions as well as water molecules. The salt (0.5 M) is required for the critical temperature to be in the range between the freezing point of the mixture (approximately -10°C) and the denaturation temperature of lysozyme (approximately 55°C).[11]

We study the temperature dependence of light scattered at an angle of 90° from the lysozyme-water mixture at several different volume fractions. As the temperature is lowered, the intensity increases continuously until a temperature, T_c, is reached at which the mixture suddenly becomes opaque with no light passing through the mixture. The inverse of the scattered light extrapolates to zero at a low temperature T_s (see Figure 1). Figure 2 shows T_c and T_s for various mixtures. Mixtures with a volume fraction of greater than 50%

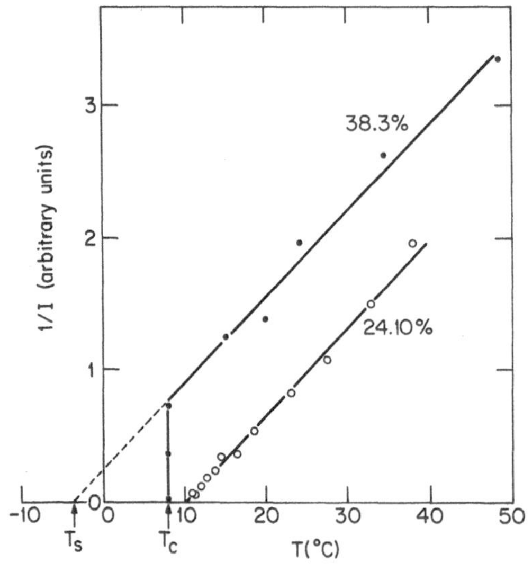

Figure 1. Reciprocal of the intensity of scattered light plotted against the temperature at a concentration very close to the critical concentration (24.1%) and at a non-critical concentration (38.3%).

in the refractive index in the lens.[1] However, only a few causes of fluctuations have been experimentally identified for the cataractous lens. Kinoshita and his collegues,[2] for example, demonstrated that the cortical opacities observed in the cataracts associated with defects in metabolism are due to the inhomogeneous swelling of lens fiber cells. Spector et al[3] and Jedziniak et al[4] found an increase in the amount of heavy molecular weight protein aggregates in the cataractous lens.

In this lecture we shall demonstrate that the phase separation of the protein-water mixture in the lens cytoplasma plays an essential role in causing the nuclear opacity in the cold cataract and the galactosemic cataract.

First, we shall see that a model solution of lysozyme-water mixture shows the characteristic behavior of phase separation and critical phenomena of binary fluid mixtures.[5] In contrast to binary fluid mixtures, however, in the phase separation of protein-water mixture, the mixture does not separate into two macroscopic phases. It separates into two domains of different concentrations with sizes of the order of 1 μm. The domains scatter light very effectively and the mixture becomes opaque.

We will then discuss the cold cataract, which is the opacity in the excised lens of young vertebrates which appears when the lens is cooled. The opacity disappears reversibly upon warming. From the striking resemblance between the development of opacity in a lysozyme-water mixture near phase separation and the development of opacity in cold cataract in the young rat lens, we conclude that the cold cataract is a manifestation of phase separation of the lens cytoplasma.[6] We shall further study how the cold cataract temperature varies with age. It will be shown that the cold cataract temperature of a new born rat is 30°C which is 5°C higher than the newborn rat's body temperature.[7] This means that the phase separation occurs in vivo in the normal rat lens. As the rat ages, the phase separation temperature decreases monotonically and at approximately 10 days of age it becomes lower than the body temperature. Beyond this point the phase separation does not occur in vivo.[7]

Finally we demonstrate that laser light scattering spectroscopy can be safely used to determine the protein diffusivity in the lens in vivo, and thus the suceptibility of the lens to the phase separation can be determined.[8]

2. CRITICAL BEHAVIOR OF A BINARY MIXTURE OF PROTEIN AND SALT WATER[5]

In this section we investigate the light scattering from a lysozyme-water mixture and the development of opacity in the mixture. There is no particular reason for us to have chosen lysozyme, except for its ready availability in large quantities and in a pure condition.

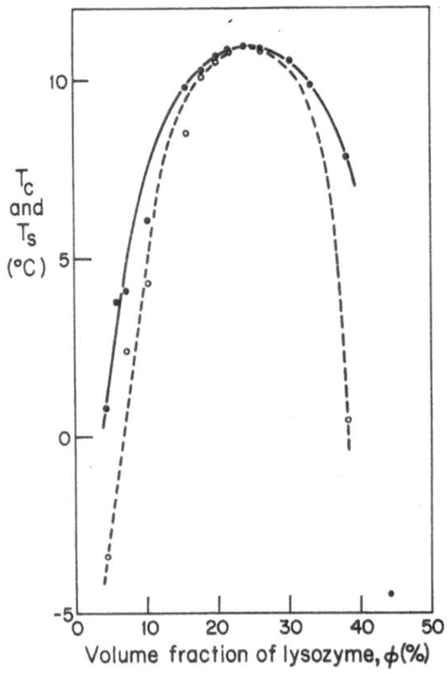

Figure 2. The coexistence curve (solid circles, the solid line indicates $|\phi_c - \phi_c^*| \propto \varepsilon^{0.41}$) and the spinodal curve (open circles, the broken line is drawn to aid the eye) of a lysozyme and salt water mixture.

were progressively more difficult to prepare. The curves of T_c and T_s correspond to the coexistence curve and to the spinodal curve, respectively, of the binary mixture. These two curves have maxima which coincide at the critical mixing point at a volume fraction, ϕ_c^*, of (33±5%) and a temperature T_c^* of (10.90±0.05°C). The coexistence curve is described by $|\phi_c - \phi_c^*| \propto \varepsilon^\beta$ with the critical exponent $\beta = 0.41 \pm 0.1$, where ϕ_c is the volume fraction of the mixture on the coexistence curve, and $\varepsilon = (T-T_c^*)/T_c^*$ is the reduced temperature. This value of the exponent β is consistent with the prediction of mean field theory.

In order to investigate the critical behavior of the osmotic isothermal compressibility κ and the long-range correlation length ξ, the turbidity of the mixture was measured along the critical isochore. There is a theoretical formula which combines the turbidity, the thermal compressibility, and the correlation length[12]

$$\tau = \frac{\pi^3}{\lambda^4}\left(\frac{\phi\partial n^2}{\partial\phi}\right)_T^2 k_B T \kappa_T \left\{\frac{2\alpha^2+2\alpha+1}{\alpha^3}\ln(1+2\alpha) - \frac{2(1+\alpha)}{\alpha^2}\right\} \tag{1}$$

where λ is the wavelength of light in the salt solution, ϕ is the volume fraction of the mixture, n is the refractive index of the mixture, k_B is the Bolzmann constant, κ_T is the osmotic compressibility, and T is the absolute temperature. Here $\alpha = 2(k_o\xi)^2$, where k_o is the wave vector of the incident light in the mixture and ξ is the long-range correlation length. For $\alpha << 1$. Equation (1) becomes

$$\tau_o = \frac{8}{3}\left(\frac{\pi^3}{\lambda^4}\right)\left(\frac{\phi\,\partial n^2}{\partial\phi}\right)_T^2 k_B T \kappa_T. \tag{2}$$

On the assumption that $\kappa_T \propto \tau_o/T$ has a simple power law dependence on ε, such a power law should be manifest in the range of large ε where $\alpha << 1$. The departure of τ/T from simple power law behavior near the critical point can be used to determine ξ.[12]

The intensity, I, of the beam from a He-Ne laser (5 mW) transmitted through the mixture (1 cm path length) was determined by direct measurement of the photocurrent of a photomultiplier tube. The reference intensity I_o was determined for the same mixture without NaCl, since such a mixture is far from its critical point. The turbidity was calculated as $\tau = -\ln(I/I_o)\ \text{cm}^{-1}$. The measured values are shown as solid circles in Figure 3. The deviation from τ_o/T is shown with the broken line. A linear extrapolation of high-temperature data gives the line for τ_o/T. The large value of ξ_0, however, leads to only a short linear region since there is an upper limit to the experimental range of ε at 10^{-1} because of the susceptibility of lysozyme to thermal denaturation. Thus the slope of the line for τ_o/T can only be determined to be in the range 0.9 - 1.1. The open circles show the calculated results for ξ. These

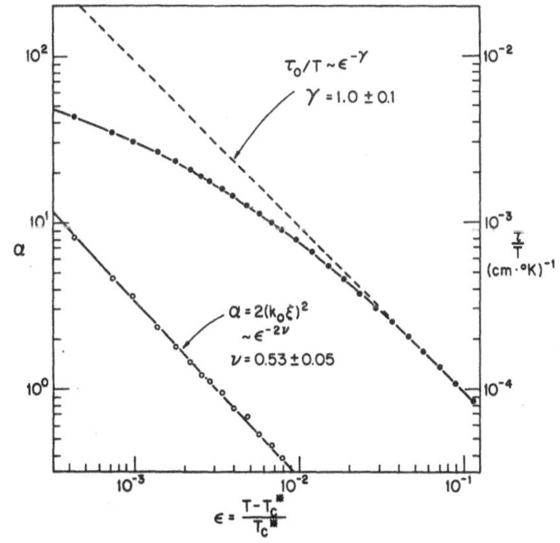

Figure 3. The log-log plot of τ/T (solid circles) and of α (open circles) versus the reduced temperature along the critical isochore. The broken line represents τ_o/T.

results are described by

$$\kappa_T = \kappa_T^o \, \varepsilon^{-\gamma} \propto \tau_o/T, \tag{3-a}$$

$$\xi = \xi_o \varepsilon^{-\nu}, \tag{3-b}$$

with $\kappa_T^o = (3.1 \pm 0.2) \times 10^{-7} \mathrm{cm\ sec}^2/\mathrm{g}$, $\gamma = 1.0 \pm 0.1$, $\xi_o = 26 \pm 5$ Å, and $\nu = 0.53 \pm 0.05$. These values of the critical exponets γ and ν are consistent with the predictions of mean-field theory rather than of the renormalization-group theory ($\gamma = 1.23$, $\nu = 0.63$).[9]

The decay rate of concentration fluctuations was measured as a Rayleigh linewidth of the scattered laser light along the critical isochore by optical-mixing spectroscopy using an argon-ion laser. All measurements were in the hydrodynamic regime ($q\xi < 0.1$, where q is the magnitude of the scattering wave vector). In this regime, the prediction of the mode-coupling theory is approximated to within 10% by

$$\Gamma/q^2 = k_B T/6\pi\eta\xi, \tag{4}$$

where η is the macroscopic shear viscosity.[14] The measured values of Γ/q^2 are shown in Figure 4. In order to check the validity of Equation (4) for this mixture, Γ/q^2 was calculated from ξ given in Equation (3-b) and η_m determined by rolling-ball viscometry. A stainless-steel microball, 0.635 mm diameter, and a glass micropipette, 1 mm diameter, were used and calibrated against glycerol-water standard solutions. Figure 5 shows the divergence of the macroscopic shear viscosity η_m at smaller values of ε in a semilog plot. The viscosity anomaly is extremely large compared to that found in binary mixtures of small molecules. The nonanomalous background is negligibly small. The broken line in Figure 4 shows the calculated curve for Γ/q^2 using this η_m. There is a discrepancy of several orders of magnitude between the measured and calculated values. This discrepancy is probably due to a difference between the nature of η which appears in Equation (4) and the nature of the measured η_m.

In this section, we have demonstrated the existence of a critical mixing point in a protein/solvent mixture. The static behavior of the concentration fluctuations is in close agreement with the prediction of the mean-field theory. The dynamic behavior, however, is not in agreement with the mode-coupling theory unless the viscosity of the mixture is properly evaluated.

3. PHASE SEPARATION OF A PROTEIN-WATER MIXTURE IN COLD CATARACT[6]

In this section we will discuss the striking resemblance between

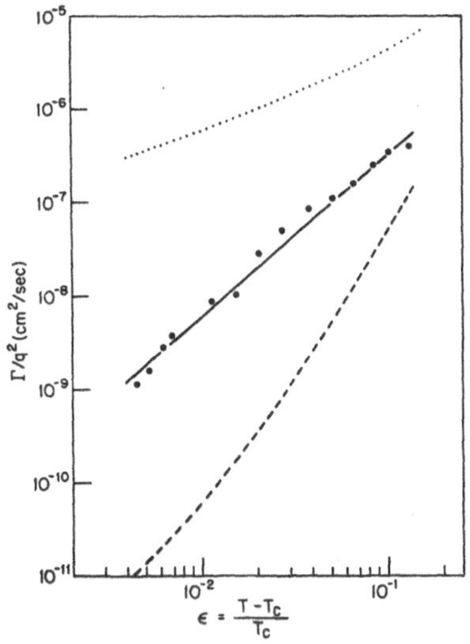

Figure 4. Log-log plots of Γ/q^2 versus the reduced temperature along the critical isochore. Solid circles indicate experimental values obtained from linewidth measurements. The broken line indicates the curve calculated from Equation (4) using independently observed values for ξ and η_m. For reference, the dotted line indicates the curve calculated from Equation (4) using independently observed values for ξ and the viscosity of the salt water.

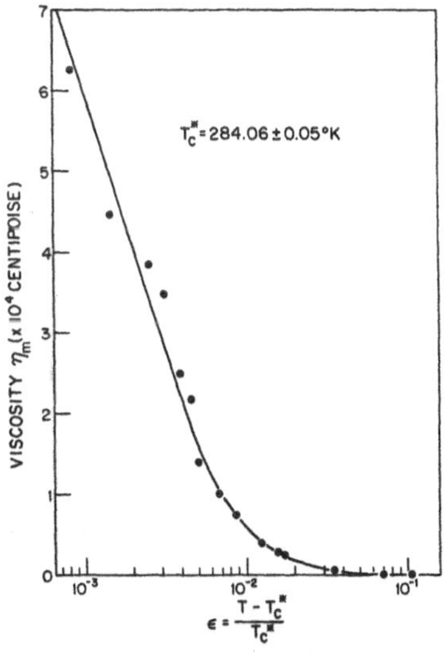

Figure 5. The macroscopic shear viscosity η_m versus the reduced temperature along the critical isochore.

the development of opacity in a lysozyme-water mixture near phase separation and the development of opacity in the cold cataract in the lens of a young rat.

Let us start by briefly looking at the structure of the lens. During the development of the eye in the embryo, the lens is induced from epidermis by nerve tissue. The fiber cells which make up the bulk of the adult lens are derived from epithelical cells. The lens contains only this one type of cell. In the equatorial region the epithelial cells differentiate and elongate into fiber cells. In these processes fiber cells are denucleated. The fiber cells are laid down throughout life (Figure 6). The cells defferentiating and elongating from epithelial cells into fiber cells start synthesizing specific lens proteins.[11] Since the lens grows continuously throughout life laying down new fiber cells at the periphery, newer cells are found toward the periphery and older cells toward the center of the lens. As a result there is a gradient of protein concentration as a function of radial position. The long, spindle-shaped fiber cells of the lens are packed in such a fashion that each fiber cell lies approximately along a concentric shell. Since there is no extensive compartmentalization of the cytoplasm of an individual cell, the protein volume fraction can be considered internally homogeneous. Thus, the lens can be thought of as a bag of protein solution with a concentration gradient from center to periphery maintained by the cell membranes.

We have observed the concentration fluctuations of the protein solution in the intact lens using laser light scattering spectroscopy. Figure 7-b shows the temperature dependence of the decay rate,

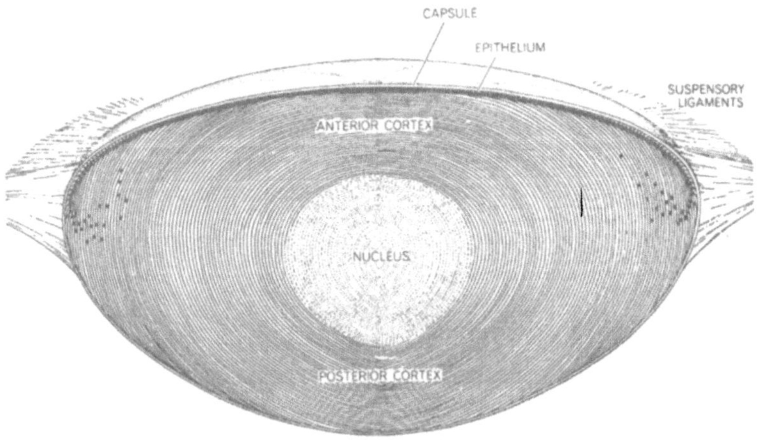

Figure 6. Cross-section of the lens.

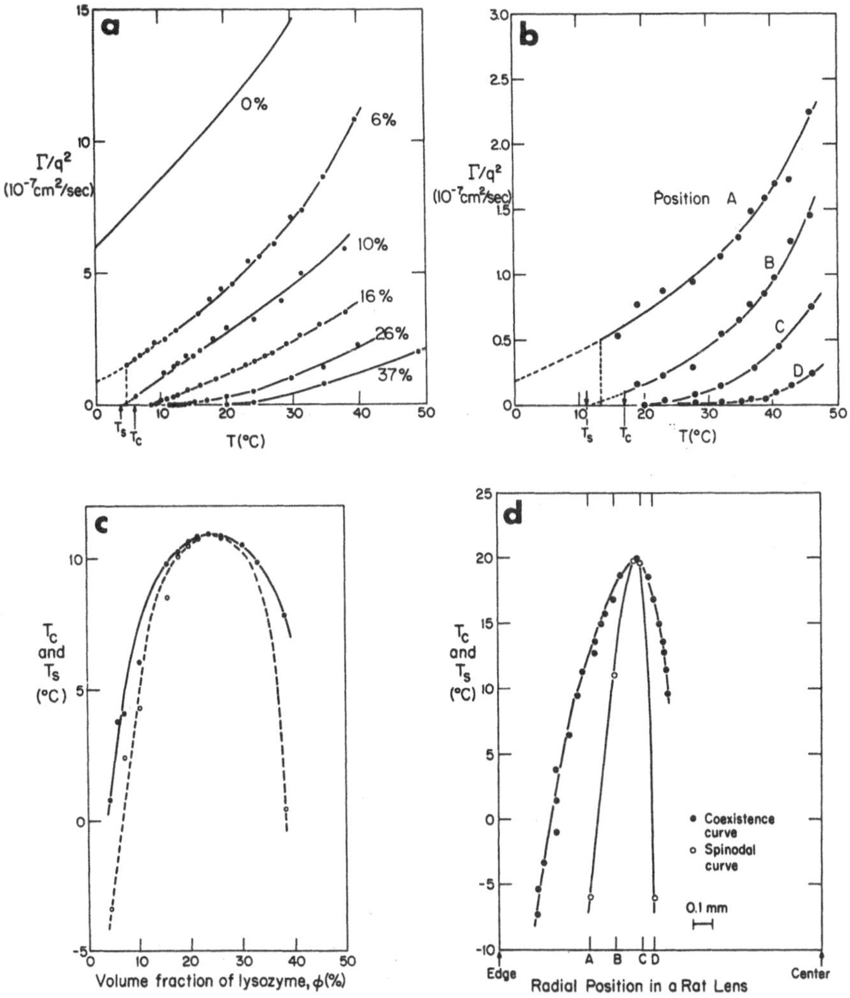

Figure 7. (a) and (b) Temperature dependence of Γ/q^2 for (a) a lysozyme-salt water mixture at several different volume fractions, and (b) a protein-water mixture in a fresh young rat lens at several different radial positions in the equatorial plane of the lens. The radius is 1.75 mm. The curve labeled 0 percent in (a) was calculated for an infinitely dilute lysozyme-salt water mixture. (c) and (d) Coexistence curve (•) and spinodal curve (o) for (c) lysozyme-salt water mixtures, and (d) protein-water mixtures in a young rate lens, determined by measurements of the intensity and the decay rate of fluctuations of laser light scattered by the mixture. Some points in the coexistence curve for the lens were measured by observing the appearance of opacity.

divided by q^2, of the concentration fluctuations of the protein solution at several positions in the equatorial plane inside the excised lens of a young albino rat of body weight 100g. The radius of the lens was 1.75 mm. The scattered light was collected from a scattering region with an approximate linear dimension of 20 μm, which is several times the diameter of a lens fiber cell. We observed quite similar behaviour as that of the lysozyme-water mixture which is shown in Figure 8-a: As the temperature is lowered Γ/q^2 diminishes, until at a certain temperature T_c, it suddenly goes to zero. The temperature extrapolated to zero decay rate corresponds to the spinodal temperature, T_s. Figure 7-d shows both T_c and T_s at several radial positions in the equatorial plane inside the excised lens.

We can now compare the temperature dependences of the decay rate, the coexistence curves and the spinodal curves for the lens and the lysozyme-water mixtures in Figures 7-a to 7-d. There is remarkable similarity between the curves in Figures 7-a and t-b, and those in Figures 7-c and 7-d. The general character of the dependence of the features of the curves in Figure 7-b and of the values of T_c and T_s (Figure 7-d) on radial position may be attributable in large part to the monotonic increase of the protein volume fraction along the radius from the periphery to the center of the lens.[16] As the temperature of the lens is lowered, a concentric shell of opacity appears at 20°C (Figure 8). The shell of opacity grows both toward the center and toward the periphery as the temperature is lowered futher. This corresponds to the widening distance between the two intersections of the isotherm with the coexistence curve as the temperature is lowered. At a given temperature, opacity due to phase separation is produced at all volume fractions between the two intersections. The presence of opacity at lower temperatures below the coexistence curve is similar to the behavior of the lysozyme-salt water mixture rather than the binary mixture of small molecules, which undergo bulk phase separation. It is possible, however, to obtain a bulk phase separation by applying a centrifugal force on the protein molecules in solution, both of lysozyme and calf lens homogenate.[17]

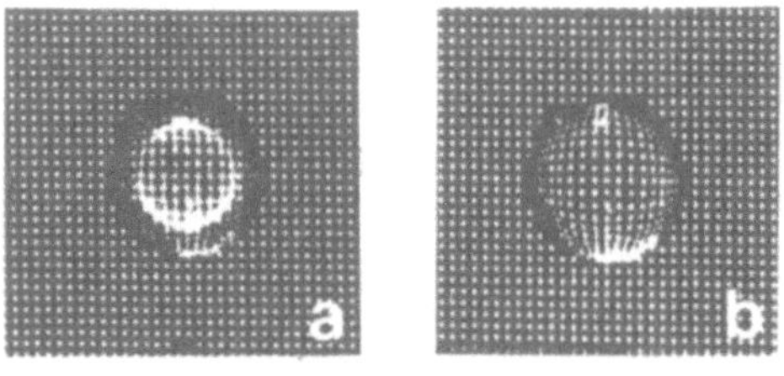

Figure 8. Excised lens of a 100-g albino rat. (a) at 12°C, showing the opacity of cold cataract in the form of a shell, and (b) at 25°C, showing the absence of the opacity. The radius of the lens is 1.75 nm.

It has been suggested by Lerman and Zigman[18] that the change of state of the protein-water mixture in cold cataract reflects a change in the conformation of the lens protein, perhaps to expose more hydrophobic regions at low temperatures. The analysis presented here shows that such conformational changes are not needed to explain the experimental findings. The critical behavior of a binary mixture can be understood theoretically by considering the free energy to include a constant mixing enthalpy term and a temperature-dependent mixing entropy term.

It is now clear that the opacity in the cold cataract is a manifestation of cytoplasmic phase separation. In the next section we shall demonstrate that the lens opacity associated with the galactosemia is indeed caused by this phase separation.

4. CYTOPLASMIC PHASE SEPARATION IN GALACTOSEMIC CATARACT[7]

The galactosemic cataract which develops in the lens of a young rat when it is fed a high galactose diet has been studied extensively, since it is an excellent model of human cataract associated with defects in sugar metabolism such as diabetes and galactosemia.[19] In this section we present evidence that the nuclear opacity in the galactosemic rat lens is a phase separation of the cytoplasm.

Female Sprague-Dawley rats of different ages were obtained from Charles River Breeding Laboratories, Wilmington, Massachusetts. The animals were given free access to tap water and Purina Rat Chow. Galactose-fed animals were given, in place of whole chow, a mixture of equal parts galactose and ground chow starting at 21 days of age. At intervals, the animals were asphyxiated in a CO_2 chamber. The lenses were then removed and placed in a temperature controlled cell filled with silicone oil. The cell temperature was regulated to an accuracy of ± 0.05°C. Each lens was viewed from the anterior axial direction through a dissecting microscope having a calibrated eyepiece.

The phase separation temperature at which the opacity appears upon cooling was determined as a function of the radial position in the lens for rats of various ages (Figure 9-a). The area under each curve of the diagram represents opacity and the remainder represents transparency. The opaque area after one day is a solid spheroid with the boundary at different temperatures at the positions indicated by the curve. As the temperature is lowered, the area of opacity first appears at the center of the lens at approximately 30°C and then grows toward the periphery. It is of interest to note that for newborn rats, the miximum phase separation temperature is higher than the ocular temperature. In fact, we observed a pinpoint opacity in the lenses of newborn rats *in vivo* by cutting open the eyelids of the rat. In older lenses, the center is clear and the opaque region is a speroidal shell with the outside boundary at the

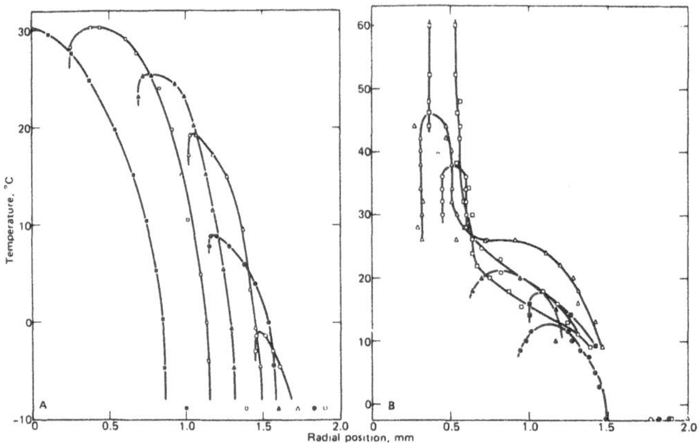

Figure 9. (a) Progressive change of the characteristics of cold cataract in lenses of normal animals during normal aging on day 1 (■), 10 (□), 18 (▲), 25 (△), 42 (•), and 56 (◦). (b) Change in lenses of galactose-fed animals at different stage of galactosemia at 34 (▲, △, □) and 35 (◦) days and in two lenses of normal animals at 31 (■) and 38 (•) days. Each curve in (a) and (b) represents the radial position of the boundary between the clear and opaque areas of the lens. The part of the diagram under each curve represents opacity and the remainder, transparency. Thus, another way of reading the diagram is to regard the curve as representing the temperature at which opacity appears at each radial position upon cooling. The symbols at the bottom of the diagrams mark the subcapsular radius of each lens. Experimental details are described in the text.

positions indicated in the figure. As the temperature is lowered, the region of opacity initially appears as a thin shell which then increases in thickness toward the periphery. It is difficult to observe the inside boundary of the shell of opacity, but observation of the cross-section of a lens which had been cut open showed that there was extremely little movement of the inside boundary of the shell toward the center as the temperature was lowered. The curves in Figure 9-a show a shift with age of the boundary away from the center of the lens and toward lower temperature.

The galactose-fed animals developed nuclear opacity in vivo at around 35 days, or after 14 days of galactose feeding (Figure 10). We emphasize that the present observations concern only the nuclear opacity and not the cortical vacuolar opacity which also developes prior to the nuclear opacity in these animals. The vacuolar opacity is not as intense as the nuclear opacity and did not interfer with our observation.[20] We have examined the progressive

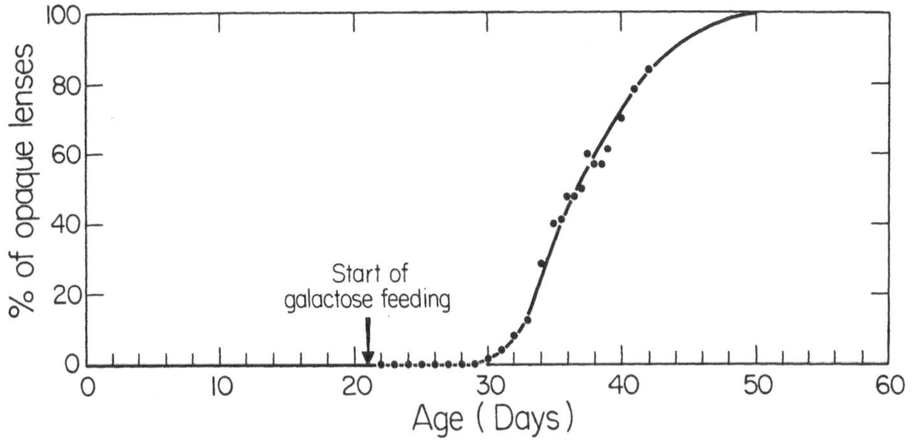

Figure 10. Percentage of galactosemic rat lenses having in vivo nuclear opacity as a function of age. The rats were fed a 50% galactose diet starting at the age of 21 days.

change with age of the characteristics of the cold cataract in the galactose-fed animals. There were no differences from the cold cataract in normal animals until around 30 days. After this date, the phase separation temperature started to rise and became progressively highter. When it became higher than ocular temperature, the opacity could then be seen in vivo. This situation is clearly seen in Figure 11, in which the maximum phase separation temperature of normal and galactosemic lenses are plotted as a function of age. We can see that in the normal lens the maximum phase separation temperature decreases monotonically with age, whereas in the galactosemic lens, it begins to increase at 30 days and then increases rapidly at 35 days, after 14 days of feeding. Figure 9b shows some typical phase diagrams of the cold cataract at various stages in the development of galatosemia. It is important to note that the phase boundary of opacity not only rises in temperature, but also moves toward the center of the lens.

We suggest that the change of the radial position of the phase boundary with age and galactosemia can be understood qualitatively by considering the dependence of the phase separation temperature on protein concentration. The lens grows by laying down long, thin fiber cells in concentric layers at the periphery of the lens. The protein concentration in the fiber cells at each radial position increases with age due to synthesis and dehydration.[16] Thus, the protein concentration increases along the radius from the periphery to the center of the lens due to the increasing age of the fiber cells.[16] Figure 9a show that except at very young ages, there is an intermediate radial position, or protein concentration, at which the phase separation temperature is highest. This type of dependence

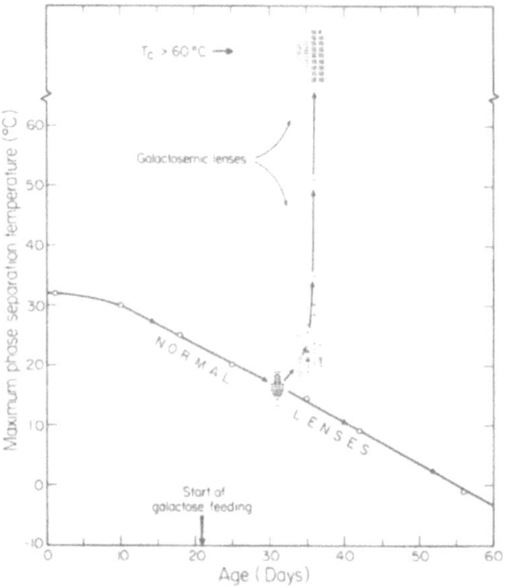

Figure 11. Maximum phase separation temperature of normal (open circles) and galactosemic (dots) rat lenses as a function of age. The lenses were heated only to 60°C. Observations at high temperatures were impossible because of the denaturation of the lens protein. The number of the rat lenses examined was 34, 62, 65, 36, and 31 for the ages of 31, 34, 35, 36, and 37 days, respectively. Each lens at the age of the 34 or 35 days is represented by a single dot. For lenses at the ages of 31, 36, or 37 days, each is represented by two connected dots so that the total number of dots, isolated or connected, is approximately the same for each day's measurement. This gives each dot approximately equal statistical weight.

of the phase separation temperature on protein concentration has been observed in lysozyme-salt water mixtures which undergo phase separation, producing opacity. The maximum phase separation temperature corresponds to the critical point. The progressive shift away from the center with age of the radial position of the critical point is due to the general increase in protein concentration throughout the lens.[16] Because of the monotonic decrease of protein concentration with radial position, the critical concentration appears at a larger radial position when the protein concentration increases at every point in the lens. It is well established that there is an osmotic swelling of the lens caused by the accumulation of dulcitol formed from galactose by lens aldose reductase.[21] The general decrease in protein concentration throughout the lens due to this swelling during the development of galactosemic cataract[22] may explain why the change in the characteristics of cold cataract during the process corresponds to a reversal of the changes which occur with aging.

The phase separation temperature depends not only on the total protein concentration, but also on other factors such as the composition of proteins and the concentration of ions.[23] Some preliminary results on normal rat lenses incubated in salt solutions indicate that there is a strong dependence of the phase separation temperature on salt concentration. The decrease of the maximum phase separation temperature with age in normal rat lenses may be related to the change of ionic strength during the dehydration process. For galactose-fed animals, the lens membrane becomes weak and eventually begins to malfunction after approximately two weeks of feeding.[24] This breakdown of the membrane leads to changes in protein concentrations and salt composition inside the lens fiber cells.[24] Indeed we have observed a strong similarity between the evolution of phase boundaries of the galactosemic lens and normal lenses incubated in saline solution. Further study is needed to identify which factors are improtant to the change of the phase separation temperature.

It is now clearly desirable to make in vivo determinations of the phase separation temperature of the cytoplasm. The difference of the phase separation temperature from the ocular temperature is a direct indicator of the susceptibility of the lens to cytoplasmic phase separation, and therefore to nuclear opacity. These determinations may be achieved by measuring the protein diffusivity, which is zero at the critical point and becomes larger when the ocular temperature is higher than the critical temperature.

5. In Vivo OBSERVATION OF PROTEIN DIFFUSIVITY[8]

In the previous section, we have seen that the nuclear opacity in the galactosemic lens can be understood in terms of the coexistence curve in the phase diagram. When the critical temperature, which is normally below body temperature, becomes higher than body temperature, the phase separation occurs in vivo and the lens becomes opaque. It is now important to measure the coexistence curve or the phase separation temperature, T_c, in vivo, since it is a direct measure of the susceptibility of the lens to the phase separation. Direct measurement of T_c is not possible because we can not change the body temperature easily. As shown in early section, however, the protein diffusivity, D, is a function of $T - T_c$. Therefore from the in vivo measurement of D we can obtain a information on $T_{body} - T_c$; the difference between the body temperature and the phase separation temperature.

In this section, we consider some of the problems encountered in carrying out in vivo studies and present results of protein diffu-

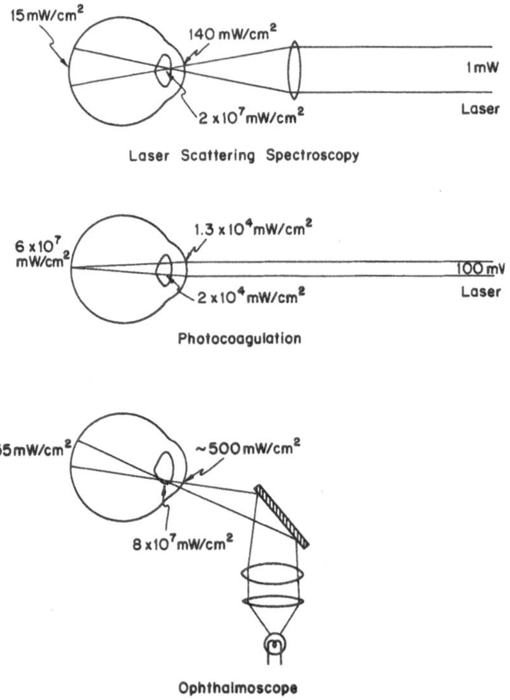

Figure 12. Light paths and radiation levels at the cornea, lens, and retina in laser light scattering spectroscopy, photocoagulation treatment, and direct ophthalmoscopic examination.

the region about the focus was reflected twice by an Abbe's prism and imaged by a lens with a 15 cm focal length on a slit in front of a photomultiplier tube. By passing only the light from the focus through a pinhole slit, it is possible to exclude all undesired scattered light. Since the image of the focus does not move appreciably when the focus is positioned at different locations inside the lens by moving the rabbit, no readjustment of the optical setup is required.

The scattered light intensity is measured by a photomultiplier tube and its temporal fluctuation is recorded in the form of an autocorrelation function. From an analysis with a minicomputer of the precise shape and the characteristic time of decay of the autocorrelation function, we can determine the spectrum of protein diffusivity in the lens.

In *in vivo* experiments, it is essential to avoid any physiological damage to the subject. This is particularly important in sivity measurements on the rabbit lens. Since satisfactory measurements could be made on rabbit lenses in vivo using levels of laser light irradiation which we consider as safe as routine ophthalmoscopy, we expect that the technique will be valuable in human studies as well. The use of this technique has several advantages over conventional methods for making in vivo observations of cataract formation:

1. Sensitivity. Changes in protein diffusivity can be observed considerably in advance of the stage when they give rise to opacity detectable by visual inspection with an ophthalmoscope or a slit-lamp microscope.

2. Quantitative characterization. Protein diffusivity provides additonal information over that supplied by measurements of the opacity[3] about the physiochemical state of lens protein. The light scattering responsible for opacity includes not only that from the protein but also that from static structural nonhomogeneities. On the other hand, the diffusivity is related to the characteristic distance over which protein molecules are correlated in their random motions, which reflects the state of reversible and irreversible association of the molecules.

In view of these advantages, we hope that this technique will be useful in refining in vivo studies of the effects of aging, various drugs, diets, and radiation on cataract formation.

A rabbit was held on a platform by wrapping its torso with a cloth jacket anchored to the platform. The head was immobilized by means of hard, contoured bag filled with beans under the chin and a metal bar pressed against its forehead. The eyelids were held open with small surgical clamps. Several drops of 0.5% tropicamide (Mydriacyl) were applied to the eye to dilate the iris, and saline was applied from time to time to maintain the cornea in good condition. It was necessary to maintain this condition only for several minutes in order to obtain one measurement.

A beam of light from a 5 mW He-Ne laser was expanded to 6 mm diameter and attenuated with a neutral density filter to 1 mW. The beam was focused by a lens of 25 mm focal length. The rabbit was placed so that the focus was positioned at the desired point within the lens (Figure 12). The precise position of the focus could be easily located by sight because of the large beam size and short focal length. This position could be changed be means of three perpendicular micrometer translators which could move the platform on which the rabbit was immobilized. The diameter of the beam at the focal point, which reflects the diffraction limit, was about 10 μm. Beyond the focal point the diameter expanded to 2.5 mm at the retina. Therefore, a high intensity was obtained at the focal point with low irradiance on the retina. The light scattered from considering use of the present method on human subjects. In this section, therefore, we shall show that the present in vivo method is as safe as routine ophthalmoscopy for all the tissues of the human as well as rabbit eye which receives the laser radiation. Figure 12 shows the optical configurations and intensities of irradiance on various tissues of the eye in photocoagulation treatment and in ophthalmoscopic examination.

Retina is the tissue most sensitive to laser radiation in the eye. The maximum permissible level of He-Ne laser light at the retina was found to be 100 mW/cm^2 for 1 hr of illumination.[25] Since the lens, the vitreous humor, and the cornea do not absorb as much light as the retina, they may be exposed to more light. In order to minimize irradiance at the retina while maximizing scattering intensity from the lens, the original laser light of 1 mW power was expanded to 6 mm diameter and focused very sharply on the measurement position in the lens. Thus, after passing through the lens, the beam was defocused quickly and the irradiance at the retina was about 15 mW/cm^2, which was well below the maximum permissible level. The irradiance is also less than that produced on the retina by direct or indirect ophthalmoscopy, which were measured by Pomerantzeff and associates[26] to be 77 and 55 mW/cm^2, respectively.

Although there are some reports on the absorption of light by the lens, there has been no decisive determination of the safety limit of laser radiation on the lens. However, we can show that our optical arrangement is quite safe to the lens by the following consideration. Weale[27] has determined the amount of light absorbed by human lenses. With visible light, he did not find any resonance absorption or special photochemical reactions in the lens. Thus, we consider only the heat effects produced by the absorption of the laser light in the lens. At the 6,328 Å wavelength of the He-Ne laser, the optical density of normal lenses from an approximately 50-year-old subject was about 0.1 to 0.3. This means that 20 to 50 percent of the light is either scattered or absorbed by the lens. The temperature increase due to this absorption was calculated by a steady-state thermal flow model to be approximately 0.3 °C. This is not sufficient to damage the lens. This result was confirmed by measurements of the diffusivity of lens proteins with laser light scattering spectroscopy. The diffusivity, D, changes approximately 5 percent at 37 °C, if the temperature increases by 1 °C. Our experiments indicate that the diffusivity does not depend on laser intensity up to at least 150 mW. This means that the temperature rise is less than 1 °C up to at least 150 mW. With a 1 mW laser beam the temperature increase is negligible.

At the cornea the irradiance is about one-fourth that of direct ophthalmoscopic radiation and 1/100 less than that of a 100 mW photocoagulation treatment. Since there are no reports of corneal damage caused by ophthalmoscopic examination or photocoagulation, the irradiance used in this method should be safe.

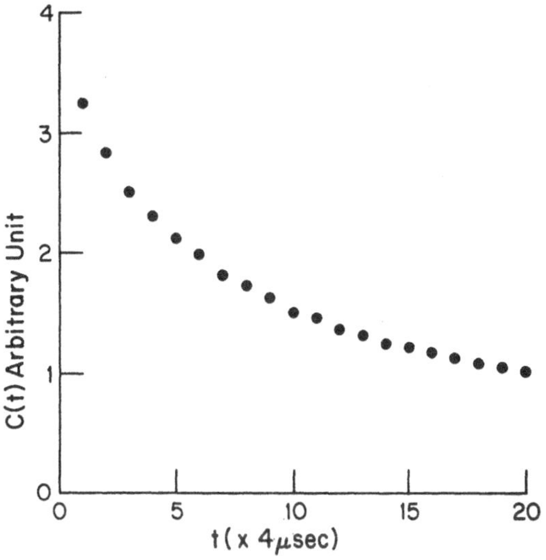

Figure 13. Correlation function of laser light intensity scattered from the in vivo lens of a 70-day-old New Zealand albino rabbit.

Since the vitreous humor and aqueous humor both consist mainly of water, the tolerance to laser radiation is higher than that of the cornea, lens, or retina.

We have discussed only the gross physical effects on various tissues of the eye expected to result from use of the method described. We have not thoroughly examined more delicate possibilities of disturbance to the eye, and further confirmation of safety by histological examination may be prudent before this method is used with human subjects.

Using the methods described above, we measured correlation functions of laser light scattered from lenses of New Zealand albino rabbits. The scattering angle was calculated to be 124 degrees (+6 and -2 degrees) by considering the geometry of the cornea, the aqueous humor, and the lens and their refractive indices. Any inaccuracies affect the absolute value of the average diffusivity but not quantities which depend on relative values, such as variance, skew, and kurtosis.[28] Figure 13 shows the measured correlation function from the lens of a 70-day-old rabbit. The measurement time was about 1 minute. Analysis of the data gives an average diffusivity of 2.9×10^{-7} cm^2/sec, variance of 121 %, skewness of 113 %, and kurtosis of 72 %.

In order to check that small eye movements do not affect the data, we excised and measured the same lens at 36°C at a 90-degree

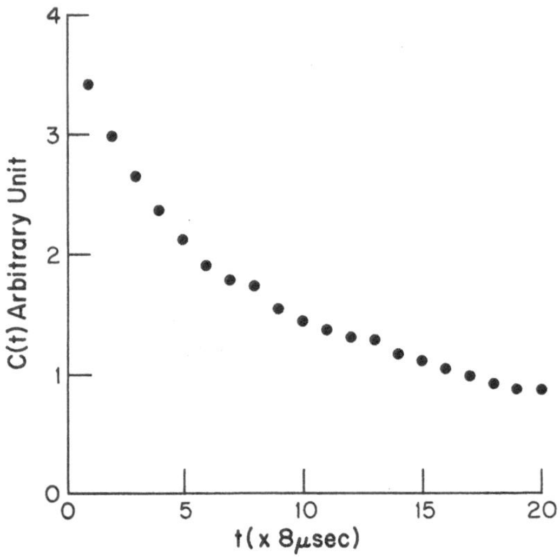

Figure 14. Correlation function of laser light intensity scattered from the excised lens of the same rabbit as in Figure 13.

scattering angle. The measured correlation function is shown in Figure 14. The data in Figures 13 and 14 are similar to each other except for the different time scales due to different scattering angles and give the same results upon cumulant analysis: an average diffusivity of 2.8×10^{-7} cm^2/sec, variance of 111%, skewness of 116%, and kurtosis of 81%.

6. CONCLUSION

In this lecture we have demonstrated that the cold cataract is a manifestation of the phase separation of the cytoplasma in the lens fiber cells. The nuclear opacity in the galactosemic rat lens appears when the phase separation temperature becomes higher than body temperature. We have some preliminary data which indicate that the opacity seen in the rat lens organ-cultured in hypoglycemic conditions, that is, in a medium of less glucose level than usual, is also due to the phase separation. This indicates that the opacity which appears in an infant with hypoglycemia may also be due to the phase separation of lens cytoplasma. It is clearly desirable to identify the cause of the rise of the phase separation temperature. Finally we have shown the feasibility of usage of laser light scattering spectroscopy for live animals. We believe that the technique will provide a powerful clinical means with which to characterize the lens opacity in the precataractous lenses of patients.

ACKNOWLEDGEMENTS

This lecture is merely a review of the work done in our laboratory at M.I.T. The people who have participated in the work are Dr. C. Ishimoto, Mr. P. Goalwin and Prof. L. T. Chylack, Jr. in addition to the authors. It is our pleasure to thank Prof. G. B. Benedek and Prof. J. Clark for their helpful discussions and advice.

This work was supported by NIH Grants EY 01696 and EY 02433, and NSF DMR 78-24185.

REFERENCES

1. G. B. Benedek, Applied Optics 10: 459 (1971).
2. J. H. Kinoshita, L. O. Meorla, and E. Dikmak, Exp. Eye Res. 1: 405 (1962).
3. A. Spector, S. Li, and J. Sigelman, Invest. Ophthalmol. 13: 795 (1974).
4. J. A. Jedzniak, J. H. Kinoshita, E. M. Yate, and G. B. Benedek, Exp. Eye Res. 20: 367 (1975).
5. C. Ishimoto and T. Tanaka, Phys. Rev. Lett. 39: 474 (1977).
6. T. Tanaka, C. Ishimoto, and L. T. Chylack, Jr., Science 197: 1010 (1977).
7. C. Ishimoto, P. W. Goalwin, S.-T. Sun, I. Nishio, and T. Tanaka, Proc. Natl. Acad. Sci. U.S.A. 76: 4414 (1979).
8. T. Tanaka and C. Ishimoto, Invest. Ophthalmol. 16: 135 (1977).
9. R. E. Dickerson and I. Geis, The Structure and Action of Proteins, (Harper and Row, New York, 1969).
10. C. Tanford and M. L. Wagner, J. Am. Chem. Soc. 76: 3331 (1954).
11. D. Nicoli and G. B. Benedek, Biopolymers 15: 2421 (1976).
12. L. D. Landau and E. M. Lifshitz, Statistical Physics, (Pergamon, Oxford, England, 1969), 2nd rev. English ed.
13. K. Kawasaki, in Phase Transitions and Critical Phenomena, edited by C. Domb and M. S. Green, (Academic, London, 1976), vol. 5A.
14. T. Tanaka, Phys. Rev. A17: 763 (1978).
15. R. W. Young and H. W. Fulhrorst, Invest. Ophthalmol. 5: 288 (1966).
16. B. Philipson, Invest. Ophthalmol. 8: 258 (1969).
17. J. I. Clark and G. B. Benedek, Biochem. and Biophys. Res. Comm. 96: 482 (1980).
18. S. Lerman and S. Zigman, Acta Ophthalmol. 45: 193 (1967).
19. R. Van Heyningen, Exp. Eye Res. 11: 415 (1971).
20. T. O. Sipple, Invest. Ophthalmol. 5: 568 (1966).
21. J. H. Kinoshita, Invest. Ophthalmol. 13: 713 (1971).
22. B. Philipson, Invest. Ophthalmol. 8: 281 (1969).
23. G. B. Benedek, J. I. Clark, E. Serrallach, C. Y. Young, L. Mengel, T. Sauke, A. Bagg, and K. Benedek, Phil. Trans. Roc. Soc. London 292: 121 (1979).

24. V. E. Kinsey and K. R. Hightower, Exp. Eye Res. 26: 521 (1978).
25. AMERICAN NATIONAL STANDARD Z136.1-1976.
26. O. Pomerantzeff, J. Govignon, and C. L. Schepens, Trans. Am. Acad. Ophthalmol. Otolaryngol. 73: 246 (1969).
27. R. A. Weale, Optical Acta 1: 107 (1954).
28. D. E. Koppel, J. Chem. Phys. 57: 4818 (1972).

STRUCTURES AND DYNAMICS OF MUSCLE CELLS AND MUSCLE PROTEINS STUDIED BY INTENSITY FLUCTUATION SPECTROSCOPY OF LASER LIGHT

S. Fujime and T. Maeda

Mitsubishi-Kasei Institute of Life Sciences
Tokyo 194

and

Y. Umazume

Department of Physiology
The Jikei University School of Medicine
Tokyo 105

CONTENTS

1. INTRODUCTION

In this article, we will describe (1) the intensity fluctuation spectroscopic (IFS) studies on dynamics of muscle thin filaments in solution, and (2) IFS studies on structural fluctuations of muscle cells during contraction. For persons who are not familiar with muscle structure and muscle proteins, we briefly introduce some current ideas about muscle.

Figure 1 is a schematic representation of the structure of skeletal (or more generally speaking, striated) muscle. (a) shows a gross anatomical feature of whole muscle. A piece of muscle (b) is called a muscle bundle, which consists of many muscle cells (or muscle fibers mf). A single fiber (c) has a diameter of 50-100 μm and consists of many myofibrils f. A fibril (d) has a diameter of the order of 1 μm and consists of many myofilaments called thin and thick filaments. Electron micrography has shown that the myofilaments pack

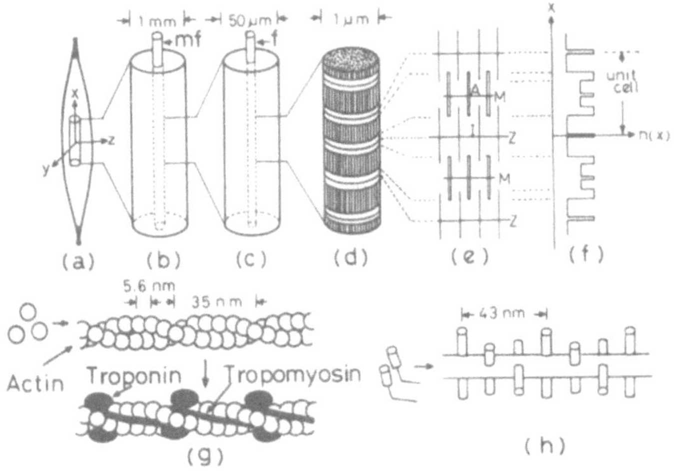

Figure 1. Schematic representation of the structure of skeletal muscle. (a) gross anatomical feature of whole muscle, (b) a piece of muscle (or muscle bundle), (c) a single fiber (or cell), (d) a single fibril, (e) ultrastructure of fibril; A: thick filament, I: thin filament, M: M-line, Z: Z-line, (f) functional form of protein concentration or refractive index n(x) along the fiber axis, (g) and (h) ultrastructures of thin and thick filaments respectively.

in the fibril forming a hexagonal lattice (e). The striations observed under an optical microscope are due to the periodic change in optical anisotropy or of the protein concentration in the fibril (f). A thin filament (g) is a fibrous actin-filament (F-actin) decorated with regulatory proteins named tropomyosin and troponin, and its total length is 1.0 µm. A thick filament (h) consists of myosin molecules and has projections, each of which is the enzymatically active portion of a myosin molecule.

Muscle contraction occurs in each sarcomere in a way that cross-bridges (projections from thick filaments) move out and combine with thin actin filaments and ATP hydrolysis occurs in a manner that chemical energy is transduced into mechanical energy. Muscle contraction and relaxation is regulated by a level of free calcium ion concentrations in the myofibrillar space. When the calcium ion concentration increases to a micromolar level, troponin molecules on thin filaments bind calcium ions and the actin-myosin interaction accompanying power strokes of cross-bridges occurs. When troponin molecules release calcium ions as a result of the decrease of the calcium ion concentration, the actin-myosin interaction is inhibited.

2. INTENSITY FLUCTUATION STUDIES OF MUSCLE PROTEINS*

The IFS measurements of muscle proteins in solution have been made for myosin[1], F-actin[2], and F-actin complexed with other muscle proteins.[3-5] The earlier works on F-actin and its complexes with other muscle proteins by Fujime and his associates were made by use of an analog spectrum analyzer. Later Carlson and his associates made the IFS measurements of F-actin and its complexes with other muscle proteins by use of a digital correlator, and gave a quite different result compared with those of Fujime et al.[6,7] Maeda and Fujime then reinvestigated the IFS spectrum of F-actin by use of a digital correlator, and obtained qualitatively similar results to those of Carlson et al.[8] Very recently, Newman and Carlson reported an interesting result on native thin filaments in solution.[9] We should like to restrict ourselves to discuss only the interpretation of the IFS spectrum of F-actin.

At first, our model on the field correlation function $g^{(1)}(\tau)$ for F-actin in solution is briefly reviewed for the later discussion.[10] As to the equation of motion of a semiflexible filament, we adopt the Harris-Hearst model[11]

$$\rho(\partial^2\vec{r}/\partial t^2) + \zeta(\partial\vec{r}/\partial t) + \varepsilon(\partial^4\vec{r}/\partial x^4) - \kappa(\partial^2\vec{r}/\partial x^2) = \vec{F}(t) \qquad (1)$$

where ρ is the linear density, ζ is the friction constant per unit length, $\vec{F}(t)$ is the fluctuating Brownian force, and $\vec{r}(x,t)$ is a space curve representing the chain ($-L/2 \leq x \leq L/2$; L being the contour length of the chain). The elastic constants have been given in terms of γ(the inverse of the statistical length) as[11]

$$\varepsilon = 3k_BT/4\gamma \quad \text{and} \quad \kappa = 3k_BT\gamma\mu(\gamma L) \qquad (2)$$

where

$$\mu(\gamma L) = [1 - (1 - e^{-2\gamma L})/2\gamma L]^{-1} \qquad (3)$$

Although some inconsistency of this model has been pointed out[12], we again adopt this model because Equation (1) is very tractable mathematically. Even when we adopt, for example, the Bixon-Zwanzig model[13], the situation may not be so different from the present one. By mode expansion

$$\vec{r}(x,t) = \sum_n \vec{q}(n,t)Q(n,x) \quad \text{and} \quad \vec{F}(t) = \sum_n \vec{B}(n,t)Q(n,x) \qquad (4)$$

* Work with T. Maeda

we have

$$\varepsilon Q(n,x)^{IV} - \kappa Q(n,x)^{II} - \lambda_n Q(n,x) = 0 \tag{5}$$

$$\rho\ddot{\vec{q}}(n,t) + \zeta\dot{\vec{q}}(n,t) + \lambda_n \vec{q}(n,t) = \vec{B}(n,t) \tag{6}$$

Equation (5) with the free-end boundary conditions

$$Q(n,x)^{II} = \varepsilon Q(n,x)^{III} - \kappa Q(n,x)^{I} = 0 \quad \text{at} \quad x = \pm L/2 \tag{7}$$

determines the eigenvalues λ_n and eigenfunctions $Q(n,x)$. Putting

$$\begin{Bmatrix} \alpha \\ i\beta \end{Bmatrix} = \left[\kappa/2\varepsilon \pm \{(\kappa/2\varepsilon)^2 + \lambda_n/\varepsilon\}^{\frac{1}{2}} \right]^{\frac{1}{2}} \tag{8}$$

where $\pm\,\alpha$ and $\pm\,i\beta$ are the roots of the characteristic equation to Equation (5), we have the following transcendental equation

$$1 - \cos(\beta L)\cosh(\alpha L) + \tfrac{1}{2}(y - 1/y)\sin(\beta L)\sinh(\alpha L) = 0 \tag{9}$$

where $y \equiv (\beta/\alpha)^3$. The nth root $\beta_n L$ of Equation (9) gives $\lambda_n = \varepsilon\beta_n^{\,4} + \kappa\beta_n^{\,2}$ or

$$3D\tau_n/L^2 = (4\gamma L)\Big/\left[(\beta_n L)^4 + (2\gamma L)^2(\beta_n L)^2\mu(\gamma L)\right] \tag{10}$$

where $D = k_B T/\zeta L$ was assumed because ζ was defined as the friction constant per unit length. The first four $\beta_n L$ and three τ_n are shown in Figure 2 as function of γL. The eigenfunctions are given by

$$Q(0,x) = (1/L)^{\frac{1}{2}} \tag{11}$$

$$Q(n,x) = \left(\frac{c_n}{L}\right)^{\frac{1}{2}}\left[\frac{\cos(\beta_n x)}{\cos(\beta_n L/2)} + \left(\frac{\beta_n}{\alpha_n}\right)^2 \frac{\cosh(\alpha_n x)}{\cosh(\alpha_n L/2)}\right] \quad \text{for even } n \tag{12}$$

$$Q(n,x) = \left(\frac{c_n}{L}\right)^{\frac{1}{2}}\left[\frac{\sin(\beta_n x)}{\sin(\beta_n L/2)} + \left(\frac{\beta_n}{\alpha_n}\right)^2 \frac{\sinh(\alpha_n x)}{\sinh(\alpha_n L/2)}\right] \quad \text{for odd } n \tag{13}$$

The eigenfunction $Q(0,x)$ belonging to the eigenvalue $\lambda_0 = 0$ represents the translational motion of the center-of-resistance of the chain. The eigenfunction $Q(1,x)$ in the stiff limit ($\gamma L \ll 1$) has the form

$$Q(1,x) = (12/L^3)^{\frac{1}{2}}x \tag{14}$$

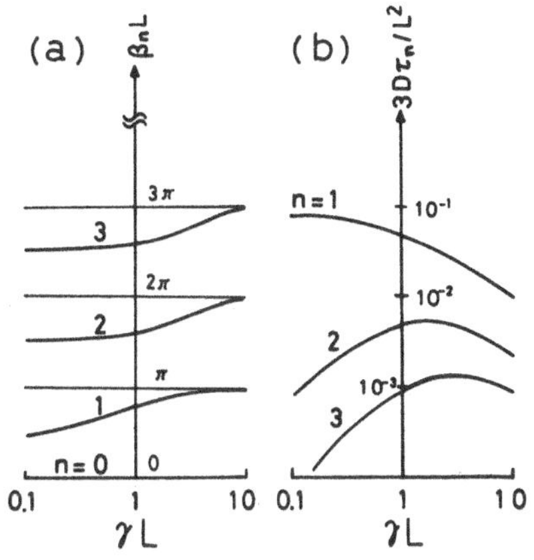

Figure 2. Graphic representation of eigenvalues and relaxation times. (a) $\beta_n L$ versus γL and (b) $3D\tau_n/L^2$ versus γL. Note that at $\gamma L << 1$, $\beta_1 L$ tends to zero but $3D\tau_1/L^2$ tends to 1/12.

which represents the rotational motion of the rod.

Under usual approximation, Equation (6) gives ($\tau = |t - t'|$)

$$\langle\vec{q}(n,t)\vec{q}(n,t')\rangle = \langle q(n,t)^2\rangle \exp(-\tau/\tau_n) \text{ and } \tau_n = \zeta/\lambda_n \quad (n\neq 0) \qquad (15)$$

The elastic energy of the chain in the present model is given by

$$\langle V\rangle = \tfrac{1}{2}\sum_n{}' \lambda_n \langle q(n,t)^2\rangle \qquad (16)$$

Thus, from equipartition of energy we have

$$\langle q(n,t)^2\rangle = 3k_B T/\lambda_n \quad (n \neq 0) \qquad (17)$$

Now the field correlation function $g^{(1)}(\tau)$ is given by

$$g^{(1)}(\tau) = G(\tau)/G(0), \quad G(\tau) = L^{-2}\iint_{-L/2}^{L/2} J(x,x',\tau)dxdx' \qquad (18)$$

$$J(x,x',\tau) = \exp(-DK^2\tau)\exp\Big[-(K^2/6)\sum_n{}'\langle q(n,t)^2\rangle \times \{Q(n,x)^2+Q(n,x')^2-2Q(n,x)Q(n,x')\exp(-\tau/\tau_n)\}\Big] \qquad (19)$$

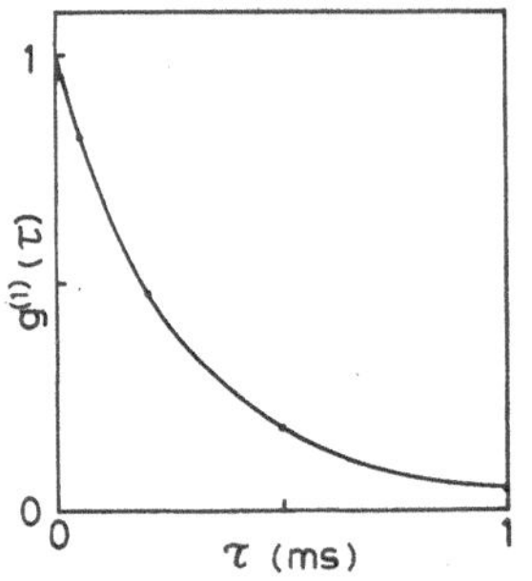

Figure 3. Correlation function for rigid rod. KL = 24 and D = 2.35×10^{-8} cm^2/s. (.) is from Equation (21) and (—) is from Equation (23).

Before computing $g^{(1)}(\tau)$ for a semiflexible chain, we consider the stiff limit of the above model. The smallest non-zero root of Equation (9) in the stiff limit is given by

$$\beta_1 L = (48\gamma L)^{1/4} \quad \text{or} \quad 3D\tau_1/L^2 = 1/12 \tag{20}$$

When we consider only the n = 0 and 1 modes, Equation (18) gives

$$g^{(1)}(\tau) = J^{-1} \sum_{\ell:\text{even}} P_\ell(KL)\exp\left[-(DK^2 + 3\ell\Theta)\tau\right] \quad \text{for } \gamma L \ll 1 \tag{21}$$

where $J = \sum P_\ell(KL)$ and

$$P_\ell(KL) = \frac{2^\ell}{\ell!}\left(\frac{1}{h}\int_0^h z^\ell e^{-z^2} dz\right)^2, \quad h = KL/2\sqrt{6} \text{ and } \Theta = D/(L^2/12) \tag{22}$$

Equation (21) should be compared with Pecora's formula[14]

$$g^{(1)}(\tau) = J^{-1} \sum_{\ell:\text{even}} B_\ell(KL)\exp\{-\left[DK^2 + \ell(\ell + 1)\Theta\right]\tau\} \tag{23}$$

where $J = \sum B_\ell(KL)$. Numerical computation suggests that both equations give very similar decay curves for large KL values (Figure 3).

To see the effect of polymer flexibility on $g^{(1)}(\tau)$, we compute Equation (18) for different values of γL at low and high scattering angles (Figure 4). In the actual computation, we took n up to 8 in Equation (19). The $g^{(1)}(\tau)$ computed with only the n = 1 (rotational) mode gives almost the same decay curves for γL = 0.1 and 0.5 (Figure 4c). This means that, for high scattering angles, the modes higher than n = 2 are essentially important for different decaying behaviors

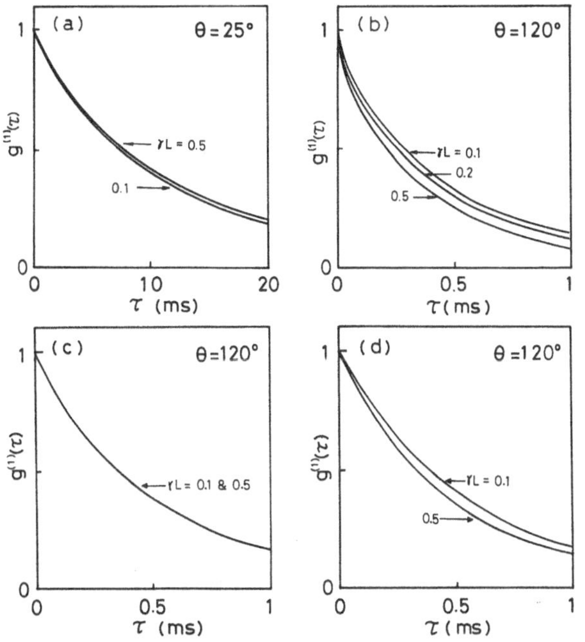

Figure 4. Correlation functions at varied flexibility parameters. (a) and (b) computation included modes n = 1 to 8. D = 1.24×10^{-8} cm^2/s, L = 1.01 μm and $K^2 = 0.549 \times 10^{10}$ cm^{-2} in (a) and 8.8×10^{10} cm^{-2} in (b). (c) computation included only n = 1 (rotational) mode. D, L, and K^2 are the same as in (b). (d) computation included modes n = 2 to 8. D = (1.24 + 0.748) $\times 10^{-8}$ cm^2/s, and L and K^2 are the same as in (b).

of $g^{(1)}(\tau)$ at different γL values (Figure 4d). [For γL = 0.5, the mode as high as n = 4 was found to appreciably contribute to $g^{(1)}(\tau)$.]

Newman and Carlson recently reported the IFS data of solution of intact thin filaments from scallop adductor muscle.[9] If the thin filament was a rigid rod, Equation (23) [and also Equation (21)] would be applicable. In such a case, $g^{(1)}(\tau)$ at the same scattering angle but at different temperatures should scale with the ratio of temperature to solvent viscosity T/η. *The low angle (10°-25°) time correlation functions of scattered light were independent of concentrations (0.08-1.3 mg/ml), scaled with the ratio T/η over a range of 4°-45°C, and yielded a value for D(5°C) = 1.24 × 10^{-8} cm^2/s which corresponded to the filament length of 1.06 μm. On the other hand, high angle (60°-150°) IFS temperature dependence did not scale with T/η and hence did not show the temperature dependence expected for rigid rods. Quantitative SDS polyacrylamide gel electrophoresis showed that at high temperature (>35°) tropomyosin completely disso-*

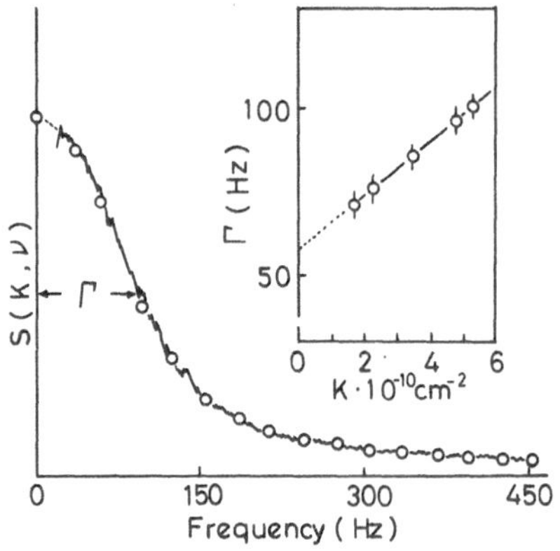

Figure 5. An example of power spectra of reconstituted F-actin. (o) is a single Lorentzian fit. Inset shows the half-height widths Γ versus K^2. F-actin at 4 mg/ml. From reference 5.

ciated from native thin filaments. They confirmed that the non-scaling at higher scattering angles was not due to any change in length distribution of filaments with temperature or to free tropomyosin in solution. They attributed the non-scaling to non-rigid motions of the filament.

Their data can be interpreted by the present model. We assume that the γL value of the thin filament increases with temperature, partly due to loosening of the structure of F-actin and partly due to weakening of F-actin/tropomyosin bonds. In fact, our own measurements by another technique[15] on the γL values of thin filaments of intact muscle have shown that $\gamma L = 0.2$ at 5°C and $\gamma L = 0.5$ at 45°C.[16] Adopting these values, we find in Figure 4a and b a quantitative agreement between their experimental and our computed $g^{(1)}(\tau)$'s.

ADDENDA We would like to discuss briefly the discrepancy between results from analog and digital methods mentioned earlier. An example of the power spectra obtained from solution of reconstituted F-actin is shown in Figure 5. When the shot noise level was determined at 5 kHz, the spectrum could well be fitted with a Lorentzian with width Γ. The inset in Figure 5 shows the Γ versus K^2 thus obtained, which suggests $\Gamma = AK^2 + B$ where A and B are constants. Since the contour length of F-actin reconstituted in vitro is very long, up to several μm, we ignored the rotational motion and obtained[10] (cf. Figure 4d)

$$g^{(1)}(\tau) = J^{-1} \sum_{M:\text{even}} P_{OM}(KL) \exp\left[-(DK^2 + M/\tau_2)\tau\right] \quad (24)$$

where $J = \sum P_{OM}(KL)$. We considered that the P_{OO} term was also too

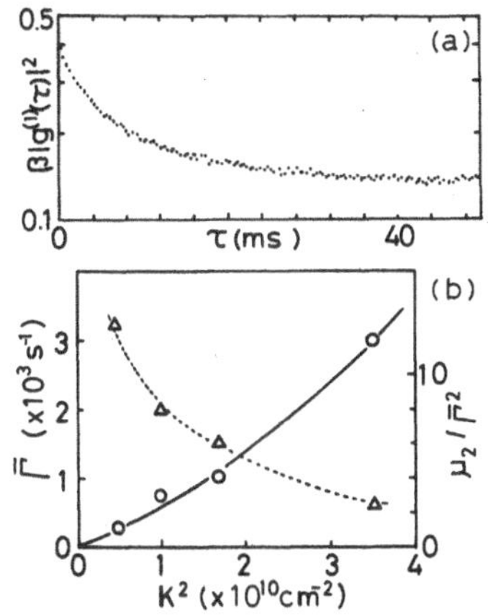

Figure 6. An example of correlation functions and the first two cumulants of reconstituted F-actin. (a) correlation function at T = 0.4 ms and scattering angle of 30°. Note that the β value at T = 1 μs was as large as 0.8. (b) $\bar{\Gamma}$ (o) and $\mu_2/\bar{\Gamma}^2$ (Δ) versus K^2. From reference 8.

narrow to be detected by our ac-coupled spectrum analyzer, and assumed that Γ was equal to the P_{OM}-weighted average of $(DK^2 + M/\tau_2)$.

On the other hand, F-actin prepared by the same procedure as above gave $|g^{(1)}(\tau)|^2$ as shown in Figure 6a, where the ordinate is in the logarithmic scale.[8] The correlation functions are highly non-exponential and have very slow decay components, much slower than 1 s^{-1}. The cumulants $\bar{\Gamma}$ and $\mu_2/\bar{\Gamma}^2$ are shown in Figure 6b, which suggests[6,8] $\bar{\Gamma} = A(K^2)K^2$, where $A(K^2)$ is an increasing function of K^2. A further study showed that the lower the actin concentrations and/or the higher the scattering angles were, the smaller the contribution from very slow components. This means that the slow components came from very slow translational movements of large gels of loosely entangled actin filaments.

The power spectra were calculated from experimental correlation functions. The Fourier transform of $|g^{(1)}(\tau)|^2$ is given by

$$S(K,\nu) = \int_0^{NT} |g^{(1)}(\tau)|^2 \cos(2\pi\nu\tau)d\tau = S_o(K,\nu)*S_w(\nu) \qquad (25)$$

where N is the total number of channels of the correlator, T the channel width, $S_o(K,\nu)$ the true power spectrum, $S_w(\nu)$ the Fourier transform of the window function defined by $G_w(\tau) = 1$ for $0 \leq \tau \leq NT$ and = 0 otherwise, and the asterisk means convolution. $S_w(\nu)$ has the width $\delta = 1/2NT$ (Hz), which corresponds to the band-width of the spec-

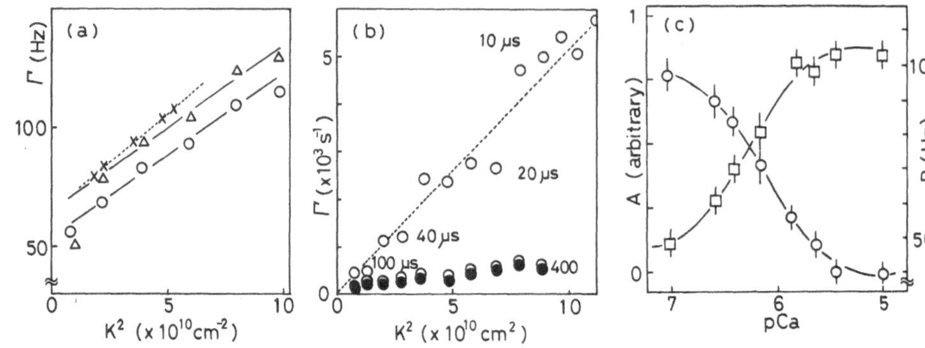

Figure 7. IFS results of reconstituted F-actin. (a) half-height widths (Γ) of power spectra. (x) is from Figure 5, (o at 1.4 mg/ml and Δ at 1 mg/ml) is from Fourier transformed spectra by use of Equation (25). (b) Γ (•) and Γ_1 (o) versus K^2, calculated by use of Equation (26). F-actin at 1 mg/ml and T specified in the figure. (c) an example of previous results showing very sensitive change in spectral widths. A complex of reconstituted thin filament and HMM at 1 mg/ml F-actin, and pCa = -(base-10 logarithm of free calcium ion concentration). A (□) and B (o) are from $\Gamma = AK^2 + B$. From reference 5.

trum analyzer. On the other hand, $\Delta = N\delta = 1/2T$ (Hz) corresponds to a frequency range scanned by the analyzer. For N = 125 and T = 0.8 ms, we have δ = 5 Hz and Δ = 625 Hz, which are very close to the previous conditions, i.e. $\delta = \pm$ 3 Hz and Δ = 400-800 Hz. In practical computation, the contribution from the very slow components to $|g^{(1)}(\tau)|^2$ observed at lower scattering angles were subtracted from the original data. This procedure was necessary to minimize the "termination error" and, at the same time, it played a role as the high-pass filter of the ac-coupled analyzer. As expected, the calculated spectra were in agreement with those obtained previously by the analog method (Figure 7a).

The correlation functions were further analyzed by assuming

$$|g^{(1)}(\tau)|^2 = P_1\exp(-\Gamma\tau)+P_2 \quad \& = Q_1\exp(-\Gamma_1\tau)+Q_2\exp(-\Gamma_2\tau) \qquad (26)$$

The results are shown in Figure 7b, where both Γ and Γ_1 (the fast component) at longer channel widths are close to the previous results.

The very slow components give very narrow spectral widths, so that the ac-coupled spectrum analyzer we used previously could not detect the narrow components. This is the most important origin of the discrepancy between results by the two methods. Further, F-actin reconstituted *in vitro* has a very broad length distribution, so that, in addition to the presence of very slow components mentioned above, experimental correlation functions depend strongly on the channel widths. The spectrum analyzer had only one band-width and hence it measured the power spectra corresponding to correlation functions at one particular channel width.

Anyway, the half-height widths of power spectra of light scattered from solutions of F-actin and of its complexes with other muscle proteins were very sensitive to the solvent conditions. An example of the previous results is shown in Figure 7c.[5] The earlier conclusion that F-actin is flexible in some cases and stiff under other conditions was confirmed later by different experimental techniques[15-18], and also by the recent study mentioned above. But, our earlier interpretation on spectral widths based on Equation (24) must be examined further, and what A and B in Figure 7c, for example, mean is still under question.

3. INTENSITY FLUCTUATION STUDIES OF MUSCLE CELLS*

Carlson and his associates first made the IFS measurements of muscle during contraction.[19-21] Fujime systematically studied the change in the intensities of optical diffraction lines from muscle fibers during contraction.[22] Combining these results, Fujime proposed a model of dynamics of sarcomere fluctuations.[23] At first this model is reviewed in a slightly modified version.

Let us consider a smooth cylinder whose radius and length (of the illuminated portion) are R and $\underline{\ell}$ respectively. Location in the cylinder is specified by the vector $\vec{j}$ whose cylindrical coordinates are denoted by r, ϕ and x (Figure 8). The scattering vector $\vec{K}$ also has cylindrical coordinates Ξ, Φ and u. The scattered field from the cylinder is given by

$$e(\vec{K}) = C\int_0^{\ell} n(x)e^{iux}dx\int_0^R\int_0^{2\pi} e^{i\Xi r\cos(\phi - \Phi)}rdrd\phi \tag{27}$$

where C is a proportionality constant (hereafter C = 1), and n(x) is the refractive index (Figure 1f). Noting that n(x) has period L (i.e. ℓ = NL, where L is the sarcomere length and N is the number of unit cells in the illuminated portion), we have the intensity expression

$$i(\vec{K}) = (\pi R^2)^2\left[2J_1(\Xi R)/\Xi R\right]^2|F(u)|^2L(u) \tag{28}$$

---

*Work with Y. Umazume

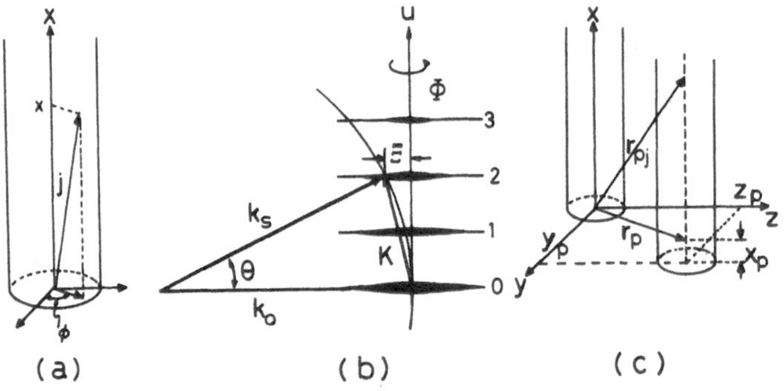

Figure 8. Geometrical relationships. (a) real space, (b) reciprocal space, and (c) definition of coordinates representing fibril arrangement.

where

$$F(u) = \int_0^L n(x)e^{iux}dx \quad \text{and} \quad L(u) = \frac{\sin^2(uLN/2)}{\sin^2(uL/2)} \tag{29}$$

and $J_1(z)$ is the Bessel function of the first order. The maxima in the intensity function occur along the u-axis at $u_n = 2\pi n/L$, where n is an integer (diffraction order index) specifying each maximum.

Now we consider the time-dependent fluctuations during contraction of the positions of scatterers in the unit cell around the lattice points. Although we consider below a simple lattice, i.e. one lattice point in a unit cell, extension to a complex lattice is straightforward. Let the position vector of the scatterer in the jth cell be

$$x_j(t) = jL + x^\circ + \delta x_j(t) \tag{30}$$

where x° is the position of the lattice point in the cell and $\delta x_j(t)$ is the fluctuation of the position of the scatterer around x°. We assume

$$<\delta x_j(t)\delta x_k(t')> = <\delta x^2>\Phi_x(\tau)\delta_{jk} \tag{31}$$

Letting $f \equiv F(u)[2\pi R^2 J_1(\Xi R)/\Xi R]$, we have

$$c(\vec{K},\tau) = <e(-\vec{K},t)e(\vec{K},t')> = f^2 <\sum_{jk}\sum e^{iu[x_j(t) - x_k(t')]}>$$

$$= f^2 e^{-2M} \sum_{jk}\sum e^{iu(j-k)L} \exp[2M\Phi_x(\tau)\delta_{jk}] \tag{32}$$

where the "Debye-Waller" factor M is defined as $M = \frac{1}{2}u^2\langle\delta x^2\rangle$. Now, Equation (32) can be evaluated as

$$c(\vec{K},\tau) = f^2 e^{-2M} L(u) + f^2 N e^{-2M}\{\exp[2M\Phi_x(\tau)] - 1\} \tag{33}$$

For $\tau = 0$ and $u = u_n \equiv 2\pi n/L$, the diffraction intensity is given by $i(u_n,\Xi) = f^2N^2\exp(-2M) + f^2N[1 - \exp(-2M)]$. The first term gives the intensity of diffraction lines and the second term the background intensity. The intensity drop by exp(-2M) is compensated by the increase of the background intensity. The field correlation function for $u \neq u_n$ (i.e. at off-line positions) is given by

$$g^{(1)}(\tau) \equiv c(\vec{K},\tau)/c(\vec{K},0) = \{\exp[2M\Phi_x(\tau)] - 1\}/\{\exp(2M) - 1\}$$

$$\simeq \Phi_x(\tau) \quad \text{for small } M \tag{34}$$

To consider the simplest model of a muscle fiber (or a bundle), we assume that the cylindrical fiber consists of many identical fibrils in near register. The fiber has a radius R_o while each fibril has a radius R ($R_o >> R$). Each fibril, if independent of each other, would diffract according to Equation (27). It is necessary, however, to take account of phase relations of the scattering by different fibrils. In the following, we consider a very simplified case in order to see the effect of fibril arrangement on the intensity. At time t, the position vector $\vec{r}_{pj}(t)$ of the scatterer in the jth unit cell of the pth fibril is assumed to be given by (Figure 8c)

$$\vec{r}_{pj}(t) = \vec{r}_p + \delta\vec{r}_p(t) + \vec{a}[jL + x^\circ + \delta x_j^p(t)] \tag{35}$$

where $\vec{r}_p = (x_p,y_p,z_p)$ is the position vector of the origin of the pth fibril relative to the reference fibril ($|x| < L$), whose fluctuations during contraction are denoted by $\delta\vec{r}_p(t) = [\xi_p(t),\eta_p(t),\zeta_p(t)]$. The $\delta x_j^p(t)$ is the fluctuation during contraction of the position of the scatterer in the jth cell of the pth fibril (cf. Equation (30)) and $\vec{a} = (1,0,0)$. If we assume for simplicity of algebra that both $\delta\vec{r}_p(t)$ and $\delta x_j^p(t)$ are Gaussian random variables with zero mean and that, for $(\mu_p,\nu_p) = \xi_p$, η_p or ζ_p,

$$\langle\mu_p(t)\nu_p(t)\rangle = \langle\mu_p(t)\delta x_j^p(t)\rangle = 0 \quad (\mu \neq \nu) \tag{36}$$

$$\langle\mu_p(t)\mu_q(t')\rangle = \langle\mu^2\rangle\Phi_\mu(\tau)\delta_{pq} \tag{37}$$

$$\langle\delta x_j^p(t)\delta x_k^q(t')\rangle = \langle\delta x^2\rangle\Phi_x(\tau)\delta_{jk}\delta_{pq} \tag{38}$$

where $\langle\xi^2\rangle \neq \langle\eta^2\rangle = \langle\zeta^2\rangle$ and $\Phi(\tau)$'s are suitably defined correlation functions, the similar mathematical procedure as in Equation (32) leads to the correlation function of scattered light from contracting muscle as

$$C(\vec{K},\tau) \equiv f^2 \sum_{pqjk}\sum\sum\sum \langle e^{i\vec{K}[\vec{r}_{pj}(t) - \vec{r}_{qk}(t')]}\rangle$$

$$= f^2 e^{-2(M_\xi+M_\eta+M)} L(u)\{L_f(uvw)+P[e^{2M_\xi\Phi_\xi(\tau)+2M_\eta\Phi_\eta(\tau)} - 1]\}$$

$$+ f^2 NPe^{-2(M_\xi+M_\eta+M)} e^{2M_\xi\Phi_\xi(\tau)+2M_\eta\Phi_\eta(\tau)} [e^{2M\Phi_x(\tau)} - 1] \tag{39}$$

where $P \equiv (R_o/R)^2$ is the total number of fibrils in a fiber and

$$M_\xi = \tfrac{1}{2}u^2\langle\xi^2\rangle, \quad M_\eta = \tfrac{1}{2}\Xi^2\langle\eta^2\rangle \quad \text{and} \quad M = \tfrac{1}{2}u^2\langle\delta x^2\rangle \tag{40}$$

and (uvw) are the Cartesian coordinates of $\vec{K}$ $[v^2 + w^2 = \Xi^2]$. The function

$$L_f(uvw) = |\sum_p e^{i(ux_p + vy_p + wz_p)}|^2 \tag{41}$$

is the "Laue function" similar to L(u), but is only statistically defined ($P \leq L_f(uvw) \leq P^2$). The radial shape factor of diffraction lines in the present model is given by $S(u_n,\Xi) = [2J_1(\Xi R)/\Xi R]^2 L_f(u_n vw)$. According to Bonner and Carlson[21], the experimental radial shape of the zero-order diffraction line during contraction is well fitted with only $[2J_1(\Xi R)/\Xi R]^2$ provided that 70% of fibrils have a diameter of 0.7 μm and 30% of fibrils have a diameter of 0.45 μm. This 2R value is comparable to or slightly smaller than the mean diameter of fibrils observed on electron micrographs. This result means, at least for contracting muscle, that $L_f(0vw)$ is a very slowly decreasing function of (vw) and, at the same time, that $L_f(uvw) \simeq P$ except for $L_f(000) = P^2$.

For $u \neq u_n$, we have from Equation (39)

$$g^{(1)}(\tau) = e^{-2M_\xi[1-\Phi_\xi(\tau)]} e^{-2M_\eta[1-\Phi_\eta(\tau)]} [e^{2M\Phi_x(\tau)} - 1]/[e^{2M} - 1] \quad (42)$$

$$\simeq \Phi_x(\tau) e^{-2M_\xi[1-\Phi_\xi(\tau)]} e^{-2M_\eta[1-\Phi_\eta(\tau)]} \quad \text{for small } M \quad (43)$$

For $u = u_n$, $L(u_n) = N^2 >> 1$. If we assume $L_f(u_n vw) = P$, we have

$$g^{(1)}(\tau) = e^{-2M_\xi[1-\Phi_\xi(\tau)]} e^{-2M_\eta[1-\Phi_\eta(\tau)]} \quad (44)$$

The correlation function $\Phi_x(\tau)$ can be calculated on a certain molecular basis. In order for the regular structure of sarcomeres to be kept, A-bands (and also I-bands) must be connected to each other with a spring-like force. We assume the viscoelastic model of a fibril shown in Figure 9. Let the friction constant of an A-band be ζ. Then the position $\delta x_j(t)$ of the jth M-line deviating from its mean position will obey the following equations of motion ($j = 1,2, \ldots, N$)

$$\zeta(\delta \dot{x}_j) - \kappa_A^\circ(\delta x_{j-1} - 2\delta x_j + \delta x_{j+1}) = F_j(t) \quad \text{for } \kappa_A^\circ << \kappa_A \quad (45)$$

$$\zeta(\delta \dot{x}_j) + \kappa_A(\delta x_j) = F_j(t) \quad \text{for } \kappa_A^\circ >> \kappa_A \quad (46)$$

The Z-line motion will also obey the same type of equations as above. In the case of Equation (45), the fluctuation is the largest at j = N/2. Experimentally speaking, this is probably not the case. Thus, we assume Equation (46). In the following, suffices j and A are omitted.

In contracting muscle, there will be a large fluctuating force F(t) due to imbalance of the actin-myosin connections in half sarcomeres of a fibril. We assume (see Figure 9) $F(t) = F^+(t) + F^-(t)$, $F^+(t) = \sigma[\bar{n} + \delta n^+(t)]$ and $F^-(t) = -\sigma[\bar{n} + \delta n^-(t)]$, where σ is the force developed by one actin-myosin connection, $\bar{n}$ the average number

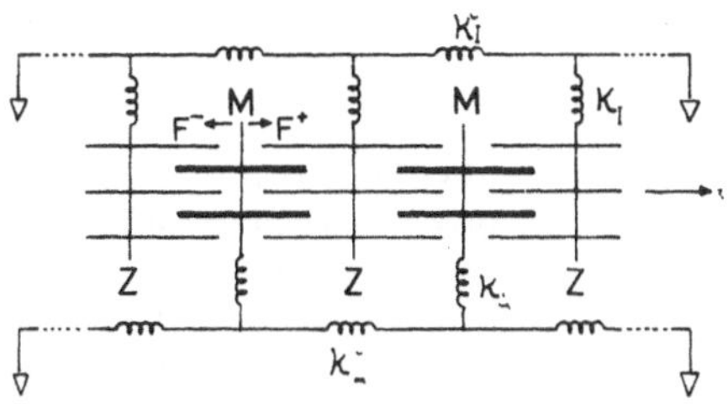

Figure 9. Viscoelastic model of myofibril.

of such connections in a half sarcomere of a fibril and $\delta n(t)$ the fluctuation in n. Then we have $\langle F^+(t)\rangle = -\langle F^-(t)\rangle = F_o/2$ (F_o is the contractile force per fibril) and $\langle F(t)F(t')\rangle = 2\sigma^2\langle\delta n(t)\delta n(t')\rangle$, where no correlation between $\delta n^+(t)$ and $\delta n^-(t)$ was assumed. We write, for simplicity of algebra, $\partial n/\partial t = (n_o - n)f - ng$, where f and g are the kinetic constants of the formation and breaking of the actin-myosin connections respectively, and n_o the total number of cross-bridges in a half sarcomere of a fibril.[24] Then, we have $\bar{n} = n_o f/(f + g)$ and $\langle\delta n(t)\delta n(t')\rangle = \bar{n}\exp(-\Gamma\tau)$, where $\Gamma \equiv f + g$ and $\langle\delta n^2\rangle = \bar{n}$ was assumed. In this model, we have $\langle F(t)F(t')\rangle = (2\sigma^2\bar{n}) \times \exp(-\Gamma\tau)$, or the power spectrum of the fluctuating force

$$G(\omega) = 4(2\sigma^2\bar{n})\Gamma/(\omega^2 + \Gamma^2) \tag{47}$$

According to a linear response theory[25], Equation (46) gives, for $F(t) = A_\omega\exp(i\omega t)$, $\delta x_\omega(t) = A_\omega\exp(i\omega t)/(i\zeta\omega + \kappa)$, or since $\langle A_\omega^* A_\omega\rangle$ is the power spectrum of the force

$$\langle\delta x^2\rangle\Phi_x(\tau) = \frac{1}{2\pi}\int_0^\infty \langle\delta x_\omega^*(t)\delta x_\omega(t')\rangle d\omega$$

$$= \frac{(2\sigma^2\bar{n})\Gamma}{\pi\zeta^2}\int_{-\infty}^{\infty}\frac{e^{i\omega\tau}}{(\omega^2 + q^2)(\omega^2 + p^2)}\,d\omega \tag{48}$$

Then the residue theorem gives

$$\Phi_x(\tau) = \frac{1}{1 - A}[e^{-q\tau} - Ae^{-q\tau/A}] \quad (p \neq q); \quad = e^{-q\tau} \quad (p = q) \tag{49}$$

where

$$\langle\delta x^2\rangle = (2\sigma^2\bar{n})/\kappa\zeta(p + q) = F_o\sigma/\kappa\zeta(p + q), \quad A = q/p \tag{50}$$

and $p = \Gamma$ and $q = \kappa/\zeta$ for $\Gamma > \kappa/\zeta$; and $p = \kappa/\zeta$ and $q = \Gamma$ for $\Gamma < \kappa/\zeta$. *The above consideration is valid for any case where equation of motion and the power spectrum of the driving force have the forms as in Equations (46) and (47) respectively.*

Correlation functions of scattered light from contracting muscle were obtained as follows. The number of photoelectron pulses $n(t_j,T)$ was registered at $T = 50$ μs and transferred to core memories of a Nova minicomputer in a DMA mode during the next registering cycle for $n(t_{j+1},T)$. Sixteen kilowords of memory were used for data accumulation area, so that the total time of one measurement was $Ta = 50\ \mu s \times 16 \times 1024 = 0.8192$ s. Muscle was electrically tetanized for 1 s and a measurement started usually at 0.18 s after the first stimulation. When one measurement finished, the minicomputer started to calculate

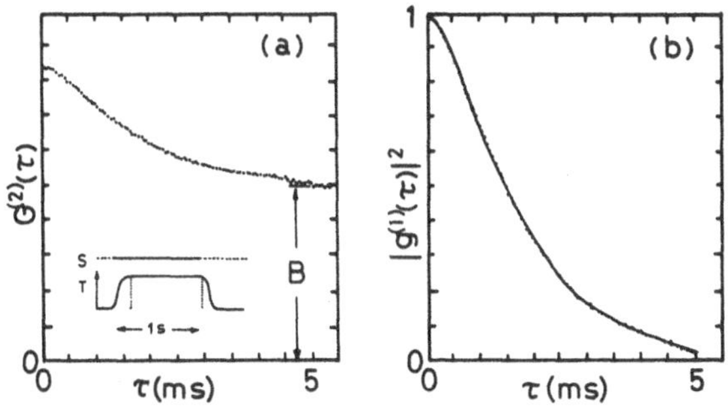

Figure 10. An example of correlation functions. (a) frog sartorius muscle during isometric contraction. On the zero-order diffraction line at the scattering angle of 25°. Inset shows electrical stimulation (S) and muscle tension (T). (b) $|g^{(1)}(\tau)|^2$ from (a). (...) is observed and (—) is least-square fitted to Equation (52). Overall behaviors of correlation functions are the same for whole sartorius muscle, bundles and single fibers (Figure 11a) of semitendinosus muscle both on the zero-order diffraction line and at off-line positions (usually between the second- and third-order diffraction lines).

a full correlation function

$$G^{(2)}(mT) = \sum_j n(t_j,T)n(t_j+mT,T) \quad \text{for } m = 1,2, \ldots , 100 \qquad (51)$$

The baseline level is given by $B = [\sum_j n(t_j,T)]^2/\sum_j 1$. Since the time required for computation of 16k data points was about 50 s, this computation finished within a waiting time for the next measurement; usually there was 5 min between successive measurements. In the case of skinned fibers or glycerinated fibers, replacement of the surrounding solution from a relaxing to contracting one initiated contraction. After every cycle of measurement for Ta s, data on core memories were transferred to disk, and calculation of $G^{(2)}(mT)$ was carried out after N times measurements.

Figure 10 shows an example of correlation functions of scattered light from contracting muscle. Overall behaviors of correlation functions are very similar to those of Carlson et al.[19-21] The characteristic points of correlation functions are

(i) The amplitude of excess correlation $\beta \equiv [G^{(2)}(0) - B]/B$ was 0.6 to 0.65 both at $u = 0$ and $u \neq u_n$. This β value was about 90% of the maximum one of our optical setup. *No heterodyning is said to occur.*

(ii) $|G^{(1)}(mT)|^2 \equiv G^{(2)}(mT) - B$ has a very characteristic be-

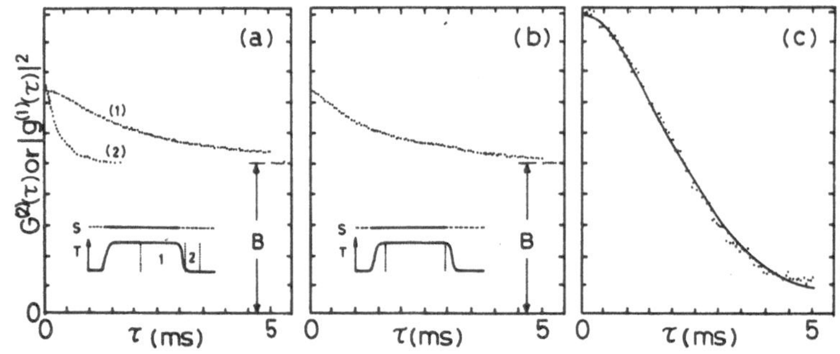

Figure 11. Example of correlation functions. (a) correlation function of a single fiber measured at u ≠ 0. (1) and (2) correspond, respectively, to correlation functions measured at time phases 1 and 2 in the tension diagram. The rapid decay curves similar to (2) were also observed during the rising phase of tension. (b) correlation function at K = 0 scattering. To avoid heterodyning, whole sartorius muscle of a big frog (thickness of about 1 mm) was intentionally used; otherwise we could observe correlation functions with large excess amplitude during only the rising and falling phases of tension, where internal shortening occurred. (c) correlation function at an off-line position from rabbit psoas muscle bundle at 7°C of contracting medium. (...) is the observed for data accumulation period of 200 ms and (—) is least-square fitted to Equation (52). In this case, we could not observe correlation functions with large excess amplitude at u = 0 (i.e. on the zero-order diffraction line). [In (a)-(c), the β values were about 0.5, which were more than 80% of the β value for solution of polystyrene latex spheres in this particular optical setup].

havior. Almost all experimental correlation functions are well expressed by

$$|G^{(1)}(mT)|^2 = G\cdot|\Phi(mT)|^2 \quad \text{and} \quad \Phi(\tau) = \frac{1}{1-A}[e^{-Q\tau} - Ae^{-Q\tau/A}] \qquad (52)$$

where G is a scale factor (Figure 10b). This was first pointed out empirically by Carlson et al.[20] It should be noted that the functional form of $\Phi(\tau)$ has been naturally derived as shown above, i.e. $\Phi_x(\tau)$.

(iii) Even at K = 0, we could observe correlation functions with a large excess amplitude characteristic of scattered light from contracting muscle (Figure 11b). This might be due to strong violation of the Rayleigh-Gans approximation (or there exist significant optical path-length fluctuations) and/or multiple scattering processes.

For u = 0, Equation (44) violates experimental facts (i) and (ii). Taking account of the fact (iii), we assume

$$g_o^{(1)}(\tau) = \Phi_o(\tau)e^{-2M_\eta[1 - \Phi_\eta(\tau)]} \quad (53)$$

where $\Phi_o(\tau)$ has the same functional form as Φ's already mentioned, and the exponential factor in Equation (53) is regarded as a modifying factor (i.e. M_η is small and/or $\Phi_\eta(\tau)$ decays very slowly) during the tension plateau.

For $u \neq u_n$, if we again assume that the exponential factors are modifying ones at tension plateau, Equation (43) is compatible with experimental facts (i) and (ii). But, logically speaking, $g^{(1)}(\tau)$ should contain $\Phi_o(\tau)$;

$$g_o^{(1)}(\tau) = \Phi_o(\tau)\Phi_x(\tau)e^{-2M_\xi[1 - \Phi_\xi(\tau)]}e^{-2M_\eta[1 - \Phi_\eta(\tau)]} \quad (54)$$

The $\Phi_o(\tau)$ corresponds to $C_A(K,\tau)$ of Bonner and Carlson.[21] We have not yet succeeded in deriving $\Phi_o(\tau)$ based on a suitable model, but we can easily suspect that $\Phi_o(\tau)$ closely relates to $\Phi_x(\tau)$.

Variations of correlation functions from measurement to measurement and from preparation to preparation probably come from the different behaviors of modifying factors in Equations (53) and (54). For simplicity of discussion, we ignore these factors for a moment. [These factors may be important in the rising and falling phases of tension development, where internal shortening occurs ((2) in Figure 11a).]

According to Carlson et al[20,21], the inverse of the half-time $\tau_{\frac{1}{2}}$, where $|G^{(1)}(\tau)|^2 = \frac{1}{2}|G^{(1)}(0)|^2$, does not depend on Ξ when u = 0, whereas it increases monotonically with u. So, the factor $\Phi_x(\tau)$ or $\{\exp[2M\Phi_x(\tau)] - 1\}/\{\exp(2M) - 1\}$ in Equation (54) is important.

An interesting fact is that when we used glycerinated fibers or skinned fibers and when we adjusted the environmental conditions such as temperature and/or the level of calcium ion concentrations so that the fibers develop a small amount of tension, we could not observe excess amplitude of correlation functions at u = 0 and $\Xi \neq 0$, whereas we did observe at $u \neq u_n$ very similar correlation functions to those of electrically (maximally) stimulated muscle (Figure 11c). As far as we have tried, we could not observe the factor $\Phi_o(\tau)$ during small tension development. The factor $\Phi_x(\tau)$ is very important.

In the following, an order-of-magnitude estimation is made. From literature values of $\sigma = 0.6 \times 10^{-6}$ dyn and $\bar{n} = 4 \times 10^4$, the assumed value of $\langle\delta x^2\rangle = (0.15\ \mu m)^2$ (see reference 22), and $\Gamma = 350\ s^{-1}$ and $\kappa/\zeta = 1940\ s^{-1}$ (A = 0.18) in Figure 10b, we have $\zeta \simeq 3 \times 10^{-3}$ dyn/cm/s and $\kappa \simeq 4$ dyn/cm. The ζ value is very close to the estimated one, 2.2×10^{-3} dyn/cm/s, from the friction constant, 2.8×10^{-6} dyn/cm/s, of one myosin filament in water multiplied by number, say 800, of thick filaments in a sarcomere of a fibril (i.e. the free draining value). On the other hand, it is not clear to what the elastic component having $\kappa \simeq 4$ dyn/cm corresponds. If our model is valid, we must assume that the elastic component (κ_A in Figure 9) is very nonlinear or the potential profile is something like a square well having a certain width, within which the M-line is very weakly bound.

4. CONCLUDING REMARK

We have seen a present status of muscle studies by IFS method. Because the system is very complex in both *in vitro* and *in vivo* experiments, quantitative investigation of IFS spectrum is rather difficult at present. Additional experimental and theoretical studies are required in order to discuss muscle physiology from IFS results.

We thank Mrs. Shuko Chiba-Nakagawa for her patient assistance.

REFERENCES

1. T. J. Herbert and F. D. Carlson, *Biopolymers* 10: 2231 (1971).
2. S. Fujime, *J. Phys. Soc. Japan* 29: 751 (1970).
3. S. Fujime and S. Ishiwata, *J. Mol. Biol.* 62: 251 (1972).
4. S. Ishiwata and S. Fujime, *J. Mol. Biol.* 68: 511 (1972).
5. S. Ishiwata, PhD thesis (Nagoya University, 1975).
6. F. D. Carlson and A. B. Fraser, *J. Mol. Biol.* 89: 273 (1974).
7. A. B. Fraser, E. Eisenberg, W. Keilley, and F. D. Carlson, *Biochemistry* 14: 2207 (1975).
8. T. Maeda and S. Fujime, *J. Phys. Soc. Japan* 42: 1983 (1977).
9. J. Newman and F. D. Carlson, *Biophys. J.* 29: 37 (1980).
10. S. Fujime and M. Maruyama, *Macromolecules* 6: 237 (1973).
11. R. A. Harris and J. E. Hearst, *J. Chem. Phys.* 44: 2595 (1966).
12. K. Soda, *J. Phys. Soc. Japan* 35: 866 (1973).
13. M. Bixon and R. Zwanzig, *J. Chem. Phys.* 68: 1896 (1978).
14. R. Pecora, *J. Chem. Phys.* 49: 4126 (1968).
15. S. Yoshino, Y. Umazume, R. Natori, S. Fujime, and S. Chiba, *Biophys. Chem.* 8: 317 (1978).
16. S. Yoshino and S. Fujime, *in* "Muscle Contraction: Its Regulatory Mechanism", S. Ebashi et al, eds., Japan Sci. Press, Tokyo/ Springer-Verlag, Berlin (1980).
17. T. Takebayashi, Y. Morita, and F. Oosawa, *Biochim. Biophys. Acta* 492: 357 (1977).
18. T. Yanagida and F. Oosawa, *J. Mol. Biol.* 126: 507 (1978).
19. F. D. Carlson, R. Bonner, and A. Fraser, *Cold Spring Harbor Symp. Quant. Biol.* 37: 389 (1972).

20. F. D. Carlson and A. Fraser, *in* "Photon Correlation and Light Beating Spectroscopy", H. Z. Cummins and E. R. Pike, eds., Plenum, New York (1973). pp. 519-538.
21. R. F. Bonner and F. D. Carlson, *J. Gen. Physiol.* 65: 555 (1975).
22. S. Fujime, *Biochim. Biophys. Acta* 379: 227 (1975).
23. S. Fujime and S. Yoshino, *Biophys. Chem.* 8: 305 (1978).
24. A. F. Huxley, *in* "Progress in Biophysics", volume 7 (1957). pp. 255-318.
25. C. Kittel, *in* "Elementary Statistical Physics", John Wiley, New York (1958). p. 147.

PART V

Contributed Papers

DETERMINATION OF LARGE MOLECULAR WEIGHTS BY FLUCTUATION SPECTROSCOPY WITH QUASIELASTIC LIGHT SCATTERING

Kohji Ohbayashi, Masaharu Minoda, and Hiroyasu Utiyama

Faculty of Integrated Arts and Sciences
Hiroshima University
Higashisenda-machi, Hiroshima 730, Japan

A new method is proposed to determine the molecular weight M of large molecules by observing the probability distribution function of the fluctuation of quasielastic light scattering. The experimental system used is depicted in Figure 1. The quasielastic light $I_s(t)$ scattered by molecules in a finite volume V is detected by a photomultiplier tube. The photocurrent is amplified, and passed through a low pass filter having a cutoff frequency of ω_c. The low pass filter is used to reduce the fluctuation of the photocurrent not associated with the number fluctuation. The probability distribution function of the photocurrent fluctuation is calculated with a correlator and then analyzed with a minicomputer.

The correlation function of the light scattered by the molecules in the finite volume V is given by[1]

$$\langle I_s(t)I_s(0)\rangle = \langle N\rangle^2\beta^2 + \langle N\rangle^2\beta^2\exp(-2Dk_s^2|t|)$$

$$+ \langle N\rangle \frac{\beta^2}{1+4D|t|/\alpha^2}\int_{-\infty}^{\infty}\frac{dx}{\pi}\left(\frac{\sin x}{x}\right)^2 \exp\left(-\frac{D|t|x^2}{\ell^2}\right) \tag{1}$$

where $\langle N\rangle$ is the mean number of molecules in V, β is the suitably chosen detection efficiency, D is the diffusion constant of the molecules, k_s is the scattering vector, 2d is the diameter of the laser beam, and 2ℓ is the length of the beam observed by the receiving optics. In Equation (1), the first term is the d.c. part, the second

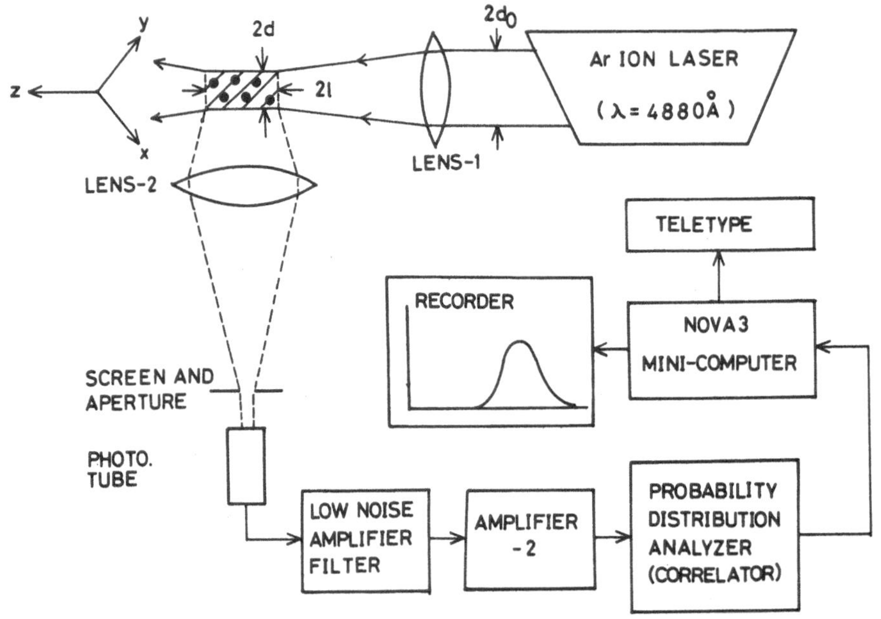

Figure 1. The block diagram of the experimental arrangement for the determination of molecular weight by fluctuation spectroscopy with quasielastic light scattering.

corresponds to the selfbeat of the quasielastic light scattering, and the last term expresses the number fluctuation. Then, the correlation function of the photocurrent fluctuation which passes through the low pass filter is given by

$$<J(0)J(t)> = <N> \frac{e^2 a\beta\omega_c}{2} \exp(-\omega_c|t|) + <N>^2 e^2 a^2 \beta^2$$

$$+ <N>^2 \frac{e^2 a^2 \beta^2 \omega_c}{2Dk_s^2} \exp(-\omega_c|t|)$$

$$+ <N> \frac{e^2 a^2 \beta^2}{1+4D|t|/d^2} \int_{\infty}^{\infty} \frac{dx}{\pi} \left(\frac{\sin x}{x}\right)^2 \exp\left(-\frac{D|t|x^2}{\ell^2}\right) \quad (2)$$

where e and a are the electronic charge and the detection efficiency, respectively. The value of ω_c was chosen so that d/ℓ^2, $D/d^2 << \omega_c << Dk_s^2$. The first term in Equation (2) is due to the shot noise.

The probability distribution function of the photocurrent fluctuation is

$$P(J) = \frac{1}{\pi\sqrt{<\delta J^2>}} \exp(-(J - <J>)^2/<\delta J^2>) \tag{3}$$

and

$$\begin{aligned} <J>^2 &= <N>^2\beta^2 e^2 a^2 \\ <\delta J^2> &= <N>\beta^2 e^2 a^2 + <N>^2\beta^2 e^2 a^2 \omega_c/2Dk_s^2 \ . \end{aligned} \tag{4}$$

The contribution of the shot noise to $<\delta J^2>$ can be neglected. If the weight concentration of the solution is $<c>$, the molecular weight is given by

$$M = <c>VN_A/<N> \tag{5}$$

where N_A is Avogadro's number. From Equations (4) and (5), we get

$$\frac{<\delta J^2>}{<J>^2} <c> = \frac{M}{VN_A} + \frac{\omega_c}{2Dk_s^2} <c> \tag{6}$$

This relation, being linear with respect to $<c>$, shows that, by plotting the quantity on the left hand side against $<c>$, we obtain M/VN_A from the ordinate intercept. Thus we may obtain the molecular weight M if V is known. It is also noted from Equation (6) that the slope of the plot is inversely proportional to D. Hence, if $\omega_c/2k_s^2$ is known, D may be estimated from the slope.

The present method is an extension of the number fluctuation method proposed by Weissman et al.[2] The advantage of the present method is that we are using a fixed cell instead of a rotating cell, which simplifies the experimental set-up and reduces the amount of the sample needed. The use of quasielastic light scattering instead of fluorescence may allow a wider application of this method, if contamination by dust is avoided.

In Equation (6), the scattering volume V and ω_c/k_s^2 are the quantities that should be determined by separate experiments. For this purpose, we observed the scattering from polystyrene latex sphere of 0.091 μm in diameter. The catalogue molecular weight and the diffusion constant are 2.50×10^8 and 4.75×10^{-8} cm^2/sec, respectively, at 20°C in water. The weight concentrations we used were $<c> = 6.97 \times 10^{-7}$, 1.29×10^{-6}, 1.75×10^{-6}, 3.74×10^{-6}, and $5.21 \times$

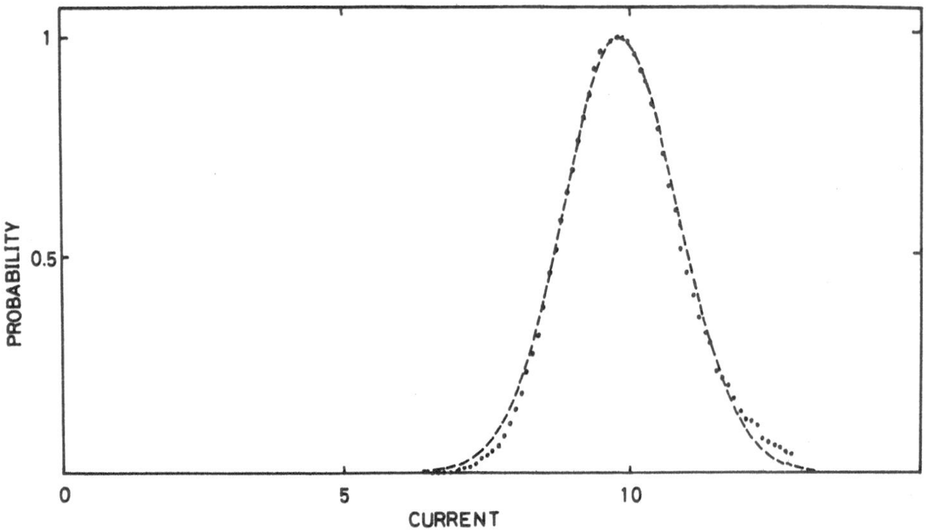

Figure 2. The probability distribution of the photocurrent fluctuation of the intensity of light scattered by the polystyrene latex spheres suspended in water. The diameter of the sphere is 0.091 μm, and the weight concentration of the solution is 1.29×10^{-6} g/cm^3. The scales are arbitrary.

10^{-6} in g/cm^3. One of the results of the probability distribution of the current fluctuation is shown in Figure 2 for the sample of $\langle c \rangle = 1.29 \times 10^{-6}$ g/cm^3. The broken curve is the best fit of Equation (3) to the experimental data (filled circles). From the fit, relative values of $\langle \delta J^2 \rangle$ and $\langle J \rangle^2$ were determined. The ratio thus determined for the present case is $\langle \delta J^2 \rangle / \langle J \rangle^2 = 0.0113$. The contribution of solvent to $\langle J \rangle$ had been subtracted before taking the ratio. About 20 measurements were made for each sample and the averages were taken. Figure 3 plots the results of $\langle c \rangle \langle \delta J^2 \rangle / \langle J \rangle^2$ against $\langle c \rangle$. The results fall on a straight line as expected from Equation (6). From the intercept with the ordinate, the scattering volume V was determined to be $V = (3.2 \pm 0.32) \times 10^{-7}$ cm^3. The slope gave a value of $\omega_c / 2k_s^2 = (7.1 \pm 0.7) \times 10^{-11}$ cm^2/sec.

Using the parameters thus determined, we measured the molecular weight of bacteriophage T4. Four samples having the concentrations $\langle c \rangle = 1.83 \times 10^{-7}$, 7.60×10^{-7}, 1.61×10^{-6}, and 1.84×10^{-6} in g/cm^3 were measured. From a similar plot, we obtained $M = (1.9 \pm 0.2) \times 10^8$. This value is in satisfactory agreement with the value $(1.925 \pm 0.066) \times 10^8$ determined by Dubin et al. using other methods.[3]

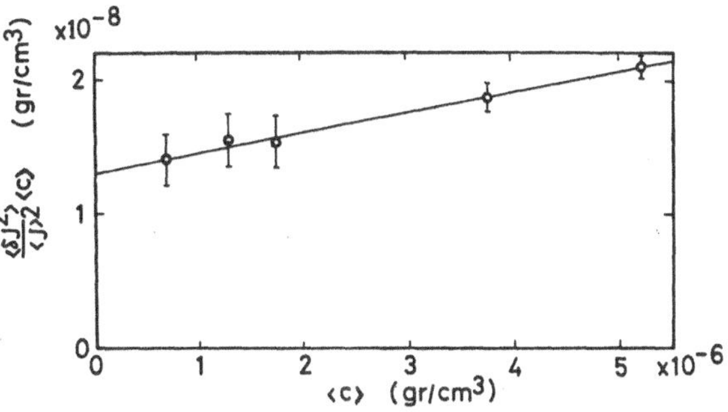

Figure 3. The plot of $\langle c\rangle\langle\delta J^2\rangle/\langle J\rangle^2$ against $\langle c\rangle$, where $\langle c\rangle$ is the mean weight concentration, $\langle\delta J^2\rangle$ is the mean square fluctuation of the photocurrent, and $\langle J\rangle$ is the mean d.c. photocurrent. From the ordinate and the slope of the best-fit straight line, we can determine the molecular weight and the diffusion constant, respectively (see text).

REFERENCES

1. K. Ohbayashi, M. Minoda, and H. Utiyama, Macromolecule vol. 14 (1981).
2. M. Weissman, H. Schindler, and G. Feher, Proc. Natl. Acad. Sci. USA 73: 2776 (1976).
3. S. B. Dubin, G. B. Benedek, F. C. Bancroft, and D. Freifelder, J. Mol. Biol. 54: 547 (1970).

NUMERICAL INVERSION OF THE LAPLACE TRANSFORM

Didier Sornette and Nicole Ostrowsky

Laboratoire de Physique de la Matière Condensée
Associé au CNRS (LA 190), Université de Nice,
Parc Valrose, 06034 NICE CEDEX, France

We wish to present a practical way to numerically reconstruct $G(\Gamma)$ from the experimentally measured correlation function $g(\tau)$, i.e. deal with polydispersity problems:

$$g(\tau) = \int_0^\infty G(\Gamma)\, e^{-\Gamma\tau}\, d\Gamma \tag{1}$$

MATHEMATICAL THEORY

Expanding $G(\Gamma)$ on the eigen functions of the Laplace transform:[1]

$$G(\Gamma) = \int_{-\infty}^{+\infty} d\omega \Psi_\omega(\Gamma)\, a_\omega \tag{2}$$

yields for $g(\tau)$:

$$g(\tau) = \int_{-\infty}^{+\infty} \lambda_\omega a_\omega\, \Psi_\omega(\tau)\, d\omega \tag{3}$$

with:

$$\lambda_\omega = \frac{\omega}{|\omega|}\sqrt{\frac{\pi}{ch(\pi\omega)}} \tag{4}$$

As McWhirter and Pike pointed out, when ω goes to infinity, λ_ω goes to zero which means that for large values of ω (say $\omega \geq \omega_{max}$) the eigen functions are "transmitted" so weakly across equation (1) that they cannot be distinguished from the noise. We must then rewrite:

$$g(\tau) = \int_{-\omega_{max}}^{\omega_{max}} \lambda_\omega \, a_\omega \, \psi_\omega(\tau) \, d\omega + \text{noise} \tag{5}$$

This amounts to band limiting $G(\Gamma)$ which becomes an approximated function $\tilde{G}(\Gamma)$. Using the explicit value of $\psi_\omega(\Gamma)$, this function can be expressed as:

$$e^{1/2 \, \mathrm{Log}\Gamma} \, \tilde{G}(\Gamma) = \int_{-\omega_{max}}^{\omega_{max}} \beta(\omega) \, e^{i\omega \, \mathrm{Log}\Gamma} d\omega \tag{6}$$

where $\beta(\omega)$ is a linear combination of a_ω and $a_{-\omega}$.

Defining a new function $\mathcal{G}(\mathrm{Log}\Gamma)$ as:

$$\mathcal{G}(\mathrm{Log}\Gamma) \, d(\mathrm{Log}\Gamma) = \tilde{G}(\Gamma) \, d\Gamma \tag{7}$$

equation (6) is conveniently rewritten as:

$$e^{-1/2 \, \mathrm{Log}\Gamma} \mathcal{G}(\mathrm{Log}\Gamma) = \int_{-\omega_{max}}^{\omega_{max}} \beta(\omega) \, e^{i\omega \mathrm{Log}\Gamma} \, d\omega \tag{8}$$

The left hand side of this equation thus appears as a band limited Fourier transform[2] for which the sampling theorem applies:
- $\mathcal{G}(\mathrm{Log}\Gamma)$ can only be known for points $\mathrm{Log}\,\Gamma_n$, $\mathrm{Log}\,\Gamma_{n+1}$, etc... separated by at least π/ω_{max}.
- $\mathcal{G}(\mathrm{Log}\Gamma)$ can be reconstructed from these points by the following interpolation formula:

$$\mathcal{G}(\mathrm{Log}\Gamma) = \sum_{n=1}^{\infty} \mathcal{G}(\mathrm{Log}\,\Gamma_n) \sqrt{\frac{\Gamma}{\Gamma_n}} \, \frac{\sin(\omega_{max} \mathrm{Log}(\Gamma/\Gamma_n))}{\omega_{max} \mathrm{Log}(\Gamma/\Gamma_n)} \tag{9}$$

PRACTICAL APPLICATIONS[3]

a) For a given noise on the experimental points, ω_{max} must be chosen so that:

$$\lambda_{\omega_{max}} \approx \text{noise},$$

that is, using equation (4):

$$\omega_{max} \approx \frac{2}{\pi} \text{Log} \frac{\sqrt{\pi}}{\text{noise}} \quad (10)$$

For example, for noise $\approx 10^{-3}$, $\omega_{max} \approx 5$.
However, we found experimentally that an a priori knowledge on the range allowed for Γ permits us to significantly raise ω_{max}.

b) The sampling of $G(\Gamma)$ is performed, using the following approximation:

$$G(\Gamma) = \sum_{n=1}^{N} a_n \delta(\Gamma - \Gamma_n) \qquad \text{with } \Gamma_n = \left[e^{n\pi/\omega_{max}}\right] \Gamma_1 \quad (11)$$

One can show that:

$$\frac{\omega_{max}}{\pi} a_n = \mathcal{G}(\text{Log}\Gamma_n) \quad (12)$$

This formula, correct in the limit ω_{max} goes to infinity, is valid within a few per thousands for $\omega \gtrsim 5$.

c) The coefficients a_n are obtained by minimizing the function:

$$\left| \sum_{j=1}^{M} g(\tau_j) - \sum_{n=1}^{N} a_n e^{-\Gamma_n \tau_j} \right|^2 \quad \text{with respect to the } a_n\text{'s.}$$

(M is the number of experimental points on the correlation function, and N the number of sampling points for $G(\Gamma)$).

d) The reconstruction of the band limited function $\mathcal{G}(\text{Log}\Gamma)$ can be performed in two ways:
- Use the interpolation formula (9)
- Sample $\mathcal{G}(\text{Log}\Gamma)$ for different sets of δ-functions slightly shifted with respect to one another.

The equivalence of these two methods was tested on a series of experimental (see Figure 1a) and simulated data (see Figures 1b and c). Note that for this latter case, the simulated distribution function shown in solid line was used to generate a correlation function to which noise was then added. The set of δ-functions and the reconstructed $\mathcal{G}(\text{Log } R)$ (dashed line) was then obtained by another operator without any information concerning the form of the original function.

In addition to having a great resolution when the Γ range is limited, this method has proven most useful in cases where there was no a priori knowledge on the shape of the distribution function $G(\Gamma)$.

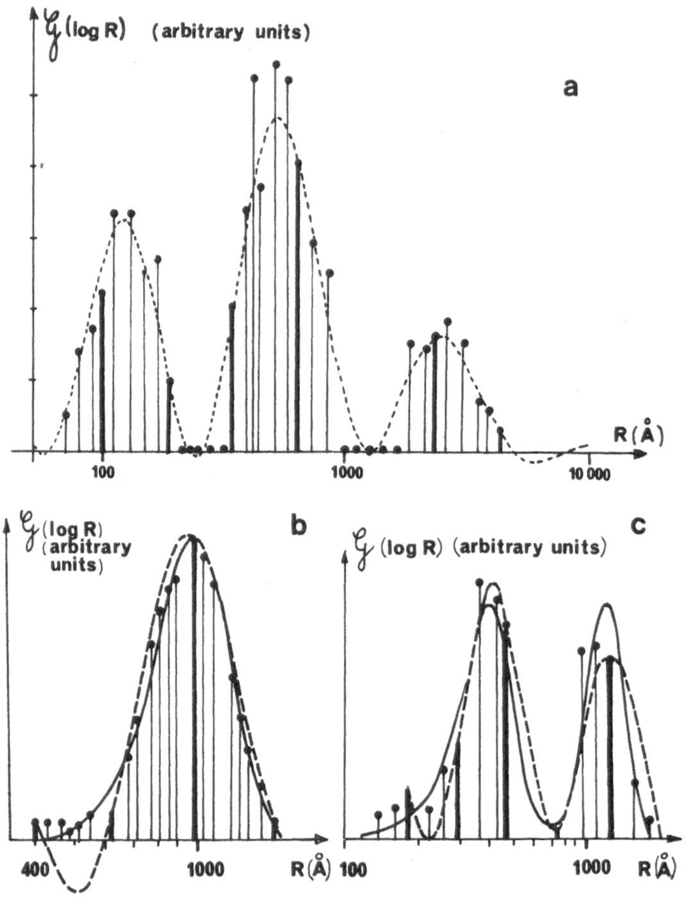

Figure 1. Reconstruction of the distribution function 𝒢(Log R) from experimentally measured (a) or simulated (b and c) correlation functions.
The dashed line was obtained from the interpolation formula (9) using for 𝒢(Log R_n) the values found with the set of δ-functions indicated by thick vertical lines. The thin lines represent the other sets of δ-functions shifted with respect to one another.

REFERENCES

1. J. Mc Whirter and E. R. Pike, J. Phys. A. Math Gen 11: 1729 (1978).
2. J. Mc Whirter, Optica Acta 27: 83 (1980).
3. N. Ostrowsky, P. Parker, E. R. Pike, and D. Sornette, to be published in Optica Acta (1981).

DEPOLARIZED RAYLEIGH SPECTRA AND HYDROGEN BONDING OF LIQUID 1,5- AND 2,4-PENTANEDIOL

G. Fytas, W. Mersch, and Th. Dorfmüller

Department of Chemistry
University of Bielefeld
4800 Bielefeld 1, West Germany

CONTENTS

1. INTRODUCTION

Low frequency depolarized light scattering (DLS) in a liquid provides information about the intrinsic anisotropy of the polarizability tensor. In the case of small non hydrogen bonded anisotropic molecules, it has been well established, that the dynamics of molecular anisotropy is reflected in the depolarized Rayleigh spectra.[1] In the case of liquids with strong intermolecular interactions, such as hydrogen bonding, the DLS may contain components due to a structural anisotropy induced by these interactions. Thus we have observed in 1,2,6-hexanetriol that the isolated molecule at high temperatures has a very low anisotropy in the low frequency region ($\sim$ 10 GHz).[2] Similar observations have been reported for liquid water.[3] Also the different dielectric relaxation behavior of five isomeric pentanediols has been related to the extent of intramolecular H-bonding. In order to examine closely these interactions, as manifested in the average molecular anisotropy, we have measured the depolarized Rayleigh spectra of 1,5- and 2,4-pentanediol in the temperature range 230 to 350K using Fabry-Perot interferometry.

2. EXPERIMENTAL

The DLS spectra were taken using a similar apparatus as described elsewhere.[4] The light source was an argon-ion laser operating at 514.5 nm and delivering ∿ 400 mW on a single mode. All the measurements were performed at a scattering angle of 90° with the incident light beam polarized parallel (H) to the scattering plane. The scattered light from the sample was spectrally analyzed by a piezo-elctrically scanned F-P interferometer (Tropel). A plate separation of 1.50 cm, corresponding to a 10.00 GHz free spectral range, was chosen throughout these experiments. The interferometer was operated with a finesse of about 60. The experimental depolarized Rayleigh spectra were fitted to a lorentzian function. After correcting for the overlap of the neighboring orders and for the instrumental function the halfwidth at half height Γ was obtained. The linewidths are reproducible to better than 10%. Depolarized Rayleigh spectra were recorded at temperatures between 230 and 350 K. We calculated the depolarized intensities by integration of the fitted lorentzian profiles within an error ± 8%. The two isomeric pentanediols were purchased from Merck (Munich, West Germany). After vacuum destillation, the two samples were filtered through a 0.45 μm Millipore filter into the optical cells with 10 x 10 square cross section.

3. RESULTS AND DISCUSSION

A. Intensity

The DLS intensity I_{VH} for the two pentanediols includes no high frequency component and is thus due to the permanent optical anisotropy of the scattering "particle". Figure 1a visualizes the temperature dependence of I_{VH}^* ($=I_{VH}/\rho$) for the two neat pentanediols. We did not correct I_{VH} for the local field because the exact form of this correction is not known but in our case the local field in the two liquids is expected to be quite similar as the indices of refraction do not differ appreciably. Firstly, we have to deal with the question: What is the origin of the scattered depolarized light observed in our experiments? The n-pentane is a very weak depolarized scatterer as compared to either of the pentanediols. In the case of independent molecular scatterers I_{VH}^* is generally independent of temperature although a marked decrease of the VH-intensity with increasing temperature to some asymptotic value has been observed in higher alkanes ($n > 16$)[5] and in some polymeric liquids.[6] This behavior is however due to the orientational intermolecular correlation, which decreases with temperature according to $T/(T - T_0)$. In our case I_{VH} falls nearly linearly and tends to vanish at high temperatures (Figure 1a). Additionally the large activation energy of ∿ 6 kcal/mole leads to the conclusion that the major contribution to the observed optical anisotropy is due to a structural anisotropy (β_{inter}) produced by intermolecular H-bonding interactions. Davidson[7] found with infrared techniques in 2,4-pentanediol almost 40%

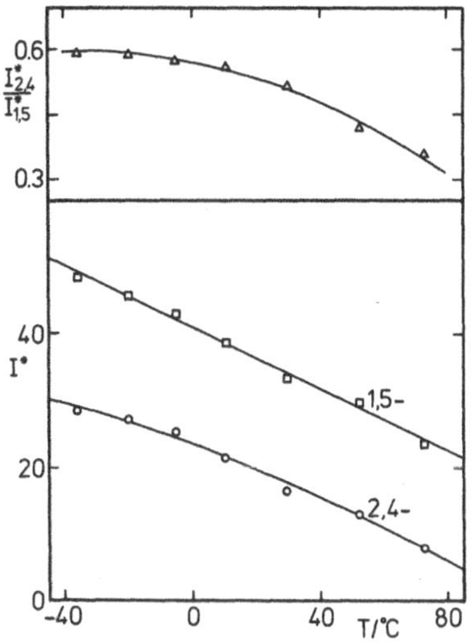

Figure 1. a) $I_{VH}^* = I_{VH}/\rho$ vs temperature and
b) intensity ratio $I_{2,4}^*/I_{1,5}^*$ vs T for 1,5- and 2,4-pentanediol

intramolecular bonded -OH groups in a very dilute solution, while 1,5-pentanediol showed no evidence of internal H-bonding. In Figure 1a we have plotted the ratio of the depolarized intensities corrected for the density as obtained for the 2,4- and 1,5-pentanediol. For temperatures between -40 and 20°C the intensity ratio is about 0,6 and decreases with increasing temperature. In order to account at least qualitatively for this result, we adopt the following simple model. The internal H-bonding in 2,4-pentanediol results in the formation of flexible hexagonal rings with a very low contribution to the optical anisotropy, so that the major part of the observed anisotropy can be assigned to an intermolecular H-bonding configuration. By neglecting static orientational correlations (F) and assuming $\beta^{1,5}_{inter} \simeq \beta^{2,4}_{inter}$ the experimental intensities ratio leads to a mole fraction $x \simeq I_{2,4}^*/I_{1,5}^* = 0.6$ of the 2,4-pentanediol engaged in intermolecular H-bonds. This number, of course, has only a qualitative meaning mainly because of the second assumption. Further evidence supporting the existence of internally H-bonding in 2,4-pentanediol as probed by DLS and giving some measure as to its extent is given in Figure 2. The curve in Figure 2 clearly demonstrates the fact that the ability of diols to form intramolecular structures decreases with increasing distance of the OH-groups. Additionally DLS measurements of solutions in CCl_4/tert-butanol mixtures have shown that $I_{2,4}^*/I_{1,5}^*$ is independent from concentration.[8] This fact supports the assumption $\beta^{2,4}_{inter} \simeq \beta^{1,5}_{inter}$ made above, which reflects a similar intermolecular structure in both pentanediols throughout the measured concentration range.

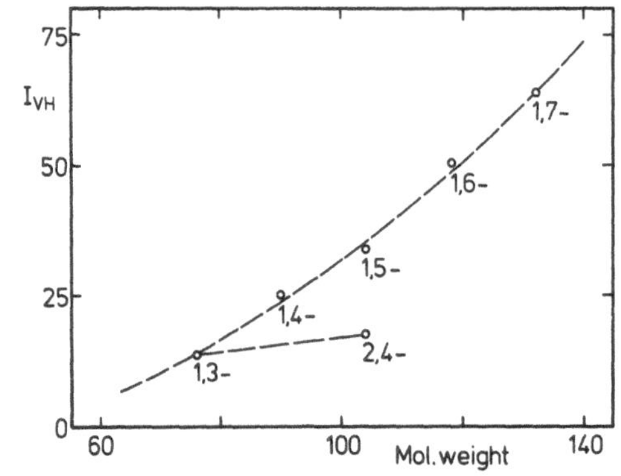

Figure 2. I_{VH} vs molecular weight for some α,ω-alkanedioles at 25°C.

B. Orientation Relaxation Times

The relaxation time $\tau (= 1/\Gamma)$ is a decreasing function of the temperature as shown in Figure 3 for bulk 1,5- and 2,4-pentanediol. We observed activation energies of 6,8 ± 0,5 kcal/mol/Ks and 5,9 ± 0,5 kcal/mol for 1,5- and 2,4-pentanediol, respectively. This rather high value of the activation energy, which corresponds roughly to two H-bonds per molecule, is an additional indication, that the depolarized Rayleigh spectrum is related to the intermolecular H-bonding in the two studied alcohols. It is worth mentioning for a comparison that the corresponding activation energy of 1,2,6-hexanetriol is 8.3 kcal/mol.[2] Figure 3 displays that, in contrast to the intensity (Figure 1), the τ values in both samples are very similar. This experimental fact is compatible with the assumption of a similar

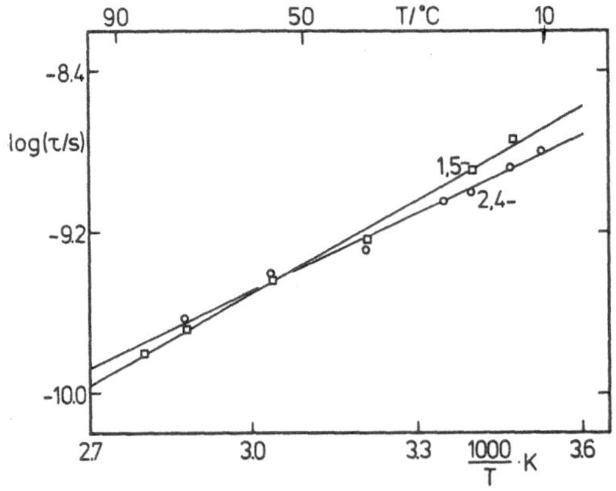

Figure 3. Semilog plot of orientations time τ vs 1/T for 1,5- and 2,4-pentanediol.

anisotropic structure in the two pentanediols. Oriéntational relaxational times can be often described by a Debye-Stokes-Einstein hydrodynamic equation of the form:

$$\tau = (1 + F) \cdot \frac{C}{k_B} \cdot \frac{\eta_s}{T} + \tau_o \qquad (1)$$

where k_B is the Boltzmann's constant, τ_o accounts for a non zero intercept and C is related to the hydrodynamic volume of the rotating moiety dependent on its shape, and the hydrodynamic boundary conditions. In Figure 4, we have plotted our DLS relaxation times versus η/T for the samples studied here. For 1,5-pentanediol the orientational time follows very well Equation 1, while a large deviation from linear behavior is observed in 2,4-pentanediol. Similar deviations have been reported for rhodamine 6G in ethylene glycol[9] for styrene in polystyrene[10] and for local side group motions in some polymers.[6] It is obvious in these cases, that the orientational motion does not experience the full frictional effects, which are contained in the macroscopic shear viscosity.

Assuming as above that the DLS technique probes in this particular case, mainly the intermolecular H-bonded network, we would expect the same shape C for both pentanediols, as it seems to be the case at high temperatures (Figure 4). A similar effect should be observed by diluting the pentanediols. We found, that the relaxation times at 20°C for 79 wt % and 53 wt % 2,4-pentanediol in CCl_4/t-

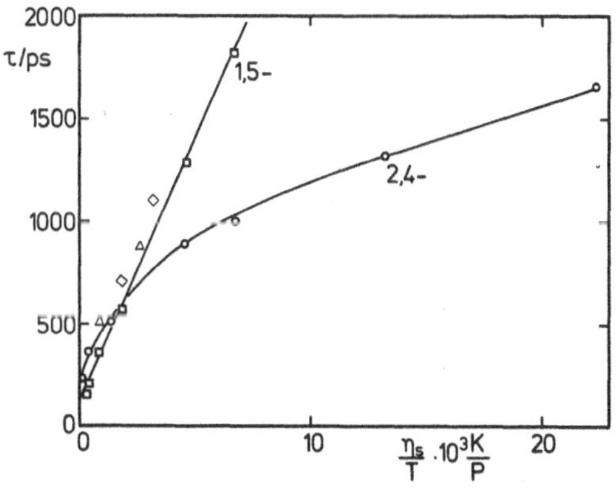

Figure 4. τ vs η/T ($\eta_s = 3.99 \cdot 10^{-6} \exp(\frac{2956}{T-62})$ for 1,5- and $\eta_s = 5.29 \cdot 10^{-6} \exp(\frac{1661}{T-165})$ for 2,4)

□ ◇ : neat and diluted 1,5- and

○ △ : neat and diluted 2,4-pentanediol.

butanol follows Equation 1 with a value of C similar to that of the 1,5-pentanediol (Figure 4). We can point out, that this exceptional behavior of 2,4-pentanediol suggests the existence of two physically distinct relaxation processes.[8] The faster one, which is exposed here, is visible with DLS-techniques, while the slower has been observed by photon correlation spectroscopy[8] and dielectric relaxation methods.[7]

REFERENCES

1. B. J. Berne and R. Pecora, Dynamic Light Scattering, Wiley-Interscience, New York, 1976.
2. Th. Dorfmüller, H. Dux, G. Fytas, and W. Mersch, J. Chem. Phys. 71: 366 (1979).
3. C. J. Montrose, J. A. Bucaro, J. Marshall-Coakley, and T. A. Litovitz, J. Chem. Phys. 60: 5025 (1974).
4. G. Fytas, Y.-H. Lin, and B. Chu, J. Chem. Phys. 74: 3131 (1981).
5. G. D. Patterson, A. D. Kennedy, and J. P. Latham, Macromolecules 10: 667 (1977).
6. Y.-H. Lin, G. Fytas, and B. Chu, J. Chem. Phys., in press (1981).
7. D. W. Davidson, Can. J. Chem. 39: 2139 (1961).
8. G. Fytas and Th. Dorfmüller, to be published.
9. T. J. Chuang and K. B. Eisenthal, Chem. Phys. Lett. 11: 368 (1971).
10. G. R. Alms, G. D. Patterson, and J. R. Stevens, J. Chem. Phys. 70: 2145 (1979).

THERMODYNAMIC PROPERTIES OF SUPERHEATED WATER

DEDUCED FROM BRILLOUIN SCATTERING MEASUREMENTS

J. Leblond and M. Hareng

Laboratoire de Résonance Magnétique
Ecole de Physique et Chimie
10 rue Vauquelin 75231 Paris, Cedex 5, France

ABSTRACT

To test any nucleation theory in liquid, it is necessary to know properties of the liquids in the metastable state. We present here measurements of the sound velocity obtained by Brillouin light scattering in superheated and supercooled water. We give a fitting of the sound velocity data versus temperature by means of a fourth degree polynomial. The isothermal compressibility as deduced from hypersound velocity is fitted by a critical law in the superheating state. Finally we deduce the specific heat at constant volume to evaluate the amount of broken hydrogen bonds.

INTRODUCTION

Thermodynamics of pure liquids predict that it is possible to observe liquid states beyond the binodal curve which is defined by the equality of the chemical potentials of the two phases. Observed in this conditions, the liquid is in a metastable state. The theory of thermal and mechanical stability predicts a limit to the existence of this metastable state defined by the spinodal curve along which $(\partial P/\partial v) = 0$. The temperature limit of stability at atmospheric pressure is estimated to be 325°C.[1] However, homogeneous theory[2] predicts that vapour cavities can form in a pure superheated liquid before the spinodal curve is reached and suggests that superheating is due to an energy barrier to vaporization which is equal to the work necessary to create the surface of the cavity. It is possible to predict the P-T states for which there is a reasonable probability that during an experiment at least one vapour cavity can overcome surface forces and attain microscopic size. The predicted maximum superheating temperature for water at atomospheric pressure is about

300°C,[3] i.e. 25°C below the temperature corresponding to the spinodal.

The knowledge of the physical properties of the metastable liquid state contributes to test the validity of the nucleation theory and to bring a better estimation of the limit of stability.

Light scattering seems to be a non-perturbative and sensitive method to obtain information on the properties of metastable water. We present measurements of hypersound velocity at atmospheric pressure obtained from a Brillouin spectra analysis.[4]

EXPERIMENTAL RESULTS

The velocity of the sound deduced from Brillouin scattering analysis is shown in Figure 1. The sound velocity, which

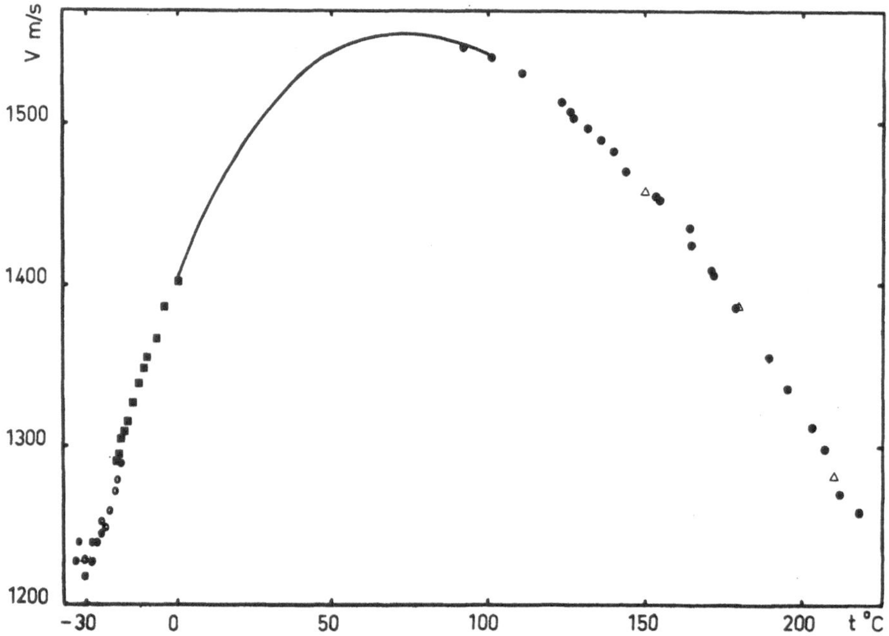

Figure 1. Sound velocity V vs temperature.
o : E. H. Trinh and R. E. Apfel[18]
▪ : O. Conde, J. Teixeira[6]
— : V. A. Del Grosso and C. W. Mader[5]
• : M. Hareng and J. Leblond[4]
Δ : B. H. Evstefeev et al[9]

is maximum at 74.2°C,[5] decreases more and more rapidly when the temperature increases beyond that maximum. We show on the same figure the velocity of the sound obtained by the same technique in the supercooled states.[6] These measurements are in good agreement with the important lot of data available in the stable state range and with recent ultrasonic measurements given par Trinh and Apfel[7] in the superheated states. However, a small dispersion of the results exists at low temperatures in the supercooled states.

We have calculated the isothermal compressibility and the specific heat at constant volume using thermodynamic relations

$$\beta_s = \frac{1}{v^2 \rho}$$

$$\beta_T = \beta_s + \frac{T\alpha^2}{\rho C_p}$$

$$C_v = C_p \frac{\beta_s}{\beta_T}$$

where the density ρ and the thermal expansivity α are given by Kell[8] up to 150°C and by Evstefeev and others up 220°C.[9] The specific heat at constant pressure C_p are given by de Haas[10] up to 100°C. At higher temperatures, we calculated the value of C_p by means of a reduction to one atmosphere from the data in the stable states of Sirota and Melbcev.[11]

The results on β_T and C_V are reported on Figures 2 and 3. It is possible to analyze the divergent behaviour of β_T versus temperature using a critical law like in critical phenomena. Such a procedure has been widely used to fit many properties of supercooled water.[12]

If $\beta_T = \beta_o \varepsilon^x$, with $\varepsilon = (T_S - T)/T_S$, we find that β_T is very well fitted in superheated states when $T_S = 315 \pm 10$°C, and $x = -0.97 \pm 0.05$. T_S appears as the stability limit and is in good agreement with the prediction given by Skripov[1] obtained by linear extrapolation of the isochore curves determined in the stable states.

The variation of C_V versus temperature is abnormal since C_V decreases when the temperature increases. According to Kauzmann[13] C_V is the sum of two terms, one is the classical contribution of the degrees of vibration and libration in liquids, the other one is due to the existence of hydrogen bonds in water.

In fact, $C_{V_{HB}}$ can be decomposed in two terms: one is due to the hydrogen bond energy, the other corresponding to structure energy. Although this structural component has not been established experi-

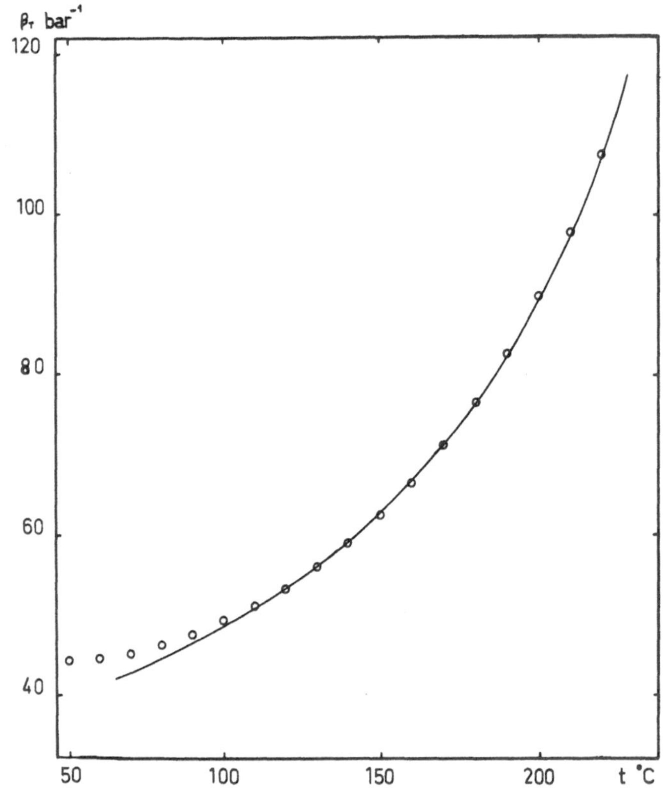

Figure 2. Isothermal compressibility. (o: our results; —— : the critical law $\beta_T = 18.5 \times 10^{-6}\ \varepsilon^{-0.97}$)

mentally by ultrasonic frequency dispersion, one can assume these effects have to be taken into account at low temperature, i.e. in supercooled states.

Figure 3 shows the experimental and theoretical contributions where we have neglected the structural effects. The theoretical part is the sum of the vibrational and hydrogen bond contributions. The vibrational one was evaluated from frequency of vibration.[14] The hydrogen bond contribution was calculated using a two state model[15] where the dependence of the number of hydrogen-bonds is in this case of the following form:

$$\log \frac{I_{HB}}{I_n} = A + \frac{B}{T}$$

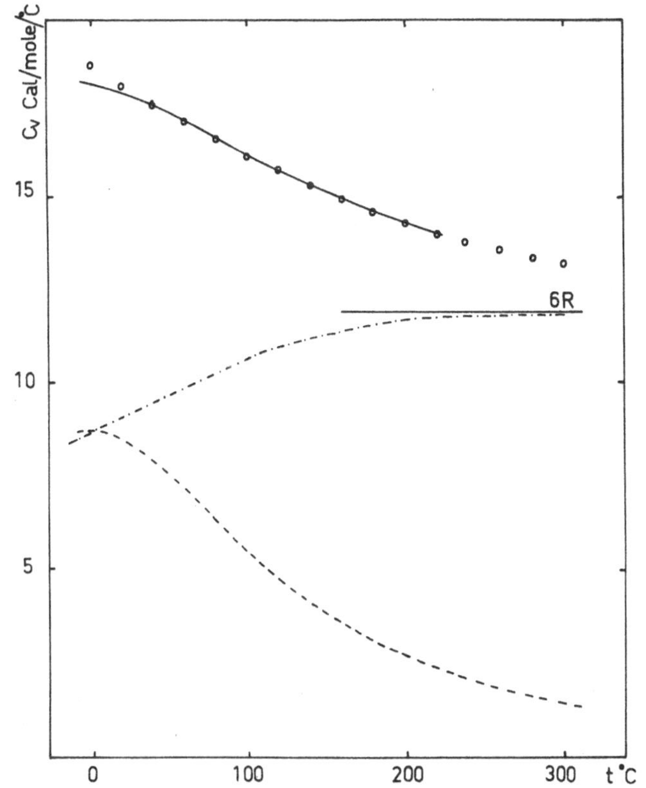

Figure 3. Specific heat at constant volume
—·—·— : component from non hydrogen-bonded molecules[14]
------ : component from hydrogen-bonded molecules[15]
o : theoretical points
——— : experimental curve

where I_{HB}/I_n represents the intensity ratio between hydrogen-bonded and non hydrogen-bonded environments. The parameters A = -1.623 and B = 546.8°K give the the best agreement. The hydrogen-bond energy deduced from B is equal to 2.5 kcal/mole which is in good agreement with results obtained with Raman[16] or infrared methods.[17]

REFERENCES

1. V. N. Chukanov and V. P. Skripov, Proceedings of the 8th International Conference on the Properties of Water and Steam. Vol. 1, p. 512 (1974).
2. M. Blander and J. L. Katz, AIChE Journal 21: 833(1975).
3. R. E. Apfel, J. Chem. Phys. 54: 62 (1971).

4. M. Hareng, and J. Leblond, J. Chem. Phys. 73: 622 (1980).
5. V. A. Del Grosso and C. W. Mader, J. Acous. Soc. Am. 52: 1442 (1972).
6. O. Conde, Thesis.
7. E. H. Trinh and R. E. Apfel, J. Chem. Phys. 69: 4245 (1978).
8. G. S. Kell, J. Chem. and Engineering Data 20: 97 (1975).
9. B. H. Evstefeev, V. P. Skripov and V. N. Chukanov, Teplofiz. Vyso. Temp. 17: 299 (1979).
10. W. J. de Haas, P. V. Com. Int. Poids Mes. 22: 85 (1950).
11. A. M. Sirota and B. K. Malbcev, Teploenergetika 9: 7 (1959).
12. R. J. Speedy, C. A. Angell, J. Chem. Phys. 65: 851 (1978).
13. D. Eisenberg, W. Kauzmann, The Structure and Properties of Water. University of Oxford Press, 1969.
14. K. S. Pitzer, Quantum Chemistry. Prentice-Hall, Englewood Cliffs, New Jersey.
15. G. E. Walrafen, J. Chem. Phys. 48: 244 (1968).
16. W. A. Senior, A. E. Verall, J. Phys. Chem. 73: 4242 (1969).
17. J. Leblond, to be published.
18. E. H. Trinh and R. E. Apfel, J. Chem. Phys. 72: 6731 (1980).

OSMOTIC COMPRESSIBILITY AND LINEWIDTH OF

(NBS-705) POLYSTYRENE IN METHYL ACETATE*

K. Abbey, K. Kubota, and B. Chu

Chemistry Department
State University of New York at Stony Brook
Long Island, New York 11794

CONTENTS

1. INTRODUCTION

Static and dynamic properties of polymer solutions have recently been investigated by laser light scattering.[1] The concentrations studied have ranged from the very dilute to the semi-dilute and concentrated regions. In our case, we have chosen to characterize polystyrene (NBS 705 standard, $M_w = 179{,}300$, $M_w/M_n = 1.07$) dissolved in methyl acetate. In this system, the upper and lower critical solution temperatures differ by only ∿125°C and the theta temperatures by only ∿70°C.[2,3] Consequently, one can monitor the change in quality of the solvent over a range of temperatures which are experimentally accessible.

2. STATIC MEASUREMENTS

Measurement of the intensity of scattered light as a function of scattering angle yields the osmotic compressiblity

*Work supported by the National Science Foundation and the U.S. Army Research Office.

$$\left(\frac{\partial \pi}{\partial C^V}\right)_{T,P} = \frac{4\pi^2 n^2}{N\lambda^4}\left(\frac{\partial n}{\partial C^V}\right)^2_{T,P} C^V RT/R_c(0) \tag{1}$$

where $R_c(0)$ is the excess Rayleigh ratio obtained from absolute excess scattered intensity extrapolated to zero scattering angle. The quantities n, C^V, N, T, P, and R are the refractive index of the solution, the concentration (g/cm^3), Avogadro's number, temperature, pressure, and gas constant. From the expression for the virial expansion in dilute solutions,

$$\left(\frac{\partial \pi}{\partial C^V}\right)_{T,P} = \frac{RT}{\bar{M}_w} + 2A_2RTC^V + \ldots \tag{2}$$

the second virial coefficient, A_2, can be determined.

For dilute solutions, scaling relations predict that the osmotic compressibility is essentially independent of concentration.[4] As the polymer concentration in solution approaches the overlap concentration, C*, the osmotic compressibility is expected to become proportional to $(C^V)^n$ where the value of the exponent n is predicted to be 5/4 in the semidilute region and 2 in the concentrated solution region.

3. DYNAMIC MEASUREMENTS

The linewidths (Γ) measured for dilute solutions where polymer coils do not overlap can be related to the macroscopic translational diffusion coefficient, $D = \Gamma/K^2$ with K being the magnitude of the momentum transfer vector. In more concentrated solutions in the semidilute region, this interpretation must be modified to include motions of the entangled polymers. The introduction of other effects can be observed in the large increases in the variance of the linewidth estimates as concentration increases.

4. EXPERIMENTAL PROCEDURES

Polystyrene (NBS 705 standard, $M_w = 179{,}300$, $M_w/M_n = 1.07$) was dissolved in benzene and freeze dried in vacuum. Methyl acetate was fractionally distilled over P_2O_5 in a nitrogen atmostphere. A stock solution (5%) was filtered through a 0.22 Millipore filter into the light scattering cells and the samples for each concentration were prepared either by evaporation or dilution. The cells were then flame sealed under vacuum.

The intensity of scattered light was measured using a modified Malvern spectrometer with a Helium-Neon laser. Acquisition of data from the light scattering spectrometer was controlled by a PDP-11 minicomputer. This involved rotation of the spectrometer turn-table, control of the digital counting of photoelectron pulses, sampling of the reference intensity in order to correct for laser fluctuations, and data analysis. Typically, the intensity was measured for thirty one-second intervals, and a preset criterion used to eliminate anomalously high or low values. The temperature was monitored and controlled (to 0.01°C at t = 30°C and t = 43°C, to 0.05° at t = 55°C and t = 70°C) by a Bayley precision temperature controller, thermistors, and an a.c. impedance bridge. The refractive index increment $\partial n/\partial C^V$ was determined using a Brice Phoenix differential refractometer. Dynamic light scattering measurements were made using both an argon-ion laser and a Helium-Neon laser. The apparatus has been described previously.[5]

5. RESULTS AND DISCUSSIONS

Figure 1 shows a log-log plot of osmotic compressiblity versus concentration for four different temperatures. At t = 30°, the temperature is below the theta temperature and in the region of chain collapse. The decrease in $(\partial\pi/\partial C^V)_{T,P}$ near C ∿ 0.04 appears to be caused by critical effects and is similar to that observed for polystyrene in transdecalin.[6] At t = 43°, the system is very near the theta temperature predicted by the Flory-Schultz plot.[3] At t = 43°, 55° and 70°, the characteristics of the curves are very similar.

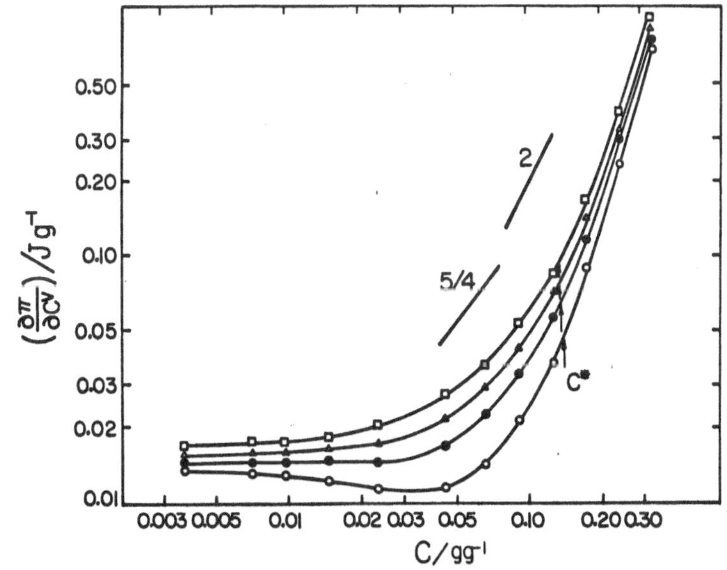

Figure 1. Log-Log plot of osmotic compressibility as a function of concentration (g/g) at 30°C (hollow circles), 43°C (filled circles), 55°C (hollow triangles) and 70°C (hollow squares). $C^* = \bar{M}_w/N\rho_s\langle r_g^2\rangle_z^{3/2}$ with ρ_s being the solvent density.

In the dilute solution region, one observes that the osmotic compressiblity is essentially independent of concentration. At higher concentrations, a gradual transition from an exponent of approximately zero to a higher value (greater than 5/4) is observed.

The values of A_2 determined from the slope of osmotic compressibility as a function of concentration extrapolated to infinite dilution are given in Figure 2. The experimentally determined values of A_2 were then used to calculate the radius of gyration at each temperature using the modified FKO equation[7-9] as expressed in Equation (18) of reference 6. We used a measured value of $\langle r_g^2\rangle_z^{1/2}$ = 13.4 nm at 43°C in the calculations and obtained $\langle r_g^2\rangle_z^{1/2}$ = 13.2, 13.6, 13.7 nm at 30°, 55°, and 70°C, respectively, reflecting the small differences between upper and lower critical solution temperatures. As the values of $\langle r_g^2\rangle_z$ were quite small, we believe that it was more appropriate to use the computed $\langle r_g^2\rangle_z$ rather than the measured ones in the scaling of C/C*.

Figure 3 shows a comparison of data reported here for polystyrene in methyl acetate and that determined for polystyrene in trans-decalin where the theta temperature is 20.5°C. Near the theta temperatures for the two systems, the values of $(\partial\pi/\partial C)_{T,P}$ are almost superimposable while larger deviations exist at higher temperatures, in particular for polystyrene in t-decalin at concentrations below C*. Such observations suggest that scaling concepts must be modified to adequately describe the solution behavior in these regions.

Results of the linewidth measurements at t = 43° and scattering angle of 90° are shown in Figure 4. For concentrations less than C*, the diffusion coefficient decreases with increasing concentration

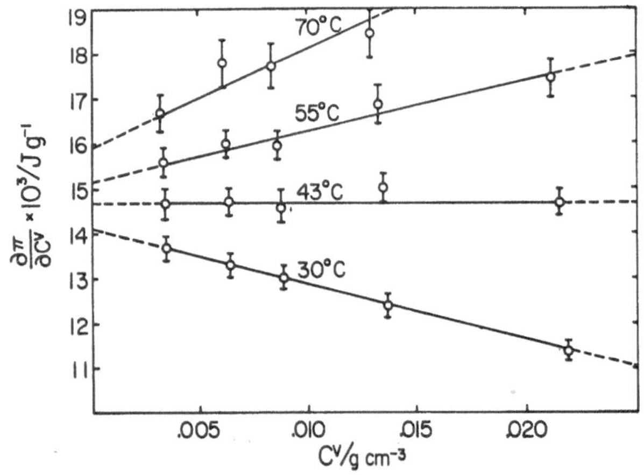

Figure 2. Plot of osmotic compressibility as a function of concentration in the dilute solution region.

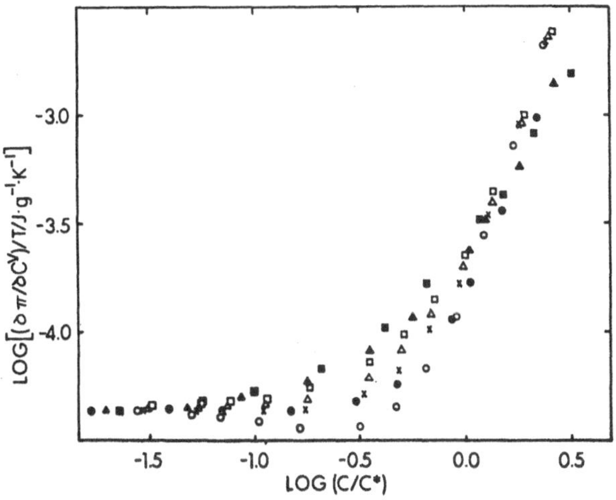

Figure 3. Log-log plot of $(\partial\pi/\partial C^V)/T$ as a function of C/C* for NBS-705 in methyl acetate at 30° (hollow circles), 43°C (crosses), 55°C (hollow triangles) and 70° (hollow squares), and NBS-705 in transdecalin at 20°C (filled circles), 30°C (filled triangles) and 40°C (filled squares).

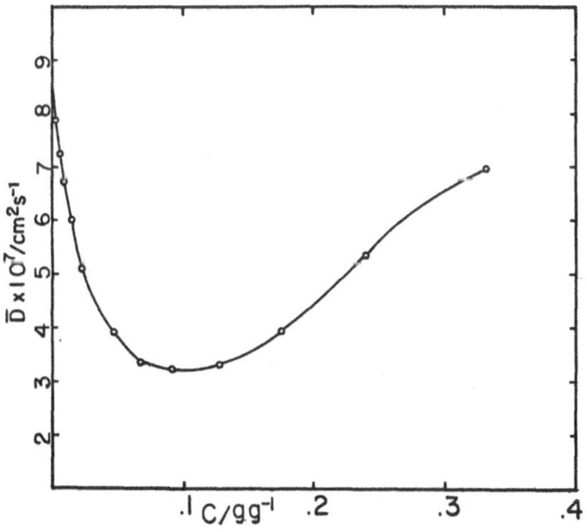

Figure 4. Plot of $\bar{D}(=\bar{\Gamma}/K^2)$ as a function of concentration at 43° and $\theta = 90°$.

as expected when the predominant motion is translational diffusion of the polymer coil. Calculation of the linewidth variance indicates a broadening of the distribution at higher concentrations and will be discussed in detail elsewhere.

REFERENCES

1. See main lectures on "Static and Dynamic Properties of Polymer Solutions" by B. Chu in this volume.
2. S. Saeki, S. Konno, N. Kuwahara, M. Nakata, and M. Kaneko, Macromolecules 7: 521 (1974).
3. S. Konno, S. Saeki, N. Kuwahara, M. Nakata, and M. Kaneko, Macromolecules 8: 799 (1975).
4. M. Daoud and G. Jannink, J. de Physique 37: 973 (1976).
5. F. C. Chen, A. Yeh, and B. Chu, J. Chem. Phys. 66: 1290 (1977).
6. T. Nose and B. Chu, Macromolecules 12: 590 (1979).
7. M. Kurata, "Modern Industrial Chemistry", Vol. 18, Asakura Publishing Co., Japan, 1975, p. 288.
8. S. Imai, Rep. Prog. Polymer Phys. Jpn. 8: 9 (1965).
9. H. Yamakawa, A. Aoki, and G. Tanaka, J. Chem. Phys. 45: 1938 (1966).

THE CHANGE IN THE HYDRODYNAMIC RADIUS OF POLYSTYRENE AROUND THE THETA TEMPERATURE

David Caroline and Michael J. Pritchard

School of Physical and Molecular Sciences
University College of North Wales
Bangor, Gwynedd LL57 2UW, UK

A descriptive feature of the scaling-law approach to polymers, as pioneered and developed by de Gennes,[1] is the subchain or 'blob' model of the polymer coil. The polymer chain is notionally divided into sections or blobs within which there is considered to be no excluded-volume interaction, thus making the configuration of the section Gaussian. Interactions take place between blobs giving rise to coil expansion (or contraction) for sufficiently long chains. The length of chain forming a blob increases as the theta point is approached, and becomes infinite there.

A simple interpretation of this picture suggests that the configuration of the overall chain length remains Gaussian for a finite temperature excursion from the theta point, and indeed the coil dimension should remain constant over this interval. Qualitatively, it can be seen that the extent of this theta region will increase with decreasing molecular weight of the polymer. Daoud and Jannink[2] have deduced the temperature-concentration diagram of polymer solutions and, for dilute solutions, they predict that the temperature range of the theta region of constant coil dimension is independent of concentration but decreases with molecular weight M as $M^{-\frac{1}{2}}$. Further, they expect a collapse region on reducing the temperature below the theta region, where the radius of gyration R_g varies as $\tau^{-1/3}$, with $\tau = (T - \Theta)/T$.

A test of these predictions has been carried out by Nierlich, Cotton and Farnoux[3] who used small-angle neutron scattering (SANS) to observe the decrease in R_g below Θ for a low molecular-weight polystyrene in deuterated cyclohexane ($\bar{M}_w = 29000$, $\bar{M}_w/\bar{M}_n < 1.2$). Between 38°C and 20°C, R_g decreased only slightly, and this was taken to indicate the extent of the theta region. Below 20°C, the experimental points were shown to vary as $\tau^{-0.32}$ as predicted for the collapse region. However, the 2% uncertainty in R_g would not preclude a smoother and continuous variation over the temperature region investigated, as predicted by mean-field theories (see, for example, Sanchez[4]) with no constant-size region.

We have studied the variation of the hydrodynamic radius R_h of polystyrene in cyclohexane from precipitation through the theta temperature (34.5°C) to about 60°C for six narrow fractions with $\bar{M}_w$ ranging from 37000 to 5.05 x 10^6 and $\bar{M}_n/\bar{M}_n$ typically < 1.06. High-precision measurements of the diffusion coefficient D in the dilute regime were made by photon-correlation spectroscopy as a function of temperature T and concentration C. The values obtained for D were reproducible to between 0.2 and 0.5%. The experimental details are described in an earlier paper.[5] It turns out that this work overlaps, in part, a recently reported investigation by Bauer and Ullman[6] who have studied the identical system over a similar molecular-weight range, but only for temperatures below the theta point.

RESULTS AND DISCUSSION

As an example, the experimental data for the sample $\bar{M}_w$ = 1.26 x 10^6 are shown in Figure 1 where D is plotted as a function of T for the four concentrations studied. A four-parameter polynomial was fitted to each set of points, thus enabling D to be obtained as a function of C as well as T. As expected for dilute solutions, D varies linearly with C at constant T: $D = D_o(1 + k_D C)$ where D_o is the value of D at infinite dilution. Values of R_h were calculated from the Stokes-Einstein relation $D_o = kT/6\pi\eta R_h$, where k is Boltzmann's constant; values of the viscosity η were taken from Landolt-Börnstein.[7]

The variation of the hydrodynamic expansion factor α_h $(=R_h/R_h^{\Theta})$ with T is shown in Figure 2. These curves were calculated from polynomial fits to the variation of D_o with T. The coil dimensions are seen to change continuously and smoothly through the theta point, and the general pattern is in good qualitative agreement with the theoretical predictions of Sanchez[4] for α_g $(=R_g/R_g^{\Theta})$. In particular, the slope of the curves, $d\alpha_h/dT$, decreases with increasing T, and the curves level off more quickly for the lower molecular weights. The slope at the theta point varies as $M^{0.42\pm0.04}$, and this favours the mean-field theory ($M^{\frac{1}{2}}$) rather than scaling-law theory which predicts zero slope. It thus appears that the blob concept is not as helpful in interpreting behaviour in the dilute regime as it is in more concentrated solutions.

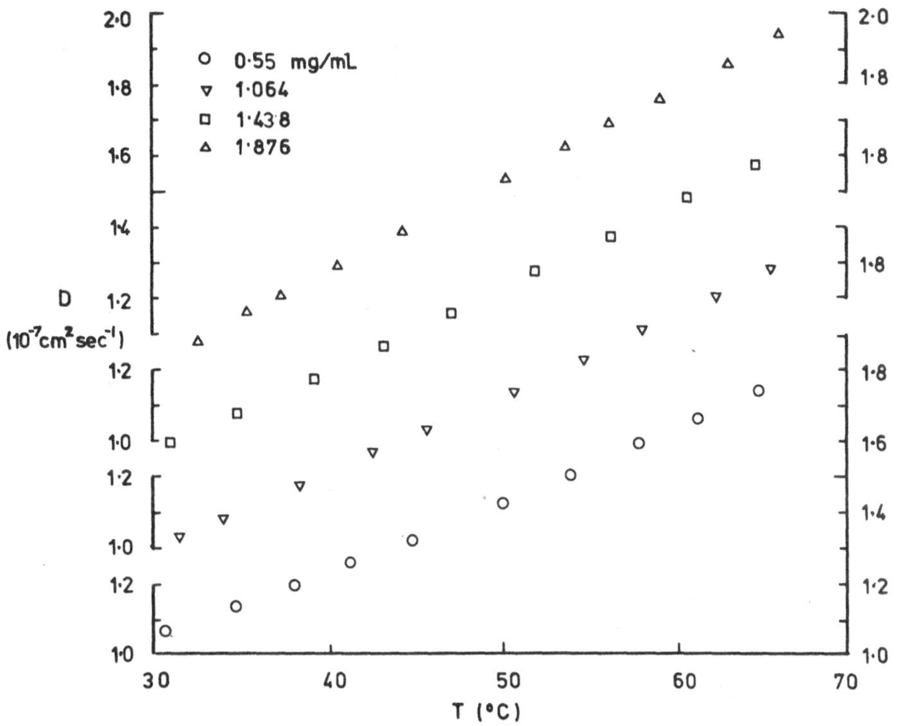

Figure 1. Variations of D with T for sample $\bar{M}_w = 1.26 \times 10^6$.

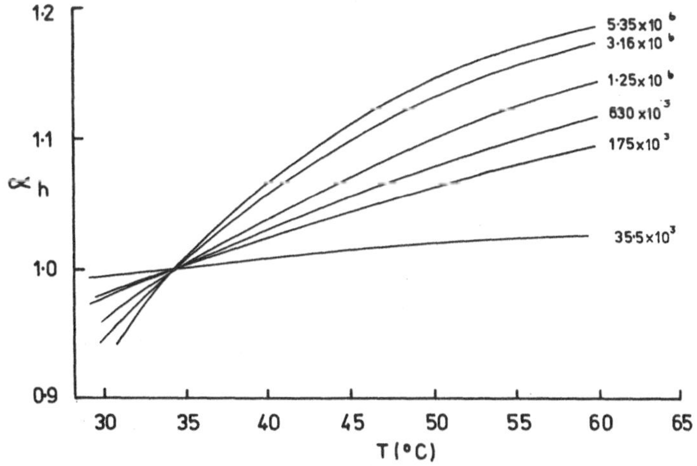

Figure 2. Variations of α_h with T.

REFERENCES

1. P. G. de Gennes, Scaling Concepts in Polymer Physics, (Cornell University Press, Ithaca, New York, 1979).
2. M. Daoud and G. Jannink, J. Phys. (Paris) 37: 973 (1976).
3. M. Nierlich, J. P. Cotton, and B. Farnoux, J. Chem. Phys. 69: 1379 (1978).
4. I. C. Sanchez, Macromolecules 12: 980 (1979).
5. M. J. Pritchard and D. Caroline, Macromolecules, vol. 13 (1980).
6. D. R. Bauer and R. Ullman, Macromolecules 13: 392 (1980).
7. Landolt-Börnstein, 6 Auflage, Band II/5a, p. 157.

MICROSCOPIC STRUCTURE OF SHEARED COLLOIDS

Bruce J. Ackerson and Noel A. Clark

Physics Department
Oklahoma State University
Stillwater, Oklahoma 74078

Physics Department
University of Colorado
Boulder, Colorado 80302

Aqueous suspensions of monodisperse, charged, plastic spheres are known to form liquid and solid structures with average particle spacing or lattice constants on the order of the wavelength of light.[1] Thus, light scattering can be used to determine both the static[1] and dynamic structure[2,3] of these model interacting particle systems. Such equilibrium structures are very delicate and can be destroyed by mild agitation, but will reform in a brief time if left unmolested. If such suspensions are subjected to a simple shear, the structure is visibly modified as observed by the light scattering intensity pattern, which is related to the spatial Fourier transform of the real space configurations.

A system, which forms a polycrystalline structure of body center cubic (bcc) "colloidal crystallites" in equilibrium, produces the sequence of scattering patterns shown in Figure 1 as the rate of shear (S) is decreased slowly from a value where the crystal structure is totally melted by the shear stress. Here the shear has been applied with the same results by either of two methods:[4,5] a concentric cylinder cell or a rocking cuvette. The samples are Dow polystyrene latex spheres of radii r = 54.5 nm or r = 117 nm in aqueous suspension (∿ 0.1 ωt %) with sufficient ion exchange resin added to remove all salt ions.

For all but one of the scattering patterns shown in Figure 1, the incident wave vector $\bar{k}_I$ is parallel to the shear ($\hat{V}$) and normal to the velocity ($\hat{V}$). The scattered wave vector $\bar{k}_s$ is sufficiently close to the forward direction that $\bar{k}$ (= $\bar{k}_I - \bar{k}_s$) lies nearly in the plane determined by $\bar{k}_v$, the unit wave vector in the direction of the velocity ($\hat{v}$), and by $\bar{k}_e$, the unit wave vector which is perpendicular to both $\bar{k}_v$ and $\bar{k}_\nabla$. Here $\bar{k}_\nabla$ is the unit wave vector in the direction of the shear ($\hat{\nabla}$). In this plane the intensity pattern gives directly the $\bar{k}$-space or reciprocal lattice amplitudes of the real space structure with little projective distortion.

Figure 1a shows a scattering pattern typical of an amorphous or liquid structure. A diffuse ring is seen; however, this pattern is of additional interest because it exhibits a shear induced distortion such that the ring is elliptical rather than circular about the laser beam (Figure 1c). This distortion is the same as that expected for simple fluids undergoing shear[6] and that seen in molecular dynamics experiments;[7] albeit, it has never been observed experimentally for pure fluids. This is because enormously high rates of shear ($\sim 10^{12}$/sec) are required to produce these effects. Colloidal particles, however, have much slower dynamics and the effect is clearly visible[5] at low rates of shear.

As the rate of shear is reduced, diffuse intensity bands appear at $k_v = 0$, $k_e = n(2\pi/a)$ with $n = \pm 1,2$ for the $\bar{k}_I \perp \hat{v}$ geometry (Figure 1e). Here $2\pi/a$ is the radius of the first Debye-Scherrer ring (Figure 1a). This $\bar{k}$-space structure indicates the presence of interparticle correlations corresponding to strings of a few particles running parallel to $\hat{v}$ (Figure 1d) spaced by a mean distance of a along $\hat{e}$ but with no interstring positional correlation along $\hat{v}$.

As S is lowered further, sharp spots appear at $k_v = 0$, $k_e = n(2\pi/b)$ and sharp lines appearing along $k_v = n(2\pi/a)$ are modulated in intensity to form rows of spots which are elongated along $\hat{e}$ (Figure 1g). The resulting pattern in the $\bar{k}_v$-$\bar{k}_e$ plane is a distorted two dimensional hexagonal close pack (2d-hcp) array with a 20% shrinkage along $\hat{e}$. We interpret this $\bar{k}$-space structure as a condensation of the strings into hcp layers which are parallel to the $\hat{e}$-$\hat{v}$ plane and stretched by 20% along $\hat{e}$ (Figure 1f, open circles). Adjacent layers along $\hat{\nabla}$ are displaced parallel to $\hat{e}$ by b/2 (Figure 1f, closed circle) which results in a reduction of intensity of odd order $k_v = 0$ reflections. This b/2 displacement and the large interstring spacing are the result of a layer "channeling down the grooves" in the hcp structure of the neighboring layers in a shear flow. These sliding 2d-hcp layers lack translational order along $\hat{v}$ which produces an experimentally observed line structure in k-space. Also visible in Figure 1g are the diffuse rings which have now become hexagonal. These arise from local hexagonal clusters which are oriented by the shear and are apparently distinct from the regions ordered into layers as evidenced by the inward displacement of the $k_v = 0$ spots

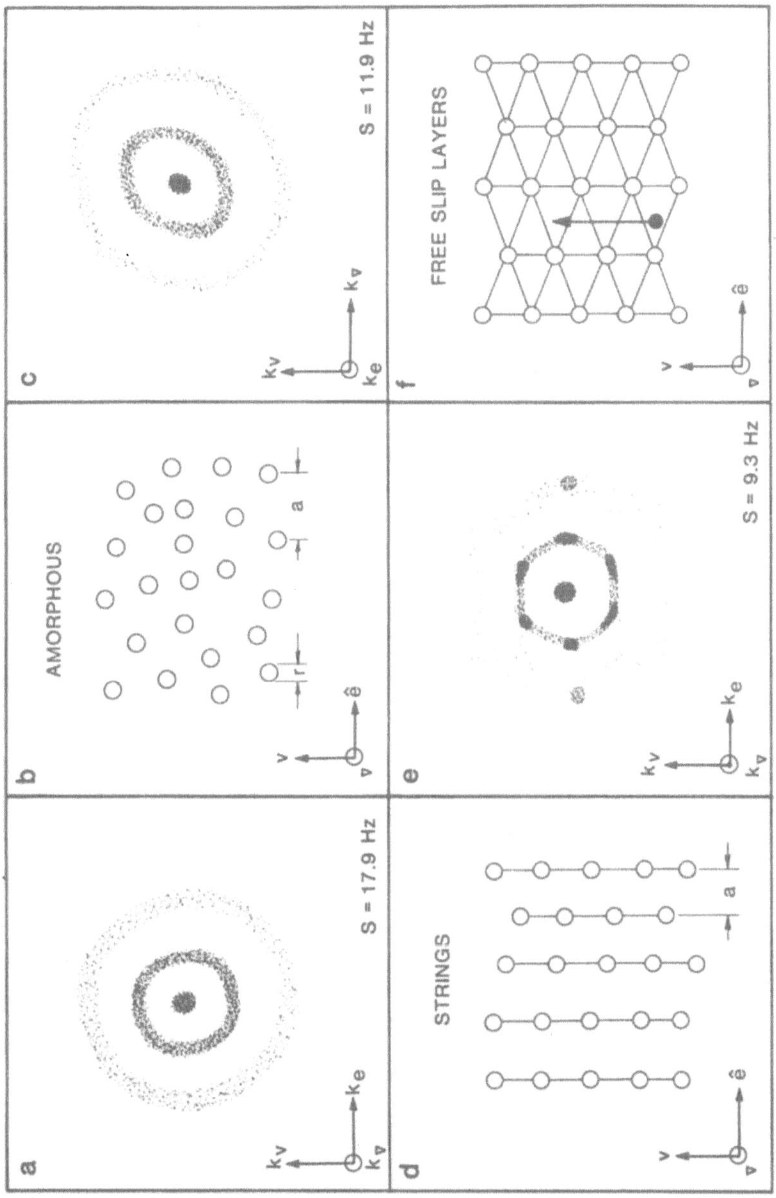
a
S = 17.9 Hz
b
AMORPHOUS
c
S = 11.9 Hz
d
STRINGS
e
S = 9.3 Hz
f
FREE SLIP LAYERS

Figure 1.

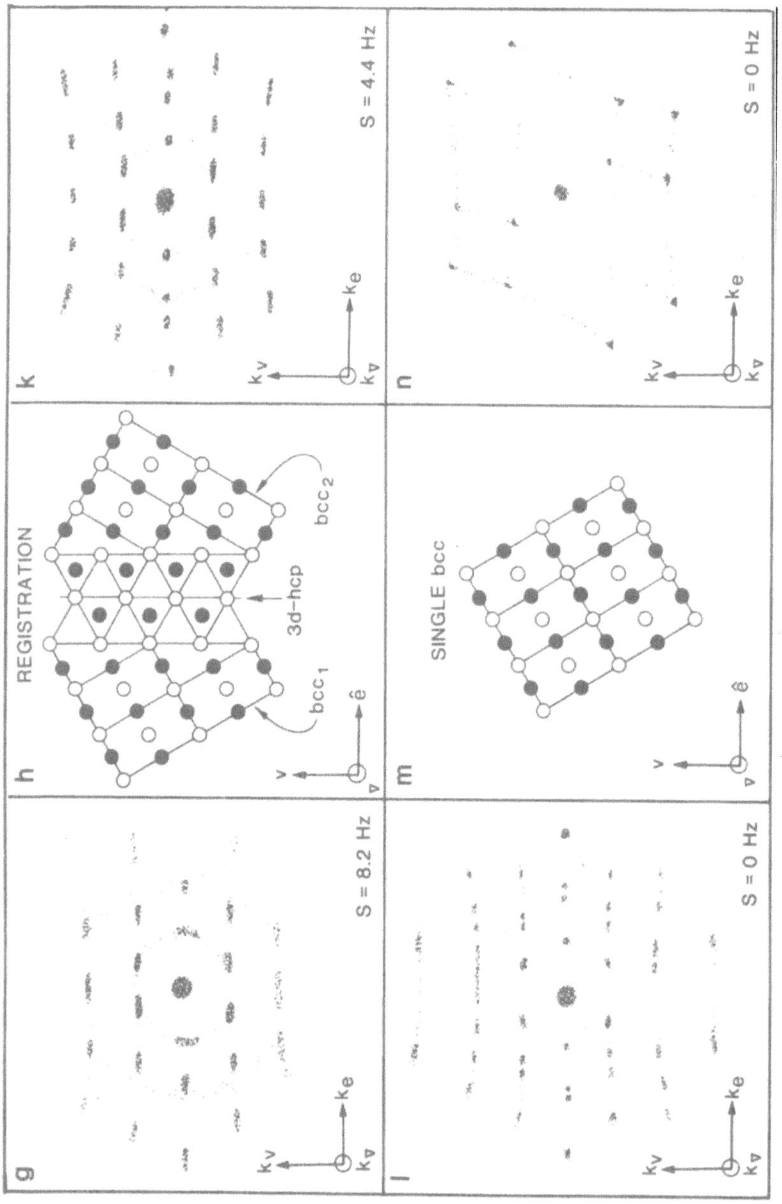

Figure 1. (continued)

from the diffuse hexagon maximum extent along k_e.

Upon further lowering the shear rate, registration of adjacent sheets along $\hat{v}$ is achieved. Because there is a 20% stretch of the 2d-hcp sheets along $\hat{e}$, registration sites for adjacent layers (closed circles in Figure 1h in bcc_1 and bcc_2 regions) lie on the centers of either of two sets of lines joining particles (open circles in Figure 1h) to form locally either of two twin related bcc structures. Particles also register at the centers of triangles formed by three particles to make a local 3d-hcp structure. Experimentally, this is heralded by the appearance of the 3d-hcp structure as evidenced by a second set of sharp $k_v = 0$ spots (Figure 1k) which are properly placed to make a k-space 2d-hcp, i.e., at $k_v = 0$, $k_e = n(2\pi/a)$. This structure persists to zero shear where the structure is a mosaic of the two bcc crystal orientations (and was verified by scattering from other crystal orientations as well). This indicates that the hcp structure is present only at finite shear rates and, therefore, must be associated with the state of intermediate translational order along $\hat{v}$: a mosaic of bcc_1, bcc_2, and hcp correlated region. At low S the bcc dominate and the 3d-hcp regions become kinks which separate regions that have formed into the different bcc orientations. Relaxation of strain associated with such a boundary favors a local 3d-hcp.

The two orientations of the bcc crystallites with respect to the shear cells continue to evolve in time in the absence of shear with the large crystals growing at the expense of the smaller crystallites.[4] After several days or weeks many large oriented crystals fill the cell. In some samples left undisturbed for several months, crystals as large as 7 mm x 5 mm x 1 mm have been found. Laser scattering from such crystals is analogous to x-ray scattering from single atomic crystals. Bragg reflections index to within a few per cent of that expected for bcc crystals. Thermal diffuse scattering[8] is observed as lines of weakly scattered light connecting the nearly visible Bragg spots in Figure 1n. Kossel lines, which indicate a high degree of ordering, are visible in some samples.

Further details including data and observations from other scattering geometries and theoretical models for the observed transitions will be published elsewhere.[9] One of us (B.J.A.) wishes to thank the Research Corporation for support of this project.

REFERENCES

1. P. A. Hiltner, and I. M. Krieger, J. Phys. Chem. 73: 2386 (1969); J. W. Goodwin, R. H. Ottewill, and A. Parentich, J. Phys. Chem. 84: 1580 (1980).
2. J. C. Brown, P. N. Pusey, J. W. Goodwin, and R. H. Ottewill, J. Phys. A8: 664 (1975).

3. D. W. Schaefer, and B. J. Ackerson, Phys. Rev. Lett. 35: 1448 (1975).
4. N. A. Clark, A. J. Hurd, and B. J. Ackerson, Nature 281: 57 (1979).
5. N. A. Clark, and B. J. Ackerson, Phys. Rev. Lett. 44: 1005 (1980).
6. H. S. Green, The Molecular Theory of Fluids, (Dover, New York, 1969). p. 135ff.
7. W. T. Ashurst, and W. G. Hoover, Phys. Rev. A11: 658 (1975).
8. R. W. James, The Optical Principles of the Diffraction of X-Rays, (Cornell University Press, 1965).
9. B. J. Ackerson, and N. A. Clark, Phys. Rev. Lett. 46: 123 (1981).

LIGHT SCATTERING STUDY OF WATER IN OIL MICROEMULSIONS

A. M. Cazabat, D. Chatenay, D. Langevin, A. Pouchelon

Laboratoire de Spectroscopie Hertzienne de l'E.N.S.
24, Rue Lhomond
75231 Paris Cedex 05

CONTENTS

1. INTRODUCTION

Microemulsions have attracted much interest since their introduction because of their applications in numerous fields of science and technology.[1] They are apparently homogeneous and transparent systems of low viscosity, containing emulsifier molecules, oil, and water. They usually behave as dispersions of very small spherical droplets of water (or oil) surrounded by emulsifier molecules, in a continuous medium containing the oil (or water). The droplet sizes range from 100 to 1000 Å in diameter.

Light scattering techniques have already been extensively applied to these systems.[2-5] They allow us to investigate not only sizes but also interaction forces in these media. Unusually strong attractive forces have been found in this way. These results will be helpful to achieve a better understanding of microemulsions stability. Indeed, contrary to the more commonly known emulsions (that are also dispersions of oil-or water-droplets into water -or oil- but with larger droplets sizes - around 1 μ), the microemulsions seem to be thermodynamically stable systems.

The analysis of light scattering data on microemulsions involves the use of theories of brownian diffusion of particles with hydrodynamic interactions. Up to the very recent years, several discrepancies could be found between the different authors. But the origin of these discrepancies seems now to be well understood.[6,7] Unfortunately, the experiments do not allow simple comparison with theory, since in most cases interparticular interactions are superpositions of repulsive (hard sphere, electrical) and attractive (Van der Waals) contributions. From this point of view, water in oil microemulsions are simple systems, because electrical forces are negligible in these media.

The characteristics of the light scattered by microemulsions are usually typical of "small" particles interacting through "small" range potentials ("small" compared to optical wavelengths): the correlation function of the scattered electric field is exponential and both the scattered intensity and the diffusion coefficient (deduced from the correlation time) are independent of the scattering angle.

In some cases however (vicinity of a gel for instance) the correlation function becomes non exponential.[8] In other cases, (vicinity of a percolation threshold), the correlation function stays exponential, but both the scattered intensity and the diffusion coefficient are found to depend on scattering angle.[9] This indicates that the attractive forces become long range forces.

In this paper, we will present a brief survey of our own recent results on microemulsions. More details could be found in refs. 5 and 9.

2. EXPERIMENTAL PROCEDURE

We studied quaternary mixtures of water, sodium dodecyl sulfate (SDS) as surfactant, butanol, pentanol or hexanol as cosurfactant, cyclohexane or toluol as oil.

Most microemulsions indeed are formed only if two different kinds of emulsifier molecules are used: surfactant + cosurfactant. This is the source of a great practical difficulty. The cosurfactant (an alcohol here) is generally not only present on the interfacial region of the droplets, but also in the oil of the continuous phase, together with small amounts of water. The droplets concentration cannot therefore be varied by simply adding oil to the microemulsion: this will change also the droplet size. In order to interpret the light scattering data it is necessary to work out a correct dilution procedure.[5] This then allows us to determine the volume fraction of the droplets.

We prepared different series of microemulsions by varying either the water to soap ratio in the droplets, either the amount of salt in water, either the oil or the alcohol. The experimental set up is described in refs. 5 and 9.

3. DATA ANALYSIS

Figures 1 and 2 show typical examples of scattered intensity I and diffusion coefficient D variations versus droplets volume fraction. They illustrate the fact that large differences can be found among the different studied mixtures.

At small volume fractions the data are analyzed in terms of virial coefficients:

$$I \propto \phi \left(\frac{\partial n}{\partial \phi}\right)^2 \frac{4}{3} \pi R^3 (1 + B\phi)^{-1}$$

$$D = D_o (1 + \alpha\phi) \qquad D_o = \frac{kT}{6\pi\eta R_H}$$

where R is the radius of the droplet, R_H its hydrodynamic radius, B the virial coefficient of the osmotic compressibility and α that of the diffusion coefficient; n is the refractive index of the microemulsion, η the viscosity of the continuous phase.

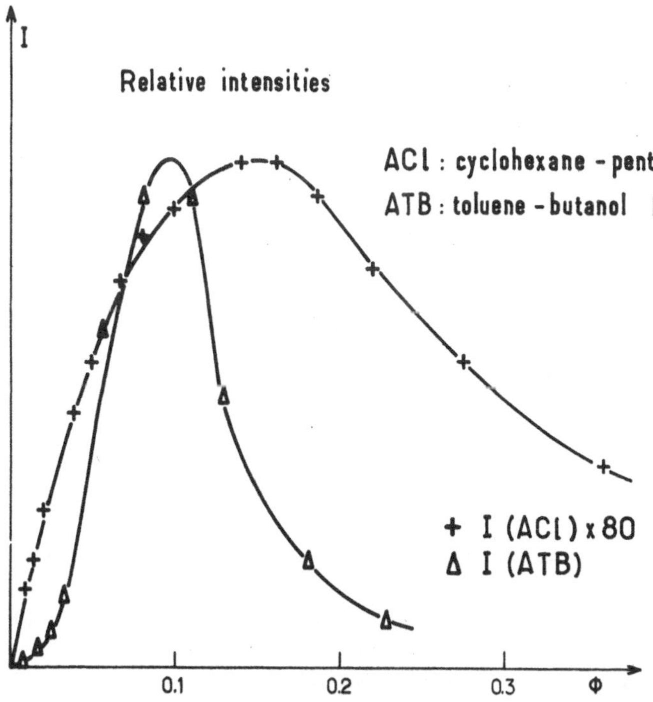

Figure 1.

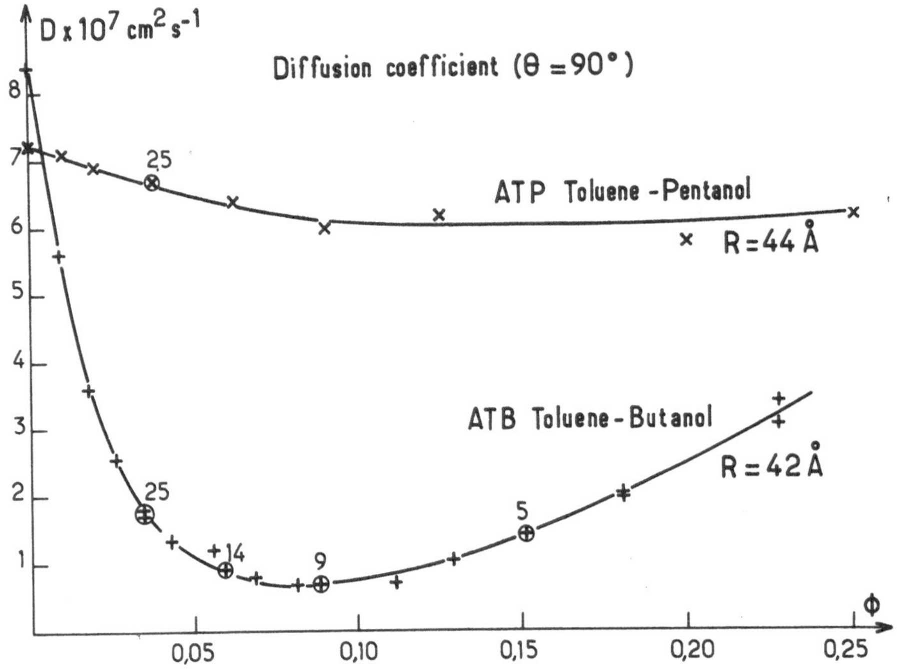

Figure 2.

The droplet sizes are determined through extrapolations to zero concentration. The slopes of ϕ/I and D with ϕ at small ϕ, $\phi \lesssim 0.1$ give the virial coefficients B and α. In most cases B and α differ largely from the hard sphere values $B_{HS} = 8$ and $\alpha_{HS} = 1.5$. The difference can be related to the strength of an attractive interaction between droplets. Numerical estimations led to the striking conclusion that this strength is about 2 orders of magnitude larger than Van der Waals interaction between the same droplets. Up to now the origin of these strong attractive interactions has been very poorly understood.

The theories of particles' brownian motion allow us to relate the virial coefficients B and α.[2,6] In our experiments the agreement is qualitatively good.

At larger volume fractions, one can still fit the intensity variation versus ϕ with expressions derived from a model by Vrij,[2] in which he assumed that the osmotic pressure is the sum of a hard sphere term and of a contribution of the attractive forces treated as a perturbation. The fit accounts well for the maximum of $I(\phi)$ at $\phi \sim 0.1$; it is usually very good up to large volume fractions $\phi \sim 0.2 - 0.3$ except in some cases, in particular when the attraction becomes very strong (case of the microemulsions ATB in Figure 1).

$D(\phi)$ shows a minimum at the volume fraction which corresponds to the maximum of $I(\phi)$: $\phi \sim 0.1$. Now D and I are given by:[6]

$$D = \frac{4}{3}\pi R^3 \frac{\partial \Pi}{\partial \phi}\frac{1}{f} \quad \text{and} \quad I \propto \phi\left(\frac{\partial n}{\partial \phi}\right)^2 \left(\frac{\partial \Pi}{\partial \phi}\right)^{-1}$$

f being the friction coefficient and $\partial\Pi/\partial\phi$ the osmotic compressibility. By forming the product $\phi x D^{-1} \times I^{-1}$ one gets a quantity proportional to the friction coefficient f. No extremum of f at $\phi \sim 0.1$ is found in this way. This means that the minimum of D can be entirely accounted by the osmotic compressibility variation. I and D are independent of scattering angle in most cases. However, in the case of microemulsions which exhibit an extremely strong variation of D versus ϕ (microemulsion ATB Figure 2) and for volume fractions ϕ close to the extrema $\phi \sim 0.1$, both D and I were found to depend on scattering angle. This is shown in Figures 3 and 4. In these systems the virial coefficients are very negative, indicating strong attractive interactions between droplets. On the other hand, a careful analysis of the autocorrelation function shows no evidence of departure from exponential behavior. Such volume fractions correspond also to the percolation threshold already observed in electrical conductivity measurements.[10] In the "stirred percolation" model proposed to interpret these measurements, it is assumed that the

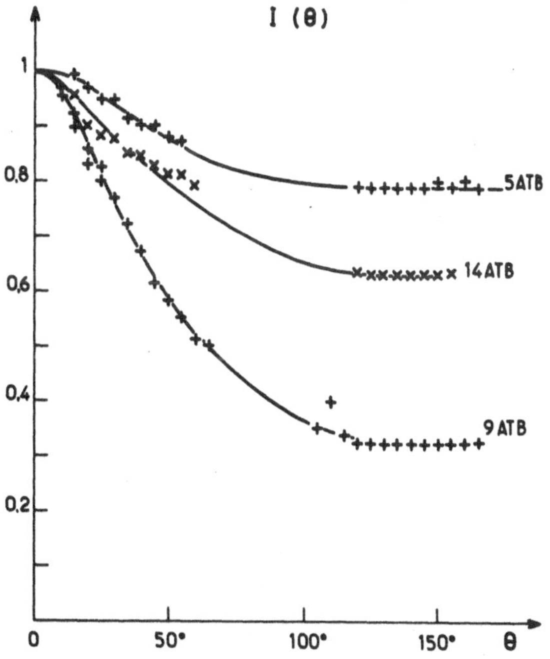

Figure 3.

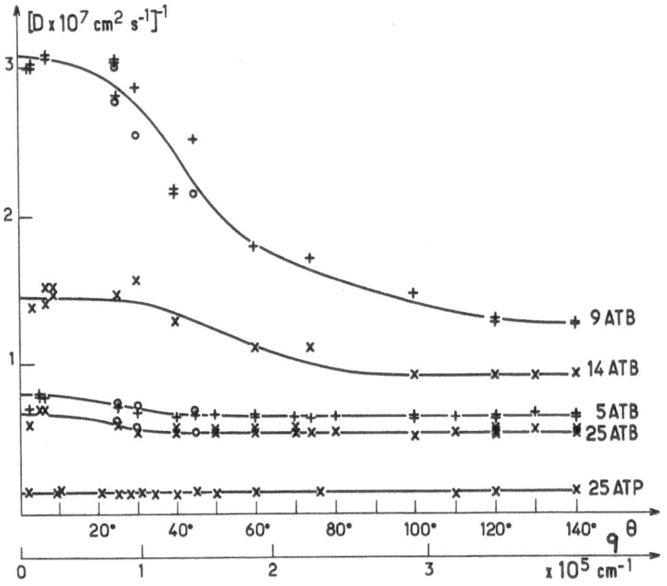

Figure 4.

droplets behave as randomly distributed hard spheres. Transient clusters are statistically formed. Above a critical volume fraction ϕ_c an infinite path appears through the droplets. The electrical conductivity increases steeply (several orders of magnitude) around ϕ_c.

Dependences of D and I versus scattering angle were already found in other systems when the correlation length becomes of the order of the optical wavelengths: fluid near a critical point, charged macromolecular solutions,[11] latex dispersions.[12,13] In the two last cases the autocorrelation function has a pronounced non-exponential behavior. Pusey[12] attributed this effect to the existence of two slow processes in the medium: diffusion of individual particles across solvent molecules and changes of the instantaneous configuration of particle neighbors. The mean interparticle spacing being large, the characteristic time τ_I of the second process is long ($\sim$ 200 μs); the correlation function decay is determined by free particle motion for $\tau < \tau_I$ and by interparticle forces for $\tau > \tau_I$. In our present problem, the mean interparticle spacing is not significantly different from particle radius: $d \sim R(4\phi^{-1})^{1/3} \sim 3.5R$ for $\phi \sim 0.1$; τ_I is very small ($\sim$ 10 ns) compared to correlation delay times. Consequently, the particle concentration can be considered as the only slow decaying property, and its correlation function will be exponential as observed.

The analogy between the present problem and critical phenomena is closer. In both cases, the correlation function for the concentration is exponential (no coupling with other hydrodynamic variables).

Near a critical point, a picture of a fluid can also be made with transient clusters of size ξ. The correlation length ξ is the range over which the concentration fluctuations are correlated and diverges at the critical point. In the percolation problem, the correlation length is the spatial size of the largest clusters and diverges also beyond the percolation threshold. However, the analogy cannot be driven too far: indeed the scattering of light by a percolating system does not usually diverge near the threshold: the important fluctuations are not in the concentrations but in the connectivity.[14]

An explanation of the variation of both scattered intensity and diffusion coefficient with scattering angle in our experiments can be given with the help of the recent theories of brownian motion of interacting particles:[6]

$$I(q) \propto \phi S(q) \qquad\qquad D(q) = D_o \frac{H(q)}{S(q)} .$$

S(q) is the structure factor and H(q) contains the hydrodynamic interactions. Here S(q) decreases when q increases. This indicates the existence of long range ($\sim$ 500 Å) attractive interactions. Unfortunately, the nature of the interaction potential is not sufficiently well known to permit a precise comparison with experimental data. The problem of hydrodynamic interactions is still more complicated. Let us mention that the experiments indicate that long range contributions are much less important that for S(q):H(q) $\sim$ cst.

4. CONCLUSION

Light scattering technique appears as a very useful method for obtaining information about sizes and interactions in microemulsions. Some general conclusions can be made over the large variety of water in oil systems that we have studied. Droplets sizes depend mostly on the ratio of water to soap as it can be expected by simply assuming that the area per polar head of soap molecules at the surface of water droplets is constant. Interactions are more closely related to the molecular organization of the interface: compact interfaces lead to hard sphere like interactions, fluid interfaces to strong attractive interactions. In the last case these interactions can become of long range character ($\sim$ 500 Å) in the vicinity of a percolation threshold.

These results can be a guide for the elaboration of a model of interaction potential, allowing achievement of a better understanding of the problem of microemulsions stability.

REFERENCES

1. L. M. Prince, in "Microemulsions, Theory and Practice", Academic Press, New York, 1978.
2. A. A. Calje, W. G. M. Atgerof, A. Vrij, in "Micellization, Solubilisation and Microemulsions", Vol. 2, Plenum Press, New York, 1977.
3. A. Graciaa et al., J. Phys. Lett. 38: 253 (1977); 39: 235 (1978).
4. M. Zulauf, H. E. Eicke, J. Phys. Chem. 83: 480 (1979).
5. A. M. Cazabat, D. Langevin, A. Pouchelon, J. Coll. Int. Sci. 73: 1 (1980); A. M. Cazabat, D. Langevin, J. Chem. Phys., 74: 3148 (1981).
6. B. J. Ackerson, J. Chem. Phys. 64: 242 (1976); 64: 684 (1978).
7. W. Hess, Light Scattering in Liquids and Macromolecular Solutions (Plenum Press, New York, 1980).
8. A. M. Bellocq et al., Optica Acta, to appear.
9. A. M. Cazabat, D. Chatenay, D. Langevin, A. Pouchelon, J. Phys. Lett. 41: 441 (1980).
10. M. Lagües, R. Ober, C. Taupin, J. Phys. Lett. 39: 487 (1978).
11. D. W. Schaefer, B. J. Berne, Phys. Rev. Lett. 32: 1170 (1974).
12. P. N. Pusey, J. Phys. A 8: 1433 (1975).
13. H. M. Fijnaut et al., Chem. Phys. Lett. 32: 351 (1978).
14. A. Coniglio, H. E. Stanley, W. Klein, Phys. Rev. Lett. 42: 518 (1979).

LIGHT SCATTERING BY OIL-WATER INTERFACES

D. Chatenay, D. Langevin, J. Meunier, and A. Pouchelon

Laboratoire de Spectroscopie Hertzienne de l'E.N.S.
24, Rue Lhomond
75231 Paris Cedex 05

CONTENTS

1. INTRODUCTION

The use of light scattering techniques in the field of liquid interfaces has been developed recently.[1] Contrary to most of the other scattering techniques (X-rays, neutrons, etc.) light scattering does not suffer from limitations due to low scattering cross sections. It also proved to be particularly useful when the more classical mechanical techniques also became limited: near critical points, in the presence of liquid crystal phases, or for soap films, polymer solutions, and monolayers on water. It is also particularly well-suited for the study of oil water interfaces of low interfacial tensions. These systems, recently discovered, are of great practical interest because they are the basis of several oil recovery processes. In these processes, the oil water interfacial tensions, typically of the order of 50 dyn/cm, are decreased to less than about 10^{-3} dyn/cm by adding small amounts of surfactant molecules.[2] Although numerous groups have observed ultralow tensions and have worked out empirical rules for the selection of convenient mixtures for each reservoir, there is little basic understanding of how ultralow tensions arise.

The low interfacial tensions are associated with unusual phase equilibria. The surfactant is not only present at the interface between water and oil, but also in both bulk phases. This can allow some water (or oil) molecules to dissolve into the oil (or water), and to form a water in oil (or oil in water) microemulsion. Phase equilibria can be modified by introducing some other chemicals besides the surfactant (alcohol as cosurfactant, salt, etc.) or by varying the temperature. For instance, if the salinity is increased, the partitioning of the surfactant changes. One generally observes at low salinity a O/W microemulsion in equilibrium with oil, and at high salinities a W/O microemulsion in equilibrium with brine. At intermediate salinities one can observe a middle phase microemulsion in equilibrium with both oil and brine. The lowest interfacial tensions are usually found in the three-phase region.

Up to now, the relationship between phase equilibria involving microemulsions and low interfacial tensions is not clearly established. It has been suggested that when the amount of surfactant at the interface is sufficiently large, the spreading pressure can be large enough to produce small values of the interfacial tensions. Another interesting possibility is the formation of a thick oil-water interface, the thickness being related to droplet sizes in the microemulsion.[3] The case of middle phase microemulsion is certainly particular since its structure seems different compared to that of usual microemulsions. It has been proposed that it may be a true molecular solution among surfactants, oil and water, or that it may have a bicontinuous structure (sponge-like structure, mixture of O/W and W/O microemulsions, lamellar structure).

A great deal of both experimental and theoretical work is still needed to elucidate these problems. In this paper, we will present our recent light scattering measurements of both interfacial and bulk properties of one of these systems.

2. EXPERIMENTAL PROCEDURE

The samples have been prepared by mixing several pure chemicals: bidistilled water + sodium chloride (brine, 46.8% by weight), toluol (oil, 47.25%), sodium dodecyl sulfate (surfactant, 1.99%), and butanol

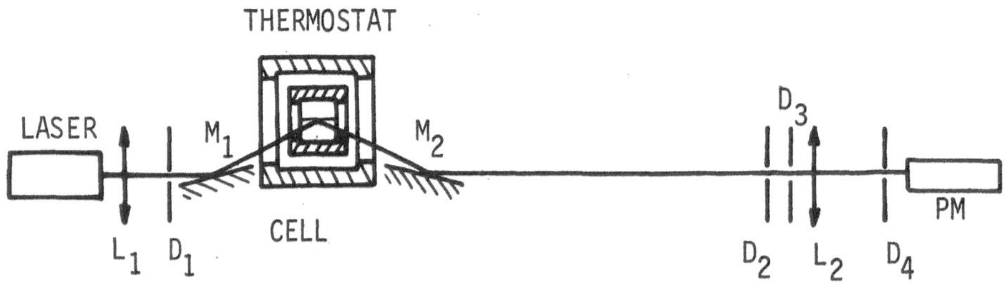

Figure 1. Set-up.

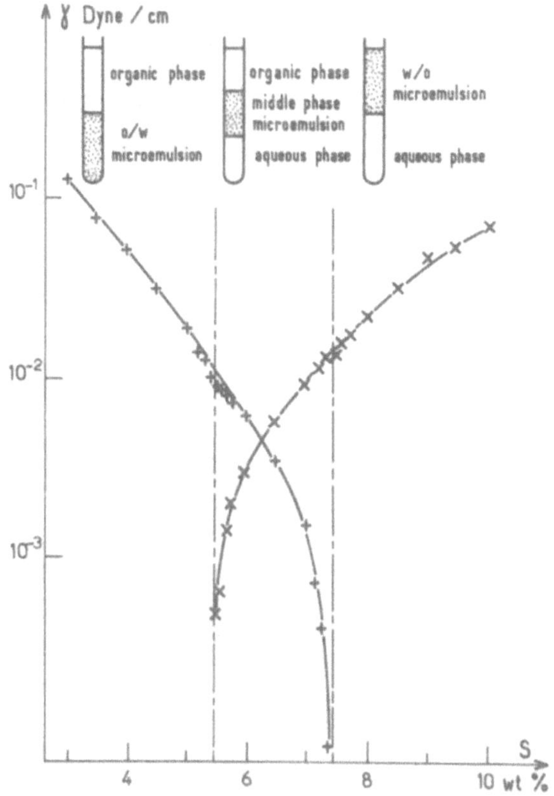

Figure 2. γ versus S.

(cosurfactant 3.96%). The temperature of phase separation is 20°C. The interfacial tensions have been measured by a light scattering technique developed in our laboratory.[4] The set-up is represented in Figure 1. The spectrum of the scattered light is recorded and fitted with a Lorentzian shape. The width is related to interfacial tension γ by

$$\Delta\nu = \frac{(\gamma + \Delta\rho g/q^2)q}{4\pi(\eta + \eta')} .$$

$\Delta\rho$ is the density difference between bulk phases, q is the scattering wave vector, and η and η' are the viscosities of the bulk phases below and above the interface. In the two-phase regions, microemulsion phases have been diluted with a continuous phase according to the procedure described in reference 5. We wanted to investigate in this way the influence of micellar concentration on the interfacial tension. We did not find the way to dilute the middle phase microemul-

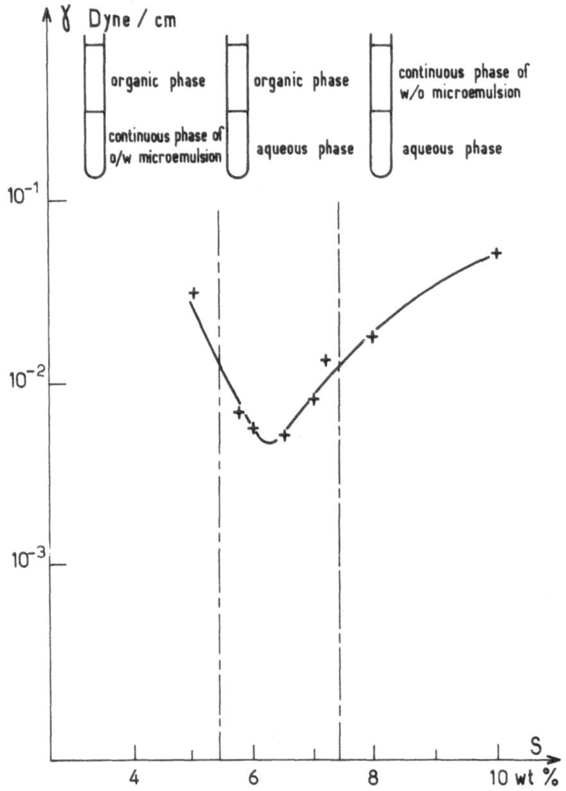

Figure 3. γ versus S.

sions in the three-phase region. This was not really surprising because this phase is expected to have a different structure.

3. RESULTS

Figure 2 represents the measurements of interfacial tensions at different salinities S. Very small values were obtained. As in most systems, interfacial tension between microemulsions and oil phase γ_{Om} decreases when salinity increases and tension between microemulsions and aqueous phases γ_{mW} increases with salinity.

In the two-phase domain interfacial tension was found to be independent of droplet concentrations in the microemulsion phase. In particular, its value was not changed when the microemulsion was replaced by its continuous phase. Moreover the composition of this continuous phase approximately coincides with the extrapolation at the corresponding salinity of the composition of either aqueous phases for O/W microemulsions or oil phases for W/O microemulsions. This led us to measure also the interfacial tension γ_{OW} between the upper oil phase and the lower aqueous phase in the three-phase domain. These results are shown in Figure 3. As expected the curve shows no discontinuities at the three-phase domain boundaries. At all other salinities, γ_{OW} stays equal to the highest interfacial tension in the initial three-phase system (Figure 2) within experimental accuracy (± 10%).

4. DISCUSSION

The dilution of a microemulsion does not change the interfacial tension when the microemulsion is in equilibrium with an aqueous or an organic phase. The microemulsion can even be replaced by its continuous phase provided this phase is saturated in surfactant monomers.[5] We can therefore conclude that in these systems the origin of low interfacial tensions is entirely due to a thin condensed surfactant layer at the interface. In the three-phase region the interfacial tension between organic and aqueous phases can be as low as $\gamma^* = 4.5 \times 10^{-3}$ dyn/cm. Again this is only possible if a thin surfactant layer forms at the interface.

The interfacial tension does not change when one replaces the aqueous phase with the middle phase if the salinity S is less than $S^* = 6.35$ wt %, and when one replaces the organic phase with the middle phase if $S > S^*$. This suggests that the interface structure is not strongly affected by this procedure and, in particular, that interfacial tensions do not depend on middle phase structure. This is essentially the same result which was obtained for the two-phase region. The origin of all these low interfacial tensions can then probably be attributed to thin surfactant layers at the interface.

The still lower interfacial tensions in the three-phase domain $\gamma < \gamma^*$ seem to be of quite different origin. They can most likely be associated with thicker interfaces, as in critical phenomena. Some evidence of the vicinity of a consolute point was indeed found close to the three-phase domain boundaries. Figure 4 shows that the intensity scattered by the microemulsion phases increases sharply. This is associated with the decrease of the diffusion coefficient (Figure 5) and the onset of angular variations of both scattered light intensity and diffusion coefficient. The existing theories of interfacial tensions cannot yet explain all the observed experimental features[6] but they can certainly be improved.

5. CONCLUSION

When a microemulsion is in equilibrium with oil and brine, the origin of the two resulting low interfacial tensions is due to quite different phenomena. One is associated with the presence of a thin surfactant layer at the interface, as when the microemulsion is in equilibrium with only oil or with only brine. The second low inter-

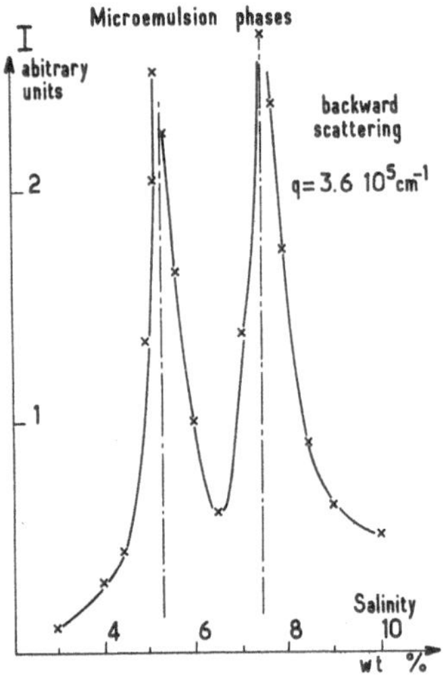

Figure 4. I versus S.

facial tension is more likely related to thicker interfaces as in critical phenomena.

Since the oil displacing efficiency during the recovery process is controlled by the largest interfacial tension value, the presence of the microemulsion is not required to achieve such a value. In fact, microemulsions are expected to act as surfactant reservoirs which keep the oil-water interfaces saturated in surfactant molecules during the recovery process.

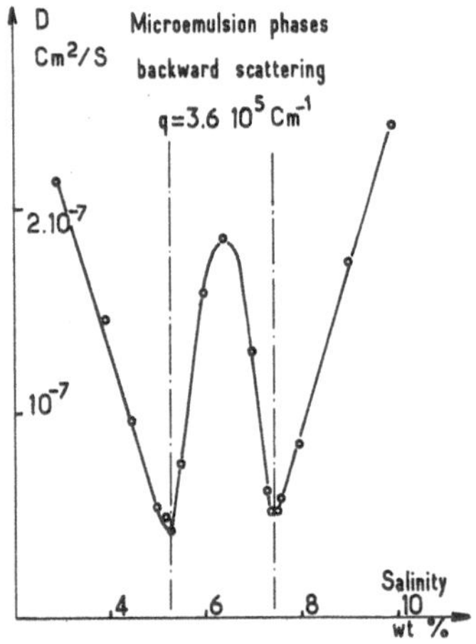

Figure 5. D versus S.

REFERENCES

1. D. Langevin, J. Meunier, in "Photon Correlation Spectroscopy and Velocimetry" (Plenum, New York, 1976)(review article).
2. W. F. Foster, J. Pet. Tech. 25: 205 (1973).
3. C. A. Miller, R. Hwan, W. J. Benton, and T. Fort, J. Coll. Int. Sci. 61: 554 (1977).
4. A. Pouchelon, J. Meunier, D. Langevin, and A. M. Cazabat, J. Phys. Lett. 41: 239 (1980).
5. A. Pouchelon, J. Meunier, D. Langevin, and D. Chatenay, to appear in J. Coll. Int. Sci.
6. A. Pouchelon, J. Meunier, D. Langevin, D. Chatenay, and A. M. Cazabat, Chem. Phys. Lett., 76: 277 (1980).

DYNAMIC LIGHT SCATTERING STUDIES OF DODECYLDIMETHYL-AMMONIUM CHLORIDE MICELLES

A. Flamberg and R. Pecora

Department of Chemistry
Stanford University
Stanford, California 94305

CONTENTS

1. INTRODUCTION

In this work, we use the dynamic light scattering technique of photon correlation spectroscopy to study the properties of micellar dodecyldimethyl-ammonium chloride ($C_{12}H_{25}(CH_3)_2NHCl$, abbreviated DDAC). This system has been extensively studied by S. Ikeda and co-workers[1] using surface tension, viscosity, static light scattering and C-13 NMR measurements. From these studies it was observed that at surfactant concentrations well above the critical micelle concentration the micelles undergo some kind of transition as the ionic strength is changed by adding increasing amounts of NaCl. From the evidence thus far obtained, a transition from spherical micelles at low salt concentrations to flexible rod-like micelles at high salt concentrations has been proposed. This theory has also been used to explain data from another well-studied micelle system, SDS,[2] but much controversy exists.[3]

By determining the diffusion coefficients of these micelles and comparing them to the measured molecular weights and radii of gyration, information about the micelle shape is obtained. If the micelles at very high NaCl concentrations are in fact long rods (predicted to be about 3000 Å in 4 M NaCl) then a structure factor effect in the autocorrelation function of the scattered light intensity should be seen.

2. EXPERIMENTAL

The cationic surfactant DDAC used was carefully prepared in the laboratory of S. Ikeda and used without further purification. Solutions were filtered with 0.2 μ Nucleopore membrane filters until they were dust-free.

The instrumental set up was as follows: The 4880 Å line of a Lexel Model 95 Argon ion laser was focused into a square cuvette in a thermostated sample block whose temperature was controlled to ±0.05°C by a water bath. After passing through the collection optics, the light was detected with an ITT FW 130 phototube and a Mech-Tronics 511 photon discriminator. The signal was then processed by a 48 channel Malvern digital autocorrelator whose output was stored on floppy disks by a Data General Nova 3 computer, which also performed certain curve fitting routines. Other fitting routines and drawings of the data were done on a Data General Eclipse 840 computer and a Versatec plotter.

3. RESULTS AND DISCUSSION

Translational diffusion coefficients are obtained by fitting the autocorrelation function to a single exponential of the form

$$C(\tau) = A + B \exp(-2k^2 D\tau)$$

where D is the translational diffusion coefficient, A and B are constants, and k is the length of the scattering vector, which is related to the experimental scattering angle θ by

$$k = \frac{4\pi n}{\lambda} \sin(\theta/2)$$

The Stokes-Einstein relation is then used to determine an apparent hydrodynamic radius. The results are then given in Table 1 along with parameters previously measured by Ikeda _et al_.[1] In Figure 1, the measured hydrodynamic radius is plotted against the radius of gyration. These points can be compared to the solid curves which represent the theoretical predictions for rods (using Broersma's relations), oblate ellipsoids (using Perrin's relations) and spheres. This type of analysis is similar to that done on SDS[4] and seems to indicate a similar rod-like shape for DDAC.

The deviation from rod-like behavior for the 4 M sample is of interest. If the micelles are truly rods, then the autocorrelation function should show a dependence on the scattering angle. The theory for scattering from rigid rods is well known[5] and predicts an autocorrelation function of the form

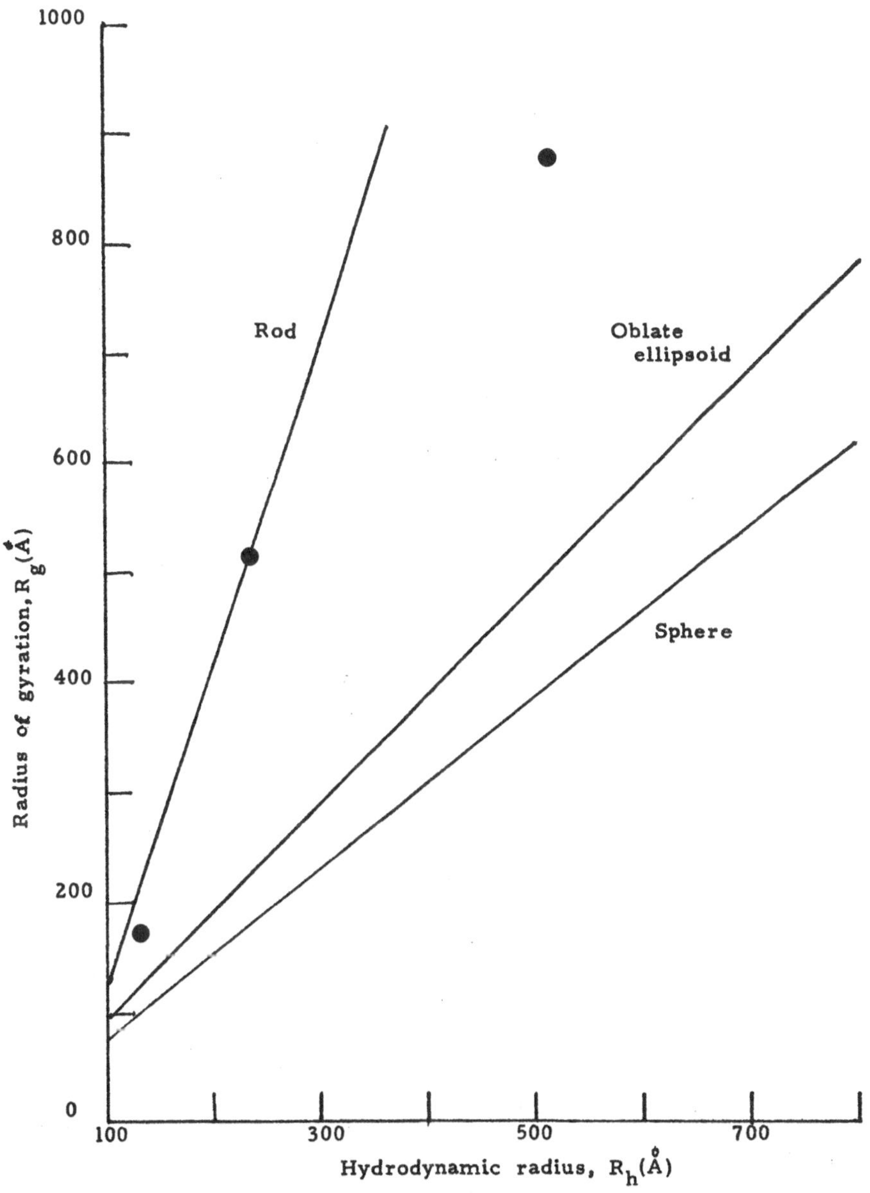

Figure 1. Plot of measured radii of gyration vs. hydrodynamic radii. The points are for DDAC in 2 M, 3 M, and 4 M NaCl.

Table 1. Hydrodynamic radius dependence on salt and surfactant concentration. The radii of gyration and the molecular weights are taken from reference 1. All data was collected at 25°C.

NaCl (M)	DDAC (g/dl)	R_h (Å)	R_g (Å)	MW
0.1	0.99	20.5		19,400
1.0	0.38	25		25,000
	0.75	25		
	2.66	37		56,000
2.0	0.88	137	174	333,000
3.0	0.22	232	515	1,200,000
4.0	0.50	508	884	2,940,000

$$C(kL,\tau) = A + [S_o(kL)\exp(-k^2D\tau) + S_1(kL)\exp(-k^2D-6\theta) + \ldots]^2 \quad 2$$

where L is the length of the rod and θ is the rotational diffusion coefficient. The amplitudes S_o and S_1 are usually much greater than the higher order amplitudes S_2, S_3, etc. (For a spherical system S_1 and all of the higher order terms would be equal to zero). The actual data is shown in Figure 2. At low scattering angle the data fits well to a single exponential, but a high scattering angle it does not. This behavior would be expected for a 3000 Å rod where $S_o >> S_1$ at low angle and $S_o \approx S_1$ at high angle. The data for the 2 M NaCl system fits well to a single exponential at both low and high scattering angles. In this system the micelles are predicted to be rods 600 Å in length. For rods of this size, theory predicts that $S_o >> S_1$ at all scattering angles. Our data is therefore consistent with this prediction.

Other effects which need to be considered are those of flexibility, polydispersity and concentration. Of these, concentration is probably the most important. If the rod model is again assumed, then the concentrations studied in this work would fall within the region discussed in the works of Doi and Edwards.[6] These concentrations lead to restricted diffusion, both rotational and translational. A slower translational diffusion coefficient would give a larger apparent hydrodynamic radius, and perhaps this could account

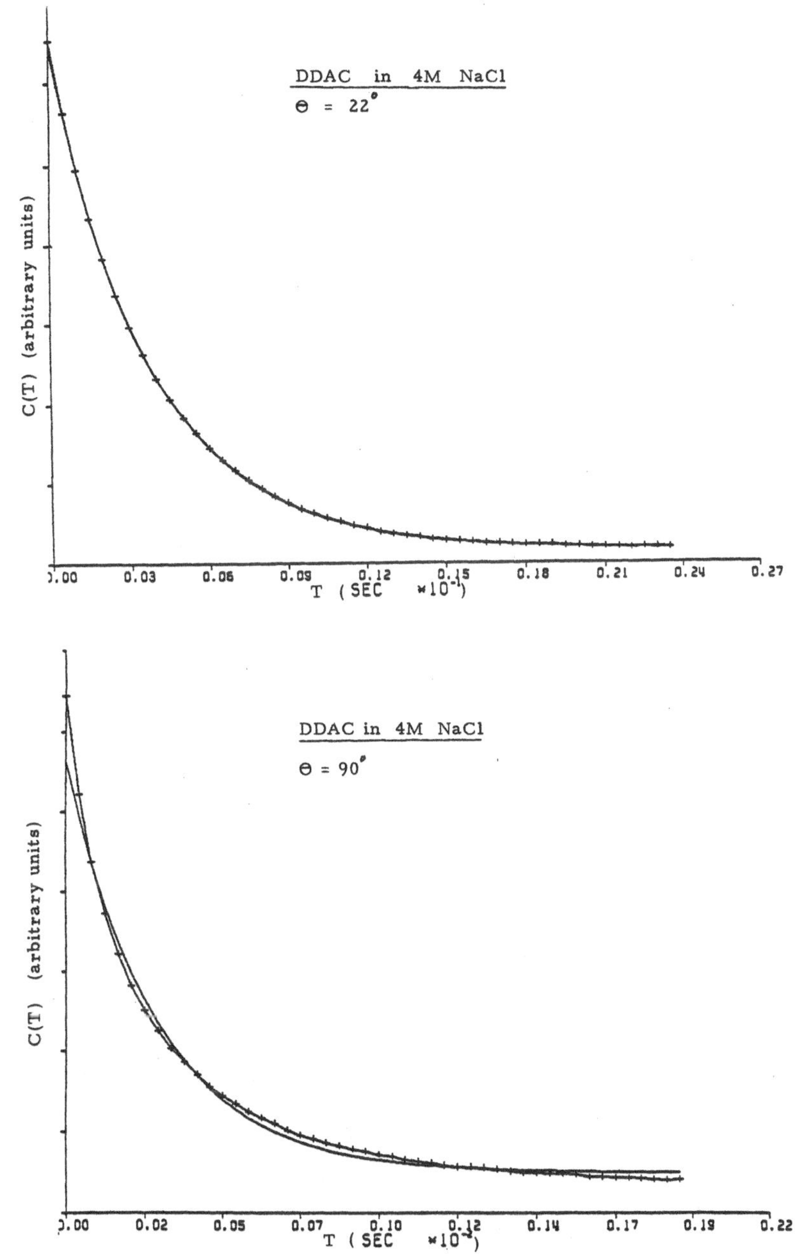

Figure 2. C(τ) in arbitrary units <u>vs</u>. τ. The data points (+) have been fit with a single exponential (solid line). The samples are DDAC micelles in 4 M NaCl, with scattering angles of θ = 22° and 90°. The data was collected at 25°C.

for the deviation of the 4 M point in Figure 1. Also, Doi and Edwards predict an additional exponential term in the light scattering correlation function of these concentrated systems. The decay time of this additional exponential term is predicted to be proportional to the scattering vector length. Therefore great care must be exercised in interpreting the data in Figure 2.

In conclusion, experimental evidence presented here is consistent with the theory of rod-like micelles. However, much more work needs to be done on this system. More experiments are presently in progress in our laboratory.

ACKNOWLEDGEMENTS

We wish to thank Professor S. Ikeda for introducing us to this problem and supplying the surfactant samples used in this work. We also wish to thank the National Science Foundation (USA) and the National Institutes of Health (USA) for support of this work.

REFERENCES

1. S. Ikeda, S. Ozeki, and M. Tsunoda, J. Colloid Interface Sci. 73: 27 (1980).
2. P. J. Missel, N. A. Mazer, G. B. Benedek, C. Y. Young, and M. C. Carey, J. Phys. Chem. 84: 1044 (1980).
3. M. Corti and V. Degiorgio, in Solution Chemistry of Surfactants, ed. by K. L. Mittel, Plenum Press, New York, 1979, p. 377.
4. C. Y. Young, P. Missel, N. A. Mazer, G. B. Benedek, and M. C. Carey, J. Phys. Chem. 82: 1375 (1978).
5. B. J. Berne and R. Pecora, Dynamic Light Scattering, Ch. 8, John Wiley, New York, 1976.
6. M. Doi and S. F. Edwards, J. C. S. Faraday II 74: 560 (1978).

CRITICAL BEHAVIOR OF A MICROEMULSION

John S. Huang and M. W. Kim

Exxon Research and Engineering Company

Linden, New Jersey 07036

We have studied the nature of the phase transition in a water-in-oil microemulsion near the upper cloud point temperature and found that the transition resembled a critical transition in fluid mixtures.

Our 3-component microemulsion contains 3% (wt./vol.) sodium di-2-ethyl hexyl sulfosuccinate, a surfactant commonly known as AOT-100, 4.76% (vol./vol.) water and the rest is n-decane. The molar ratio [H_2O]/[AOT] is roughly 41. The solubilization of water by the dialkyl sodium sulfosuccinates in hydrocarbon phase has been studied by many authors,[1,2] and in the concentration range of interest, the microemulsion would have a structure that consists of 100Å size water swollen micelles "dissolved" in the continuous decane phase. The ternary phase diagram of AOT, water and decane is illustrated in Figure 1, the solid line separates the 1-phase region from the 2-phase region near the H_2O-decane base at 23°C. This phase boundary moves up as the temperature is raised, as shown by the broken line at 35°C. The composition of our test sample is marked by the cross (+). At 36.01°C, the phase boundary would pass through that point and the microemulsion becomes unstable at this temperature called the cloud point temperature, Tc, above which the system will split into two micellar phases separated by an interface.

Recently, Zulauf and Eicke have studied extensively the structure of the AOT-water-hydrocarbon systems.[3] Senatra and Guibilaro[4] have reported on a study on the dielectric properties of a water-hexanol-hexadecane system at constant temperature (20°C) as a function of the water concentration. They have found that both the real and imaginary parts of the dielectric constant diverge at certain water concentration (62.5% wt./wt.) and they describe this behavior as "following a critical phenomena pattern".

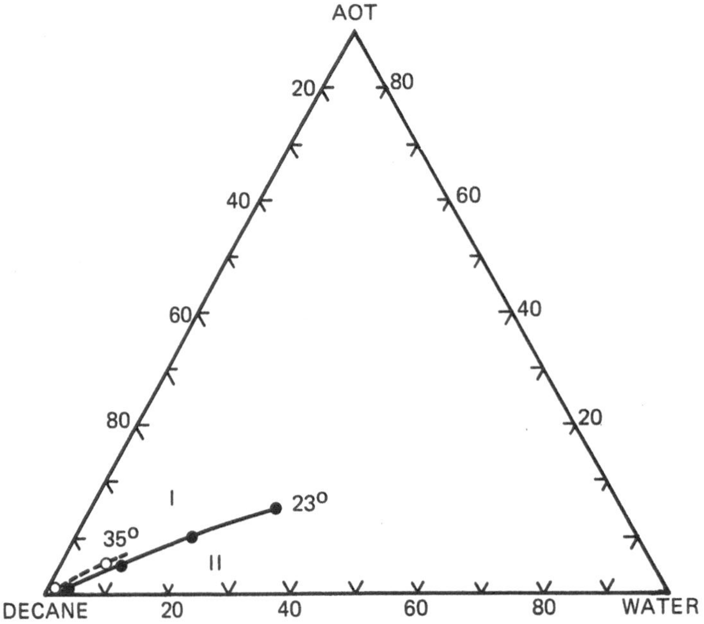

Figure 1. The ternary phase diagram of AOT, water and n-decane. The solid line divides the 1-phase and 2-phase region at 23°C, the broken line denotes the same phase boundary at 35°C. The + indicates the test sample composition.

In this study, we report dynamic light scattering measurements of the temperature dependence of the correlation lengths (static and dynamic) as well as the osmotic compressibility near the cloud point temperature, Tc. The overall composition is held constant.

The standard dynamic laser light scattering techniques and the photon-photon correlation spectroscopy were employed to measure the linewidth of the homodyne spectra as well as the scattering intensity as a function of the scattering angle and temperature (Figure 2). The sample cell is a sealed ampule made of precision bored nmr tubes of 10 mm O.D. AOT was twice recrystalized from methanol, and n-decane, a 99% + gold label product from Aldrich Chemical Co. was used without further treatment. Water was twice distilled from deionized water in our laboratory. Over 600 data were taken, the results are summarized as follows:

(1) The scattering intensity was found to obey the Ornstein Zernike-Debye relation even in the immediate neighborhood of the transition temperature. A static correlation length Rg can be deduced from the slope of the best straight line fit to the inverse

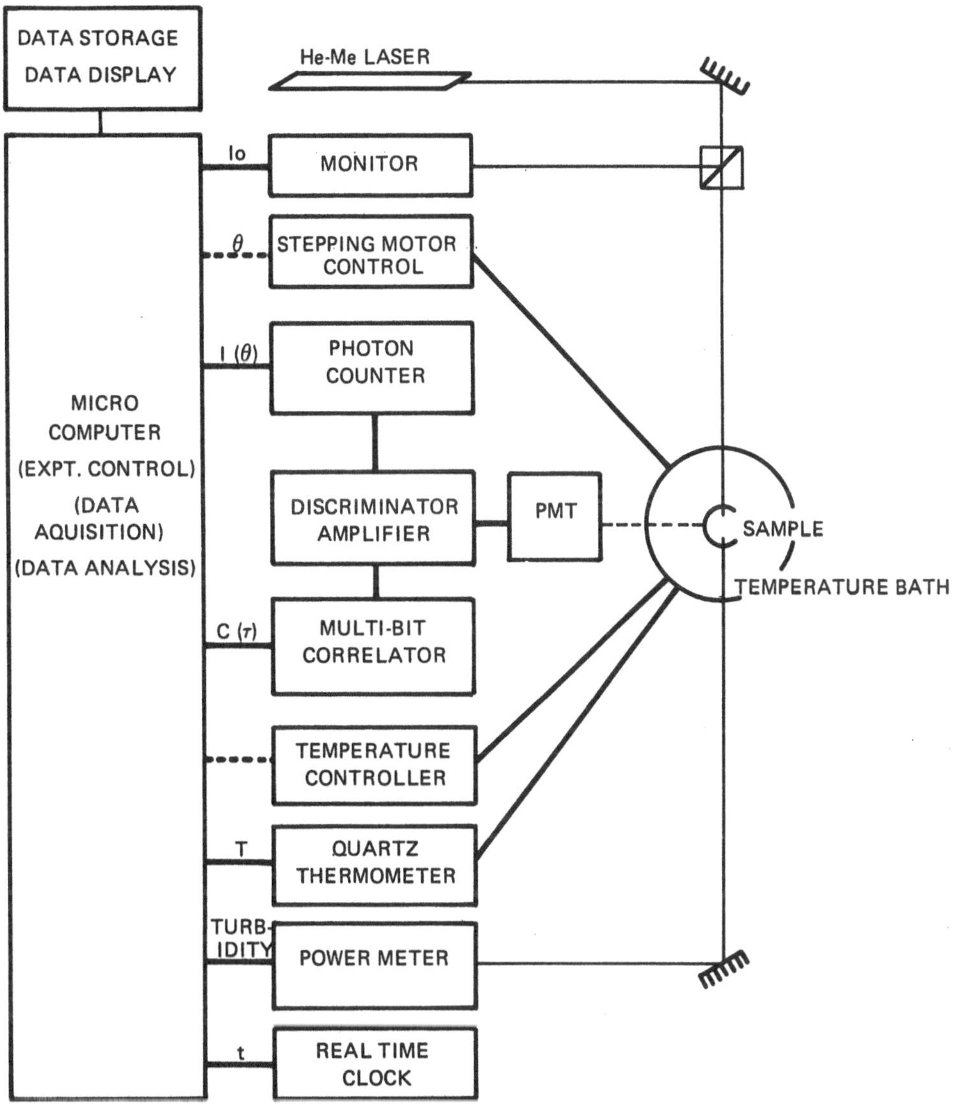

Figure 2. Experimental set-up of the light scattering experiment. The light source is a 15 mw He-Ne laser, the correlator is a Melvin multibit correlator, the microcomputer is a Hewlett Packard HP 9845B system.

scattering intensity $I(k)^{-1}$ versus k^2 plot as shown in Figure 3. The scattering vector

$$k = \frac{4\pi n}{\lambda} \sin \theta/2$$

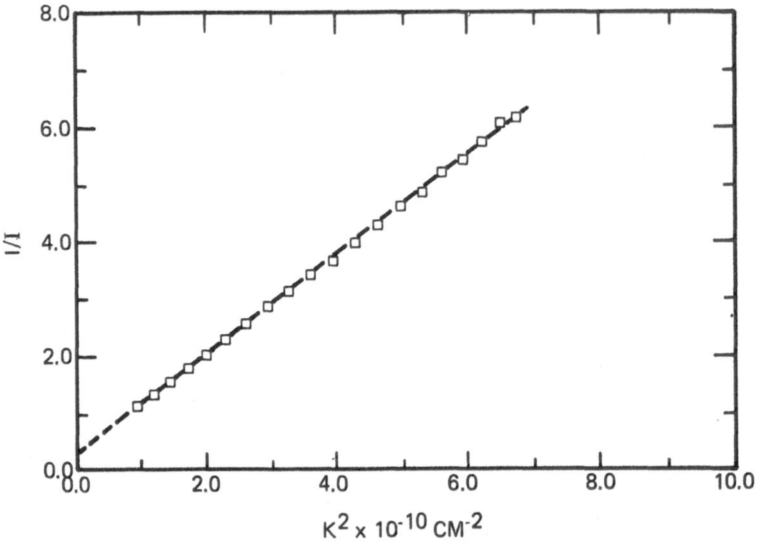

Figure 3. The inverse intensity $I(k)^{-1}$ is plotted against k^2 near Tc (35.18°C).

is expressed in terms of the index of refraction n, the wavelength of light in vacuo λ, and the scattering angle θ. The extrapolated forward scattering intensity I(0), a quantity proportional to the osmotic compressibility of the microemulsion droplets, is seen to diverge at the cloud point Tc = 36.01°C. The log-log plot for the forward scatterer intensity vs. the relative temperature Tc - T is shown in Figure 4, from the slope of the best straight line fit we

$$I(0) \sim (Tc - T)^{-\gamma}$$

where γ = 1.22. ($\sim$ reads, goes as, or, is proportional to.) A value in good agreement with the corresponding critical indices as measured in most critical binary fluids. The static correlation length also diverges at Tc, more about this in the following section.

(2) From the linewidth measurement, Γ, we can deduce a dynamic correlation length. Since we know that the scattering is mainly due to the swollen micelles, we could at least in the temperature range far from the cloud point, deduce a meaningful hydrodynamic radius R_h by using the Stokes-Einstein relation

$$R_h = \frac{k_B T}{6\pi\eta D} \quad (1)$$

where $D = \frac{\Gamma}{k^2}$ is the mutual diffusion coefficient and η is the sol-

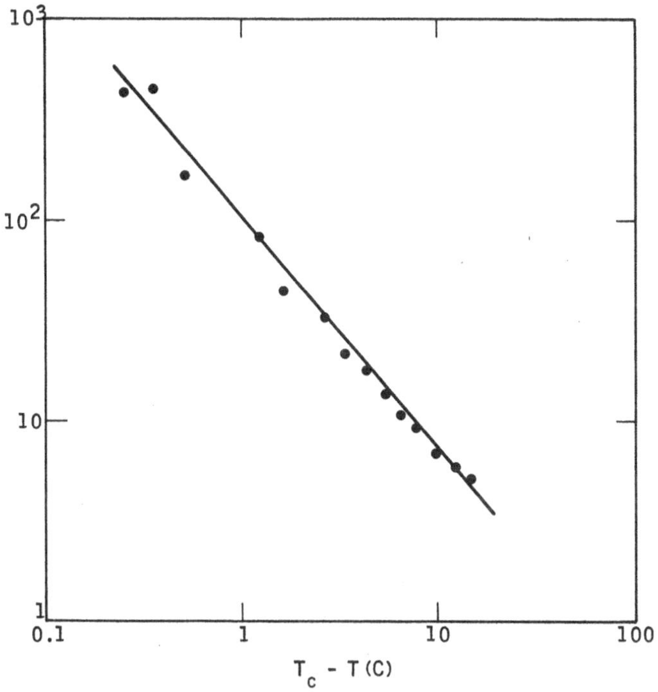

Figure 4. The forward extrapolated scattering intensity is plotted versus the relative temperature Tc - T on a log-log scale Tc = 36.01°C is the cloud point temperature.

vent (n-decane) viscosity. The value for this hydrodynamic radius compares very well with the radius of gyration for a spherical droplet as measured by the angular dependence of the scattering intensity.

In the neighborhood of the Tc (Tc - T < 3°C), Rh_o begins to increase dramatically from 100 Å to several thousand Å. Furthermore, the apparent value for diffusion constant begins to show a temperature dependence (where ξ is a length comparable to the hydrodynamic radius),

$$\frac{\Gamma}{k^2} = D_o(1 + \xi^2 k^2) \quad (2)$$

and a single exponential fit to the intensity correlation function deteriorates rapidly near Tc. This together with the scattering intensity results suggests a poly-dispersed, highly distorted droplet system.[5] However, the extrapolated value for $\Gamma/k^2 = D_o$ at forward angle (k = 0) would still be a measure for the average center of mass diffusion constant of the droplet system. Figure 5 depicts the critical slowing down of the scattering intensity fluctuations near Tc. It should be noted that the extrapolated temperature at which $D_o = 0$ and $I(0)^{-1} = 0$ are the same and it is undistinguishable from the visual cloud point temperature Tc. The hydrodynamic radius R_h of the scatterer is seen to diverge at Tc, and the divergence can be expressed by

$$R_h \sim \left(\frac{T_c - T}{T_c}\right)^{-\nu}$$

where $\nu = 0.75$ in the 1 phase region and $\nu' = 0.68$ in the 2 phase region. It is worth noting that the prefactor for the divergence for Tc - T (1-phase) case is roughly twice that in the 2-phase region, again resembling the critical binary mixtures.

Finally, the linewidth data taken from the above mentioned sample and several other samples with somewhat different concentration apparaently obey the dynamic scaling behavior as predicted for the critical transition by Kawasaki[6] and by Ferrell.[7] It should be noted that the criticality of the cloud point transition occurs in other microemulsion systems including the oil-in-water type. We intend to report fully the results of this study in the near future.

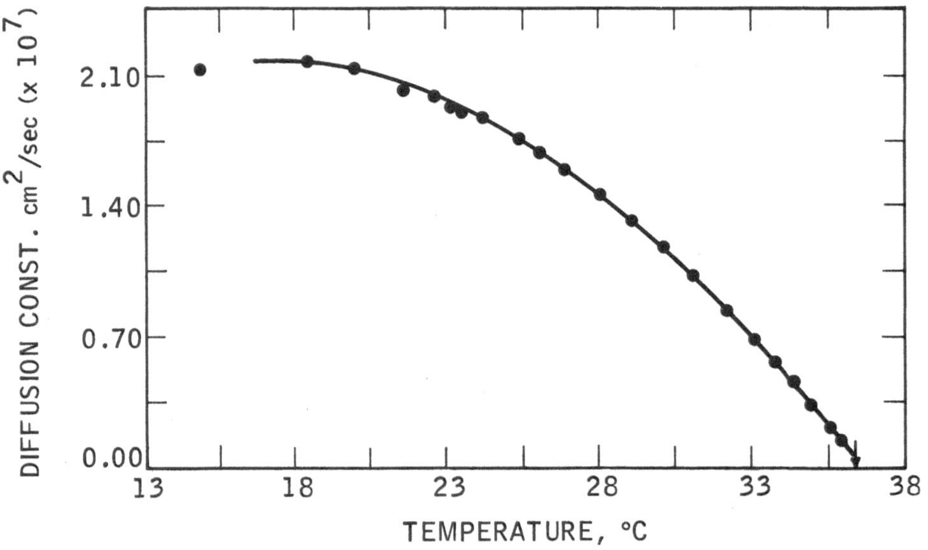

Figure 5. The critical slowing down of the composition fluctuation is illustrated by the vanishing of the forward extrapolated linewidth data $D_o = \Gamma/k^2$ at k = 0 as a function of temperature.

REFERENCES

1. P. Ekwall, L. Mandell, and K. Fontell, J. Coll. and Interface. Sci. 33: 215 (1970).
2. S. G. Frank, Y. H. Shaw, and N. C. Li, J. Phys. Chem. 77: 239 (1973).
3. M. Zulauf and H.-Friedrich Eicke, J. Phys. Chem. 83: 480 (1979).
4. D. Senatra and G. Guibilaro, J. Colloid and Interface Sci. 67: 448 (1978).
5. H. W. Huang, J. Chem. Phys. 73: 1084 (1980); H. W. Huang, F. Y. Chen, private communications.
6. K. Kawasaki, Phys. Rev. 150: 291 (1966); Phys. Lett. A30: 325 (1969); Phys. Rev. A1: 1750 (1970).
7. R. A. Ferrell, Phys. Rev. Lett. 24: 1169 (1970).

THE SPHERE-ROD TRANSITION IN SDS MICELLES

P. J. Missel and G. B. Benedek

Department of Physics
Massachusetts Institute of Technology
Cambridge, MA 02139
U.S.A.

ABSTRACT

We have employed laser light scattering methods to investigate the size, shape, and polydispersity of alkyl sulfate micelles in aqueous NaCl solutions. Our studies have demonstrated the existence of a continuous transition from roughly spherical aggregates (R ~ 25 Å) to polydisperse rod-like structures whose mean size increases with increasing NaCl, decreasing temperature and increasing detergent concentration. We present here a brief description of a thermodynamic model for this sphere-to-rod transition. In this model, detergent molecules are assumed to form a distribution of micelles with aggregation numbers $n = n_o, n_o+1, \ldots, n_o+k$, where n_o corresponds to the minimum spherical micelle and n_o+k corresponds to a prolate spherocylinder with K molecules in the cylindrical portion between hemispherical endcaps each having $n_o/2$ molecules. The theory describes the micellar size distribution using two parameters: $(X-X_B)$, the amount of detergent in micellar form, and K, which reflects the difference in chemical potential between a molecule in the spherical and cylindrical portions of the micelle. We first present experimental data which tests the ability of our theory to predict successfully the dependence of the mean hydrodynamic radius $\bar{R}_h$ on detergent concentration and the shape of the micellar size distribution. New data on the influence of chain length, counterion size and the addition of urea are discussed.

REFERENCES

P. J. Missel, N. A. Mazer, G. B. Benedek, and M. C. Carey, Thermodynamics of the sphere-to-rod transition in alkyl sulfate micelles,

to appear in "Solution Behavior of Surfactants," from the symposium held at Clarkson College in June, 1980 at the Tenth Northeast Regional Meeting of the American Chemical Society, K. L. Mittal and E. J. Fendler, eds. (Plenum, New York, due in 1981).
N. A. Mazer, M. C. Carey, and G. B. Benedek, in "Micellization, Solubilization, and Microemulsions," vol. 1, K. L. Mittal, ed. (Plenum, New York, 1977), p. 359.
N. A. Mazer, G. B. Benedek, and M. C. Carey, J. Phys. Chem. 80: 1975 (1976).
C. Y. Young, P. J. Missel, N. A. Mazer, G. B. Benedek. and M. C. Carey, J. Phys. Chem. 82: 1375 (1978).
P. J. Missel, N. A. Mazer, G. B. Benedek, C. Y. Young, and M. C. Carey, J. Phys. Chem. 84: 1044 (1980).

LIGHT SCATTERING STUDY OF THE PHYSICAL STRUCTURE OF HUMAN CERVICAL MUCUS

A. M. Cazabat

Laboratoire de Spectroscopie Hertzienne de l'E.N.S.
24, Rue Lhomond
75231 Paris Cedex 05

B. Volochine

Orme des Merisiers
CEN Saclay 91190
Gif Sur Yvette

J. M. Kunstmann and F. C. Chretien

Laboratoire d'Histologie
Hôpital de Bicêtre
94270 Kremlin Bicêtre

1. INTRODUCTION

We present here a study, performed by light-scattering techniques, of dynamical properties of a biological medium, the human cervical mucus. This medium exhibits a hydrogel-like structure (weak linkages[1]) containing up to 98% water. The solid phase is a tridimensional network, the meshes of which, for the studied samples, had a mean size equal to 5 μm (determined by scanning electron microscopy[2]) (Figure 1). Rheological experiments[3] have shown the strong non-newtonian character of this medium. The distribution of relaxation times τ, determined by viscoelastic measurements[4], is very large.

Furthermore, a nuclear magnetic resonance study[3,5] seems to show that in this medium there exist huge thermal oscillations. Although

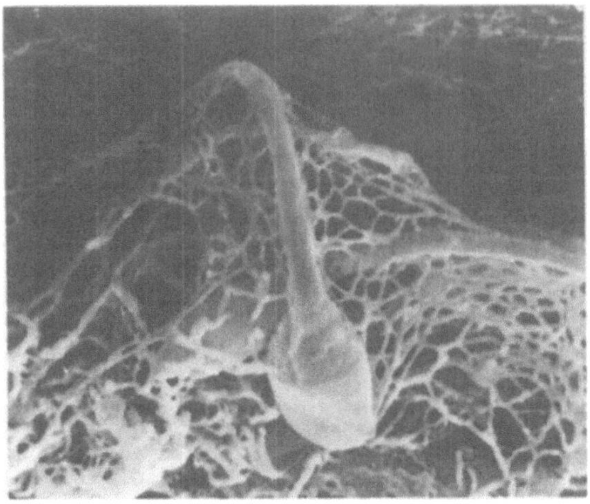

Figure 1. Example of human cervical mucus in non-ovulatory period. The magnification is equal to 11000.

all this information seems to be coherent with a model of a hydrogel, it has been strongly criticized.[6] The authors of this paper disclaimed competence for obtaining valid data on the true spatial organization and the behavior of cervical mucus infrastructure, from both nuclear magnetic resonance and scanning electron microscopy, which was deemed to be "largely artifactitious".[6] Their arguments did not appear to us to be convincing because they are in disagreement with actual results on hydrogels and entangled molecules.[7-10]

For this reason we decided to perform, by means of a non-disturbing method (light scattering), a thorough study of this medium which is very sensitive to mechanical stresses (the restoring time of broken linkages is very large[4]).

2. EXPERIMENTAL DEVICE

The experimental set-up (Figure 2), described in previous papers[11,12], is a classical two-beam, heterodyne device with variable scattering angle and adjustable heterodyne efficiency. We always worked with a heterodyne efficiency at least equal to 20 which is a value necessary for light-scattering experiments with gels.[13] The pinholes were selected to give one coherence area upon the photomultiplier. The signals were analyzed either by a classical real time wave analyzer or by an analog correlator.

3. RESULTS

The experimental spectra range continuously up to several kilohertz and are not reducible to a sum of two Lorentzian lines as it

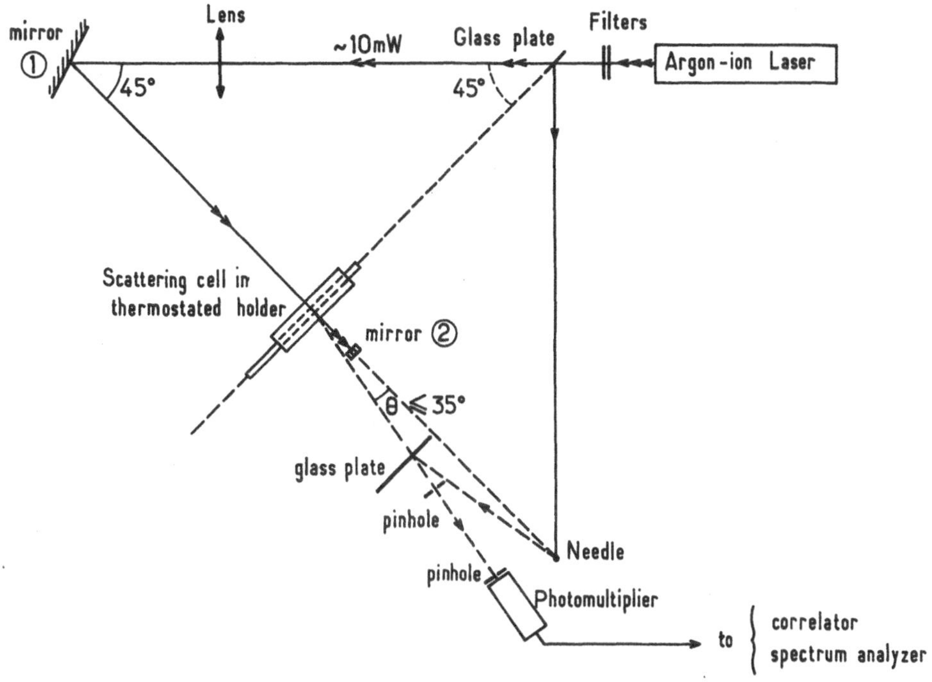

Figure 2. Experimental device (heterodyne technique).

was previously pointed out.[6] Figures 3 and 4 give, in log-log plot, an example of spectra obtained at different scattering angles on the same sample. The limit behavior of spectra in function of frequency ω is given by $\omega^{-(\beta + 1)}$ where β is a parameter such that $0 < \beta < 1$. The spectrum shown in Figure 3 is in good agreement with a model of hydrogel. At this scattering angle, the investigation domain has dimensions far less than the size of the meshes so the experimental method used is sensitive to all the local modes of the network and not to collective behavior of the network in its whole (the latter would give in gels[13] a simple Lorentzian line).

Figure 5 shows the results of a detailed angular study. Each experimental point is the mean value of data obtained from the 47 studied sample. The dispersion of experimental data is roughly equal to 20% from one sample to another.

This study shows that the spectrum is the sum of two contributions. The first one does not depend on q (with $q = (4\pi n/\lambda)\sin \theta/2$) and is well apparent for a scattering angle less than 15°. The second one, which appears for $\theta > 15°$, is approximately q^2-dependent. These two contributions are well-separated for $\theta > 35°$. These results exclude the possibility of a medium described by a model of "entangled macromolecules" since, in such a medium, all the modes are q-dependent.[7-9,14] On the contrary, these results are in good agreement with a model of a hydrogel with weak linkages.

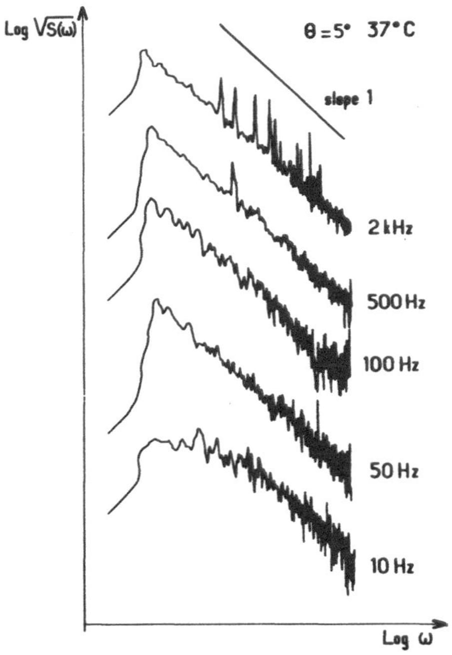

Figure 3. Log-log plot of the amplitude spectrum at $\theta = 5°$ for different frequency ranges of analysis.

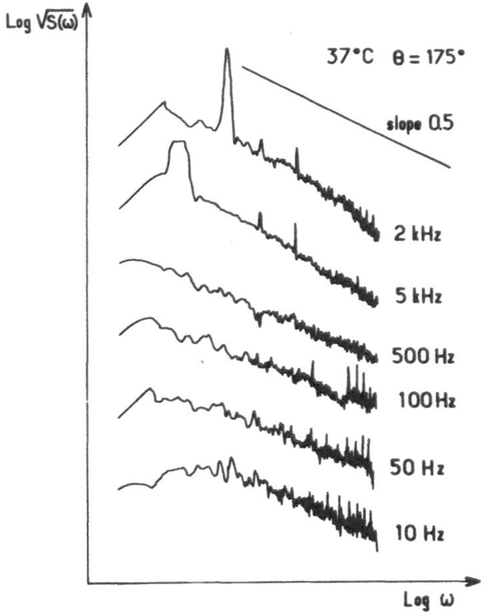

Figure 4. Log-log plot of the amplitude spectrum at $\theta = 175°$ for different frequency ranges of analysis.

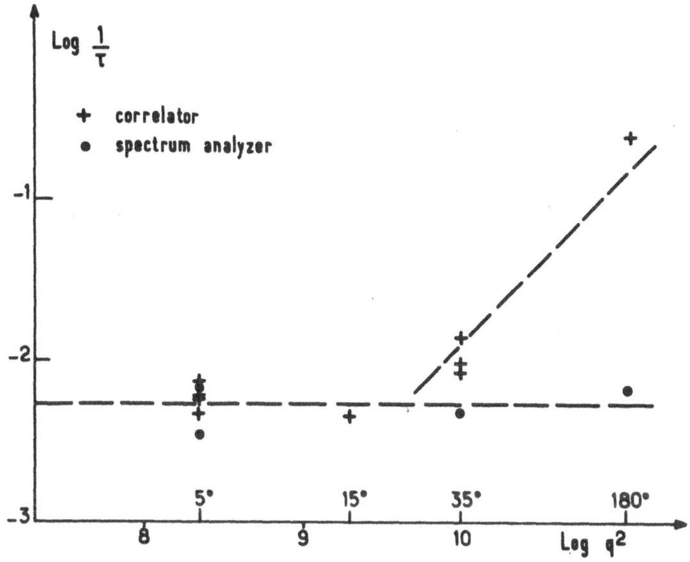

Figure 5. Log-log plot of τ^{-1} versus q^2.

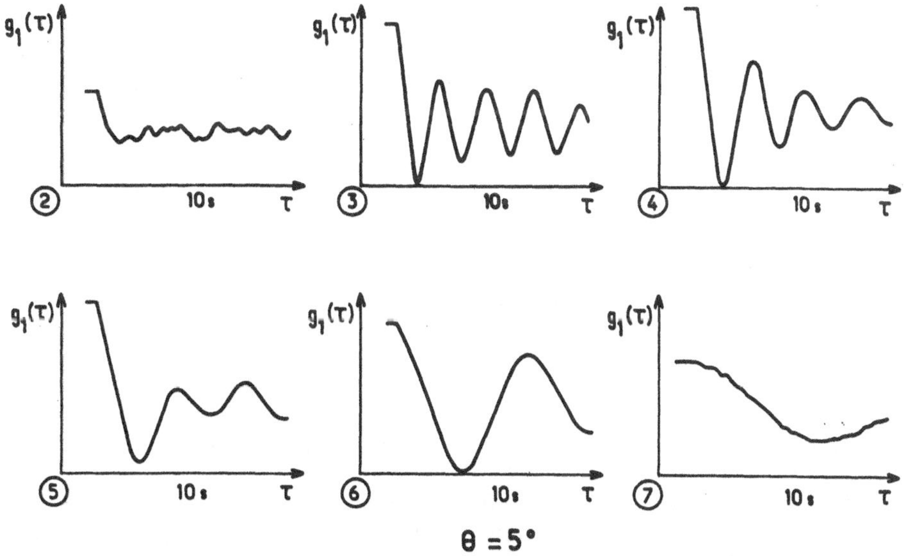

Figure 6. Example of strong oscillations affecting the temporal decay function at $\theta = 5°$ (the time interval between two consecutive plots is equal to $\sim$ 30 seconds).

The contribution which is not q-dependent describes the structural relaxation of linkages[15] and the contribution which is q^2-dependent describes the vibrational modes of the network.[10,13,16,17]

After very weak thermal or mechanical perturbation of the sample, we observe, after a delay time of an order of 30s, very strong oscillations (Figure 6) in the temporal decay function. It corresponds to propagating modes; this can only be explained by the non-newtonian properties of this medium (such behavior is known to occur in non-newtonian fluids[18]). The perturbation induces a relatively rapid oscillation (period $\sim$ 1 s) which, in turn, induces a slower one, and so on. This effect is very strong in the forward direction but its angular dependence is not clear (it is probably strongly coupled with structural relaxation).

REFERENCES

1. P. G. De Gennes, cours Collège de France, 1978-79.
2. F. C. Chretien, J. Gyn. Obst. Biol. Rep. 3: 711 (1974); 5: 313 (1976).
3. E. Odeblad, Act. Obst. Gyn. Scand. 47 (Suppl. 1): 59 (1968).
4. N. Eliezer, Biorheology 11: 61 (1974).
5. E. Odeblad, "Female Infertility," P. Keller Karger, ed. (Bâle, 1977).
6. W. Lee, R. J. Blandau, and P. Verdugo, "Workshop Conf. Rottach.," Egern V. Insler and G. Bettendorf, eds. (1977), p. 68.
7. E. Dubois-Violette and P. G. De Gennes, Physics 3: 181 (1967).
8. P. G. De Gennes, Macromolecules 9: 587 (1976).
9. M. Adam and M. Delsanti, Macromolecules 10: 1229 (1977); J. de Phys. Lettres 38: L271 (1977).
10. E. Geissler and A. M. Hecht, J. Chem. Phys. 65: 103 (1976).
11. M. Dubois, P. Jouannet, P. Bergé, and G. David, Nature 252 (1974); M. Dubois and P. Bergé, Rev. Phys. Appl. 8: 89 (1973).
12. B. Volochine, P. Bergé, and I. Laguës, Phys. Rev. Lett. 25: 1414 (1970); B. Volochine and J. Boscq Rolland, Rev. Phys. Appl. 14: 391 (1979).
13. T. Tanaka, L. O. Hocker, and G. B. Benedek, J. Chem. Phys. 59: 5151 (1973); S. J. Candau, C. Y. Young, and T. Tanaka, J. Chem. Phys. 70: 4694 (1979).
14. M. Adam, M. Delsanti, and G. Pouyet, J. de Phys. Lettres 40: L435 (1979).
15. N. Ostrowsky, in "Photon Correlation and Light Beating Spectroscopy," H. Z. Cummins and R. E. Pike, eds. (Plenum, New York, 1974).
16. A. M. Hecht and E. Geissler, J. de Phys. 39: 631 (1978).
17. S. J. Candau, C. Y. Young, and T. Tanaka, J. Chem. Phys. 70: 4694 (1979).
18. Colloque de la SFP, Discussions, Bordeaux, 1979.

CONFORMATIONAL CHANGE OF PROTEIN/SDS COMPLEX IN LOW IONIC STRENGTH SOLUTION - A STUDY BY DYNAMIC LIGHT SCATTERING

B. Herpigny, S. H. Chen, R. E. Tanner, and C. K. Rha

Massachusetts Institute of Technology

Cambridge, Massachusetts 02139

Proteins are macromolecules which contain a polypeptide backbone (strongly hydrophilic) and many side chains which are generally hydrophobic. In order to minimize its energy a protein molecule in an aqueous solution tends to form a globular structure which buries as many hydrophobic side chains in the interior as possible. But a complete removal of the hydrophobic side chains from their contact with water is impossible and these hydrophobic patches (if sufficiently large) can constitute some binding sites for other hydrocarbon or amphiphilic molecules. The detergents are known to be good protein denaturing agents. When these detergent molecules attach themselves to the binding sites and neutralize the hydrophobic effect, it becomes energetically favorable for some new interior hydrophobic side chains to expose themselves to the water molecules and thus form some new binding sites. This kind of cooperative effect has been discussed by Reynolds and Tanford.[1]

The interest of detergent such as Sodium Dodecyl sulfate (SDS)/ protein complex stems from the conjecture that SDS/protein complex may be an idealized model system for biological membranes in which there are amphiphilic molecules, primarily phospholipids, in interactions with many membrane proteins.

The purpose of the dynamic light scattering is to study the kinetics of conformational change of protein/SDS complex as a function of weight ratio of the bound SDS to protein in the solution. The photon correlation functions were measured at a series of temperatures ranging from 12 to 48°C at 3 different scattering vector values (q) in solutions with 2% BSA (Bovine Serum Albumin) or OA (Cvalbumin). The detergent used was SDS at a very low ionic strength, I = 0.025M,

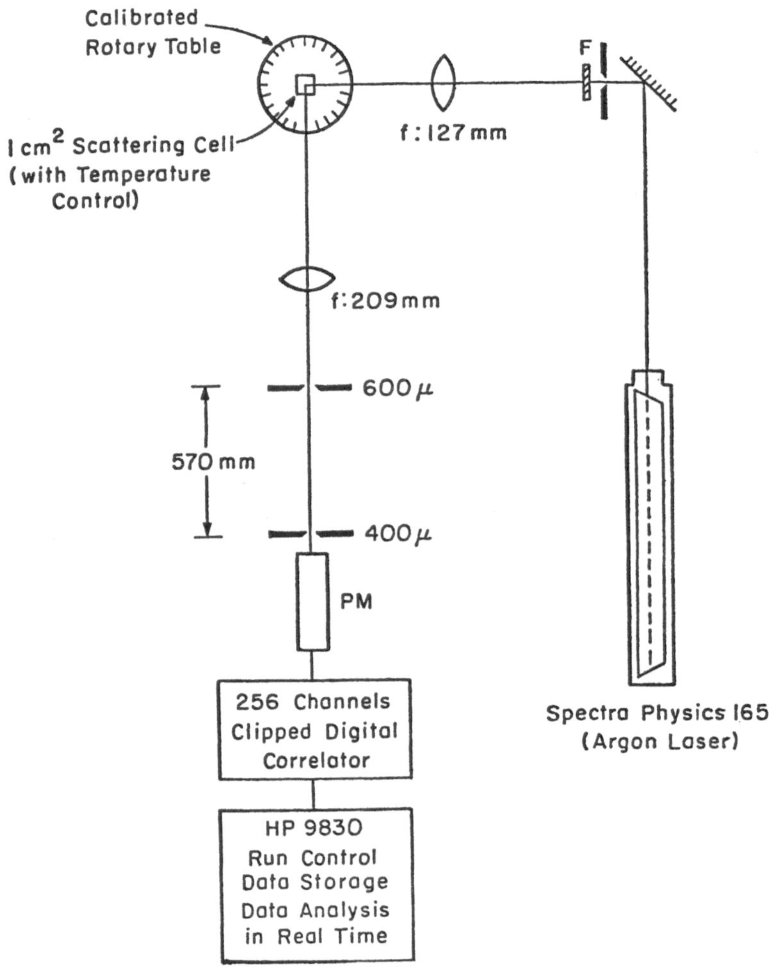

Figure 1. Schematic diagram of experimental set-up.

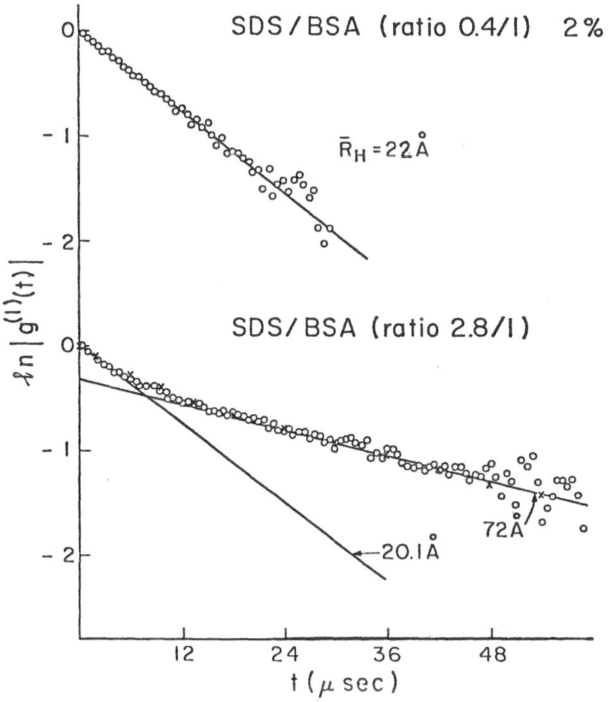

Figure 2. Correlation function for mixture of SDS/BSA in the buffer solution at two weight ratios SDS(gr)/BSA(gr) = .4/1 and SDS(gr)/BSA(gr) = 2.8/1.
The cumulant analysis shows that at weight ratio .4/1, we see essentially one exponential indicating existence of particles of an effective hydrodynamic radius of 22 Å, identified to be an average between the protein size and the micellar size. On the other hand at ratio 2.8/1, the cumulant analysis shows existence of mixture of two distinct sizes of 20 Å and 72 Å, respectively micelles and protein/SDS complexes.

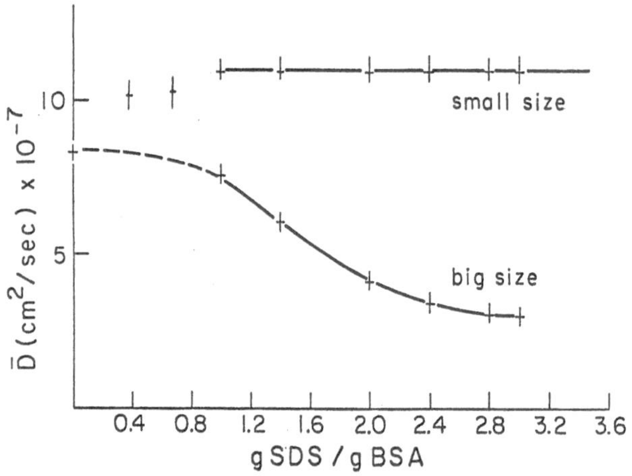

Figure 3. Effective diffusion constants for the mixture of SDS/BSA. The smaller size is identified to be the same as the measured micellar size obtained from pure SDS and remains constant as a function of SDS/BSA ratio. The bigger size, presumably representing the BSA/SDS complex, grows bigger and saturates above 2.8/1 ratio.

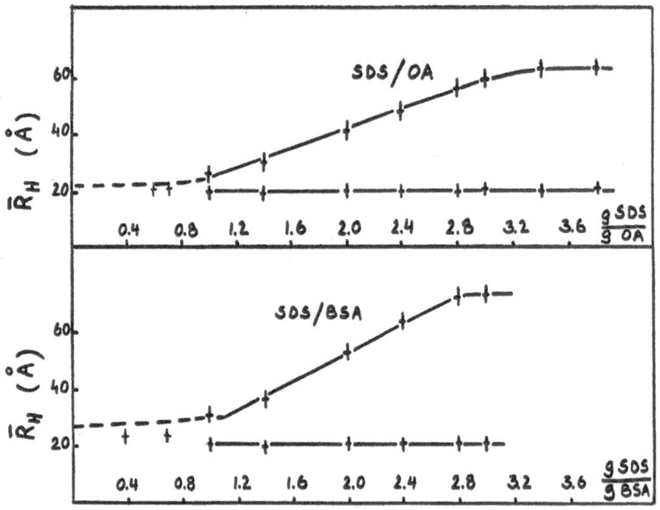

Figure 4. Same plot as Figure 3 but in terms of effective hydrodynamic radius. This graph shows that size of the complex increases slowly initially till a ratio of approximately 1/1 and then goes linearly till a saturation level at a ratio of about 2.8/1 for both proteins (BSA and OA).

with phosphate buffer (NaH_2PO_4-Na_2HPO_4) and concentrations above the critical micellar concentration. The spectrometer was a standard photon-correlation set-up with a 256 channel correlator and its associated computer.

RESULTS

- Dynamic light scattering detects the presence of micelles in equilibrium with SDS/protein complexes in the solutions.
- As the ratio of SDS/protein is increased, the size of micelles remains constant while the protein/SDS complex linearly increases in size and reaches a saturation.
- The hydrodynamic shape of the complex at the maximum binding has not been determined in these experiments. This should be possible with additional measurements of the rotational diffusion coefficient by depolarized dynamic light scattering. With the present experiments, it is not possible to distinguish between a rod shape (as proposed by Reynolds and Tanford[2]) and an ellipsoidal shape (as proposed by Tanner and Rha[3]) inferred from a recent viscosity measurement.
- A complete version of the present work will be published elsewhere.[4]

ACKNOWLEDGEMENTS

B. Herpigny gratefully acknowledges financial support from I.R.S.I.A. (Belgium) during her stay at M.I.T.

REFERENCES

1. C. Tanford, The Hydrophobic Effect, 2nd edition, John Wiley, New York, 1980.
2. J. A. Reynolds and C. Tanford, Proc. National Acad. Sci. 66: 1002 (1970).
3. R. Tanner and C. K. Rha, "Hydrophobic Effect on the Intrinsic Viscosity of Globular Proteins", paper presented at 9th international Congress on Rheology, Naples (1980).
4. R. E. Tanner, B. Herpigny, S. H. Chen, C. K. Rha, preprint entitled "Conformational Change of Protein/SDS Complex in Solution - A Study by Dynamic Light Scattering", to be published in J. Chem. Phys. (1981).

HEAT AGGREGATION OF HUMAN IgG

A DYNAMIC LIGHT SCATTERING STUDY

T. Jossang and J. Feder

Institute of Physics
Box 1048
University of Oslo, Norway

and

E. Rosenqvist

National Institute of Public Health
Postuttak, Oslo, Norway

CONTENTS

1. INTRODUCTION

The thermal stability of immunoglobins is a critical parameter both to experimental immunologists and clinicians responsible for administration of antibody preparations. In diagnostic immunology heating of serum at 56C for 30 minutes has become a well established method of inactivating complement in immunological assays. This procedure usually does not influence the antibody activity or the other main biological properties of the immunoglobins. In contrast, heating of IgG at 63°C for 30 minutes is a widely used method to produce soluble IgG aggregates. Such aggregates possess many biological properties which make them suitable for use as controls in studies concerning soluble antigen-antibody complexes.[1,2] The structure of the heat aggregates and the mechanism by which they are formed has been studied by several methods. For the analysis of the hydrodynamic properties, ultracentrifugation and gelfiltration have been the most used techniques,[3,4,5] and for the study of the biological properties of the soluble aggregates, complement fixation methods has been shown to be very sensitive.[6] By dynamic light scattering we measure the diffusion coefficients of the components in a solution of IgG. In this way we can follow the process of aggregation under different conditions.

2. EXPERIMENTAL

Monomeric IgG was prepared from pooled human immunoglobins (Gammaglobulin Kabi 16%, AB Kabi, Stockholm, Sweden), by gelfiltration on a Sephacryl S 300 Superfine column. Gammaglobulin Kabi contains at least 97% Immunoglobins, mainly IgG. The gelfiltration column was 2.5 cm in diameter and 92 cm long. Samples of 4 ml Gammaglobulin Kabi (80 mg/ml) were eluted by a Tris-HCl buffer pH 7.6 (0.05M Tris-HCl, 0.2M NaCl, 0.2nM EDTA, 0.02% NaN_3). From the monomeric fractions, samples were collected and passed through a Millipore filter (0.22 µm) into thoroughly cleaned cylindrical glass-cells. Two series of identical samples with different concentrations (4.1 and 15.4 mg/ml) were prepared. The more concentrated series was prepared by ultrafiltration in collodium bags. The protein concentration was calculated from OD-measurements ($E_{1\,cm}^{1\%}$ = 14.0 at 280 nm). A solution of monomeric ferritin molecules, for use as a reference standard, was prepared by gelfiltration of horse spleen ferritin (Koch-Light) on a Sepharose 6B column in the same Tris-HCl buffer as above.

Autocorrelation functions of the intensity fluctuations of laser light scattered at different angles were measured using a Malvern spectrometer (RR102) and a 128 channel digital single clipped correlator built after the design of S. H. Chen et. al.[7] We control the correlator using a M6800 microprocessor system for data collection, continuous display and accumulation of the correlation function. The microprocessor is in turn controlled by a laboratory mini computer (NORD 10S) which reads the observed spectra onto mass storage and analyzes the results. The computer continuously controls, via a CAMAC interface system (KINETIC systems GPIB 3388), the temperature (termistor read by HP 3455A digital voltmeter) and measures the intensity by a HP 5328 frequency counter. With this system we obtain the overall stability required for the very long aggregation experiments - up to 60 hours and 150 different spectra for one temperature. The laser used was a Spectra Physics 124B He-Ne and the Malvern spectrometer was equipped with a RR127 Photomultiplier tube.

The data presented are the mean diffusion constants $D = \Gamma/Q^2$, or equivalent mean hydrodynamic radius, $R = kT/6\pi\eta D$.[8] The inverse time constants Γ, were obtained by a single exponential fit to each correlation function. The radii as a function of time (Figure 2) were all obtained at the same scattering vector $Q = (4\pi n/\lambda)\sin(\theta/2)$ for $\theta = 50°$. Here n is the refractive index of the solution filling the scattering volume and λ is the wavelength in vacuum for the incident light. For the viscosity η of the solution we used the experimentally determined temperature dependent viscosity for pure water.[9] Since η is very strongly temperature dependent, varying by a factor of 2 from 20° to 56°C, we have performed calibration experiments for η in the actual temperature regime to insure correct interpretation of the data. The hydrodynamic radius for the spherical and temperature stable protein molecule ferritin was measured by the same procedure as outlined above. At room temperature we found $R = 6.9 \pm .1$ nm for 26 angles between 30 deg. and 130 deg. In the temperature range up to 56°C, the thermal expansion coefficient was measured to be $k = 0.01$ nm/°C. This very low thermal expansion coefficient is consistent with the known stability of ferritin.

3. RESULTS AND DISCUSSION

In Figure 1, is shown a gelfiltration curve for Gammaglobulin

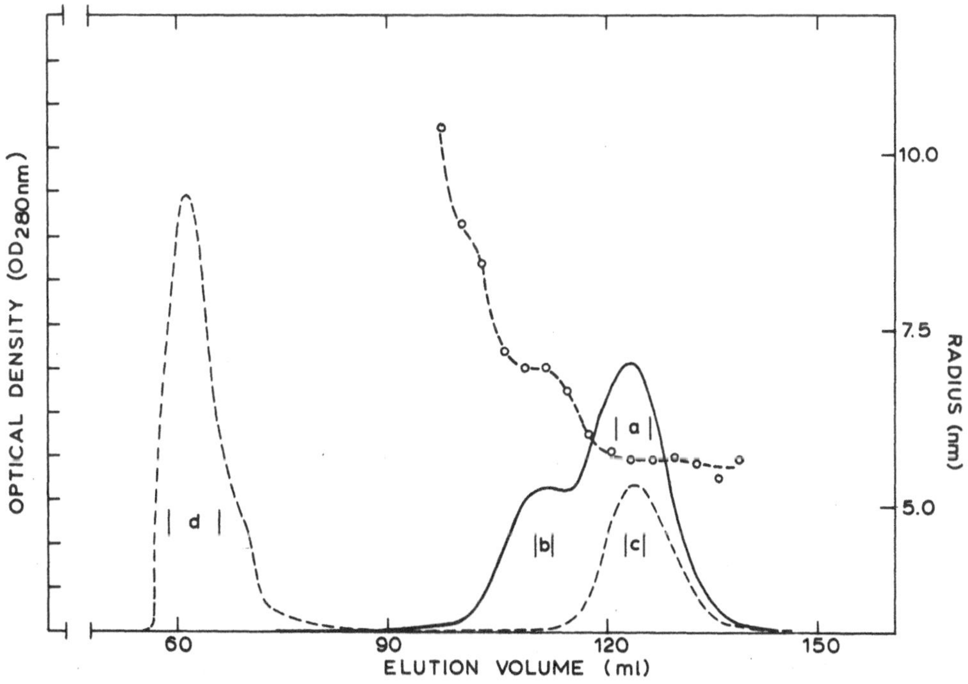

Figure 1. Gelfiltration on Biogel A5M of normal human Gammaglobulin Kabi. (———), and monomer fraction after heating at 62°C for 6 hours (----). The measured radii of the fractions of normal Gammaglobulin are given by (o). Column size 1.6 x 83 cm.

Kabi together with the radii observed for the fractions. The radius of the monomeric fraction (a) was 5.5 nm for 4.1 mg/ml, and 5.8 nm for 15.4 mg/ml. This discrepancy is probably caused by the concentration procedure. Dimers from fraction (b), were pooled and purified by another gelfiltration. The observed radius for dimers was R = 7.5 nm.

In Figure 2, we see that as a function of time the observed radius R increases as expected for an aggregating system. The rate of aggregation shows a dramatic temperature dependence. At 39° C there is no observable aggregation even after 60 hours. This is in disagreement with the observations by James et. al.,[10] who reported formation of up to 20% IgG dimers after 20 hours at 37C. Such dimerization, if present, would be easily detected by our method. Significant aggregation starts at 47°C, and increases quickly with temperature. At 62°C, aggregation proceeds so fast that stable experimental conditions are difficult to obtain. The small steps in the R(t) curve for 62°C, are caused by changes in the clipping level and in sample time during the experiment.

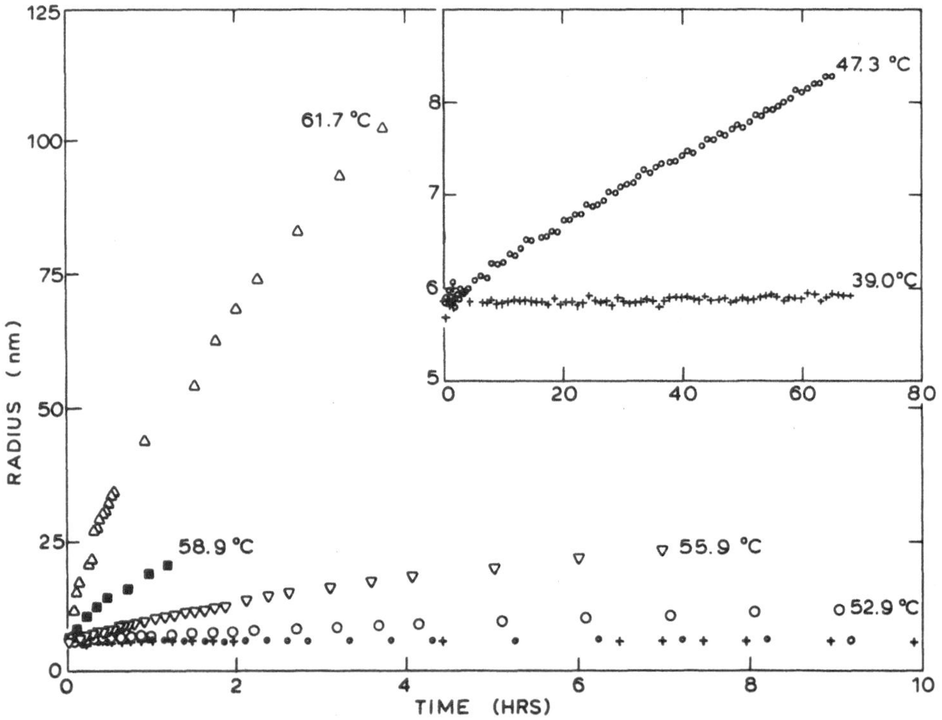

Figure 2. Observed radius of heat aggregating IgG (15.4 mg/ml), as function of time at various temperatures.

Each of the R(t) curves can be fitted by the function:

$$R = R(0)(1 + \gamma t + \beta t^2)$$

yielding the initial growth rates γ as function of temperature. In Figure 3 it is seen that γ has a temperature dependence of the form

$$\gamma = \exp[\alpha(T - T_o)]$$

with T_o = 54.8°C for 4.1 mg/ml and 53.3°C for 15.4 mg/ml. At the characteristic temperature T_o the growth rate is 1 nm/hour. The coefficient α is equal to 0.51 in units of ln(nm/deg/hour) for both concentrations. Having established that the initial growth rate has an exponential temperature dependence, we may plot the results as shown in Figure 4. By plotting the radius as a function of (γt) we have a very satisfactory data collapse, indicating that not only the initial but also the later stages of aggregation scale with the initial growth rate.

After the heat aggregation experiments we fractionated the samples by gelfiltration with a typical result shown in Figure 1. The aggregated samples had a monomer fraction (c), with R = 5.8 nm, and a fraction of aggregates (d), with R in the range 160 - 210 nm. We conclude that in the later stages of heat aggregation

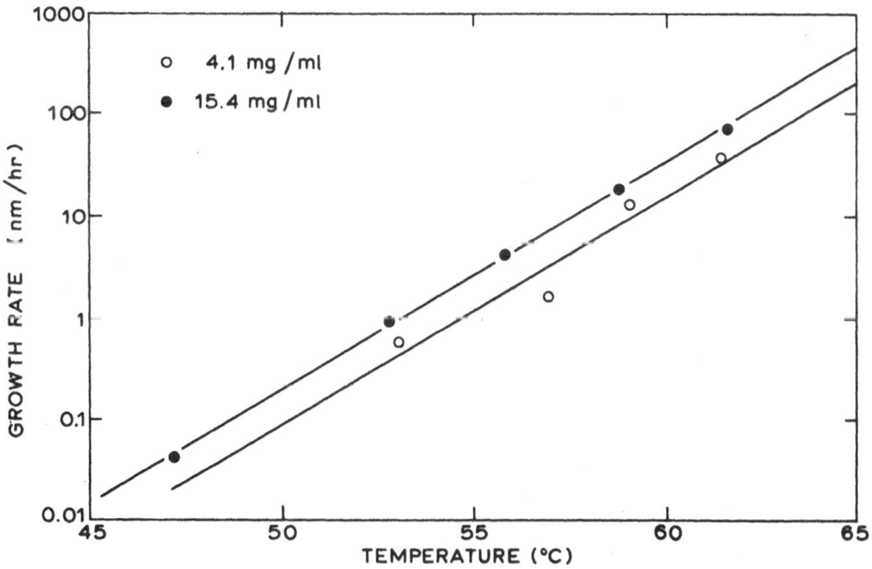

Figure 3. Initial growth rate of IgG aggregates as function of temperature.

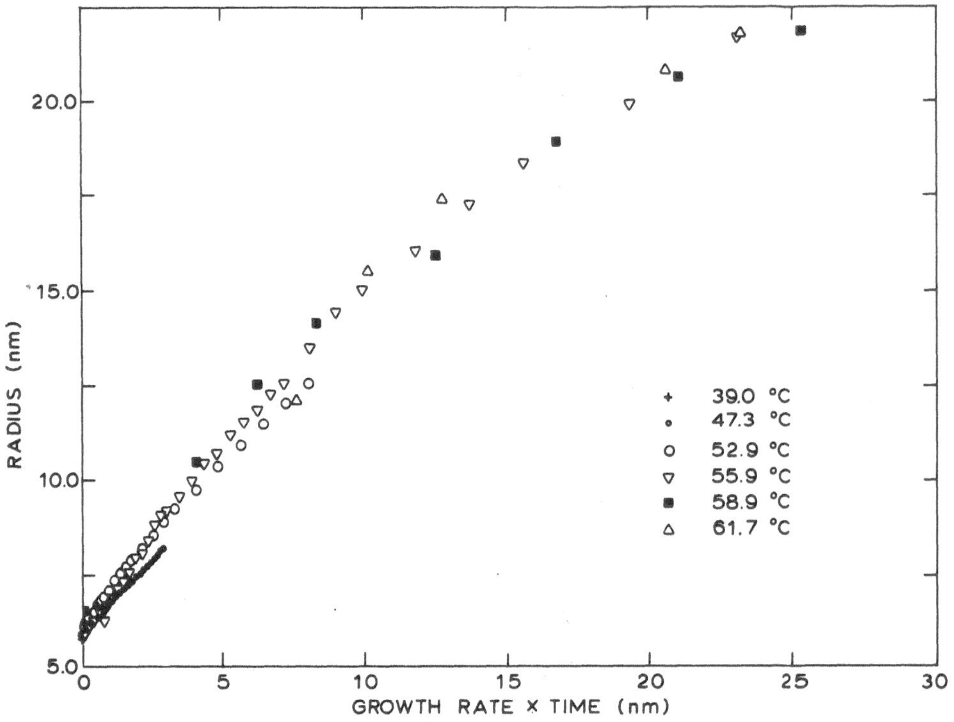

Figure 4. Observed radius of heat aggregating IgG (15.4 mg/ml), as function of x = initial growth rate x time.

proceeds by the growth of the larger complexes at the expense of the remaining monomers, and not by the formation of new aggregates. This is consistent with the observations by Knutson et. al.,[11] who found that after 10 min. at 62°C significant amounts of small aggregates were formed, and that these increase in size with time.

Further work is in progress to obtain the growth rate as a function of concentration in order to reveal the kinetics of the thermal aggregation of IgG.

REFERENCES

1. K. Ishizaka, Progr. Allergy 7: 32 (1963).
2. H. Nielsen and S. E. Svehag, Acta Path. Microbiol. Scand. C84: 261 (1976).
3. W. Augener and H. M. Grey, J. Immunol. 105: 1024 (1970).
4. R. H. Kauffmann, L. A. van Es, and R. Daha, J. Immunol. Meth. 31: 11 (1979).
5. P. Johns and D. R. Stanworth, J. Immunol. Meth. 10: 231 (1976).

6. R. D. Soltis, D. Hasz, M. J. Morris, and I. D. Wilson, Immunol. 36: 37 (1976).
7. S. H. Chen, W. B. Veldkamp, and C. C. Lai, Rev. Sci. Instrum. 46: 1356 (1975).
8. H. Z. Cummins, in Photon Correlation and Light Beating Spectroscopy, Plenum Press, New York, 1974, p. 285.
9. Handbook of Chemistry and Physics, 56th ed., CRC Press, Inc., 1976. p. F-49.
10. K. James, C. S. Henney, and D. R. Stanworth, Nature 202: 563 (1964).
11. D. W. Knutson, A. Kijlstra, J. Lentz, and L. A. van Es, Immunol. Com. 8: 337 (1979).

QUASIELASTIC LIGHT SCATTERING STUDIES

OF MICROTUBULE ASSEMBLY IN VITRO

G. R. Palmer[†] and D. B. Sattelle

A. R. C. Unit
Department of Zoology
Downing Street, Cambridge CB2 3EJ, U.K.

CONTENTS

1. INTRODUCTION

Quasi-elastic light scattering (QELS) is a rapid, non-invasive technique for the measurement of macromolecular hydrodynamic radii and size distribution (polydispersity).[1,2] This method has recently been applied to the study of the self-assembly of supramolecular structures.[3,4,5] Its usefulness in studying such heterogeneous systems is based on:

(a) the capacity to probe the system at equilibrium in a non-perturbative manner;
(b) a very high sensitivity to the appearance of polymeric forms in the presence of high concentrations of monomer;
(c) the ability to obtain quantitative data without prior physical separation of the components.

In this report the properties and assembly of beef brain tubulin into microtubules in vitro are characterized by QELS. Microtubules are hollow, cylindrical organelles (∿ 25nm in diameter) made up of linear protofilaments of αβ heterodimers of tubulin (MW of heterodimer = 110,000).[6] Assembly-competent tubuline solutions are prepared from brain homogenates by successive cycles of assembly and disassembly by subjecting the material to temperature changes

[†]NATO (Canada) Postdoctoral Fellow

in the presence of guanine nucleotides and magnesium ions. This
material contains a number of microtubule associated proteins (MAPS)
including HMW 1 (MW 300,000)[7] and HMW 2 (MW 280,000)[7] and the tau
proteins (MW 58-66,000).[8] Microtubules are labile polymers implicated
in a variety of cellular functions including cell division, transport,
maintenance of cell shape and movement. The kinetics of in vitro
microtubule assembly which exhibits many of the properties of in
vivo assembly has been investigated largely by means of turbidity[9,10]
and viscometry.[9] QELS provides additional information on the sizes
of the components of tubulin solutions prior to and during microtubule assembly.

2. MATERIALS AND METHODS

Assembly-competent tubulin was extracted from beef brain and
purified through three cycles of assembly and disassembly, a procedure based on the purification scheme of Shelanski et al.[11] Polymerization steps at 37°C were carried out in 3M glycerol, whereas
depolymerization steps at 4°C were performed in 1M glycerol, thereby
promoting assembly and disassembly. All buffers were filtered
through Amicon UM10 filters to remove contaminants > 10,000 MW.
Light scattering measurements were made in 1M glycerol. All protein
concentrations were measured by the Coomassie brilliant blue G
assay.[12]

A spectrometer set to a scattering angle (Θ) of 90° and in which
rapid temperature changes of the sample cells could be achieved,
was used throughout these experiments. The 488nm line from an Argon-
Ion (Spectra-Physics 164) laser illuminated the tubulin samples.
Photocount autocorrelation analysis of the scattered laser light
was performed using a Langley-Ford correlator. Since the assembly
competent tubulin solutions were polydisperse, all correlation functions were analysed by the method of cumulants.[13] Fitting data to
a third order polynomial provided a z-averaged diffusion coefficient
($\bar{D}$) and in addition a polydispersity index ($\mu_2/\bar{\Gamma}^2$). All $\bar{D}$ measurements were standardized to water at 20°C.

The diffusion coefficients obtained in this way provide a direct
measure of average hydrodynamic particle size, whereas the polydispersity index provides a measure of the width of the distribution
of particles sizes. The total scattered intensity (I) at Θ = 90
was also recorded to provide a measurement comparable to turbidity,
which for long microtubules appears insensitive to length and is
proportional only to the mass of the tubule.[10]

3. RESULTS

Depolymerized, assembly-competent, beef brain tubulin was allowed
to equilibrate for 10-15 min. to successively higher temperatures.
Changes in $\bar{D}$, $\mu_s/\bar{\Gamma}^2$ and I were recorded. As shown in Figure 1, at
a temperature in the region of 20-25°C polymerization of the tubulin

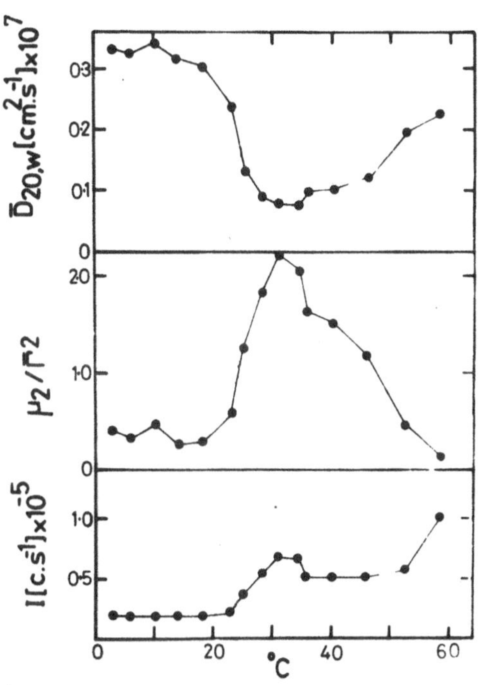

Figure 1. Laser diffusion measurements on beef brain tubulin (1.1 mg/ml) at a range of temperatures. Each point is the average of three measurements made at a fixed sample time (τ = 20μs). Values for $\bar{D}_{20,w}$ and $\mu_2/\bar{\Gamma}^2$ are not extrapolated to $\bar{\Gamma}\tau_{max}$ = 0. The sample was equilibrated (10-15 min) at each new temperature prior to collection of light scattering data.

was detected by a sharp change in all three parameters. The decrease in $\bar{D}$ was accompanied by an increase in both $\mu_2/\bar{\Gamma}^2$ and I. From 32-45°C, $\bar{D}$, $\mu_2/\bar{\Gamma}^2$ and I varied only slightly, whereas over the range 45-60°C substantial, irreversible changes were noted in all three parameters which probably indicates irreversible thermal denaturation.

Rapid temperature jump experiments were also performed on depolymerized, assembly-competent, beef brain tubulin. Following the elevation of the temperature from 4°C to 37°C, a rapid decrease in $\bar{D}$ was accompanied by an increase in $\mu_2/\bar{\Gamma}^2$ and I (Figure 2).

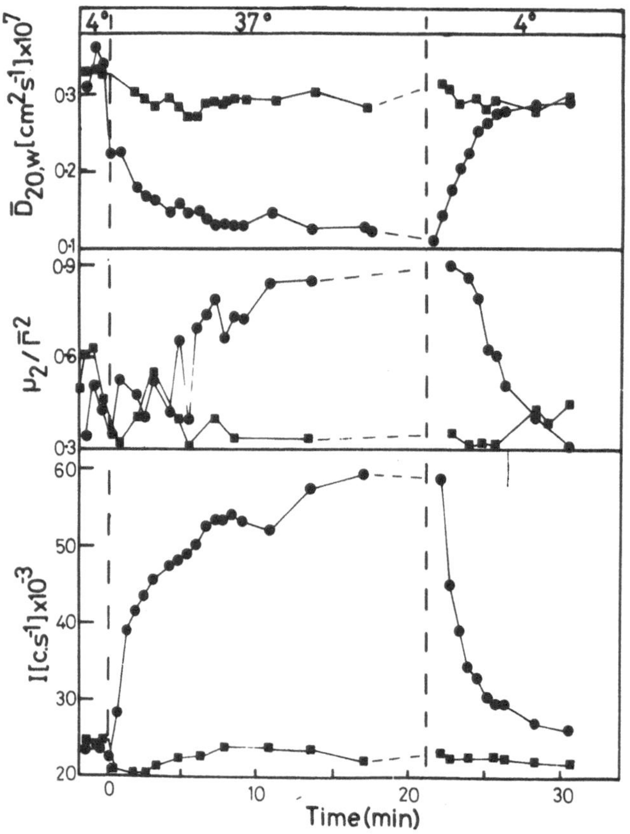

Figure 2. Laser diffusion measurements of beef brain tubulin (1.1 mg/ml) obtained during assembly and disassembly induced by temperature jumps. Broken lines indicate the time at which the temperature jump was applied. (●) Beef brain tubulin; (■) beef brain tubulin in the presence of 100 μM colchicine.

After 21.5 min at the elevated temperature, the sample was rapidly returned to 4°C and $\bar{D}$, $\mu_2/\bar{\Gamma}^2$ and I all returned to values close to the initial measurements taken prior to assembly. Using the same experimental protocol, the inhibitory effects of the alkaloid colchicine (100 μM) on polymerization were detected (Figure 2), indicating the potential of QELS for quantitative assessment of the interactions of pharmacological agents with microtubule assembly.

Table I provides a comparison of the present diffusion coefficient measurements with previous QELS studies on vertebrate tubulins from a variety of sources. The values determined from this present study are extrapolations of $\bar{D}$ to $\bar{\Gamma}\tau_{max} = 0$. This was done in order to take into account the dependence of $\bar{\Gamma}$ on sample time (τ) resulting from sample polydispersity.

Table 1. Comparison of laser-diffusion measurements on brain tubulin from a variety of vertebrate tissues

Source	Preparation	Diffusion coefficient ($\times 10^7$ cm^2 S^{-1})	Reference
Beef	4°C (3 cycles)	0.55	present study
Pig	4°C < 60S	0.61	14
	4°C < 10S	3.33	14,15
Dogfish	4°C (2 cycles)	3.00	16

3. DISCUSSION

Quasi-elastic light scattering (QELS) has proved a useful tool for investigating the steady-state solution properties of disassembled tubules as well as a means of following the rapid assembly of microtubules. At 20-25°C polymerization of beef tubulin is detected. The temperature at which tubulin polymerization is initiated *in vitro* may be characteristic of the tubulin source. It is of interest for example that material from a cold-blooded vertebrate polymerizes optimally at 18-21°C.

The diffusion coefficient (extrapolated to $\bar{\Gamma}\tau_{max} = 0$) for beef brain, assembly-competent tubulin purified through three cycles of assembly and disassembly was $0.55 \pm 0.04 \times 10^{-7}$ cm^2 S^{-1} a value considerably lower than the value of 4.7×10^{-7} cm^2 S^{-1} predicted by the Svedberg equation for the tubulin dimer.[15] This discrepancy may be attributed to the presence of the 36S single-layered, double rings detected in tubulin samples by electron microscopy and sedimentation analysis. Our data for beef tubulin is in reasonable agreement with the value of 0.61×10^{-7} cm^2 S^{-1} for a pig brain tubulin solution following centrifugation to remove particles larger than 60S.[14,15] Diffusion coefficients much closer to the value expected for a tubulin dimer have been reported. For example, centrifugation of pig brain tubulin to remove material larger than 10S[14,15] results in a measured diffusion coefficient of 3.33×10^{-7} cm^2 S^{-1}. Also, dogfish (*Squalus acanthus*) brain tubulin purified through two cycles of assembly yields a diffusion coefficient of 3.0×10^{-7} cm^2 S^{-1}. Dogfish tubulin prepared in this way is known to be largely devoid of MAPS and rings.[17] Pig brain tubulin centrifuged to remove particles greater than 10S should also be devoid of rings.

Thus laser diffusion measurements on a variety of tubulin preparations provide quantitative data that is consistent with earlier electron microscopy and sedimentation analysis on the same material. QELS is also able to monitor the rapid polymerization of microtubules. Further kinetic studies employing QELS should provide detailed information on the composition and mechanism of assembly of microtubules.

REFERENCES

1. B. Chu, Laser Light Scattering, Academic Press, New York (1974).
2. B. J. Berne and R. Pecora, Dynamic Light Scattering, John Wiley, New York (1976).
3. L. Hocker, J. Krupp, and G. B. Benedek, Biopolymers 12: 1677-1678 (1973).
4. G. P. Agarwal, J. G. Gallagher, K. C. Aune, and C. D. Armeniades, Biochem. 16: 1865-1870 (1977).
5. G. R. Palmer and O. G. Fritz, Biopolymers 18: 1659-1672 (1979).
6. P. Dustin, Microtubules, Springer-Verlag, Berlin (1978).
7. W. Herzog and K. Weber, Eur. J. Biochem. 92: 1-8 (1978).
8. R. P. Frigon and S. N. Timasheff, Biochemistry 14: 4559-4566 (1975).
9. G. G. Borisy, F. M. Marcum, J. B. Olmsted, D. B. Murphy, and K. A. Johnson, Ann. N.Y. Acad. Sci. 253: 107-132 (1975).
10. F. Gaskin, C. R. Cantor, and M. L. Shelanski, J. Mol. Biol. 89: 737-758 (1974).
11. M. L. Shelanski, F. Gaskin, and C. R. Cantor, Proc. Natn. Acad. Sci. U.S.A. 70: 765-768 (1973).
12. H. M. Bradford, Anal. Biochem. 72: 248-254 (1976).
13. D. E. Koppel, J. Chem. Phys. 57: 4814-4820 (1972).
14. F. Gaskin and J. S. Gethner, in Cell Motility, ed. by R. Goldman, T. Pollard and J. Rosenbaum, Cold Spring Harbour Laboratory (1976).
15. J. S. Gethner, G. W. Flynn, B. J. Berne, and F. Gaskin, Biochemistry 16: 5776-5781 (1977).
16. D. B. Sattelle, G. M. Langford, and K. H. Langley, in Photon Correlation Spectroscopy and Velocimetry, ed. by H. Z. Cummins and E. R. Pike (1976).
17. G. M. Langford, Exp. Cell Res. 111: 139-151 (1978).

PHOTON-CORRELATION SPECTROSCOPY OF CENTRIFUGED SOLUTIONS. APPLICATION TO *ARTEMIA* RIBOSOME SUBUNITS.

Paul Nieuwenhuysen and Julius Clauwaert

Department of Cell Biology
University of Antwerp
2610 Wilrijk, Belgium

CONTENTS

1. INTRODUCTION

Quasi-elastic light scattering can be used to probe the dynamics of macromolecules in solution (see the review by Bloomfield and Lim, 1978, and papers in this volume). In our laboratory, photon-count autocorrelation spectroscopy of scattered laser light is mostly used to measure the translational diffusion coefficient of biological macromolecules as a part of their physical-chemical characterization. In particular, we have studied the ribosomes isolated from an eukaryote, the brine shrimp *Artemia* (Nieuwenhuysen and Clauwaert, 1978). An extension of this work to the ribosomal subunits is hampered by their high tendency to aggregate. The aggregation was first avoided as far as possible, then photon-correlation spectroscopy was performed on the monomeric subunits directly after they were separated from their eventual aggregates by sedimentation in a transparent centrifugation tube.

2. EXPERIMENTAL PROCEDURES

A. Preparation of Ribosomal Particles

Ribosomes were isolated from cryptobiotic, undeveloped embryos (cysts) of *Artemia* from San Francisco Bay (Nieuwenhuysen and Clauwaert, 1980, and ref. therein). The large and small ribosome subunits were dissociated and separated by sucrose gradient centrifugation in 20 mM Hepes/KOH buffer (pH 7.5), containing 11 mM Mg acetate, 400 mM KCl, and 1 mM dithiothreitol, using a zonal rotor. The obtained subunits tend to aggregate. This hampers the application of light, X-ray and neutron scattering. Therefore aggregation was avoided as much as possible. High KCl concentration inhibits aggregation, but converts the subunits to slower sedimenting particles (Nieuwenhuysen and Clauwaert, 1980). As a compromise, our subunit-buffer contained 20 mM Hepes or Triethanolamine buffer (pH 7.5), 10 mM Mg acetate, 250 mM KCl, and 1 mM dithiothreitol.

B. Photon-Correlation Spectroscopy (of Centrifuged Solutions)

The diffusion coefficients of the various ribosomal particles have been determined by single-clipped photon-count autocorrelation spectroscopy. This technique is based on the efficient and accurate analysis of the intensity fluctuations of laser light scattered by a solution of macromolecules in Brownian motion. In the case of a dilute and monodisperse solution of macromolecules, which are small compared to the wavelength of light, the normalized correlation function g(iT) decays exponentially with a decay time proportional to the searched diffusion coefficient D:

$$g(iT) - 1 = A \exp(-2 DK^2 iT) \tag{1}$$

where i is the channel number, T the chosen sample time, A the amplitude, and K the modulus of the scattering vector:

$$K = (4\pi n/\lambda_0) \sin(\theta/2) \tag{2}$$

with n being the refractive index of the solution, λ_0 the wavelength of the light in vacuo, and θ the scattering angle.

Our spectrometer was placed in a thermostated room. A beam of light (λ_0 = 4888 Å) from an intensity-stabilized Coherent Radiation argon-ion laser was focused in the investigated solution. The scattered light passed an interference filter and was detected with an ITT FW130 photomultiplier positioned on a Malvern goniometer. The light scattering intensity was monitored with a Malvern photon count rate meter and recorded on paper, to detect eventual instabilities or scattering by dust particles in the solution, and to check if the intensity of light scattering by benzene, water, and the solution varied with θ as expected from theory. The single-clipped autocorrelation function of the photon-counts was then built up in a Malvern K 7023 24-channel correlator, and displayed on an oscilloscope. The data were recorded on paper tape for subsequent computer analysis by different curve fitting procedures; the decay time and small deviations of monoexponential decay of each spectrum of several series of measurements at different angles were all taken into account to yield finally a value for the searched diffusion coefficient (Nieuwenhuysen, 1978). The reliability of our apparatus and experimental procedures has been proved by measurements on 'monodisperse' polystyrene latex spheres, on the small spherical bacteriophage MS2, on the complete ribosomes from Artemia, etc. (Nieuwenhuysen and Clauwaert, 1980, and ref. therein). This work, however, could not be simply extended to the study of ribosomal subunits because these aggregate. Quasi-elastic light scattering of a mixture of the monomers and their aggregates would only yield their z-average diffusion coefficient, from which no accurate value for the diffusion coefficient of the monomers can be derived. Therefore photon-correlation spectroscopy was performed of laser light scattered by the monomeric subunits directly after they were centrifuged in various ways to separate them from their aggregates and from dust particles. Transparent, cellulose-nitrate centrifugation tubes were used in a Beckman SW40 swing-out rotor. After centrifugation they were put in a specially constructed holder and used directly as scattering cells: the monomers were located by their laser light scattering intensity and photon-correlation spectroscopy yielded their diffusion coefficient (Nieuwenhuysen and Clauwaert, 1980). In the experiments involving rate-zonal sedimentation of ribosomal particles in a density gradient, the refractive index n and viscosity η in the gradient must be known accurately. Indeed, calculating the diffusion coefficient D of the macromolecules in the scattering volume from the decay times of their photon-correlation spectra involves the factor n^2 (Equations 1, 2), and the reduction of D to standard conditions ($D_{20,w}$) involves the factor η. Figure 1 shows that these factors and thus their uncertainty is smaller when CsCl or ^2H_2O is used to create a density gradient, instead of the more conventional glycerol or sucrose. The case of CsCl is detailed elsewhere (Nieuwenhuysen and Clauwaert, 1980); another advantage is that the viscosity of an aqueous CsCl solution below room temperature does not increase but decrease with concentration, which favors a good separation by sedimentation in a CsCl density gradient.

Subunits in diluted solution (1 A_{260}/ml) were fixed for 20 hr with 1% neutralized formaldehyde, dialysed against subunit-buffer, concentrated by ultrafiltration (using Amicon stirred cells with XM300 membranes), loaded on 3-11% CsCl gradients in subunit-buffer without KCl, and centrifuged at 40,000 rpm near 0°C to sediment the zone of monomers close to the bottom while aggregates were pelleted. When ^2H_2O was used to create a density gradient, the subunits had not to be fixed: top fractions of subunits were dialysed against

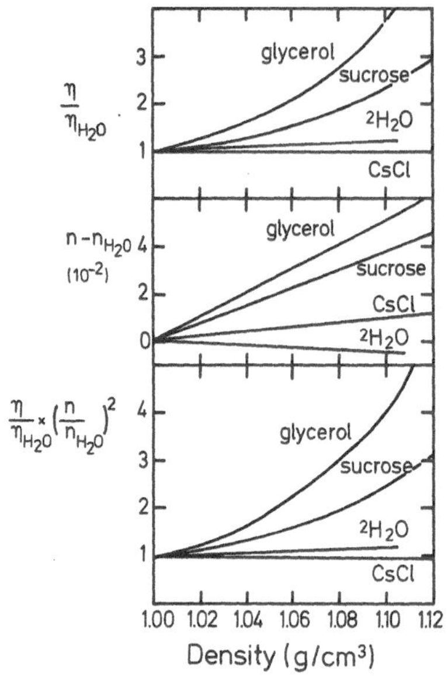

Figure 1. Properties of aqueous solutions of density gradient media, which are relevant to quasi-elastic light scattering. The viscosity η and refractive index n as a function of the density, and their combination in the factor required to obtain the standard diffusion coefficient of macromolecules sedimented in a density gradient from their quasi-elastic light scattering spectra. The data for glycerol, sucrose and CsCl were taken from CRC Handbook of Chemistry and Physics; the viscosity of H_2O - ^2H_2O mixtures was taken from Baker (1936), and their refractive index was measured.

subunit-buffer, concentrated in dialysis tubing with dry Sephadex, loaded on a 20-100% ^2H_2O gradient in subunit-buffer and centrifuged as above. After light scattering, the density gradients were fractionated to verify the location of subunits by their uv-absorbance, as well as the concentration distribution of the gradient-medium by refractive index measurements.

3. RESULTS

The diffusion coefficient of the monomeric ribosome subunits was determined by photon-correlation spectroscopy directly after they were separated from their aggregates by centrifugation. In the most simple experiments, tubes were filled with a ribosomal solution and centrifuged at 40,000 rpm, long enough to pellet aggregates, but short enough to leave monomers in solution near the bottom (dashed line in Figure 2). After fractionation, uv-absorbance spectroscopy

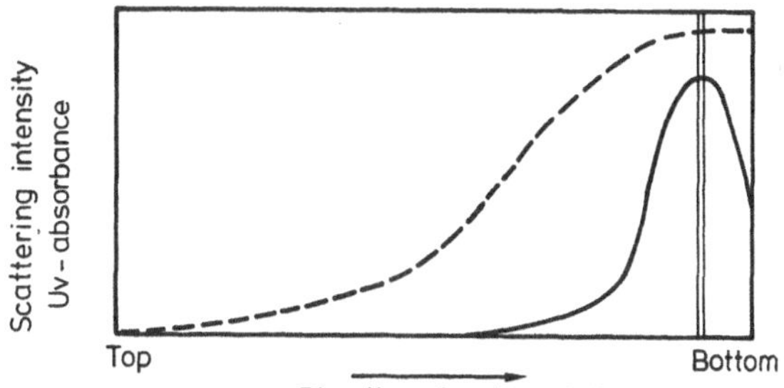

Figure 2. Centrifugation of ribosomal subunits before quasi-elastic light scattering.
Typical concentration distributions are shown of ribosomal particles in the centrifugation tube after sedimentation, as determined from their light scattering intensity and uv-absorbance: (---) after centrifugation of a homogeneous solution to pellet aggregates and to keep a layer of monomeric ribosome subunits near the bottom, (——) after band sedimentation of a ribosomal solution in a density gradient of CsCl or ^2H_2O. The vertical lines indicate the region in the gradients suited for quasi-elastic light scattering.

revealed a continuous decrease of the ratio A_{260}/A_{280} towards the top of the gradient; this indicates that the subunits lost some proteins. This phenomenon also occurred when subunits were pelleted to concentrate them for measurements of their density increment and partial specific volume. It has also been reported that slime mold ribosomes gradually lost their proteins when they were centrifuged repeatedly (Ramagopal an Ennis, 1980). More satisfactory results were obtained by rate-zonal sedimentation of ribosome subunits in density-gradients to pellet eventual aggregates, bringing the intact, monomeric subunits in a zone near the bottom of the centrifugation tube (full line in Figure 2). This yielded with fixed large subunits in CsCl, $D^\circ_{20,w} = (1.58 \pm 0.03) \times 10^{-7}$ cm^2/s (Nieuwenhuysen and Clauwaert, 1980), and with unfixed large subunits in ^2H_2O, $D^\circ_{20,w} = (1.61 \pm 0.03) \times 10^{-7}$ cm^2/s. With fixed small subunits in CsCl, $D^\circ_{20,w} = (1.80 \pm 0.04) \times 10^{-7}$ cm^2/s was found, but a much lower value in ^2H_2O. It seems relevant here that neutron scattering experiments with contrast variation have indicated that ^2H_2O does not affect the shape of the large ribosome subunits from the bacterium Escherichia coli, but that ^2H_2O induces aggregation of the small subunits (Serduyk et al., 1979).

Besides their diffusion coefficient other physicochemical characteristics of the ribosomal particles have also been measured and derived from the basic measured values. Due to space limitations no details can be given here; some methodology has been described earlier (Nieuwenhuysen and Clauwaert, 1978, 1980). Our data obtained before mid 1980 have been listed in Table 1.

Table 1. Summary of the physicochemical properties of ribosomal particles from *Artemia*.

	Complete ribosomes	Large Subunits	Small Subunits
Buoyant density in CsCl (g/cm^3)	1.551±0.004	1.567±0.004	1.519±0.004
RNA content (% by weight)	49±2	52±2	43±2
$A_{260,\ 1\ cm}^{1\ mg/ml}$	12.1±0.4	12.8±0.4	10.5±0.4
A_{260}/A_{280}	1.92±0.02	1.94±0.02	1.87±0.02
Partial specific volume, $\bar{v}°$ (cm^3/g)	0.63±0.02	0.62±0.02	0.65±0.02
Sedimentation coefficient, $s^{o}_{20,w}$ (10^{-13} s)	81±1	59±1	38±1
Diffusion coefficient, $D^{o}_{20,w}$ (10^{-7} cm^2/s)	1.41±0.02	1.61±0.03	1.80±0.04
Molecular weight, M (10^6)	3.85±0.2	2.45±0.2	1.45±0.1
Total molecular weight of RNA(10^6)	1.9±0.1	1.28±0.08	0.63±0.04
proteins(10^6)	2.0±0.1	1.2±0.1	0.83±0.08
Hydrodynamic radius, R_h (Å)	152±2	133±3	119±3
Maximal hydrodynamic volume, $V_{h,max}$ (10^6 $Å^3$)	14.7±0.6	9.9±0.6	7.1±0.5
Dry-particle volume, V_d (10^6 Å)	4.1±0.3	2.5±0.2	1.6±0.2
Dry-particle radius, R_d (Å)	99±3	84±3	72±3
Frictional ratio, R_h/R_d	1.54±0.04	1.58±0.05	1.66±0.06
Maximal hydrodynamic solvation,			
δ_{max} (cm^3 solvent/g ribosome)	1.7±0.1	1.9±0.2	2.3±0.2
F_{max} (cm^3 solvent/cm^3 ribosome)	0.72±0.03	0.75±0.03	0.78±0.03

4. DISCUSSION

The results seem consistent: the molecular weight of the complete ribosome is equal to within experimental error to the sum of the molecular weights of its two subunits, and the same is true for the 'dry particle' volumes. The values for the buoyant densities in CsCl, for the RNA contents, and for the sedimentation coefficients fall in the ranges of values published for ribosomal particles from other eukaryotes (see the compilations in Nieuwenhuysen and Clauwaert, 1980). A comparison with the characteristics of the corresponding ribosomal particles from the bacterium E. coli (Koppel, 1974, and the ref. cited therein) shows mainly how much smaller the latter are, while their frictional ratios are comparable, which reflects their similarity in morphology and solvation. For both kinds of ribosomal particles the measured frictional ratio is higher for the small than for the large subunit and lowest for the complete ribosome, i.e. both subunits associated; this agrees with their morphology as revealed by electron microscopy: the elongated small subunit and the spheroidal large subunit with protrusions associate to form the more globular ribosome.

ACKNOWLEDGEMENTS

Mr. H. Backhovens and Mrs. D. Bauwens provided technical assistance. P. N. is Senior Research Assistant of the National Fund for Scientific Research (Belgium) (N.F.W.)./F.N.R.S.). This research was partially supported by the N.F.W.O.

REFERENCES

W. N. Baker, J. Chem. Phys. 4: 294 (1936).
V. E. Bloomfield and K. Lim, Methods Enzymol. 48: 415 (1978).
D. E. Koppel, Biochemistry 13: 2712 (1974).
P. Nieuwenhuysen, Macromolecules 11: 832 (1978).
P. Nieuwenhuysen and J. Clauwaert, Biochemistry 17: 4260 (1978).
P. Nieuwenhuysen and J. Clauwaert, in The Brine Shrimp Artemia, Vol. 2., Physiology, Biochemistry, Molecular Biology, ed. by G. Persoone, P. Sorgeloos, O. Roels, and E. Jaspers, Universa Press, Wetteren, Belgium (1980), p. 491.
S. Ramagopal and H. L. Ennis, Eur. J. Biochem. 105: 245 (1980).
I. N. Serdyuk, A. K. Grenader, and G. Zaccai, J. Mol. Biol. 135: 691 (1979).

QUASIELASTIC LIGHT SCATTERING AND ELECTROPHORETIC LIGHT SCATTERING STUDIES ON MONONUCLEOSOMES AND POLYNUCLEOSOMES

Kenneth S. Schmitz and Nambi Parthasarathy

Department of Chemistry
University of Missouri - Kansas City
Kansas City, Missouri 64110

Chromatin is a nucleohistone complex which exhibits a repeat unit structure. The repeat unit, or nucleosome, is composed of approximately 200 base pairs of DNA wrapped about the surface of an octameric histone core, the latter containing two copies each of the histones H2A, H2B, H3, and H4. Approximately 145 base pairs of DNA that are in intimate contact with the histone core are resistant to nuclease digestion. This region of the nucleosome is called the "core particle". The remaining DNA which is nuclease accessible is referred to as "linker DNA".

We have studied mononucleosome and polynucleosome preparations as a function of ionic strength and temperature using the techniques of sedimentation velocity, quasielastic light scattering (QLS), and electrophoretic light scattering (ELS). The applied electric fields in the ELS experiments were either static or varied sinusoidally in time. The apparent diffusion coefficient D_{app} was assumed to have the functional form[1]

$$D_{app} = D/G(K,0) \tag{1}$$

where D is the "true" diffusion coefficient and G(K,0) is the static structure factor which results for a system of interacting particles. The spectral density $S(K,\omega-\omega_o)$ in the case of a sinusoidal electric field is of the form[2]

$$S(K,\omega-\omega_o) = B\int_{-\infty}^{+\infty} \langle c(K,0)^* c(K,0)\rangle \exp(-D_{app}K^2t)\exp(z(y-1))\exp(i(\omega-\omega_o)t)dt \tag{2}$$

where B is a constant, $\langle c(K,0)^*c(K,0)\rangle$ is the amplitude of the concentration fluctuation, $z = \mu EK_x/2\pi\nu'$, and $y = \exp(i2\pi\nu' t)$, μ is the electrophoretic mobility, E is the magnitude of the applied electric field, K_x is the x-component of the scattering vector K, and ν' is the frequency (cps) of the driving electric field. Expansion of $\exp(z(y-1))$ leads to the following form for the amplitude of the peak centered at the driving frequency ν',

$$A_1(\nu') = \sum_{n=1}^{\infty} z^n/(n!) = (\mu EK_x/2\pi\nu')\exp(-\mu EK_x/2\pi\nu') \quad (3)$$

for $\nu' > \mu EK_x/2\pi$. A plot of $\ln(\nu' A_1(\nu'))$ vs. $1/\nu'$ has a slope $-\mu EK_x/2\pi$. This method of determining μ is called the amplitude modulation method.

RESULTS

The QLS data were analyzed as a single exponential function with a baseline, where D_{app} was defined at each scattering angle by the expression

$$D_{app} = 1/2\tau K^2 \quad (4)$$

where τ is the relaxation time. D_{app} was found to be strongly dependent on the scattering vector K and in many cases plots of D_{app} vs. K exhibited oscillations. As illustrated in Figure 1A, $D_{app}(\Theta)/D_{app}(90°)$ vs. K plots appear to follow $1/G_H(K,0)$ vs. K plots, where $G_H(K,0)$ is the static structure factor for a linear array of Brownian particles held in a lattice structure by a harmonic potential as derived by Ackerson,[1]

$$G_H(K,0) = \frac{1 - \exp(-K^2/2\alpha\beta)}{1 - 2\cos(Ka)\exp(-K^2/4\alpha\beta) + \exp(-K^2/2\alpha\beta)} \quad (5)$$

where α is the harmonic coupling constant, $\beta = 1/kT$, and a is the interparticle spacing. In general, D_{app} vs. K plots appeared to oscillate with increasing amplitude as the temperature was raised from 10°C to 35°C for polynucleosomes in ionic strength buffers of 30mM and above. As the temperature was raised above 35°C, D_{app} became less dependent on K as the temperature was raised from 10°C to 60°C for ionic strengths below 10mM. ELS experiments with an applied static field suggest the apparent charge on the mono- and poly- nucleosomes increase as the ionic strength is lowered. As shown in Figure 1B, the elctrophoretic mobility for mononucleosomes increases from 0.5×10^{-4} cm^2/Vs to 1.2×10^{-4} cm^2/Vs as the ionic strength is lowered from 11mM to 1mM. Also shown in Figure 1B is

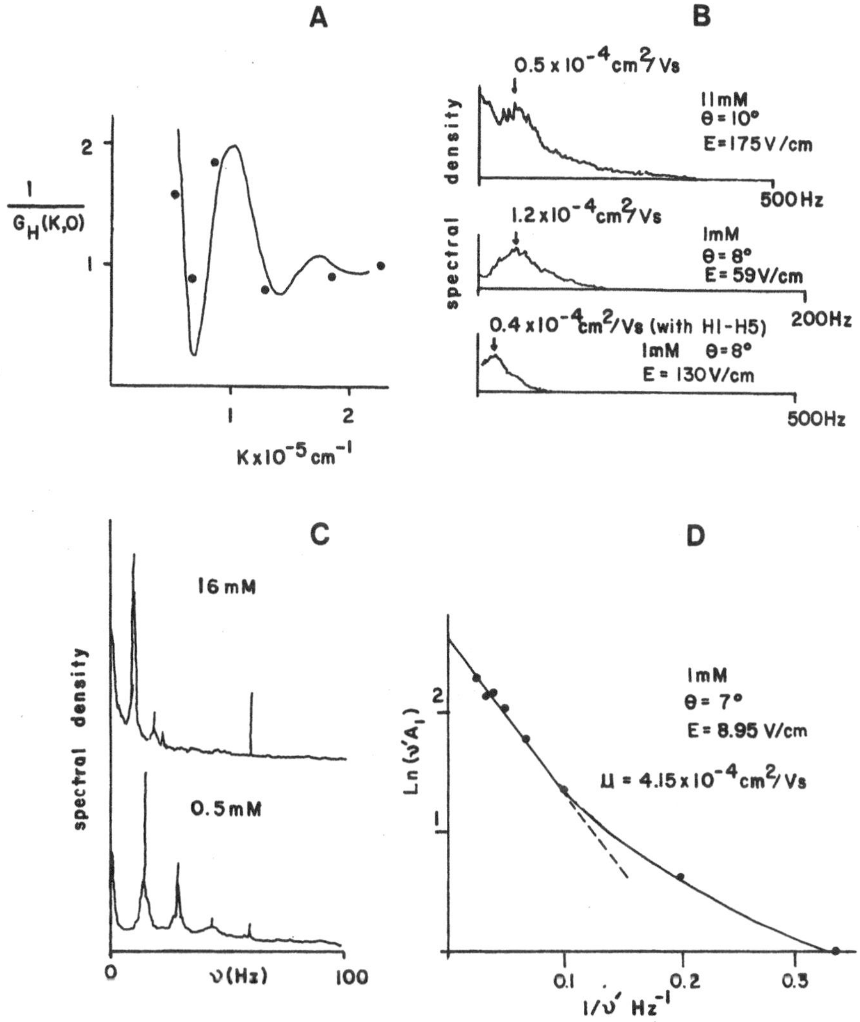

Figure 1. Experimental Results on the Mononucleosome and Polynucleosome Systems.

<u>Figure 1A:</u> If it is assumed that G(K,0) = 1 at the scattering angle of 90 degrees, then the ratio $D_{app}(\Theta)/D_{app}(90)$ should be equal to 1/G(K,0) in accordance with Equation 1. This ratio for Fraction F68 at 10.4°C in 1mM cacodylate is illustrated above along with the theoretical curve in the harmonic model approximation (cf. Equation 5) for an interparticle distance $a = 9 \times 10^{-5}$ cm and a coupling constant $4\alpha\beta = 1 \times 10^{10}$ cm^{-2}.

Figure 1B: Electrophoretic light scattering experiments using a static electric field were performed on mononucleosomes in 11mM and 1mM ionic strength buffers. As illustrated above, the apparent electrophoretic mobility goes from 0.5×10^{-4} cm^2/Vs in 11mM to 1.2×10^{-4} cm^2/Vs in 1mM ionic strength buffer thus suggesting a net increase in the apparent change on the mononucleosome without the histones H1 and H5. Mononucleosomes with the external histones have an apparent electrophoretic mobility of 0.4×10^{-4} cm^2/Vs in 1mM ionic strength buffer.

Figure 1C: Electrophoretic light scattering experiments using a sinusoidal electric field were performed on polynucleosomes in 16mM and 0.5mM ionic strength buffer solutions. The number and location of the peaks depend on the quantities $\mu EK_x/2\pi\nu'$ and ν', respectively, in accordance with Equation 2. As shown above, a very slow, large amplitude relaxation process results for the polynucleosome system in going from 16mM to 0.5mM ionic strength buffer systems. We believe this is due to interparticle interactions and the formation of local domains of ordered structure.

Figure 1D: The amplitude modulation method (cf. Equation 3) is illustrated above for mononucleosomes in 1mM ionic strength buffer solution. The relative amplitude d/d' of the peak at the driving frequency ν' was used in the calculations in the above plot, where d is the magnitude of the peak and d' is the magnitude of the internal reference at 60 Hz (cf. Figure 1C) due to the laser background.[2] The electrophoretic mobility obtained from the slope in the high frequency region was 4.15×10^{-4} cm^2/Vs, which is closer to the value for sodium ions than mononucleosomes using the static electric field method (cf. Figure 1B).

the Doppler-shifted spectrum for mononucleosomes containing the externally bound histones H1 and H5. The electrophoretic mobility for the latter system is 0.4×10^{-4} cm^2/Vs. ELS experiments performed with a sinusoidal electric field appear to obey Equations 2 and 3 as illustrated in Figure 1C and 1D. As shown in Figure 1C, a second, very long-lived relaxation time appears as the ionic strength is lowered from 16mM to 0.5mM cacodylate. The electrophoretic mobility for mononucleosomes in 1mM cacodylate as calculated by the amplitude modulation method (4.15×10^{-4} cm^2/Vs) is closer to that of sodium ions (4.62×10^{-4} cm^2/Vs) than to the electrophoretic mobility obtained from the static electric field experiment (cf. Figure 1B).

DISCUSSION

Due to the relatively high ionic strength buffers used in the present study, the interparticle interaction as inferred from D_{app} vs. K plots does not originate from direct interactions between particles. We propose that the origin lies in charge fluctuations within the polynucleosome structure and subsequent effects on the small ion distribution about the polyion. That is, spontaneous fluctuations of DNA-core histone and DNA-external histone (H1 and H5) contacts cause a pulsating distortion of the ion cloud in the vicinity of these contacts. This mechanism can explain the difference in the temperature profiles of D_{app} vs. K for ionic strength buffers above and below 30mM. For the higher ionic strengths, both DNA-core histone and DNA-external histone fluctuations contribute to the total charge fluctuation within the nucleosome. As the temperature is raised, the amplitude of the fluctuation associated with the more labile external histones increases thus leading to more domains of ordered structure within the solution. In the lower ionic strength solutions, however, the external histones are more tightly bound to the DNA and thus the amplitudes of these fluctuations are not as large as the temperature is raised to 35°C. The thermal energy is sufficient, therefore, to disrupt the postulated ordered domains. Continual increase in the temperature above 50°C for all ionic strengths leads to aggregation, as inferred from the lower D_{app} values, and loss of ordered domains, as inferred from the weaker dependence of D_{app} on K. Involvement of small ion distribution in the proposed mechanism is an extension of the electrolyte dissipation theory proposed by Schurr[3] and is somewhat supported experimentally by the ELS data which suggest the electrophoretic mobility obtained by the amplitude modulation method is closer to that of sodium ions than that obtained for mononucleosomes when a static electric field is applied.

ACKNOWLEDGEMENTS

We thank Ms. Jacqueline C. Kent and Ms. Jennifer Gauntt for technical assistance in the preparation of the chromatin samples. This research was supported in part by grants from the National Institutes of Health (USPH GM24346), the National Science Foundation (NSF PCM 7905895), and the American Cancer Society (NP332).

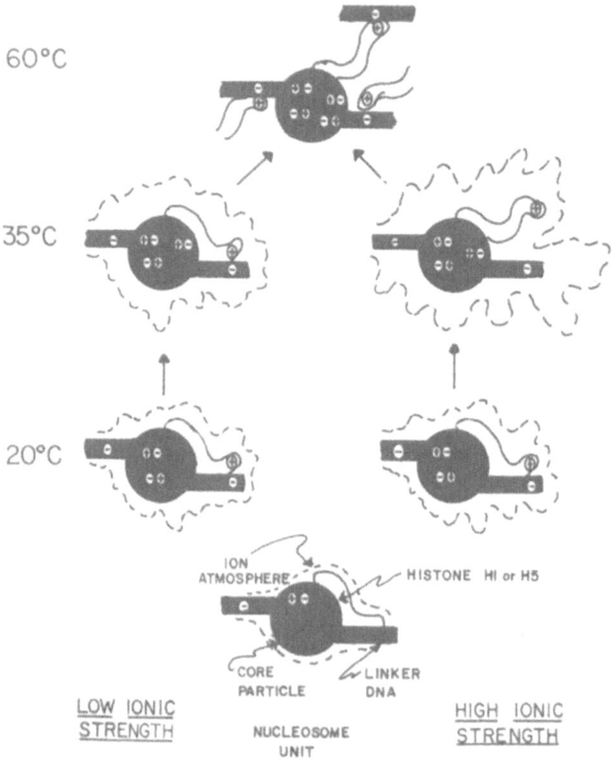

Figure 2

Proposed Mechanisms Leading to Localized "Ordered Domains" in the Chromatin Solutions.

Fluctuations of the contact points between the histones and the DNA within the core particle structure (balls in the above diagram) and the linker DNA regions (rods in the above diagram) result in fluctuations in the small ion concentration about the chromatin particle. It is further assumed that one end of the external histones H1 and H5 is very labile, hence relatively large fluctuations in the charge density should result in high ionic strength buffers as compared to the low ionic strength buffers (cf. the 35°C illustration above). These large fluctuations lead to large distortions in the ion atmosphere and hence interparticle coupling if the ion atmospheres of neighboring molecules overlap. As the temperature is increased for all ionic strength solutions examined in the present study, the number of cross-link contacts by the external histones becomes sufficient to cause aggregation and eventual precipitation. High ionic strength refers to 30mM and above while low ionic strength refers to 10mM and below.

REFERENCES

1. B. J. Ackerson, J. Chem. Phys. 64: 242 (1976).
2. K. S. Schmitz, Chem. Phys. Letters 63: 259 (1979).
3. J. M. Schurr, Chem. Phys. 45: 119 (1980).

ELECTROPHORETIC LIGHT SCATTERING OF RED CELL MEMBRANE VESICLES: A STUDY OF THE ELECTRICAL CHARGES OF THE CYTOPLASMIC MEMBRANE SURFACE

W. S. Yen, R. W. Mercer, B. R. Ware, and P. B. Dunham

Departments of Chemistry and Biology
Syracuse University
Syracuse, New York 13210

The membrane of human red blood cells is structurally asymmetric in that most, if not all, of its constituent molecules are oriented in such a manner that different groups are exposed at each membrane surface. Carbohydrates are restricted to the external surface and peripheral proteins to the cytoplasmic surface. Integral proteins which span the membrane are always asymmetrically arranged. Of the major phospholipids, the ones at the outer surface are mainly the choline phosphatides, while the inwardly facing ones are amino lipids.[1,2,3]

Under physiological conditions the external surface of the red cell membrane is negatively charged due to carbohydrates (N-acetyl neuraminic acid) and proteins.[4] Recent preliminary studies have suggested that the cytoplasmic surface is also negatively charged.[5] We have undertaken to characterize further the species responsible for the electrical charge of the cytoplasmic surface of the red cell membrane. Inside-out vesicles (IOVs) prepared from red cell membranes were examined using electrophoretic light scattering (ELS).[6,7] In this technique, the velocity distribution of the charged particles under the influence of an external electric field can be determined by analyzing the frequency spectrum of the interference between the incident and the Doppler shifted light. From the velocities, the electrophoretic mobilities and the surface charge of the particles can be inferred. Electrophoretic light scattering is particularly suited for a study of the surface of the vesicles (spheres, 0.8 μm average diameter) since no visual observation is required (as it is in conventional microelectrophoresis). We report here preliminary results in which we demonstrate the potential of studying the properties of the cytoplasmic surface of the membrane using this laser light scattering approach.

We have prepared IOVs by the method of Steck.[8] We routinely determined the relative sidedness of a preparation of vesicles by measuring the activity of cholinesterase, an enzyme restricted to the external surface of red cells.[8] Our preparations of IOVs were generally 75-85% inside out.

Figure 1 shows an ELS spectrum of suspensions of vesicles and of intact red cells. The horizontal frequency axis (Doppler shift) may also be considered to be an electrophoretic mobility axis. The spectrum of the vesicles is reproducible in that the smaller peak with highter mobility (60 Hz in this case) was always observed. Figure 1 shows that the mobility of this fraction of vesicles corresponds exactly to that of intact cells. We conclude tentatively that the more numerous vesicles with slightly lower mobility (53 Hz) are inside-out, while the smaller peak with higher mobility is due to ROVs. The fraction of vesicles which is inside-out, determined by cholinesterase activity (80%), is consistent with this conclusion.

In an attempt to confirm this conclusion, we treated a suspension of vesicles with neuraminidase, an enzyme which reduces the charge of intact cells (and presumably of ROVs) by cleaving N-acetyl neuraminic acid, a negatively charged specie. Neuraminidase should be expected to have no effect on the charge of IOVs. Figure 2 shows

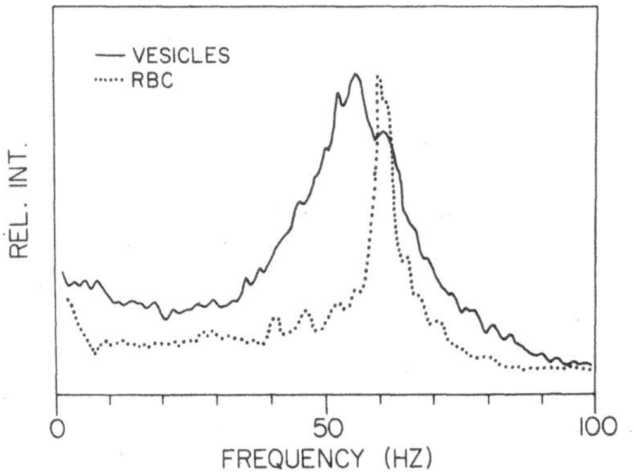

Figure 1. ELS spectra of intact red blood cells (RBC) and vesicles. Each spectrum is a plot of the scattered light intensity as a function of Doppler shift frequency. For both spectra the suspending aqueous medium contained 25 mM NaCl, 2.5 mM Hepes (N-2-hydroxyethylpiperazine-N'-2-ethane sulfonic acid) buffered to pH 7.4, and sufficient sorbitol to maintain isotonicity with intact cells (270 mOsM). The vesicle suspension was diluted to 1 μg protein per ml of suspension; cells were suspended at 10^5/ml. For both spectra the scattering angle was 36.6°, the electric field strength was 28 V/cm, and the temperature was 22°C.

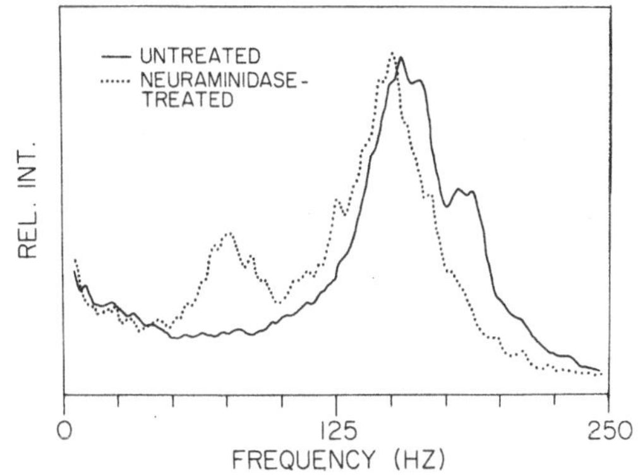

Figure 2. ELS spectra of vesicles of mixed sidedness before and after treatment with neuraminidase. Vesicles were suspended in a solution containing 20 mM NaCl, 5 mM KCl, 1 mM $MgCl_2$, and 2.5 mM tris-glycyl glycine buffered to pH 7.4. Vesicles were incubated with neuraminidase (type V, Sigma Chemical Co.) at 5 units/ml, 37°C, for 60 min, followed by 100-fold dilution with the suspension medium. For both spectra the scattering angle was 36.6°, the electric field strength was 63 V/cm, and the temperature was 22°C.

the result of such an experiment. In the untreated (normal) vesicles the two peaks are clearly distinguishable, with about 20% of the vesicles having higher electrophoretic mobility (at 180 Hz), and most of the vesicles having a Doppler frequency of 150 Hz. Treatment of the vesicles with neuraminidase had little or no effect on the mobility of the major population of vesicles, represented by the peak at 150 Hz, thus supporting our conclusion that these vesicles are IOVs. However the peak at the higher frequency was eliminated and a new peak appeared at 75 Hz, which is consistent with the conclusion that vesicles of higher mobility are ROVs since the N-acetyl neuraminic acid is known to be responsible for greater than 60% of the electrical charge of the external surface of red cells.

We conclude that ELS is an ideal technique with which to distinguish IOVs from ROVs. We expect this new approach to be valuable for the study of the electrical charge of the cytoplasmic surface of the cell membrane of red cells. We are currently identifying the groups responsible for this charge.

REFERENCES

1. M. S. Bretscher, Science 181: 622 (1973).
2. J. E. Rothman and J. Lenard, Science 195: 743 (1977).

3. L. D. Bergelson and L. I. Barsukov, Science 197: 224 (1977).
4. G. V. F. Seaman, in The Red Blood Cell, vol. 2, 2nd edition, ed. by D. M. Surgenor, Academic Press, New York, 1975.
5. H. G. Heidrich and G. Leutner, Eur. J. Biochem. 41: 37 (1974).
6. B. R. Ware and W. H. Flygare, Chem. Phys. Lett. 12: 81 (1971).
7. B. R. Ware, in New Applications of Lasers to Chemistry, ed. by G. M. Hieftje, ACS Symposium Series No. 85, 1978.
8. T. L. Steck, in Methods in Membrane Research, vol. 2, ed. by E. D. Korn, Plenum, New York, 1974.

NEUTRON AND LIGHT SCATTERING STUDIES OF BROME MOSAIC VIRUS AND ITS PROTEIN

M. Zulauf, M. Cuillel, and B. Jacrot

EMBL Outstation
85X, F-38041 Grenoble Cédex

CONTENTS

1. INTRODUCTION

Brome Mosaic Virus (BMV; MW 4.6 10^6) is a small spherical plant virus of 133 Å radius which possesses a divided genome of single-stranded RNA (MW about 10^6) contained in a protein capsid which is composed of 180 copies of a single polypeptide of MW 20,300. The proteins are located at quasiequivalent crystallographic sites on an icosahedron of triangulation number T = 3: twelve pentamers form the vertices and twenty hexamers the faces of the icosahedron. The surface structure is occasionally visible on electron micrographs of negatively stained virus, and has been analyzed to 25 Å by image reconstruction of two-dimensional arrays[1] of the closely related Cowpea Chlorotic Mottle Virus (Figure 1; Figure 2 shows a schematic view of the central protein arrangement visible in Figure 1).

In the absence of RNA, at pH 7, 0.5M KCl, BMV protein exists solubilized in small molecular weight subunits, probably dimers (see below); lowering the pH to 5 induces a crystallization of the protein in empty capsids, again with icosahedral symmetry T = 3. This implies that the protein monomer is capable of several, somewhat different bondings, according to the different local curvature around the hexamer and pentamer vertices. The molecular mechanism for the symmetry-breaking is investigated by SANS of the native virus, and by photon correlation and neutron scattering of the solubilized protein.

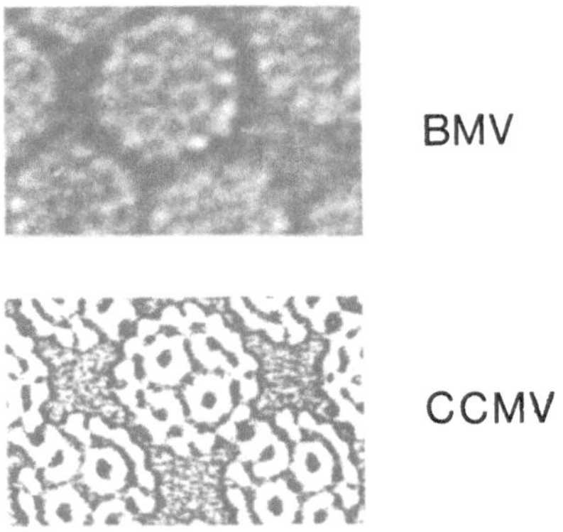

Figure 1.

2. SMALL ANGLE NEUTRON SCATTERING[2]

Spectra obtained from concentrated samples of native BMV in buffers with 5 different ratios of D_2O/H_2O are shown in Figure 3. The formal resolution is 37 Å (Q_{max} = 0.17 $Å^{-1}$). In 40% D_2O, the scattering density of the protein is virtually the same as that of the solvent, and the signal stems from the inner RNA core of the virus. In 68%, the RNA is matched, and the scattering is due to the outer protein shell. In 100%, all scattering densities are negative with respect to the solvent, in 0% they are positive, and in 55% the protein appears with negative contrast, the RNA with positive.

This data is analyzed in terms of a spherical shell model:[3,4,5] the virus is depicted as a sphere subdivided into an _a priori_ unknown

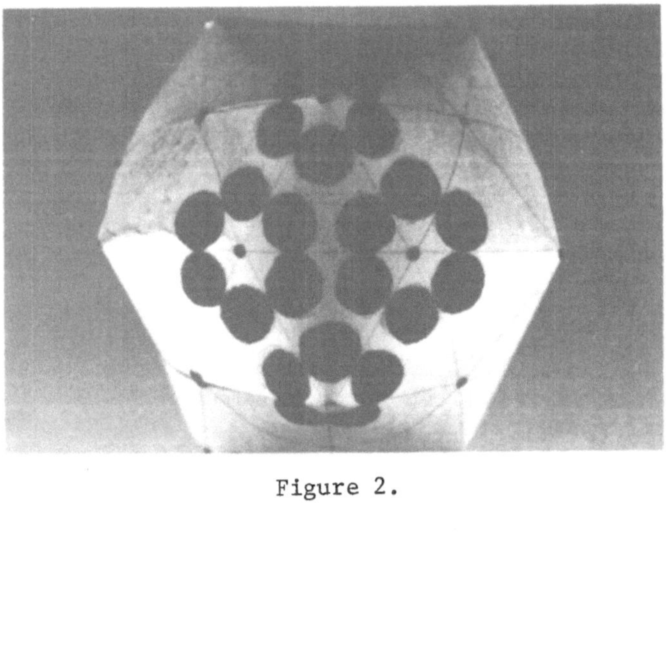

Figure 2.

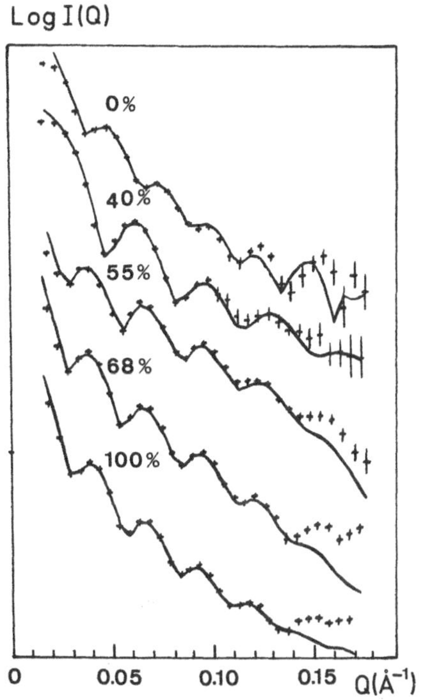

Figure 3.

number of concentric shells with various radii and scattering densities. Each shell is thought to define a compartment of uniform composition of biological material (protein, nucleic acid, water, or a mixture). The scattered intensity is then calculated, smeared according to the 9% wavelength spread of the neutrons and the effects of finite camera apertures, and compared with the experimental data. Any combination of the model parameters may be adjusted by least squares fitting. Note that since the values of the scattered intensity can be determined on an absolute scale, the densities are known on an absolute scale and the material can be identified.[5] The following strategy is applied to obtain a model that fits all 5 spectra simultaneously:

(1) The radii of the shells are the same in all 5 contrasts (assumption of homogeneous deuteration; labile H atoms will exchange).

(2) The densities of the shells are required to vary linearly with the contrast (assumption of equilibrium proton exchange after extensive dialysis).

(3) Only the model with the minimum number of shells that satisfies (1) is considered. Deviations between this model and the data are then attributed to non-spherical scattering contributions.

Result: Assuming that BMV has a central hole (in electron microscopy, certain stains color the center of the virus) we find it necessary to use 6 shells to fit the data in the range $0.03 < Q < 0.14$ Å^{-1} (the forward direction shows interparticle interferences). The large number of shells is partially due to the finding that the RNA density decreases gradually to the inside of the virions. This variation needs 4 shells to be described properly by the model. The fits are shown in Figure 3, the radii and relative scattering densities in Figures 4 and 5.

Starting from the outside of the virus, we have first a shell 32 Å thick that comprises 95% of the viral protein. The excess scattering density of that shell is $1.84\ 10^{-14}$ cm/Å^3; densities of compact proteins are usually around $2.35 - 2.40\ 10^{-14}$ cm/Å^3. Next to the inside is a 6 Å thick shell with very little protein and much solvent. The next shell of 10 Å thickness contains 10% protein and 90% RNA. Further to the inside we find no protein and decreasing RNA density.

We confirm, with improved resolution, the previous finding[4] that the capsid of BMV is not very densely packed with protein. The clustering observed by electron microscopy is likely to extend

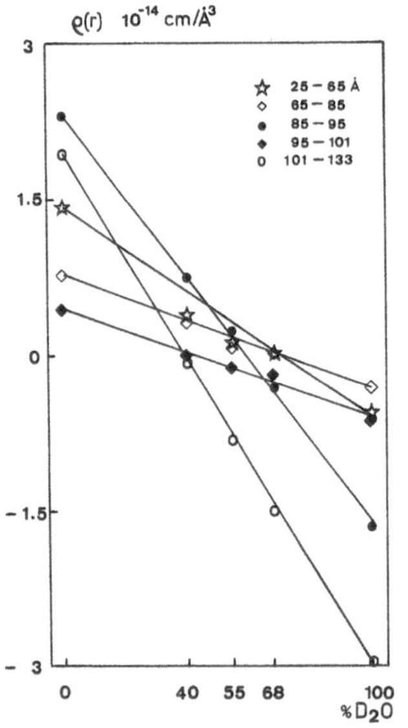

Figure 4.

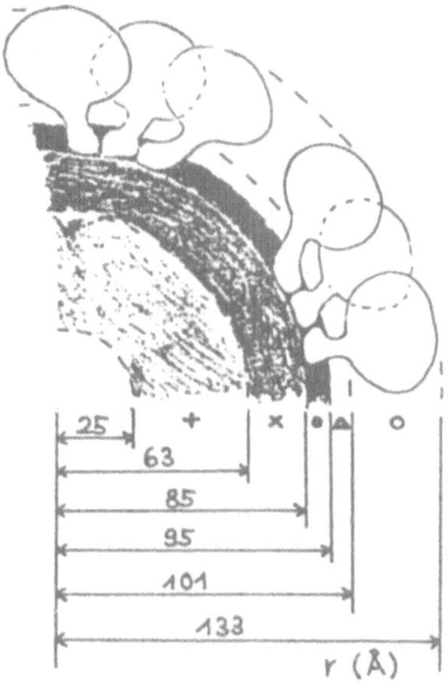

Figure 5.

through the whole capsid. The systematic underestimation of the intensity by the model for wave vectors > 0.14 $Å^{-1}$ could be due to non-spherical scattering contributions arising from the protein clustering.

From this, the following low-resolution picture of the shape of the protein monomer arises: it has a globular compact part of about 30 Å in diameter and a thin, probably flexible arm of 20-25 Å length with which it protrudes into the RNA. A schematic view of the virus showing a central section through adjacent hexamers is shown in Fig. 5.

3. PHOTON CORRELATION AND NEUTRON SCATTERING

The correlation spectra obtained from solutions of BMV protein at pH 7, 0.5M KCl, are not single exponentials. Analysis in terms of two exponents leads to two relaxations that correspond to particles with apparent hydrodynamic radii of 28 ± 2 Å (68% of the scattered intensity) and 80 ± 10 Å (Figure 6). These values are independent of the sampling time, the scattering angle and the protein concentration. Combining the diffusion coefficient of the fast species, (7.55 ± 0.55) 10^{-7} cm^2/s, with the observed single component sedimentation velocity, (2.50 ± 0.25) S, leads to a molecular weight of 29,600, which is neither the monomer nor the dimer of the protein. The observed z-average of the diffusion coefficient, (6.04 ± 0.07) 10^{-7} cm^2/s, combines to a weight of 36,800. It is interesting that in sedimentation equilibrium centrifugation, a single component with a weight of 40,000 appears.

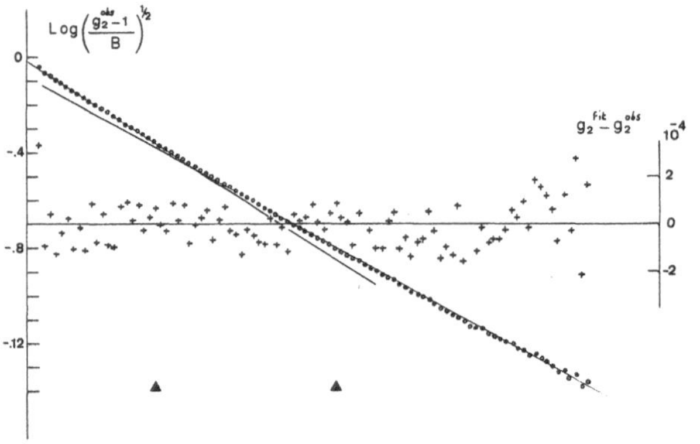

Figure 6.

Neutron scattering off the solubilized protein leads to Guinier plots that show marked curvature in the forward direction (Figure 7). In D_2O, the radius of gyration deduced from the flat part of the plot at higher angles is 26 ± 2 Å. The intensity extrapolated from this region to zero scattering angle corresponds[6] to a molecular weight of 28,200, in good agreement with the weight of the fast component found by photon correlation. The residual intensity can again be analyzed in terms of a single scattering species; the corresponding radius of gyration is around 43 Å.

We conclude that both the correlation and the neutron data cannot be interpreted easily in terms of polydispersity of the BMV protein in solution. The observation of a single boundary in sedimentation velocity experiments and the dimer weight in sedimentation equilibrium renders polydispersity also unlikely. A possible interpretation of the data could be that the protein is solubilized as a

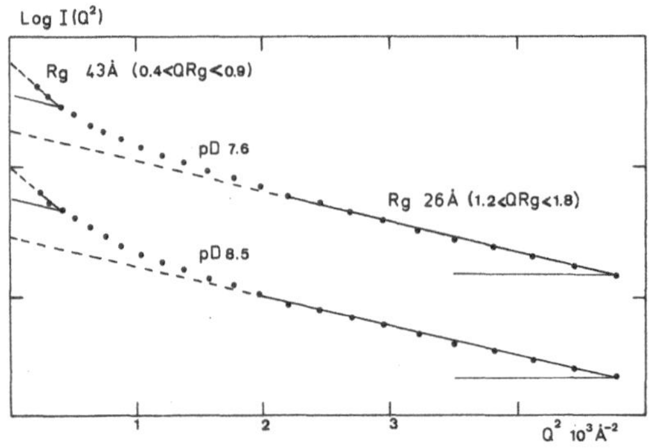

Figure 7.

dimer, in which the two rigid globular parts of each monomer are linked by means of the thin flexible arms with which we found the protein shell of the whole virus anchored in the RNA. The scattering unit is then basically of dumb-bell shape, with a certain probability distribution for the separation between the globular parts. Unfortunately there is no worked-out theory of diffusion of such an object.

REFERENCES

1. A. C. Steven, R. C. Smith, and M. Horne, J. Ultrastructure Res. 64: 63 (1978).
2. G. Zaccai, this volume.
   B. Jacrot, Rep. Prog. Phys. 39: 911 (1976).
3. D. Schneider, M. Zulauf, R. Schäfer, and R. M. Franklin, J. Mol. Biol. 124: 97 (1978).
4. C. Chauvin, P. Pfeiffer, J. Witz, and B. Jacrot, Virology 88: 138 (1978).
5. C. Chauvin, J. Witz, and B. Jacrot, J. Mol. Biol. 124: 641 (1978).
6. B. Jacrot and G. Zaccai, Biopolymers, in press.

SCALING OF HALFWIDTHS WITH SCATTERING VECTOR FOR AUTOCORRELATION FUNCTIONS FROM NORMALLY SWIMMING BOVINE SPERMATOZOA

T. Craig and F. R. Hallett

Physics Department
University of Guelph
Guelph, Ontario, Canada

CONTENTS

1. INTRODUCTION

The bull spermatozoon is modelled as a ellipse of semi-axes 9.0 μm x 2.3 μm x 0.5 μm both coated and uncoated. The trajectories of the normal cells are assumed to be helical with the overall direction of motion along the long axis of the cell.

2. THEORY

In order to calculate the electric field autocorrelation function for non-point particles in the Rayleigh-Gans-Debye approximation (RGD) it is necessary to evaluate the expression below (Equation 1) as shown by Craig et al. (1979)

$$g^{(1)}(\tau) = N\langle e^{i\vec{k}\cdot[\vec{r}(\tau) - \vec{r}(o)]} e^{i\vec{k}\cdot[\vec{R}(\tau) - \vec{R}(o)]} A(\vec{k},\tau)A^*(\vec{k},o)\rangle \quad (1)$$

and $A(\vec{k},\tau)$ is the dynamic form factor. It has been shown before (Craig et al., 1979) that;

$$A(\vec{k},\tau) = 3j_1(K)/K \quad (2)$$

where j_1 is the first order spherical Bessel function and

$$K^2 = (\vec{k}\cdot\vec{a})^2 + (\vec{k}\cdot\vec{b})^2 + (\vec{k}\cdot\vec{c})^2 \qquad (3)$$

where a, b and c are the lengths of the semi-axes of the scatterer.

The normal cell undergoes a helical trajectory which is determined by vt as well as setting,

$$\psi = \psi_o + \omega\tau \qquad (4)$$

where ω is the angular frequency and ψ is the instantaneous phase around the helix. The cell appears to crawl around the helix, i.e. the "c" axis is tangent to the trajectory. This is described by a tilting of angle α about the "a" axis. As well, the cell exhibits a "duck-footed" behavior. This is shown as a tilt β about the "b" axis. These angles are all needed to determine the dot products in Equations 1 and 4. They are found to be about 57° and 12° respectively for real spermatozoa.

It is possible to model a coat on the particle by using the method of Chen et al. (1977). A similar treatment is done here.

If the distributions of the form;

$$P(x) = xe^{\frac{-2x}{\bar{x}}} \qquad (5)$$

are used for both speed and frequency the final form of the autocorrelation function is (Craig et al., 1979);

$$g^{(1)}(\tau) = N'R\int_o^1 d\nu \sum_n (1+iq_\nu)^{-2}(1+ip_n)^{-2}|B_n|^2 \qquad (6)$$

where

$$q_\nu = \tfrac{1}{2}k\nu\bar{v}\tau \qquad (7)$$

$$p_n = \tfrac{1}{2}n\bar{\omega}\tau \qquad (8)$$

$$B_n = \frac{1}{2\pi}\int d\psi e^{-in\psi} e^{i\vec{k}\cdot\vec{R}(\tau)} A(\vec{k},\tau) \qquad (9)$$

R denotes the real part and $\nu = \cos\theta$.

Here θ is the angle between the scattering vector and the overall direction of motion.

3. SCALING

As shown by Craig et al., the theoretical functions based on Equation 6 and the data for normal cells agree well.

The parameters of the model functions which have most effect on the half widths are the length of the "c" axis and the frequency of rotation. The frequency of rotation cannot vary greatly from 10 Hz although the length of the "c" axis was considered more adjustable since it was not obvious how much of the midpiece contributed to the scattering power of the cell. The frequency was only used to "fine tune" the value of the half width in order to match it to that of the experimental function. If any major changes were required the length of the scatterer was varied.

Figure 1 shows the scaling of the half widths of both autocorrelation data and theoretical calculations using Equation 6. The error bars on the data points indicate the standard error of the mean for the half widths at that scattering angle. Here the frequency was adjusted to 11.6 Hz to make the theory match the average of the data at 15°. It is clear that although the general features of the two are similar the absolute values of the half widths of the model functions are always smaller.

The three curves in Figure 2 correspond to the data, a particle with a coat with frequency (11.1 Hz) matched to the data at a scattering angle of 15° and a particle with a coat matched at 20°. The model with a coat ($n_1 = 1.42$, $n_2 = 1.35$, thickness ≈ 0.27 μm) lies consistently below the data. The inclusion of the coat makes the agreement somewhat better than without a coat especially at $\sin(\theta/2) \simeq 0.26$. This thickness although too thick for an actual membrane may be realistic for the heavy acrosome which covers the end of the head. This indicates that the peak in this region is probably due to the presence of a coat on the particle. The curve is not very sensitive to the thickness of the coat. Varying this parameter does not affect the position of the peak appreciably. The height of this peak is quite sensitive to the value of n_2; lowering n_2 increases the height. This is due to the fact that lowering the value of n_2 makes the inner medium more closely resemble the environment thereby making the coat more "prominent" from a light scattering standpoint. This further emphasizes the fact that this peak is due to the shell.

The second curve in Figure 2 shows a scaling curve similar to the first except that the length of the particle, "c", was adjusted to make the data and theory correspond at 20°. This curve shows somewhat better agreement with the data as far as absolute value

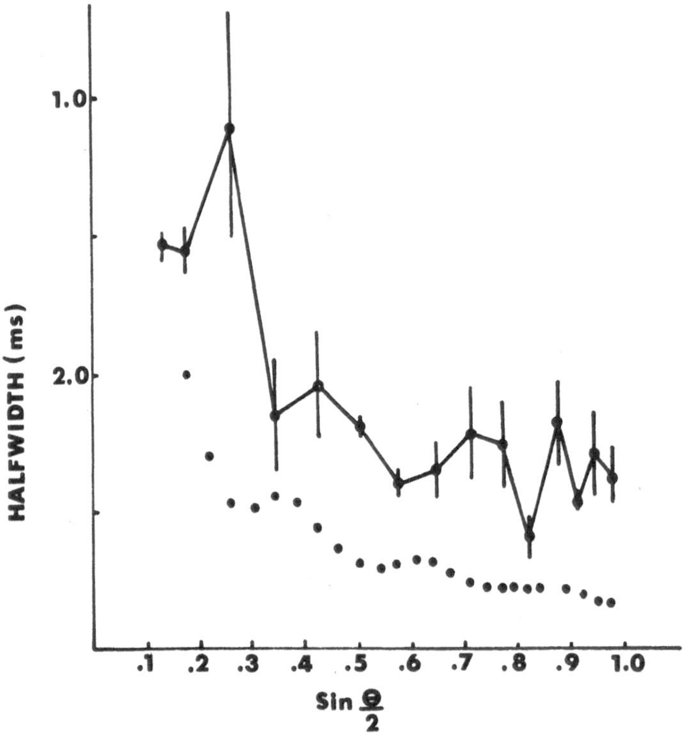

Figure 1. Halfwidth scaling curves for data from normally swimming spermatozoa (•-•-•) and for theoretical functions calculated from Equation 6 (...).

is concerned. Although there is a larger discrepancy between the data and theory at 15° it should be noted that the slope in the theory is so steep at this point that a slight error in the measurement of the scattering angle could strongly influence the agreement.

4. CONCLUSIONS

For these cells the form factor appears to have been chosen correctly as far as the "a" axis is concerned. This is shown by the correspondence of the peaks in the scaling curve. However, there is somewhat of an ambiguity as far as the length of the cell and rotational frequency is concerned. This is to be expected since as shown by Craig et al. (1979), these two parameters affect the width of the autocorrelation functions as a product. The proper interpretations appears to be that from a light scattering point of view, the length of the cell is about 9.0 μm since this gives the proper frequency to match the data at 15°. The differences

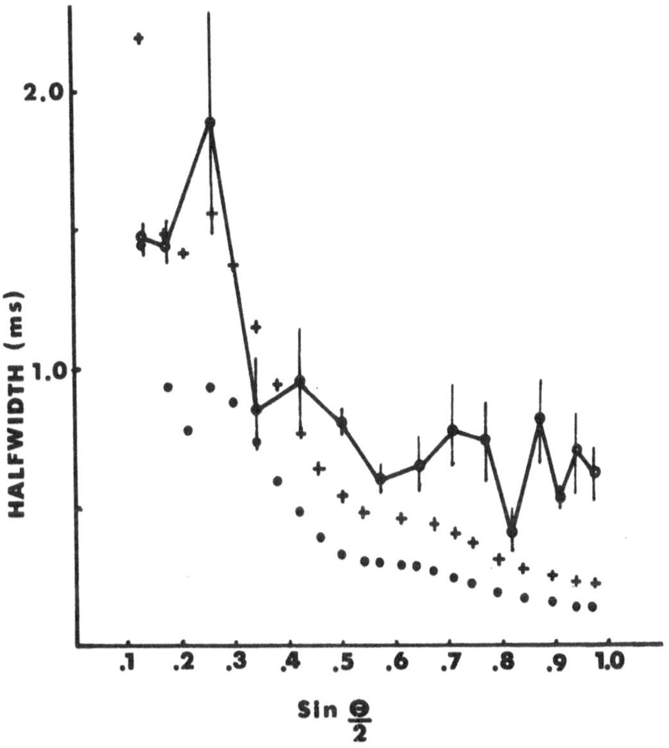

Figure 2. Halfwidth scaling curves for data from normally swimming spermatozoa (-0-) and for theoretical functions calculated from Equation 6 with a coat and matched to data at 15° (...) and a coat and matched at 20° (+++).

then between the data and the theory at higher angles may be a problem with the choice of trajectory or it may be that the form factor is in some way $\vec{k}$ dependent and a more complex form factor is in order.

REFERENCES

1. S.-H. Chen, M. Holz, and P. Tartaglia, "Quasi-elastic light scattering from structured particles", Appl. Opt. 16: 187 (1977).
2. T. Craig, F. R. Hallett, and B. Nickel, "Quasi-elastic light scattering spectra of swimming spermatozoa: rotational and translational effects", Biophys. J. 28: 457 (1979).
3. F. R. Hallett, T. Craig, and J. Marsh, "Swimming speed distributions of bull spermatozoa as determined by quasi-elastic light scattering", Biophys. J. 21: 203 (1978).

LIGHT SCATTERING STUDIES OF CATARACT

M. Delaye*, J. I. Clark, G. B. Benedek

13-2014 Department of Physics,
Massachusetts Institute of Technology
Cambridge, Massachusetts 02139

The early stages of cataract development are characterized by a progressive opacification which is due to increased light scattering within the lens. Our purpose is to establish the physicochemical basis of this scattering. We have chosen the reversible phase transition cataract in the nucleus of the calf lens[1] as a model system. In this cataract the nuclear opacification increases as the temperature decreases. We studied the development of the opacity in both the intact lens and the isolated cytoplasm.

To obtain the size and motion of the elements responsible for reversible opacification, we combine static and dynamic light scattering experiments. The setup shown in Figure 1 allows us to perform these two kinds of measurements simultaneously. If the chopper is stopped the laser beam can be focussed into the sample along a fixed direction (M_3M_1 for example). Under this condition, we use the setup for quasielastic light scattering (QLS) experiments where the photomultiplier tube (PMT) detects the intensity fluctuations in the scattered light at θ and the correlator provides the correlation function, $C(\tau)$, of the photocurrent fluctuations. $C(\tau)$ is defined by the equation:

$$C(\tau) = \langle i(t) \cdot i(t + \tau) \rangle \qquad (1)$$

where $i(t)$ is the photocurrent which is directly proportional to the

---

*Present Address:
Laboratoire de Physique du Solide Bat. 510, Universite de Paris-Sud
91405 Orsay, France

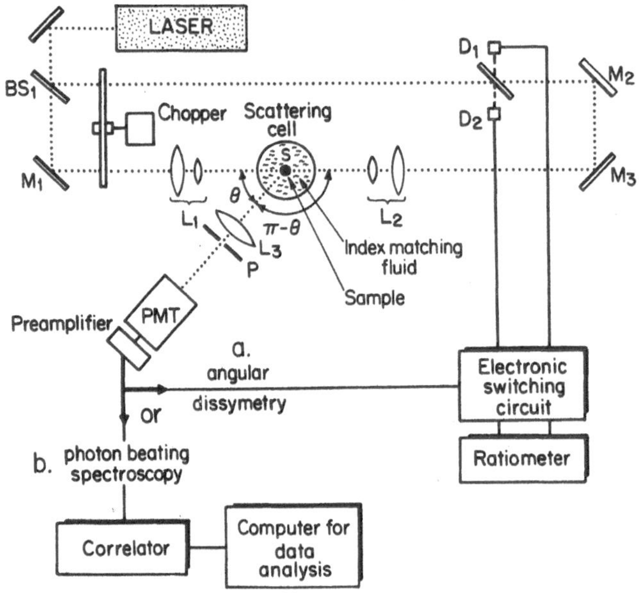

Figure 1. Block-diagram of the light-scattering setup.

intensity of the scattered light. If the chopper is turned on, the direction of the incident light is switched every 5ms so that the phototube alternately detects the intensity $i(\theta)$ and $i(\pi - \theta)$, and θ can be varied from 90° to 40°. This instrumentation allows us to use one sample for the measurement of the angular dissymetry $d(\theta) = i(\theta)/i(\pi-\theta)$ of the scattered light, which provides information on the size and shape of the scatterers, and the correlation function, $C(t)$, which provides information on the motion of the scatterers.

We will first present the results of the analysis of the correlation function of the scattered light measured by QLS. The inset in Figure 2 shows the correlation function $C(\tau)$ obtained, in the range $\tau = 4$ to 200μs, from the light scattered by the isolated cell cytoplasm at $\theta = 90°$ and $T = 25°C$. This correlation function looks reasonably exponential and by fitting $C(\tau)$ to a single exponential with an adjustable baseline, we can deduce a relaxation time τ_1. When the scattering angle θ is varied, τ_1 shows an $1/q^2$ dependence, where $q = (4\pi\eta/\lambda)\ \sin(\theta/2)$, which is the amplitude of the scattering wave-vector. The relaxation time τ_1 thus describes a diffusive motion, which we will interpret as the Brownian motion of scattering units of radius, R_1, which are present in the cytoplasm. In a first approximation, R_1 will be deduced from the Einstein formula:

$$D_1 = 1/\tau_1 q^2 = kT/6\pi\eta R_1 \qquad (2)$$

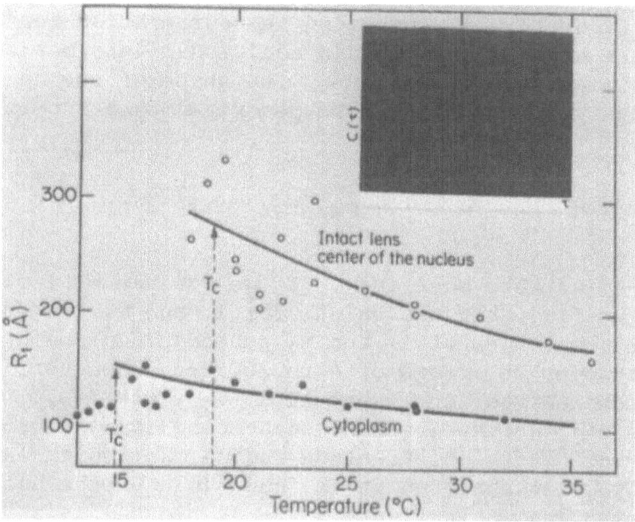

Figure 2. Effective radius of the small scatterers, R_1, as a function of temperature in the isolated cytoplasm and in the center of the intact lens. The inset shows the correlation function $C(\tau)$ for $4 < \tau < 200\mu sec$ as obtained from the light scattered at $\theta = 90°$ and $T = 25°C$ by the cytoplasm.

where k is the Boltzmann constant and T the absolute temperature. η, which is the local viscosity of the cytoplasmic fluid, will be taken as equal to the viscosity of water. The values of R_1, which are calculated following this procedure, are plotted as a function of temperature in the lower part of Figure 2. The effective radius of these cytoplasmic scattering units is roughly 140Å and R_1 does not vary significantly with temperature. The correlation function of the light scattered by the center of the nucleus is qualitatively similar to the correlation function of light scattered by the cytoplasm. However, we see from the plot of R_1 versus temperature that the scatterers in the nucleus have a radius, R_1, which is approximately 210Å at 25°C and varies significantly with temperature.* If one takes into account the high concentration of the cytoplasm (≃ 40% protein by weight) one realizes that the values of Figure 2 should be reduced by 25% to get the real radius of the scatterers.[3]

When the correlation function $C(\tau)$ is studied over a broader time range (τ = 40μsec to 2msec), it displays a shape reported in

*For both experimental systems, T_{cat} represents the "cold-cataract-temperature" for which 10% of the incident light is scattered.

the inset of Figure 3. By analyzing the correlation function over this broad range of τ, we can obtain additional information about the scattering elements in the lens. The shape of the correlation function in Figure 2 (inset) can be described by a two-exponential fit:

$$C(\tau) = [i_1 \exp(-\tau/\tau_1) + i_2 \exp(-\tau/\tau_2)]^2 + B \tag{3}$$

where B is an adjustable baseline*. If we analyze $C(\tau)$ according to formula (3), we find that the values for i_1 and τ_1 are in good agreement with the values of i_1 and τ_1 determined using a single exponential fit over the range of 4 to 200μsec. However, the analysis of the second part of the expression ($i_2 \exp(-\tau/\tau_2)$) of Equation (3) permits us to identify a second scattering component. This component corresponds to an effective radius, R_2, which is several thousand angstroms in size and it is found in both the nucleus and the isolated cytoplasm. In Figure 3, R_2 is plotted as a function

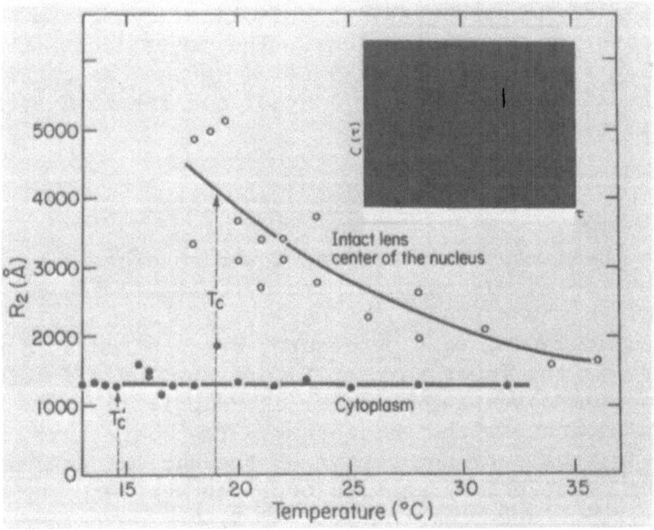

Figure 3. Effective radius of the large scatterers, R_2, as a function of temperature in the isolated cytoplasm and in the center of the intact lens. The inset shows the correlation function $C(\tau)$ for 40μsec < τ < 2msec as obtained from the light scattered at $\theta = 90°$ and $T = 25°C$ by the cytoplasm.

*Note: Actually when $C(\tau)$ is analyzed over a time range even broader, one realizes that the baseline shows itself a slight temperature dependence. We believe that this very slow component describes internal structural relaxations in the viscous cytoplasm.

of temperature in the cytoplasm and in the intact nucleus. In the cytoplasm, R_2 shows very little change with temperature (lower plot). In the intact nucleus (upper plot) R_2 increases from 2000Å at 36°C when the lens is clear, to 5000Å at 19°C, when the cataract develops in the nucleus.

From the QLS measurements one can also deduce the temperature variation of the intensities i_1 and i_2 scattered respectively by the small and large scatterers identified above. These data, which are reported in Figure 4, represent the behavior of i_1 and i_2 in both nuclear cytoplasm and the intact lens. Figure 4 also shows that the intensity, i_1, increases slightly as the temperature is increased from 15°C, where the cytoplasm is turbid, to higher temperatures where the cytoplasm is transparent. Thus the small scatterers cannot be responsible for the cataractous opacification. In contrast, the intensity, i_2, scattered by the large scattering elements increases in the same way as the total scattered intensity. This indicates that the large elements are the cause of the turbidity of a cataractous lens.

We complemented these dynamic measurements by a careful study of the angular dissymetry $d(\theta)=i(\theta)/i(\pi-\theta)$ as a function of cosθ for various temperatures. The data obtained on the cytoplasmic homogenate are reported in Figure 5. At low temperatures ($T \simeq 15$ C), when the cytoplasm is turbid, the dissymetry $d(\theta)$ reaches very high values. This shows that the scattering of light at low temperature is

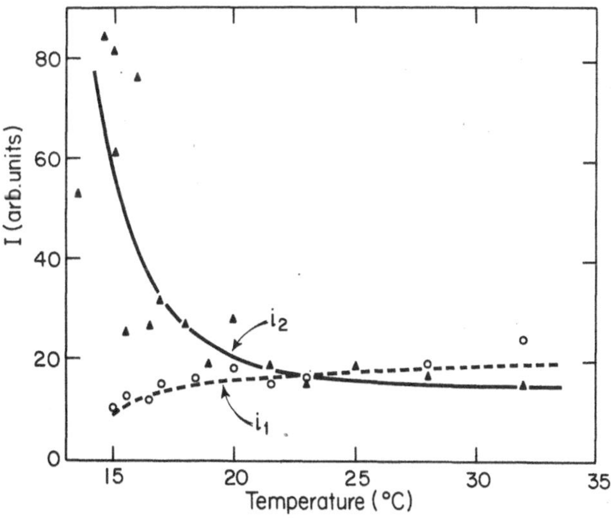

Figure 4. Intensities i_1 (small scatterers) and i_2 (large scatterers) scattered from isolated cytoplasm at $\theta = 90°$ as a function of temperature.

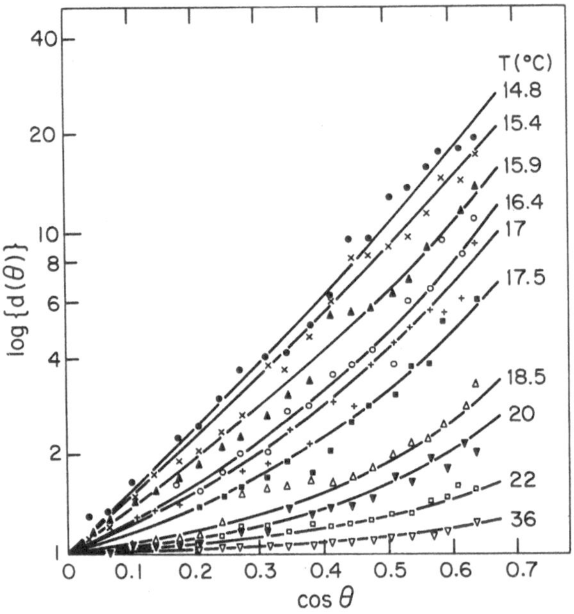

Figure 5. Angular dissymetry d(θ) in the nuclear cytoplasm plotted as a function of cosθ for various temperatures. The vertical scale is logarithmic. The solid lines correspond to a theoretical calculation of the dissymetry using a continuous bimodal size distribution of the scatterers.

dominated by the contribution of large scatterers which have a radius of a few thousand Angstroms. At high temperatures however, the scattering pattern becomes almost isotropic, showing that the main contribution in the scattered light is now due to small scattering units, which have a radius smaller than 200Å. A detailed study of the shapes of the curves log d(θ) vs. cosθ is reported in Reference 4. This angular dissymetry (AD) data suggests that there are two kinds of scatterers: small ones and large ones. It also implies that the relative proportion of large scatterers decreases when T is increased above T_{cat}. These conclusions are in perfect agreement with the QLS data. We can even reach a quantitative agreement between AD and QLS data, if we assume that the large scattering elements have a broad asymmetric size distribution and an average value of R ≃ 1300Å.

SUMMARY

We identified both in the isolated and in the native cytoplasm of the calf lens nucleus, two kinds of scattering elements: small ones, of one hundred Angstroms radius and large ones whose size is distributed around 1300Å. The small scatterers, which are probably proteic structures, are not responsible for the cataractous opacification. On the contrary, the large elements are causing the turbidity associated with the cold-cataract, since their proportion increases when the temperature is lowered. These large elements can be associated with a phase separation in the cytoplasm[4,5] and can be considered as droplets of high proteic concentration nucleating in the bulk of the lens cell cytoplasm.

ACKNOWLEDGEMENT

Supported by Grant #2-R01 EY 00797-09 from the National Institutes of Health and Grant #81-01 from the Whitaker Health Sciences Fund.

REFERENCES

1. S. Zigman and S. Lerman, Exp. Eye Res. 4: 24 (1965).
2. J. I. Clark, M. Delaye, P. Hammer, L. Mengel, and G. B. Benedek, Exp. Eye Res., to be published (1981).
3. G. D. J. Phillies, G. B. Benedek, and N. A. Mazer, J. Chem. Phys. 65: 1883 (1976).
4. M. Delaye, J. I. Clark, and G. B. Benedek, accepted by Biophysical Journal.
5. J. I. Clark and G. B. Benedek, Biochem. Biophys. Res. Comm. 95: 482 (1980).

LASER LIGHT SCATTERING ON BACTERIAL COLONIES AS A POSSIBLE TOOL FOR RAPID IDENTIFICATION*

Tibor Illéni

UNEP-UNESCO-ICRO Microbial Resource Centre (MIRCEN)
Department of Microbiological Engineering
Karolinska Institutet
104 01 Stockholm 60, Sweden

ABSTRACT

A rapid and simple method was developed for bacterial colony recognition by using the polarised laser light diffraction pattern of the colony structure. The diffraction patterns were photographed and classified according to their degree of symmetry. A correlation between the symmetry groups of diffraction patterns and the taxonomical identity of some eubacterial colonies was studied.

THE METHOD

A thin layer (approximately 2 mm) of nutrient agar gel (5 ml of Difco's) was prepared in a disposable plastic petri dish (100 by 15 mm).

The following species were studied:

1.	Escherichia coli 0112 ab:B 13	(+)
2.	Escherichia coli 113-3	ATCC 11105
3.	Klebsiella pneumoniae	(++)
4.	Micrococcus luteus	ATCC 9341
5.	Proteus rettgeri	(++)
6.	Salmonella IS9	(++)
7.	Serratia marcescens	(++)
8.	Staphylococcus albus	(++)

---

*This investigation was supported by research grant DNR 80-3571 from the Swedish Board for Technical Development.

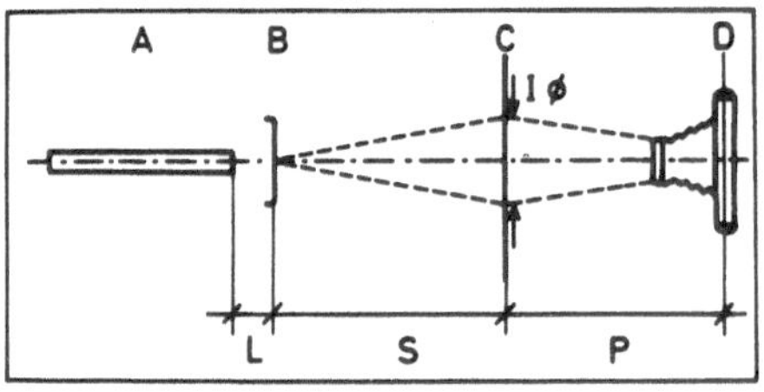

Figure 1. Experimental set-up for generation and photography of transmitted laser light diffraction patterns. A: He-Ne laser; B: petri dish with colonies; C: semi-transparent screen for visualization of the scattering patterns; D: Polaroid roll film camera; L: distance between light source and specimen (constant, L = 70 mm); S: distance between specimen and image (variable); I: size of the image adjusted to a constant (40 mm diameter) by variation of S; P: distance between image and photographic plate (constant, P = 280 mm).

9.	Staphylococcus aureus	(++)
10.	Streptococcus faecalis	ATCC 9790
11.	Streptococcus pyogenes	M2224, 7 (++)

(+) = National Bacteriological Laboratory, Stockholm.
(++) = Karolinska Institutet, Stockholm.

For the inoculum preparation, the cells were grown on agar slants and then suspended in saline. The suspensions were then further diluted with saline to a concentration at which single, well-separated colonies were produced on the petri plates.

After 18 hours of incubation at 37°C, five well-separated and undisturbed colonies were selected and their diameters measured with an ocular micrometer. The chosen colonies were then investigated with the instrument shown in Figure 1.

RESULTS

The diffraction patterns could be arranged into four groups:

I. The pattern consists of symmetric circles as shown in Figure 2.
II. The pattern consists of non-circular rings as shown in Figure 3.
III. The pattern consists of non-circular rings with star-like "rays" extending from a more or less luminous and complicated middle area as shown in Figure 4.
IV. The pattern has an amorphous asymmetric structure as shown in Figure 5.

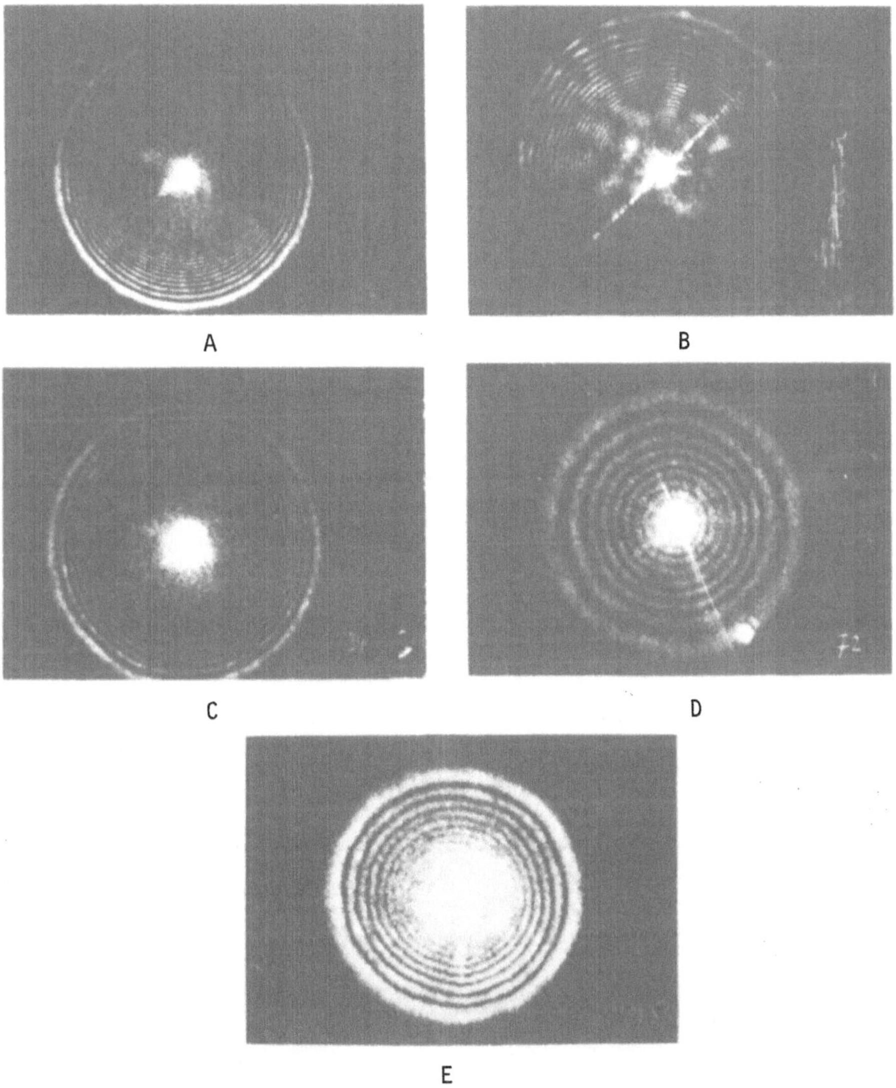

Figure 2. Laser diffraction patterns through colonies of A: Staphylococcus aureus; B: Micrococcus luteus; C: Staphylococcus albus; D: Streptococcus faecalis; and E: Streptococcus pyogenes. The colony diameters were less than 0.80 mm.

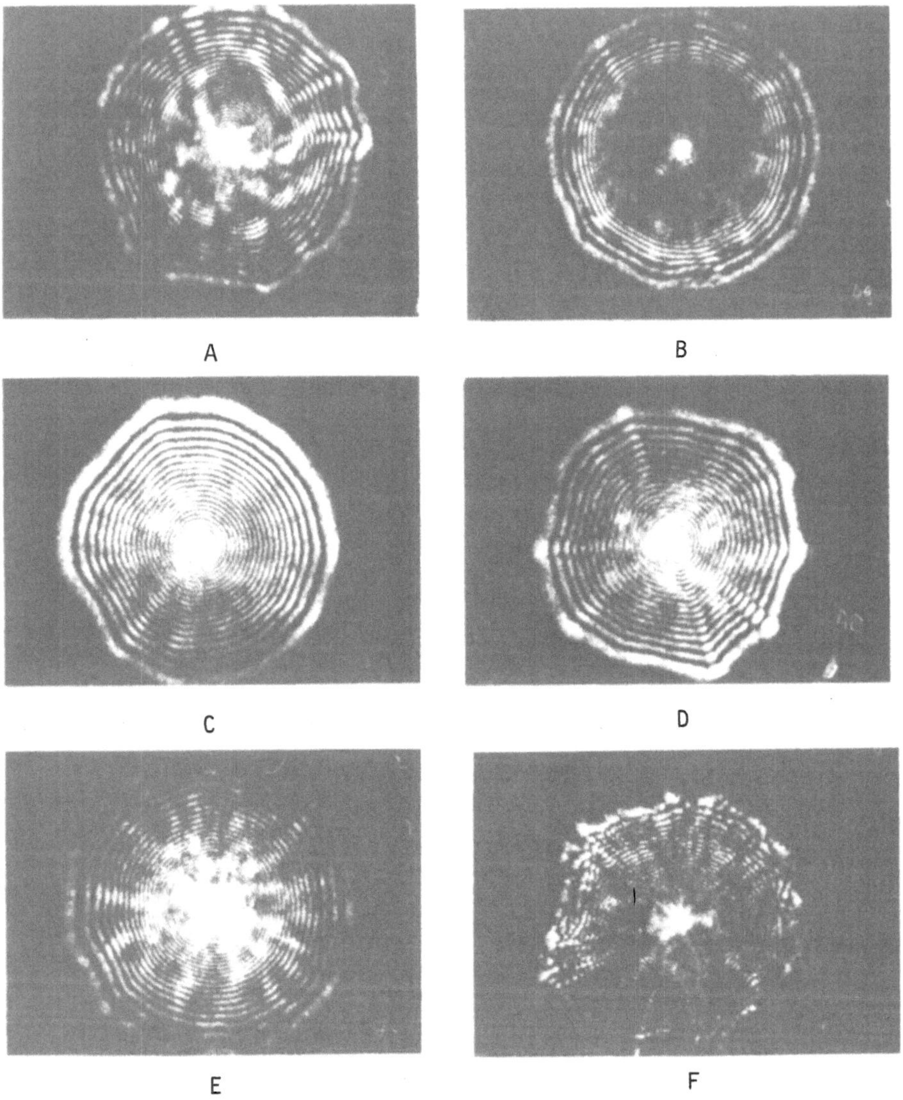

Figure 3. Laser diffraction patterns through colonies of A: Serratia marcescens; B: Escherichia coli 0112; C: Klebsiella pneumoniae; D: Salmonella IS9; E: Proteus rettgeri; and F: Escherichia coli 113. The colony diameters were less than 0.80 mm.

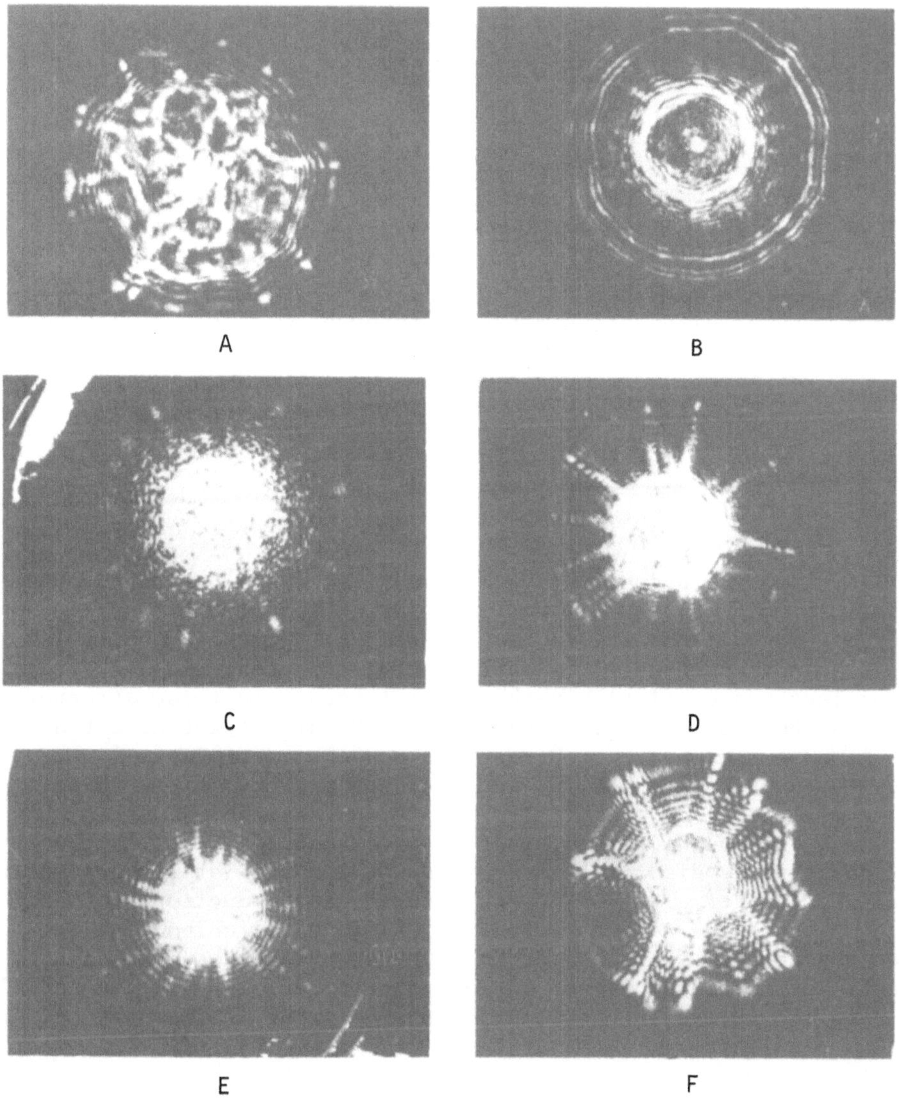

Figure 4. Laser diffraction patterns through colonies of A: <u>Serratia marcescens</u>; B: <u>Escherichia coli 0112</u>; C: <u>Klebsiella pneumoniae</u>; D: <u>Salmonella IS9</u>; E: <u>Proteus rettgeri</u>; and F: <u>Streptococcus pyogenes</u>. The colony diameters were between 0.80 - 1.70 mm.

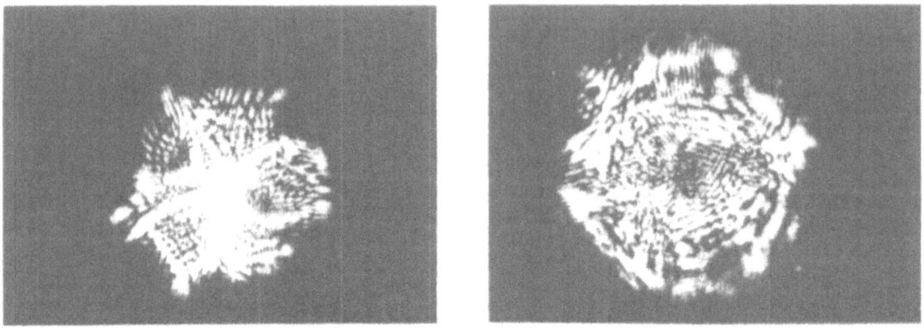

A B

Figure 5. Laser diffraction patterns through colonies of A: Streptococcus pyogenes; and B: Escherichia coli 113. The colony diameters were over 1.70 mm.

DISCUSSION

The measuring system which was used (Figure 1) served as a tool for evaluating the potential use of laser diffraction in rapid colony identification. According to the experimental findings, bacterial colonies can be considered as optical gratings consisting of several more or less systematically-packed cell layers. In these structures there must be some "more transparent spots" which together act as a grating. When the highly coherent and monochromatic laser light passes through those spots, an interference pattern is generated.

The technique described shows that transmitted laser light diffraction patterns disclose differences between similar-looking microcolonies and offer a promising approach to data reduction relevant to the morphological characterisation of microbial colonies. The method has obvious applications in connection with the detection of variants in a homogeneous population of colonies. It might eventually lead even to pattern recognition methods allowing the identification of unknown colonies.

A QUASI-ELASTIC LIGHT SCATTERING AND CINEMATOGRAPHIC INVESTIGATION OF MOTILE *Chlamydomonas reinhardtii*

T. J. Racey and F. R. Hallett

Interdepartmental Biophysics Group
Department of Physics, University of Guelph
Guelph, Ontario, Canada N1G 2W1

CONTENTS

1. INTRODUCTION

In this presentation motile *Chlamydomonas reinhardtii* cells are shown to yield functions whose decay times are inversely related to the translational speed of the cell. This conclusion is based on a comparison of experimentally determined functions obtained over a range of scattering angles with functions calculated using the assumption that the cells behave as translating Rayleigh-Gans-Debye (R-G-D) particles. The swimming parameters obtained from the light scattering experiments were consistent with those obtained from high speed cinematography.

2. THEORY

If one assumes a spherical shape for the *C. reinhardtii* cell (this is consistent with the linear scaling curve shown in Figure 3) then the autocorrelation function can be written as

$$g^{(1)}(\tau) = \int_0^\infty dv \int_{-1}^{1} d\cos\theta \; P_s(v) e^{ikr\cos\theta} \qquad (1)$$

where θ is the angle between k and r.

The strategy in this study was to develop a suitable expression for $P_s(v)$ from high speed cinematographic analysis. This function thus determined was used in conjunction with Equation 1 to calculate electric field autocorrelation functions over a range of accessible scattering angles. These calculated functions were then compared to the measured functions.

3. EXPERIMENTAL PROCEDURE

The culture of C. reinhardtii (wild type ATCC#18798) was acquired from the American Type Culture Collection. The cells were grown in medium 277 as described by ATCC at 25°C. They were exposed to fifteen hours of light daily under a broad spectrum GROLUX light.

The samples used were examined about four hours into their daily cycle of light. A consistent level of activity in the cells was strived for by spinning the cells down, resuspending them in fresh medium, and then spinning them down an hour later and then allowing them to sit. The cells examined were removed from the upper portion of the centrifuge tube, which resulted in a fairly consistent level of activity among the different cell cultures. The cells were prepared the same way for both light scattering and cinematography.

4. RESULTS AND MODELS

A detailed trajectory of a cell obtained from high speed cinematography revealed a pattern such as Figure 2. The measured average progressive swimming speed histogram, obtained from cinematography, (Figure 1) was modelled using the equation,

$$P_s(v) = \frac{128}{3} \frac{v^3}{\bar{v}^4} e^{\frac{-4v}{\bar{v}}} \qquad (2)$$

The scaling curve of the half-width times (HWT) of the experimental autocorrelation function is shown in Figure 3.

Two models were developed to match the experimental results.

Two speed distribution model with a stationary function

This model incorporated the forward and backward motion of Figure 2 with a swimming speed distribution of the form,

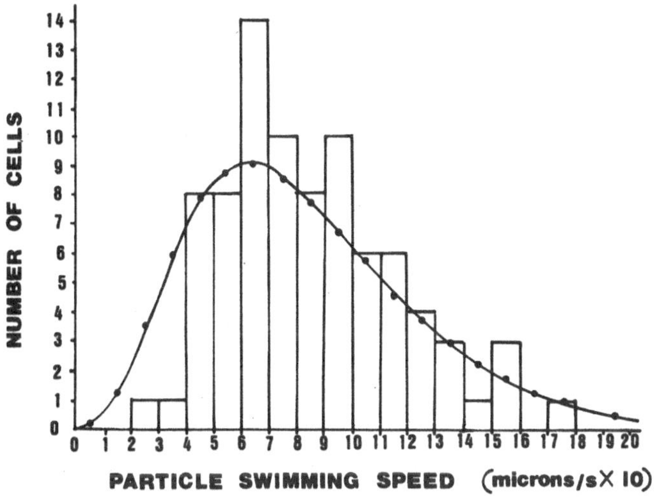

Figure 1. Speed distribution histogram of the C. reinhardtii cells. The superimposed function is

$$P_s(v) = \frac{128}{3\bar{v}^{-4}} v^3 e^{-4v/\bar{v}}$$

with $\bar{v}$ = 84 μm/s and v_{mp} = 63 μm/s.

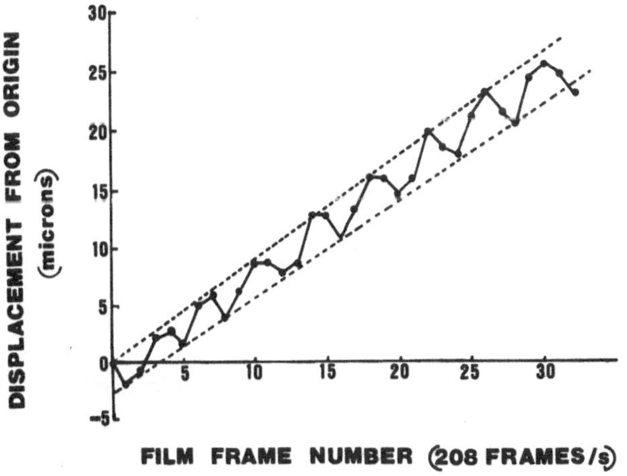

Figure 2. Characteristic motion of a C. reinhardtii cell.

$$P_s(v) = \frac{128}{3}\left(X_2 \frac{v_2^{\,3}}{\bar{v}_2^{\,4}} e^{\frac{-4v_2}{\bar{v}_2}} + X_1 \frac{v_1^{\,3}}{\bar{v}_1^{\,4}} e^{\frac{-4v_1}{\bar{v}_1}}\right) + \text{stationary function} \qquad (3)$$

where X_2 is the fraction of cells moving quickly with a speed distribution V_2 (forward) and X_1 is the fraction of cells moving less quickly with a speed distribution V_1 (backwards). The stationary function was a polynomial given by

$$S.F. = (1-(X_2+X_1))(1-k^2\tau(-.30209+k^2\tau(-.27928(.45634-k^2\tau)(.33518))))) \qquad (4)$$

This form of $P_s(v)$ was used in Equation 1 with $r = v\tau$.

Individual autocorrelation functions with scattering angles between 15° to 150° were fitted with a least squares fit of Equation 1 to give values of X_2 and X_1 that were consistent with the cinematographic data of Figure 1. An example of the fit is shown in Figure 4.

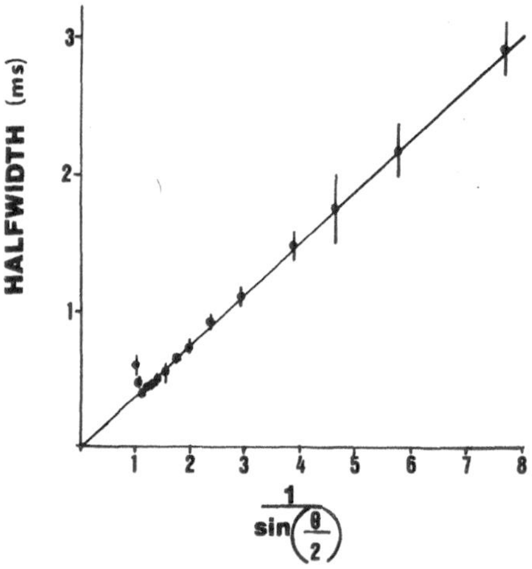

Figure 3. The half width time (ms) of the experimental electric field autocorrelation function scaled against $\frac{1}{\sin(\theta/2)}$, θ = the scattering angle.
The solid represents a least squares fit to all except the first two points.

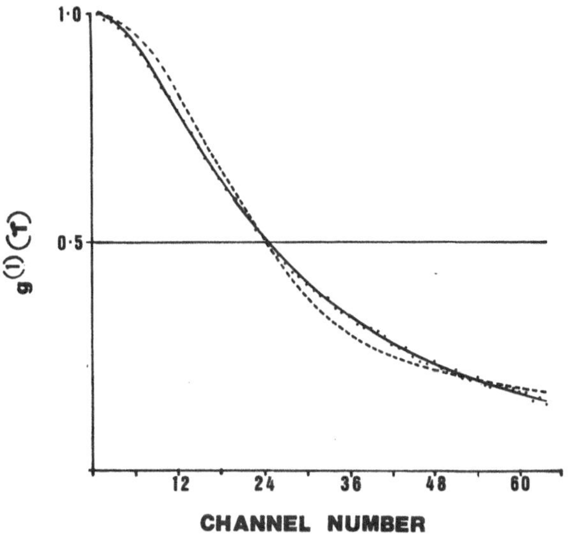

Figure 4. An autocorrelation function at $\theta = 15°$. The solid line is from the two speed model. The dashed line is the sine function model.

Sine Function Model

The trajectory of Figure 2 was considered to be a ramp function with a superimposed oscillation. Equation 2 was combined with Equation 1 with

$$r(\tau) = v\tau + A(\sin(\omega\tau + q) - \sin q) \qquad (5)$$

where A is the amplitude of the oscillation and q is the phase at the origin. This gives a reasonable fit to the individual autocorrelation functions. (i.e. Figure 4)

5. SUMMARY

The best results for the parameters obtained in this investigation are shown in Table 1.

Both models, as described, 1) incorporated the forward and reverse motion that is observed cinematographically. 2) matched the individual autocorrelation functions reasonably well and 3) match the HWT of the experimental autocorrelation functions and 4) scale similarly to the data in Figure 3.

The sine function model models the 'average' cell quite well but it is not as good as the two speed distribution model in matching

Table 1. A Summary of the Parameters Obtained in this Investigation

Average Progressive Speed: $\bar{v}$	
RACEY, HALLETT	
cinematography: long time average, 85 cells	84 μm/s
cinematography: detailed study, '12' cells	124 ± 32 μm/s
two speed model for 'average' cell $N_f = 0.47$, $N_b = 0.38$, $N_s = 0.15$	85 μm/s
two speed model for '12' cells $N_f = 0.57$, $N_b = 0.32$, $N_s = 0.11$	117 μm/s
sine function model for 'average' cell $\bar{\nu} = 34$ Hz, $A = 1.29$ μm	84 μm/s
Beat frequency: $\bar{\nu}$	
RACEY, HALLETT	≃ 34 Hz

individual autocorrelation functions. The two speed distribution model requires more calibration from cinematography to interpret the results of the X_1 and X_2 values determined from the fits. Together the two models would seem to provide a complementary framework for the interpretation of the light scattering data.

SOLUTION OF A NON-LINEAR FOKKER-PLANCK EQUATION GOVERNING BACTERIAL CHEMOTACTIC BAND FORMATION AND PROPAGATION BY STOCHASTIC COMPUTER SIMULATION METHOD

Piero Tartaglia

Institute of Physics
Faculty of Engineering
University of Rome

and

Sow-Hsin Chen

Nuclear Engineering Department
Massachusetts Institute of Technology

Chemotaxis refers to a manifestation by certain strains of bacteria with the ability to sense the presence of chemical gradients of nutrients in liquid media. The chemotactic response of E. Coli to oxygen has been extensively studied by Adler and his co-workers in recent years.[1] The simplest demonstration of the phenomenon is to inject suitable numbers of the bacteria into the bottom of a tube containing oxygen saturated motility buffer solution. A sharp band of the bacteria is soon formed and the band migrates slowly upward because the bacteria collectively seek a certain optimum oxygen concentration within a traveling oxygen gradient created by their own metabolism. Mathematical theory of this striking phenomenon of band formation and propagation was formulated by Keller and Segel[2] and the consequence of the theory was quantitatively tested experimentally by a light scattering experiment of Holz and Chen.[3] The Keller-Segel model consists of a pair of coupled non-linear Fokker-Planck equations involving the bacterial density $b(z,t)$ and the substrate concentration $c(z,t)$. Solution of these coupled equations generally requires a numerical procedure. This short note is intended to outline a technique we are currently developing to study numerically the transition from the transcient (band formation) to steady state (band propagation) solution. Readers are referred to Reference 3 for detail background information and notations.

The Fokker-Planck equation for the conditional probability $P(\bar{z},\bar{t})$ of a bacterium to be in the position $\bar{z}$ at time $\bar{t}$, given that it was in a certain position at time $\bar{t} = 0$, can be easily transformed into a stochastic differential equation for the position itself of the bacterium $\bar{z}(\bar{t})$, i.e.

$$d\bar{z}(\bar{t}) = \bar{v}_c d\bar{t} + \sqrt{2}\, dW(\bar{t}) \tag{1}$$

where $dW(\bar{t})$ is a stochastic Wiener process with average and correlation function

$$\langle dW(\bar{t})\rangle = 0, \qquad \langle dW(\bar{t})dW(\bar{t}')\rangle = \delta(\bar{t} - \bar{t}')d\bar{t}d\bar{t}' \tag{2}$$

The drift velocity $\bar{v}_c$ is related to the substrate concentration by the relation

$$\bar{v}_c = \frac{\bar{\delta}}{\bar{c}}\frac{\partial \bar{c}}{\partial \bar{z}} \tag{3}$$

The initial condition for the stochastic differential equation is assumed to be $\bar{z}(0) = 0$, that is a δ-function initial distribution for the bacteria.

When there is no infinite barrier, the bacteria will undergo an overdamped motion in the range $(-\infty, \infty)$ of the $\bar{z}$ axis under the action of the systematic force $\bar{v}_c$ and the stochastic force $dW/d\bar{t}$ (which is a white noise). In the present case, since the motion is bounded at $\bar{z} = 0$ because of the boundary condition of no incoming flux, one has to use the reflection principle of Désiré André[4] and introduce a reflecting barrier at the origin. This procedure is equivalent to using a new process which is equivalent to using a new process which is $\bar{z}(\bar{t})$ for $\bar{z} \geq 0$ and $-\bar{z}(\bar{t})$ for $\bar{z} < 0$.

Since the diffusion of the substrate is slow it is assumed to be zero, so that the rate of change of the nutrient concentration only depends on the consumption rate which is proportional to the bacterial density

$$\frac{\partial \bar{c}}{\partial \bar{t}} = \bar{P}(\bar{z},\bar{t}) \tag{4}$$

with the initial condition $\bar{c}(\bar{z},0) = 1$ and no flux boundary conditions.

Equations 1 and 4 constitute a set of coupled equations which do not have an analytic solution, except in the stationary state, so that it is necessary to solve them numerically.[5,6] One chooses a time interval $\Delta\bar{t}$ and a space interval $\Delta\bar{z}$ such that $\bar{t}_n = n\Delta\bar{t}$ ($n = 0,1,2,\ldots$) $\bar{z}_i = i\Delta\bar{z}$ ($i = 0,1,2,\ldots$) and Equation 1 ca be written as

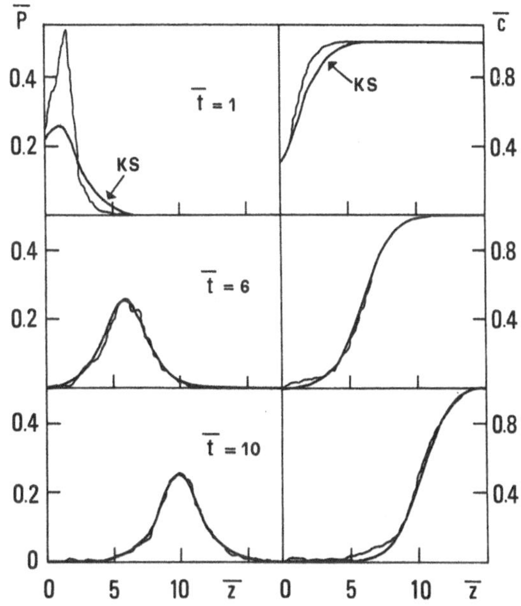

Figure 1. The bacterial distribution $\bar{P}(\bar{z},\bar{t})$ and substrate concentration distribution $\bar{c}(\bar{z},\bar{t})$ at different stages of development. Both the positional and time variables are in non-dimensional scaled quantities. It should be noted that the computer simulation method gives the same results as the Keller-Segel analytical solutions in the steady state regime as it should. In the transcient regime ($\bar{t}$ = 1) it gives however more accurate solutions.

$$\bar{z}(\bar{t}_{n+1}) = \bar{z}(\bar{t}_n) + \int_{\bar{t}_n}^{\bar{t}_{n+1}} d\bar{t}\ \bar{v}_c(\bar{x}(\bar{t}),\bar{t}) + \sqrt{2}\ W_n \tag{5}$$

The term which gives the systematic evolution is then approximated by

$$\int_{\bar{t}_n}^{\bar{t}_{n+1}} d\bar{t}\ \bar{v}_c(\bar{x}(\bar{t}),\bar{t}) \simeq \bar{v}_c(\bar{x}(\bar{t}_n),\bar{t}_n)\Delta\bar{t} \tag{6}$$

while the variable

$$W_n \equiv \int_{\bar{t}_n}^{\bar{t}_{n+1}} dW(\bar{t}) \tag{7}$$

is easily seen to be a gaussian random variable with

$$\langle W_n\rangle = 0, \qquad \langle W_n^2\rangle = \langle \int_{\bar{t}_n}^{\bar{t}_{n+1}} dW(\bar{t}) \int_{\bar{t}_n}^{\bar{t}_{n+1}} dW(\bar{t}')\rangle = \Delta t \tag{8}$$

The first step is a purely diffusive motion starting at $\bar{z} = 0$, since for $\bar{t} = 0$, $\bar{v}_c = 0$. Equation 5 gives the motion of each individual bacterium, repeating the procedure for many trajectories one gets the hystogram of $P(\bar{z},\bar{t}_n)$ using intervals of width $\Delta\bar{z}$. Once $P(\bar{z},\bar{t}_n)$ is known $\bar{c}(\bar{z},\bar{t}_{n+1})$ can be obtained from the difference equation

$$\bar{c}(\bar{z},\bar{t}_{n+1}) \simeq c(\bar{z},\bar{t}_n) - P(\bar{z},\bar{t}_n)\Delta t \tag{9}$$

The drift $\bar{v}_c(\bar{z},\bar{t}_{n+1})$ is then calculated by numerical differentiation and allows the next time step according to Equation 5.

At each step the averages $\langle\bar{z}(\bar{t})\rangle$ and $\langle\bar{z}^2(\bar{t})\rangle - \langle\bar{z}(\bar{t})\rangle^2$ are also computed as sums over the trajectories, and can then be compared with the Keller-Segel solution for the average position of the bacteria $\langle\bar{z}(\bar{t})\rangle = \bar{t}$ and $\langle\bar{z}^2(\bar{t})\rangle - \langle\bar{z}(\bar{t})\rangle^2 = \langle\bar{\xi}^2\rangle$ for the width of their distribution, for large values of $\bar{t}$.

In the case $\bar{\delta} = 2$, illustrated in the figures, we used $\Delta\bar{t} = 0.1$ ($\simeq$ 10s for the real time), $\Delta\bar{z} = 0.25$ ($\simeq$ 0.025 mm for the real displacement) and 1000 trajectories. The stationary value of $\langle\bar{\xi}^2\rangle$ is $\pi^2/3$.

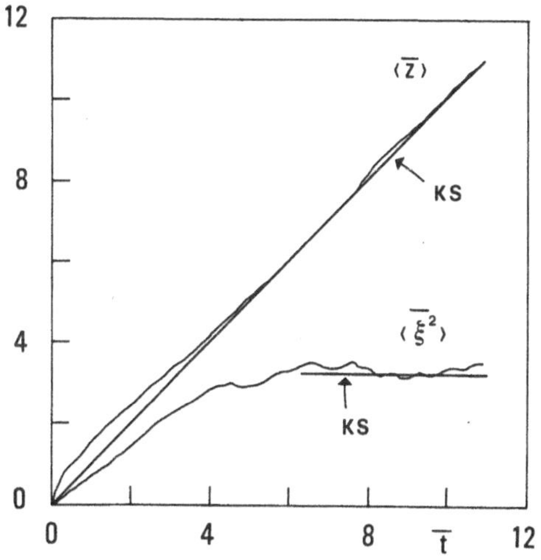

Figure 2. Plot of the first and second moments of the bacterial distribution as a function of time. It is seen that the bacterial band propagates with a uniform speed and maintains a constant width after the transcient period.

REFERENCES

1. J. Adler, "Chemotaxis in bacteria", Science 153: 708 (1966).
2. E. F. Keller and L. A. Segel, J. Theor. Biol. 30: 235 (1971).
3. M. Holz and S. H. Chen, Biophys. J. 26: 243 (1979).
4. A. Papoulis, "Probability, Random Variables, and Stochastic Processes", McGraw-Hill, New York, 1965, p. 505.
5. N. J. Rao, J. D. Borkwankar, and D. Ramkrishna, SIAM J. Control 12: 124 (1974).
6. F. de Pasquale, P. Tartaglia, and P. Tombesi, Nuovo Cimento Lett. 28: 141 (1980).

A RUN-TWIDDLE MODEL FOR THE SELF CORRELATION FUNCTION OF BACTERIA IN CHEMOTACTIC BANDS

Paul C. Wang and Sow-Hsin Chen

Nuclear Engineering Department
Massachusetts Institute of Technology
Cambridge, Massachusetts 02139

CONTENTS

1. INTRODUCTION

Microscopic motions of individual Escherichia coli in liquid chemotactic media have been extensively studied by stroboscopic photomicrographs[1] and especially by tracking microscope[2]. The motions are shown to be characterized by sequential repetitions of either a "running" state in which a bacterium moves in a fairly straight path or a "twiddle" state in which the bacterium jigs around locally. Both the stroboscopic photomicrograph and tracking microscope techniques work well only when bacterial density in solution is relatively low. For studying bacterial motion in a chemotactic band, which has a very high bacterial density (10^8/ml), quasi-elastic light scattering technique is more suitable.[3,4]

Traditionally, calculation of the self correlation functions takes into account only the contribution from the running state.[3,4] However, in the chemotactic band case, bacteria spend a significant portion of time in the twiddle state; therefore, it is necessary to include the twiddle motions in the self correlation function. It has been pointed out from previous works[5-11] that in order to eliminate contributions from wobbling effects and structural and nonspherical shape effects from the correlation function one should perform scattering experiments at small angles. We thus undertook a series of small angle light scattering from bacteria forming a chemotactic band with the wave-vector pointing in the direction of the band propagation and in a direction perpendicular to it. We formulated a stochastic theory of bacterial motions taking into account explicitly pertinent features of the run-twiddle model. The resulting self correlation function for a bacterium was derived as

$$F_s(Q,t) = \frac{\tau_1}{\tau_1 + \tau_2} \exp\left(-\frac{t}{\tau_1}\right) \exp\left(-Q^2 L_1^2 \left[\frac{t}{T_1} - 1 + \exp\left(-\frac{t}{T_1}\right)\right]\right)$$

$$+ \frac{\tau_2}{\tau_1 + \tau_2} \exp\left(-\frac{t}{\tau_2}\right) \exp\left(-\tfrac{1}{2} Q^2 V_2^2 t^2\right)$$

$$+ \frac{2}{\tau_1 + \tau_2} \int_0^t dt_1 \exp\left(-\frac{(t - t_1)}{\tau_1}\right) \exp\left(-Q^2 L_1^2 \left[\frac{(t - t_1)}{T_1} - 1 + \exp\left(-\frac{(t - t_1)}{T_1}\right)\right]\right) \exp\left(-\frac{t_1}{\tau_2}\right) \exp\left(-\tfrac{1}{2} Q^2 V_2^2 t_1^2\right) \quad (1)$$

where

τ_1: average duration of twiddle

τ_2: average duration of run

L_1: average small step length of a bacterium in twiddle state

T_1: average duration of the small step

V_2: average running speed

$\vec{Q}$: scattering vector.

The first term in Equation (1) is the contribution from the bacteria which twiddle all the time during time t (observation time). The second term is the contribution from bacteria which maintain the running state all the time. The third term is the contribution from bacteria which change state once, either from run to twiddle or from twiddle to run, during t. When the small step length L_1 is small compared with Q^{-1} (i.e. $QL_1 << 1$), the twiddle correlation function can be replaced by a simpler exponential form with an effective diffusion constant D ($=L_1^2/T_1$), i.e.

$$\text{First term} = \frac{\tau_1}{\tau_1 + \tau_2} \exp\left(-\frac{t}{\tau_1}\right) \exp\left(-DQ^2 t\right) \quad (2)$$

2. EXPERIMENT AND RESULT

The way to prepare bacteria culture and optical set up were the same as in our previous experiment.[12] Small scattering angles were used (θ=5.8° in the parallel case and 6.9° in the perpendicular case). Wavelength of the laser beam is 6328Å. The self correlation function in the parallel direction was measured right after the perpendicular case.

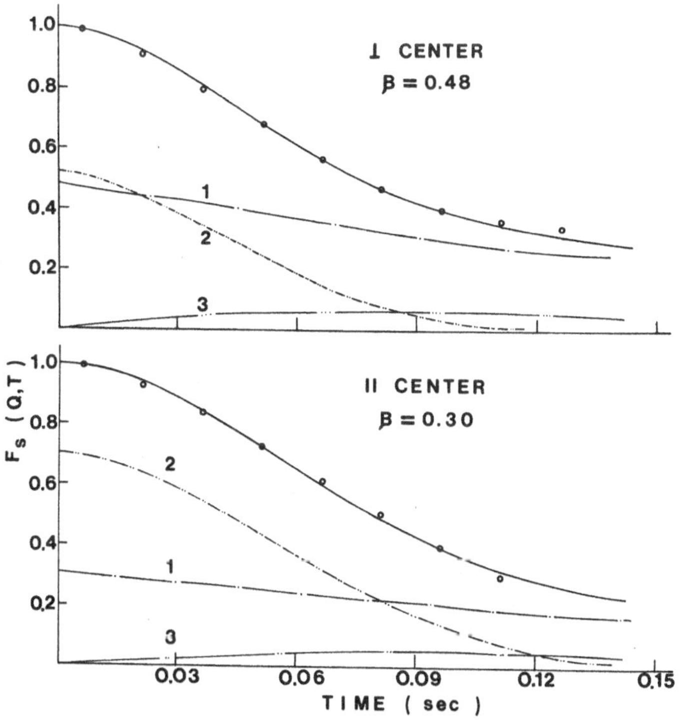

Figure 1. Self correlation functions of bacteria in the center part of the migrating band for Q perpendicular and parallel, respectively, to the direction of oxygen gradient. Circles are the experimental points and solid lines are theoretical prediction of the run-twiddle model. The numbered dash lines are contributions from each term in Equation (1).

Table I.

	D ($\frac{\mu m^2}{sec}$)	τ_1 (sec)	τ_2 (sec)	V_2 ($\frac{\mu m}{sec}$)	β*
Perpendicular Case	4.74	0.43	0.47	29	0.48
Parallel Case	4.74	0.43	1.00	27	0.30

$*\beta \equiv \frac{\tau_1}{\tau_1 + \tau_2}$ = fraction of bacteria in the twiddle state

In Figure 1, there are two self correlation functions. One is from the front part of the band with $\vec{Q}$ in the parallel direction, and the other is also from the front part of the band with $\vec{Q}$ in the perpendicular direction. The solid lines are theoretical calculations according to Equation (1). The best fit parameters are given in Table I. The dotted lines indicate separate contributions from each term in the self correlation function. The first term is from the twiddle motion which behaves like a simple exponential function. The second term is from the running motion, which behaves like a Gaussian function. The third term starts with a value at t = 0, increases to about 0.1 at the tail of the curve, and eventually dies out. The dots in these figures are the experimental data. The agreement between theory and experiment is excellent.

By choosing the direction of the band propagation to be the z-direction, we can write the definition of the two correlation functions measured as

$$F_{\parallel}(Q,t) = \langle e^{iQZ(t)} \rangle \tag{3}$$

$$F_{\perp}(Q,t) = \langle e^{iQX(t)} \rangle \tag{4}$$

We can show[12] from Equations (3) and (4) with Equations (1) and (2) that the average displacements $\langle Z^2(t) \rangle$ and $\langle X^2(t) \rangle$ in the parallel and perpendicular directions are respectively

$$\langle Z^2(t) \rangle \text{ or } \langle X^2(t) \rangle = \frac{\tau_1}{\tau_1 + \tau_2}(2Dt)\exp\left(-\frac{t}{\tau_1}\right) + \frac{\tau}{\tau_1 + \tau_2}(V_2^2 t^2)\exp\left(-\frac{t}{\tau_2}\right)$$

$$+ \frac{2}{\tau_1+\tau_2} \int_o^t dt_1 [2D(t-t_1) + V_2^2 t_1^2] \exp\left(- \frac{(t-t_1)}{\tau_1}\right) \exp\left(- \frac{t_1}{\tau_2}\right)$$

Figure 2 shows the mean square displacements in both directions as a function of time using parameters in Table I.

3. CONCLUSION

For a typical correlation function recorded in the perpendicular direction, there are about one hundred channels of reliable data points. These data are sufficient to determine a unique set of parameters D, τ_1, τ_2, and V_2 in Equation (1). However, bacterial velocity distribution may be shifted towards the chemotactic direction (+$\hat{Z}$-direction) in the parallel case; the correlation function tends to

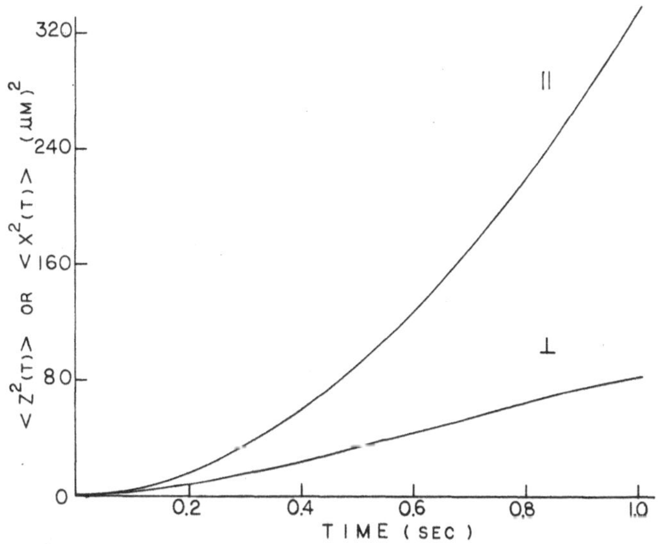

Figure 2. Mean square displacements of a bacterium parallel and perpendicular to the direction of oxygen concentration gradient as a function of time computed from Equation (5) using parameters given in Table I. Notice the asymmetry in the two directions which is due mainly to the difference in β values. This illustrates clearly the microscopic mechanism of chemotaxis.

become oscillatory. As a consequence, the background subtraction becomes more difficult. This causes the tail part of the correlation function to become less reliable and we generally omit these data points in the fitting procedure. We have thus less data points to fit in the parallel case and as a result we obtain non-uniqueness of parameters.

The procedure of finding a set of best fit parameters is as follows. First, we adjust D and V_2 so as to fit the front part of the curve. We then adjust τ_1 and τ_2 individually to match the rear part of the data. In the perpendicular case, usually the curve is smoother and more reliable. The set of the best fit parameters is unique. In the parallel case, there are fewer data points available and it was found that as long as we keep β constant, the absolute values of τ_1 and τ_2 do not make too much difference in the fitting. In order to select one particular set of τ_1 and τ_2 in the parallel case, we can use τ_1 from the perpendicular analysis. This is consistent with the tracking microscopic observation[2]; the twiddle time in these two cases is the same. τ_2 can then be determined from β and τ_1.

From Figure 2, we see that even though the running velocity is a little slower in the parallel direction, a bacterium travels twice as long in the parallel direction as in the perpendicular direction. This difference is due to different β values in these two cases. This is exactly the reason why the chemotactic bands migrate along the favorable chemical direction.

ACKNOWLEDGEMENTS

This research is supported by the National Science Foundation.

REFERENCES

1. R. M. MacNab and D. E. Koshland, Jr., The gradient sensing mechanism in bacterial chemotaxis, Proc. Nat'l. Acad. Sci. U.S.A. 60: 2509 (1972).
2. H. C. Berg and D. A. Brown, Chemotaxis in Escherichia coli analyzed by three-dimensional tracking, Nature (Lond.) 239: 500 (1972).
3. R. Nossal and S. H. Chen, Laser measurements of chemotactic response of bacteria, Opt. Comm. 5: 117 (1972).
4. M. Holz and S. H. Chen, Quasi-elastic light scattering from migrating chemotactic bands of Escherichia Coli, Biophy. J. 23: 15 (1978).
5. G. B. Stock and F. C. Carlson, Photon autocorrelation spectra of wobbling and translating bacteria, in "Symposium on Swimming and Flying in Nature," T. Y-T. Wu, C. J. Brokaw, and C. Brenen, eds. (Plenum, New York, 1974), pp. 57-68.
6. J. P. Boon, R. Nossal, and S. H. Chen, Light scattering spectrum due to wiggling motions of bacteria, Biophy. J. 14: 847 (1974).

7. M. Holz and S. H. Chen, Rotational-translational models for interpretation of quasi-elastic light scattering spectra of motile bacteria, Appl. Opt. 17: 3197 (1978).
8. B. J. Bern and R. Nossal, Inelastic light scattering by large structural particles, Biophy. J. 14: 865 (1974).
9. S. H. Chen, M. Holz, and P. Tartaglia, Quasi-elastic light scattering from structured particles, Appl. Opt. 16: 187 (1977).
10. M. Holz and S. H. Chen, Structural effects in quasi-elastic light scattering from motile bacteria of E. coli, Appl. Opt. 17: 1930 (1978).
11. G. B. Stock, The measurement of bacterial translation by photon correlation spectroscopy, Biophy. J. 22: 79 (1978).
12. P. C. Wang and S. H. Chen, Quasi-elastic light scattering from migrating chemotactic bands of Escherichia coli II: Analysis of anisotropic bacterial motions, submitted to Biophy. J. (1981).

QUASI-ELASTIC LIGHT SCATTERING

STUDIES OF MUSCULAR CONTRACTION

Francis D. Carlson and Richard C. Haskell

Thomas C. Jenkins Department of Biophysics
Johns Hopkins University
Baltimore, Maryland 21218

For the past quarter of a century studies of the molecular mechanism of muscular contraction have been guided by the sliding filament model of striated muscle. According to this model myosin cross-bridges projecting from the myosin thick filaments interact with the actin thin filaments of the muscle cell in such a way that each cross-bridge produces an impulsive contractile force between the interdigitating thick and thin filaments. Each thick filament contains a few hundred cross-bridges which attach to the thin filament independently and asynchronously so that collectively they produce an average contractile force that depends linearly on the extent to which the thick and thin filaments overlap. Much of what is known about the molecular dynamics of the cross-bridge cycle is derived from low angle x-ray diffraction studies of contracting muscle and is subject to the limitations inherent in space-time average measurements on a dynamic process. There is currently an active search for more direct evidence to corroborate the independent cycling of cross-bridges.[1,2] We have been using quasi-elastic light scattering to obtain evidence for cross-bridge cycling.[3,4] Our most recent results obtained on single muscle fibers, are free of many of the complicating factors present in our early work on whole muscle and it is these new results that are presented here.

The dominant light scattering structures in striated muscle are assemblages of the contractile proteins. The fact that these proteins are organized into thick and thin filaments endows them with a considerably higher scattering power than the dissolved sarcoplasmic proteins. The myosin cross-bridges which project from the myosin thick filament and are presumed to move during cycling represent about 35% of the contractile protein and so it is reasonable to expect that dynamic light scattering studies on muscle might provide information about the structural dynamics of these moving components of the myosin molecule. However, cross-bridge motions are expected to be of the order of 0.01 μm, while the reciprocal scattering vector in our light scattering experiment is 0.1 μm and the intensity fluctuations arising from such small aplitude cross-bridge motion would be less than 1% and thus scarcely detectable. On the other hand, independent cycling of a few hundred cross-bridges could lead to fluctuating imbalances in the forces acting upon thick and thin filaments or myofibular A and I bands. Fluctuating force imbalances could produce displacements as large as the reciprocal of the scattering vector if the structures which constrain the thick and thin filaments into A and I bands are sufficiently flexible. Under these conditions measurement of the normalized intensity autocorrelation function might give information about the dynamics of cross-bridge cycling.

We have measured the intensity autocorrelation function, $g^{(2)}(\tau)$, of light quasi-elastically scattered from single semi-tendinosis muscles of the frog during the plateau phase of an isometric contraction.[4] A comparison of data taken during the two successive halves of a tetanus indicated that the intensity fluctuations of light scattered from a contracting muscle fiber arise from a quasi-stationary random process. Further, the statistics of the scattered field were, to a good approximation Gaussian. In a series of 18 experiments the average value of $g^{(2)}(\tau)-1$ was 0.58 and the average value of the effective half-time of $g^{(2)}(\tau)$, $\tau_{½}$, was 73 ms. In resting muscle the amplitude of the autocorrelation function was 0.01 indicating virtually no structural fluctuations of the thick and thin filaments.

Comparison of scattering data obtained with the muscle fiber axis oriented in the scattering plane and normal to that plane conclusively established that the velocity of scattering elements was almost exclusively along the fiber axis. For fibers oriented with the fiber axis in the scattering plane the reciprocal of $\tau_{½}$ (decay rate of $g^{(2)}(\tau)$) varied approximately linearly with the projection of the scattering vector on the fiber axis. These results are consistent with the interpretation that the scattering elements in the muscle fiber possess relative velocities along the fiber axis which are constant for time intervals longer than the field coherence time.

If the motion of the scattering elements along the fiber axis is due only to a slow elongation of the sarcomeres in the region under observation then $\tau_{½}$ should depend upon the length of the region observed in a calculable way. Further, if the sarcomeres are elongating there should be a corresponding gradual displacement of their first order diffraction maximum. Experiments showed that under conditions where there was no change in sarcomere spacing as indicated by the lack of movement of the first order diffraction maximum, relative motions associated with fiber elongation did not make a major contribution to the decay of $g^{(2)}(\tau)$.

If the scattering elements are constrained to axial displacements less than the reciprocal of the projection of the scattering vector on the fiber axis, the amplitude of $g^{(2)}(\tau)$ will be much less than it would be for unconstrained motion. The amplitude of $g^{(2)}(\tau)$ was measured over a wide range of scattering angles and the data were corrected for a bias that occurs in experiments of short duration. In fibers exhibiting no elongation of the sacromeres during a tetanus, because of the variability of the data, the possibility cannot be excluded that the motion responsible for the decay of $g^{(2)}(\tau)$ is constrained to rms relative displacements as small as 0.3 μm.

Measurements of the amplitude and $\tau_{1/2}$ of $g^{(2)}(\tau)$ were made as a function of sarcomere length. During the plateau phase of tension both the amplitude and $\tau_{1/2}$ $g^{(2)}(\tau)$ were essentially independent of the sarcomere length. However, during the initial "creep phase" of a tetanus at longer sarcomere lengths, $\tau_{1/2}$ was significantly less than it was during the tension plateau.

These results can be summarized as follows. In a 200 micron long region observed near the center of a single muscle fiber during the tension plateau of an isometric tetanus no significant fraction of the scattering material possesses relative velocities greater than 3 μm/s accompanied by displacements greater than 0.05 μm. The dependence of $\tau_{1/2}$ on scattering angle and orientation of the fiber indicates that the decay of $g^{(2)}(\tau)$ is due primarily to relative motion of the scattering material along the fiber axis. The rms value for these relative axial velocities was 1-2 μm/s and the large amplitude of $g^{(2)}(\tau)$ observed at small scattering angles excludes the possibility that the motion responsible for the decay of $g^{(2)}(\tau)$ was constrained to relative displacements with an rms value less than r equal to 0.1 μm. These results are consistent with an interpretation which has large groups of myofibular sarcomere moving axially relative to other large groups of myofibular sarcomeres with relative axial velocities of 1-2 μm/s and constant sarcomere lengths. Careful microscopic observation under high magnification reveals such relative axial motions between groups of myofibular sarcomeres in different regions of the muscle fiber with little or no change in sarcomere length. Our results do not exclude structural fluctuations at the level of the myofibular sarcomeres due to independent cross-bridge cycling and force imbalances. However, such fluctuations if they exist must be limited to relative velocities less than 3 μm/s between thick and thin filaments and displacements less than 0.05 μm. If these structural fluctuations contributed to the decay of $g^{(2)}(\tau)$ one would expect a change in the amplitude and decay rate of $g^{(2)}(\tau)$ when there is no overlap of the thick and thin filaments. We have found both the amplitude and $\tau_{1/2}$ to be roughly independent of the sarcomere length, making it unlikely that there are any significant fluctuations at the myofibrillar sarcomere level.

Our failure to find any strong evidence for structural fluctuations in contracting muscle has several possible explanations. It may be that the asynchronously cycling cross-bridge model of contraction is incorrect. Also, it is possible that the A and I band structures of muscle possess sufficient structural rigidity to prevent any relative displacements of thick filaments due to fluctuations in the number of cross-bridges acting. It is also possible that during contraction cross-bridges are attached to the thin filaments for a large fraction of their cycle time and thus endow the structure with enough rigidity to make relative displacements unresolveable. Finally, the upper limit 3 μm/s for relative velocities corresponds to the relative velocity of thick and thin filaments during unloaded shortening of a fiber, as a consequence, smaller relative velocities, less than 1-2 μm/s would be masked by this large scale motion and thus rendered undetectable.

If, as is currently thought, the cross-bridge is not spherically symmetrical and its orientation changes during the cycle, there should be fluctuations in the birefringence of the muscle fiber associated with the orientational changes. Studies in our laboratory now in progress are directed toward the measurement of birefringence fluctuations in contracting muscle.

REFERENCES

1. J. Borejdo and M. F. Morales, Biophys. J. 20: 315 (1977).
2. J. Borejdo, S. Putnam, and M. F. Morales, Proc. Nat. Acad. Sci. U.S.A. 76: 6346 (1979).
3. R. F. Bonner and F. D. Carlson, J. Gen. Physiol. 65: 555 (1975).
4. R. F. Haskell and R. D. Carlson, Biophys. J. 33: 39 (1981).

PARTICIPANTS

Dr. K.M. Abbey	Dept. of Chemistry, SUNY at Stony Brook Stony Brook, NY 11794 USA (P)
Dr. B.J. Ackerson	Physics Dept., Oklahoma State University Stillwater, OK 74078 USA (P)
Mr. H.A. Aljuwair	Nuclear Engineering Dept., MIT, 24-209 Cambridge, MA 02139 (P)
Dr. B. Bedwell	Chemical Engineering Dept., University of Michigan, Ann Arbor, MI 48109 USA (P)
Prof. G.B. Benedek	Physics Dept., MIT, Cambridge, MA 02139 USA (L)
Dr. C.M. Berke	Polaroid Corp., Research Division 750 Main St., 4J Cambridge, MA 02139 USA (P)
Dr. D. Beysens	C.E.N. Saclay, SPSRM, BP. No. 2, F91190 Gif-Sur-Yvette, France (L)
Dr. M.H. Birnboim	Mechanical Engineering Dept., Rensselaer Polytech Institute, Troy, NY 12181 USA (P)
Dr. R. Bonner	Biomedical Engineering & Inst. Branch Building 13, Room 3W13 National Institutes of Health Bethesda, MD 20014 USA (L)
Dr. J.P. Boon	Service de Chime Physique II, Campus Plain ULB, Code Postal #231 Boulevard du Triomphe, B-1050 Bruxelles, Belgium (L)

L: Lecturer P: Participant

Dr. H.C. Burstyn	Physics Dept., University of Maryland College Park, MD 29742 USA (P)
Prof. F.D. Carlson	Biophysics Dept., Johns Hopkins University Baltimore, MD 21218 USA (P)
Dr. D. Caroline	School of Physics & Molecular Sciences, University College of North Wales Bangor, Gwyneed, LL57 2UW, United Kingdom (P)
Dr. P. Chabrat	Lab. d'Optiques Molec., University of Bordeaux I, 351 Cours de la Liberation, 33405 Talence, France (P)
Dr. D. Chatenay	Lab. Sepc. Hertzienne de I'ENS, 24 rue Lhomond, 75231 Paris cedex 05 France (P)
Dr. F.P. Chen	Polaroid Corp., Research Division, 750 Main Street, 4J Cambridge, MA 02139 USA (P)
Prof. S.H. Chen	Nuclear Engineering Dept., MIT, 24-209 Cambridge, MA 02139 USA (Director, Organizer, L)
Prof. T.K. Chou	Dept. of Chemistry, Beijing University, Beijing, China (P)
Prof. B. Chu	Dept. of Chemistry, SUNY at Stony Brook Stony Brook, NY 11794 USA (Organizer, L)
Dr. C. Cohen	School of Chemical Engineering, Cornell University, Ithaca, NY 14853 USA (P)
Dr. M. Corti	CISE, P.O. Box 3986 20100 Milan, Italy (L)
Mr. T. Craig	Dept. of Physics, University of Guelph, Guelph, Ontario, Canada (P)
Prof. L.S. Dai	Dept. of Physics, Fudan University Shanghai, China (P)

Prof. V. Degiorgio	CISE, Casella Postale 3986 20100 Milano, Italy (L)
Dr. M. Delaye	Physics Dept., MIT, 13-2018 Cambridge, MA 02139 USA (P)
Mr. A. Dinapoli	Chemistry Dept., SUNY at Stony Brook Stony Brook, CY 11794 USA (P)
Ms. M. Djabourov	Lab. de Physique Thermique, Ecole de Physique et Chimie, 10 rue Vauquelin, 75231 Paris cedex 05, France (P)
Dr. B. Farnoux	CEA, Centre D'Etudes Nucleaires de Saclay, BP No. 2, 91 Gif-Sur-Yvette, France (L)
Dr. A.R. Faruqi	MRC Lab. of Molec. Biol., University Medical School, Hills Road Cambridge CB220H England (L)
Dr. G.T. Feke	Retina Foundation, 20 Staniford Street Boston, MA 02144 USA (P)
Mr. A. Flamberg	Chemistry Dept., Stanford University Stanford, CA 94305 USA (P)
Dr. J.R. Ford	Chemistry Dept., University of Massachusetts Amherst, MA 01003 USA (P)
Dr. S. Fujime	Mitsubishi-Kasei Inst. of Life Sciences, 11 Minamiooya, Machida-Shi, Tokyo, Japan (L)
Dr. G. Fytas	Physical Chemistry, University of Bielefeld, 48 Bielefeld, West Germany (P)
Dr. A. Ganz	Exxon Production Research Co., P.O. Box 2189 Houston, TX 77001 USA (P)
Dr. M. Giglio	CISE, P.O. Box 3986 20100 Milan, Italy (L)
Prof. W.I. Goldburg	Physics Dept., University of Pittsburgh, Pittsburgh, PA 15260 USA (L)

Prof. R. Graham	Universitat Essen-Gesamthochschule Fachbereich Physik, D-4300, Essen 1, West Germany (L)
Dr. E. Gulari	Wayne State University, Chemical Engineering Dept., Detroit, MI 48202 (P)
Dr. D.E. Guveli	School of Pharmacy, University of Bradford, Bradford West Yorkshire BD7 1DP, England (P)
Dr. R.S. Hall	Polaroid Corp., Waltham, MA 02154 USA (P)
Dr. F.R. Hallett	Dept. of Physics, University of Guelph, Guelph, Ontario, Canada (L)
Dr. J. Hayter	Inst. Max von Laue-Paul Langevin, Avenue des Martyrs, 156 X-38042 Grenoble Cedex, France (L)
Prof. M.J. He	Chemistry Dept., Fudan University, Shanghai, China (P)
Ms. B. Herpigny	Service de Chimie Physique II, Campus Plaine ULB, Code Postal #231 Boulevard du Triomphe, B-1050 Bruxelles, Belgium (P)
Dr. J.S. Huang	Exxon Research & Engineering Co., P.O. Box 45 Linden, NJ 07036 USA (P)
Dr. T. Illeni	Dept. of Bact. Engineering, Karolinska Inst., S-104 01, Stockholm 60, Sweden (P)
Prof. C.S. Johnson	Chemistry Dept., University of North Carolina, Chapel Hill, NC 27514 USA (P)
Dr. T. Jossang	Box 1048, University of Oslo Oslo 3, Norway (P)
Dr. W.G. Jung	Konigsbergerstr. 1, 7500 Karlsruhe, West Germany (P)

Prof. K. Kawasaki	Dept. of Physics, Kyushu University Kukouka 812, Japan (L)
Dr. W.C. Koehler	Solid State Division, Oak Ridge National Laboratory, Box X Oak Ridge, TN 37830 USA (L)
Mr. M. Kotlarchyk	Nuclear Engineering Dept., MIT, 24-209 Cambridge, MA 02139 USA (P)
Dr. K. Kubota	Chemistry Dept., SUNY at Stony Brook Stony Brook, NY 11794 (P)
Dr. A. Kumar	Physics & Astronomy Dept., University of Pittsburgh, Pittsburgh, PA 15260 (P)
Dr. J. Leblond	Lab. de Res. Magnetique, Universite de Paris VI, 10 rue Vauquelin, 75231 Paris cedex 05, France (P)
Mr. D-C. Lee	Chemistry Dept., SUNY at Stony Brook Stony Brook, NY 11794 (P)
Dr. W.I. Lee	Center for Bioengineering, Mail Stop F1-20, University of Washington Seattle, WA 98195 USA (L)
Dr. L. Letamendia	CNRS, University of Bordeaux I, 351 Cours de la Liberation, 33405 Talence, France (P)
Dr. J.S. Lin	Oak Ridge National Laboratory, P.O. Box X Oak Ridge, TN 37830 USA (L)
Dr. Y-H. Lin	Chemistry Dept., SUNY at Stony Brook Stony Brook, NY 11794 (P)
Dr. J.A. Marqusee	Chemistry Dept., MIT Cambridge, MA 02139 USA (P)
Mr. P. Missel	Physics Dept., MIT, 13-2018 Cambridge, MA 02139 USA (P)
Prof. R.C. Mockler	Physics Dept., University of Colorado Boulder, CO 80309 USA (P)

Dr. D.F. Nicoli	Physics Dept., University of California at Santa Barbara Santa Barbara, CA 93106 USA (L)
Dr. P. Nieuwenhuysen	Dept. of Cell Biology, Univ. Instelling Antwerpen, Universiteitsplein 1, 2610 Wilrijk, Belgium (P)
Prof. T. Nose	Dept. of Polymer Chemistry, Tokyo Institute of Technology Ookayama, Meguro-ku, Tokyo, Japan (P)
Dr. R. Nossal	Building 12A, Rm. 2007, National Institutes of Health Bethesda, MD 20205 USA (Organizer, L)
Dr. K. Ohbayashi	Hiroshima University Hiroshima 730, Japan (P)
Dr. C.J. Oliver	Royal Signals & Radar Estab., St. Andrews Road, Great Malvern, Worcs. WR143PS, England (L)
Dr. A. Onuki	Physics Dept., Kyushu University Fukuoka 812, Japan (P)
Prof. I. Oppenheim	Chemistry Dept., MIT Cambridge, MA 02139 USA (L)
Dr. N. Ostrowsky	Lab. de Physique de la Matiere Condensee, University of Nice, Parc Valrose, 06034 Nice cedex, France (L)
Dr. G.R. Palmer	Zoology Dept., University of Cambridge, Downing St., Cambridge CB23EJ England (P)
Dr. A. Patkowski	Institute of Physics, A. Mickiewicz University Poznan, Poland (P)
Prof. R. Pecora	Dept. of Chemistry, Stanford University, Stanford, CA 94305 USA (L)
Dr. E.R. Pike	Royal Signals & Radar Estab., St. Andrews Road, Great Malvern, Worc. WR143PS, England (L)

Dr. B.G. Pinsky	Biophysics Graduate Group, University of California Dept. of Applied Science Davis, CA 95616 USA (P)
Mr. J.W. Pope	Chemistry Dept., SUNY at Stony Brook Stony Brook, NY 11794 USA (P)
Dr. B. Quentrec	Lab. de Physique de la Matiere Condensee, Univ. de Nice, Parc Valrose, 06034 Nice cedex, France (P)
Mr. T. Racey	Dept. of Physics, University of Guelph, Guelph, Ontario, Canada (P)
Dr. M. Rawiso	Inst. Laue-Langevin, Centre de Tri 156X 38042 Grenoble cedex, France (P)
Prof. J. Rouch	Lab. d'Optique Moleculaire, Univ. de Bordeaux I, 351 Cours de la Liberation, 33405 Talence, France (P)
Dr. J. Schelten	Inst. fur Festkorperforschung, Postfach 1913, D-517 Julich 1, West Germany (L)
Dr. K.S. Schmitz	Chemistry Dept., University of Missouri at Kansas City, 5100 Rockhill Road, Kansas City, MO 64110 USA (P)
Mr. D. Sornette	Univ. de Nice, Parc Valrose 06034, Nice cedex, France (P)
Mr. D. Statman	Chemistry Dept., SUNY at Stony Brook Stony Brook, NY 11794 USA (P)
Prof. G. Summerfield	Nuclear Engineering Dept., University of Michigan Ann Arbor, MI 48109 USA (P)
Mr. G. Swislow	Physics Dept., MIT, 13-2013 Cambridge, MA 02139 USA (P)
Dr. T. Tanaka	Physics Dept., MIT, 13-2009 Cambridge, MA 02139 USA (L)
Dr. P. Tartaglia	Inst. of Physics, Faculty of Engineering, University of Rome Rome, Italy (P)

Dr. J. Teixeira	Ecole de Physique et Chimie, 75231 Paris cedex 05, France (P)
Dr. C. Vaucamps	Univ. of Bordeaux, 351 Cours de la Liberation, 33405 Talance, France (P)
Prof. C-H. Wang	Chemistry Dept., University of Utah, Salt Lake City, UT 84112 USA (P)
Prof. C.W. Wang	Chemistry Dept., Beijing University, Beijing PROC, China (P)
Mr. P.C. Wang	Nuclear Engineering Dept., MIT, 24-209 Cambridge, MA 02139
Prof. D. Woermann	Inst. of Physical Chemistry, University of Koln, D-500 Koeln Luxembergerstr. 116, West Germany (P)
Mr. Q. Yan	Inst. of Physics, Academia Sinica Beijing PROC. China (P)
Dr. W.S. Yen	Chemistry Dept., Syracuse University, Syracuse, NY 13210 USA (P)
Prof. Q. Ying	Inst. of Chemistry, Academia Sinica, Beijing PROC, China (P)
Dr. K.W.R. Yoo	Physics Dept., University of Massachusetts Amherst, MA 01003 (P)
Dr. C. Young	Dept. of Physics, Brandeis University Waltham, MA 02254 USA (P)
Prof. H. Yu	Dept. of Chemistry, University of Wisconsin Madison, WI 53706 USA (L)
Dr. G. Zaccai	Inst. Laue-Langevin 156X, 38042 Grenoble cedex, France (L)
Dr. G. Zalczer	Dept. of Chemistry, SUNY at Stony Brook Stony Brook, NY 11794 (P)
Ms. B.E. Zebrowski	Chemistry Dept., Stanford University Stanford, CA 94305 USA (P)
Dr. M. Zulauf	CENG, LMA, BP 85 Centre de Tri, 38041 Grenoble cedex, France (P)

INDEX